Imaging from Cells to Animals *In Vivo*

Series in Cellular and Clinical Imaging

The purpose of the series is to promote education and new research using cellular and clinical imaging techniques across a broad spectrum of disciplines. The series emphasizes practical aspects, with each volume focusing on a particular theme that may cross various imaging modalities. Each title covers basic to advanced imaging techniques as well as detailed discussion dealing with interpretations of the studies. The series provides cohesive, complete, and state-of-the-art cross-modality overviews of the most important and timely areas within cellular and clinical imaging.

Super-Resolution Imaging in Biomedicine
edited by Alberto Diaspro, Marc A. M. J. van Zandvoort

Imaging in Photodynamic Therapy
edited by Michael R. Hamblin, Yingying Huang

Optical Probes in Biology
edited by Jin Zhang, Sohum Mehta, Carsten Schultz

Coherent Raman Scattering Microscopy
edited by Ji-Xin Cheng, Xiaoliang Sunney Xie

Natural Biomarkers for Cellular Metabolism
Biology, Techniques, and Applications
edited by Vladimir V. Ghukasyan, Ahmed A. Heikal

Handbook of Neurophotonics
edited by Francesco S. Pavone, Shy Shoham

Imaging from Cells to Animals In Vivo
edited by Margarida Barroso, Xavier Intes

For more information about this series, please visit: [https://www.crcpress.com/Series-in-Cellular-and-Clinical-Imaging/book-series/CRCSERCELCLI]

Imaging from Cells to Animals *In Vivo*

Edited by
Margarida M. Barroso and Xavier Intes

CRC Press
Taylor & Francis Group
Boca Raton London New York

CRC Press is an imprint of the
Taylor & Francis Group, an **informa** business

MATLAB® is a trademark of The MathWorks, Inc. and is used with permission. The MathWorks does not warrant the accuracy of the text or exercises in this book. This book's use or discussion of MATLAB® software or related products does not constitute endorsement or sponsorship by The MathWorks of a particular pedagogical approach or particular use of the MATLAB® software.

First edition published 2021
by CRC Press
6000 Broken Sound Parkway NW, Suite 300, Boca Raton, FL 33487-2742

and by CRC Press
2 Park Square, Milton Park, Abingdon, Oxon, OX14 4RN

© 2021 Taylor & Francis Group, LLC

CRC Press is an imprint of Taylor & Francis Group, LLC

Library of Congress Cataloging-in-Publication Data

Names: Barroso, Margarida, editor. | Intes, Xavier, editor.
Title: Imaging from cells to animals in vivo / edited by Margarida M.
Barroso, Department of Molecular and Cellular Physiology, Albany Medical
College, Xavier Intes, Biomedical Engineering Department and Co-Director
of the Center for Modeling, Simulation and Imaging for Medicine
(CeMSIM), at Rensselaer Polytechnic Institute.
Description: First edition. | Boca Raton : CRC Press, 2020. | Series:
Series in cellular and clinical imaging | Includes bibliographical
references and index.
Identifiers: LCCN 2020023039 (print) | LCCN 2020023040 (ebook) | ISBN
9781138041097 (hardback) | ISBN 9781315174662 (ebook)
Subjects: LCSH: Imaging systems in biology. | Imaging systems in medicine.
| Diagnostic imaging--Digital techniques.
Classification: LCC R857.O6 I458 2020 (print) | LCC R857.O6 (ebook) | DDC
616.07/54--dc23
LC record available at https://lccn.loc.gov/2020023039
LC ebook record available at https://lccn.loc.gov/2020023040

ISBN: 9781138041097 (hbk)
ISBN: 9781315174662 (ebk)

Typeset in Times
by Deanta Global Publishing Services, Chennai, India

Contents

Section I   Overview of Imaging Methods and Instrumentation

Section II   Imaging Cellular Behavior

Section III   Whole-Organ and Whole-Organism Imaging

Preface

This book was conceived with the goal of providing a complete and up-to-date summary on cellular processes in biology visualized by a wide variety of techniques at multiscale *in vitro*, *ex vivo,* and *in vivo*. Our focus is to offer readers an overview of the most important current imaging techniques that allow for detailed investigation of cellular processes (**Section I**). We have also included in-depth descriptions of how these state-of-the-art imaging techniques can be used to address biological processes in cells (**Section II**) and whole tissues and organs within living organisms (**Section III**). Specific areas of coverage include cell metabolism, receptor-ligand interactions, membrane trafficking, cell signaling, cell migration, and cell adhesion. We have also incorporated chapters that deal with cellular processes investigated by intravital microscopy as well as optical and molecular imaging in living organisms such as mice and zebrafish.

The main goal of cell biology is to study cellular processes in whole tissues within living organisms. Technical limitations have restricted cell biology to work *ex vivo* and *in vitro,* using 2D/3D cultured cells, tissues, and organoids. Crucial technological breakthroughs leading to development of intravital microscopy and optical imaging have enabled researchers to begin to analyze cell biological processes *in vivo*. This book offers a single, coherent source of information on imaging techniques used to investigate protein function *in vivo* and continue to advance the work of imaging of cells and tissues in their native environment. We hope that cell biologists, imaging experts, and researchers interested in the molecular and cellular biology of diseases will find this book useful to help encourage future cross-disciplinary collaborations.

MATLAB® is a registered trademark of The MathWorks, Inc. For product information, please contact:

The MathWorks, Inc.
3 Apple Hill Drive
Natick, MA 01760-2098 USA
Tel: 508 647 7000
Fax: 508-647-7001
E-mail: info@mathworks.com
Web: www.mathworks.com

The Editors

Margarida M. Barroso is a Professor in the Department of Molecular and Cellular Physiology, Albany Medical College in Albany, New York. She received her Ph.D. in Genetics from the University of Lisbon/Gulbenkian Institute of Sciences in Portugal and was a postdoctoral fellow at the Department of Molecular Biology, Princeton University, New Jersey. She is a faculty instructor in several international imaging courses and has two issued patents on Förster resonance energy transfer (FRET) imaging technology. Dr. Barroso belongs to the following scientific societies: SPIE - The International Society of Optics and Photonics, American Society of Cell Biology (ASCB), Biophysical Society and is currently the Past-President of the Histochemical Society. She has published more than 40 papers in peer-reviewed journals and acts as a reviewer for several internationally recognized journals as well as for US and international research funding institutions.

Dr. Barroso's research goal is to accelerate preclinical drug discovery by developing novel imaging assays to screen and optimize the delivery of targeted anticancer drugs. She is also interested in the regulation of membrane trafficking pathways and of receptor-mediated cholesterol and iron transport *in vitro* and *in vivo*. Dr. Barroso's diverse expertise integrates basic cell biology with methodological advances in imaging technology to position her research group as a major force in the visualization, quantitation, and optimization of drug delivery into cancer cells using receptor-targeted approaches.

Xavier Intes is a Professor in the Biomedical Engineering Department and Co-Director of the BioImaging Center at Rensselaer Polytechnic Institute in Troy, New York. Dr. Intes received his Ph.D. in physics from Université de Bretagne Occidentale, France. He was a postdoctoral fellow at the University of Pennsylvania, Philadelphia, Pennsylvania under the mentorship of Britton Chance and Arjun Yodh. Dr. Intes was the Chief Scientist of Advanced Research Technologies Inc. (Montreal, Quebec) and oversaw the development of two commercial time-resolved tomographic optical imaging platforms: Optix® and SoftScan®. He has been a faculty member of Rensselaer Polytechnic Institute since 2006. He is a recipient of the National Science Foundation (NSF) CAREER award and a Fellow of the American Institute for Medical and Biological Engineering (AIMBE).

Dr. Intes' research interests are on the application of optical techniques for biomedical imaging in preclinical and clinical settings. His research concentration is on functional imaging of the breast and brain, fusion with other modalities and fluorescence molecular imaging. The goal of his laboratory is to develop quantitative thick-tissue optical imaging platforms by focusing on three main areas: (a) design of new optical tomographic imaging instrumentation; (b) developing new reconstruction algorithms for quantitative volumetric imaging; and (c) investigating optimal experimental and theoretical parameters for functional, molecular, and dynamical optical imaging.

List of Contributors

Maísa Mota Antunes
Department of Morphology
Universidade Federal de Minas Gerais
Belo Horizonte, Brazil

Lisa Beckmann
Department of Biomedical Engineering
Northwestern University
Evanston, Illinois

Stephen A. Boppart
Beckman Institute for Advanced Science and Technology
University of Illinois at Urbana-Champaign
Urbana, Illinois

Jose Javier Bravo-Cordero
Department of Medicine
The Tisch Cancer Institute
New York, New York

Amelia Brumbaugh
BioTech Inc.
Knoxville, Tennessee

Ruofan Cao
The W.M. Keck Center for Cellular Imaging
University of Virginia

and

Department of Biology
University of Virginia
Charlottesville, Virginia

Raquel Carvalho-Gontijo
Department of Morphology
Universidade Federal de Minas Gerais
Belo Horizonte, Brazil

Jenu Varghese Chacko
Department of Biomedical Engineering
University of Wisconsin-Madison
Madison, Wisconsin

Dan Close
BioTech Inc.
Knoxville, Tennessee

John Condeelis
Department of Anatomy and Structural Biology
Albert Einstein College of Medicine

and

Gruss Lipper Biophotonics Center
Albert Einstein College of Medicine

and

Department of Surgery
Albert Einstein College of Medicine
Bronx, New York

Michael Conway
BioTech Inc.
Knoxville, Tennessee

Claudia Crocini
Department of Molecular, Cellular, and Developmental
    Biology
University of Colorado Boulder

and

BioFrontiers Institute
University of Colorado Boulder
Boulder, Colorado

Siân Culley
Medical Research Council Laboratory for Molecular Biology
University College London

and

The Francis Crick Institute
London, United Kingdom

Richard N. Day
Department of Cellular and Integrative Physiology
Indiana University School of Medicine
Indianapolis, Indiana

Érika de Carvalho
Department of Morphology
Universidade Federal de Minas Gerais
Belo Horizonte, Brazil

Roshan Dsouza
Beckman Institute for Advanced Science and Technology
University of Illinois at Urbana-Champaign
Urbana, Illinois

Kenneth W. Dunn
Department of Medicine, Division of Nephrology
Indiana University Medical Center
Indianapolis, Indiana

Kevin W. Eliceiri
Department of Biomedical Engineering
University of Wisconsin-Madison

and

Department of Medical Physics
University of Wisconsin-Madison
Madison, Wisconsin

David Entenberg
Department of Anatomy and Structural Biology
Albert Einstein College of Medicine

and

Gruss Lipper Biophotonics Center
Albert Einstein College of Medicine
Bronx, New York

Maria Alice Freitas-Lopes
Department of Morphology
Universidade Federal de Minas Gerais
Belo Horizonte, Brazil

Masahiro Fukuda
Program in Neuroscience and Behavioral Disorders
Duke-NUS Medical School
Singapore

and

International Research Center for Medical Sciences (IRCMS)
Kumamoto University
Kumamoto, Japan

Alena I. Gavrina
Privolzhsky Research Medical University

and

Lobachevsky State University of Nizhny Novgorod
Nizhny Novgorod, Russia

Jacky G. Goetz
Université de Strasbourg
Fédération de Médecine Translationnelle de
    Strasbourg (FMTS)
Strasbourg, France

Ricardo Henriques
Medical Research Council Laboratory for Molecular Biology
University College London

and

The Francis Crick Institute
London, United Kingdom

Michael J. Hickey
Centre for Inflammatory Diseases
Monash University
Melbourne, Australia

Hajime Hirase
Laboratory for Neuron-Glia Circuitry
RIKEN Center for Brain Science
Saitama, Japan

Burkhard Höckendorf
Howard Hughes Medical Institute
Ashburn, Virginia

Song Hu
Department of Biomedical Engineering
University of Virginia
Charlottesville, Virginia

Philipp J. Keller
Howard Hughes Medical Institute
Ashburn, Virginia

Anand T. N. Kumar
Massachusetts General Hospital
Harvard Medical School
Boston, Massachusetts

Romain F. Laine
Medical Research Council Laboratory for
    Molecular Biology
University College London

and

The Francis Crick Institute
London, United Kingdom

Irina V. Larina
Baylor College of Medicine
Houston, Texas

Dawei Li
Department of Biomedical Engineering
Johns Hopkins University
Baltimore, Maryland

Xingde Li
Department of Biomedical Engineering
Johns Hopkins University
Baltimore, Maryland

Maria M. Lukina
Privolzhsky Research Medical University

and

Lobachevsky State University of Nizhny Novgorod
Nizhny Novgorod, Russia

Julie S. Di Martino
Department of Medicine
The Tisch Cancer Institute
New York, New York

Aleksandra V. Meleshina
Privolzhsky Research Medical University
Nizhny Novgorod, Russia

Gustavo Batista Menezes
Department of Morphology
Universidade Federal de Minas Gerais
Belo Horizonte, Brazil

Chandrani Mondal
Department of Medicine
The Tisch Cancer Institute
New York, New York

M. Ursula Norman
Centre for Inflammatory Diseases
Monash University
Melbourne, Australia

Maja H. Oktay
Department of Anatomy and Structural Biology
Albert Einstein College of Medicine

and

Gruss Lipper Biophotonics Center
Albert Einstein College of Medicine

and

Department of Pathology
Albert Einstein College of Medicine
Bronx, New York

Katsuya Ozawa
Laboratory for Neuron-Glia Circuitry
RIKEN Center for Brain Science
Saitama, Japan

Mehmet S. Ozturk
Electrical and Electronics Engineering Department
Karadeniz Technical University
Trabzon, Turkey

Hyeon-Cheol Park
Department of Biomedical Engineering
Johns Hopkins University
Baltimore, Maryland

Antonio Peixoto
Institut de Pharmacologie et Biologie Structurale

and

Université de Toulouse
Toulouse, France

Pedro Matos Pereira
Medical Research Council Laboratory for
    Molecular Biology
University College London

and

The Francis Crick Institute
London, United Kingdom

Ammasi Periasamy
The W.M. Keck Center for Cellular Imaging
University of Virginia

and

Department of Biology
University of Virginia

and

Department of Biomedical Engineering
University of Virginia
Charlottesville, Virginia

Mario Perro
Roche Pharma Research and Early Development
Roche Innovation Center
Zürich, Switzerland

Robert Prevedel
European Molecular Biology Laboratory
Heidelberg, Germany

Christopher A. Reissaus
Department of Pediatrics
Indiana University School of Medicine
Indianapolis, Indiana

Steven Ripp
Center for Environmental Biotechnology
The University of Tennessee
Knoxville, Tennessee

Leonardo Sacconi
European Laboratory for Non-Linear Spectroscopy
Florence, Italy

and

National Institute of Optics (INO-CNR)
Florence, Italy

and

Institute for Experimental Cardiovascular Medicine
University Freiburg
Freiburg, Germany

Md Abdul Kader Sagar
Department of Biomedical Engineering
University of Wisconsin-Madison
Madison, Wisconsin

Marina V. Shirmanova
Privolzhsky Research Medical University
Nizhny Novgorod, Russia

Karsten H. Siller
University of Virginia
Charlottesville, Virginia

Brian T. Soetikno
Department of Biomedical Engineering
Northwestern University
Evanston, Illinois

Ilya V. Turchin
Institute of Applied Physics of the Russian Academy
    of Sciences
Nizhny Novgorod, Russia

Tianxiong Wang
Department of Biomedical Engineering
University of Virginia
Charlottesville, Virginia

Horst Wallrabe
The W.M. Keck Center for Cellular Imaging
University of Virginia

and

Department of Biology
University of Virginia
Charlottesville, Virginia

Yinan Wan
Howard Hughes Medical Institute
Ashburn, Virginia

Shang Wang
Baylor College of Medicine
Houston, Texas

Tingting Xu
Center for Environmental Biotechnology
The University of Tennessee
Knoxville, Tennessee

Xinwen Yao
Department of Biomedical Engineering
Johns Hopkins University
Baltimore, Maryland

Anna Young
Center for Environmental Biotechnology
The University of Tennessee
Knoxville, Tennessee

Wu Yuan
Department of Biomedical Engineering
Johns Hopkins University
Baltimore, Maryland

Diana V. Yuzhakova
Privolzhsky Research Medical University
Nizhny Novgorod, Russia

Elena V. Zagaynova
Privolzhsky Research Medical University
Nizhny Novgorod, Russia

Hao F. Zhang
Department of Biomedical Engineering
Northwestern University
Evanston, Illinois

Section I

Overview of Imaging Methods and Instrumentation

1

Fluorescence Microscopy Techniques

Mehmet S. Ozturk and Robert Prevedel

CONTENTS

1.1 Introduction

Fluorescence microscopy has been a quintessential and enabling tool for life scientists for more than a century and, as such, a source of major discoveries in biology. These days, advanced imaging techniques based on fluorescence dyes and proteins are routinely used to visualize subtle processes in living cells at a subcellular resolution and often with molecular specificity. These microscopes have changed our visual perception as well as our understanding of cellular mechanisms. They allow us to learn from direct observation and thereby come to a mechanistic understanding of how complex, multicellular living systems operate. In this chapter, we will introduce the reader to the basic principles in fluorescence microscopy, starting with the photophysics of fluorescence molecules. Here, we are purposely focusing on fluorescent markers suitable for *in vivo* imaging and discussing their relevant parameters, which influence their function and respective applications. The remainder of this chapter, then, is dedicated to a technical discussion and review of various fluorescence microscopy techniques that are relevant for *in vivo* studies.

1.2 Principles of Fluorescence Imaging

1.2.1 Photophysics of Fluorescence

A fluorescence molecule has to fulfill many requirements in order to be most useful for a particular *in vivo* study. Thus, it is helpful to understand the basic principles underlying fluorescence signal generation. In this section, we will introduce the mechanism and parameters relevant for fluorescent imaging. While today there exists an extensive library of available fluorophores with well-documented parameters, environmental conditions can often affect those parameters positively or negatively. Therefore, we will also briefly discuss and, where possible, quantify these effects.

Inherent Parameters of Fluorophores

Quantum Electronic Behavior: Fluorescence is an incoherent light emission phenomenon that stems from the quantum mechanical behavior of electrons within a molecule. Even in the case of a coherent (light) source excitation, the resulting emission will be incoherent, i.e. not in phase with the excitation light, because of the vibrational relaxation within the molecule (Wang, Lihong, and Wu, 2007).

The underlying mechanism works as follows. An incoming (excitation) photon is absorbed by the molecule, residing in the ground state (S_0). Thus, one of the electrons makes a transition to an *excited state* (S_0–S_1). The excited electron dwells in the excited state for an average time (*fluorescence lifetime*), during which it experiences a series of *nonradiative transitions* (e.g. internal conversion, intersystem crossing, etc.), before spontaneously returning to the S_0 state (S_1–S_0) by *radiative transition*. The energy difference between the excited and ground state is radiated in the form of visible/near infrared (VIS/NIR) light (Figure 1.1).

Because some of the energy is effectively lost to vibrational relaxation of the molecule, the fluorescence emission results in less energetic, i.e. longer wavelength (red-shifted), photons. This energy difference is called *Stokes shift* (Figure 1.1b). Note that the excitation and emission spectra often display some symmetry. This "mirror spectrum" is due to the similarity of vibrational levels for the ground state (S_0) and the excited state (S_1) (Sauer, Hofkens, and Enderlein, 2011).

The excitation rate (k_{ex}), i.e. the rate at which electrons make a transition from the ground state to the excited state, is directly proportional to the excitation light intensity (I_{ex}) and the molecular absorption cross section (σ), $k_{ex} = I_{ex}\sigma$. To illustrate the fluorescence working principle quantitatively, let us assume a generic fluorophore with a molecular cross section (σ) of 3×10^{-16} cm^2. If our molecule is illuminated by a 1mW light source (560 nm) with a beam radius of 0.5 μm, the power density will be ~1.3×10^5 W/cm^2, which corresponds to 3.1×10^{23} photons/cm^{2}sec (according to P = $\hbar c/\lambda$). Therefore, the excitation rate of a single molecule within the focus will be $k_{ex} = 9.58 \times 10^7$ photons/sec. All the excited electrons will experience a de-excitation at a rate of k_{dex}, which is the sum of radiative (k_r) and nonradiative (k_{nr}) de-excitation processes, $k_{dex} = k_r + k_{nr}$. Next, we establish a relationship between the fraction of excited state electrons (x) and de-excited electrons ($1-x$), which is $k_{ex}(1-x) = k_{dex}x$ (e.g. $x = 0.75$) (Panula, 2003). Then, k_{dex} becomes 3.2×10^7 photons/sec. The aim is here to identify the radiative relaxation rate (k_r), for which we need

to know the so-called quantum yield (Φ). The quantum yield is the ratio of electrons that undergo radiative relaxation (k_r) compared to all excited electrons undergoing both radiative and nonradiative de-excitation processes ($k_r + k_{nr}$),

$$\Phi = \frac{k_r}{k_r + k_{nr}} = \frac{k_r}{k_{dex}}.$$

For simplicity, let us assume a quantum yield (Φ) of 1/3, which results in a radiative relaxation rate k_r of $1{,}07 \times 10^7$ photons/sec, approximately.

The expected fluorescence intensity (I_f) of a single molecule at the detector further depends on the overall efficiency of the imaging system, η (a composite metric that includes light collection efficiency of optics and the quantum efficiency of the detector), or $I_f = k_{dex}\Phi\eta$. A realistic value for η is ~10%, which leaves us with 1.07×10^6 photon/sec of fluorescence from a molecule. Typical exposure times of detectors span from μsec for a confocal or multiphoton microscope to msec for camera-based microscopes such as a wide-field or light sheet; therefore, this fluorescence yield will result in between 1 photon (exposure time 1 μsec) to 10^3 photons (exposure time 1 ms). Eventually, the total number of photons will depend on the number of molecules that are simultaneously excited by the focused light. For our illustration purpose, let us assume a fluorophore concentration of 1 μM, which translates to ~2,520 molecules inside a spherical focal volume of 1 μm radius ($M = N \times n/V$), where M is the molar concentration, N is the Avogadro's Number (6.02×10^{23}), n is the amount of mol, and V is the volume (1L = 10^{15} μL). Thus, we end up with a range of 10^3–10^6 photons per detector pixel in our example calculation, which also resembles typical values for bright fluorescence samples in practice. When pushing this to lower values, care must be taken to ensure that the remaining photon count is large enough to exceed the noise floor of the detection device in order to result in an appropriate signal-to-noise ratio sufficient for further image processing.

The photon budget of fluorophores is generally limited, so every effort needs to be taken to ensure efficient excitation and collection of the emitted photons, as both affect the signal

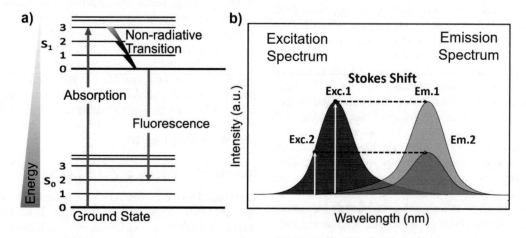

FIGURE 1.1 (a) Joblonski diagram of quantum electronic behavior of electrons, indicating the various stages of the fluorescence cycle. (b) Illustration of excitation and emission spectra. Excitation at the peak of the absorption spectra (Exc.1) leads to optimal emission (Em.1), excitations away from the peak (Exc.2) result in correspondingly less emission intensity. Note that the spectra display a mirror-like symmetry.

quality (i.e. signal to background ratio). To illustrate these effects, let us examine how signal quality is dependent on excitation intensity. Here, the rates of excitation and fluorescence emission are important, in particular their ratio (Panula, 2003):

$$N = k_{ex} / \left(k_{ex} + k_r \right)$$

When the excitation and fluorescence emission rates are equal ($k_{ex} = k_r$), 50% of all fluorophores would be in the excited state at any given time (assuming a quantum efficiency of fluorescence of 1). Increasing the power of the excitation light by, e.g. three-fold, ($k_{ex} = 3 \times k_r$) would yield $N = 0.75 k_r$, or a 50% increase of the fluorescence intensity (from 0.50 to 0.75). However, in determining signal quality, one must also consider any increase in background signals. Assuming a realistic case in which background is comprised of $B = n_c + n_B \times k_{ex}$, i.e. a constant noise term n_c (e.g. 0.1) and an excitation dependent background term n_B (e.g. 0.1), increasing the excitation three-fold would yield a two-fold increase in background (from 0.20 to 0.40). Hence, the signal to background ratio would effectively drop by 25% (from 2.5 to 1.88), leading to an overall worse signal quality. This effect is further illustrated in Figure 1.2. Moreover, the sample would be subject to unnecessary exposure of excitation light, which in turn may lead to photobleaching.

The main cause of photobleaching in practice is the over-excitation and thus saturation of the fluorophores (Widengren et al., 2007), leading to an exponential decay of fluorescence intensity with time. Coming back to our above calculation, let us assume 2,520 fluorescent molecules in our focal volume together with a radiative relaxation rate of ~$1 \times 10^7 \ s^{-1}$. Because most endogenous fluorophores, such as GFP, are limited to a maximum average number of fluorescence cycle of 10^4 before photobleaching, all fluorescence molecules inside our focus would photobleach within approximately 2.5 sec of constant illumination ($2520 \times 10^4 / 10^7$). While this seems short

on first sight, most microscopes scan the illumination light over the sample in one way or another, thus only revisiting the same focal spot for a small fraction of the total imaging time. Therefore, in most *in vivo* imaging settings, the photobleaching time will be considerably longer. Overall, as we have shown, it is important to avoid fluorescence signal saturation and photobleaching, as both negatively affect the achievable imaging time and signal quality.

Fluorescence Lifetime: The fluorescence lifetime, i.e. average duration a molecule spends in the excited state before it experiences radiative relaxation, is of critical importance because it defines the shortest time scale in which dynamic events can be imaged. The fluorescence lifetime (τ) of a fluorophore is described theoretically as

$$\tau = \frac{\Phi}{k_r} = \frac{1}{k_{dex}}.$$

Experimentally, the fluorescence lifetime obeys an exponential decay of the form $I(t) = I_0 e^{-t/\tau}$, where τ is the time until the emission has fallen from the initial intensity, $I(0) = I_0$, down to $I(\tau) = I_0 / e$. For many fluorophores, τ is in the range of 1–10 nsec. Yet, because every fluorophore often possesses a unique and different decay constant, this can be used to separate fluorophores with an otherwise identical or strongly overlapping emission spectrum (for details, see Chapter 10: Fluorescence Lifetime). One of the widely used techniques to measure the fluorescence lifetime is through "time-correlated single photon counting (TCSPC)" (Pian et al., 2017), which measures the distribution of photon arrival times after a short (<psec) excitation pulse. The lifetime can be found by fitting an exponential decay curve to the photon distribution. An alternative and often faster way to calculate the lifetime is through phasor analysis (Chen et al., 2017), which involves transforming the photon arrival distribution to the Fourier domain in which unmixing complex spectra of overlapping fluorophores is computationally more straightforward.

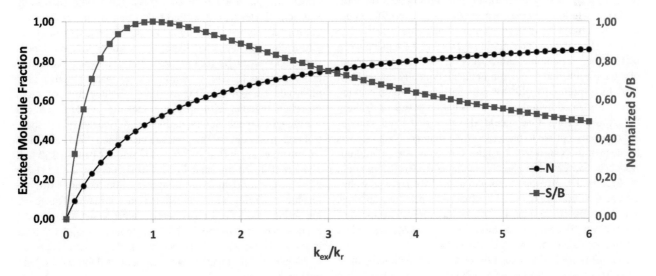

FIGURE 1.2 The ratio between the rates of excitation and fluorescence emission is plotted against the fraction of excited molecule and signal-to-background ratio (S/B). The signal-to-background ratio has its optimum value when half of the total molecules are excited. Further increase in excited molecules (through increasing the excitation power), effectively reduces the signal quality. Here for the calculating the background, $n_c = n_B = 0.1$ was assumed.

1.2.2 Effects of the Host Environment on Fluorescent Molecules

Several environmental factors can affect the properties of a fluorescent molecule such as pH, electric potential, or the temperature. Although it seems disadvantageous to have such sensitivity toward environmental factors, as we will see further below, those factors can be exploited to understand the cellular dynamics or to discern healthy and diseased tissues.

Another aspect of environmental effects is the natural removal of exogenous nontargeted fluorescence dyes from the hosting body through metabolism. This is called *systemic clearing,* which is a practical constraint in *in vivo* imaging experiments because it dictates the time window during which the fluorophores can actually be imaged before they disappear. Here, two contradicting aspects must be balanced. On the one hand, the foreign molecule should be removable from the body as soon as possible to prevent toxic effects. On the other hand, the clearing process should allow enough time for the imaging experiment. In clinical environments or translational experiments, fluorescent agents are often evaluated by their accumulation and clearing time, e.g. in tumor imaging (Ozturk et al., 2014) or photodynamic therapy (Escobedo et al., 2010).

1.2.3 Effects of Fluorescent Molecules on Host Environment

The most deleterious effect of fluorophores, particularly exogenous ones, is toxicity and other related photodamage mechanisms. Those mechanisms impose certain restrictions on the excitation power and illumination duration to maintain the sample integrity and fluorescence signal level. In the following, different mechanisms leading to photodamage will be discussed, namely photobleaching, phototoxicity, and photothermal effects.

Photobleaching is an irreversible, photon-induced chemical process that causes the molecule to go into a nonfluorescent state, typically a long-lived triplet state. From there, the molecule can interact with the environment as well as other molecules, which can result in covalent modifications. The average number of excitation and emission cycles that the molecule can undergo before photobleaching occurs depends on the structure of the molecule and its environment, but typically ranges from $\sim 10^4$ for proteins to $\sim 10^8$ for nanoparticles. Photobleaching can be mitigated by reducing the excitation power or limiting the exposure time. One should also note that fixed samples are more susceptible to photobleaching than aqueous samples because of the motility of molecules. A technique called fluorescence recovery after photobleaching (FRAP) (Erami et al., 2016) is used to investigate molecular mobility. Here, any photobleached fluorophores will be replaced by new fluorophores, and the fluorescence intensity will be recovered. The speed of this recovery depends on the mobility of the molecules and its environmental conditions (e.g. diffusion).

Phototoxicity is the generation of harmful chemicals, most notably reactive oxygen species (ROS) that are created as a result of laser radiation on fluorescent or nonfluorescent molecules (Débarre et al., 2014; Boas et al., 2011). If the perturbation is large enough, it will exceed the repair mechanisms of the cell, and toxic effects, such as increased intracellular calcium levels, membrane destruction, and eventually apoptotic cell death, will take place. In the presence of fluorescent molecules, ROS damage the molecular structure of the fluorophore and cause irreversible *photobleaching.* In general, these effects depend on the proximity as well as concentration and diffusion of ROS (Fischer et al., 2011). In *photodynamic therapy* (PDT), these mechanisms are used to their advantage to clear the tissue from malignant, e.g. cancer, cells by administering specialized therapy agents (e.g. ALA-PPIX, HPPH) (Ormond and Freeman, 2013). Phototoxic effects can be mitigated by keeping excitation light intensities low, either by reducing the power (continuous-wave laser) or repetition rate (pulsed laser source). Furthermore, the choice and concentration of the dye or fluorophore and its localization inside the cell are key factors that influence the severity of the photodamage. Finally, photodamage can also stem from nonchemical effects (heat, thermoelastic stress, ionization, etc.) that do not originate from fluorophores. For example, heating through one-photon absorption of the excitation light by the tissue can cause thermal damage.

1.2.4 Fluorophore Types

Fluorescent agents are commercially available in different forms and often highly specialized based on the desired application. In the following, we focus on fluorescent markers that are suitable for *in vivo* imaging; however, for further details, we refer the reader to the extensive documentation (*Molecular Probes Handbook, A Guide to Fluorescent Probes and Labeling Technologies,* Life Technologies, 2010). Two critical parameters to consider for fluorescence imaging *in vivo,* i.e. in deeper tissue, are scattering and absorption. Scattering is a phenomenon that induces deviations from straight photon because of localized nonuniformities in the medium – eventually, this results in blurring of the acquired image when wide-field detectors, such as cameras, are used. Scattering effects are generally reduced exponentially with increasing wavelength. Hence, red-shifted light offers deeper penetration compared to blue-shifted light in tissue (Ntziachristos, 2010). Absorption, on the other hand, has a nonmonotonic dependence on wavelength (see Figure 1.3). In biological tissue, the absorption coefficient is dominated by water (Figure 1.3a) alongside hemoglobin (Figure 1.3b), as they are the main absorbers together with lipids and melanin. According to the wavelength dependent attenuation length, four favorable spectral regions for optical imaging can be defined (Shi et al., 2016; Tromberg et al., 2000): I) 700–1,000 nm, II) 1,100–1,350 nm, III) 1,600–1,870 nm, and IV) 2,100–2,300 nm. I and II are used mostly for deep-brain imaging, while the latter two are recently emerging with compatible light sources and detectors becoming available. Consequently, fluorophores that can be excited and emit in the NIR have become a major focus of research (Hong et al., 2014), as they enable deeper penetration of excitation light together with less loss of fluorescence light emanating from the tissue.

Fluorophores in tissue can be classified depending on whether they are naturally occurring inside the tissue (*endogenous*) or whether they have to be administered externally

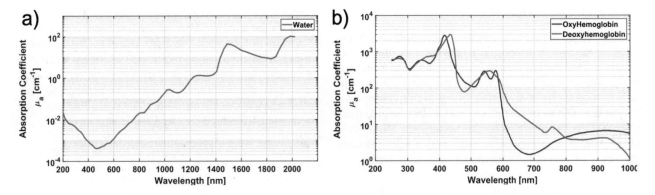

FIGURE 1.3 (a) Absorption coefficient of water within 200–2,000 nm. (b) Absorption coefficient of oxyhemoglobin and deoxyhemoglobin within 200–1,000 nm.

(*exogenous*). Depending on the application, endogenous fluorophores may be desired, e.g. when measuring metabolic activity; however, it can also be a background source of fluorescence against the actual signal, particularly when recording light from exogenous fluorophores. Naturally, care has to be taken in the latter case, especially in regard to choosing a suitable exogenous fluorophore in order to avoid overlap with any endogenous fluorophores or autofluorescence, which mostly resides on the blue side of the spectrum, 300–450 nm for absorption and 300–550 nm for emission. Consequently, appropriate filters need to be employed to prevent contamination of the desired signal.

1.2.5 Endogenous Fluorophores (EnF)

The interest in endogenous fluorophores dates back to the early twentie century due to the observed correlation between autofluorescence signal changes and the progression of malignant diseases (Policard, 1924). Since then, fluorescent proteins became a focus of research for both structural and metabolic imaging. Cellular metabolism and intracellular redox states can be visualized with nicotinamide dinucleotide and flavin adenine dinucleotide (NAD, FAD). They can be found in oxidized or reduced states (e.g. NAD$^+$, NADH, respectively) and relate to energy consumption within cells (Ying, 2008). NADP on the other hand plays a key role in biosynthetic pathways. The FAD/NADH fluorescence ratio can be used to quantify the metabolic activity of cells (redox ratio). A decrease in redox ratio indicates an increase in metabolic activity, which is one of the characteristics of cancer cells (Valeur Bernard and Berberan-Santos, 2013). NADH and NADPH have similar fluorescence excitation and emission spectra, 340 $\mp$ 30 nm and 460 $\mp$ 50 nm, respectively. The overlapping spectrum makes it difficult to differentiate the origin of the signals, while their distinct fluorescence lifetime facilitates separation by fluorescence lifetime imaging (Blacker et al., 2014).

Structural proteins, collagen, and elastin, which are the main components of connective tissue in animal bodies, are another source of autofluorescence. The orientation of collagen is critical for understanding the structural integrity of the tissue (Balu and Tromberg, 2015). Here, second harmonic generation (SHG), which is excited by multiphoton absorption (see Section 3.3), offers a label-free visualization of the collagen matrix. It

has been shown that autofluorescence signal actually decreases in collagen because of the remodeling of stroma during carcinogenesis in the cervix (Sokolov et al., 2002). Another study reported autofluorescence intensity variations depending on the disease, e.g. where a benign inflammation shows a decrease and a dysplasia shows an increase in autofluorescence in comparison to a healthy tissue (Pavlova et al., 2008).

1.2.6 Exogenous Fluorophores (ExF)

Exogenous fluorophores (ExF) are the workhorse of modern biology, as they enable one to visualize and monitor a wide range of inter- and intracellular activities, up to the domain of whole-body imaging of small animals and clinical studies (Panula, 2003). Live-tissue imaging often requires the use of multiple fluorophores in order to visualize different structures and functionalities at the same time. In this situation, having a large library of fluorophores available is essential to avoid spectral overlap. ExF constitute a wide range of contrast agents, from organic dyes to genetically encoded proteins (gene-reporters) (Ozturk et al., 2013) and to nanostructures (nanocrystals/quantum dots, carbon nanotubes, etc.).

1.2.6.1 Fluorescent Dyes

Fluorescent dyes offer a wide spectral coverage, from UV to VIS (400–650 nm) to NIR (700–1,000 nm) regime. These dyes include cyanines, fluoresceins, and rhodamines as well as the widely used Alexa and Atto dye families. Those dyes can be either used for nontargeted imaging or for targeted imaging by labeling biomolecules. To date, only a handful of dyes were approved for clinical use in humans (Nguyen and Tsien, 2013). Indocyanine green (ICG, Ex/Em: 788/813), fluorescein sodium (Ex/Em: 460/515), and methylene blue (Ex/Em: 660/692) (Vansevičiūtė et al., 2015) are the best representatives of this group because of their wide use and minimal phototoxicity. Especially fluorophores with spectra close to the NIR are advantageous for avoiding autofluorescence contamination, which was featured in an *ex vivo* study (Golijanin et al., 2016) in which ICG identified carcinoma in organs with high specificity and sensitivity. Similarly, in an *in vivo* study, Methylene blue delineated nuclei and gland formation within rectal mucosae via gastrointestinal endoscopy (Ichimasa et al., 2014).

1.2.6.2 Quantum Dots

Quantum dots (QD) are semiconductor nanoparticles that have emerged recently in fluorescence imaging. QD offer broad absorption spectra and distinct, narrow-band emission, that are tunable by changing the size of the nanoparticle (typically 2–10 nm) (Jin et al., 2008). Owing to this unique optical property, QD make multiplexed (multicolor) imaging more practical than its organic/inorganic counterparts. Multiplexed imaging with conventional fluorophores usually requires individual excitation light source for each fluorophore, and because many fluorophores possess broad emission spectrums, they often overlap strongly with each other. The unique optical properties of QD stem from the toxic semiconductor core that is composed of e.g. cadmium selenide, lead selenide, or indium arsenide. Thus, the core has to be passivated to be applicable in biological studies (Walling, Novak, and Shepard, 2009). Another unique property is the large area to volume ratio, which carries the potential for multiple functionalization and multimodal imaging such as fluorescence-positron emission tomography and fluorescence-magnetic resonance imaging (Michalet et al., 2005). However, phototoxic effects still limit the usefulness of many nanoparticles for *in vivo* imaging. For example, cadmium-based QDs release free cadmium, which in turn reacts with common biological oxidants that are present in many inflammatory tissues (Hilderbrand and Weissleder, 2010). Still, this is a highly active research area, and future work will show how to best mitigate these detrimental effects (Duffner et al., 2005; Michalet et al., 2005).

1.2.6.3 Fluorescence Indicators

E×F can also be functionalized to bind in a highly specific and quantifiable manner to a given target, while their spectra can be chosen such as to enable separation of different targets at the same time. Often, indicators demonstrate a change in their characteristic features, i.e. their lifetime, spectral response, or intensity, when bound to the target. Another class of fluorophores, which is co-localized with their target, are called *labels* or *trackers*. Like fluorescent indicators, they, too, experience a change in their characteristic parameters, such as a shift in the spectrum, change in the intensity, or change in the fluorescence lifetime.

pH indicators bring forth image contrast because of a pH change in their environment. One exemplary indicator (a modified fluorescein, bis-carboxyethyl-carboxyfluorescein [BCECF]) has both pH-sensitive and pH-insensitive parts, which enables ratiometric imaging. Others, such as 3,6-dihydroxy phthalonitrile and various naphthofluorescein derivatives, yield a pH dependent shift in their emission spectra (Urano et al., 2009). Recent research is actively pushing the emission spectrum toward the NIR for deep-tissue imaging (Myochin et al., 2011).

Calcium (Ca^{+2}) has a vital role in neurotransmission, muscle contraction, and many enzymatic reactions. Hence, Ca^{+2} indicators are used in various *in vivo* imaging studies, such as visualizing neuronal activity (Prevedel et al., 2014), heart imaging (Tallini et al., 2006), and stimulation by temperature (Ermakova et al., 2017). Recent studies reported red-shifted chimera of Ca^{+2} indicators (Zhao et al., 2011), which are essential for studies requiring large penetration depth, e.g. visualizing activity in deeper brain regions, such as the hippocampus.

Fatty acids are an important source of energy when they are metabolized; therefore, their indicators deliver important clues for cell metabolic activity, which have implications ranging from cancer cell characteristics (Yu et al., 2014; Kowada, Maeda, and Kikuchi, 2015) to diabetic studies (Wen et al., 2011).

Voltage is a direct measurement of cellular electric potential that is enabled by voltage indicators. Genetically encoded voltage indicators are proteins that target the membrane and whose fluorescent signal change depending on voltage potential. Widely used proteins are ArcLight (Cao et al., 2013), ASAP1 (St-Pierre et al., 2014), and ASAP2f (Yang et al., 2016), with new indicators constantly emerging (Gong et al., 2015). Applications mostly focus on neuronal activity recordings, as the flow of electrical signals *in vivo* deliver critical insight for the neuronal information processing and complex function of neuronal networks.

To summarize, readily available fluorescent molecules are serving the ever-expanding variety of purposes in imaging, ranging from imaging structural features to metabolic activities and to targeted high-specificity imaging. Although many currently available fluorophores meet those requirements, translating their use into *in vivo* settings often face additional challenges, especially in regard to potential phototoxic effects. In the future, designing fluorophores that operate in the NIR optical window while minimizing their toxicity to an acceptable level will be of high importance, but the lengthy and complex nature of fluorophore development will likely take time to realize.

1.3 Principles of Microscopy

Now that we have covered the underlying physical principles of fluorescence and given an overview of commonly used fluorophores and their use for *in vivo* imaging, let us now turn to the instrumentation that enables fluorescence microscopy. We will first lay the foundation by discussing the basic optical principle of a microscope and parameters that influence its performance. Then, we will introduce standard microscopy techniques that are suitable for *in vivo* imaging. Here, we chose to restrict our technical review to major techniques, such as confocal, light sheet, and multiphoton microscopy.

1.3.1 Image Formation and Magnification

The simplest way to form an image is to use a thin lens. If an object is placed at distance S_0 away from the lens, then an image is formed at distance S_1 at the back of the lens (see Figure 1.4). The location where the image is formed can vary between infinity and the focal distance (f) of the lens, depending on the object distance (S_0). This relation can be written as:

$$\frac{1}{S_0} + \frac{1}{S_1} = \frac{1}{f}$$

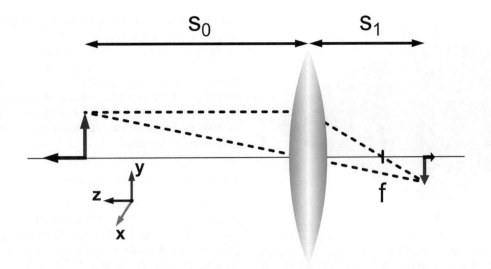

FIGURE 1.4 Thin lens imaging schematics, showing the magnification on x and y (lateral) direction as well as z (axial) direction. Here f denotes the focal length of the lens.

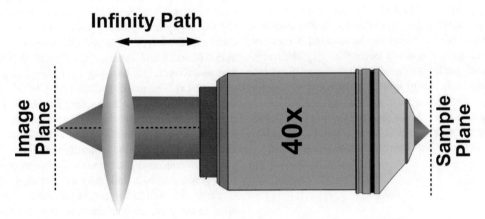

FIGURE 1.5 A standard microscope comprises of an objective lens and a tube lens. The designed magnification can be realized by placing the appropriate tube lens in tandem with the objective lens.

while the effective magnification in the lateral (x–y) plane (M_L) as well as along the axial (z) direction (M_A) can be defined with the following equations:

$$M_L = -\frac{S_1}{S_0}$$

$$M_A = -\left[\frac{S_1}{S_0}\right]^2 = -M^2$$

The negative sign here points to the fact that the image is inverted in respect to the object plane. Standard microscopes these days typically consist of an *objective* lens in combination with an imaging lens (*tube lens*) (see Figure 1.5). The spacing between the lenses is commonly referred to as the *infinity* path, where all the light from the sample is collimated. This allows the convenient introduction of various other optics (beamsplitters, filters, etc.). In order to realize the desired magnification of the objective (e.g. 40×), a tube lens of certain focal length has to be used, depending on the manufacturer of the objective (e.g. 165 mm for Zeiss, 200 mm for Nikon, Olympus, etc.).

1.3.2 Numerical Aperture and Spatial Resolution

The *numerical aperture* (NA) of a lens quantifies the light collection and focusing ability of a lens and thus has a direct effect on the resolution of the system. The NA is characterized by the angular span of the light collection cone (θ) with respect to the focal point, f, (Figure 1.6) and is defined as $NA = n\sin\theta$, where n is the refractive index of the medium that the lens is in contact with.

Spatial resolution in microscopy refers to the ability of the imaging modality to differentiate two objects close to each other, i.e. the minimum distance of objects in the lateral as well as axial dimension. Because of diffraction, even an infinitely small, point-like object will yield an object of a finite size at the image plane of a perfect microscope. The spatial pattern of such an object is the so-called *Airy* disk, whose characteristic size is defined as the radius of the first dark ring of this Airy disk in the focal plane:

$$r_{airy} = 0.61\frac{\lambda}{NA_{obj}}$$

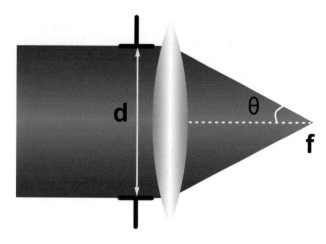

FIGURE 1.6 Numerical aperture (NA) characterizes the range of angles θ over which the system can accept or emit light and is closely related to the resolution of an imaging system. Here d is the aperture size, and f denotes the focal length of the lens.

where NA_{obj} is the numerical aperture of the objective and λ is the wavelength of the light. According to the Rayleigh criterion, two equally bright objects can be resolved if they are separated by a distance d equal or larger than r_{airy}. Naturally, this definition only holds for a perfectly aligned optical system free from aberration. In reality, optical aberrations can slightly deteriorate the spatial resolution of an imaging system.

Similar considerations hold for the axial resolution, which is measured along the optical axis (z-axis). Here, the minimum distance between two objects that can still be separated is:

$$z_{min} = \frac{2\lambda n}{NA_{obj}^2}$$

where n is the refractive index of the medium. It is important to note that the axial resolution is inversely proportional to the square of the numerical aperture of the system and thus results in substantially lower value compared to the lateral resolution. Figure 1.7 shows these dependencies in plots of the so-called point spread functions (PSF), i.e. the response of an imaging system to a point-like object.

1.3.3 Fluorescence Microscopy Techniques for *In Vivo* Studies

In vivo samples bring several challenges to the domain of microscopy. On the optical side, this is exemplified by nontransparent and scattering tissue types, and different techniques were developed to address these issues. Here, we will give a brief overview of standard techniques more suitable for thin and minimally scattering tissues before introducing multiphoton microscopy techniques that are advantageous for deep-tissue imaging.

1.3.3.1 Deconvolution

Deconvolution is a computational method that achieves superior contrast by removing deleterious contributions from out-of-focus light. Because an image is formed through the convolution of the signal in the sample with the PSF of the imaging system, knowledge of the PSF enables deconvolution, i.e. to reverse the convolution process, and thus to computationally remove the out-of-focus light. The PSF can be derived either experimentally or theoretically by knowing the optical parameters of the microscope. Deconvolution delivers a much clearer (crisper) 3D image. It is, in fact, more of a tool than a microscopy technique and can easily be integrated with different imaging modalities

1.3.3.2 Confocal Microscopy

A confocal microscopy (CM), first patented by Marvin Minsky in 1957 (Minsky, 1961), utilizes a pinhole to physically reject any out-of-focus light from hitting the detector (Figure 1.8a). It features superb optical resolution, particularly, in the axial direction and, therefore, has found extensive use in the scientific community. In the context of *in vivo* imaging, it is mostly appreciated for its optical sectioning capability, i.e. to generate optical slices (in the x–y plane) of an intact tissue analogous to histology but without sectioning the tissue physically. In the past, CM has been employed for different tissue types and biological questions ranging from skin (González and Tannou, 2002) to eye (Jalbert et al., 2003) and even intravital imaging of internal organs (Marques et al., 2015). The main limitation of CM is that it is limited to the so-called ballistic regime, in which the majority of the light remains unscattered. Typically, this is only tens of microns deep in mammalian tissue (Smithpeter et al., 1998). With CM being a point scanning technique, it can become very time-consuming to acquire 3D images of large fields-of-view (Figure 1.8b–c) and bears a high risk of photobleaching (Figure 1.8d). While faster CM techniques have been developed in the past, such as the *spinning disk* CM (Quinn and DeLeo, 2014; Patel, Rajadhyaksha, and Dimarzio, 2011), the main limitation for *in vivo* imaging studies still remains in the limited-depth penetration.

1.3.3.3 Light Sheet Microscopy

Light sheet microscopy (LSM) utilizes an orthogonal geometry of illumination and detection planes (Figure 1.8e). Typically, in LSM, a thin light sheet (thickness of a few micrometers) is generated that enables optical sectioning (Figure 1.8f). Subsequently, a wide-field detection, e.g. using a camera, at an orthogonal direction enables highly multiplexed and, therefore, much faster (100–1,000 times) imaging in comparison to CM (Figure 1.8g). The "selected plane illumination" feature of LSM drastically reduces the volume under potential risk of photobleaching, which is another important advantage of LSM compared to CM (Figure 1.8h). Mounting of specimens, such as a zebrafish embryo, drosophila larva, or mouse embryo, is often nontrivial because of the unusual geometry and the lack of sample space. Often samples are placed in gels, such as agarose, for stabilization. Recently, more advanced versions of LSM have appeared that allow illuminating and imaging the sample from multiple views, thus improving contrast and resolution (de Medeiros et al., 2015).

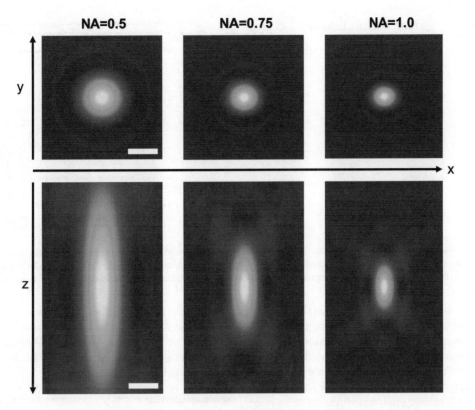

FIGURE 1.7 Cross-sectional views from different simulated point spread functions (PSF). The PSF shows a stronger dependence on numerical aperture (NA) on the axial (bottom panel) compared to the lateral (top panel) dimensions. Scale bar = 0.5 µm.

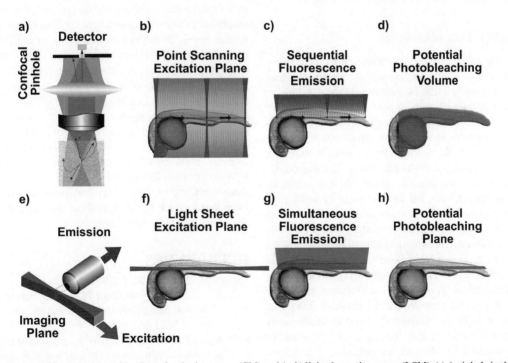

FIGURE 1.8 Basic working principles of (a–d) confocal microscopy (CM) and (e–h) light sheet microscopy (LSM). (a) A pinhole in the image plane of CM enables rejection of out of focus light, thus achieving high (axial) resolution and contrast. (b) In conventional CM, the sample is excited point by point and (c) fluorescence emission is collected sequentially. (d) This configuration is prone to photobleaching as the excitation light illuminates the sample along the entire light cone. (e) An orthogonal configuration of the illumination and detection path minimizes the generation of out-of-focus light, thus maximizing signal to background ratio. (f) A sheet of light excites only the region of interest from which fluorescence is later collected. (g) Simultaneous detection of fluorescence is enabled by wide-field detectors such as cameras. (h) Because of the confined excitation, the potential photobleaching area is minimized.

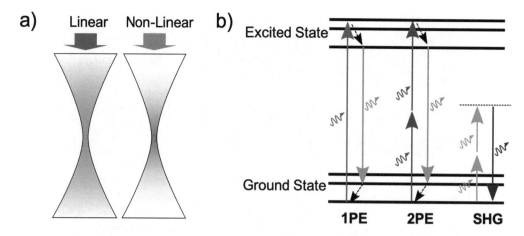

FIGURE 1.9   (a) Nonlinear excitation leads to a more confined region of fluorescence generation. In one-photon-based imaging methods, such as confocal imaging, fluorescence is excited along the entire cone of the focus, while in nonlinear microscopy the higher order dependence leads to an effectively smaller excitation spot size. (b) Schematic Jablonski diagram, showing two-photon absorption (2PA) as well as second-harmonic generation (SHG).

Because LSM is based on wide-field detection (i.e. scientific cameras), it works best in highly transparent and little-scattering objects (i.e. naturally transparent or chemically cleared). Hence, *in vivo* imaging applications have focused on samples such as mouse, zebrafish, and drosophila embryos with an emphasis on embryonic development (Keller et al., 2008). Chemical clearing is widely used, e.g. structural brain imaging, in which LSM is used to achieve a highly detailed map of neuronal wiring *ex vivo* (Ahrens et al., 2013).

However, scattering remains a major source of contamination for the fluorescence signal for deep-tissue imaging.

1.3.3.4 Multiphoton Microscopy

All of the techniques mentioned above rely on ballistic propagation of light, and thus light scattering is detrimental for those modalities. This severely limits the usefulness of these techniques to thin sections or cultured cells *in vitro* and superficial layers of tissue *in vivo*. Multiphoton microscopy (MPM), on the other hand, relies on a nonlinear light-matter interaction to generate the fluorescence signal, thus making it less sensitive to scattering. Over the past three decades, such nonlinear optical microscopy techniques have been continuously refined and engineered for improved performance. Since the seminal paper by Denk et al. (1990), two-photon microscopy in combination with *in vivo* labeling strategies have led to a rapidly expanding field with vastly different applications in live-tissue and animal imaging. Typical applications include mouse brain (Kerr and Denk, 2008), skin (Laiho et al., 2005), heart (Vinegoni et al., 2015), kidney (Dunn, Sutton, and Sandoval, 2012), and lymph nodes (Liu et al., 2015).

The fact that more than one photon contributes to the signal generation in multiphoton microscopy leads to qualitatively new properties because of the "higher order" light-matter interaction, in which the fluorophore excitation depends on the square (or cube, etc.) of the incident light intensity. This effectively results in a tighter confinement of the signal (fluorescence) generation and practically leads to focal volumes on the order of femtoliters close to the optical diffraction limit (see Figure 1.9a). In order to excite a molecule, two (or more) photons need to impinge near simultaneously (within ~0.5fs;

see Figure 1.9b); therefore, the efficiency of multiphoton absorptions depends crucially on the physical properties of the molecule as well as the spatial and temporal density of the excitation light. Further details on multiphoton microscopy are discussed in Chapter 22 of this book.

1.4 Summary

In this chapter, we have introduced the reader to the basics and principles of fluorescence imaging, putting particular emphasis on the underlying photophysical properties of common fluorophores as well as different types of fluorophores used in life science today. To aid in the discussions, example calculations regarding typical fluorescence emission and excitation power considerations in fluorescence imaging were included, as they form a critical aspect of *in vivo* imaging, affecting both image quality and sample viability. Furthermore, major concepts of fluorescence microscopy techniques were discussed, including image formation and spatial resolution, and common imaging technologies that are compatible with imaging in small, living animals, such as confocal, light sheet, and multiphoton microscopy were introduced. We would also like to refer the readers to Chapters 20 and 22 of this book, which will focus in more detail on the use of light sheet and multiphoton microscopy for *in vivo* imaging, respectively.

REFERENCES

Ahrens, M. B., Orger, M. B., Robson, D. N., Li, J. M., and Keller, P. J. (2013). Whole-brain functional imaging at cellular resolution using light-sheet microscopy. *Nature Methods*, *10*(5): 413–20. doi: 10.1038/nmeth.2434.

Balu, M., and Tromberg, B. J. (2015). Multiphoton microscopy for non-invasive optical biopsy of human skin. *Romanian Journal of Clinical and Experimental Dermatology*, *2*(3): 160–66.

Blacker, T. S., Mann, Z. F., Gale, J. E., Ziegler, M., Bain, A. J., Szabadkai, G., and Duchen, M. R. (2014). Separating NADH and NADPH fluorescence in live cells and tissues using FLIM. *Nature Communications*, *5*(May): 1–9. doi: 10.1038/ncomms4936.

Boas, D. A., Pitris, C., and Ramanujam, N. (Eds.) (2011). *Handbook of biomedical optics*. Boca Raton, Florida: CRC Press.

Cao, G., Platisa, J., Pieribone, V. A., Raccuglia, D., Kunst, M., and Nitabach, M. N. (2013). Genetically targeted optical electrophysiology in intact neural circuits. *Cell*, *154*(4): 904–13. doi: 10.1016/j.cell.2013.07.027.

Chen, S., Sinsuebphon, N., Barroso, M., Intes, X., and Michalet, X. (2017). AlliGator: A phasor computational platform for fast in vivo lifetime analysis. *Optics in the Life Sciences Congress*, OmTu2D.2. Optical Society of America. doi: 10.1364/OMP.2017.OmTu2D.2.

Débarre, D., Olivier, N., Supatto, W., and Beaurepaire, E. (2014). Mitigating phototoxicity during multiphoton microscopy of live drosophila embryos in the 1.0–1.2 μm wavelength range. *PLOS ONE*, *9*(8): e104250. doi: 10.1371/journal.pone.0104250.

Denk, W., Strickler, J. H., and Webb, W. W. (1990). Two-photon laser scanning fluorescence microscopy. *Science*, *248*(4951): 73–6.

Duffner, F., Ritz, R., Freudenstein, D., Weller, M., Dietz, K., and Wessels, J. (2005). Specific intensity imaging for glioblastoma and neural cell cultures with 5-aminolevulinic acid-derived protoporphyrin IX. *Journal of Neuro-Oncology*, *71*(2): 107–11. doi: 10.1007/s11060-004-9603-2.

Dunn, K. W., Sutton, T. A., and Sandoval, R. M. (2012). Live-animal imaging of renal function by multiphoton microscopy. *Current Protocols in Cytometry*, *347*: 12.9.1–12.9.18. Hoboken, NJ: John Wiley & Sons, Inc. doi: 10.1002/0471142956.cy1209s62.

Erami, Z., Herrmann, D., Warren, S. C., Nobis, M., McGhee, E. J., Lucas, M. C., Leung, W., et al. (2016). Intravital FRAP imaging using an E-Cadherin-GFP mouse reveals disease- and drug-dependent dynamic regulation of cell-cell junctions in live tissue. *Cell Reports*, *14*(1): 152–67. doi: 10.1016/j.celrep.2015.12.020.

Ermakova, Y. G., Lanin, A. A., Fedotov, I. V., Roshchin, M., Kelmanson, I. V., Kulik, D., Bogdanova, Y. A., et al. (2017). Thermogenetic neurostimulation with single-cell resolution. *Nature Communications*, *8*(May): 15362. doi: 10.1038/ncomms15362.

Escobedo, J. O., Rusin, O., Lim, S., and Strongin, R. M. (2010). NIR dyes for bioimaging applications. *Current Opinion in Chemical Biology*, *14*(1): 64–70. doi: 10.1016/j.cbpa.2009.10.022.

Fischer, R. S., Wu, Y., Kanchanawong, P., Shroff, H., and Waterman, C. M. (2011). Microscopy in 3D: A biologist's toolbox. *Trends in Cell Biology*, *21*(12): 682–91. doi: 10.1016/j.tcb.2011.09.008.

Golijanin, J., Amin, A., Moshnikova, A., Brito, J. M., Tran, T. Y., Adochite, R. C., Andreev, G. O., et al. (2016). Targeted imaging of urothelium carcinoma in human bladders by an ICG PHLIP peptide ex vivo. *Proceedings of the National Academy of Sciences of the United States of America*, *113*(42): 11829–34. doi: 10.1073/pnas.1610472113.

Gong, Y., Huang, C., Li, J. Z., Grewe, B. F., Zhang, Y., Eismann, S., and Schnitzer, M. J. (2015). High-speed recording of neural spikes in awake mice and flies with a fluorescent voltage sensor. *Science*, *350*(6266): 1361–66. doi: 10.1017/CBO9781107415324.004.

González, S., and Tannous, Z. (2002). Real-time, in vivo confocal reflectance microscopy of basal cell carcinoma. *Journal of the American Academy of Dermatology*, *47*(6): 869–74. doi: 10.1067/mjd.2002.124690.

Hilderbrand, S. A., and Weissleder, R. (2010). Near-infrared fluorescence: Application to in vivo molecular imaging. *Current Opinion in Chemical Biology*, *14*(1): 71–9. doi: 10.1016/j.cbpa.2009.09.029.

Hong, G., Diao, S., Chang, J., Antaris, A. L., Chen, C., Zhang, B., Zhao, S., et al. (2014). Through-skull fluorescence imaging of the brain in a new near-infrared window. *Nature Photonics*, *8*(9): 723–30. doi: 10.1038/nphoton.2014.166.

Ichimasa, K., Ei Kudo, S., Mori, Y., Wakamura, K., Ikehara, N., Kutsukawa, M., Takeda, K., et al. (2014). Double staining with crystal violet and methylene blue is appropriate for colonic endocytoscopy: An in vivo prospective pilot study. *Digestive Endoscopy*, *26*(3): 403–8. doi: 10.1111/den.12164.

Jalbert, I., Stapleton, F., Papas, E., Sweeney, D. F., and Coroneo, M. (2003). In vivo confocal microscopy of the human cornea. *The British Journal of Ophthalmology*, *87*(2): 225–36. doi: 10.1136/bjo.87.2.225.

Jin, T., Fujii, F., Komai, Y., Seki, J., Seiyama, A., and Yoshioka, Y. (2008). Preparation and characterization of highly fluorescent, glutathione-coated near infrared quantum dots for in vivo fluorescence imaging. *International Journal of Molecular Sciences*, *9*(10): 2044–61. doi: 10.3390/ijms9102044.

Keller, P. J., Schmidt, A. D., Wittbrodt, J., and Stelzer, E. H. K. (2008). Reconstruction of zebrafish early embryonic development by scanned light sheet microscopy. *Science*, *322*(5904): 1065–69. doi: 10.1126/science.1162493.

Kerr, J. N. D., and Denk, W. (2008). Imaging in vivo: Watching the brain in action. *Nature Reviews. Neuroscience*, *9*(3): 195–205. doi: 10.1038/nrn2338.

Kowada, T., Maeda, H., and Kikuchi, K. (2015). "BODIPY-based probes for the fluorescence imaging of biomolecules in living cells. *Chemical Society Reviews*, *44*(14): 4953–72. doi: 10.1039/C5CS00030K.

Laiho, L. H., Pelet, S., Hancewicz, T. M., Kaplan, P. D., and So, P. T. C. (2005). Two-photon 3-D mapping of ex vivo human skin endogenous fluorescence species based on fluorescence emission spectra. *Journal of Biomedical Optics*, *10*(2): 024016. doi: 10.1117/1.1891370.

Liu, Zhiduo, Gerner M. Y., N. Van Panhuys, A. G. Levine, A. Y. Rudensky, and R. N. Germain. (2015). Immune homeostasis enforced by co-localized effector and regulatory t cells. *Nature*, *528*(7581): 225–30. doi: 10.1038/nature16169.

Marques, P. E., Antunes, M. M., David, B. A., Pereira, R. V., Teixeira, M. M., and Menezes, G. B. (2015). Imaging liver biology in vivo using conventional confocal microscopy. *Nature Protocols*, *10*(2): 258–68. doi: 10.1038/nprot.2015.006.

de Medeiros, G., Norlin, N., Gunther, S., Albert, M., Panavaite, L., Fiuza, U. M., Peri, F., et al. (2015). Confocal multiview light-sheet microscopy. *Nature Communications*, *6*: 8881. doi: 10.1038/ncomms9881.

Michalet, X., Pinaud, F. F., Bentolila, L. A., Tsay, J. M., Doose, S., Li, J. J., Sundaresan, G., et al. (2005). Quantum dots for live cells, in vivo imaging, and diagnostics. *Science (New York, NY)*, *307*(5709): 538–44. doi: 10.1126/science.1104274.

Minsky, M. (1961). Microscopy apparatus, issued 1961. https://www.google.com/patents/US3013467.

Myochin, T., Kiyose, K., Hanaoka, K., Kojima, H., Terai, T., and Nagano, T. (2011). Rational design of ratiometric near-infrared fluorescent PH probes with various PKa values, based on aminocyanine. *Journal of the American Chemical Society, 133*(10): 3401–9. doi: 10.1021/ja1063058.

Nguyen, Q. T., and Tsien, R. Y. (2013). Fluorescence-guided surgery with live molecular navigation—A new cutting edge. *Nature Reviews. Cancer, 13*(9): 653–62. doi: 10.1038/nrc3566.

Ntziachristos, V. (2010). Going deeper than microscopy: The optical imaging frontier in biology. *Nature Methods, 7*(8): 603–14. doi: 10.1038/nmeth.1483.

Ormond, A., and Freeman, H. (2013). Dye sensitizers for photodynamic therapy. *Materials, 6*(3): 817–40. doi: 10.3390/ma6030817.

Ozturk, M. S., Lee, V. K., Zhao, L., Dai, G., and Intes, X. (2013). Mesoscopic fluorescence molecular tomography of reporter genes in bioprinted thick tissue. *Journal of Biomedical Optics, 18*(10): 100501. doi: 10.1117/1.JBO.18.10.100501.

Ozturk, M. S., Rohrbach, D. J., Sunar, U., and Intes, X. (2014). Mesoscopic fluorescence tomography of a photosensitizer (HPPH) 3D biodistribution in skin cancer. *Academic Radiology, 21*(2): 271–80. doi: 10.1016/j.acra.2013.11.009.

Panula, P. A. J. (2003). Handbook of biological confocal microscopy. *Journal of Chemical Neuroanatomy, 25*(3): 228–29. doi: 10.1016/S0891-0618(03)00007-3.

Patel, Y. G., Rajadhyaksha, M., and Dimarzio, C. A. (2011). Optimization of pupil design for point-scanning and line-scanning confocal microscopy. *Biomedical Optics Express, 2*(8): 2231–42. doi: 10.1364/BOE.2.002231.

Pavlova, I., Williams, M., El-Naggar, A., Richards-Kortum, R., and Gillenwater, A. (2008). Understanding the biological basis of autofluorescence imaging for oral cancer detection: High-resolution fluorescence microscopy in viable tissue. *Clinical Cancer Research, 14*(8): 2396–404. doi: 10.1158/1078-0432.CCR-07-1609.

Pian, Q., Yao, R., Sinsuebphon, N., and Intes, X. (2017). Compressive hyperspectral time-resolved wide-field fluorescence lifetime imaging. *Nature Photonics, 11*(7): 411–14. doi: 10.1038/nphoton.2017.82.

Policard, A. (1924). Study on the aspects offered by experimental tumors examined in the light of wood. *CR Social Biology, 91*: 1423–24.

Prevedel, R., Yoon, Y. G., Hoffmann, M., Pak, N., Wetzstein, G., Kato, S., Schrödel, T., et al. (2014). Simultaneous whole-animal 3D imaging of neuronal activity using light-field microscopy. *Nature Methods, 11*(7): 727–30. doi: 10.1038/nmeth.2964.

Quinn, M. T., and DeLeo, F. R. (Eds.) (2014). *Neutrophil Methods and Protocols, 1124. Methods in Molecular Biology.* Totowa, NJ: Humana Press. doi: 10.1007/978-1-62703-845-4.

Sauer, M., Hofkens, J., and Enderlein, J. (2011). *Handbook of fluorescence spectroscopy and imaging.* Weinheim, Germany: Wiley-VCH Verlag GmbH & Co. KGaA. doi: 10.1002/9783527633500.

Shi, L., Sordillo, L. A., Rodríguez-Contreras, A., and Alfano, R. (2016). Transmission in near-infrared optical windows for deep brain imaging. *Journal of Biophotonics, 9*(1–2): 38–43. doi: 10.1002/jbio.201500192.

Smithpeter, C. L., Dunn, A. K., Welch, A. J., and Richards-Kortum, R. (1998). Penetration depth limits of in vivo confocal reflectance imaging. *Applied Optics, 37*(13): 2749–54. doi: 10.1364/ao.37.002749.

Sokolov, K., Galvan, R., Myakov, A., Lacy, A., Lotan, R., and Richards-Kortum, R. (2002). Realistic three-dimensional epithelial tissue phantoms for biomedical optics. *Journal of Biomedical Optics, 7*(1): 148. doi: 10.1117/1.1427052.

St-Pierre, F., Marshall, J. D., Yang, Y., Gong, Y., Schnitzer, M. J., and Lin, M. Z. (2014). High-fidelity optical reporting of neuronal electrical activity with an ultrafast fluorescent voltage sensor. *Nature Neuroscience, 17*(6): 884–89. doi: 10.1038/nn.3709.

Tallini, Y. N., Ohkura, M., Choi, B.-R., Ji, G., Imoto, K., Doran, R., Lee, J., et al. (2006). Imaging cellular signals in the heart in vivo: Cardiac expression of the high-signal Ca2+ indicator GCaMP2. *Proceedings of the National Academy of Sciences of the United States of America, 103*(12): 4753–58. doi: 10.1073/pnas.0509378103.

Tromberg, B. J., Sha, N., Lanning, R., Cerussi, A., Espinoza, J., Pham, T., Svaasand, L., and Butler, J. (2000). Non-invasive in vivo characterization of breast tumors using photon migration spectroscopy. *Neoplasia, 2*(1): 26–40. doi: 10.1038/sj.neo.7900082.

Urano, Y., Asanuma, D., Hama, Y., Koyama, Y., Barrett, T., Kamiya, M., Nagano, T., et al. (2009). Selective molecular imaging of viable cancer cells with PH-activatable fluorescence probes. *Nature Medicine, 15*(1): 104–9. doi: 10.1038/nm.1854.

Valeur, B., and Berberan-Santos, M. N. (2013). *Molecular fluorescence: Principles and applications* (2nd ed.). Wiley-VCH, New York.

Vansevičiūtė, R., Venius, J., Žukovskaja, O., Kanopienė, D., Letautienė, S., and Rotomskis, R. (2015) 5-aminolevulinic-acid-based fluorescence spectroscopy and conventional colposcopy for in vivo detection of cervical pre-malignancy. *BMC Women's Health, 15*(1): 35. doi: 10.1186/s12905-015-0191-4.

Vinegoni, C., Lee, S., Aguirre, A. D., and Weissleder, R. (2015). New techniques for motion-artifact-free in vivo cardiac microscopy. *Frontiers in Physiology, 6*(May): 147. doi: 10.3389/fphys.2015.00147.

Walling, M. A., Novak, J. A., and Shepard, J. R. E. (2009). Quantum dots for live cell and in vivo imaging. *International Journal of Molecular Sciences, 10*(2): 441–91. doi: 10.3390/ijms10020441.

Wang, L. V., and Wu, H. (2007). *Biomedical Optics: Principles and Imaging.* Wiley-Interscience, New York.

Wen, H., Gris, D., Lei, Y., Jha, S., Zhang, L., Huang, M. T. H., Brickey, W. J., and Ting, J. P.-Y. (2011). Fatty acid–induced NLRP3-ASC inflammasome activation interferes with insulin signaling. *Nature Immunology, 12*(5): 408–15. doi: 10.1038/ni.2022.

Widengren, J., Chmyrov, A., Eggeling, C., Löfdahl, P. Å., and Seidel, C. A. M. (2007). Strategies to improve photostabilities in ultrasensitive fluorescence spectroscopy. *Journal of Physical Chemistry. Part A, 111*(3): 429–40. doi: 10.1021/jp0646325.

Wiederschain, G.Y. (2010). *The molecular probes handbook, a guide to fluorescent probes and labeling technologies* (11th ed.). Biochemistry, Moscow: 76: 1276.

Yang, H. H. H., St-Pierre, F., Sun, X., Ding, X., Michael, Z. Z., and Clandinin, T. R. (2016). Subcellular imaging of voltage and calcium signals reveals neural processing in vivo. *Cell*, *166*(1): 245–57. doi: 10.1016/j.cell.2016.05.031.

Ying, W. (2008). NAD+/NADH and NADP /NADPH in cellular functions and cell death: Regulation and biological consequences. *Antioxidants & Redox Signaling*, *10*(2): 179–206. doi: 10.1089/ars.2007.1672.

Yu, S., Levi, L., Casadesus, G., Kunos, G., and Noy, N. (2014). Fatty acid-binding protein 5 (Fabp5) regulates cognitive function both by decreasing anandamide levels and by activating the nuclear receptor peroxisome proliferatoractivated receptor α/β (PPARα/β) in the brain. *Journal of Biological Chemistry*, *289*(18): 12748–58. doi: 10.1074/jbc.M114.559062.

Zhao, Y., Araki, S., Wu, J., Teramoto, T., Chang, Y.-F., Nakano, M., Abdelfattah, A. S., et al. (2011). An expanded palette of genetically encoded Ca2+ indicators. *Science*, *333*(6051): 1888–91. doi: 10.1126/science.1208592.

2

Intravital Microscopy

Mario Perro, Jacky G. Goetz, and Antonio Peixoto

CONTENTS

Our understanding of cell biology has significantly progressed thanks to the development of *in vitro* and *ex vivo* technological approaches. However, even the most sophisticated of these approaches fails to faithfully reproduce the complex cellular and molecular environment of physiological and pathophysiological processes that occur *in vivo*. Hence, the relevance of *in vitro* or *ex vivo* observations needs to be constantly evaluated and refined in living organisms. A key experimental approach to perform this "reality check" is intravital microscopy (IVM) because it allows high resolution visualization of cellular behavior, cell fates, and molecular events in living organisms. The practice of IVM was initiated centuries ago and involved the use of bright-field microscopy in relatively optically transparent tissues (van Leeuwenhoek and Hoole, 1800; Wagner, 1839; Metchnikoff, 1893). Today, IVM is largely based on fluorescence detection and allows *in vivo* imaging of multiple cell types simultaneously in

optically dense tissues over extended periods of time and at high resolutions. Progress was possible because of key developments in the instrumentation, such as laser excitation sources (single and multiphoton), high-performance detectors, chemical and genetic probes for labeling different cell types, and surgical techniques to render deep tissues accessible for IVM. These technological developments have rendered IVM within reach of many researchers and students in a variety of fields of research. Bearing this in mind, this chapter is aimed at researchers or students who are interested in adding IVM to their technological toolbox as a valuable complement to other approaches. For this purpose, we will describe the current state of the art, the challenges, limitations, and future perspectives to push the limits of this technique. Even though our focus will be mainly in IVM for small rodents, the vast majority of the principles described are applicable to other *in vivo* animal models.

2.1 Instrumentation

IVM can be performed using a variety of light microscopy techniques such as wide-field (bright-field or epi-fluorescence), laser scanning confocal, multiphoton, spinning disk and selective single plane illumination microscopy (SPIM). Currently, there is no "one fits all" microscopy technique for IVM. Instead, several criteria, such as optical transparency of the tissue, the spatial and temporal resolution for observation, phototoxicity, and the depth of imaging, need to be taken in consideration in order to choose the most appropriate technique (Figure 2.1). Below, we will describe the main points to consider for the choice of IVM microscopy technique for each specimen/project.

2.1.1 Optically Transparent vs. Opaque Specimens

Both bright-field and SPIM techniques deliver very low peak light intensity dose to the specimen during imaging because they rely on the low-power light sources (bright-field) or a light sheet that excites only a thin optical section of the sample (SPIM) (Figure 2.2). Accordingly, bright-field and SPIM techniques are associated with very low levels of phototoxicity and allow long-term imaging of sensitive specimens. However, these techniques are most suitable for optically transparent specimens (zebrafish and drosophila embryos) because propagation of the light source through the specimen is strictly required for proper detection of the signal. Indeed, opaque structures can act as a barrier to light as it travels across the tissue plane, thus occluding other structures of the specimen. Even though specimen clearing techniques have been developed to circumvent these problems, these are not applicable to IVM because they rely on treatments that are incompatible with cell/tissue viability. In transparent specimens, bright-field microscopy and SPIM are extremely powerful techniques and allow long-term imaging of whole but small organisms (Krzic et al., 2012). In order to perform IVM studies in opaque specimens, other techniques such as epi-fluorescence, laser scanning confocal, multiphoton, and spinning disk are more appropriate. Epi-fluorescence microscopy relies on a low peak light intensity illumination of all focal planes simultaneously

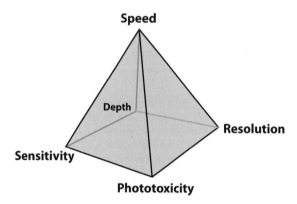

FIGURE 2.1 Intravital microscopy conundrum. The ideal IVM technique does not exist, thus favoring one parameter, for instance speed, will eventually result in poor performance in sensitivity or the size of the 3D volume that is sampled over time. Hence, the choice on the IVM technique will rely on the definition of the parameters that are most important for a given project/scientific question.

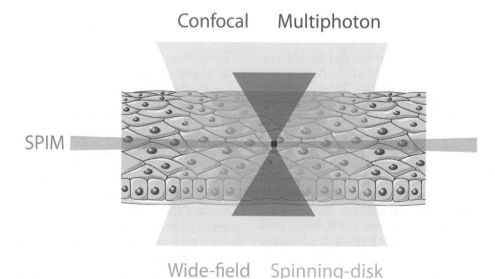

FIGURE 2.2 Light microscopy techniques applied in intravital microscopy. Illustration of the illumination pattern of confocal (red), multiphoton (brown), wide-field (gray), single plane illumination (SPIM) and spinning disk confocal (orange) microscopy of a tissue.

(wide-field setup) and the detection of fluorescence that is reflected through the specimen. Accordingly, fluorescence signals originate from the entire volume of the specimen, which degrades the resolution of details in the objective focal plane. This feature renders this technique inappropriate for thick specimens (>2 μm) that generate a high degree of emitted fluorescence. In contrast, confocal and spinning disk microscopies use high-power light sources (e.g. continuous laser) to illuminate the specimen and detect the reflected fluorescence (Figure 2.2). One of the main advantages of these techniques is that emitted fluorescence outside the focal plane is excluded by the use of a pinhole, thus providing the capacity of optical sectioning of the specimen. Hence, controlled movement in the z-axis allows the acquisition of sequential optical sections and generation of three-dimensional (3D) images of the specimen. Both confocal and spinning disk microscopy deliver high peak light intensity into the specimen that allow imaging of thick specimens (<100 μm). However, spinning disk microscopy is more appropriate for prolonged imaging sessions, as it is less prone to phototoxicity because illumination of the specimen is performed at multiple points simultaneously. Multiphoton microscopy relies on the use of high peak intensity light derived from pulsed (femtoseconds) lasers to illuminate the specimen, and fluorescence is only generated upon the simultaneous absorption (within a few femtoseconds) of two or more low-energy photons. Indeed, photon density of the light pulses decreases steeply outside the focal plane, which prevents fluorescence from being generated elsewhere and dispenses the use of a pinhole. Therefore, multiphoton microscopy is less prone to phototoxicity than confocal microscopy because fluorescent activation only occurs at the focal plane (a few femtoliters) (Figure 2.2). An additional advantage of the multiphoton microscope is that it uses longer wavelengths (near infrared and infrared) that are less scattered by tissue and reach deeper areas, thus allowing imaging of thicker specimens (>100 μm) (Table 2.1).

2.1.2 Spatial Resolution

According to Abbe's theory, lateral resolution$_{x,y}$ = $\lambda/2NA$ and axial resolution$_z$ = $2\lambda/NA^2$, where λ is the wavelength of the excitation light and the numerical aperture (NA) of the lens.

Because each IVM technique makes use of lenses with different NA and employs different wavelengths for excitation light, their spatial resolution will differ considerably. Hence, the best choice of IVM technique will depend on the level of detail that is required to visualize the specimen (higher for subcellular imaging and lower for tissue/cellular imaging). In terms of wavelength, the use of laser sources with longer wavelengths for multiphoton IVM, although advantageous for opaque specimens, results in degraded resolution when compared to SPIM, confocal, spinning disk, and wide-field microscopy. In terms of objectives, most IVM techniques make use of long working distance lenses in order to allow imaging of deep areas of the specimen. However, these lenses have relatively low NA, which result in a poor axial resolution but an intact lateral resolution. Among these techniques, SPIM provides theoretically better axial resolution than epi-fluorescence and multiphoton microscopy, and, at NA < 0.8, it provides better axial resolution than confocal microscopy (Engelbrecht and Stelzer, 2006).

Currently, the maximal spatial resolution of IVM techniques remains limited by the diffraction limit: 200–250 nm in x- and y-axes and 500 nm in z-axis. Various super-resolution techniques such as stimulated emission depletion (STED), structured illumination microscopy (SIM), and photoactivated localization microscopy/stochastic optical reconstruction microscopy (PALM/STORM) have been developed to break the diffraction limit (Liu et al., 2015). However, when STED and multiphoton SIM were applied *in vivo* to thick specimens, these have only provided improved resolution in superficial areas of the specimen (<25 μm) (Berning et al., 2012; Ingaramo et al., 2014). In contrast, multiphoton IVM combined with spatial modulation of excitation light has been shown to significantly improve axial resolution and lateral resolution to a lower extent while allowing deep imaging (<110 μm) (Andresen et al., 2012). Spatial resolution is also limited in thick specimens commonly used for IVM because they are prone to scattering both excitation and emission light, thus creating a depth-dependent degradation of the spatial resolution due to the aberration of the point spread function (PSF) (de Grauw et al., 1999; Niesner et al., 2007; Herz et al., 2010). These aberrations can be corrected using postacquisition deconvolution through the blind estimation of the PSF, for

TABLE 2.1

Advantages and Limitations of Intravital Microscopy Techniques

Technique	Excitation	Light Source	Advantages	Limitations
Wide-field	Single photon	Mercury lamp	Low toxicity Low costs Fast acquisition	Limited depth Out of focus light
Confocal	Single photon	Continuous laser	High spatial resolution	Limited depth High phototoxicity Slow acquisition
Spinning disk	Single photon	Continuous laser	Fast acquisition Low phototoxicity	Limited depth
SPIM	Single photon	Continuous laser	Fast acquisition Low phototoxicity	Limited compatibility of samples
Mulitphoton	2/3 photon	Pulsed laser	Low phototoxicity Extended depth	Slow acquisition

instance using the Huyguens® or ImageJ software. However, 3D deconvolution is more efficient when the PSF aberrations caused by the microscope setup and scattering of the specimen are integrated (Ghosh and Preza, 2015). In this regard, the integration of adaptive optics in the microscope setup can mitigate specimen-induced aberrations through its capacity to dynamically compensate for this effect while imaging (Ji et al., 2012). However, PSF aberrations are heterogeneous in the specimen and require time-consuming corrections that will slow down the acquisition rate of images.

2.1.3 Temporal Resolution

Biological processes differ considerably in their temporal dynamics and can range from extremely fast molecular events occurring in a cell (range of µm/s) to slower migration of a cell within a tissue (range of µm/min). In order to choose the appropriate IVM technique, it is important to realize that the speed at which images are collected is strictly dependent on the thickness of the 3D volume to be imaged, the desired resolution, and the microscope setup. Indeed, the collection of large volumes to increase depth and the collection of a large number of optical sections to increase resolution slows down the sampling rate of the 3D volume. As a consequence, the confidence of cellular/molecular tracking of *in vivo* events is significantly decreased when depth and resolution is privileged. The physical design of the microscope also has significant impact on the temporal resolution of IVM. For instance, laser line scanning IVM techniques, such as confocal and multiphoton microscopy, are slower in collecting images (up to four frames/s at 512×512pix) because the movement of galvanometer scanning mirrors by servo motors that direct the laser beam to each x–y position in field of view are limited by inertia. The use of resonant galvanometer scanning mirrors that vibrate at a high and fixed frequency (up to 8 kilohertz) can improve temporal resolution (up to 30fps at 512×512) (Kirkpatrick et al., 2012). However, short laser pixel dwell times associated with this type of scan often require image averaging in order to improve contrast, thus reducing the advantage of this type of technology for weak signals. However, the combination of resonant galvanometer scanners with high-sensitivity detection achieved with gallium arsenide phosphide (GaAsp) or hybrid photodetectors can partially overcome this limitation. Similarly, the temporal resolution of SPIM is dictated by the mechanical movement of the light sheet through the specimen, thus resulting in limited image acquisition rates. To improve the speed of acquisition of SPIM, simultaneous collection of multiple views of the specimen (Tomer et al., 2012) or implementation of multiphoton excitation (Truong et al., 2011) have been developed, although the applications of these developments remain limited to optical transparent specimens. Finally, spinning disk and fluorescent wide-field techniques offer the best temporal resolution (up to 200fps in spinning disk microscopy) because of the rapid illumination of the specimen and detection of the emitted signal at multiple points simultaneously.

2.1.4 Depth

Imaging depth is dictated mainly by the IVM technology used and the transparency of the tissue. Among IVM techniques, multiphoton microscopy excels at deep imaging, enabling up to 1.6 mm deep into the optically opaque tissues. In order to further improve imaging depth, different strategies can be used for scattering specimens in multiphoton IVM. For instance, higher laser intensities can be employed to compensate for the scattering of photons as the focus is driven deeper into the specimen. However, this strategy has limits because, at high laser intensities, out-of-focus fluorescence can be generated, in particular in superficial regions of the specimen, that can obscure the features of interest and limit imaging depth (Theer and Denk, 2006). Alternatively, higher wavelengths ($\approx$1280 nm) can be used to improve imaging depths up to 1.6 mm in brain tissue (Kobat et al., 2011). However, wavelengths longer than 1300 nm will eventually be absorbed by water molecules present in the specimen and generate heat that can compromise tissue physiology (Kobat et al., 2009). Another barrier for deep imaging is the scattering of emitted photons that considerably reduce the number of events detected by laser pulse, thus reducing imaging depth. Currently, the incorporation in the microscope setup of GaAsP detectors that have higher quantum efficiency (50%) than alkali detectors (25%) can improve photon sensibility and, therefore, imaging depth. Another approach consists in incorporating photon counting capacity in the microscope setup in order to better distinguish the scarce emitted photons from background electronic noise of the detectors (Driscoll et al., 2011). Still on the side of the microscope setup, adaptive optics can also be incorporated to correct for optical aberrations that become larger as the focus is driven deeper into the tissue (Tuohy and Podoleanu, 2010; Ji et al., 2012).

Besides the characteristics of microscope setup, extended imaging depth (up to 2000 µm) can be achieved through approaches at the level of the specimen itself. For instance, the use of gradient refractive index (GRIN) lenses and microprisms, which require the surgical insertion of optical elements directly in living animals, reach deep areas of a given tissue (Engelbrecht et al., 2008; Chia and Levene, 2009). (Figure 2.3). Conventional GRIN lenses use a gradient in the refractive index of glass to bend and focus light that acts as a relay of excitation and emission light to and from the specimen (Figure 2.3). GRIN lenses are 0.35 to 2 mm in diameter, have low NA (up to 0.6), and have optical aberrations that significantly degrade optical resolution. However, new approaches that couple GRIN lenses with high-NA plano-convex lenses improve both NA (up to 0.85) and optical resolution to the level of a water-immersion objective (Barretto et al., 2009). Currently ultraslim objectives with a diameter down to 1.3 mm are commercially available but also have low NA. Alternatively, the surgical insertion of right-angle glass prisms (microprism) in living animals can also provide access to deep areas of tissue. Indeed, reflective coating of the hypotenuse of these microprisms bends excitation and emission light, thus allowing one to visualize the surrounding tissue (Chia and Levene, 2009) (Figure 2.3). So far, these approaches have yielded significant results in brain and kidney imaging in mice and even skeletal muscle in humans (Li and Yu, 2008; Llewellyn et al., 2008; Barretto et al., 2011). Even though the use of GRIN and microprisms extend the depth of imaging beyond the capacity of the microscope setup, it remains an invasive technique, thus care should be taken to distinguish the biological findings from the potential bias introduced by the surgical procedure.

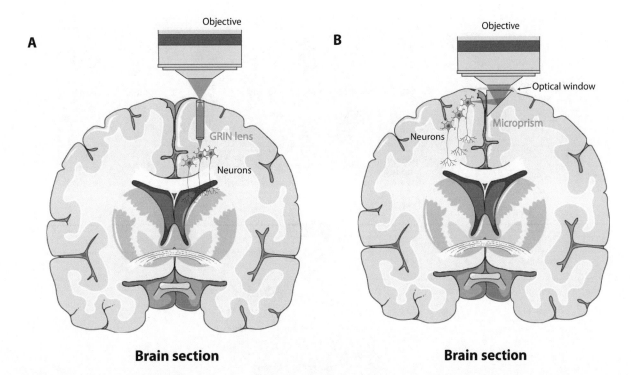

FIGURE 2.3 Surgical approaches to improve depth of brain intravital microscopy approaches. Gradient refractive index (GRIN) lenses and microprisms are optical elements that can be surgically inserted in living animals to reach deep areas of the brain. These optical elements can act as relay lenses of the excitation/emission light (GRIN lens) A or as mirrors that reflect excitation and emission light and allow visualization of the surrounding tissues (microprisms) B. Brain section and neuron images were obtained from Servier Medical Art images (http://smart.servier.com/) and are used according to the Creative Commons License 3.0 (https://creativecommons.org/licenses/by/3.0/).

2.2 Surgical Preparation

The appropriate use of IVM requires a good understanding of tissue physiology and anatomy to choose the best surgical solution to visualize the organ of interest with minimal perturbation in blood and lymph flow or innervation. To illustrate this point, lymphocyte motility ceases immediately after cardiac arrest, even in optimal temperature conditions. The majority of the organs, besides the skin surface, require surgery with different levels of invasiveness to gain optical access to the tissue. Importantly, highly invasive techniques may compromise tissue physiology and will compromise longitudinal observations. Furthermore, exposed organs need to be kept constantly hydrated, otherwise risking cell/tissue viability. In order to allow long-term and longitudinal IVM studies, optical windows have been designed to gain optical access. These optical windows are composed of transparent glass windows that can be surgically implanted in different areas, such as dorsal skin, brain and abdominal cavity (Vajkoczy and Menger, 1994; Drew et al., 2010; Ritsma et al., 2013). Below we describe a variety of surgical techniques to visualize different organs by IVM in the mouse model (Figure 2.4).

2.2.1 Digestive System

The digestive system is comprised of several key organs, such as the liver, pancreas, small intestine, and tongue, that have been probed by IVM. The pancreas, liver, and intestine can be exteriorized after a midline incision to gain optical access (Hosoe et al., 2004; Coppieters et al., 2010; Marques et al., 2015). Once the organs are exteriorized, these are immobilized on a microscope stage where humidity and temperature are controlled. For IVM of the tongue, no surgery is required, and the organ is simply pulled-out of the oral cavity and immobilized on a suction holder (Choi et al., 2015). Alternatively, optical windows or a GRIN lens can be implanted to perform IVM of the liver and small and large intestine (Kim et al., 2010; Ritsma et al., 2013; Ritsma et al., 2014). Of note, the use of atropine, antagonist of muscarinic acetylcholine receptor in the parasympathetic nervous system, can reduce peristalsis, thus increasing the stability of the intestine IVM preparation (Choe et al., 2015). So far, these techniques have been instrumental for studying autoimmune reactions in type I diabetes (Coppieters et al., 2012), granuloma formation in response to *Mycobacterium bovis*-BCG infection (Egen et al., 2008), liver metastasis (Ritsma et al., 2014), lymphocyte migration into Peyer's patches (Imai et al., 2008), colorectal cancer (Jung et al., 2017), delivery of luminal antigens to small intestine dendritic cells (McDole et al., 2012), and the functional activity of taste cells (Choi et al., 2015).

2.2.2 Respiratory and Cardiovascular System

The challenges to perform IVM in the heart and lung are three: These are vital organs that require extreme care during surgical preparation, access to the thoracic cage requires mechanical ventilation, and these organs constitute the major sources of motion

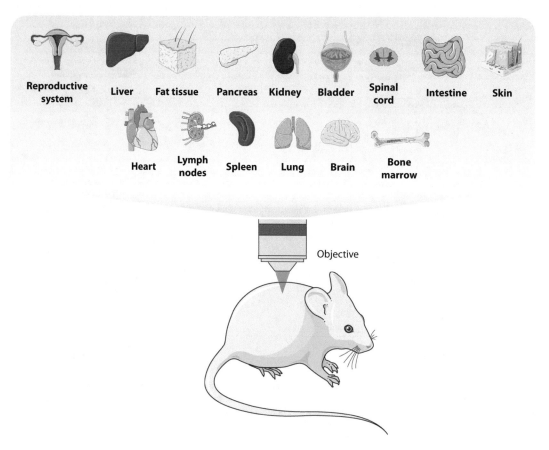

FIGURE 2.4 Several systems and organs that have been explored by intravital microscopy techniques in the mouse model. All images, except objective image, were obtained from Servier Medical Art images (http://smart.servier.com/) and are used according to the Creative Commons License 3.0 (https://creativecommons.org/licenses/by/3.0/).

artifacts in IVM preparations. For IVM of the heart, several methods have been employed to stabilize this organ. These rely on the externalization of the organ and fixation either by sutures (Chilian and Layne, 1990), by compression with a coverslip (Li et al., 2012), or by attachment to a mechanical or suction-based stabilizer (Lee et al., 2012; Vinegoni et al., 2012). All of the above require mechanical ventilation, as proper intrathoracic pressure is no longer maintained during surgery. Alternatively, surgical techniques for longitudinal studies have been developed that rely on the surgical implantation of GRIN lenses (Jung et al., 2013). Similarly, several techniques have been developed for the lung IVM that rely on the establishment of a thoracic window by a thoracotomy (incision in the pleural space of the chest), that may or may not involve resection of some ribs but requires mechanical ventilation. Once the thoracic window is established, the lung is stabilized using either Vetbound glue to bond the organ to a cover glass on a microscope stage (Kreisel et al., 2010), a suction-based stabilizer equipped with a cover glass (Looney et al., 2011), clamping of the bronchus to stop movement (Hasegawa et al., 2010), or a transparent membrane to seal the thoracic window, while ensuring contact with the organ (Tabuchi et al., 2008). In addition, longitudinal IVM studies of the lung, indeed, a longitudinal lung IVM approach has recently been published, but the surgical implantation of a GRIN lens equipped with a suction-based stabilizer holds some promise. Altogether, these techniques have allowed fundamental

studies of the mechanisms involved in the recruitment of neutrophils during pulmonary inflammation due to transplantation (Kreisel et al., 2010), ischemia reperfusion in the heart (Li et al., 2012), the immune-modulatory role of a subset of alveolar macrophages (Westphalen et al., 2014), the establishment of pioneer metastatic cells in the lung (Headley et al., 2016; Entenberg et al., 2018), and recruitment of monocytes during acute myocardial infarction (Jung et al., 2013).

2.2.3 Reproductive System

Currently, several organs of the reproductive system such as the prostate, uterus, placenta, and ovary have been explored by IVM. For all these organs, IVM requires the exposure of the organ by a small incision that is made to the skin and peritoneum of the animal. After externalization, the prostate can be stabilized by bonding it to a cover glass (Ghosh et al., 2010), and the ovary can be stabilized by using insect pins to secure surrounding tissues, such as fat pad and connective tissue, without affecting blood flow (Migone et al., 2013). In contrast, the uterus and placenta require a more elaborate preparation. For uterus IVM, one of the uterine horns is stabilized with an aluminum foil band, embedded in low meting agarose, and drenched in caffeine solution to prevent muscle contractions without affecting blood flow (Zenclussen et al., 2013). For placental IVM, the uterine wall is sectioned to reveal individual amniotic sacs, and the respective placenta is stabilized using an aluminum foil clip (Zenclussen et al., 2012). All of the above techniques require maintaining proper humidity and temperature of the exposed organs. Alternatively, ovary IVM can be performed by surgical implantation of an optical window for longitudinal studies (Bochner et al., 2015). So far, these techniques have allowed studies of the mechanisms involved in ovulation (Migone et al., 2016), progression and response to treatment of ovarian cancer (Zhong et al., 2009), *Plasmodium*-infected erythrocyte sequestration in the placenta (de Moraes et al., 2013), and visualization of uterine mast cells (Schmerse et al., 2014).

2.2.4 Urinary System

The physiology of few organs of the urinary system, such as the kidney and bladder, have been investigated using IVM techniques. For these organs, as for most internal organs, IVM preparation requires their externalization by a laparotomy (surgical incision through the abdominal wall to gain access into the abdominal cavity). The stabilization of the kidney can be performed by placing the organ on a coverslip-bottomed heated chamber bathed in normal saline (Dunn et al., 2002) or by using a custom holder arm (Dunn et al., 2012). Importantly, longitudinal and deep observations of the kidney can be performed using this surgical technique and individual glomerulus revisited over time (Hackl et al., 2013; Schuh et al. 2015). As for bladder IVM, after externalization, the organ is emptied of urine and replaced with Phosphate-buffered saline (PBS) by a catheter inserted transurethrally. Next, the bladder is immobilized using a custom-made suction-stabilized imaging window equipped with a cover glass (Sano et al., 2016) or by positioning it against a cover glass (Kowalewska et al., 2011). Alternatively, the use of a GRIN lens has shown potential to be used in kidney and bladder IVM (Kim et al., 2012). Altogether, IVM of the kidney and bladder have allowed studies of mechanisms of glomerular injury and disease (Burford et al., 2014), kidney function (Schiessl et al., 2016), and leukocyte behavior in the glomerular microvasculature (Devi et al., 2013) and bladder (Kowalewska et al., 2011).

2.2.5 Nervous System

Among the organs of the nervous system, such as the brain and spinal cord, the brain has been largely explored by IVM techniques for its importance but also, owing to its optical properties, because it allows deep imaging up to 1.6 mm in an intact brain (Kobat et al., 2009). For brain IVM, several techniques have been developed to access the cortex for longitudinal studies: an open-skull window that requires a craniotomy (removal of a part of the skull bone) that is sealed using a cover glass, a thinned-skull window in which the skull-bone thickness is significantly reduced (Holtmaat et al., 2009) or a reinforced-skull window in which skull bone is thinned and reinforced using a cover glass (Drew et al., 2010). For stabilization, the animal is usually placed in a stereotaxic frame, and the head is secure with ear bars. Among these techniques, the open-skull window is very prone to inflammation that requires use of anti-inflammatory treatment, whereas the reinforced-skull

preparation does not cause inflammation. In order to reach deeper areas of the brain, the surgical implantation of GRIN lenses or microprisms are a solution, but they are invasive techniques (Barretto et al., 2009; Chia and Levene, 2009). For spinal cord IVM, a midline incision in the skin of the back and retraction of the of the paraventral muscles is required to expose the vertebra laminae. After exposure, a laminectomy is performed to remove the back of the vertebra that covers the spinal canal, and the dura matter is opened over the spinal cord (Kerschensteiner et al., 2005). Alternatively, a chronic spinal cord window has been developed for longitudinal studies (Farrar et al., 2012). So far, these techniques have been instrumental to study the activity in dendrites of single neurons (Grutzendler et al., 2002), immune cell recruitment into auto-immune central nervous system lesions (Bartholomaus et al., 2009), immune-mediated vascular injury during acute viral meningitis (Kim et al., 2009), experimental cerebral malaria pathogenesis (Pai et al., 2012), the dynamics of axon regeneration after spinal cord injury (Kerschensteiner et al., 2005), and longitudinal tracking of glioma and brain metastasis (Kienast et al., 2010; Osswald et al., 2015).

2.2.6 Immune System

Several secondary lymphoid organs, such as the spleen and bone marrow as well as the inguinal, popliteal, and cervical mesenteric lymph nodes (LN), have been observed by IVM. For IVM of LNs, the surgical preparation is dependent on their location. For the popliteal LN, a midline vertical incision in midcalf is performed at the level of the popliteal fossa, and the lymph node is separated from the surrounding tissue while taking care to spare blood and lymph vessels. Stabilization of this preparation is performed by fixing the leg, tail, and/or spine of the animal (Mempel et al., 2004; Liou et al., 2012). For the inguinal LN IVM preparation, the LN is surgically exposed by creating a skin flap that is stabilized by compressing an O-ring surrounding the organ with a Plexiglas support (Miller et al., 2003) or by fixing traction sutures inserted in the skin flap to a Plexiglas stage (von Andrian, 1996). For the cervical lymph node, an incision is performed at the ventrolateral level of the neck to create a skin flap, and the fat tissue covering the submandibular gland is split and turned outward to expose the LN (Schramm et al., 2006). For IVM of the mesenteric LN and spleen, a laparotomy is performed to externalize the organs, and stabilization is achieved by bonding them to a modified Petri-dish equipped with a cover glass (Grayson et al., 2003) or by compression with a cover glass using a spring-loaded platform (Arnon et al., 2013). Recently, an optical window for the inguinal LN has been developed for longitudinal studies (Meijer et al., 2017). To perform cranial bone marrow IVM, a midline incision is performed in the scalp and the frontoparietal skull. A plastic ring is inserted to spread the skin, and stabilization of the preparation is achieved by placing the animal in a stereotaxic holder (Mazo et al., 1998). Alternatively, a cranial window can be used for longitudinal cranial or femur bone marrow IVM (Chen et al., 2016; Le et al., 2017). The use of IVM techniques to investigate the inner working of the immune system have been influential in defining the multistep cascade of events necessary for leukocyte trafficking to the secondary lymphoid organs (Butcher, 1991; Springer, 1994), T cell immune responses in LNs (Miller et al., 2003), immune responses against *Listeria* in the spleen (Waite et al., 2011) or against vaccinia and vesicular stomatitis virus (Hickman et al., 2008), *Toxoplasma* in the LNs (Chtanova et al., 2008), mobilization of neutrophils from the bone marrow (Kohler et al., 2011), and engraftment of tumor cells in the bone marrow (Sipkins et al., 2005).

2.2.7 Other Tissues

Several other tissues have been visualized by IVM techniques such as muscle, skin, eye, knee joint, bone, mammary glands, and adipose tissue. The skin is the most accessible tissue for IVM and does not require surgery unless access to the subcutaneous side is necessary. A common strategy for this is the use of the skin-flap technique (Miller et al., 2003) or the surgical implantation of a dorsal optical window for longitudinal studies (Sipkins et al., 2005). Muscle IVM can also be observed using a dorsal optical window or by surgical implantation of a GRIN lens (Barretto et al., 2009). The IVM of the eye is a significant challenge because of ocular movement but does not require surgery. In this technique, immobilization is performed by a custom-made corneal clamp combined with a cranial holder (Steven et al., 2011). For knee joint IVM, a small incision is made distal to the patella tendon, the fascia is removed, and the tendon is sectioned and removed. Stabilization of the preparation is achieved by fixation of the extremity of the leg in silicon and placement of a cover glass on the knee capsule (Veihelmann et al., 1998). For mammary gland IVM, a skin flap can be used similarly to the inguinal LN IVM (Miller et al., 2003) or an optical window can be surgically implanted for longitudinal studies (Alieva et al., 2014). For IVM of adipose tissue, a skin-flap technique can be used to observe subcutaneous fat or, for abdominal fat, a laparotomy is performed for externalization, and stabilization is achieved by compressing the tissue on a custom-made stage (Cho et al., 2014). Altogether, these techniques allowed the description of the dynamics of mammary stem cells (Scheele et al., 2017), inflammation of the adipose tissue (Cho et al., 2014), immune response to HSV-1 in the skin (Ariotti et al., 2012), corneal lymphangiogenesis (Kang et al., 2016), and murine arthritis (Veihelmann et al., 2002).

2.3 Motion Artifact Control

One of the main challenges of using IVM is minimizing periodic and random motion artifacts derived from respiratory, cardiac, and muscle activity. For this, several techniques are employed to stabilize the tissue under observation, but these are frequently not sufficient, thus further increasing the rejection rate of imaging sets and reducing the throughput of IVM techniques. A simple solution relies on frame rejection in sets of images with motion artifacts (Soulet et al., 2013). However, this strategy affects the quality of the tracking of cell behavior and can be performed only when sufficient frames are acquired during an imaging session. Image registration can also compensate for motion artifacts by the realignment of images and rejection of unmatched components (Liebling et al., 2005; Greenberg and Kerr, 2009), but these approaches work best when motion artifacts are mainly on an x–y axis. Other approaches based on image processing that rely on distortion models can be effective in eliminating motion artifacts (Vercauteren et al., 2006; Greenberg and Kerr, 2009). Alternatively, the implementation of active image acquisition modes can eliminate motion artifacts that are not resolved by physical stabilization. For instance, the acquisition of multiple images while recording the electrocardiograph (ECG) cycle and phase of the respiratory cycle, followed by an image reconstruction algorithm that retains only images at precise points in time, allows one to eliminate the vast majority of the motion artifacts (Lee et al., 2012). A similar approach can be envisioned in which image acquisition is only triggered at precise points in time of the ECG and respiratory cycle. Other active approaches rely on the modification of the imaging setup. For instance, the integration of a system that continuously monitors the position of the tissue and a feedback system to control the position of the objective to maintain constant distance can eliminate many of the motion artifacts in spinal cord IVM (Laffray et al., 2011). Alternatively, imaged-based correction can be applied to IVM, allowing the removal of physiological motion from *in vivo* (and even clinical) functional imaging data (Warren et al. eLife, 2018)

2.4 Anesthesia Control

The choice and dosage of anesthesia is a critical step for the success of IVM. The effective anesthesia for IVM must avoid hypotension and respiratory depression because they negatively impact tissue physiology. Furthermore, proper dosage is critical because anesthetics' potency varies among animal models and may even vary among strains within the same model, such as for mice (Sonner et al., 2000). Anesthetics can be delivered by injection or by inhalation, and the route of delivery can be chosen according to the experimental design. The most common injectable anesthesia cocktail for small rodents consists of ketamine and xylazine, either complemented with acepromazine or completely substituted by tribromoethanol (Germain et al., 2006). A bolus of these anesthetics cocktails can provide approximately 30–45 min of deep sedation, as these are rapidly metabolized and eliminated, thus allowing only a short imaging interval. For longer imaging sessions, it is necessary to constantly administer anesthetics to compensate their loss, either through multiple injections or an intravenous/subcutaneous line coupled with a perfusion pump. An alternative approach is the use of aerosolized delivered anesthetics like the halogenated ethers isoflurane and sevoflurane. These anesthetics allow fast and stable narcotic effects with little side effect, as they are well tolerated by the small rodents and big mammals. In addition, new generation anesthetic dispensers allow long-term, precise control of the concentration of oxygen and anesthetic that reaches the animal. Therefore, aerosolized delivered anesthetics is preferred for long-term imaging sessions.

Besides hypotension and respiratory depression, anesthetics may have a direct impact on cell functions. For instance, isoflurane is an allosteric inhibitor of lymphocyte function-associated antigen 1 (LFA1), an important molecule necessary for immune cells' recruitment to tissues (Dustin and Springer, 1988; Zhang et al., 2009). Indeed, isoflurane has been shown to block neutrophil recruitment to the skin in a cutaneous Arthus reaction model (Carbo et al., 2013). Therefore, it is important to account for the potential impact of anesthesia in the biological observations under study.

2.5 Temperature Control

Anesthesia is accompanied by a global reduction in metabolism that decreases body temperature and may compromise both cell function and animal survival. In addition, the externalization of the target organ lowers local tissue temperature. Hence, to maintain proper cell function and animal survival, body temperature must be tightly regulated and monitored. For instance, lymphocyte motility is optimal between 37–38°C, but a small decrease in temperature below 35°C severely impairs their migratory behavior. Proper temperature can be achieved by using a coil perfused with a heated liquid that surrounds the organs of interest; however, systemic temperature and, therefore, blood pressure are not maintained. Other approaches rely on maintaining body heat with a heating pad but cannot control the local temperature in the exposed organ. So far, the best strategy consists in using a thermoregulated imaging chamber surrounding the microscope stage that ensures the maintenance of the entire animal's body at a physiological temperature, therefore controlling both local and systemic temperature. Because, when in contact with liquid where the organs are immersed, immersion objectives frequently act as a heat sink and decrease sample temperature, and it is sometimes necessary to use a heating collar on the objectives to prevent this phenomenon. However, the use of a thermoregulated imaging chamber comes with the risk of animal dehydration that will decrease blood flow and negatively impact animal survival. In order to maintain proper hydration, fluids need to be administered manually or using a perfusion pump by intravenous or intraperitoneal injection during the imaging session. For extended periods of imaging (>6 h), a source of energy should also be added to the hydration fluid, e.g. sucrose. In general, monitoring of vital signs, such as body temperature, oxygen levels in blood, and heart rate during IVM, can provide important feedback for success in long-term imaging sessions.

2.6 Fluorescent Labeling

Because the majority of current IVM techniques rely on fluorescence, it is critical to identify a suitable fluorescent labeling approach for each tissue/cell. Many fluorescent labeling strategies can be used for IVM, and these rely either on chemical labeling with organic or inorganic dyes, antibody and plant-lectins, or genetic labeling to investigate many aspects of cell structure and function. Here, we focus on the labeling techniques that are used in IVM.

2.6.1 Antibodies and Lectins

Currently, a specific cell type or tissue-specific molecules can be fluorescently labeled using either antigen-specific antibodies or plant-derived lectins that have affinity for different carbohydrate moieties in glycoconjugates. Lectins conjugated with fluorochromes are extremely useful labeling tools that have a well-defined staining pattern and can be used to highlight tissue structures, such as blood vessels or nerves (Wong et al., 2010; KleinJan et al., 2014; Kataoka et al., 2016), and several types of lectins are commercially available. As far as for antibodies, these can be conjugated to a variety of different fluorochromes and used to label specific cell types *in vivo* for IVM, such as blood and lymphatics vessels and lymphocytes (Cummings et al., 2008; Egeblad et al., 2008; McElroy et al., 2009; Moreau et al., 2012). However, the presence of an Fc portion on some of these antibodies may be associated with nonspecific labeling because many immune cells *in vivo* have a specific receptor for this portion of antibodies. To solve this problem, fluorescent labeled Fab fragments of antibodies are now available. Recently, alpaca-derived antibody fragments have been developed that have significant advantages over traditional antibodies and hold great potential for IVM techniques. Indeed, single-chain camelid antibody fragments lack an Fc portion and are much smaller (~15 kDa) than immunoglobulins (~150 kDa) that improve tissue penetration and improve specificity (Rashidian et al., 2015). Finally, fluorochromes for antibody and lectin labeling should be bright and stable within a broad range of pH, which are properties that can be found in several Alexa dyes. Nanocrystal labeling of antibodies can also been used for *in vivo* targeting (Akerman et al., 2002; Gao et al., 2004; Kim et al., 2004). Altogether, targeting of specific molecules *in vivo* is a promising strategy but should be performed with care as antibody/lectin may either block or enhance cellular functions and potentially bias the interpretation of the IVM observations.

2.6.2 Organic Dyes

A common strategy to fluorescently label cells for IVM is their incubation with chemical dyes *ex vivo* followed by adoptive transfer into recipient animals. Some of these dyes, such as Cell Trackers™ and boron-dipyrromethene (BODIPY), have been used in multiple studies, as they are able to penetrate cells and, after being metabolized, will become reactive to amine or thiol motifs of cellular components (Bousso et al., 2002; Miller et al., 2002; Stoll et al., 2002; Mempel et al., 2004; Miller et al., 2004). Others are lipophilic and bind to cell membrane lipid bilayers such as PKH family (PKH2, PKH67, PKH3 and PKH26) and carbocyanine dyes (DiO, DiL, DiD and DiR) or have high affinity for double stranded DNA (Hoescht 33242 and DRAQ5). Even though these dyes provide very bright labeling of cells and exist with a variety of emission spectra, they are prone to cytotoxicity when used at high concentrations. For instance, high concentrations of Hoescht 33242 have been shown to inhibit cell proliferation (Parish, 1999) and DRAQ5 may inhibit protein binding to DNA (Mari et al., 2010). In addition, these dyes are often diluted after each cell division, which may hinder long-term imaging of labeled cells after adoptive transfer. Still, dyes such as CFSE that are consistently halved upon each cell division can be used to track cell proliferation *in vivo*. Moreover, a common strategy to identify artifacts induced by labeling with dyes is to switch the labeling agent and verify whether the biological observations are similar.

2.6.3 Inorganic Dyes: Quantum Dots

Quantum dots (QDs) are semiconductor nanocrystals that possess high photostability and quantum yield, a narrow emission spectrum, and a large excitation cross section (Alivisatos et al., 1996; Michalet et al., 2005). Given these features, QDs are commonly used for highlighting the vasculature upon intravenous injection (Larson et al., 2003). The time of persistence in the circulation varies according to each QD. For example, addition of polyethylene glycol (PEGylation) groups enhance the stability of the QDs, reducing their accumulation in the liver and bone marrow and extending their persistence in the circulation (Ballou et al., 2004). One of the main limitations of QDs is the limited potential to deliver such labeling into the cell of interest. So far, the only way to perform whole-cell labeling is to use microinjection, electroporation, or naturally occurring phagocytosis (Dubertret et al., 2002; Jaiswal et al., 2003; Mattheakis et al., 2004), but these approaches have the potential of affecting cell and organ function.

2.6.4 Fluorescent Protein

So far, the best strategy to fluorescently label cells *in vivo* for IVM relies in the genetic manipulation of the genome to integrate fluorescent reporter proteins, which are not cytotoxic and allow long-term tracking in bacteria to yeast to vertebrates, such as zebrafish and mice. The first fluorescent protein (FP), green fluorescent protein (GFP), was cloned from *Aequorea victoria* jellyfish, and it revolutionized the field of cell biology (Prasher et al., 1992; Heim et al., 1994). Subsequently, several other FPs have been discovered that have a wide range of emission/excitation spectra (Shaner et al., 2005). The applications range from the expression of single FP (Tsien, 1998), to fusion proteins, Forster resonance energy transfer (FRET) reporter systems or photoactivable proteins (Zhang et al., 2002; Bevan, 2004; Giepmans et al., 2006; Matheu et al., 2007; Victora et al., 2010). In order to be expressed in the cell of interest for long periods of time, FP encoding genes need to be inserted into the target cell genome. One possibility is to transfer the FP gene by DNA transfection or retroviral or lentiviral transduction directly into cells (Wan et al., 2000; Belland et al., 2004; Perro et al., 2010; Marangoni et al., 2013). This strategy is appropriate for a wide variety of cell types and very efficient in dividing cells. In contrast, gene delivery into quiescent immune cells, such as primary T lymphocytes, is inefficient. For immune cells, an alternate remains the delivery of FP encoding genes into bone marrow progenitors that are more amenable to viral transduction, followed by adoptive transfer into recipient animals. Upon transfer, these progenitors will give rise to differentiation of mature cells stably expressing the FP of interest (Cyster and Schwab, 2012). Importantly, this delivery strategy of FP encoding genes can be applied to stem

cells or embryos and can give rise to novel transgenic animal lines in shorter periods of time than conventional transgenesis. Alternatively, germ-line manipulation of the genome by knock-in, bacterial artificial chromosome (BAC), or transgene technology can be used to fluorescently label cells/tissues.

The expression of the FP encoding genes may be controlled by a strong and ubiquitous promoter, such as the hybrid CMV-chicken beta actin, or by a tissue/cell specific promoter, as in the case of CD11c-YFP mice (Lindquist et al., 2004). The expression of fluorescent protein can also be driven by a gene promoter of a particular cellular function, such as granzyme B, IFNγ, and IL4, providing functional information. In addition, the use of photoactivable FPs allow one to specifically track individual cells among a heterogeneous population of cells over time, thereby increasing resolution for studying cellular dynamics (Victora et al., 2010). Even though most FP reporter lines have faithful expression of the gene of interest, promoter leakiness may result in labeling unwanted cell populations. For instance, in the CD11c-YFP transgenic mice, both DCs, alveolar macrophages, and some subset of CD8 T cells are fluorescent. To obviate this problem, imaging algorithms can be used to distinguish cells with the same fluorescent marker by their shape, thus allowing the analysis of these populations separately (Thornton et al., 2012). Recent developments in intravital imaging now allow the usage of biosensors that can be either endogenously expressed (Nobis et al., 2017; Conway et al., 2018) in mouse models or xenografted (when expressed in tumor cells) (Warren et al., 2018). When combined with FLIM-FRET imaging on organs exposed using abdominal windows, such imaging now reaches accurate assessment of important signaling molecules such as RhoGTPases (Nobis et al., 2017), which can be expressed in various cellular types. In addition, one can now interrogate the behavior of tumors and monitor therapeutic response using biosensors of the PI3K-AKT pathway (Conway et al., 2018). Similar approaches can be used for addressing the binding of powerful drugs to its target and for assessing subsequent cell killing in a tumor context (Sparks et al., 2018).

2.7 Pitfalls and Shortcomings

IVM is a powerful technique that allows the researcher to eavesdrop into the inner working of cells and tissues. Indeed, the visualization of cellular behavior *in vivo* allows one to perform both a "reality check" as well as reveal unexpected phenomenon, such as the dynamics of migratory behavior of lymphocytes in lymphoid organs (Miller et al., 2003). Despite significant progress in the instrumentation, surgical preparations and cell labeling strategies, IVM remains a complex and low throughput technique. Indeed, most IVM techniques require surgery that is inevitably accompanied by some degree of inflammation and tissue damage, with the exception of superficial skin IVM. Hence, IVM may induce biases in the experimental results that need to be considered during data analysis. Furthermore, the regions that are imaged are frequently small, which makes the task of imaging similar regions in successive experiments difficult. Some of these shortcomings can be addressed by monitoring the state of the sample by including probes for blood and/or lymph flow, using tissue autofluorescence and second (SHG) and third harmonic generation (THG) signals, or using a system of coordinates in the optical window in order to allow longitudinal studies for the same region over extended periods of time. Another approach is to include experimental and control cells in the same imaging volume, which is very powerful for separating uncontrolled differences from experimental manipulation. Of course, all of these strategies require multiple color acquisitions that are not necessarily compatible with the microscope configuration and restricts the number of colors for experimental cells. Another aspect to consider is that IVM permits one to visualize what is labeled while all else (cells, structural components, vessels, cytokines, etc.) is not visible. Hence, any cellular/molecular behavior observed is very likely to be conditioned by this densely and heterogeneous packed tissue environment. Finally, despite considerable improvements in imaging depth regions beyond 2 mm are not possible to investigate by IVM. To circumvent this limitation, thick vibratome sections of living tissue can be performed, but it induces tissue damage, eliminates blood and lymph flow as well innervation with impacts on cell function that are hard to anticipate.

2.8 Toward High Resolution: IVM Correlative Light and Electron Microscopy (CLEM)

One of the major pitfalls of IVM is the lower resolution it provides in comparison to SIM, STED, or PALM/STORM microscopy, which prevents the obtainment of accurate subcellular information. The recent development of intravital correlative light and electron microscopy (CLEM), which combines the advantages of IVM and electron microscopy, gives access to ultrastructural information on tissues previously visualized by IVM (Karreman et al., 2016). This correlative approach relies on the identification or creation of landmarks during IVM in order to retrieve the region of interest prior to resin embedding and electron microscopy analysis. These landmarks can be anatomical features; artificial markers, such as exogenous probes (QDs, polymer beads, photo-oxidable fluorophores, or peroxidases); or features generated by tissue branding using a near-infrared laser during IVM (Bishop et al., 2011; Luckner et al., 2018) (Sparks et al., 2018). The use of microscopic X-ray computed tomography of the resin-embedded sample can provide significant improvement in the accuracy and speed of the registration to retrieve the region of interest, providing a significant increase in throughput (Karreman et al., 2016). So far, this approach has allowed one to visualize the dynamics of astrocytic coverage during dendritic lifetime, interrogate its ultrastructure at high resolution using a label-free CLEM method (Luckner et al., 2018), and follow extravasation and metastasis of tumor cells at very high resolution (Follain et al., 2018).

2.9 Future Developments

Currently, IVM techniques allow us to visualize a few of the components within a given tissue while the majority remain unseen, thus not capturing the entire complexity of biological phenomenon. In order to assess complexity, new reporter lines that include several stromal cells and structural components need to be developed. These novel lines should also allow more than simple cell labeling but provide means to visualize cell function by using biological sensors (e.g. FRET or FP fusion proteins) in signaling pathways. Of course, the instrumentation also needs to be developed in order to simultaneously capture an increasing number of signals. In parallel, instrumentation should be able to distinguish signals from FP or dyes with close emission spectra by means of spectral umixing during acquisition. Multicolor imaging should be performed at faster rates in order to capture fast events to improve the resolution of cell tracking and with improved sensitivity to detect faint signals. So far, super-resolution techniques have failed to improve resolution of IVM beyond the 25–40 µm, thus efforts at the level of instrumentation and labeling strategies should be made. In order to extend IVM to the molecular levels, new approaches of multiphoton Raman spectroscopy should be further developed in order to characterize the molecular complexity and their dynamics in living animals (Holtom et al., 2001).

In summary, IVM has made it possible to achieve major breakthroughs in different fields of research. Excitingly, the future still holds great promises, especially regarding future instrumentation developments. Indeed, we can envision that IVM will be extended to the level of molecular imaging, getting a step closer to visualizing the full complexity of cell events in living animals.

REFERENCES

Akerman, M. E., Chan, W. C., Laakkonen, P., Bhatia, S. N., and Ruoslahti, E. (2002). Nanocrystal targeting in vivo. *Proc Natl Acad Sci U S A*, *99*(20): 12617–12621.

Alieva, M., Ritsma, L., Giedt, R. J., Weissleder, R., and van Rheenen, J. (2014). Imaging windows for long-term intravital imaging: General overview and technical insights. *Intravital*, *3*(2): e29917.

Alivisatos, A. P., Johnsson, K. P., Peng, X., Wilson, T. E., Loweth, C. J., Bruchez, M. P., Jr., and Schultz, P. G. (1996). Organization of 'nanocrystal molecules' using DNA. *Nature*, *382*(6592): 609–611.

Andresen, V., Pollok, K., Rinnenthal, J. L., Oehme, L., Gunther, R., Spiecker, H., Radbruch, H., Gerhard, J., Sporbert, A., Cseresnyes, Z., Hauser, A. E., and Niesner, R. (2012). High-resolution intravital microscopy. *PLOS ONE*, *7*(12): e50915.

Ariotti, S., Beltman, J. B., Chodaczek, G., Hoekstra, M. E., van Beek, A. E., Gomez-Eerland, R., Ritsma, L., van Rheenen, J., Maree, A. F., Zal, T., de Boer, R. J., Haanen, J. B., and Schumacher, T. N. (2012). Tissue-resident memory CD8+ T cells continuously patrol skin epithelia to quickly recognize local antigen. *Proc Natl Acad Sci U S A*, *109*(48): 19739–19744.

Arnon, T. I., Horton, R. M., Grigorova, I. L., and Cyster, J. G. (2013). Visualization of splenic marginal zone B-cell shuttling and follicular B-cell egress. *Nature*, *493*(7434): 684–688.

Ballou, B., Lagerholm, B. C., Ernst, L. A., Bruchez, M. P., and Waggoner, A. S. (2004). Noninvasive imaging of quantum dots in mice. *Bioconjug Chem*, *15*(1): 79–86.

Barretto, R. P., Ko, T. H., Jung, J. C., Wang, T. J., Capps, G., Waters, A. C., Ziv, Y., Attardo, A., Recht, L., and Schnitzer, M. J. (2011). Time-lapse imaging of disease progression in deep brain areas using fluorescence microendoscopy. *Nat Med*, *17*(2): 223–228.

Barretto, R. P., Messerschmidt, B., and Schnitzer, M. J. (2009). In vivo fluorescence imaging with high-resolution microlenses. *Nat Methods*, *6*(7): 511–512.

Bartholomaus, I., Kawakami, N., Odoardi, F., Schlager, C., Miljkovic, D., Ellwart, J.W., Klinkert, W. E., Flugel-Koch, C., Issekutz, T. B., Wekerle, H., and Flugel, A. (2009). Effector T cell interactions with meningeal vascular structures in nascent autoimmune CNS lesions. *Nature*, *462*(7269): 94–98.

Belland, R., Ojcius, D. M., and Byrne, G. I. (2004). Chlamydia. *Nat Rev Microbiol*, *2*(7): 530–531.

Berning, S., Willig, K. I., Steffens, H., Dibaj, P., and Hell, S. W. (2012). Nanoscopy in a living mouse brain. *Science*, *335*(6068): 551.

Bevan, M. J. (2004). Helping the CD8(+) T-cell response. *Nat Rev Immunol*, *4*(8): 595–602.

Bishop, D., Nikic, I., Brinkoetter, M., Knecht, S., Potz, S., Kerschensteiner, M., and Misgeld, T. (2011). Near-infrared branding efficiently correlates light and electron microscopy. *Nat Methods*, *8*(7): 568–570.

Bochner, F., Fellus-Alyagor, L., Kalchenko, V., Shinar, S., and Neeman, M. (2015). A novel intravital imaging window for longitudinal microscopy of the mouse ovary. *Sci Rep*, *5*: 12446.

Bousso, P., Bhakta, N. R., Lewis, R. S., and Robey, E. (2002). Dynamics of thymocyte-stromal cell interactions visualized by two-photon microscopy. *Science*, *296*(5574): 1876–1880.

Burford, J. L., Villanueva, K., Lam, L., Riquier-Brison, A., Hackl, M. J., Pippin, J., Shankland, S. J., and Peti-Peterdi, J. (2014). Intravital imaging of podocyte calcium in glomerular injury and disease. *J Clin Invest*, *124*(5): 2050–2058.

Butcher, E. C. (1991). Leukocyte-endothelial cell recognition: Three (or more) steps to specificity and diversity. *Cell 67*(6): 1033–1036.

Carbo, C., Yuki, K., Demers, M., Wagner, D. D., and Shimaoka, M. (2013). Isoflurane inhibits neutrophil recruitment in the cutaneous Arthus reaction model. *J Anesth*, *27*(2): 261–268.

Chen, Y., Maeda, A., Bu, J., and DaCosta, R. (2016). Femur window chamber model for in vivo cell tracking in the murine bone marrow. *J Vis Exp*. (113):e54205

Chia, T. H., and Levene, M. J. (2009). Microprisms for in vivo multilayer cortical imaging. *J Neurophysiol*, *102*(2): 1310–1314.

Chilian, W. M., and Layne, S. M. (1990). Coronary microvascular responses to reductions in perfusion pressure. Evidence for persistent arteriolar vasomotor tone during coronary hypoperfusion. *Circ Res*, *66*(5): 1227–1238.

Cho, K. W., Morris, D. L., DelProposto, J. L., Geletka, L., Zamarron, B., Martinez-Santibanez, G., Meyer, K. A., Singer, K., O'Rourke, R. W., and Lumeng, C. N. (2014). An MHC II-dependent activation loop between adipose tissue macrophages and CD4+ T cells controls obesity-induced inflammation. *Cell Rep*, *9*(2): 605–617.

Choe, K., Jang, J. Y., Park, I., Kim, Y., Ahn, S., Park, D. Y., Hong, Y. K., Alitalo, K., Koh, G. Y., and Kim, P. (2015). Intravital imaging of intestinal lacteals unveils lipid drainage through contractility. *J Clin Invest*, *125*(11): 4042–4052.

Choi, M., Lee, W. M., and Yun, S. H. (2015). Intravital microscopic interrogation of peripheral taste sensation. *Sci Rep*, *5*: 8661.

Chtanova, T., Schaeffer, M., Han, S. J., van Dooren, G. G., Nollmann, M., Herzmark, P., Chan, S. W., Satija, H., Camfield, K., Aaron, H., Striepen, B., and Robey, E. A. (2008). Dynamics of neutrophil migration in lymph nodes during infection. *Immunity*, *29*(3): 487–496.

Conway JRW, Warren SC, Herrmann D, Murphy KJ, Cazet AS, Vennin C, Shearer RF, Killen MJ, Magenau A, Mélénec P, Pinese M, Nobis M, Zaratzian A, Boulghourjian A, Da Silva AM, Del Monte-Nieto G, Adam ASA, Harvey RP, Haigh JJ, Wang Y, Croucher DR, Sansom OJ, Pajic M, Caldon CE, Morton JP, Timpson P. (2018). Intravital imaging to monitor therapeutic response in moving hypoxic regions resistant to PI3K pathway targeting in pancreatic cancer. *Cell Rep*, *23*(11): 3312–3326

Coppieters, K., Amirian, N., and von Herrath, M. (2012). Intravital imaging of CTLs killing islet cells in diabetic mice. *J Clin Invest*, *122*(1): 119–131.

Coppieters, K., Martinic, M. M., Kiosses, W. B., Amirian, N., and von Herrath, M. (2010). A novel technique for the in vivo imaging of autoimmune diabetes development in the pancreas by two-photon microscopy. *PLOS ONE*, *5*(12): e15732.

Cummings, R. J., Mitra, S., Lord, E. M., and Foster, T. H. (2008). Antibody-labeled fluorescence imaging of dendritic cell populations in vivo. *J Biomed Opt*, *13*(4): 044041.

Cyster, J. G., and Schwab, S. R. (2012). Sphingosine-1-phosphate and lymphocyte egress from lymphoid organs. *Annu Rev Immunol*, *30*: 69–94.

de Grauw, C. J., Vroom, J. M., van der Voort, H. T., and Gerritsen, H. C. (1999). Imaging properties in two-photon excitation microscopy and effects of refractive-index mismatch in thick specimens. *Appl Opt*, *38*(28): 5995–6003.

de Moraes, L. V., Tadokoro, C. E., Gomez-Conde, I., Olivieri, D. N., and Penha-Goncalves, C. (2013). Intravital placenta imaging reveals microcirculatory dynamics impact on sequestration and phagocytosis of *Plasmodium*-infected erythrocytes. *PLOS Pathog*, *9*(1): e1003154.

Devi, S., Li, A., Westhorpe, C. L., Lo, C. Y., Abeynaike, L. D., Snelgrove, S. L., Hall, P., Ooi, J. D., Sobey, C. G., Kitching, A. R., and Hickey, M. J. (2013). Multiphoton imaging reveals a new leukocyte recruitment paradigm in the glomerulus. *Nat Med*, *19*(1): 107–112.

Drew, P. J., Shih, A. Y., Driscoll, J. D., Knutsen, P. M., Blinder, P., Davalos, D., Akassoglou, K., Tsai, P. S., and Kleinfeld, D. (2010). Chronic optical access through a polished and reinforced thinned skull. *Nat Methods*, *7*(12): 981–984.

Driscoll, J. D., Shih, A. Y., Iyengar, S., Field, J. J., White, G. A., Squier, J.A., Cauwenberghs, G., and Kleinfeld, D. (2011). Photon counting, censor corrections, and lifetime imaging for improved detection in two-photon microscopy. *J Neurophysiol*, *105*(6): 3106–3113.

Dubertret, B., Skourides, P., Norris, D. J., Noireaux, V., Brivanlou, A. H., and Libchaber, A. (2002). In vivo imaging of quantum dots encapsulated in phospholipid micelles. *Science*, *298*(5599): 1759–1762.

Dunn, K. W., Sandoval, R. M., Kelly, K. J., Dagher, P. C., Tanner, G. A., Atkinson, S.J., Bacallao, R. L., and Molitoris, B. A. (2002). Functional studies of the kidney of living animals using multicolor two-photon microscopy. *Am J Physiol Cell Physiol*, *283*(3): C905–916.

Dunn, K. W., Sutton, T. A., and Sandoval, R. M. (2012). Live-animal imaging of renal function by multiphoton microscopy. *Curr Protoc Cytom*, Chapter 12: 62:12.9.1–12.9.18.

Dustin, M. L., and Springer, T. A. (1988). Lymphocyte function-associated antigen-1 (LFA-1) interaction with intercellular adhesion molecule-1 (ICAM-1) is one of at least three mechanisms for lymphocyte adhesion to cultured endothelial cells. *J Cell Biol*, *107*(1): 321–331.

Egeblad, M., Ewald, A. J., Askautrud, H. A., Truitt, M. L., Welm, B. E., Bainbridge, E., Peeters, G., Krummel, M. F., and Werb, Z. (2008). Visualizing stromal cell dynamics in different tumor microenvironments by spinning disk confocal microscopy. *Dis Model Mech*, *1*(2–3): 155–167; discussion 165.

Egen, J. G., Rothfuchs, A. G., Feng, C. G., Winter, N., Sher, A., and Germain, R. N. (2008). Macrophage and t cell dynamics during the development and disintegration of mycobacterial granulomas. *Immunity*, *28*(2): 271–284.

Engelbrecht, C. J., Johnston, R. S., Seibel, E. J., and Helmchen, F. (2008). Ultra-compact fiber-optic two-photon microscope for functional fluorescence imaging in vivo. *Opt Express*, *16*(8): 5556–5564.

Engelbrecht, C. J., and Stelzer, E. H. (2006). Resolution enhancement in a light-sheet-based microscope (SPIM). *Opt Lett*, *31*(10): 1477–1479.

Entenberg, D., Voiculescu, S., Guo, P., Borriello, L., Wang, Y., Karagiannis, G. S., Jones, J., Baccay, F., Oktay, M., and Condeelis, J. (2018). A permanent window for the murine lung enables high-resolution imaging of cancer metastasis. *Nat Methods*, *15*(1): 73–80.

Farrar, M. J., Bernstein, I. M., Schlafer, D. H., Cleland, T. A., Fetcho, J. R., and Schaffer, C. B. (2012). Chronic in vivo imaging in the mouse spinal cord using an implanted chamber. *Nat Methods*, *9*(3): 297–302.

Follain G, Osmani N, Azevedo AS, Allio G, Mercier L, Karreman MA, Solecki G, Garcia Leòn MJ, Lefebvre O, Fekonja N, Hille C, Chabannes V, Dollé G, Metivet T, Hovsepian F, Prudhomme C, Pichot A, Paul N, Carapito R, Bahram S, Ruthensteiner B, Kemmling A, Siemonsen S, Schneider T, Fiehler J, Glatzel M, Winkler F, Schwab Y, Pantel K, Harlepp S, Goetz JG. (2018) Hemodynamic forces tune the arrest, adhesion, and extravasation of circulating tumor cells. *Dev Cell*, *45*(1): 33–52:e12.

Gao, X., Cui, Y., Levenson, R. M., Chung, L. W., and Nie, S. (2004). In vivo cancer targeting and imaging with semiconductor quantum dots. *Nat Biotechnol*, *22*(8): 969–976.

Germain, R. N., Miller, M. J., Dustin, M. L., and Nussenzweig, M. C. (2006). Dynamic imaging of the immune system: Progress, pitfalls and promise. *Nat Rev Immunol*, *6*(7): 497–507.

Ghosh, S., and Preza, C. (2015). Fluorescence microscopy point spread function model accounting for aberrations due to refractive index variability within a specimen. *J Biomed Opt*, *20*(7): 75003.

Ghosh, S. K., Kim, P., Zhang, X. A., Yun, S. H., Moore, A., Lippard, S. J., and Medarova, Z. (2010). A novel imaging approach for early detection of prostate cancer based on endogenous zinc sensing. *Cancer Res*, *70*(15): 6119–6127.

Giepmans, B. N., Adams, S. R., Ellisman, M. H., and Tsien, R. Y. (2006). The fluorescent toolbox for assessing protein location and function. *Science*, *312*(5771): 217–224.

Grayson, M. H., Hotchkiss, R. S., Karl, I. E., Holtzman, M. J., and Chaplin, D. D. (2003). Intravital microscopy comparing t lymphocyte trafficking to the spleen and the mesenteric lymph node. *Am J Physiol Heart Circ Physiol*, *284*(6): H2213–2226.

Greenberg, D. S., and Kerr, J. N. (2009). Automated correction of fast motion artifacts for two-photon imaging of awake animals. *J Neurosci Methods*, *176*(1): 1–15.

Grutzendler, J., Kasthuri, N., and Gan, W. B. (2002). Long-term dendritic spine stability in the adult cortex. *Nature*, *420*(6917): 812–816.

Hackl, M. J., Burford, J. L., Villanueva, K., Lam, L., Susztak, K., Schermer, B., Benzing, T., and Peti-Peterdi, J. (2013). Tracking the fate of glomerular epithelial cells in vivo using serial multiphoton imaging in new mouse models with fluorescent lineage tags. *Nat Med*, *19*(12): 1661–1666.

Hasegawa, A., Hayashi, K., Kishimoto, H., Yang, M., Tofukuji, S., Suzuki, K., Nakajima, H., Hoffman, R. M., Shirai, M., and Nakayama, T. (2010). Color-coded real-time cellular imaging of lung t-lymphocyte accumulation and focus formation in a mouse asthma model. *J Allergy Clin Immunol*, *125*(2): 461–468: e466.

Headley, M. B., Bins, A., Nip, A., Roberts, E. W., Looney, M. R., Gerard, A., and Krummel, M. F. (2016). Visualization of immediate immune responses to pioneer metastatic cells in the lung. *Nature*, *531*(7595): 513–517.

Heim, R., Prasher, D. C., and Tsien, R. Y. (1994). Wavelength mutations and posttranslational autoxidation of green fluorescent protein. *Proc Natl Acad Sci U S A*, *91*(26): 12501–12504.

Herz, J., Siffrin, V., Hauser, A. E., Brandt, A. U., Leuenberger, T., Radbruch, H., Zipp, F., and Niesner, R. A. (2010). Expanding two-photon intravital microscopy to the infrared by means of optical parametric oscillator. *Biophys J*, *98*(4): 715–723.

Hickman, H. D., Takeda, K., Skon, C. N., Murray, F. R., Hensley, S. E., Loomis, J., Barber, G. N., Bennink, J. R., and Yewdell, J. W. (2008). Direct priming of antiviral CD8+ t cells in the peripheral interfollicular region of lymph nodes. *Nat Immunol*, *9*(2): 155–165.

Holtmaat, A., Bonhoeffer, T., Chow, D. K., Chuckowree, J., De Paola, V., Hofer, S. B., Hubener, M., Keck, T., Knott, G., Lee, W. C., Mostany, R., Mrsic-Flogel, T. D., Nedivi, E., Portera-Cailliau, C., Svoboda, K., Trachtenberg, J. T., and Wilbrecht, L. (2009). Long-term, high-resolution imaging in the mouse neocortex through a chronic cranial window. *Nat Protoc*, *4*(8): 1128–1144.

Holtom, G. R., Thrall, B. D., Chin, B. Y., Wiley, H. S., and Colson, S. D. (2001). Achieving molecular selectivity in imaging using multiphoton Raman spectroscopy techniques. *Traffic*, *2*(11): 781–788.

Hosoe, N., Miura, S., Watanabe, C., Tsuzuki, Y., Hokari, R., Oyama, T., Fujiyama, Y., Nagata, H., and Ishii, H. (2004). Demonstration of functional role of TECK/CCL25 in t lymphocyte-endothelium interaction in inflamed and uninflamed intestinal mucosa. *Am J Physiol Gastrointest Liver Physiol*, *286*(3): G458–466.

Imai, Y., Park, E. J., Peer, D., Peixoto, A., Cheng, G., von Andrian, U. H., Carman, C. V., and Shimaoka, M. (2008). Genetic perturbation of the putative cytoplasmic membrane-proximal salt bridge aberrantly activates alpha(4) integrins. *Blood*, *112*(13): 5007–5015.

Ingaramo, M., York, A. G., Wawrzusin, P., Milberg, O., Hong, A., Weigert, R., Shroff, H., Patterson, and G. H. (2014). Two-photon excitation improves multifocal structured illumination microscopy in thick scattering tissue. *Proc Natl Acad Sci U S A*, *111*(14): 5254–5259.

Jaiswal, J. K., Mattoussi, H., Mauro, J. M., and Simon, S. M. (2003). Long-term multiple color imaging of live cells using quantum dot bioconjugates. *Nat Biotechnol*, *21*(1): 47–51.

Ji, N., Sato, T. R., and Betzig, E. (2012). Characterization and adaptive optical correction of aberrations during in vivo imaging in the mouse cortex. *Proc Natl Acad Sci U S A*, *109*(1): 22–27.

Jung, K., Heishi, T., Khan, O. F., Kowalski, P. S., Incio, J., Rahbari, N. N., Chung, E., Clark, J. W., Willett, C. G., Luster, A. D., Yun, S. H., Langer, R., Anderson, D. G., Padera, T. P., Jain, R. K., and Fukumura, D. (2017). Ly6Clo monocytes drive immunosuppression and confer resistance to anti-VEGFR2 cancer therapy. *J Clin Invest*, *127*(8): 3039–3051.

Jung, K., Kim, P., Leuschner, F., Gorbatov, R., Kim, J. K., Ueno, T., Nahrendorf, M., and Yun, S. H. (2013). Endoscopic time-lapse imaging of immune cells in infarcted mouse hearts. *Circ Res*, *112*(6): 891–899.

Kang, G. J., Ecoiffier, T., Truong, T., Yuen, D., Li, G., Lee, N., Zhang, L., and Chen, L. (2016). Intravital imaging reveals dynamics of lymphangiogenesis and valvulogenesis. *Sci Rep*, 6: 19459.

Karreman, M. A., Hyenne, V., Schwab, Y., and Goetz, J. G. (2016). Intravital correlative microscopy: Imaging life at the nanoscale. *Trends Cell Biol*, *26*(11): 848–863.

Kataoka, H., Ushiyama, A., Kawakami, H., Akimoto, Y., Matsubara, S., and Iijima, T. (2016) Fluorescent imaging of endothelial glycocalyx layer with wheat germ agglutinin using intravital microscopy. *Microsc Res Tech*, *79*(1): 31–37.

Kerschensteiner, M., Schwab, M. E., Lichtman, J. W., and Misgeld, T. (2005). In vivo imaging of axonal degeneration and regeneration in the injured spinal cord. *Nat Med*, *11*(5): 572–577.

Kienast, Y., von Baumgarten, L., Fuhrmann, M., Klinkert, W. E., Goldbrunner, R., Herms, J., and Winkler, F. (2010). Real-time imaging reveals the single steps of brain metastasis formation. *Nat Med*, *16*(1): 116–122.

Kim, J. K., Lee, W. M., Kim, P., Choi, M., Jung, K., Kim, S., and Yun, S. H. (2012). Fabrication and operation of GRIN probes for in vivo fluorescence cellular imaging of internal organs in small animals. *Nat Protoc*, *7*(8): 1456–1469.

Kim, J. V., Kang, S. S., Dustin, M. L., and McGavern, D. B. (2009). Myelomonocytic cell recruitment causes fatal CNS vascular injury during acute viral meningitis. *Nature*, *457*(7226): 191–195.

Kim, P., Chung, E., Yamashita, H., Hung, K. E., Mizoguchi, A., Kucherlapati, R., Fukumura, D., Jain, R. K., and Yun, S. H. (2010). In vivo wide-area cellular imaging by side-view endomicroscopy. *Nat Methods*, *7*(4): 303–305.

Kim, S., Lim, Y. T., Soltesz, E. G., De Grand, A. M., Lee, J., Nakayama, A., Parker, J. A., Mihaljevic, T., Laurence, R. G., Dor, D. M., Cohn, L. H., Bawendi, M. G., and Frangioni, J. V. (2004). Near-infrared fluorescent type II quantum dots for sentinel lymph node mapping. *Nat Biotechnol*, *22*(1): 93–97.

Kirkpatrick, N. D., Chung, E., Cook, D. C., Han, X., Gruionu, G., Liao, S., Munn, L. L., Padera, T. P., Fukumura, D., and Jain, R. K. (2012). Video-rate resonant scanning multiphoton microscopy: An emerging technique for intravital imaging of the tumor microenvironment. *IntraVital*, *1*(1):60–68.

KleinJan, G. H., Buckle, T., van Willigen, D. M., van Oosterom, M. N., Spa, S. J., Kloosterboer, H. E., and van Leeuwen, F. W. (2014). Fluorescent lectins for local in vivo visualization of peripheral nerves. *Molecules*, *19*(7): 9876–9892.

Kobat, D., Durst, M. E., Nishimura, N., Wong, A. W., Schaffer, C. B., and Xu, C. (2009). Deep tissue multiphoton microscopy using longer wavelength excitation. *Opt Express*, *17*(16): 13354–13364.

Kobat, D., Horton, N. G., and Xu, C. (2011). In vivo two-photon microscopy to 1.6-mm depth in mouse cortex. *J Biomed Opt*, *16*(10): 106014.

Kohler, A., De Filippo, K., Hasenberg, M., van den Brandt, C., Nye, E., Hosking, M. P., Lane, T. E., Mann, L., Ransohoff, R. M., Hauser, A. E., Winter, O., Schraven, B., Geiger, H., Hogg, N., and Gunzer, M. (2011). G-CSF-mediated thrombopoietin release triggers neutrophil motility and mobilization from bone marrow via induction of Cxcr2 ligands. *Blood*, *117*(16): 4349–4357.

Kowalewska, P. M., Burrows, L. L., and Fox-Robichaud, A. E. (2011). Intravital microscopy of the murine urinary bladder microcirculation. *Microcirculation*, *18*(8): 613–622.

Kreisel, D., Nava, R. G., Li, W., Zinselmeyer, B. H., Wang, B., Lai, J., Pless, R., Gelman, A. E., Krupnick, A. S., and Miller, M. J. (2010). In vivo two-photon imaging reveals monocyte-dependent neutrophil extravasation during pulmonary inflammation. *Proc Natl Acad Sci U S A*, *107*(42): 18073–18078.

Krzic, U., Gunther, S., Saunders, T. E., Streichan, S. J., and Hufnagel, L. (2012). Multiview light-sheet microscope for rapid in toto imaging. *Nat Methods*, *9*(7): 730–733.

Laffray, S., Pages, S., Dufour, H., De Koninck, P., De Koninck, Y., and Cote, D. (2011). Adaptive movement compensation for in vivo imaging of fast cellular dynamics within a moving tissue. *PLOS ONE*, *6*(5): e19928.

Larson, D. R., Zipfel, W. R., Williams, R. M., Clark, S. W., Bruchez, M. P., Wise, F. W., and Webb, W. W. (2003). Water-soluble quantum dots for multiphoton fluorescence imaging in vivo. *Science*, *300*(5624): 1434–1436.

Le, V. H., Lee, S., Lee, S., Wang, T., Hyuk Jang, W., Yoon, Y., Kwon, S., Kim, H., Lee, S. W., and Hean Kim, K. (2017). In vivo longitudinal visualization of bone marrow engraftment process in mouse calvaria using two-photon microscopy. *Sci Rep*, 7: 44097.

Lee, S., Vinegoni, C., Feruglio, P. F., Fexon, L., Gorbatov, R., Pivoravov, M., Sbarbati, A., Nahrendorf, M., and Weissleder, R. (2012). Real-time in vivo imaging of the beating mouse heart at microscopic resolution. *Nat Commun*, 3: 1054.

Li, W., Nava, R. G., Bribriesco, A. C., Zinselmeyer, B. H., Spahn, J. H., Gelman, A. E., Krupnick, A. S., Miller, M. J., and Kreisel, D. (2012). Intravital 2-photon imaging of leukocyte trafficking in beating heart. *J Clin Invest*, *122*(7): 2499–2508.

Li, X., and Yu, W. (2008). Deep tissue microscopic imaging of the kidney with a gradient-index lens system. *Opt Commun*, *281*(7): 1833–1840.

Liebling, M., Forouhar, A. S., Gharib, M., Fraser, S. E., and Dickinson, M. E. (2005). Four-dimensional cardiac imaging in living embryos via postacquisition synchronization of nongated slice sequences. *J Biomed Opt*, *10*(5): 054001.

Lindquist, R. L., Shakhar, G., Dudziak, D., Wardemann, H., Eisenreich, T., Dustin, M. L., and Nussenzweig, M. C. (2004). Visualizing dendritic cell networks in vivo. *Nat Immunol*, *5*(12): 1243–1250.

Liou, H. L., Myers, J. T., Barkauskas, D. S., and Huang, A. Y. (2012). Intravital imaging of the mouse popliteal lymph node. *J Vis Exp.* (60): e3720

Liu, Z., Lavis, L. D., and Betzig, E. (2015). Imaging live-cell dynamics and structure at the single-molecule level. *Mol Cell, 58*(4): 644–659.

Llewellyn, M. E., Barretto, R. P., Delp, S. L., and Schnitzer, M. J. (2008). Minimally invasive high-speed imaging of sarcomere contractile dynamics in mice and humans. *Nature, 454*(7205): 784–788.

Looney, M. R., Thornton, E. E., Sen, D., Lamm, W. J., Glenny, R. W., and Krummel, M. F. (2011). Stabilized imaging of immune surveillance in the mouse lung. *Nat Methods, 8*(1): 91–96.

Luckner, M., Burgold, S., Filser, S., Scheungrab, M., Niyaz, Y., Hummel, E., Wanner, G., and Herms, J. (2018). Label-free 3D-CLEM using endogenous tissue landmarks. *iScience, 6*: 92–101.

Marangoni, F., Murooka, T. T., Manzo, T., Kim, E. Y., Carrizosa, E., Elpek, N. M., and Mempel, T. R. (2013). The transcription factor NFAT exhibits signal memory during serial t cell interactions with antigen-presenting cells. *Immunity, 38*(2): 237–249.

Mari, P. O., Verbiest, V., Sabbioneda, S., Gourdin, A. M., Wijgers, N., Dinant, C., Lehmann, A. R., Vermeulen, W., and Giglia-Mari, G. (2010). Influence of the live cell DNA marker DRAQ5 on chromatin-associated processes. *DNA Repair (Amst), 9*(7): 848–855.

Marques, P. E., Antunes, M. M., David, B. A., Pereira, R. V., Teixeira, M. M., and Menezes, G. B. (2015). Imaging liver biology in vivo using conventional confocal microscopy. *Nat Protoc, 10*(2): 258–268.

Matheu, M. P., Deane, J. A., Parker, I., Fruman, D. A., and Cahalan, M. D. (2007). Class IA phosphoinositide 3-kinase modulates basal lymphocyte motility in the lymph node. *J Immunol, 179*(4): 2261–2269.

Mattheakis, L. C., Dias, J. M., Choi, Y. J., Gong, J., Bruchez, M. P., Liu, J., and Wang, E. (2004). Optical coding of mammalian cells using semiconductor quantum dots. *Anal Biochem, 327*(2): 200–208.

Mazo, I. B., Gutierrez-Ramos, J. C., Frenette, P. S., Hynes, R. O., Wagner, D. D., and von Andrian, U. H. (1998). Hematopoietic progenitor cell rolling in bone marrow microvessels: Parallel contributions by endothelial selectins and vascular cell adhesion molecule 1. *J Exp Med, 188*(3): 465–474.

McDole, J. R., Wheeler, L. W., McDonald, K. G., Wang, B., Konjufca, V., Knoop, K. A., Newberry, R. D., and Miller, M. J. (2012). Goblet cells deliver luminal antigen to CD103+ dendritic cells in the small intestine. *Nature, 483*(7389): 345–349.

McElroy, M., Hayashi, K., Garmy-Susini, B., Kaushal, S., Varner, J. A., Moossa, A. R., Hoffman, R. M., and Bouvet, M. (2009). Fluorescent LYVE-1 antibody to image dynamically lymphatic trafficking of cancer cells in vivo. *J Surg Res, 151*(1): 68–73.

Meijer, E. F. J., Jeong, H. S., Pereira, E. R., Ruggieri, T. A., Blatter, C., Vakoc, B. J., and Padera, T. P. (2017). Murine chronic lymph node window for longitudinal intravital lymph node imaging. *Nat Protoc, 12*(8): 1513–1520.

Mempel, T. R., Henrickson, S. E., and Von Andrian, U. H. (2004). T-cell priming by dendritic cells in lymph nodes occurs in three distinct phases. *Nature, 427*(6970): 154–159.

Metchnikoff, E. (1893). *Lectures on the comparative pathology of inflammation delivered at the Pasteur Institute in 1891.* London Kegan Paul, Trench, Trübner.

Michalet, X., Pinaud, F. F., Bentolila, L. A., Tsay, J. M., Doose, S., Li, J. J., Sundaresan, G., Wu, A. M., Gambhir, S. S., and Weiss, S. (2005). Quantum dots for live cells, in vivo imaging, and diagnostics. *Science, 307*(5709): 538–544.

Migone, F. F., Cowan, R. G., Williams, R. M., Gorse, K. J., Zipfel, W. R., and Quirk, S. M. (2016). In vivo imaging reveals an essential role of vasoconstriction in rupture of the ovarian follicle at ovulation. *Proc Natl Acad Sci U S A, 113*(8): 2294–2299.

Migone, F. F., Cowan, R. G., Williams, R. M., Zipfel, W. R., and Quirk, S. M. (2013). Multiphoton microscopy as a tool to study ovarian vasculature in vivo. *IntraVital, 2*(1): e24334.

Miller, M. J., Wei, S. H., Cahalan, M. D., and Parker, I. (2003). Autonomous t cell trafficking examined in vivo with intravital two-photon microscopy. *Proc Natl Acad Sci U S A, 100*(5): 2604–2609.

Miller, M. J., Safrina, O., Parker, I., and Cahalan, M. D. (2004). Imaging the single cell dynamics of CD4+ t cell activation by dendritic cells in lymph nodes. *J Exp Med, 200*(7): 847–856.

Miller, M. J., Wei, S. H., Parker, I., and Cahalan, M. D. (2002). Two-photon imaging of lymphocyte motility and antigen response in intact lymph node. *Science, 296*(5574): 1869–1873.

Moreau, H. D., Lemaitre, F., Terriac, E., Azar, G., Piel, M., Lennon-Dumenil, A. M., and Bousso, P. (2012). Dynamic in situ cytometry uncovers T cell receptor signaling during immunological synapses and kinapses in vivo. *Immunity, 37*(2): 351–363.

Niesner, R., Andresen, V., Neumann, J., Spiecker, H., and Gunzer, M. (2007). The power of single and multibeam two-photon microscopy for high-resolution and high-speed deep tissue and intravital imaging. *Biophys J, 93*(7): 2519–2529.

Nobis M, Herrmann D, Warren SC, Kadir S, Leung W, Killen M, Magenau A, Stevenson D, Lucas MC, Reischmann N, Vennin C, Conway JRW, Boulghourjian A, Zaratzian A, Law AM, Gallego-Ortega D, Ormandy CJ, Walters SN, Grey ST, Bailey J, Chtanova T, Quinn JMW, Baldock PA, Croucher PI, Schwarz JP, Mrowinska A, Zhang L, Herzog H, Masedunskas A, Hardeman EC, Gunning PW, Del Monte-Nieto G, Harvey RP, Samuel MS, Pajic M, McGhee EJ, Johnsson AE, Sansom OJ, Welch HCE, Morton JP, Strathdee D, Anderson KI, Timpson P (2017). A RhoA-FRET biosensor mouse for intravital imaging in normal tissue homeostasis and disease contexts. *Cell Rep, 21*(1): 274–288.

Osswald M, Jung E, Sahm F, Solecki G, Venkataramani V, Blaes J, Weil S, Horstmann H, Wiestler B, Syed M, Huang L, Ratliff M, Karimian Jazi K, Kurz FT, Schmenger T, Lemke D, Gömmel M, Pauli M, Liao Y, Häring P, Pusch S, Herl V, Steinhäuser C, Krunic D, Jarahian M, Miletic H, Berghoff AS, Griesbeck O, Kalamakis G, Garaschuk O, Preusser M, Weiss S, Liu H, Heiland S, Platten M, Huber PE, Kuner T, von Deimling A, Wick W, Winkler F. (2015). Brain tumour cells interconnect to a functional and resistant network. *Nature, 528*(7580): 93–98.

Pai, S., Danne, K. J., Qin, J., Cavanagh, L. L., Smith, A., Hickey, M. J., and Weninger, W. (2012). Visualizing leukocyte trafficking in the living brain with 2-photon intravital microscopy. *Front Cell Neurosci, 6*: 67.

Parish, C. R. (1999). Fluorescent dyes for lymphocyte migration and proliferation studies. *Immunol Cell Biol*, *77*(6): 499–508.

Perro, M., Tsang, J., Xue, S. A., Escors, D., Cesco-Gaspere, M., Pospori, C., Gao, L., Hart, D., Collins, M., Stauss, H., and Morris, E. C. (2010). Generation of multi-functional antigen-specific human t-cells by lentiviral TCR gene transfer. *Gene Ther*, *17*(6): 721–732.

Prasher, D. C., Eckenrode, V. K., Ward, W. W., Prendergast, F. G., and Cormier, M. J. (1992). Primary structure of the Aequorea victoria green-fluorescent protein. *Gene*, *111*(2): 229–233.

Rashidian, M., Keliher, E. J., Bilate, A. M., Duarte, J. N., Wojtkiewicz, G. R., Jacobsen, J. T., Cragnolini, J., Swee, L. K., Victora, G. D., Weissleder, R., and Ploegh, H. L. (2015). Noninvasive imaging of immune responses. *Proc Natl Acad Sci U S A*, *112*(19): 6146–6151.

Ritsma, L., Ellenbroek, S. I. J., Zomer, A., Snippert, H. J., de Sauvage, F. J., Simons, B. D., Clevers, H., and van Rheenen, J. (2014). Intestinal crypt homeostasis revealed at single-stem-cell level by in vivo live imaging. *Nature*, *507*(7492): 362–365.

Ritsma, L., Steller, E. J., Ellenbroek, S. I., Kranenburg, O., Borel Rinkes, I. H., and van Rheenen, J. (2013). Surgical implantation of an abdominal imaging window for intravital microscopy. *Nat Protoc*, *8*(3): 583–594.

Sano, T., Kobayashi, T., Negoro, H., Sengiku, A., Hiratsuka, T., Kamioka, Y., Liou, L. S., and Ogawa, O., Matsuda, M. (2016). Intravital imaging of mouse urothelium reveals activation of extracellular signal-regulated kinase by stretch-induced intravesical release of ATP. *Physiol Rep*, *4*(21): e13033.

Scheele, C. L., Hannezo, E., Muraro, M. J., Zomer, A., Langedijk, N. S., van Oudenaarden, A., Simons, B. D., and van Rheenen, J. (2017). Identity and dynamics of mammary stem cells during branching morphogenesis. *Nature*, *542*(7641): 313–317.

Schiessl, I. M., Hammer, A., Kattler, V., Gess, B., Theilig, F., Witzgall, R., and Castrop, H. (2016). Intravital imaging reveals angiotensin II-induced transcytosis of albumin by podocytes. *J Am Soc Nephrol*, *27*(3): 731–744.

Schmerse, F., Woidacki, K., Riek-Burchardt, M., Reichardt, P., Roers, A., Tadokoro, C., and Zenclussen, A. C. (2014). In vivo visualization of uterine mast cells by two-photon microscopy. *Reproduction*, *147*(6): 781–788.

Schramm, R., Schafers, H. J., Harder, Y., Schmits, R., Thorlacius, H., and Menger, M. D. (2006). The cervical lymph node preparation: A novel approach to study lymphocyte homing by intravital microscopy. *Inflamm Res*, *55*(4): 160–167.

Schuh, C. D., Haenni, D., Craigie, E., Ziegler, U., Weber, B., Devuyst, O., and Hall, A. M. (2015). Long wavelength multiphoton excitation is advantageous for intravital kidney imaging. *Kidney Int*, *89*(3): 712–719.

Shaner, N. C., Steinbach, P. A., and Tsien, R. Y. (2005). A guide to choosing fluorescent proteins. *Nat Methods*, *2*(12): 905–909.

Sipkins, D. A., Wei, X., Wu, J. W., Runnels, J. M., Cote, D., Means, T. K., Luster, A. D., Scadden, D. T., and Lin, C. P. (2005). In vivo imaging of specialized bone marrow endothelial microdomains for tumour engraftment. *Nature*, *435*(7044): 969–973.

Sonner, J. M., Gong, D., and Eger, E. I., 2nd. (2000). Naturally occurring variability in anesthetic potency among inbred mouse strains. *Anesth Analg*, *91*(3): 720–726.

Soulet, D., Pare, A., Coste, J., and Lacroix, S. (2013). Automated filtering of intrinsic movement artifacts during two-photon intravital microscopy. *PLOS ONE*, *8*(1): e53942.

Sparks, H., Kondo, H., Hooper, S., Munro, I., Kennedy, G., Dunsby, C., French, P., and Sahai, E. (2018). Heterogeneity in tumor chromatin-doxorubicin binding revealed by in vivo fluorescence lifetime imaging confocal endomicroscopy. *Nat Commun*, *9*(1): 2662.

Springer, T. A. (1994). Traffic signals for lymphocyte recirculation and leukocyte emigration: The multistep paradigm. *Cell*, *76*(2): 301–314.

Steven, P., Bock, F., Huttmann, G., and Cursiefen, C. (2011). Intravital two-photon microscopy of immune cell dynamics in corneal lymphatic vessels. *PLOS ONE*, *6*(10): e26253.

Stoll, S., Delon, J., Brotz, T. M., and Germain, R. N. (2002). Dynamic imaging of t cell-dendritic cell interactions in lymph nodes. *Science*, *296*(5574): 1873–1876.

Tabuchi, A., Mertens, M., Kuppe, H., Pries, A. R., and Kuebler, W. M. (2008). Intravital microscopy of the murine pulmonary microcirculation. *J Appl Physiol 1985*, *104*(2): 338–346.

Theer, P., and Denk, W. (2006). On the fundamental imaging-depth limit in two-photon microscopy. *J Opt Soc Am A Opt Image Sci Vis.*, *23*(12): 3139–3149.

Thornton, E. E., Looney, M. R., Bose, O., Sen, D., Sheppard, D., Locksley, R., Huang, X., and Krummel, M. F. (2012). Spatiotemporally separated antigen uptake by alveolar dendritic cells and airway presentation to t cells in the lung. *J Exp Med*, *209*(6): 1183–1199.

Tomer, R., Khairy, K., Amat, F., and Keller, P. J. (2012). Quantitative high-speed imaging of entire developing embryos with simultaneous multiview light-sheet microscopy. *Nat Methods*, *9*(7): 755–763.

Truong, T. V., Supatto, W., Koos, D. S., Choi, J. M., and Fraser, S. E. (2011). Deep and fast live imaging with two-photon scanned light-sheet microscopy. *Nat Methods*, *8*(9): 757–760.

Tsien, R. Y. (1998). The green fluorescent protein. *Annu Rev Biochem*, *67*: 509–544.

Tuohy, S., and Podoleanu, A. G. (2010). Depth-resolved wavefront aberrations using a coherence-gated Shack-Hartmann wavefront sensor. *Opt Express*, *18*(4): 3458–3476.

Vajkoczy, P., and Menger, M. D. (1994). New model for the study of the microcirculation of islet grafts in hairless and nude mice. *Transplant Proc*, *26*(2): 687.

van Leeuwenhoek, A., and Hoole, S. (1800). *The Select Works of Antony Van Leeuwenhoek, Containing His Microscopical Discoveries in Many of the Works of Nature*: London Whittingham and Arliss.

Veihelmann, A., Hofbauer, A., Krombach, F., Dorger, M., Maier, M., Refior, H. J., and Messmer, K. (2002). Differential function of nitric oxide in murine antigen-induced arthritis. *Rheumatol Oxf Engl*, *41*(5): 509–517.

Veihelmann, A., Szczesny, G., Nolte, D., Krombach, F., Refior, H. J., and Messmer, K. (1998). A novel model for the study of synovial microcirculation in the mouse knee joint in vivo. *Res Exp Med (Berl)*, *198*(1): 43–54.

Vercauteren, T., Perchant, A., Malandain, G., Pennec, X., and Ayache, N. (2006). Robust mosaicing with correction of motion distortions and tissue deformations for in vivo fibered microscopy. *Med Image Anal, 10*(5): 673–692.

Victora, G. D., Schwickert, T. A., Fooksman, D. R., Kamphorst, A. O., Meyer-Hermann, M., Dustin, M. L., and Nussenzweig, M. C. (2010). Germinal center dynamics revealed by multiphoton microscopy with a photoactivatable fluorescent reporter. *Cell, 143*(4): 592–605.

Vinegoni, C., Lee, S., Gorbatov, R., and Weissleder, R. (2012). Motion compensation using a suctioning stabilizer for intravital microscopy. *IntraVital, 1*(2): 115–121.

von Andrian, U. H. (1996). Intravital microscopy of the peripheral lymph node microcirculation in mice. *Microcirculation, 3*(3): 287–300.

Wagner, R. (1839). Icones physiologicae; tabulae physiologiam et geneseos historiam illustrantes. Lipsiae: Voss.

Waite, J. C., Leiner, I., Lauer, P., Rae, C. S., Barbet, G., Zheng, H., Portnoy, D. A., Pamer, E. G., and Dustin, M. L. (2011). Dynamic imaging of the effector immune response to listeria infection in vivo. *PLOS Pathog, 7*(3): e1001326.

Wan, Y. Y., Leon, R. P., Marks, R., Cham, C. M., Schaack, J., Gajewski, T. F., and DeGregori, J. (2000). Transgenic expression of the coxsackie/adenovirus receptor enables adenoviral-mediated gene delivery in naive t cells. *Proc Natl Acad Sci U S A, 97*(25): 13784–13789.

Warren, S. C., Nobis, M., Magenau, A., Mohammed, Y. H., Herrmann, D., Moran, I., Vennin, C., Conway, J. R., Melenec, P., Cox, T. R., Wang, Y., Morton, J. P., Welch, H. C., Strathdee, D., Anderson, K. I., Phan, T. G., Roberts, M. S., and Timpson, P. (2018). Removing physiological motion from intravital and clinical functional imaging data. *eLife* 7:e35800.

Westphalen, K., Gusarova, G. A., Islam, M. N., Subramanian, M., Cohen, T. S., Prince, A.S., and Bhattacharya, J. (2014). Sessile alveolar macrophages communicate with alveolar epithelium to modulate immunity. *Nature, 506*(7489): 503–506.

Wong, S. E., Winbanks, C. E., Samuel, C. S., and Hewitson, T. D. (2010). Lectin histochemistry for light and electron microscopy. *Methods Mol Biol, 611*: 103–114.

Zenclussen, A. C., Olivieri, D. N., Dustin, M. L., and Tadokoro, C. E. (2012) In vivo multiphoton microscopy technique to reveal the physiology of the mouse placenta. *Am J Reprod Immunol, 68*(3): 271–278.

Zenclussen, A. C., Olivieri, D. N., Dustin, M. L., and Tadokoro, C. E. (2013). In vivo multiphoton microscopy technique to reveal the physiology of the mouse uterus. *Am J Reprod Immunol, 69*(3): 281–289.

Zhang, H., Astrof, N. S., Liu, J. H., Wang, J. H., and Shimaoka, M. (2009). Crystal structure of isoflurane bound to integrin LFA-1 supports a unified mechanism of volatile anesthetic action in the immune and central nervous systems. *FASEB J, 23*(8): 2735–2740.

Zhang, J., Campbell, R.E., Ting, A. Y., and Tsien, R.Y. (2002). Creating new fluorescent probes for cell biology. *Nat Rev Mol Cell Biol, 3*(12): 906–918.

Zhong, W., Celli, J. P., Rizvi, I., Mai, Z., Spring, B. Q., Yun, S. H., and Hasan, T. (2009). In vivo high-resolution fluorescence microendoscopy for ovarian cancer detection and treatment monitoring. *Br J Cancer, 101*(12): 2015–2022.

3

An Introduction to Live-Cell Super-Resolution Imaging

Siân Culley, Pedro Matos Pereira, Romain F. Laine, and Ricardo Henriques

CONTENTS

Fluorescence microscopy has been a crucial tool in the advancement of modern cell biology because of its noninvasive nature, compatibility with imaging live samples, and molecule-specific labeling tools. However, the resolving power of conventional fluorescence microscopy is limited to ~250–300 nm. To resolve cellular structures on a smaller size scale than this, researchers have typically relied on electron microscopy, which can provide insight into structures on a nanometer scale. While electron microscopy continues to be a valuable tool for investigating fine intracellular structures, incompatibility with live samples and its limited labeling capabilities remain obstacles to studying dynamic phenomena with high confidence in molecular identities. This has led to the development of super-resolution microscopy methods, which were developed in the early 2000s. Super-resolution microscopy bridges the resolution gap between conventional fluorescence microscopy and electron microscopy while retaining the advantages associated with light microscopy. This chapter provides a brief overview of commonly used super-resolution microscopy techniques, their applications to live-cell imaging, and future directions for this family of techniques.

3.1 The Diffraction Limit

The resolution of light microscopy has a fundamental limit. This is commonly referred to as the diffraction limit, as it arises from the diffraction of light that occurs as it passes through an aperture. In the case of a microscope, the aperture is the microscope objective whose physical size is related to the numerical aperture (NA) of the microscope by Equation 3.1 (Figure 3.1a):

$$NA = n\sin\theta \approx \frac{nD}{2f} \tag{3.1}$$

where n is the refractive index of the medium between the lens and the imaged object, D is the effective diameter of the lens aperture, and f is its focal length (distance between the lens and an in-focus object). After a point source of light undergoes diffraction through a microscope objective, it reaches the observer, not as an infinitesimally small point, but, rather, a blurred diffraction pattern called the point spread function (PSF). The three-dimensional (3D) shape of the PSF depends on both the physical properties of the microscope optics and the wavelength of the light (Figure 3.1b).

The size of the PSF compared to the size of subcellular structures being imaged imposes a limit on the structures that can be accurately resolved using fluorescence microscopy. This is referred to as the diffraction limit and was first formalized by Ernst Abbe as:

$$\Delta d_{x,y} \approx \frac{\lambda}{2NA} \tag{3.2}$$

where $\Delta d_{x,y}$ is the smallest resolvable distance between two light-emitting objects in the lateral dimension. Figure 3.1c shows how closely separated points become indistinguishable

because of the diffraction limit. The axial resolution of a microscope is considerably poorer than its lateral resolution and is given by Equation 3.3 (Inoue, 2006).

$$\Delta d_z = \frac{2\lambda n}{NA^2} \tag{3.3}$$

Thus, the axial resolution is worse by a factor of $4n/NA$, and resolutions are typically on the order of >1,000 nm. Equations 3.2 and 3.3 describe the best possible resolution achievable in an ideal imaging system; in reality, optical aberrations within the microscope optics and scattering of light induced by the sample degrade resolution further.

3.2 Super-Resolution Microscopy Techniques

Super-resolution microscopy refers to any technique based on optical microscopy that is capable of resolving object separations smaller than allowed by Abbe's diffraction limit. There are three main techniques within the field: structured illumination microscopy (SIM), stimulated emission depletion (STED) microscopy, and single molecule localization microscopy (SMLM). Table 3.1 summarizes some key features of these methods. The features described in Table 3.1 are achievable only if the user pays due attention to sample preparation and imaging conditions. These are outlined in Table 3.2.

3.2.1 Structured Illumination Microscopy

SIM is a widefield super-resolution technique that uses the mathematical analysis of a sequence of images capturing a sample fluorescently excited by a patterned illumination to retrieve subdiffraction limit features. It was first described by Mats Gustafsson in 2000 (Gustafsson, 2000).

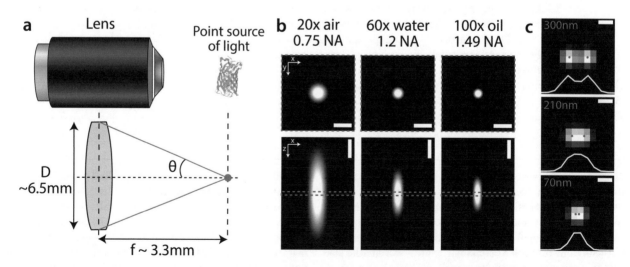

FIGURE 3.1 The diffraction limit. The resolution of fluorescence microscopy is dependent on diffraction through the optics of the microscope and varies with wavelength and properties of the objective. (a) Simplified diagram of the geometry of imaging a point source of light through a microscope objective (not to scale). (b) Simulated point spread functions (PSFs) for three different microscope objectives for imaging at $\lambda = 500$nm in the lateral (top row) and axial (bottom row) dimensions. The images in the top row correspond to the focus of the beam, highlighted with a dashed red box on the axial images. Scale bars = 500nm. Simulations were performed using PSF Generator (Kirshner et al., 2013). (c) Simulated microscope images of two point sources of light (red points) separated by different separations for a system where $\Delta d_{x,y} = 210$nm. Yellow lines indicate line profiles across center of image. Scale bars = 250nm.

TABLE 3.1

Features of Routine Applications of Commonly Used Super-Resolution Microscopy Techniques

	SIM	STED	SMLM
Type of microscope	Widefield	Confocal	Widefield
Lateral resolution	~150 nm	~50 nm	~30 nm
Axial resolution	~300 nm	~100 nm	~50 nm
Temporal resolution	~1 s	~100 ms	>10 s
Optical complexity	High	High	Low
Computational requirements	High	Low	High
Multicolor compatibility	≤ 4 colors	≤ 2 colors	≤ 2 colors
Sensitivity to sample preparation	Low	Medium	High

Values in this table represent a typical "best-case" for each method, although these can be improved using custom hardware, software, and reagents/fluorophores.

TABLE 3.2

Key Parameters to Optimize or Calibrate for Each Super-Resolution Technique

	SIM	STED	SMLM
Acquisition	Brightness of the sample and photobleaching rate during a single SIM acquisition. Correction collar (if available) of the objective to minimize spherical aberrations. Refractive index matching of immersion medium.	Choice of good STED dye(s), dense labeling, sample embedding. Depletion power, number of line and frame accumulations, gating position (for gated-STED only).	Choice of reducing agent and oxygen scavenging buffer composition (for dSTORM). Choice of dyes with appropriate photoswitching properties. Illumination and reactivation laser intensities. Mechanical stability and fiducial markers for drift correction. High labeling density required.
Reconstruction	Deconvolution parameters	None, but deconvolution is frequently applied for image restoration.	Method of localization and visualization (and their respective parameters). Pixel size. Applying appropriate thresholding of molecule localizations to reject poor fitting. Drift correction method.

When two patterns are overlaid, interference between the two results in a new emergent pattern known as moiré fringes (Figure 3.2). Relative shifts between the two patterns leads to the emergence of new moiré fringes (Figure 3.2b); as a result these fringes contain information about both underlying patterns. In SIM, the illumination field is patterned, usually in a striped or sinusoidal distribution. When the fluorescently labeled structure in the sample, which can also be thought of as a pattern, is excited with patterned illumination, the resulting collected fluorescence will thus contain moiré fringes. As one pattern is known (the illumination), computational analysis can disentangle the data from the moiré fringes in the acquired image to provide higher resolution information of the unknown pattern (the structure being imaged).

Here, it is useful to discuss the principle behind SIM in Fourier space. Typical microscopy images are a representation of the spatial information contained within the data; the Fourier transform displays the same data but in the frequency domain. The Fourier transform of a widefield fluorescence microscopy image is shown in Figure 3.3a, and it can be seen that the frequency information is contained within a central circular distribution. The radius of this circle is $k_0 = 1/\Delta d$, where Δd is the lateral resolution limit as described in Equation 3.2. Within this representation, smaller structures (i.e. higher

frequencies) lie further toward the edge of the circle, whereas larger structures (i.e. lower frequencies) are toward the center.

In SIM, the striped illumination pattern replicates the frequency distribution of the imaged structure into two additional circles within Fourier space, which represent the moiré fringes (Figure 3.3b). The positions of these new circles from the center depend on the spatial frequency of the illumination pattern (k_1) and the angle at which this pattern is incident on the sample. If the fluorophore pattern within the sample has a spatial frequency k, then moiré fringes will be created at the difference frequency ($k - k_1$). This difference frequency relocates higher frequency information (i.e. subdiffraction limit structures) into the range of frequencies observable by the microscope (i.e. the k_0 circle).

To isolate the contribution of the fluorophore pattern from the illumination pattern, the illumination is shifted at different orientations and phases; this series of images is then processed to reconstruct the super-resolution image (Figure 3.3c). This super-resolution image has a maximum frequency of ($k_0 + k_1$). However, the spatial frequency of the illumination field (k_1) is also governed by the diffraction limit, so the highest frequency within the illumination pattern is limited to k_0. This limits the highest observable image frequency in SIM to ~$2k_0$, i.e. a doubling in the resolution.

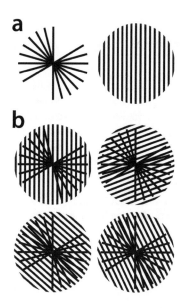

FIGURE 3.2 Moiré fringes are generated when two patterns (a) are overlaid (b). These fringes change according to the relative displacement and orientation of the two patterns.

3.2.1.1 Axial Resolution

The implementation of SIM described above increases only the lateral resolution; however, it is also possible to improve the resolution in the axial dimension by varying the illumination pattern such that rather than a striped field, a 3D checkerboard-like pattern is produced with periodic variations in the axial direction as well as the lateral. This implementation of SIM is known as three-beam SIM (for this reason, two-dimensional (2D) SIM is often referred to as two-beam SIM) (Gustafsson et al., 2008). Again, the increase in axial resolution is limited to a two-fold improvement over the diffraction limit (as described in Equation 3.3).

3.2.1.2 Temporal Resolution

The factors affecting the temporal resolution of a SIM image are the camera exposure time, the number of different grating rotations (usually either three or five), and the number of phase shifts per rotation (usually five). Depending on the optics used to generate the illumination pattern, there may also be a time delay introduced as the illumination pattern is changed. For example, for 30ms exposure, the minimum acquisition time for a raw SIM data set comprising three rotations and five phases would theoretically be 450ms. However, when testing on a commercial SIM microscope (Zeiss Elyra PS.1), the acquisition time was in fact 1.41s. It is generally accepted that the best temporal resolution of SIM is of the order of 1 second for a single color image on a commercial system. It is important to note that while the raw data acquisition can be performed at this rate the reconstructed super-resolution image is typically not immediately available for visualization, as the reconstruction is computationally intensive.

3.2.1.3 Hardware and Sample Preparation

SIM typically requires a widefield microscope body and a laser illumination source with the appropriate optics for

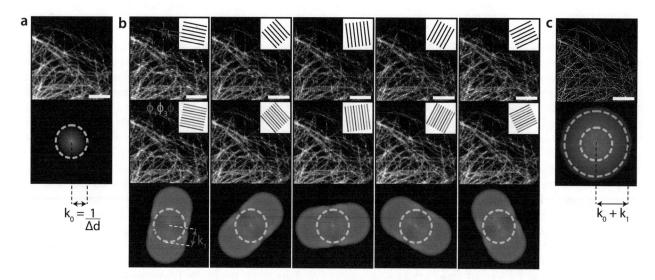

FIGURE 3.3 Structured illumination microscopy. SIM involves taking images of a labeled structure with patterned illumination and then reconstructing these into a single super-resolution image. (a) A widefield image of microtubules labeled with Alexa Fluor 488 (top) can be represented in the frequency domain by plotting its Fourier transform (bottom). The frequency information in the Fourier transform is contained within a circle whose radius is the reciprocal of the diffraction limit. (b) The same field of view is imaged using a striped illumination pattern at different rotation angles and phases. Top: Moiré patterns generated by imaging the sample at five different illumination rotations as illustrated in the top right corners. Middle: A single SIM acquisition requires that each rotation is also shifted to several different phases (typically five). Here the moiré fringes from three of the five phases are overlaid for each angle in red, green, and blue. Bottom: The Fourier transforms of the moiré fringe images now contain frequency information beyond that available in (a). (c) Computationally reconstructed SIM image (top) and its Fourier transform (bottom). The Fourier transform contains twice the frequency information of its widefield equivalent. Scale bars = 5µm.

beam shaping and filtering to produce the illumination pattern. The illumination pattern is usually generated with a spatial light modulator such as a liquid crystal device. One advantage of SIM is that it is compatible with any wavelength of illumination; however, it should be noted that, as the best resolution achievable with SIM is $\Delta d/2$, a longer wavelength of illumination results in poorer resolution than with shorter wavelengths. Scientific complementary metal oxide semiconductor (sCMOS) cameras are commonly used as detectors for SIM, as they have a large chip capable of imaging a large field of view and can also image at a high frame rate.

Commercial systems for SIM imaging are available from Cytiva (DeltaVision OMX), Nikon (N-SIM), and Zeiss (Elyra). These platforms also include their own proprietary software for reconstruction of the super-resolution data sets. However, it is also fairly straightforward to build a SIM path for an existing microscope. There are useful step-by-step guides available for this (Young, Ströhl, and Kaminski, 2016; Lu-Walther et al., 2015). An ImageJ/Fiji (Schneider, Rasband, and Eliceiri, 2012; Schindelin et al., 2012) plugin is also available for open-source reconstruction of SIM data acquired using both home-built and commercial equipment (Müller et al., 2016).

The main considerations in fluorophore choice for SIM are the same as for any laser-based widefield microscopy: good signal-to-noise ratio (SNR) and resilience to photobleaching. Fluorophores that do not meet these two criteria will lead to the generation of poor-quality images containing artifacts. Aside from fluorophore choice, SIM does not have any specific imaging buffer or sample mounting requirements, apart from needing attention to minimize refractive index mismatch in the optical path (see Table 3.2).

3.2.1.4 Examples of Live-Cell SIM

SIM has been used extensively for live-cell super-resolution microscopy for a variety of cell biology applications, which is testament to its ease of use and low phototoxicity. Fiolka et al. (2012) provided an early example showcasing multicolor live-cell 3D SIM imaging for a range of biological structures, such as mitochondria, cytoskeleton, and clathrin-coated pits. SIM has also been used for quantification of actin and membrane reorganisation at the immunological synapse (Ashdown et al., 2014, 2017), visualization of vesicular trafficking in the neuronal growth cone (Nozumi et al., 2017), and uncovering dynamic interactions between lysosomes and mitochondria (Han et al., 2017).

3.2.1.5 Limitations of SIM

The major limitation of SIM is that it fundamentally cannot attain the same resolutions as STED and SMLM techniques. Insufficient SNR in the raw images can also yield artificial structures, such as its common honeycomb-like patterning. Because SIM relies on the generation and the measurement of subtle optical fringes within the sample, any optical aberrations or distortions of the signal lead to artifactual reconstructions and poor image quality. This limits the applicability of SIM approaches in thick samples, e.g. in tissue or in animal

models, and, realistically, most 2D SIM applications are most successful when combined with total internal reflection fluorescence (TIRF).

3.2.1.6 Other Microscopy Implementations Based on the SIM Principle

A number of techniques based on similar principles to SIM have been developed. A subset of these specifically exploits the concept of photon reassignment in point-scanning confocal microscopy in so-called image scanning microscopy (ISM) methods (Sheppard, 1988; Sheppard, Mehta, and Heintzmann, 2013; Müller and Enderlein, 2010). The ISM principle has also been parallelized for increased imaging speed (York et al., 2012). These techniques are reviewed in further detail in Ströhl and Kaminski (2016). Another form of photon reassignment capable of achieving a 1.7-fold resolution increase in both the lateral and axial dimensions uses a novel "Airyscan" detector acting as a collection of small-diameter pinholes (Huff, 2015); this is commercially available from Zeiss. Another variant, known as "instant SIM" (iSIM), uses advanced optical paths to generate images with $\sqrt{2}$-fold improved resolution without the need for any computational reconstruction (York et al., 2013; Guo et al., 2018). A commercial implementation of iSIM is available from VisiTech International. Other related SIM variants, such as multifocal SIM (York et al., 2012) and rescan confocal microscopy (De Luca et al., 2013) are also available commercially (including from confocal.nl and Olympus).

Nonlinear methods also exist for increasing the resolution of SIM beyond its 2-fold limit. In theory, the resolution of SIM can be increased by increasing the illumination intensity such that fluorophore excitation becomes saturated, leading to the generation of additional harmonics (i.e. higher multiples of k_0) (Gustafsson, 2005). However, the laser illuminations required for this are so large that they are highly incompatible with live cell imaging. Alternatively, photoswitchable fluorescent proteins can be used in conjunction with patterned illumination to achieve a >2-fold improvement in resolution (Li et al., 2015).

3.2.2 Stimulated Emission Depletion Microscopy

STED microscopy is a laser-scanning confocal technique that uses an additional laser beam to deplete fluorescence from the periphery of the excitation volume and confine the emitted fluorescence to a subdiffraction limit region. The theory of STED microscopy was first described by Stefan Hell and Jan Wichmann (1994), and the first practical demonstration was performed in 2000 on yeast and *E. coli* (Klar et al., 2000).

In confocal microscopy a diffraction-limited spot of laser light is scanned across a region of interest in a labeled sample, and the resulting fluorescence is collected simultaneously at each point to build up an image pixel-by-pixel. This is in contrast to widefield techniques in which the whole sample is illuminated simultaneously. In STED microscopy, in addition to this "excitation" beam, which promotes the fluorophore from the ground state into the excited state, a second so-called "depletion" beam also scans the sample

(Figure 3.4a). The wavelength of the depletion beam is chosen to coincide with the emission spectrum of the fluorophore but not the excitation spectrum (Figure 3.4b). As a result, the depletion beam causes stimulated emission. This is a resonant process that returns fluorophores to the ground state via a transition matching the energy of the depletion beam photons. If the flux of depletion photons incident upon an excited fluorophore is large enough (i.e. the depletion beam has a sufficiently high intensity), then the fluorophore will return to the ground state preferentially via stimulated emission rather than by fluorescence. During stimulated emission, as the fluorophores are driven to de-excite through the transition dictated by the depletion beam, this excess energy will be released in the form of a photon of the same wavelength as the depletion beam.

STED microscopy exploits stimulated emission by shaping the depletion beam into an annular, or "doughnut" shape (Figure 3.4c). The doughnut beam is overlapped with the excitation beam, and both beams scan the sample simultaneously. Beneath the intense regions of the doughnut beam, stimulated emission prevents fluorescence from occurring, whereas in the dark hole at the center of the doughnut, fluorescence can occur without inhibition. In this way, STED microscopy reduces the size of the fluorescence PSF to the size of the subdiffraction limit hole in the center of the depletion beam (Figure 3.4c).

Resolution in STED microscopy scales with the size of the hole in the center of the doughnut; a smaller hole will generate a smaller fluorescent spot and therefore a higher resolution. The size of this hole is proportional to the efficiency of stimulated emission and therefore is dependent upon the intensity of the depletion beam (Figure 3.4d). This is described by a modification to Abbe's equation, as shown in Equation 3.4.

$$\Delta d_{x,y}^{\text{STED}} = \frac{\lambda}{2\text{NA}\sqrt{1 + \dfrac{I}{I_{\text{sat}}}}} \qquad (3.4)$$

Here I is the peak intensity of the doughnut beam and I_{sat} is the "saturation intensity" defined as the laser intensity required to reduce the fluorescence emission of a fluorophore to 50% of its undepleted intensity (Harke et al., 2008). It should be noted that I_{sat} depends on not just the identity of the fluorophore, but also the depletion wavelength, fluorophore microenvironment, and polarisation effects among other factors; therefore, it is not a straightforward value to ascertain reliably. Resolutions in STED microscopy acquired with high-intensity lasers can reach ~50–60 nm.

The spatial resolution of STED microscopy can also be improved via temporal "gating" of the detected fluorescence (gated-STED) (Vicidomini et al., 2011). If an excitation beam that is pulsed in time at high frequency (e.g. 80 MHz) is used in conjunction with a doughnut shaped depletion beam, the probability that a detected photon originated from the central hole (as opposed to unwanted bleed-through from the intense part of the doughnut beam) increases with time from the excitation pulse on a nanosecond timescale. Therefore, resolution can be increased by rejecting any "early" fluorescence – typically the first few nanoseconds after the fluorophore being excited – and creating the image from only later-arriving photons

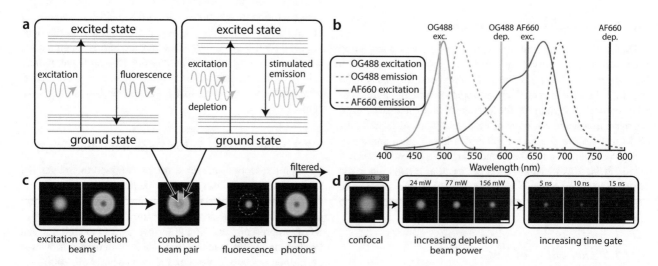

FIGURE 3.4 STED microscopy uses the photophysics of stimulated emission and beam shaping to achieve super-resolution images. (a) Energy level diagrams showing fluorescence (left) and stimulated emission (right) processes. Each wavy line represents a single photon. (b) The excitation and emission spectra of two commonly used STED dyes, Oregon Green 488 (OG488) and Alexa Fluor 660 (AF660), with appropriate choices of excitation (exc.) and depletion (dep.) laser lines. (c) Schematic showing the 2D shapes of the excitation and emission beams and the resultant light produced from this beam pair. The green-dashed line in the "detected fluorescence" panel indicates the perimeter of the PSF if the excitation beam alone was used. (d) Resolution in STED microscopy can be improved by increasing the power of the depletion beam and, in the case of gated-STED, increasing the time gate delay. Images are simulations based on the PSF sizes and count rates from Vicidomini et al. (2011) in which a single nitrogen vacancy was imaged with STED and gated-STED to obtain the super-resolution PSF. Power for the time gate panel is constant at 156 mW; all images are on same intensity scale. Scale bars = 200nm.

(Figure 3.4d). Again, the optimum time gate will depend on the properties of the specific fluorophore being imaged, namely its fluorescence lifetime.

3.2.2.1 Axial Resolution

As STED microscopy is based on confocal microscopy, images are optically sectioned. However, the depletion beam as described above increases only the resolution in the lateral dimension and provides no other increase in axial resolution. To gain an increase in the axial resolution, an additional depletion beam is required to suppress fluorescence from above and below the focus, again via stimulated emission. As a result, dedicated 3D-STED microscopes are capable of achieving axial resolutions <100 nm but only when the additional optics for the axial depletion beam are present (Hein, Willig, and Hell, 2008).

3.2.2.2 Temporal Resolution

As the physical process of stimulated emission occurs on the same timescale as fluorescence, i.e. nanoseconds, this is much faster than the typical pixel dwell time in laser scanning microscopy; therefore, the underlying physics does not limit the acquisition speed (unlike with SMLM, see next section). Aside from the increase in the time taken to image a region of interest because of the smaller pixel sizes, in theory, the temporal resolution of STED microscopy is the same as that of confocal microscopy (Galbraith and Galbraith, 2011). However in practice, as far fewer photons are collected in STED microscopy, pixel dwell times are often increased, or multiple frames are averaged to increase the SNR of the resultant images (Wegel et al., 2016).

3.2.2.3 Hardware and Sample Preparation

STED microscopy requires a microscope with scanning mirrors for fast laser scanning and at least one highly sensitive point detector, such as a photomultiplier tube, avalanche photodiode, or hybrid photodetector. The major modifications to convert a conventional confocal microscope into a STED microscope are the depletion laser, beam-shaping optics, and additional filter sets. For gated-STED, the excitation laser must be pulsed, and photon counting electronics are required (Vicidomini et al., 2011).

Researchers in optical physics and spectroscopy labs often opt to build custom STED microscopes for greater accessibility to optical components for a more flexible system. However, building a STED microscope is not a trivial matter and requires sufficient expertise to align and maintain the microscope. An excellent guide to building a STED microscope has been published by Wu et al. (Wu et al., 2015). Leica Microsystems and Abberior Instruments also manufacture commercial STED microscopes with 3D capability and multiple depletion laser wavelengths.

In theory, STED microscopy does not require special fluorophores or sample preparation, as all fluorescent molecules are physically capable of undergoing stimulated emission.

However, there are some features of fluorophores that will impact upon the quality and resolution of the super-resolution image. The probability of stimulated emission occurring follows approximately the same distribution as the emission spectrum of the fluorophore (Vicidomini et al., 2012), so there must be an overlap between the emission spectrum of the dye and the wavelength of the depletion laser. On the other hand, overlap between the excitation spectrum of the fluorophore and the depletion wavelength must be minimal. Additionally, depletion lasers need to provide sufficient power for high resolution (see Equation 3.4); therefore, the depletion laser(s) of a STED microscope will often limit the choice of fluorophores that can be used, and it is highly desirable to use fluorophores that have well-separated excitation and emission spectra (Sednev, Belov, and Hell, 2015). Other desirable photophysical properties of fluorophores for STED microscopy are good photostability, resistance to bleaching, and long fluorescence lifetimes. Care must be taken during sample preparation to minimize the presence of any autofluorescent species within the mounting media, as the intense depletion beam may cause unexpected signal from these species that is not seen when using the excitation beam alone. The mounting medium Vectashield, for example, should be avoided, as it exhibits autofluorescence throughout most of the visible spectrum (Olivier et al., 2013); Mowiol is a popular alternative for STED microscopy (Lau et al., 2012).

3.2.2.4 Case Studies of Live-Cell STED Microscopy

STED microscopy for live-cell two-color imaging has been demonstrated for a range of intracellular structures (Bottanelli et al., 2016). Following this work, a further live-cell study revealed the role that the GTPase ARF1 plays in tubule formation in the Golgi apparatus (Bottanelli et al., 2017). A field that has particularly benefited from live-cell STED imaging is neuroscience; for example, in studying axon morphology and synaptic architecture (Tønnesen et al., 2014; Chéreau et al., 2017; Wegner et al., 2018). Indeed, the first live-cell STED microscopy study was investigating synaptotagmin clustering in living neurons (Willig et al., 2006), and STED microscopy is the only technique that has been used to see periodic cytoskeleton patterning in living neurons (D'Este et al., 2015).

3.2.2.5 Limitations of STED Microscopy

In practice, STED only works for a limited range of dyes with high photostability, and achieving a good SNR often requires the use of immersion media to improve the stability of the fluorescence (Blom and Widengren, 2017). These aspects limit its use in live-cell microscopy. The performance of commercial STED microscopes for live-cell imaging also appears to be limited at this time, as all of the example studies listed above used custom-built microscopes. This could be because day-to-day maintenance is required to preserve the precise beam alignment necessary to ensure sufficient power throughput of the depletion beam and accurate overlap of the excitation/depletion beam pair.

3.2.2.6 Other Microscopy Techniques Based on the Principles of STED

The idea of depleting fluorescence emission around a diffraction-limited spot in order to shrink the emission point-spread function and improve resolution can be exploited by using reversibly switchable fluorescent proteins (rsFPs) as fluorescent markers. Here, instead of using a strong depletion laser, a doughnut-shaped "off"-switching laser can be used to drive rsFPs into an "off" state. This method, called reversible saturable optical linear fluorescence transitions (RESOLFT), requires much lower laser power and, therefore, is more compatible with live imaging than standard STED (Hofmann et al., 2005). Both STED and RESOLFT techniques can be parallelized in order to speed up the acquisition times (Chmyrov et al., 2013; Yang et al. 2013).

3.2.3 Single Molecule Localization Microscopy

SMLM refers to a family of widefield super-resolution techniques in which photoswitchable/photoactivatable fluorophores are used to isolate fluorescence from individual single molecules such that they can be mapped with nanometer precision. SMLM was originally described independently as stochastic optical reconstruction microscopy (STORM) (Rust, Bates, and Zhuang, 2006), photoactivated localization microscopy (PALM) (Betzig et al., 2006), and fluorescence photoactivation localization microscopy (FPALM) (Hess, Girirajan, and Mason, 2006).

The central principle of SMLM is that, rather than acquiring one single image containing the simultaneous emission from all molecules (as is the case in conventional widefield imaging), a large number of images are acquired, each containing a small random subset of emitting "on" molecules. It follows that, in each acquired image, it is necessary that the vast majority of fluorophores reside in a temporary "off" state. Each image will contain a different subset of emitting fluorophores; if enough images are acquired, all of the fluorophores labeling the structure of interest eventually will be sampled. If the emitting molecules are separated by distances $>>\Delta d_{x,y}$ (i.e. the image is "sparse") the center of each molecule can be calculated with very high accuracy and thus used to generate the final super-resolution image (Figure 3.5a).

In order to acquire sparse images, the fluorophore, laser properties, and imaging microenvironment must be carefully selected. The original STORM implementation exploited activator-reporter pairs of dye molecules (e.g. Cy3–Cy5) acting as molecular switches. The activator, in the presence of appropriate wavelength illumination, allows for the reporter to transition between off and on states. Sparsely activated reporter molecules in the on state are then imaged. However, a more recent method, called direct STORM (dSTORM), bypasses the requirement for imaged fluorophores to have an activator molecule (Heilemann et al., 2008). In dSTORM, switching

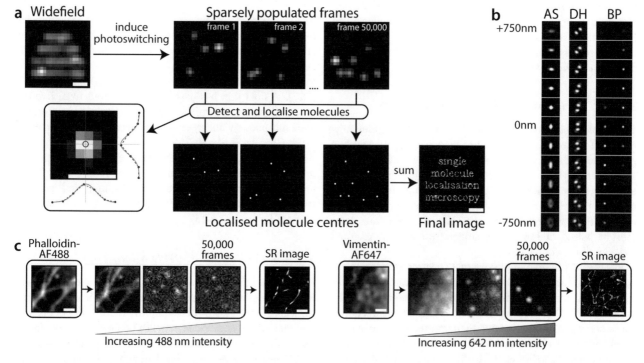

FIGURE 3.5 Single molecule localization microscopy. (a) Schematic of an SMLM acquisition workflow for a simulated structure. Inset: localizing the center (black circle) of a single molecule by Gaussian fitting. Scale bars = 500nm. (b) Different methods for encoding the axial position of a fluorophore in SMLM techniques. AS = astigmatism, DH = double helix, BP = biplane. Data is from the z-stack calibration sets available for the SMLM software benchmarking 2016 challenge ("Http://Bigwww.Epfl.Ch/Smlm/Challenge2016/Index.Html?P=datasets," n.d.). (c) Fluorophore choice can strongly impact data quality in SMLM, as demonstrated for dSTORM data of dual-labeled fixed samples imaged with a Zeiss Elyra PS.1 in blinking buffer. Left: Actin labeled with Alexa Fluor 488 displays sparse blinking but at very poor signal-to-noise ratio when laser intensity is increased to its maximum. Right: Vimentin labeled with Alexa Fluor 647 displays sparse blinking with good signal-to-noise ratio as laser intensity is increased, leading to a good reconstructed super-resolution (SR) image. Scale bars = 1μm.

behavior is induced in conventional dyes using high-intensity illumination and redox reactions to cycle in and out of the off state. The on state, in this case, is the singlet state of the molecule as would be imaged in conventional microscopy.

PALM techniques use photoswitchable fluorescent proteins rather than dye molecules. There are two main classes of fluorescent protein suitable for PALM imaging: proteins such as rsGFP, which has an emitting on state and dark off state, and proteins such as Kaede (Ando et al., 2002) or mEos (Wiedenmann et al., 2004) for which the on and off states are both emissive but emit at different wavelengths (e.g. a red-emitting on state and green-emitting off state). In both classes, transitions between on and off states are induced using activating laser illumination at a wavelength distinct to that used for imaging (Shcherbakova et al., 2014).

Generation of the final super-resolution image requires computational processing to localize the centers of the individual emitting molecules. This analysis typically involves initial detection of individual molecules followed by high-accuracy fitting of PSF-like functions to determine the center of the PSF and, therefore, the localization of the underlying fluorescent molecule. The precision of molecule localization (and hence resolution) is largely determined by the number of photons emitted from the molecule, N, as described in Equation 3.5 (Thompson, Larson, and Webb, 2002).

$$\Delta d_{x,y}^{SMLM} \approx \frac{\lambda}{2NA\sqrt{N}} \qquad (3.5)$$

However, many other factors impact the final reconstructed image resolution. These include the sparsity of imaged molecules within each frame, the total number of molecules imaged, and the underlying labeling density on the sample. There is a wealth of algorithms available for performing this analysis, as reviewed in Sage et al. (2015) of varying levels of complexity and usability. Two recommended software packages for performing this analysis are QuickPALM (Henriques et al., 2010) and ThunderSTORM (Ovesný et al., 2014), both of which are available as free plugins for ImageJ/Fiji.

3.2.3.1 Axial resolution

In the SMLM approach, in order to encode information about the axial position of a fluorophore, additional hardware is required. The three most common methods for achieving this are astigmatism (Huang et al., 2008), biplane imaging (Juette et al., 2008), and double-helix PSF engineering (Pavani et al., 2009) (Figure 3.5b). The astigmatism method involves inserting a cylindrical lens into the detection path of a SMLM microscope. This introduces elongation of the PSF along the y-axis for fluorophores below the focal plane, no change in the PSF shape for in-focus molecules, and elongation of the PSF along the x-axis for fluorophores above the focal plane. The degree of elongation is proportional to the distance from the focal plane, and this can be measured, calibrated, and used to infer the axial position of a fluorophore with 50–60 nm resolution within a depth of ~1 μm. Biplane imaging requires detection of two different focal planes simultaneously, either using two halves of the same camera chip or using two separate cameras.

The two resulting image stacks of the same lateral field of view are then analyzed, with the differences between the fitted widths of the PSFs for the individual molecules used to infer the z-position. The axial resolution of biplane imaging is <100 nm but works across a wider depth of ~2 μm. Double-helix 3D SMLM uses beam-shaping optics in the detection path to transform the PSF into a double-helical shape. As a result, the fluorescence from an individual molecule appears on the camera not as a 2D Gaussian distribution, but as two lobes. The relative orientation of these two lobes encodes the axial location of the molecule, and the lateral location of the molecule is the central minimum between the two lobes. As a result, double-helix SMLM can yield axial resolutions of 20–60 nm (Carr et al., 2017) across a few microns. However, it should be noted that reconstruction approaches for 3D SMLM need to be adapted to take into account the changes in PSF shape induced to encode the axial position of the molecule.

3.2.3.2 Temporal Resolution

The temporal resolution of SMLM is severely limited by the required acquisition time for building up a large enough number of localizations and generating a complete image of the underlying structure. In standard SMLM experiments, typically 10,000–100,000 frames are required to generate a single super-resolution image. Coupled with the 10–50 ms exposure time of cameras used for SMLM acquisitions, this yields a temporal resolution of 2–90 minutes. This poor temporal resolution is one of the features of SMLM addressed by next-generation analytics (discussed below).

3.2.3.3 Hardware and Sample Preparation

The hardware requirements for SMLM are simple compared to SIM and STED, but SMLM requires much closer attention to sample preparation. The only hardware requirements for SMLM imaging are a widefield microscope with a high-NA objective, a sensitive camera (such as an electron multiplying charge coupled device [EMCCD]), and sufficiently powerful lasers to induce photoswitching behavior in the fluorophores and obtain a high SNR. As a result, home-built SMLM systems are straightforward, popular, and relatively cheap (Ma et al., 2017; Kanchanawong et al., 2010; Kwakwa et al., 2016). There are also commercial systems available, some of which offer additional features, such as optics for achieving 3D super-resolution, including the Nikon N-STORM, Zeiss Elyra, and the Oxford Nanoimaging Nanoimager.

The most important sample preparation requirement for SMLM is a fluorophore that can undergo photoswitching. For the approaches described above, important features of fluorophores include: the fraction of time that the fluorophore spends in its on and off states (its "blinking statistics"), the quantum yield of the fluorophore and the wavelength(s) required for activation and/or excitation of the fluorophore. Figure 3.5c shows the behavior of two different fluorophores for dSTORM at different illumination intensities. In-depth studies of the behavior and properties of photoswitchable fluorophores have been published for organic dye molecules (Lehmann et al., 2015; Wang et al., 2014) and fluorescent proteins (Siyuan Wang et al., 2014;

Pennacchietti, Gould, and Hess, 2017). Buffering is also a key component in maintaining the photophysical properties of the fluorophores, especially for dSTORM. Again, different fluorophores have different buffering requirements (Dempsey et al., 2011; Olivier et al., 2013).

Because of the high illumination intensities applied during SMLM imaging and the single-molecule sensitivity of the system, the presence of unwanted fluorescent species in the sample must be minimized. These species include the usual suspects, such as phenol red present in many cell culture media, but also dust and dirt on the coverslip, which frequently display blinking behavior when subjected to high laser intensities. The latter can be avoided by careful cleaning of the coverslips and slides used for sample preparation (Pereira, Almada, and Henriques, 2015).

3.2.3.4 Case Studies of Live-Cell SMLM

The major modality for live-cell SMLM has been PALM, as the use of fluorescent proteins allows for less invasive labeling of intracellular structures compared to methods required for labeling with dyes. Photoswitching mechanisms in PALM also typically require lower intensity irradiation to induce transitions to the off state compared to methods such as dSTORM. Examples of live-cell PALM include studying assembly of the FtsZ division ring in *E. coli* (Fu et al., 2010), the role of actin in hemagglutinin membrane clustering (Gudheti et al., 2013), the distribution of different phosphoinositides within the plasma membrane (Ji et al., 2015) and dynamic evolution of focal adhesion complexes (Shroff et al., 2008). Live-cell STORM imaging has been used to image clathrin-coated pits and the transferrin receptor (Jones et al., 2011) and various intracellular organelles through inducing photoswitching in commonly used probes such as Mitotracker (Shim et al., 2012). Photoactivatable fluorescent proteins were originally used for single-particle tracking studies (SPT) (Yu, 2016), and, as such, this dynamic information can be combined with the spatial accuracy of PALM in the sptPALM technique (Manley et al. 2008).

3.2.3.5 Limitations of SMLM

As SMLM approaches require a large number of camera frames with sparse emitting fluorescent molecules, long acquisition times are typically necessary (2–90 min). As many intracellular processes occur on shorter timescales than this, movement of the imaged structure will manifest as motion blur within the reconstructed image, and fine details will be lost.

The laser requirements for SMLM imaging are also limiting for live-cell microscopy. This is because the photophysical mechanisms underlying transitions between on and off states frequently rely upon high photon fluxes (Figure 3.5c) or ultraviolet (UV) wavelengths, especially for synthetic fluorophores. Both of these imaging conditions are highly phototoxic to cells (Wäldchen et al., 2015) and, when combined with the long acquisition times discussed above, can no longer be considered as noninvasive. Furthermore, the

buffers used for dSTORM imaging are typically cytotoxic, as they include reducing agents and oxygen-scavenging systems.

Care must also be exercised within the computational post-processing of SMLM data sets in order to avoid manifestation of artifacts in the images, such as interstructure merging and molecule mislocalization (Burgert et al., 2015).

3.2.3.6 Promising Approaches for SMLM

To circumvent the requirement for dependence on high-intensity/UV-wavelength illumination for achieving sparsely distributed emitters, a technique called PAINT (point-accumulation for imaging in nanoscale topography) has been developed (Sharonov and Hochstrasser, 2006). On/Off transitions in PAINT are achieved by transient binding and unbinding of free-floating fluorophores in solution to the target structure of interest. Here, the on state corresponds to when the fluorophore is bound to the structure and the off state is the diffusing fluorophore in the media. A popular implementation is DNA-PAINT, whereby fluorophores are conjugated to short ssDNA strands that transiently bind and unbind to a complementary strand conjugated to the target structure of interest (Jungmann et al., 2014; Schnitzbauer et al., 2017).

3.3 Emerging Techniques for Live-Cell Super-Resolution Microscopy

In the last part of this chapter, we will focus on strategies that have been developed specifically to enable live-cell super-resolution microscopy. Of the techniques described above, only SIM is routinely capable of imaging living samples with minimal photodamage, yet this has the limitations of a complex optical setup and only modest improvements in resolution. There are three main fronts of development for enabling better live imaging at near-molecular resolution: new microscope architectures, advanced analytics, and the design of novel fluorophores.

3.3.1 Hardware Developments

A common limiting factor to resolution in both conventional fluorescence microscopy and super-resolution microscopy is the numerical aperture of the objective (Equations 3.2–3.5). Zeiss makes a 1.57 NA objective, and Olympus has recently produced a 1.7 NA objective (commonly used "high" numerical aperature objectives are normally 1.4–1.45 NA). Thus, increased resolution can be instantly obtained by using this objective in a super-resolution microscope. For example, the 1.7 NA objective has been combined with fast multicolor SIM imaging to achieve sub-100 nm resolution for live-cell imaging of actomyosin dynamics and focal adhesions (Li et al., 2015). However, it should be noted that this is not necessarily a cost-effective method for improving resolution, as these objectives require specialized immersion media and coverslips.

There are also hardware developments aimed at increasing the throughput of live-cell SMLM. For example, these include using sCMOS cameras for larger fields of view and rapid frame rates (Almada, Culley, and Henriques, 2015; Huang et al., 2013). Another method for increasing throughput is using automated microfluidics for online fixation at the microscope, allowing for transitions between live cells at low laser intensities, and fixed-cell super-resolution imaging at high intensities (Almada et al., 2019).

Volumetric fluorescence imaging techniques such as light sheet microscopy are becoming increasingly widespread because of their capabilities for 3D imaging and, importantly, lower illumination doses to the sample. As a result, light sheet techniques are an attractive area for further developing live-cell super-resolution imaging. For example, 150nm lateral and 280nm axial resolutions have been achieved using lattice light sheet microscopy, albeit in a very specialized optical setup (Chen et al., 2014). Further developments in light sheet microscopy for super-resolution are reviewed in Girkin and Carvalho (2018).

Other hardware improvements for live-cell super-resolution microscopy center on adaptive illumination to decrease the light dose to the sample. The basic premise of adaptive illumination is that only regions containing the structure of interest are illuminated at high intensity, with the laser intensity attenuated in nonfluorophore-containing regions of the field of view. This approach has been demonstrated in both SIM (Chakrova et al., 2016) and STED (Staudt et al., 2011; Heine et al., 2017). While neither technique has yet been demonstrated in living samples, previous diffraction-limited studies using adaptive illumination have shown a marked increase in cell viability (Hoebe et al., 2007). Adaptive illumination STED has now also been commercially implemented by Abberior Instruments. Another strategy for decreasing harmful light dose is to use LED illumination rather than laser illumination, which has been recently implemented for SIM (Pospíšil et al., 2018).

3.3.2 Analytical developments

A large number of analytical developments have centered on algorithms for SMLM techniques. While the most obvious avenue for decreasing phototoxicity in SMLM imaging is to decrease the illumination laser intensity, this has the result of increasing the number of on-state molecules within a single frame, and, as such, the emission of these molecules overlaps. Such so-called "high density" data sets cause problems for the molecule localization algorithms typically used for SMLM data, leading to artifactual images. In order to enable accurate reconstructions of high-density data sets, a number of novel analytical approaches have been developed. In contrast to algorithms relying on sparse data sets, which identify single-emitting molecules and localize their centers using fitting, high-density algorithms examine features of the image such as the temporal statistics of the fluorescence or apply spatial transforms to the image. Examples of some of these techniques are summarized in Table 3.3.

TABLE 3.3

Examples of Algorithms for High-Density SMLM Data Sets

Algorithm	Resolution	Raw data requirements	Usability
SRRF: Super-resolution radial fluctuations (N. Gustafsson et al. 2016; Culley, Tosheva, et al. 2018)	Lateral: 60–150 nm Axial: No improvement vs. diffraction-limited Temporal: ~1 second	≥100 frames per super-resolution image. Exposure times ~10–50 ms per frame for best temporal resolution. No specific fluorophore requirements. Demonstrated to work with most conventional fluorescence microscopes.	Available as a plugin for ImageJ/Fiji. Analysis is GPU-enabled for speed. Moderate number of parameters to adjust.
SOFI: Super-resolution optical fluctuation imaging (Dertinger et al. 2009; Geissbuehler et al. 2012, 2014)	Lateral: 110 nm Axial: 500 nm (with additional optics for multiplane imaging) Temporal: 0.6–5 seconds	≥200 frames per super-resolution image. Exposure times ~3–20 ms per frame for best temporal resolution. Fluorophores must display and switch between discrete emission states e.g. on- and off-states	Available as a standalone Matlab toolbox ("Https://Documents.Epfl.Ch/Users/l/Le/Leuteneg/Www/BalancedSOFI/Index.Html," n.d.) or in the Localizer software for IgorPro/Matlab (Dedecker et al. 2012). Analysis is fast. Small number of parameters to adjust.
3B: Bayesian analysis of bleaching and blinking (Cox et al. 2011; Rosten, Jones, and Cox 2013)	Lateral: 50 nm Axial: No improvement vs. diffraction-limited Temporal: 4 seconds	≥200 frames per super-resolution image. Fluorophore should exhibit on-/off- switching and bleaching characteristics. Demonstrated with both laser illumination and xenon arc lamp illumination.	Available as a plugin for Fiji/ImageJ. Algorithm requires ~6 hours to produce an image for a small (60×60 pixels × 300 frames) data set. A cluster implementation is available (Hu et al. 2013) but is no longer maintained.

Other algorithms include deconSTORM (Mukamel, Babcock, and Zhuang, 2012), csSTORM (Zhu et al., 2012), FALCON (Min et al., 2015) and MUSICAL (Agarwal and Macháň, 2016), but use of these is less widespread. Resolutions are as reported for live cells in the publications.

Another area of development is the burgeoning application of machine learning and deep learning to fluorescence microscopy data. This involves training a model with pairs of raw input and corresponding high-accuracy output images. Once fully trained, the model can then be applied to input data alone and thus generate high-quality output images. There are several examples of this approach in super-resolution microscopy. For example, deep learning has been applied to accept sparse SMLM raw data sets and output super-resolution renderings, hence bypassing conventional localization algorithms (Nehme et al., 2018). Deep-learning based techniques can reconstruct super-resolution images with fewer frames than typically used in SMLM (Ouyang et al., 2018). A broader application of machine learning to fluorescence imaging is content-aware image restoration (CARE) (Weigert et al., 2018), in which, in addition to improving SNRs of images and restoring isotropic 3D resolution from under-sampled data sets, extraction of super-resolution images from widefield images has also been achieved. Deep-learning techniques must however come with a word of caution: The models are typically only capable of constructing image features that they have been trained on and can be prone to biases and are not necessarily suitable for structural discoveries.

3.3.3 Labeling Developments

In fluorescence microscopy, the protein or structure of interest is usually nonfluorescent. Therefore, a fluorescent molecule (fluorescent protein or synthetic fluorophore, either on its own or linked to a probe, such as an antibody for example) is attached to the target of interest and used as a proxy for its localization. Making sure this assumption is true is fundamental for all insights provided by fluorescence microscopy, although this is often a challenge. In super-resolution microscopy for example, the veracity of this assumption may be disputable. The nanometer scale resolutions achievable often highlight defects of the labeling approach chosen that otherwise would be hidden within the ~250–300 nm diffraction limit. This is a consequence of resolutions smaller than the distance between fluorophore and target of interest (Ries et al., 2012; Laine et al., 2015), incomplete labeling of target structures (Durisic et al., 2014; Burgert et al., 2015; Lau et al., 2012), and function and/or localization defects resulting from the labeling strategy used (Lelek et al., 2012; Hammond et al., 2010). Additionally, one common aspect to all super-resolution approaches is the need for high SNR. In SIM, insufficient SNRs can lead to artifactual reconstructions; in STED most of the fluorescence is suppressed, and undepleted molecules must emit sufficient numbers of photons; in SMLM, resolution scales with the number of photons emitted per single molecule event (Equation 3.5). Obviously, the choice of the right probe and fluorophore to use is of critical importance and, consequently, a research field developing at a considerable pace. Hence, rather than providing an in-depth description of the multiple options available for labeling, we have grouped the larger families of labeling approaches and provide pros and cons for each one in Table 3.4. We aim for the reader to have a general view of the options available and a comprehensive list of further reading material to explore any useful aspect.

To avoid an overly simplistic view of this field, it is worth pointing out that all the labeling strategies described in Table 3.4 are complementary, each having pros and cons for any given application. This complementary nature allows researchers to explore the spatial and temporal dynamics of biological systems in multiple ways, obtaining multiple degrees of information. Hence, the labeling strategies and their validity are, and will continue to be, enticing challenges for researchers developing new labeling methods and exploring the most stimulating biological questions. These labeling strategies will evolve in parallel fueled by advances in molecular genetics, biochemistry, and chemistry as well as hardware development breakthroughs and progress in the analytical methods available to analyze live-cell super-resolution data.

3.4 Super-Resolution Data Evaluation

While super-resolution microscopy techniques are becoming increasingly commonplace, it is important to note that they are prone to artifacts if sample preparation (Pereira et al. 2018), labeling, imaging, and reconstruction protocols are not appropriately followed. Furthermore, the optics and computation required for generating super-resolution reconstructions often yield nonlinear images (when compared to their diffraction-limited equivalents), which makes assessment of image quality more challenging. There are now several different methods for assessing the resolution and quality of super-resolution data, summarized in Table 3.5. It is easy to assume that very high resolution is an indicator of "good" super-resolution imaging; however, it has been shown that image resolution does not necessarily correlate with image quality and fidelity (Culley et al., 2018).

3.5 Conclusion

Super-resolution microscopy techniques provide a valuable bridge between the ultrahigh resolution of electron microscopy and the noninvasive methodologies of conventional fluorescence microscopy. Although these techniques have become well-established and their potential impact for cell biology research recognized through the awarding of the 2014 Nobel Prize in Chemistry to three pioneers in the field, applications to live-cell imaging are still in their infancy. The development of novel hardware, fluorescent labels, and analytical methods will help translate super-resolution imaging into live-cell applications that hold the promise of uncovering cellular dynamics with unprecedented detail.

TABLE 3.4

Overview of Available Labeling Approaches for Live-Cell Super-Resolution Microscopy

	Labeling approach	Advantages	Disadvantages	Exciting prospects	Refs
High-affinity probes	Antibodies (Ab) • ~150 kDa • ~10 nm Fab fragments • ~50 kDa • ~5 nm Single-domain Ab (sdAb, nanobodies); Affimers (Adhirons); Affibodies; DARPins; Aptamers (DNA/RNA based) • ~15 kDa • ~2 nm	No need for genetic encoding (if fusion functionality is an issue). Wide variety of epitopes and affinities. Can be combined with multiple synthetic fluorophores. Multiple size ranges. Probe libraries for easy selection (e.g. phage display).	Not membrane permeable unless combined with transitory membrane disruption (e.g. cell squeezing, toxins). Smaller probes are still not available for all Ab targets. Linker error, may mask epitopes or induce clustering (Ab). Mostly for external epitopes, difficult for intracellular live-cell imaging. Indirect target detection.	High number of small probes available (to match classical Ab). Small probes against encoded targets (e.g. BC2tag/BC2 sdAb, GFP/RFP nanobody). Combination with extremely robust approaches (DNA-PAINT, FRET-PAINT).	Jungmann et al, 2014; Ries et al, 2012; Laine et al, 2015; Traenkle & Rothbauer, 2017; Schlichthaerle et al, 2018; Ståhl et al, 2017; Lavis 2017; Jungmann et al, 2010; Iinuma et al, 2014; Auer et al, 118; Schueder, Lara-Gutiérrez et al, 2017; Schueder, Strauss et al, 2017; Kollmannsperger et al, 2016; Canton et al, 2013; Lambert et al, 1990; Teng et al, 2016; Braun et al, 2016; Bird et al, 1988; Pleiner et al, 2017; Opazo et al, 2012; Bedford et al, 2017; Pleiner et al, 2015; Mikhaylova et al, 2015; Platonova et al, 2015
Fluorescent Proteins (FPs)	Classical (e.g. mScarlet), Photoactivatable (e.g. PA-mCH), Photoconvertible (e.g. mEos3.2), Photoswitchable (e.g. rsKame) • ~30 kDa • ~3 nm	Genetic encoding. Easy live-cell imaging. Tight control of fusion position (C-, N- or internal). Flexibility for multiple techniques (e.g. SR, FRET, FRAP). Small linker error and no artificial clustering (if monomer). Very reproducible (e.g. stable cell lines). Direct target detection.	Laborious selection protocol for specific applications. Low(er) photon-budget/brightness. Protein fusion functionality issues. Needs to be genetically encoded. Limited variety of wavelengths (with identical high quality).	Combining with CRISPR for endogenous expression levels. Primed conversion for live-cell SMLM. Brighter and more photostable new FPs. FPs with properties adapted for specific techniques. Biosensors in combination with SR.	Gustafsson, 2000; Hein, Willig & Hell 2008; Hoffmann et al, 2005; Shcherbakova al, 2014; Gustafsson et al, 2016; Khan al, 2017; Ratz et al, 2015; Shaner, Steinbach & Tsien, 2005; Chudakov, Lukyanov & Lukyanov 2005; Wiedenmann, Oswald & Nienhaus 2009; Kremers et al, 2011; Rego et al, 2012; Nägerl et al, 2008; Tønnesen et al, 2011; Zhang et al, 2016; Hense et al, 2015; Grotjohann et al, 2012; Shaner et al, 2013; Tiwari et al, 2015; Brakemann et al, 2011; Zhang et al, 2015; Uno et al, 2015; Dempsey et al, 2015; Turkowyd et al, 2017; Mo et al, 2017; Wang et al, 2017; Richardson et al, 2017; Kaberniuk et al, 2017; Mishina et al, 2015; Sanford & Palmer, 2017; Ni, Mehta & Zhang, 2017; Bajar et al, 2016; Nienhaus & Nienhaus, 2014; Zhang et al, 2012; Hostettler et al, 2017; Rosenbloom et al, 2014; Bindels et al, 2016

(Continued)

TABLE 3.4 (CONTINUED)

Overview of Available Labeling Approaches for Live-Cell Super-Resolution Microscopy

	Labeling approach	Advantages	Disadvantages	Exciting prospects	Refs
Synthetic Fluorophores	Classical conjugation (e.g. NHS ester). Site specific labeling: Click chemistry (e.g. Tetrazine). Enzymatic (e.g. Sortase/LPXTG motif). Self-labeling systems: CLIP-tag, SNAP-tag • ~20 kDa • ~3 nm Halo-tag • ~30 kDa • ~5 nm TMP-eDHFR-tag • ~18 kDa • ~3 nm Small tags (e.g. HIS-tag/TrisNTA, FlAsH/ReAsH, Versatile Interacting Peptide, CoilY). Off-the-shelf dyes (e.g. Mitotracker, SiRActin).	Wide variety of fluorophore properties (e.g. wavelength). Can be combined with high affinity probes. Genetic encoding is possible (for cell-permeable dyes). Genetic-encoded options allow for the same target to be localized with different fluorophores. Off-the-shelf options to target specific cellular structures. Fluorogenic options. High(er) photon-budget/ brightness. Easier targeted design for specific applications	Selected option may not be membrane permeable or available for selected tagging system. Great variety of options may be daunting - difficult to access the best option for a specific problem. Frequently have to be combined with a tag or high-affinity probe.	Click Chemistry combined with genetic-code expansion and CRISPR. Fluorogenic dyes. Abundance of technique specific options. Janelia Fluor® dyes.	Shcherbakova et al, 2014; Shim et al, 2012; Wäldchen et al, 2015; Durisic et al, 2014; Lelek et al, 2012; Lavis, 2017a; Kollmannsperger et al, 2016; Khan et al, 2017; Ratz et al, 2015; Nienhaus & Nienhaus, 2014; Sednev, Belov & Hell, 2015; Dempsey et al, 2011; Grimm et al, 2017; Mateos-Gill et al, 2016; Li, Tebo & Gautier, 2017; Uttamapinant et al, 2015; Thompson et al, 2017; Lukinavičius et al, 2014; Lavis, 2017b; Gautier et al, 2008; Griffin, Adams & Tsien, 1998; Keppler et al, 2003; Los et al, 2008; Wombacher et al, 2010; Butkevich et al, 2018; Song et al, 2017; Wang, Song & Xiao, 2017; Leng et al, 2017; Lukinavičius et al, 2016; Liu et al, 2014; Prifti et al, 2014; Butkevich et al, 2017; Lukinavičius et al, 2013; Clark et al, 2016; Vreja et al, 2015; Uttamapinant et al, 2010; Liu et al, 2014; Cohen, Thompson & Ting, 2011; Howarth et al, 2005; Zane et al, 2017; Nikić et al, 2016; Sakin et al, 2017; Schvartz et al, 2017; Lukinavičius et al, 2015; Uno et al, 2014; Uno et al, 2017; Takakura et al, 2017; Wang et al, 2017; Thompson et al, 2017

TABLE 3.5

Different Methods for Assessment of Super-Resolution Images

Method	Assessed feature	Applicable modalities	Summary
Measuring width/separation of structures	Resolution	SIM, STED, SMLM	Image resolution is determined by measuring the width of a line profile drawn through a structure of interest or the distance between peaks in a line profile drawn across closely separated structures. This is simple but crude. The choice of structure to measure is subjective and is not robust against techniques that may cause artificial sharpening (namely SMLM).
Fourier Ring Correlation (FRC)	Resolution	Most robust with SMLM (Nieuwenhuizen et al. 2013), also reported for STED (Tortarolo et al. 2018)	The super-resolution image is divided into two images of the same structure, each containing half of the image information (e.g. in SMLM by creating one super-resolution image from odd-number acquired frames and another from even-number acquired frames). These images are then correlated in Fourier space to give a global estimate of image resolution. This method is unbiased from the user's perspective, but some image features, such as punctate structures, can skew results.
SIMCheck (Ball et al. 2015)	Image quality, resolution	SIM	A plugin for ImageJ/Fiji for assessing the quality of raw and reconstructed SIM images. The manuscript provides useful guidance on how quality metrics can be used to improve SIM imaging.
Localization reclassification (Fox-Roberts et al. 2017)	Image quality	SMLM	A Matlab toolbox that presents the user with localizations from an inputted SMLM data set. The user classifies these as accurate or not and the algorithm then uses this information to determine the local quality within the reconstructed image.
SQUIRREL (Culley, Albrecht, et al. 2018)	Image quality, resolution	SIM, STED, SMLM	A plugin for ImageJ/Fiji that examines a super-resolution image and the diffraction-limited equivalent of the same field-of-view to map image artifacts and provide quality metrics. The plugin also features a method for local mapping of resolution based on FRC.

REFERENCES

Agarwal, Krishna, and Radek Macháň. 2016. "Multiple Signal Classification Algorithm for Super-Resolution Fluorescence Microscopy." *Nature Communications* 7 (December): 13752. doi: 10.1038/ncomms13752

Almada, Pedro, Siân Culley, and Ricardo Henriques. 2015. "PALM and STORM: Into Large Fields and High-Throughput Microscopy with SCMOS Detectors." *Methods* 88 (October): 109–21. doi: 10.1016/J. YMETH.2015.06.004

Almada, Pedro, Pedro Pereira, Siân Culley, Ghislaine Caillol, Fanny Boroni-Rueda, Christina L. Dix, Romain F. Laine, et al. 2019. "Automating Multimodal Microscopy with NanoJ-Fluidics." *Nature Communications* 10 (1223): 1–9. doi: 10.1038/s41467-019-09231-9

Ando, Ryoko, Hiroshi Hama, Miki Yamamoto-Hino, Hideaki Mizuno, and Atsushi Miyawaki. 2002. "An Optical Marker Based on the UV-Induced Green-to-Red Photoconversion of a Fluorescent Protein." *Proceedings of the National Academy of Sciences of the United States of America* 99 (20): 12651–56. doi: 10.1073/pnas.202320599

Ashdown, George W., Andrew Cope, Paul W. Wiseman, and Dylan M. Owen. 2014. "Molecular Flow Quantified beyond the Diffraction Limit by Spatiotemporal Image Correlation of Structured Illumination Microscopy Data." *Biophysical Journal* 107 (9): L21–23. doi: 10.1016/j. bpj.2014.09.018

Ashdown, George W., Garth L. Burn, David J. Williamson, Elvis Pandžić, Ruby Peters, Michael Holden, Helge Ewers, et al. 2017. "Live-Cell Super-Resolution Reveals F-Actin and Plasma Membrane Dynamics at the T Cell Synapse." *Biophysical Journal* 112 (8): 1703–13. doi: 10.1016/j.bpj.2017.01.038

Auer, Alexander, Maximilian T. Strauss, Thomas Schlichthaerle, and Ralf Jungmann. 2017. "Fast, Background-Free DNA-PAINT Imaging Using FRET-Based Probes." *Nano Letters* 17 (10): 6428–34. doi: 10.1021/acs.nanolett.7b03425

Bajar, Bryce T., Emily S. Wang, Shu Zhang, Michael Z. Lin, and Jun Chu. 2016. "A Guide to Fluorescent Protein FRET Pairs." *Sensors (Basel, Switzerland)* 16 (9): 1–24. doi: 10.3390/s16091488

Ball, Graeme, Justin Demmerle, Rainer Kaufmann, Ilan Davis, Ian M. Dobbie, and Lothar Schermelleh. 2015. "SIMcheck: A Toolbox for Successful Super-Resolution Structured Illumination Microscopy." *Scientific Reports* 5 (1): 15915. doi: 10.1038/srep15915

Bedford, R., C. Tiede, R. Hughes, A. Curd, M. J. McPherson, Michelle Peckham, and Darren C. Tomlinson. 2017. "Alternative Reagents to Antibodies in Imaging Applications." *Biophysical Reviews* 9: 299–308. doi: 10.1007/s12551-017-0278-2

Betzig, Eric, George H. Patterson, Rachid Sougrat, O. Wolf Lindwasser, Scott Olenych, Juan S. Bonifacino, Michael W. Davidson, et al. 2006. "Imaging Intracellular Fluorescent Proteins at Nanometer Resolution." *Science (New York, N.Y.)* 313 (5793): 1642–45. doi: 10.1126/science.1127344

Bindels, Daphne S., Lindsay Haarbosch, Laura Van Weeren, Marten Postma, Katrin E. Wiese, Marieke Mastop, Sylvain Aumonier, et al. 2016. "MScarlet: A Bright Monomeric Red Fluorescent Protein for Cellular Imaging." *Nature Methods* 14 (1): 53–6. doi: 10.1038/nmeth.4074

Bird, Robert E., Karl D. Hardman, James W. Jacobson, Syd Johnson, Bennett M. Kaufman, Shwu M. Lee, Timothy Lee, et al. 1988. "Single-Chain Antigen-Binding Proteins." *Science (New York, N.Y.)* 242 (4877): 423–26. doi: 10.1126/science.3140379

Blom, Hans, and Jerker Widengren. 2017. "Stimulated Emission Depletion Microscopy." *Chemical Reviews* 117 (11): 7377–427. doi: 10.1021/acs.chemrev.6b00653

Bottanelli, Francesca, Nicole Kilian, Andreas M. Ernst, Felix Rivera-Molina, Lena K. Schroeder, Emil B. Kromann, et al. 2017. "A Novel Physiological Role for ARF1 in the Formation of Bidirectional Tubules from the Golgi." *Molecular Biology of the Cell* 28 (12): 1676–87. doi: 10.1091/mbc.E16-12-0863

Bottanelli, Francesca, Emil B. Kromann, Edward S. Allgeyer, Roman S. Erdmann, Stephanie Wood Baguley, George Sirinakis, Alanna Schepartz, et al. 2016. "Two-Colour Live-Cell Nanoscale Imaging of Intracellular Targets." *Nature Communications* 7 (March): 10778. doi: 10.1038/ncomms10778

Brakemann, Tanja, Andre C. Stiel, Gert Weber, Martin Andresen, Ilaria Testa, Tim Grotjohann, Marcel Leutenegger, et al. 2011. "A Reversibly Photoswitchable GFP-like Protein with Fluorescence Excitation Decoupled from Switching." *Nature Biotechnology* 29 (10): 942–50. doi: 10.1038/nbt.1952

Braun, Michael B., Bjoern Traenkle, Philipp A. Koch, Felix Emele, Frederik Weiss, Oliver Poetz, Thilo Stehle, et al. 2016. "Peptides in Headlock–A Novel High-Affinity and Versatile Peptide-Binding Nanobody for Proteomics and Microscopy." *Scientific Reports* 6 (October 2015): 19211. doi: 10.1038/srep19211

Burgert, Anne, Sebastian Letschert, Sören Doose, and Markus Sauer. 2015. "Artifacts in Single-Molecule Localization Microscopy." *Histochemistry and Cell Biology* 144 (2): 123–31. doi: 10.1007/s00418-015-1340-4

Butkevich, Alexey N., Vladimir N. Belov, Kirill Kolmakov, Viktor V. Sokolov, Heydar Shojaei, Sven C. Sidenstein, Dirk Kamin, et al. 2017. "Hydroxylated Fluorescent Dyes for Live-Cell Labeling: Synthesis, Spectra and Super-Resolution STED." *Chemistry (Weinheim an Der Bergstrasse, Germany)* 23 (50): 12114–9. doi: 10.1002/chem.201701216

Butkevich, Alexey N., Haisen T. A., Michael Ratz, Stefan Stold, Stefan Jakobs, Vladimir N. Belov, and Stefan W. Hell. 2018. "Two-Color 810 Nm STED Nanoscopy of Living Cells with Endogenous SNAP-Tagged Fusion Proteins." *ACS Chemical Biology.* doi: 10.1021/acschembio.7b00616

Canton, Irene, Marzia Massignani, Nisa Patikarnmonthon, Luca Chierico, James Robertson, Stephen A. Renshaw, Nicholas J. Warren, et al. 2013. "Fully Synthetic Polymer Vesicles for Intracellular Delivery of Antibodies in Live Cells." *FASEB Journal* 27 (1): 98–108. doi: 10.1096/fj.12-212183

Carr, Alexander R., Aleks Ponjavic, Srinjan Basu, James McColl, Ana Mafalda Santos, Simon Davis, Ernest D. Laue, David Klenerman, et al. 2017. "Three-Dimensional Super-Resolution in Eukaryotic Cells Using the Double-Helix Point Spread Function." *Biophysical Journal* 112 (7): 1444–54. doi: 10.1016/J.BPJ.2017.02.023

Chakrova, Nadya, Alicia Soler Canton, Christophe Danelon, Sjoerd Stallinga, and Bernd Rieger. 2016. "Adaptive Illumination Reduces Photobleaching in Structured Illumination Microscopy." *Biomedical Optics Express* 7 (10): 4263–74. doi: 10.1364/BOE.7.004263

Chen, B.-C., Wesley R. Legant, K. Wang, Lin Shao, Daniel E. Milkie, Michael W. Davidson, Chris Janetopoulos, et al. 2014. "Lattice Light-Sheet Microscopy: Imaging Molecules to Embryos at High Spatiotemporal Resolution." *Science* 346 (6208): 1257998. doi: 10.1126/science.1257998

Chéreau, Ronan, G. Ezequiel Saraceno, Julie Angibaud, Daniel Cattaert, and U. Valentin Nägerl. 2017. "Superresolution Imaging Reveals Activity-Dependent Plasticity of Axon Morphology Linked to Changes in Action Potential Conduction Velocity." *Proceedings of the National Academy of Sciences of the United States of America* 114 (6): 1401–6. doi: 10.1073/pnas.1607541114

Chmyrov, Andriy, Jan Keller, Tim Grotjohann, Michael Ratz, Elisa d'Este, Stefan Jakobs, Christian Eggeling, et al. 2013. "Nanoscopy with More than 100,000 'Doughnuts.'" *Nature Methods* 10 (8): 737–40. doi: 10.1038/nmeth.2556

Chudakov, Dmitriy M., Sergey Lukyanov, and Konstantin A. Lukyanov. 2005. "Fluorescent Proteins as a Toolkit for in Vivo Imaging." *Trends in Biotechnology* 23 (12): 605–13. doi: 10.1016/j.tibtech.2005.10.005

Clark, Spencer A., Vijay Singh, Daniel Vega Mendoza, William Margolin, and Eric T. Kool. 2016. "Light-Up 'Channel Dyes' for Haloalkane-Based Protein Labeling in Vitro and in Bacterial Cells." *Bioconjugate Chemistry* 27 (12): 2839–43. doi: 10.1021/acs.bioconjchem.6b00613

Cohen, Justin D., Samuel Thompson, and Alice Y. Ting. 2011. "Structure-Guided Engineering of a Pacific Blue Fluorophore Ligase for Specific Protein Imaging in Living Cells." *Biochemistry* 50 (38): 8221–25. doi: 10.1021/bi201037r

Cox, Susan, Edward Rosten, James Monypenny, Tijana Jovanovic-Talisman, Dylan T. Burnette, Jennifer, Lippincott-Schwartz, et al. 2011. "Bayesian Localization Microscopy Reveals Nanoscale Podosome Dynamics." *Nature Methods* 9 (2): 195–200. doi: 10.1038/nmeth.1812

Culley, Siân, David Albrecht, Caron Jacobs, Pedro Matos Pereira, Christophe Leterrier, Jason Mercer, and Ricardo Henriques. 2018. "Quantitative Mapping and Minimization of Super-Resolution Optical Imaging Artifacts." *Nature Methods* 15 (4): 263–66. doi: 10.1038/nmeth.4605

Culley, Siân, Kalina L. Tosheva, Pedro Matos Pereira, and Ricardo Henriques. 2018. "SRRF: Universal Live-Cell Super-Resolution Microscopy." *The International Journal of Biochemistry & Cell Biology* 101 (August): 74–9. doi: 10.1016/J.BIOCEL.2018.05.014

D'Este, Elisa, Dirk Kamin, Fabian Göttfert, Ahmed El-Hady, and Stefan W. Hell. 2015. "STED Nanoscopy Reveals the Ubiquity of Subcortical Cytoskeleton Periodicity in Living Neurons." *Cell Reports* 10 (8): 1246–51. doi: 10.1016/J.CELREP.2015.02.007

Dedecker, Peter, Sam Duwé, Robert K. Neely, and Jin Zhang. 2012. "Localizer: Fast, Accurate, Open-Source, and Modular Software Package for Superresolution Microscopy." *Journal of Biomedical Optics* 17 (12): 126008. doi: 10.1117/1.JBO.17.12.126008

Dempsey, Graham T., Joshua C. Vaughan, Kok Hao Chen, Mark Bates, and Xiaowei Zhuang. 2011. "Evaluation of Fluorophores for Optimal Performance in Localization-Based Super-Resolution Imaging." *Nature Methods* 8 (12): 1027–36. doi: 10.1038/nmeth.1768

Dempsey, William P., Lada Georgieva, Patrick M. Helbling, Ali Y. Sonay, Thai V. Truong, Michel Haffner, and Periklis Pantazis. 2015. "In Vivo Single-Cell Labeling by Confined Primed Conversion." *Nature Methods* 12 (7): 645–48. doi: 10.1038/nmeth.3405

Dertinger, T., R. Colyer, G. Iyer, S. Weiss, and J. Enderlein. 2009. "Fast, Background-Free, 3D Super-Resolution Optical Fluctuation Imaging (SOFI)." *Proceedings of the National Academy of Sciences* 106 (52): 22287–92. doi: 10.1073/pnas.0907866106

Durisic, Nela, Lara Laparra-Cuervo, Angel Sandoval-Álvarez, Joseph Steven Borbely, and Melike Lakadamyali. 2014. "Single-Molecule Evaluation of Fluorescent Protein Photoactivation Efficiency Using an in Vivo Nanotemplate." *Nature Methods* 11 (2): 156–62. doi: 10.1038/nmeth.2784

Fiolka, Reto, Lin Shao, E. Hesper Rego, Michael W. Davidson, and Mats G. L. Gustafsson. 2012. "Time-Lapse Two-Color 3D Imaging of Live Cells with Doubled Resolution Using Structured Illumination." *Proceedings of the National Academy of Sciences of the United States of America* 109 (14): 5311–15. doi: 10.1073/pnas.1119262109

Fox-Roberts, Patrick, Richard Marsh, Karin Pfisterer, Asier Jayo, Maddy Parsons, and Susan Cox. 2017. "Local Dimensionality Determines Imaging Speed in Localization Microscopy." *Nature Communications* 8 (January): 13558. doi: 10.1038/ncomms13558

Fu, Guo, Tao Huang, Jackson Buss, Carla Coltharp, Zach Hensel, and Jie Xiao. 2010. "In Vivo Structure of the E. Coli FtsZ-Ring Revealed by Photoactivated Localization Microscopy (PALM)." Edited by Michael Polymenis. *PLoS ONE* 5 (9): e12680. doi: 10.1371/journal.pone.0012680

Galbraith, Catherine G., and James A. Galbraith. 2011. "Super-Resolution Microscopy at a Glance." *Journal of Cell Science* 124 (Pt 10): 1607–11. doi: 10.1242/jcs.080085

Gautier, Arnaud, Alexandre Juillerat, Christian Heinis, Ivan Reis Corrêa, Maik Kindermann, Florent Beaufils, and Kai Johnsson. 2008. "An Engineered Protein Tag for Multiprotein Labeling in Living Cells." *Chemistry and Biology* 15 (2): 128–36. doi: 10.1016/j.chembiol.2008.01.007

Geissbuehler, Stefan, Noelia L. Bocchio, Claudio Dellagiacoma, Corinne Berclaz, Marcel Leutenegger, and Theo Lasser. 2012. "Mapping Molecular Statistics with Balanced Super-Resolution Optical Fluctuation Imaging (BSOFI)." *Optical Nanoscopy* 1 (1): 4. doi: 10.1186/2192-2853-1-4

Geissbuehler, Stefan, Azat Sharipov, Aurélien Godinat, Noelia L. Bocchio, Patrick A. Sandoz, Anja Huss, Nickels A. Jensen, et al. 2014. "Live-Cell Multiplane Three-Dimensional Super-Resolution Optical Fluctuation Imaging." *Nature Communications* 5: 1–7. doi: 10.1038/ncomms6830

Girkin, J. M., and M. T. Carvalho. 2018. "The Light-Sheet Microscopy Revolution." *Journal of Optics* 20 (5): 053002. doi: 10.1088/2040-8986/aab58a

Griffin, B. Albert, Stephen R. Adams, and Roger Y. Tsien. 1998. "Specific Covalent Labeling of Recombinant Protein Molecules inside Live Cells." *Science (New York, N.Y.)* 281 (5374): 269–72. doi: 10.1126/science.281.5374.269

Grimm, Jonathan B., Anand K. Muthusamy, Yajie Liang, Timothy A. Brown, William C. Lemon, Ronak Patel, Rongwen Lu, et al. 2017. "A General Method to Fine-Tune Fluorophores for Live-Cell and in Vivo Imaging." *Nature Methods* 14 (10): 987–94. doi: 10.1038/nmeth.4403

Grotjohann, Tim, Ilaria Testa, Matthias Reuss, Tanja Brakemann, Christian Eggeling, Stefan W. Hell, and Stefan Jakobs. 2012. "RsEGFP2 Enables Fast RESOLFT Nanoscopy of Living Cells." *ELife* 2012 (1): 1–14. doi: 10.7554/eLife.00248

Gudheti, Manasa V., Nikki M. Curthoys, Travis J. Gould, Dahan Kim, Mudalige S. Gunewardene, Kristin A. Gabor, Julie A. Gosse, Carol H. Kim, Joshua Zimmerberg, and Samuel T. Hess. 2013. "Actin Mediates the Nanoscale Membrane Organization of the Clustered Membrane Protein Influenza Hemagglutinin." *Biophysical Journal* 104 (10): 2182–92. doi: 10.1016/J.BPJ.2013.03.054

Guo, Min, Panagiotis Chandris, John Paul Giannini, Adam J. Trexler, Robert Fischer, Jiji Chen, Harshad D. Vishwasrao, et al. 2018. "Single-Shot Super-Resolution Total Internal Reflection Fluorescence Microscopy." *Nature Methods* 15 (6): 425–28. doi: 10.1038/s41592-018-0004-4

Gustafsson, M. G. 2000. "Surpassing the Lateral Resolution Limit by a Factor of Two Using Structured Illumination Microscopy." *Journal of Microscopy* 198 (Pt 2): 82–7. doi: 10.1046/j.1365-2818.2000.00710.x

Gustafsson, Mats G. L., Lin Shao, Peter M. Carlton, C. J. Rachel Wang, Inna N. Golubovskaya, W. Zacheus Cande, David A. Agard, and John W. Sedat. 2008. "Three-Dimensional Resolution Doubling in Wide-Field Fluorescence Microscopy by Structured Illumination." *Biophysical Journal* 94 (12): 4957–70. doi: 10.1529/biophysj.107.120345

Gustafsson, Mats G. L. 2005. "Nonlinear Structured-Illumination Microscopy: Wide-Field Fluorescence Imaging with Theoretically Unlimited Resolution." *Proceedings of the National Academy of Sciences of the United States of America* 102 (37): 13081–86. doi: 10.1073/pnas.0406877102

Gustafsson, Nils, Siân Culley, George Ashdown, Dylan M. Owen, Pedro Matos Pereira, and Ricardo Henriques. 2016. "Fast Live-Cell Conventional Fluorophore Nanoscopy with ImageJ through Super-Resolution Radial Fluctuations." *Nature Communications* 7 (12471): 1–9. doi: 10.1038/ncomms12471

Hammond, Jennetta W., T. Lynne Blasius, Virupakshi Soppina, Dawen Cai, and Kristen J. Verhey. 2010. "Autoinhibition of the Kinesin-2 Motor KIF17 via Dual Intramolecular Mechanisms." *Journal of Cell Biology* 189 (6): 1013–25. doi: 10.1083/jcb.201001057

Han, Yubing, Meihua Li, Fengwu Qiu, Meng Zhang, and Yu-Hui Zhang. 2017. "Cell-Permeable Organic Fluorescent Probes for Live-Cell Long-Term Super-Resolution Imaging Reveal Lysosome-Mitochondrion Interactions." *Nature Communications* 8 (1): 1307. doi: 10.1038/s41467-017-01503-6

Harke, Benjamin, Jan Keller, Chaitanya K. Ullal, Volker Westphal, Andreas Schönle, and Stefan W. Hell. 2008. "Resolution Scaling in STED Microscopy." *Optics Express* 16 (6): 4154. doi: 10.1364/OE.16.004154

Heilemann, Mike, Sebastian van de Linde, Mark Schüttpelz, Robert Kasper, Britta Seefeldt, Anindita Mukherjee, Philip Tinnefeld, and Markus Sauer. 2008. "Subdiffraction-Resolution Fluorescence Imaging with Conventional

Fluorescent Probes." *Angewandte Chemie (International Ed. in English)* 47 (33): 6172–76. doi: 10.1002/anie.200802376

Hein, Birka, Katrin I. Willig, and Stefan W. Hell. 2008. "Stimulated Emission Depletion (STED) Nanoscopy of a Fluorescent Protein-Labeled Organelle inside a Living Cell." *Proceedings of the National Academy of Sciences* 105 (38): 14271–76. doi: 10.1073/pnas.0807705105

Heine, Jörn, Matthias Reuss, Benjamin Harke, Elisa D'Este, Steffen J. Sahl, and Stefan W. Hell. 2017. "Adaptive-Illumination STED Nanoscopy." *Proceedings of the National Academy of Sciences of the United States of America* 114 (37): 9797–802. doi: 10.1073/pnas.1708304114

Henriques, Ricardo, Mickael Lelek, Eugenio F. Fornasiero, Flavia Valtorta, Christophe Zimmer, and Musa M. Mhlanga. 2010. "QuickPALM: 3D Real-Time Photoactivation Nanoscopy Image Processing in ImageJ." *Nature Methods* 7 (5): 339–40. doi: 10.1038/nmeth0510-339

Hense, Anika, Benedikt Prunsche, Peng Gao, Yuji Ishitsuka, Karin Nienhaus, and G. Ulrich Nienhaus. 2015. "Monomeric Garnet, a Far-Red Fluorescent Protein for Live-Cell STED Imaging." *Scientific Reports* 5: 1–10. doi: 10.1038/srep18006

Hess, Samuel T., Thanu P. K. Girirajan, and Michael D. Mason. 2006. "Ultra-High Resolution Imaging by Fluorescence Photoactivation Localization Microscopy." *Biophysical Journal* 91 (11): 4258–72. doi: 10.1529/BIOPHYSJ.106.091116

Hoebe, R. A., C. H. Van Oven, T. W. J. Gadella, P. B. Dhonukshe, C. J. F. Van Noorden, and E. M. M. Manders. 2007. "Controlled Light-Exposure Microscopy Reduces Photobleaching and Phototoxicity in Fluorescence Live-Cell Imaging." *Nature Biotechnology* 25 (2): 249–53. doi: 10.1038/nbt1278.

Hofmann, Michael, Christian Eggeling, Stefan Jakobs, and Stefan W. Hell. 2005. "Breaking the Diffraction Barrier in Fluorescence Microscopy at Low Light Intensities by Using Reversibly Photoswitchable Proteins." *Proceedings of the National Academy of Sciences* 102 (49): 17565–69. doi: 10.1073/pnas.0506010102

Hostettler, Lola, Laura Grundy, Stéphanie Käser-Pébernard, Chantal Wicky, William R. Schafer, and Dominique A. Glauser. 2017. "The Bright Fluorescent Protein MNeonGreen Facilitates Protein Expression Analysis *In Vivo.*" *G3: Genes|Genomes|Genetics* 7 (2): 607–15. doi: 10.1534/g3.116.038133.

Howarth, Mark, Keizo Takao, Yasunori Hayashi, and Alice Y. Ting. 2005. "Targeting Quantum Dots to Surface Proteins in Living Cells with Biotin Ligase." *Proceedings of the National Academy of Sciences* 102 (21): 7583–88. doi: 10.1073/pnas.0503125102

"Http://Bigwww.Epfl.Ch/Smlm/Challenge2016/Index.Html?P= datasets." n.d.

"Https://Documents.Epfl.Ch/Users/l/Le/Leuteneg/Www/BalancedSOFI/Index.Html." n.d.

Hu, Ying S., Xiaolin Nan, Prabuddha Sengupta, Jennifer Lippincott-Schwartz, and Hu Cang. 2013. "Accelerating 3B Single-Molecule Super-Resolution Microscopy with Cloud Computing." *Nature Methods* 10 (2): 96–7. doi: 10.1038/nmeth.2335

Huang, Bo, Wenqin Wang, Mark Bates, and Xiaowei Zhuang. 2008. "Three-Dimensional Super-Resolution Imaging by Stochastic Optical Reconstruction Microscopy." *Science (New York, N.Y.)* 319 (5864): 810–13. doi: 10.1126/science.1153529

Huang, Fang, Tobias M. P. Hartwich, Felix E. Rivera-Molina, Yu Lin, Whitney C. Duim, Jane J. Long, Pradeep D. Uchil, et al. 2013. "Video-Rate Nanoscopy Using SCMOS Camera-Specific Single-Molecule Localization Algorithms." *Nature Methods* 10 (7): 653–58. doi: 10.1038/nmeth.2488

Huff, Joseph. 2015. "The Airyscan Detector from ZEISS: Confocal Imaging with Improved Signal-to-Noise Ratio and Super-Resolution." *Nature Methods* 12 (12): i–ii. doi: 10.1038/nmeth.f.388

Iinuma, Ryosuke, Yonggang Ke, Ralf Jungmann, Thomas Schlichthaerle, Johannes B. Woehrstein, and Peng Yin. 2014. "Polyhedra Self-Assembled from DNA Tripods and Characterized with 3D DNA-PAINT." *Science* 344 (6179): 65–9. doi: 10.1126/science.1250944

Inoue, Shinya. 2006. "Foundations of Confocal Scanned Imaging in Light Microscopy." In *Handbook of Biological Confocal Microscopy*. Edited by James B. Pawley (3rd edition). New York, US: Springer Science.

Ji, Chen, Yongdeng Zhang, Pingyong Xu, Tao Xu, and Xuelin Lou. 2015. "Nanoscale Landscape of Phosphoinositides Revealed by Specific Pleckstrin Homology (PH) Domains Using Single-Molecule Superresolution Imaging in the Plasma Membrane." *The Journal of Biological Chemistry* 290 (45): 26978–93. doi: 10.1074/jbc.M115.663013

Jones, Sara A., Sang-Hee Shim, Jiang He, and Xiaowei Zhuang. 2011. "Fast, Three-Dimensional Super-Resolution Imaging of Live Cells." *Nature Methods* 8 (6): 499–505. doi: 10.1038/nmeth.1605

Juette, Manuel F., Travis J. Gould, Mark D. Lessard, Michael J. Mlodzianoski, Bhupendra S. Nagpure, Brian T. Bennett, Samuel T. Hess, and Joerg Bewersdorf. 2008. "Three-Dimensional Sub–100 Nm Resolution Fluorescence Microscopy of Thick Samples." *Nature Methods* 5 (6): 527–29. doi: 10.1038/nmeth.1211

Jungmann, Ralf, Maier S. Avendaño, Johannes B. Woehrstein, Mingjie Dai, William M. Shih, and Peng Yin. 2014. "Multiplexed 3D Cellular Super-Resolution Imaging with DNA-PAINT and Exchange-PAINT." *Nature Methods* 11 (3): 313–18. doi: 10.1038/nmeth.2835

Jungmann, Ralf, Christian Steinhauer, Max Scheible, Anton Kuzyk, Philip Tinnefeld, and Friedrich C. Simmel. 2010. "Single-Molecule Kinetics and Super-Resolution Microscopy by Fluorescence Imaging of Transient Binding on DNA Origami." *Nano Letters* 10 (11): 4756–61. doi: 10.1021/nl103427w

Kaberniuk, Andrii A., Nicholas C. Morano, Vladislav V. Verkhusha, and Erik Lee Snapp. 2017. "MoxDendra2: An Inert Photoswitchable Protein for Oxidizing Environments." *Chemical Communications* 53 (13): 2106–9. doi: 10.1039/C6CC09997A.

Kanchanawong, Pakorn, Gleb Shtengel, Ana M. Pasapera, Ericka B. Ramko, Michael W. Davidson, Harald F. Hess, and Clare M. Waterman. 2010. "Nanoscale Architecture of Integrin-Based Cell Adhesions." *Nature* 468 (7323): 580–4. doi: 10.1038/nature09621

Keppler, Antje, Susanne Gendreizig, Thomas Gronemeyer, Horst Pick, Horst Vogel, and Kai Johnsson. 2003. "A General Method for the Covalent Labeling of Fusion Proteins with Small Molecules in Vivo." *Nature Biotechnology* 21 (1): 86–9. doi: 10.1038/nbt765

Khan, Abdullah O., Victoria A. Simms, Jeremy A. Pike, Steven G. Thomas, and Neil V. Morgan. 2017. "CRISPR-Cas9 Mediated Labelling Allows for Single Molecule Imaging and Resolution." *Scientific Reports* 7 (1): 8450. doi: 10.1038/s41598-017-08493-x

Kirshner, Hagai, Franois Aguet, Daniel Sage, and Michael Unser. 2013. "3-D PSF Fitting for Fluorescence Microscopy: Implementation and Localization Application." *Journal of Microscopy* 249 (1): 13–25. doi: 10.1111/j.1365-2818.2012.03675.x.

Klar, Thomas A., Stefan Jakobs, Marcus Dyba, Alexander Egner, and Stefan W. Hell. 2000. "Fluorescence Microscopy with Diffraction Resolution Barrier Broken by Stimulated Emission." *Proceedings of the National Academy of Sciences of the United States of America* 97 (15): 8206–10. doi: 10.1073/PNAS.97.15.8206

Kollmannsperger, Alina, Armon Sharei, Anika Raulf, Mike Heilemann, Robert Langer, Klavs F. Jensen, Ralph Wieneke, and Robert Tampé. 2016. "Live-Cell Protein Labelling with Nanometre Precision by Cell Squeezing." *Nature Communications* 7: 10372. doi: 10.1038/ncomms10372

Kremers, Gert-Jan, Sarah G. Gilbert, Paula J. Cranfill, Michael W. Davidson, and David W. Piston. 2011. "Fluorescent Proteins at a Glance." *Journal of Cell Science* 124 (15): 2676. doi: 10.1242/jcs.095059

Kwakwa, Kwasi, Alexander Savell, Timothy Davies, Ian Munro, Simona Parrinello, Marco A. Purbhoo, Chris Dunsby, Mark A. A. Neil, and Paul M. W. French. 2016. "EasySTORM: A Robust, Lower-Cost Approach to Localisation and TIRF Microscopy." *Journal of Biophotonics* 9 (9): 948–57. doi: 10.1002/jbio.201500324.

Laine, Romain F., Anna Albecka, Sebastian Van De Linde, Eric J. Rees, Colin M. Crump, and Clemens F. Kaminski. 2015. "Structural Analysis of Herpes Simplex Virus by Optical Super-Resolution Imaging." *Nature Communications* 6: 1–10. doi: 10.1038/ncomms6980.

Lambert, Helene, Roumen Pankov, Johanne Gauthier, and Ronald Hancock. 1990. "Electroporation-Mediated Uptake of Proteins into Mammalian Cells." *Biochemistry and Cell Biology = Biochimie et Biologie Cellulaire* 68 (4): 729–34.

Lau, Lana, Yin Loon Loon Lee, Steffen J. J. Sahl, Tim Stearns, and W. E. E. Moerner. 2012. "STED Microscopy with Optimized Labeling Density Reveals 9-Fold Arrangement of a Centriole Protein." *Biophysical Journal* 102 (12): 2926–35. doi: 10.1016/j.bpj.2012.05.015

Lavis, Luke D. 2017a. "Chemistry Is Dead. Long Live Chemistry!" *Biochemistry* 56 (39): 5165–70. doi: 10.1021/acs.biochem.7b00529

Lavis, Luke D. 2017b. "Teaching Old Dyes New Tricks: Biological Probes Built from Fluoresceins and Rhodamines." *Annual Review of Biochemistry* 86 (1): 825–43. doi: 10.1146/annurev-biochem-061516-044839

Lehmann, Martin, Gregor Lichtner, Haider Klenz, and Jan Schmoranzer. 2016. "Novel Organic Dyes for Multicolor Localization-Based Super-Resolution Microscopy." *Journal of Biophotonics* 9 (1–2): 161–70. doi: 10.1002/jbio.201500119

Lelek, Mickaël, Francesca Di Nunzio, Ricardo Henriques, Pierre Charneau, Nathalie Arhel, and Christophe Zimmer. 2012. "Superresolution Imaging of HIV in Infected Cells with FlAsH-PALM." *Proceedings of the National Academy of Sciences of the United States of America* 109 (22): 8564–69. doi: 10.1073/pnas.1013267109

Leng, Shuang, Qinglong Qiao, Lu Miao, Wuguo Deng, Jingnan Cui, and Zhaochao Xu. 2017. "A Wash-Free SNAP-Tag Fluorogenic Probe Based on the Additive Effects of Quencher Release and Environmental Sensitivity." *Chemical Communications (Cambridge, England)* 53 (48): 6448–51. doi: 10.1039/c7cc01483j

Li, Chenge, Alison G. Tebo, and Arnaud Gautier. 2017. "Fluorogenic Labeling Strategies for Biological Imaging." *International Journal of Molecular Sciences* 18 (7): 1473. doi: 10.3390/ijms18071473

Li, D., L. Shao, B.-C. Chen, X. Zhang, M. Zhang, B. Moses, D. E. Milkie, et al. 2015. "Extended-Resolution Structured Illumination Imaging of Endocytic and Cytoskeletal Dynamics." *Science* 349 (6251): aab3500. doi: 10.1126/science.aab3500.

Liu, Daniel S., Lucas G. Nivón, Florian Richter, Peter J. Goldman, Thomas J. Deerinck, Jennifer Z. Yao, Douglas Richardson, et al. 2014. "Computational Design of a Red Fluorophore Ligase for Site-Specific Protein Labeling in Living Cells." *Proceedings of the National Academy of Sciences of the United States of America* 111 (43): E4551–9. doi: 10.1073/pnas.1404736111

Liu, Tao Kai, Pei Ying Hsieh, Yu De Zhuang, Chi Yang Hsia, Chi Ling Huang, Hsiu Ping Lai, Hung Sheung Lin, I. Chia Chen, Hsin Yun Hsu, and Kui Thong Tan. 2014. "A Rapid SNAP-Tag Fluorogenic Probe Based on an Environment-Sensitive Fluorophore for No-Wash Live Cell Imaging." *ACS Chemical Biology* 9 (10): 2359–65. doi: 10.1021/cb500502n

Liu, Yajun, and Jian Qiu Wu. 2016. "Cytokinesis: Going Super-Resolution in Live Cells." *Current Biology* 26 (21): R1150–52. doi: 10.1016/j.cub.2016.09.026

Los, Georgyi V., Lance P. Encell, Mark G. McDougall, Danette D. Hartzell, Natasha Karassina, Chad Zimprich, Monika G. Wood, et al. 2008. "HaloTag: A Novel Protein Labeling Technology for Cell Imaging and Protein Analysis." *ACS Chemical Biology* 3 (6): 373–82. doi: 10.1021/cb800025k

Lu-Walther, Hui-Wen, Martin Kielhorn, Ronny Förster, Aurélie Jost, Kai Wicker, and Rainer Heintzmann. 2015. "FastSIM: A Practical Implementation of Fast Structured Illumination Microscopy." *Methods and Applications in Fluorescence* 3 (1): 014001. doi: 10.1088/2050-6120/3/1/014001

Luca, Giulia M. R. De, Ronald M. P. Breedijk, Rick A. J. Brandt, Christiaan H. C. Zeelenberg, Babette E. de Jong, Wendy Timmermans, Leila Nahidi Azar, Ron A. Hoebe, Sjoerd Stallinga, and Erik M. M. Manders. 2013. "Re-Scan Confocal Microscopy: Scanning Twice for Better Resolution." *Biomedical Optics Express* 4 (11): 2644–56. doi: 10.1364/BOE.4.002644

Lukinavičius, Gražvydas, Claudia Blaukopf, Elias Pershagen, Alberto Schena, Luc Reymond, Emmanuel Derivery, Marcos Gonzalez-Gaitan, et al. 2015. "SiR-Hoechst Is a Far-Red DNA Stain for Live-Cell Nanoscopy." *Nature Communications* 6: 1–7. doi: 10.1038/ncomms9497

Lukinavičius, Gražvydas, Luc Reymond, Elisa D'Este, Anastasiya Masharina, Fabian Göttfert, Haisen Ta, Angelika Güther, et al. 2014. "Fluorogenic Probes for Live-Cell Imaging of the Cytoskeleton." *Nature Methods* 11 (7): 731–33. doi: 10.1038/nmeth.2972

Lukinavičius, Gražvydas, Luc Reymond, Keitaro Umezawa, Olivier Sallin, Elisa D'Este, Fabian Göttfert, Haisen Ta, Stefan W. Hell, Yasuteru Urano, and Kai Johnsson. 2016. "Fluorogenic Probes for Multicolor Imaging in Living Cells." *Journal of the American Chemical Society* 138 (30): 9365–68. doi: 10.1021/jacs.6b04782

Lukinavičius, Gražvydas, Keitaro Umezawa, Nicolas Olivier, Alf Honigmann, Guoying Yang, Tilman Plass, Veronika Mueller, et al. 2013. "A Near-Infrared Fluorophore for Live-Cell Super-Resolution Microscopy of Cellular Proteins." *Nature Chemistry* 5 (2): 132–39. doi: 10.1038/nchem.1546

Ma, Hongqiang, Rao Fu, Jianquan Xu, and Yang Liu. 2017. "A Simple and Cost-Effective Setup for Super-Resolution Localization Microscopy." *Scientific Reports* 7 (1): 1542. doi: 10.1038/s41598-017-01606-6

Manley, Suliana, Jennifer M. Gillette, George H. Patterson, Hari Shroff, Harald F. Hess, Eric Betzig, and Jennifer Lippincott-Schwartz. 2008. "High-Density Mapping of Single-Molecule Trajectories with Photoactivated Localization Microscopy." *Nature Methods* 5 (2): 155–57. doi: 10.1038/nmeth.1176

Mateos-Gil, Pablo, Sebastian Letschert, Sören Doose, and Markus Sauer. 2016. "Super-Resolution Imaging of Plasma Membrane Proteins with Click Chemistry." *Frontiers in Cell and Developmental Biology* 4 (September): 98. doi: 10.3389/fcell.2016.00098

Mikhaylova, Marina, Bas M. C. Cloin, Kieran Finan, Robert van den Berg, Jalmar Teeuw, Marta M. Kijanka, Mikolaj Sokolowski, et al. 2015. "Resolving Bundled Microtubules Using Anti-Tubulin Nanobodies." *Nature Communications* 6 (May): 7933. doi: 10.1038/ncomms8933

Min, Junhong, Cédric Vonesch, Hagai Kirshner, Lina Carlini, Nicolas Olivier, Seamus Holden, Suliana Manley, Jong Chul Ye, and Michael Unser. 2015. "FALCON: Fast and Unbiased Reconstruction of High-Density Super-Resolution Microscopy Data." *Scientific Reports* 4 (1): 4577. doi: 10.1038/srep04577

Mishina, Natalia M., Alexander S. Mishin, Yury Belyaev, Ekaterina A. Bogdanova, Sergey Lukyanov, Carsten Schultz, and Vsevolod V. Belousov. 2015. "Live-Cell STED Microscopy with Genetically Encoded Biosensor." *Nano Letters* 15 (5): 2928–32. doi: 10.1021/nl504710z

Mo, Gary C. H., Brian Ross, Fabian Hertel, Premashis Manna, Xinxing Yang, Eric Greenwald, Chris Booth, et al. 2017. "Genetically Encoded Biosensors for Visualizing Live-Cell Biochemical Activity at Super-Resolution." *Nature Methods* 14 (4): 427–34. doi: 10.1038/nmeth.4221

Mukamel, Eran A., Hazen Babcock, and Xiaowei Zhuang. 2012. "Statistical Deconvolution for Superresolution Fluorescence Microscopy." *Biophysical Journal* 102 (10): 2391–400. doi: 10.1016/j.bpj.2012.03.070

Müller, Claus B., and Jörg Enderlein. 2010. "Image Scanning Microscopy." *Physical Review Letters* 104 (19): 198101. doi: 10.1103/PhysRevLett.104.198101

Müller, Marcel, Viola Mönkemöller, Simon Hennig, Wolfgang Hübner, and Thomas Huser. 2016. "Open-Source Image Reconstruction of Super-Resolution Structured Illumination Microscopy Data in ImageJ." *Nature Communications* 7: 10980. doi: 10.1038/ncomms10980

Nägerl, U. Valentin, Katrin I. Willig, Birka Hein, Stefan W. Hell, and Tobias Bonhoeffer. 2008. "Live-Cell Imaging of Dendritic Spines by STED Microscopy." *Proceedings of the National Academy of Sciences of the United States of America* 105 (48): 18982–87. doi: 10.1073/pnas.0810028105

Nehme, Elias, Lucien E. Weiss, Tomer Michaeli, and Yoav Shechtman. 2018. "Deep-STORM: Super-Resolution Single-Molecule Microscopy by Deep Learning." *Optica* 5 (4): 458–64.

Ni, Qiang, Sohum Mehta, and Jin Zhang. 2017. "Live-Cell Imaging of Cell Signaling Using Genetically Encoded Fluorescent Reporters." *The FEBS Journal* 285: 203-219 June, 1–17. doi: 10.1111/febs.14134

Nienhaus, Karin, and G. Ulrich Nienhaus. 2014. "Fluorescent Proteins for Live-Cell Imaging with Super-Resolution." *Chemical Society Reviews* 43 (4): 1088–106. doi: 10.1039/C3CS60171D

Nieuwenhuizen, Robert P. J., Keith a Lidke, Mark Bates, Daniela Leyton Puig, David Grünwald, Sjoerd Stallinga, and Bernd Rieger. 2013. "Measuring Image Resolution in Optical Nanoscopy." *Nature Methods* 10 (6): 557–62. doi: 10.1038/nmeth.2448

Nikić, Ivana, Gemma Estrada Girona, Jun Hee Kang, Giulia Paci, Sofya Mikhaleva, Christine Koehler, Nataliia V. Shymanska, et al. 2016. "Debugging Eukaryotic Genetic Code Expansion for Site-Specific Click-PAINT Super-Resolution Microscopy." *Angewandte Chemie (International Ed. in English)* 55 (52): 16172–76. doi: 10.1002/anie.201608284

Nikić, Ivana, and Edward A. Lemke. 2015. "Genetic Code Expansion Enabled Site-Specific Dual-Color Protein Labeling: Superresolution Microscopy and Beyond." *Current Opinion in Chemical Biology* 28: 164–73. doi: 10.1016/j.cbpa.2015.07.021

Nozumi, Motohiro, Fubito Nakatsu, Kaoru Katoh, and Michihiro Igarashi. 2017. "Coordinated Movement of Vesicles and Actin Bundles during Nerve Growth Revealed by Superresolution Microscopy." *Cell Reports* 18 (9): 2203–16. doi: 10.1016/J.CELREP.2017.02.008

Olivier, Nicolas, Debora Keller, Pierre Gönczy, and Suliana Manley. 2013. "Resolution Doubling in 3D-STORM Imaging through Improved Buffers." Edited by Markus Sauer. *PLoS ONE* 8 (7): e69004. doi: 10.1371/journal.pone.0069004

Olivier, Nicolas, Debora Keller, Vinoth Sundar Rajan, Pierre Gönczy, and Suliana Manley. 2013. "Simple Buffers for 3D STORM Microscopy." *Biomedical Optics Express* 4 (6): 885–99. doi: 10.1364/BOE.4.000885

Opazo, Felipe, Matthew Levy, Michelle Byrom, Christina Schäfer, Claudia Geisler, Teja W. Groemer, Andrew D. Ellington, and Silvio O. Rizzoli. 2012. "Aptamers as Potential Tools for Super-Resolution Microscopy." *Nature Methods* 9 (10): 938–39. doi: 10.1038/nmeth.2179

Ouyang, Wei, Andrey Aristov, Mickaël Lelek, Xian Hao, and Christophe Zimmer. 2018. "Deep Learning Massively Accelerates Super-Resolution Localization Microscopy." *Nature Biotechnology* 36 (February): 460–68. doi: 10.1038/nbt.4106

Ovesný, Martin, Pavel Křížek, Josef Borkovec, Zdeněk Švindrych, and Guy M. Hagen. 2014. "ThunderSTORM: A Comprehensive ImageJ Plug-in for PALM and STORM Data Analysis and Super-Resolution Imaging." *Bioinformatics* 30 (16): 2389–90. doi: 10.1093/bioinformatics/btu202

Pavani, Sri Rama Prasanna, Michael A. Thompson, Julie S. Biteen, Samuel J. Lord, Na Liu, Robert J. Twieg, Rafael Piestun, and W. E. Moerner. 2009. "Three-Dimensional, Single-Molecule Fluorescence Imaging beyond the Diffraction Limit by Using a Double-Helix Point Spread Function." *Proceedings of the National Academy of Sciences of the United States of America* 106 (9): 2995–99. doi: 10.1073/pnas.0900245106

Pennacchietti, Francesca, Travis J. Gould, and Samuel T. Hess. 2017. "The Role of Probe Photophysics in Localization-Based Superresolution Microscopy." *Biophysical Journal* 113 (9): 2037–54. doi: 10.1016/j.bpj.2017.08.054

Pereira, Pedro M., David Albrecht, Caron Jacobs, Mark Marsh, Jason Mercer, and Ricardo Henriques. 2018. "Fix Your Membrane Receptor Imaging: Actin Cytoskeleton and CD4 Membrane Organization Disruption by Chemical Fixation." *BioRxiv*, October, 450635. doi: 10.1101/450635

Pereira, Pedro M., Pedro Almada, and Ricardo Henriques. 2015. "High-Content 3D Multicolor Super-Resolution Localization Microscopy." *Methods in Cell Biology* 125: 95–117. doi: 10.1016/bs.mcb.2014.10.004

Platonova, Evgenia, Christian M. Winterflood, and Helge Ewers. 2015. "A Simple Method for GFP- and RFP-Based Dual Color Single-Molecule Localization Microscopy." *ACS Chemical Biology* 10 (6): 1411–16. doi: 10.1021/acschembio.5b00046

Pleiner, Tino, Mark Bates, and Dirk Görlich. 2017. "A Toolbox of Anti-Mouse and Anti-Rabbit IgG Secondary Nanobodies." *The Journal of Cell Biology* 217 (3): 1143–1154. doi: 10.1083/jcb.201709115

Pleiner, Tino, Mark Bates, Sergei Trakhanov, Chung Tien Lee, Jan Erik Schliep, Hema Chug, Marc Böhning, et al. 2015. "Nanobodies: Site-Specific Labeling for Super-Resolution Imaging, Rapid Epitope- Mapping and Native Protein Complex Isolation." *ELife* 4 (December 2015): e11349. doi: 10.7554/eLife.11349

Pospíšil, Jakub, Tomáš Lukeš, Justin Bendesky, Karel Fliegel, Kathrin Spendier, and Guy M. Hagen. 2018. "Imaging Tissues and Cells beyond the Diffraction Limit with Structured Illumination Microscopy and Bayesian Image Reconstruction." *GigaScience* 8 (1): 1-12 doi: 10.1093/gigascience/giy126

Prifti Efthymia, Luc Reymond, Miwa Umebayashi, Ruud Hovius, Howard Riezman, and Kai Johnsson. 2014. "A Fluorogenic Probe for Snap-Tagged Plasma Membrane Proteins Based on the Solvatochromic Molecule Nile Red." *ACS Chemical Biology* 9 (3): 606–12. doi: 10.1021/cb400819c

Ratz, Michael, Ilaria Testa, Stefan W. Hell, and Stefan Jakobs. 2015. "CRISPR/Cas9-Mediated Endogenous Protein Tagging for RESOLFT Super-Resolution Microscopy of Living Human Cells." *Scientific Reports* 5: 1–6. doi: 10.1038/srep09592

Rego, E. H., L. Shao, J. J. Macklin, L. Winoto, G. A. Johansson, N. Kamps-Hughes, M. W. Davidson, and M. G. L. Gustafsson. 2012. "Nonlinear Structured-Illumination Microscopy with a Photoswitchable Protein Reveals Cellular Structures at 50-Nm Resolution." *Proceedings of the National Academy of Sciences* 109 (3): E135–43. doi: 10.1073/pnas.1107547108

Richardson, Douglas S., Carola Gregor, Franziska R. Winter, Nicolai T. Urban, Steffen J. Sahl, Katrin I. Willig, and Stefan W. Hell. 2017. "SRpHi Ratiometric PH Biosensors for Super-Resolution Microscopy." *Nature Communications* 8 (1). doi: 10.1038/s41467-017-00606-4

Ries, Jonas, Charlotte Kaplan, Evgenia Platonova, Hadi Eghlidi, and Helge Ewers. 2012. "A Simple, Versatile Method for GFP-Based Super-Resolution Microscopy via Nanobodies." *Nature Methods* 9 (6): 582–84. doi: 10.1038/nmeth.1991

Rosenbloom, Alyssa B., Sang-Hyuk Lee, Milton To, Antony Lee, Jae Yen Shin, and Carlos Bustamante. 2014. "Optimized Two-Color Super Resolution Imaging of Drp1 during Mitochondrial Fission with a Slow-Switching Dronpa Variant." *Proceedings of the National Academy of Sciences* 111 (36): 13093–98. doi: 10.1073/pnas.1320044111

Rosten, Edward, Gareth E. Jones, and Susan Cox. 2013. "ImageJ Plug-in for Bayesian Analysis of Blinking and Bleaching." *Nature Methods* 10 (2): 97–98. doi: 10.1038/nmeth.2342

Rust, Michael J., Mark Bates, and Xiaowei Zhuang. 2006. "Sub-Diffraction-Limit Imaging by Stochastic Optical Reconstruction Microscopy (STORM)." *Nature Methods* 3 (10): 793–95. doi: 10.1038/nmeth929

Sage, Daniel, Hagai Kirshner, Thomas Pengo, Nico Stuurman, Junhong Min, Suliana Manley, and Michael Unser. 2015. "Quantitative Evaluation of Software Packages for Single-Molecule Localization Microscopy." *Nature Methods* 12 (8): 717–24. doi: 10.1038/nmeth.3442

Sakin, Volkan, Janina Hanne, Jessica Dunder, Maria Anders-Össwein, Vibor Laketa, Ivana Nikić, Hans Georg Kräusslich, et al. 2017. "A Versatile Tool for Live-Cell Imaging and Super-Resolution Nanoscopy Studies of HIV-1 Env Distribution and Mobility." *Cell Chemical Biology* 24 (5): 635–45.e5. doi: 10.1016/j.chembiol.2017.04.007

Sanford, Lynn, and Amy Palmer. 2017. "Recent Advances in Development of Genetically Encoded Fluorescent Sensors." *Methods in Enzymology* 589: 1–49. doi: 10.1016/bs.mie.2017.01.019

Schindelin, Johannes, Ignacio Arganda-Carreras, Erwin Frise, Verena Kaynig, Mark Longair, Tobias Pietzsch, Stephan Preibisch, et al. 2012. "Fiji: An Open-Source Platform for Biological-Image Analysis." *Nature Methods* 9 (7): 676–82. doi: 10.1038/nmeth.2019

Schlichthaerle, Thomas, Alexandra Eklund, Florian Schueder, Maximilian Strauss, Christian Tiede, Alistair Curd, Jonas Ries, Michelle Peckham, Darren Tomlinson, and Ralf Jungmann. 2018. "Site-Specific Labeling of Affimers for DNA-PAINT Microscopy." *Angewandte Chemie (International Ed. in English)* 57: 11060–3. doi: 10.1002/anie.201804020

Schneider, Caroline A., Wayne S. Rasband, and Kevin W. Eliceiri. 2012. "NIH Image to ImageJ: 25 Years of Image Analysis." *Nature Methods* 9 (7): 671–75. doi: 10.1038/nmeth.2089.

Schnitzbauer, Joerg, Maximilian T. Strauss, Thomas Schlichthaerle, Florian Schueder, and Ralf Jungmann. 2017. "Super-Resolution Microscopy with DNA-PAINT." *Nature Protocols* 12 (6): 1198–228. doi: 10.1038/nprot.2017.024

Schueder, Florian, Juanita Lara-Gutiérrez, Brian J. Beliveau, Sinem K. Saka, Hiroshi M. Sasaki, Johannes B. Woehrstein, Maximilian T. Strauss, et al. 2017. "Multiplexed 3D Super-Resolution Imaging of Whole Cells Using Spinning Disk Confocal Microscopy and DNA-PAINT." *Nature Communications* 8 (1): 2090. doi: 10.1038/s41467-017-02028-8.

Schueder, Florian, Maximilian T. Strauss, David Hoerl, Joerg Schnitzbauer, Thomas Schlichthaerle, Sebastian Strauss, Peng Yin, et al. 2017. "Universal Super-Resolution Multiplexing by DNA Exchange." *Angewandte Chemie - International Edition* 56 (14): 4052–55. doi: 10.1002/anie.201611729

Schvartz, Tomer, Noa Aloush, Inna Goliand, Inbar Segal, Dikla Nachmias, Eyal Arbely, and Natalie Elia. 2017. "Direct Fluorescent-Dye Labeling of α-Tubulin in Mammalian Cells for Live Cell and Superresolution Imaging." *Molecular Biology of the Cell* 28 (21): 2747–56. doi: 10.1091/mbc.E17-03-0161

Sednev, Maksim V., Vladimir N. Belov, and Stefan W. Hell. 2015. "Fluorescent Dyes with Large Stokes Shifts for Super-Resolution Optical Microscopy of Biological Objects: A Review." *Methods and Applications in Fluorescence* 3 (4): 042004. doi: 10.1088/2050-6120/3/4/042004

Shaner, Nathan C., Gerard G. Lambert, Andrew Chammas, Yuhui Ni, Paula J. Cranfill, Michelle A. Baird, Brittney R. Sell, et al. 2013. "A Bright Monomeric Green Fluorescent Protein Derived from Branchiostoma Lanceolatum." *Nature Methods* 10 (5): 407–9. doi: 10.1038/nmeth.2413

Shaner, Nathan C., Paul A. Steinbach, and Roger Y. Tsien. 2005. "A Guide to Choosing Fluorescent Proteins." *Nature Methods* 2 (12): 905–9. doi: 10.1038/nmeth819

Sharonov, Alexey, and Robin M. Hochstrasser. 2006. "Wide-Field Subdiffraction Imaging by Accumulated Binding of Diffusing Probes." *Proceedings of the National Academy of Sciences of the United States of America* 103 (50): 18911–16. doi: 10.1073/pnas.0609643104

Shcherbakova, Daria M., Prabuddha Sengupta, Jennifer Lippincott-Schwartz, and Vladislav V. Verkhusha. 2014. "Photocontrollable Fluorescent Proteins for Superresolution Imaging." *Annual Review of Biophysics* 43 (1): 303–29. doi: 10.1146/annurev-biophys-051013-022836

Sheppard, C. J. R. 1988. "Super-Resolution in Confocal Imaging." *Optik (Stuttgart)* 80: 53–4.

Sheppard, Colin J. R., Shalin B. Mehta, and Rainer Heintzmann. 2013. "Superresolution by Image Scanning Microscopy Using Pixel Reassignment." *Optics Letters* 38 (15): 2889. doi: 10.1364/OL.38.002889

Shim, Sang-Hee, Chenglong Xia, Guisheng Zhong, Hazen P. Babcock, Joshua C. Vaughan, Bo Huang, Xun Wang, et al. 2012. "Super-Resolution Fluorescence Imaging of Organelles in Live Cells with Photoswitchable Membrane Probes." *Proceedings of the National Academy of Sciences* 109 (35): 13978–83. doi: 10.1073/pnas.1201882109

Shroff, Hari, Catherine G. Galbraith, James A. Galbraith, and Eric Betzig. 2008. "Live-Cell Photoactivated Localization Microscopy of Nanoscale Adhesion Dynamics." *Nature Methods* 5 (5): 417–23. doi: 10.1038/nmeth.1202

Song, Xinbo, Hui Bian, Chao Wang, Mingyu Hu, Ning Li, and Yi Xiao. 2017. "Development and Applications of a Near-Infrared Dye-Benzylguanine Conjugate to Specifically Label SNAP-Tagged Proteins." *Organic & Biomolecular Chemistry* 15 (38): 8091–101. doi: 10.1039/c7ob01698k

Ståhl, Stefan, Torbjörn Gräslund, Amelie Eriksson Karlström, Fredrik Y. Frejd, Per Åke Nygren, and John Löfblom. 2017. "Affibody Molecules in Biotechnological and Medical Applications." *Trends in Biotechnology* 35 (8): 691–712. doi: 10.1016/j.tibtech.2017.04.007.

Staudt, Thorsten, Andreas Engler, Eva Rittweger, Benjamin Harke, Johann Engelhardt, and Stefan W. Hell. 2011. "Far-Field Optical Nanoscopy with Reduced Number of State Transition Cycles." *Optics Express* 19 (6): 5644. doi: 10.1364/OE.19.005644

Ströhl, Florian, and Clemens F. Kaminski. 2016. "Frontiers in Structured Illumination Microscopy." *Optica* 3 (6): 667. doi: 10.1364/OPTICA.3.000667

Takakura, Hideo, Yongdeng Zhang, Roman S. Erdmann, Alexander D. Thompson, Yu Lin, Brian McNellis, Felix Rivera-Molina, et al. 2017. "Long Time-Lapse Nanoscopy with Spontaneously Blinking Membrane Probes." *Nature Biotechnology* 35 (8): 773–80. doi: 10.1038/nbt.3876

Teng, Kai Wen, Yuji Ishitsuka, Pin Ren, Yeoan Youn, Xiang Deng, Pinghua Ge, Andrew S. Belmont, et al. 2016. "Labeling Proteins inside Living Cells Using External Fluorophores for Microscopy." *ELife* 5 (December 2016): 1–13. doi: 10.7554/eLife.20378.

Thompson, Alexander D., Joerg Bewersdorf, Derek Toomre, and Alanna Schepartz. 2017. "HIDE Probes: A New Toolkit for Visualizing Organelle Dynamics, Longer and at Super-Resolution." *Biochemistry* 56 (39): 5194–201. doi: 10.1021/acs.biochem.7b00545

Thompson, Alexander D., Mitchell H. Omar, Felix Rivera-Molina, Zhiqun Xi, Anthony J. Koleske, Derek K. Toomre, et al. 2017. "Long-Term Live-Cell STED Nanoscopy of Primary and Cultured Cells with the Plasma Membrane HIDE Probe DiI-SiR." *Angewandte Chemie - International Edition* 56 (35): 10408–12. doi: 10.1002/anie.201704783

Thompson, Russell E., Daniel R. Larson, and Watt W. Webb. 2002. "Precise Nanometer Localization Analysis for Individual Fluorescent Probes." *Biophysical Journal* 82 (5): 2775–83. doi: 10.1016/S0006-3495(02)75618-X

Tiwari, Dhermendra K., Yoshiyuki Arai, Masahito Yamanaka, Tomoki Matsuda, Masakazu Agetsuma, Masahiro Nakano, et al. 2015. "A Fast- and Positively Photoswitchable Fluorescent Protein for Ultralow-Laser-Power RESOLFT Nanoscopy." *Nature Methods* 12 (6): 515–18. doi: 10.1038/nmeth.3362

Tønnesen, Jan, Gergely Katona, Balázs Rózsa, and U. Valentin Nägerl. 2014. "Spine Neck Plasticity Regulates Compartmentalization of Synapses." *Nature Neuroscience* 17 (5): 678–85. doi: 10.1038/nn.3682

Tønnesen, Jan, Fabien Nadrigny, Katrin I. Willig, Roland Wedlich-Söldner, and U. Valentin Nägerl. 2011. "Two-Color STED Microscopy of Living Synapses Using a Single Laser-Beam Pair." *Biophysical Journal* 101 (10): 2545–52. doi: 10.1016/j.bpj.2011.10.011

Tortarolo, Giorgio, Marco Castello, Alberto Diaspro, Sami Koho, and Giuseppe Vicidomini. 2018. "Evaluating Image Resolution in Stimulated Emission Depletion Microscopy." *Optica* 5 (1): 32. doi: 10.1364/OPTICA.5.000032

Traenkle, Bjoern, and Ulrich Rothbauer. 2017. "Under the Microscope: Single-Domain Antibodies for Live-Cell Imaging and Super-Resolution Microscopy." *Frontiers in Immunology* 8 (August): 1–8. doi: 10.3389/fimmu.2017.01030

Turkowyd, Bartosz, Alexander Balinovic, David Virant, Haruko G. Gölz Carnero, Fabienne Caldana, Marc Endesfelder, Dominique Bourgeois, et al. 2017. "A General Mechanism of Photoconversion of Green-to-Red Fluorescent Proteins Based on Blue and Infrared Light Reduces Phototoxicity in Live-Cell Single-Molecule Imaging." *Angewandte Chemie - International Edition* 56 (38): 11634–39. doi: 10.1002/anie.201702870

Uno, Shin-nosuke, Mako Kamiya, Akihiko Morozumi, and Yasuteru Urano. 2017. "A Green-Light-Emitting, Spontaneously Blinking Fluorophore Based on Intramolecular Spirocyclization for Dual-Colour Super-Resolution Imaging." *Chemical Communications (Cambridge, England)* 54 (1): 102–5. doi: 10.1039/c7cc07783a

Uno, Shin-nosuke, Mako Kamiya, Toshitada Yoshihara, Ko Sugawara, Kohki Okabe, Mehmet C. Tarhan, Hiroyuki Fujita, et al. 2014. "A Spontaneously Blinking Fluorophore Based on Intramolecular Spirocyclization for Live-Cell Super-Resolution Imaging." *Nature Chemistry* 6 (8): 681–89. doi: 10.1038/nchem.2002

Uno, Shin Nosuke, Dhermendra K. Tiwari, Mako Kamiya, Yoshiyuki Arai, Takeharu Nagai, and Yasuteru Urano. 2015. "A Guide to Use Photocontrollable Fluorescent Proteins and Synthetic Smart Fluorophores for Nanoscopy." *Microscopy* 64 (4): 263–77. doi: 10.1093/jmicro/dfv037

Uttamapinant, Chayasith, Jonathan D. Howe, Kathrin Lang, Václav Beránek, Lloyd Davis, Mohan Mahesh, Nicholas P. Barry, et al. 2015. "Genetic Code Expansion Enables Live-Cell and Super-Resolution Imaging of Site-Specifically Labeled Cellular Proteins." *Journal of the American Chemical Society* 137 (14): 4602–5. doi: 10.1021/ja512838z

Uttamapinant, Chayasith, Katharine A. White, Hemanta Baruah, Samuel Thompson, Marta Fernández-Suárez, Sujiet Puthenveetil, and Alice Y. Ting. 2010. "A Fluorophore Ligase for Site-Specific Protein Labeling inside Living Cells." *Proceedings of the National Academy of Sciences of the United States of America* 107 (24): 10914–19. doi: 10.1073/pnas.0914067107

Vicidomini, Giuseppe, Gael Moneron, Christian Eggeling, Eva Rittweger, and Stefan W. Hell. 2012. "STED with Wavelengths Closer to the Emission Maximum." *Optics Express* 20 (5): 5225. doi: 10.1364/OE.20.005225

Vicidomini, Giuseppe, Gael Moneron, Kyu Y. Han, Volker Westphal, Haisen Ta, Matthias Reuss, Johann Engelhardt, Christian Eggeling, et al. 2011. "Sharper Low-Power STED Nanoscopy by Time Gating." *Nature Methods* 8 (7): 571–3. doi: 10.1038/nmeth.1624

Vreja, Ingrid C., Ivana Nikić, Fabian Göttfert, Mark Bates, Katharina Kröhnert, Tiago F. Outeiro, Stefan W. Hell, Edward A. Lemke, and Silvio O. Rizzoli. 2015. "Super-Resolution Microscopy of Clickable Amino Acids Reveals the Effects of Fluorescent Protein Tagging on Protein Assemblies." *ACS Nano* 9 (11): 11034–41. doi: 10.1021/acsnano.5b04434

Wäldchen, Sina, Julian Lehmann, Teresa Klein, Sebastian van de Linde, and Markus Sauer. 2015. "Light-Induced Cell Damage in Live-Cell Super-Resolution Microscopy." *Scientific Reports* 5: 15348. doi: 10.1038/srep15348

Wang, Chao, Xinbo Song, and Yi Xiao. 2017. "SNAP-Tag-Based Subcellular Protein Labeling and Fluorescent Imaging with Naphthalimides." *Chembiochem: A European Journal of Chemical Biology* 18 (17): 1762–9. doi: 10.1002/cbic.201700161

Wang, Chenguang, Masayasu Taki, Yoshikatsu Sato, Aiko Fukazawa, Tetsuya Higashiyama, and Shigehiro Yamaguchi. 2017. "Super-Photostable Phosphole-Based Dye for Multiple-Acquisition Stimulated Emission Depletion Imaging." *Journal of the American Chemical Society* 139 (30): 10374–81. doi: 10.1021/jacs.7b04418

Wang, Sheng, Miao Ding, Xuanze Chen, Lei Chang, and Yujie Sun. 2017. "Development of Bimolecular Fluorescence Complementation Using RsEGFP2 for Detection and Super-Resolution Imaging of Protein-Protein Interactions in Live Cells." *Biomedical Optics Express* 8 (6): 3119–31. doi: 10.1364/BOE.8.003119

Wang, Siyuan, Jeffrey R. Moffitt, Graham T. Dempsey, X. Sunney Xie, and Xiaowei Zhuang. 2014. "Characterization and Development of Photoactivatable Fluorescent Proteins for Single-Molecule-Based Superresolution Imaging." *Proceedings of the National Academy of Sciences of the United States of America* 111 (23): 8452–7. doi: 10.1073/pnas.1406593111

Wegel, Eva, Antonia Göhler, B. Christoffer Lagerholm, Alan Wainman, Stephan Uphoff, Rainer Kaufmann, and Ian M. Dobbie. 2016. "Imaging Cellular Structures in Super-Resolution with SIM, STED and Localisation Microscopy: A Practical Comparison." *Scientific Reports* 6 (1): 27290. doi: 10.1038/srep27290

Wegner, Waja, Alexander C. Mott, Seth G. N. Grant, Heinz Steffens, and Katrin I. Willig. 2018. "In Vivo STED Microscopy Visualizes PSD95 Sub-Structures and Morphological Changes over Several Hours in the Mouse Visual Cortex." *Scientific Reports* 8 (1): 219. doi: 10.1038/s41598-017-18640-z

Weigert, Martin, Uwe Schmidt, Tobias Boothe, Andreas Müller, Alexandr Dibrov, Akanksha Jain, Benjamin Wilhelm, et al. 2018. "Content-Aware Image Restoration: Pushing the Limits of Fluorescence Microscopy." *Nature Methods* 15 (12): 1090–7. doi: 10.1038/s41592-018-0216-7

Wichmann, Jan, and Stefan W. Hell. 1994. "Breaking the Diffraction Resolution Limit by Stimulated Emission: Stimulated-Emission-Depletion Fluorescence Microscopy." *Optics Letters* 19 (11): 780–2. doi: 10.1364/OL.19.000780

Wiedenmann, Jörg, Sergey Ivanchenko, Franz Oswald, Florian Schmitt, Carlheinz Röcker, Anya Salih, Klaus-Dieter Spindler, et al. 2004. "EosFP, a Fluorescent Marker Protein with UV-Inducible Green-to-Red Fluorescence Conversion." *Proceedings of the National Academy of Sciences of the United States of America* 101 (45): 15905–10. doi: 10.1073/pnas.0403668101

Wiedenmann, Jörg, Franz Oswald, and Gerd Ulrich Nienhaus. 2009. "Fluorescent Proteins for Live Cell Imaging: Opportunities, Limitations, and Challenges." *IUBMB Life* 61 (11): 1029–42. doi: 10.1002/iub.256

Willig, Katrin I., Silvio O. Rizzoli, Volker Westphal, Reinhard Jahn, and Stefan W. Hell. 2006. "STED Microscopy Reveals That Synaptotagmin Remains Clustered after Synaptic Vesicle Exocytosis." *Nature* 440 (7086): 935–9. doi: 10.1038/nature04592

Wombacher, Richard, Meike Heidbreder, Sebastian van de Linde, Michael P. Sheetz, Mike Heilemann, Virginia W. Cornish, and Markus Sauer. 2010. "Live-Cell Super-Resolution Imaging with Trimethoprim Conjugates." *Nature Methods* 7 (9): 717–9. doi: 10.1038/nmeth.1489

Wu Yong, Wu Xundong, Toro Ligia, and Stefani Enrico. 2015. "Resonant-scanning dual-color STED microscopy with ultrafast photon counting: A concise guide," https://dx.doi.org/10.1016%2Fj.ymeth.2015.06.01

Yang, Bin, Frédéric Przybilla, Michael Mestre, Jean-Baptiste Trebbia, and Brahim Lounis. 2013. "Massive Parallelization of STED Nanoscopy Using Optical Lattices." *Optics Express* 22 (5): 5581. doi: 10.1364/OE.22.005581

York, Andrew G., Panagiotis Chandris, Damian Dalle Nogare, Jeffrey Head, Peter Wawrzusin, Robert S. Fischer, Ajay Chitnis, and Hari Shroff. 2013. "Instant Super-Resolution Imaging in Live Cells and Embryos via Analog Image Processing." *Nature Methods* 10 (11): 1122–6. doi: 10.1038/nmeth.2687

York, Andrew G., Sapun H. Parekh, Damian Dalle Nogare, Robert S. Fischer, Kelsey Temprine, Marina Mione, Ajay B. Chitnis, et al. 2012. "Resolution Doubling in Live, Multicellular Organisms via Multifocal Structured Illumination Microscopy." *Nature Methods* 9 (7): 749–54. doi: 10.1038/nmeth.2025

Young, Laurence J., Florian Ströhl, and Clemens F. Kaminski. 2016. "A Guide to Structured Illumination TIRF Microscopy at High Speed with Multiple Colors." *Journal of Visualized Experiments*, no. 111 (May): e53988–e53988. doi: 10.3791/53988

Yu, Ji. 2016. "Single-Molecule Studies in Live Cells." *Annual Review of Physical Chemistry* 67 (1): 565–85. doi: 10.1146/annurev-physchem-040215-112451.

Zane, Hannah K., Julia K. Doh, Caroline A. Enns, and Kimberly E. Beatty. 2017. "Versatile Interacting Peptide (VIP) Tags for Labeling Proteins with Bright Chemical Reporters." *ChemBioChem* 18 (5): 470–4. doi: 10.1002/cbic.201600627

Zhang, Mingshu, Hao Chang, Yongdeng Zhang, Junwei Yu, Lijie Wu, Wei Ji, Juanjuan Chen, et al. 2012. "Rational Design of True Monomeric and Bright Photoactivatable Fluorescent Proteins." *Nature Methods* 9 (7): 727–9. doi: 10.1038/nmeth.2021

Zhang, Xi, Xuanze Chen, Zhiping Zeng, Mingshu Zhang, Yujie Sun, Peng Xi, Jianxin Peng, et al. 2015. "Development of a Reversibly Switchable Fluorescent Protein for Super-Resolution Optical Fluctuation Imaging (SOFI)." *ACS Nano* 9 (3): 2659–67. doi: 10.1021/nn5064387.

Zhang, Xi, Mingshu Zhang, Dong Li, Wenting He, Jianxin Peng, Eric Betzig, and Pingyong Xu. 2016. "Highly Photostable, Reversibly Photoswitchable Fluorescent Protein with High Contrast Ratio for Live-Cell Superresolution Microscopy." *Proceedings of the National Academy of Sciences* 113 (37): 10364–9. doi: 10.1073/pnas.1611038113.

Zhu, Lei, Wei Zhang, Daniel Elnatan, and Bo Huang. 2012. "Faster STORM Using Compressed Sensing." *Nature Methods* 9 (7): 721–3. doi: 10.1038/nmeth.1978.

Zhu, Xuekai, Lei Wang, Rongzhi Liu, Barry Flutter, Shenghua Li, Jie Ding, Hua Tao, et al. 2010. "COMBODY: One-Domain Antibody Multimer with Improved Avidity." *Immunology and Cell Biology* 88 (6): 667–75. doi: 10.1038/icb.2010.21.

4

Endoscopic Optical Coherence Tomography: Technologies and Applications

Dawei Li, Hyeon-Cheol Park, Wu Yuan, Xinwen Yao, and Xingde Li

CONTENTS

4.1 Introduction

Optical coherence tomography (OCT) is a nondestructive optical sectioning technique that generates three-dimensional (3D) image volumes of biological tissues with a microscopic scale resolution at high speed (Hee et al., 1995). Since its invention 25 years ago, OCT has been widely used in various biomedical applications. It serves as a gold standard for retinal imaging in clinical practice (Cao and Tey, 2015); it has been used to diagnose skin lesions (Ulrich et al., 2016; Tearney et al., 1996); and it aids in the diagnosis of disease inside the body, such as lesions, which initiate from and beneath superficial layers of internal organs. The current gold standard for lesion diagnosis in internal organs is video endoscopy with biopsy. Compared to that technique, OCT endoscopy (Rollins et al., 1999) has two advantages: (1) OCT provides volumetric images of the tissue beneath the epithelial surface where early lesions often reside, and (2) OCT provides microscale histomorphology with the tissue intact. Various OCT endoscopes have been developed and are widely used in different lumens (Rollins et al., 1999;

Sivak et al., 2000; Xi et al., 2014; Tran et al., 2004; Tsai et al., 2013; Liang et al., 2015; Wang et al., 2015; Strathman et al., 2015; Lee et al., 2016; Liu et al., 2004; Moon et al., 2010; Kim et al., 2010; Li et al., 2000).

This chapter will provide an overview of the design of OCT endoscopes and their applications. It begins with the general structure of OCT endoscopes, which includes the optical design, scanning methods, and engineering considerations for extremely small or large lumens. Then it reviews the applications of OCT endoscopes from small animal study to clinical applications.

4.2 Design of OCT Endoscope

An OCT endoscope is for delivering, focusing, and scanning the OCT light on a suspicious area, collecting the back-reflected signal, and then transmitting it back to the detector. In general, a single-mode fiber (SMF) delivers the infrared OCT light beam to the distal part of the endoscope. The distal part consists of optics, which focus the beam on the tissue. The focused beam is spun along azimuthal direction to obtain a cross-sectional image. Application requirements decide the scanning methods. The distal optics also collect the back-reflected signal, and a signal is delivered back through the same SMF fiber to the detection and image processing module. For protection, the entire fiber is encased in a torque coil, and the micro-optics are often housed within a metal guard. When imaging, a transparent plastic sheath encases the OCT endoscope to protect it from direct contact with body fluids. Therefore, the endoscope can be conveniently disinfected for reuse. In the following paragraphs, we will talk about the design consideration in detail.

4.2.1 Micro-Optics

The optics module in the OCT sample arm focuses the beam in a desirable spot size (i.e. lateral resolution) at a given working distance (i.e. the distance between the last surface of the optics to the focal point). An OCT endoscope is a specific sample arm with size constraints in both diameter and length. Its small diameter permits the OCT endoscope to pass through the channel of clinical endoscopes such as gastroscopes or bronchoscopes. Its short, rigid length permits the OCT probe to access the organ path or the Y-shape entry port of a clinical endoscope. In practice, micro-optics design is often based on gradient index (GRIN) lens (see Figure 4.1A). The design includes a glass rod for adjusting the numerical aperture (NA) (by expanding the beam from the fiber) and a GRIN lens for focusing. Zemax (or other optical simulation software) often

help determine the proper length of the glass rod and GRIN lens to achieve the desired lateral resolution and working distance.

During the fabrication, back reflection from each optics' surface should be minimized. A solution is to angle-cleave (~8°) the end surface of the fiber and polish the entry surface of the glass rod with the same angle. While remaining parallel, the separation between the surfaces should be precisely adjusted to reach a desired working distance and beam shape. Then, the glass rod is cemented with the GRIN lens by ultraviolet (UV) glue. For a side-viewing endoscope, a beam reflector, avoiding a 45° tilting angle, deflects the light out of the sheath to mitigate specular back reflection from the sheath or sample surface. The tilting angle is usually slightly larger than 47° or slightly smaller than 43°.

4.2.2 Scanning Methods

OCT endoscopes can be divided into side-viewing endoscopes (see Figure 4.1B and D) and forward-viewing endoscopes (see Figure 4.1E and F) based on the beam direction on the tissue with respect to the longitudinal axis of the probe. A side-viewing endoscope suits for surveying a large area of a luminal organ, while a forward-viewing endoscope is better suited for image guidance.

4.2.2.1 Side-View Scanning

A side-viewing OCT endoscope can be scanned either from the proximal end (see Figure 4.1B) or at the distal

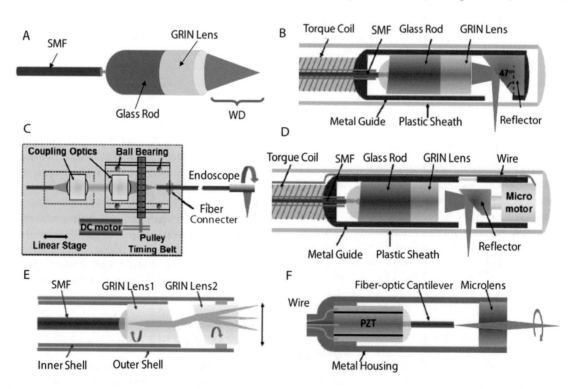

FIGURE 4.1 Schematics of A Optics design of a standard OCT endoscope; B Side-view OCT endoscope (proximal scanning); C Fiber-optic rotary joint for proximal scanning. 3D imaging is performed by pulling back the rotating endoscope; D Side-view OCT endoscope with a micromotor for scanning in distal; E Pairs-angle-rotation-scanning forward-viewing OCT endoscope; F Forward-viewing OCT endoscope with fiber-optic scanning; WD: working distance; PZT: Lead zirconate titanate; SMF: single-mode fiber.

end (see Figure 4.1D). For proximal scanning, a fiber-optic rotary joint couples light to the endoscope and rotates it (see Figure 4.1V) (Sivak et al., 2000; Xi et al., 2014; Liang et al., 2016; Tearney et al., 1997; Fujimoto et al., 1999; Bouma et al., 2000; Jang et al., 2002; Yang et al., 2005; Chen et al., 2007; Herz et al., 2004), and the torque coil allows to transfer the rotational forces along the endoscope. Proximal scanning endoscopes are widely used; however, they have two defects: (1) rotation-torque induces changes in the refractive index of the fiber that delivers the OCT signal and thus distorts imaging quality, and (2) high imaging speed is difficult to achieve. A distal scanning scheme can solve these two problems. For distal scanning, a micromotor is placed at the distal end of the endoscope. The motor rotates the microreflector mounted on its shaft to deflect and scan the light beam (see Figure 4.1D) (Tsai et al., 2013; Liang et al., 2015; Li et al., 2012; Wang et al., 2013; Oltmanns et al., 2016; Wu et al., 2006). Distal scanning endoscopes achieve higher imaging speed easily. Recently, a distal-scanning OCT endoscope with a 4,000 frame per second (fps) frame rate was demonstrated for intravascular "heartbeat" OCT imaging *in vivo* (Strathman et al., 2015). However, micromotors are usually expensive and fragile, which are major drawbacks for a distal-scanning probe. The second challenge is that micromotors (diameter is usually above 1 mm) limit the size of the endoscopes.

4.2.2.2 Forward-View Scanning

Forward-viewing endoscopes are more desired in image-guided surgery or biopsy. However, forward-beam scanning inside a flexible, compact endoscopic catheter is more challenging. Forward-scanners include paired-angle-rotation scanners (PARS), microelectromechanical systems (MEMS) lens scanners, and fiber-optic scanners.

PARS consists of a pair of angle-polished GRIN lenses and realizes forward-beam scanning by rotating both lenses in the opposite direction at the same angular speed (see Figure 4.1E) (Pan et al., 2001). Stable scanning is challenging because it needs precise alignment, such as concentricity of two GRIN lenses, and stable separation during the rotation. MEMS-based beam scanners offer a stable and controllable scanning pattern through electrostatic, electromagnetic, or electrothermal actuation (Jung et al., 2006; Strathman et al., 2013; Sun et al., 2010). This method provides high speed with low energy consumption. However, the MEMS method folds the beam by scanning mirror, resulting in a relatively larger probe diameter (Jung et al., 2006; Strathman et al., 2013; Sergeev et al., 1997).

Fiber-optic scanning provides another forward-scanning mechanism. Light delivering fiber is cantilevered on a mechanical actuator and directly scanned in the forward direction (see Figure 4.1f). The concept was first demonstrated with an electromagnetic actuator with bulk size and low speed (Sergeev et al., 1997). In 2004, a faster and more compact fiber-optic scanner was demonstrated based on a tubular piezoelectric (lead zirconate titane (PZT)) scanner (Liu et al., 2004). A fiber is cantilevered at the distal end of a tubular PZT with quartered electrodes, and the PZT operates at the resonant frequency of the fiber cantilever. Resonant scanning allows affordable scan amplitude at low operational voltages as well as high

speed (Schulz-Hildebrandt et al., 2018). The scanning speed can conveniently reach few a kilohertz by tuning the resonant frequency of the fiber (Schulz-Hildebrandt et al., 2018). Most recently, fiber-optic scanning endoscopes demonstrate imaging speed of 6 volumes/second with very compact size (1.6 mm outer diameter and 13.5 mm rigid length) (Scolaro et al., 2012).

4.2.3 Miniaturized OCT Endoscopes

Miniature flexible catheters (diameter less than 1 mm) can be easily delivered through narrow luminal sections (such as esophageal strictures) and minimize potential trauma within a small luminal area (such as small airways or arteries). Miniaturized solid endoscopes (OCT needles) minimize invasiveness when punching in solid tissues/organs for imaging (Lorenser et al., 2011; Villiger et al., 2016; Li et al., 2000; Wu et al., 2010; Tan et al., 2012). OCT needles can also integrate with a biopsy needle to provide a "first look" of suspicious tissues before the tissue biopsy (Tan et al., 2012).

As introduced above, GRIN design has been widely used for endoscopic OCT (Villiger et al., 2016; Wu et al., 2010), however, it is challenging to fabricate miniaturized catheters. An alternative is all-fiber monolithic probe design, in which the GRIN-lens-based distal optics is replaced by an angle-polished fiber tip (see Figure 4.2A–C) (Lorenser et al., 2011; Villiger et al., 2016). The design shrinks the diameter of the catheter easily but suffers from astigmatism, which degrades image quality in the transverse direction. Fortunately, an ellipsoidal ball lens can effectively reduce the astigmatism while staying small (Yuan et al., 2017; Vakoc et al., 2007). Along with the advantage of astigmatism, the ball lens design can also compensate for chromatic dispersion at 800 nm (Vakoc et al., 2007). As displayed in Figure 4.2D and E, a 2-meter long flexible OCT catheter with a ball lens design demonstrated ultrahigh resolution (i.e. an axial resolution ~2.4 μm in air) and ultrasmall form factor (i.e. an overall outer diameter of ~520 μm) in imaging a mouse colon, rat esophagus, and sheep small airways (Vakoc et al., 2007).

4.2.4 OCT Endoscopes for Imaging Large Lumens

OCT endoscopes face two challenges when imaging large lumen organs: (1) keeping the endoscope at the center of the organ; and (2) controlling the long imaging distance. Thus, balloon OCT endoscope (Fu et al., 2006; Xi et al, 2009; Gora et al., 2013) and OCT tethered capsule (Li et al., 2017) have been developed.

4.2.4.1 Balloon OCT Endoscope

As shown in Figure 4.3, an OCT balloon endoscope consists of an inner lumen and an outer lumen. The inner lumen holds an OCT endoscope for imaging, and the outer lumen stabilizes the endoscope at the center of the organ. The outer lumen includes a balloon, which is not inflated before imaging. During imaging, the balloon is inflated, then the endoscope is rotated and pulled back and forth inside the inner lumen. As the balloon endoscope should go through the working channel of a standard endoscope, the size of the OCT endoscope is limited. The size limitation increases the difficulty in precisely controlling

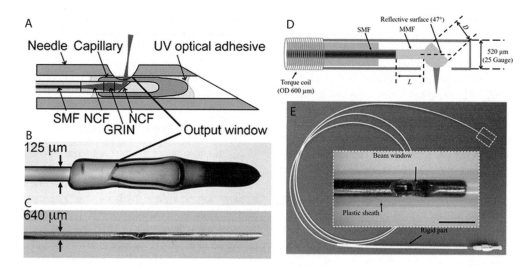

FIGURE 4.2 A Schematic of the GRIN fiber-based OCT imaging needle design. B Bright-field image of the total internal reflection (TIR) probe in a fused glass microcapillary. C Photo of the fully assembled side-viewing OCT needle. D The design of super-achromatic microprobe operating at 800 nm. E Photograph of the flexible microprobe of ~2 meters in length encased within a transparent plastic sheath of a 1 mm OD. Inset: zoomed view of the microprobe distal end boxed with yellow-dashed lines. Scale bar is 1 mm. SMF: single-mode fiber, MMF: multimode fiber, OD: outer diameter, NCF: no-core fiber. Panels A–C adapted with permission from Lorenser et al. (2011). Panels D and E adapted from Vakoc et al. (2007).

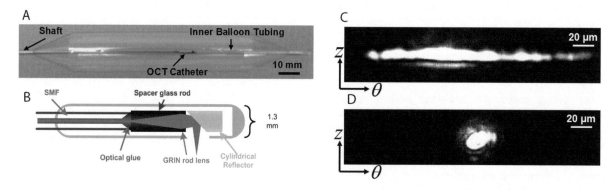

FIGURE 4.3 A Schematic of a double-lumen OCT balloon endoscope; B Schematic of optical structure inside the balloon OCT endoscope; Focused spots before C and after D astigmatism correction. Panels C and D adapted with permission from Gora et al. (2013).

the long imaging distance, which is usually larger than 10 mm. In this case, if using the popular GRIN optics design, the GRIN lens in use should be very thin (a few hundred microns) and its thickness should be precisely controlled—which is particularly challenging. An alternative is to change the GRIN lens and glass rod to two microlenses (Xi et al, 2009; Gora et al., 2013). Still, the distance between the two lenses needs to be accurately controlled.

Larger working distance also increases astigmatism and the light beam diverges severely along the azimuthal direction. For a typical balloon catheter (imaging distance is ~11 mm, and the diameter of protection tube is ~1.5 mm), the image resolution along the azimuthal direction degrades by about 40 fold (Gora et al., 2013). One solution is changing the flat reflector into a cylindrical one to precompensate the beam (Gora et al., 2013).

4.2.4.2 Capsule

OCT balloon has been demonstrated its capability of imaging large lumen organs stably, however, it still requires guidance of a clinical endoscope system, and the patient needs to be

sedated during the imaging. It is impractical for disease screening. Tethered OCT capsules (see Figure 4.4A) help overcome this hurdle (Li et al., 2017). The capsule can be swallowed directly with the help of natural peristaltic force to image the entire esophagus without sedating the patients and special settings. The 3D imaging data can be acquired when the capsule is being swallowed or when the capsule is mechanically pulled backward by the tether. Determining a suitable size for the capsule (particularly the diameter) will keep it in contact with the majority of esophagus epithelium without much folding. The size of the first tethered capsule (see Figure 4.4B) is 12.8 mm × 24.8 mm (diameter × length). It can image 20 fps with a resolution of 30 μm × 7 μm (lateral × axial in tissue). The distal scanning method enabled the OCT capsule to image at 250 fps with a resolution of 26 μm × 8.5 μm (lateral × axial in tissue) to achieve OCT *en face* image (see Figure 4.4C) (Wang et al., 2015). For structural changes at an early stage, an ultrahigh resolution is highly desirable. Recently, an ultrahigh resolution OCT tethered capsule (see Figure 4.4D), working at 800 nm, achieved an imaging resolution of 15 μm × 1.8 μm (lateral × axial in tissue) (Castro and Kraft, 2006).

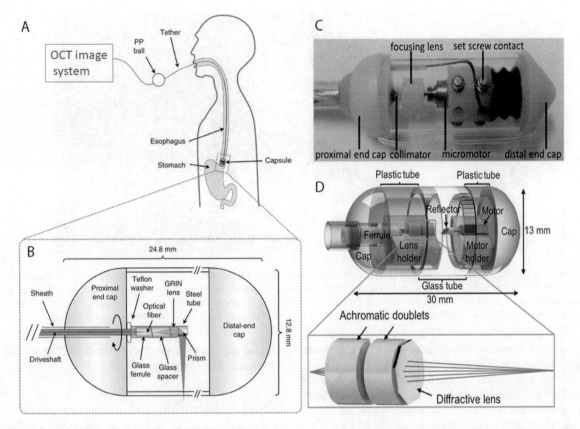

FIGURE 4.4 A Overview of a tethered OCT capsule: The PP ball and tether allow control of the capsule's location and provide a means for extracting the capsule from the subject when the procedure is complete. B Schematic of a tethered proximal-scanning OCT capsule endoscope. C Schematic of a tethered distal-scanning OCT capsule D Schematic a tethered distal-scanning OCT capsule. (Panels A and B adapted with permission from Ref Li et al. 2017.) (Panel C adapted from Wang et al. 2015.) (Panel D adapted with permission from Castro and Kraft 2008.)

4.3 From Small Animal Study to Clinical Application

4.3.1 Small Animal Study

Endoscopic OCT technology mainly targets clinical applications, and small animal studies, especially rodent models, are the starting point. Benefiting from the micro-optics design, endoscopic OCT can image small lumens of rodents, such as the gastrointestinal (GI) tract, bladder, and colon with micrometer-level image resolution.

Guinea pig models have been widely used for GI studies, such as esophageal acid including gastroesophageal reflux (GER) and airway inflammation (Li et al., 2019), in which epithelium is damaged. Cross-sectional and 3D images by endoscopic OCT help monitor the esophageal layer and structural changing. Figure 4.5A demonstrates a representative cross-sectional *in vivo* OCT image of a guinea pig esophagus in which all the layers can be clearly identified according to its histology (see Figure 4.5B) (Tran, 2004)). Statistics analysis for an eosinophilic esophagitis (EOE) model showed that the top two layers become thicker (Zhang et al., 2017; Vandamme, 2014).

Rodent models are widely used for modeling diseases such as cancer, inflammation, and hepatic steatosis resistance (Yu et al., 2016). Endoscopic OCT can distinguish not only layer information of inner lumens, but also structures such as crypt and colonic mucosa (see Figure 4.5C and D) Vakoc et al., 2007). Such ability facilitates it as a tool to monitor colon diseases such as inflammatory bowel disease and intestinal metaplasia (Pan et al., 2003) and bladder diseases such as bladder cancer (Wang et al., 2005; Adams et al., 2016).

4.3.2 OCT Imaging in Airways

Endobronchial OCT is an ideal tool for examining preneoplastic changes in an airway wall and alveoli with a resolution of 2–10 μm and an imaging depth of 1–2.5 mm in tissue, which benefits minimal-invasive and rapid volumetric study of the pulmonary tract from the trachea to terminal airways *in vivo*.

4.3.2.1 Obstructive Lung Disease

Obstructive lung disease (OLD) includes asthma, bronchiectasis, and chronic obstructive pulmonary disease (COPD). It affects more than 50 million Americans and leads to more than $60 billion in healthcare costs per year. Changes in airway structure and function of the airway smooth muscle (ASM) lead to severe airway narrowing and asthma attacks. By leveraging the intrinsic birefringence properties of ASM, orientation resolved OCT (OR-OCT) has been demonstrated to provide extra contrast to differentiate ASM from surrounding

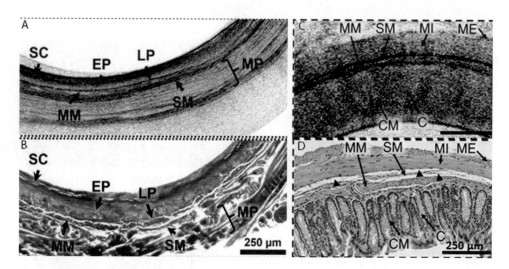

FIGURE 4.5 A Representative 2D *in vivo* OCT image of guinea pig esophagus and B its corresponding histology micrograph. C Representative *in vivo* OCT image of rat colon and D corresponding hemotoxylin and eosin (H&E) histology. SC: stratum corneum, EP: epithelium, LP: lamina propria, MM: muscularis mucosae, SM: submucosa, MP: muscularis propria, C: crypt, CM: colonic mucosa, MI: muscularis interna, ME: muscularis externa. (Panels A and B adapted with permission from Tran et al. 2004.) (Panels C and D adapted with permission from Vakoc et al. 2007.)

tissues in airways with a lumen diameter of 2–7 mm, as shown in Figure 4.6A and B (Thiboutot et al., 2018).

Several diseases occur in small airways (with diameter <2 mm): COPD, which leads to the narrowing of the terminal bronchioles and emphysematous destruction of the alveoli (Mizner, 2011), airflow obstruction (McDonough et al., 2011; Hogg et al., 1968), asthma (Hogg et al., 1968; Ding et al., 2016) and so on. Compared with large airways (with diameter >2 mm), small airways have thinner airway walls (i.e. thickness of 100s μm) and smooth muscle layers (i.e. the thickness <20 μm), which requires small OCT catheters with high resolving capability. A flexible OCT microprobe with an ultrahigh resolution of ~2.4 μm (in air) and an outer diameter of ~520 μm have been developed to study the small airway wall remodeling *in vivo*. It enabled clear delineation of fine tissue microstructures of small airways (see Figure 4.6C–E) (Vakoc et al., 2007) and provided an important pathological feature for COPD and asthma (Hariri et al., 2018).

4.3.2.2 Pulmonary Fibrosis

Pulmonary fibrosis (PF) causes scarring in the lungs. The most common type of PF is idiopathic pulmonary fibrosis (IPF), with 50,000 newly diagnosed cases each year. High-resolution computed tomography (HRCT) is usually used to diagnose IPF by identifying the characteristic usual interstitial pneumonitis (UIP) pattern of honeycombing (with size <2–3 mm); however, only 50% of cases can be diagnosed because of limited resolution, and the inclusive cases require an additional surgical biopsy for microscopic evaluation, which leads to high morbidity/mortality risks in patients. Recently, Hariri et al. (2018) reported the potential of endobronchial OCT to assess and diagnose IPF by accurately detecting microscopic honeycombing and other UIP features in patients with nondiagnostic HRCT (see Figure 4.7) (McGuire, 2016). Along with its high resolution, endobronchial OCT is also capable of rapidly imaging significant large tissue volumes, which provides a

potential method to greatly reduce the possibility of sampling errors suffered by lung biopsy.

4.3.2.3 Lung Cancer

Lung cancer causes the most cancer-related deaths in men and the second in women (after breast cancer) in worldwide (Tsuboi et al., 2005). Improved screening efficacy and optimized treatment outcomes increase survival rates, which needs new imaging method to enable detection and correct phenotype in each stage of lung cancers. OCT images can distinguish *in situ* and invasive lung carcinoma from normal bronchus by examining the integrity of airway wall microstructures (see Figure 4.8) (Lam et al., 2008). Following this study, Lam et al. found that endobronchial OCT-guided autofluorescence bronchoscopy is a useful tool for differentiating normal airway tissues from diseased tissues (including hyperplasia, metaplasia, dysplasia, and carcinoma *in situ*) based on a progressive increase in the epithelial thickness (Souza and Spechler, 2005), and it helped to guide the biopsy to increase the diagnostic yield (Souza and Spechler, 2005).

4.3.3 OCT Imaging in the Gastrointestinal Tract

4.3.3.1 Esophagus

The most investigated application of OCT in the gastrointestinal tract is the diagnosis of Barrett's esophagus (BE), which undergoes neoplastic transformation to esophageal adenocarcinoma (EAC). EAC is the most common form of esophageal cancer in the US and Europe, and squamous cell cancer (SCC), another type of esophageal cancer, is more common in the other areas, notably China and Africa.

4.3.3.1.1 Esophageal Adenocarcinoma

EAC, usually detected at a late stage, results in a dismal 5-year survival rate of 15–20% (Bouma and Tearney, 1999). The surveillance of BE helps to determine dysplasia or early

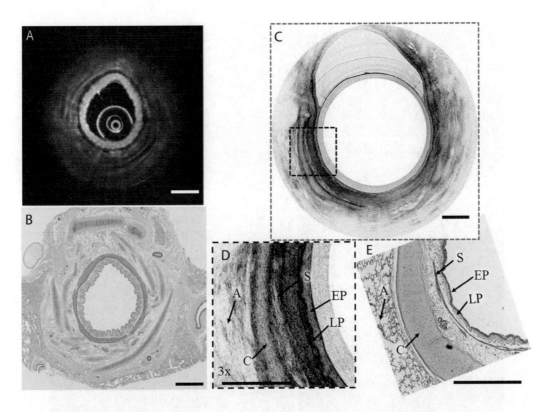

FIGURE 4.6 A OR-OCT image depicting the cross section of a canine bronchiole. B Corresponding histology match stained with aSMA antibody. Scale bars = 1 mm. C *In vivo* 2D image of the sheep small airway. D 3× zoomed view of the area boxed with red-dashed line in C. E Corresponding H&E histology. A: alveoli, ASM: airway smooth muscle, BV: blood vessel, C: cartilage, EP: epithelium, LP: lamina propria. Scale bar = 250 µm. (Panels A and B adapted with permission from Thiboutot et al. 2018.) (Panels C–E adapted from Vakoc et al. 2007.)

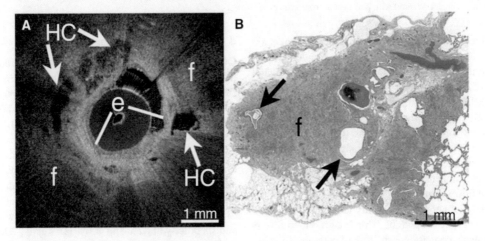

FIGURE 4.7 A *In vivo* endobronchial OCT accurately identified features of UIP/IPF, including multifocal microscopic honeycombing (HC, arrows) and peripheral destructive fibrosis (labeled f) beyond the bronchiolar epithelium (labeled e). B Subsequent surgical lung biopsies confirmed the presence of peripheral fibrosis (labeled f) and microscopic honeycombing (arrows). (Figure adapted with permission from McGuire 2016.)

stage adenocarcinoma. Treatment at an early stage dramatically increases the survival rate. Endoscopy with biopsy is the current standard of care. However, this method cannot readily identify dysplasia, as the biopsy based on the protocol in use frequently misses the most severe disease foci.

The early OCT imaging for BE used either linear driveshaft (Yang et al., 2003; Poneros et al., 2001) or fiber scanning probes (Cobb et al., 2005), which is inserted into the accessory ports of endoscopes. OCT images presented the architectural morphology of squamous mucosa (Poneros, 2004; Evans and

Nishioka, 2005; Haggitt, 1994) and helped diagnose dysplasia (see Figure 4.9A–C) (Montgomery et al., 2001; Adler et al., 2009). However, the probes could only image limited discrete locations until the invention of swept-source OCT (SS-OCT) (Liang et al., 2015; Herz et al., 2004; Yun et al., 2006) and balloon catheters (Suter et al., 2006; Suter et al., 2014). The SS-OCT system provides high speed imaging and the balloon stabilizes the OCT imaging at the center. These two inventions enables (1) comprehensive helical imaging of the entire organ rapidly (see Figure 4.9D and E), (2) implementation of

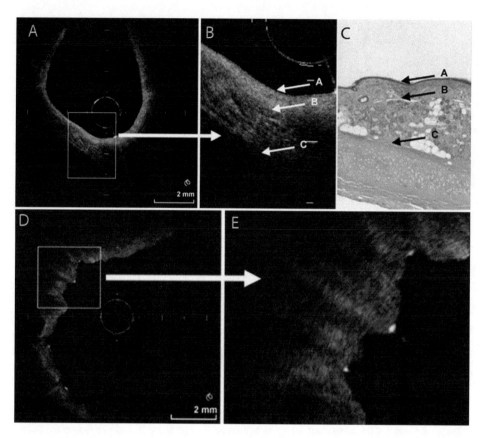

FIGURE 4.8 OCT image (A, B) and histological finding (C) of normal bronchus. A: epithelium, B: smooth muscle layer, and C: cartilage layer. (D) OCT image of squamous cell carcinoma and the enlarged view (E). The tumor in (E) is characterized by unevenly distributed high backscattering and loss of layer structure. (Figure adapted with permission from Lam et al. 2008.)

OCT-targeted biopsy with superficial laser cautery (Tsai et al., 2012), (3) assessment of the efficiency of radio-frequency ablation (RFA) treatment (Yun et al., 2006; Tsai et al., 2014), and (4) attainment of detailed structures of the micro-vasculatures (Liang et al., 2015; Liang et al., 2016; Lee et al., 2017) in the superficial layers of the esophagus, which provides additional information for disease diagnosis (Gora et al., 2013). All of these improve surveillance accuracy of BE. However, it remains inconvenient for screening. The inconvenience motivates the development of tethered capsules that unsedated patients can swallow (Wang et al., 2015; Li et al., 2017; Castro and Kraft, 2008; Gora et al., 2016), only taking about 5 minutes to image the entire esophagus. This imaging modality is preferable over endoscopy for nearly 90% of patients (Li et al., 2017; Hatta et al., 2012), which suggests that it may become an important new clinical screening tool.

4.3.3.1.2 Squamous Cell Cancer

Endoscopic ultrasound is the standard of care for early SCC staging with 80% accuracy for superficial and 70% for more invasive lesions (Jackie et al., 2000). Recently, OCT criteria, established for staging SCC, showed good accuracy for differentiating cancer invasion (Jackie et al., 2000; Hatta et al., 2010; Zuccaro et al., 2001). The accuracy of OCT for distinguishing superficial SCC limited to epithelium lamina propria was 95% and 85% for muscularis mucosa/submucosa invading lesions. The accuracy was significantly higher than ultrasound.

4.3.3.1.3 Stomach and Duodenum

OCT has been used in several clinical studies to image the stomach and the first portion of intestine (duodenum). Preliminary results showed that OCT was a promising tool to differentiate normal stomach tissue (Xi et al., 2014; Jang et al., 2002; Zhang et al., 2009; Hopper et al., 2008) from cancerous tissue (Cobb et al., 2005) (see Figure 4.10).

Celiac disease in the intestine often requires a biopsy to diagnose. Because the disease can be patchy, biopsies often come back negative even after positive serology. Studies showed that OCT could image large regions of the duodenum to assess villous blunting (Masci et al., 2009) with a sensitivity and specificity of 82% and 100% (Das et al., 2001). It provides a way to diagnose intestinal disease *in vivo*.

4.3.3.1.4 Large Intestine (Colon)

Identifying and diagnosing colon polyps is another potential application of OCT in the GI tract. To date, OCT can present microscopic architecture of the human colon *in vivo* (Sivak et al., 2000; Xi et al., 2014; Zagaynova et al., 2008) and differentiate benign from neoplastic polyps. Therefore, it has been used for colorectal cancer and inflammatory bowel disease (IBD) study. For colorectal cancer, OCT endoscopes can distinguish adenomas in colorectal cancer tissue from hyperplastic polyps and nonpolypoid normal tissue by the decrease in light scattering and more disorganization (Hopper et al., 2008; Shen et al., 2004). For IBD, an OCT endoscope

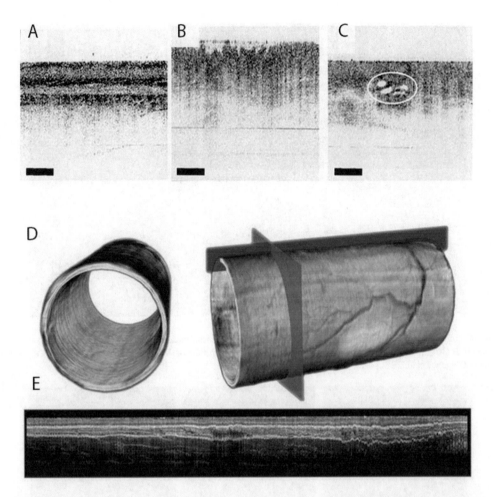

FIGURE 4.9 A OCT image of normal squamous epithelium (scale bar=500 μm) shows a five-layered appearance (from top to bottom: epithelium, lamina propria, muscularis mucosa, submucosa, and muscularis propria). B OCT image of BE with an irregular mucosal surface and absence of a layered structure. C OCT image of BE with submucosal glands (circled). D The volumetric OCT presented in arbitrary orientations and perspectives. E Longitudinal cross section through esophageal wall at the location in Panel D (inverted with epithelium at the top; dimensions: 45 mm horizontal, 2.6 mm vertical). (Panels A–C adapted with permission from Poneros 2004.) (Panels D and E adapted with permission from Suter et al. 2008.)

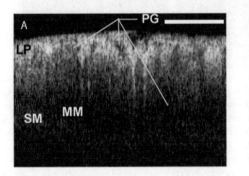
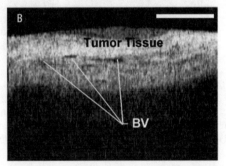

FIGURE 4.10 OCT image of healthy pyloroantral portion *in vivo* A and of exophytic tumor in the stomach B with increased vascularization. PG: pyloric glands, LP: lamia propria, SM: submucosa, MM: muscularis mucosa, BV: blood vessels. Scale bar = 500 μm. (Figure adapted from Cobb et al. 2005.)

can distinguish Crohn's disease from ulcerative colitis with a sensitivity of 90% and specificity of 83.3% by detecting transmural inflammation (Figure 4.11) (Consolo et al., 2008; Draganov et al., 2012).

4.3.3.1.5 Biliary/Pancreatic Tract

The main application of microscopic imaging in the bile duct is to identify the nature of bile duct strictures, which may be primary cancer (cholangiocarcinoma), or other inflammation caused by biliary stones. Current detection methods include endoscopic retrograde cholangiopancreatography (ECRP) and biliary endoscopy with brush biopsy (Arvanitakis et al., 2009). These methods are inadequate because they cannot detect the pathology below the surface. OCT is a valuable adjunct technology to ERCP for distinguishing the nature of indeterminate biliary strictures. Predominantly pilot clinical OCT studies

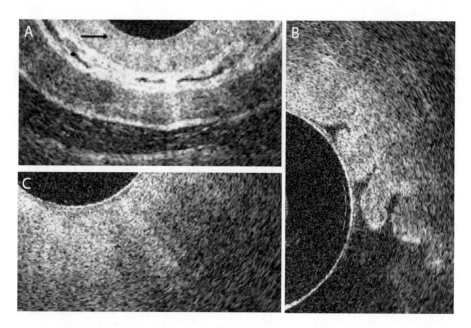

FIGURE 4.11 A OCT of normal colon; epithelium (black arrow); muscularis mucosa (black arrowhead); submucosa (black arrow); muscularis propria (black arrowhead); B OCT image of normal ileum with the peculiar pattern of intestinal villi; C Active colonic Crohn's disease. (Figure adapted with permission from Draganov et al. 2012.)

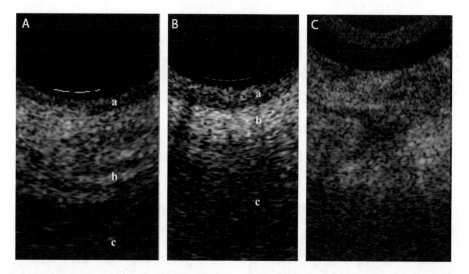

FIGURE 4.12 A OCT image from the main pancreatic duct (MPD) wall stricture in the EUS-documented normal pancreas; B OCT image from a segmental stricture of MPD in chronic pancreatitis; C OCT image from a segmental stricture of MPD in a patient with adenocarcinoma. A: a superficial hyporeflective band corresponding to the single layer of epithelial cells surrounding the MPD; B: an intermediate hyperreflective layer corresponding to the connective-fibromuscular layer surrounding the epithelium; C: an outer hyporeflective layer corresponding to the connective and acinar structure close to the duct. (Figure adapted with permission from Seitz et al. 2001.)

suggested different types of strictures have distinct microscopic morphologic features (see Figure 4.12) (Testoni et al., 2007; Seltz et al., 2001; Poneros et al., 2002; Yabushita et al., 2002).

4.3.4 OCT Imaging in the Cardiovascular System

Cardiovascular imaging, especially coronary artery imaging, is one of the clinical applications that OCT first showed its usefulness. With its superior resolution, intravascular OCT enables detailed visualization of microstructures in the coronary wall in a 3D fashion at high speed. These microstructures, such as lipids (Liu et al., 2011), calcific nodules (Liu et al., 2011), cholesterol crystals (Tearney et al., 2006; Nadkami

et al., 2007), smooth muscle/collagen (van der Sjide et al., 2016; Di Vito et al., 2015), fibrous caps (Yang et al., 2005), and macrophages (Gora et al., 2017), can be used to discriminate different types of plaques from each other (Figure 4.13) (Mehanna et al., 2011). Moreover, OCT catheters may also lend insights into the evaluation of coronary stent implantation (Bouma et al., 2003; Li et al., 2013). Increasingly, intravascular OCT is applied in clinics to facilitate the diagnosis of atherosclerotic lesions as well as to guide surgical procedures such as percutaneous coronary intervention (PCI).

OCT provides enough resolution for accurate detection of neointimal hyperplasia (NIH) observed after a drug-eluting stent, which usually exceeds the capabilities of intravascular

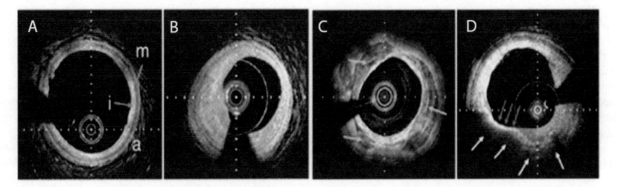

FIGURE 4.13 OCT images of human coronary plaques *in vivo*. A Artery wall with intimal hyperplasia. B Fibrous plaque showing a thickened intima. C Calcific plaque demonstrating a heterogeneous, signal poor region (white arrows) with clearly demarcated borders. D OCT fibroatheroma, showing a signal poor region with poorly defined borders (white arrows), consistent with lipid, and overlying tissue, known as the fibrous cap (red arrows). Tick marks in Panels A, B, and D: 250 μm. Tick marks in C: 500 μm. (Figure adapted from Mehanna et al. 2011.)

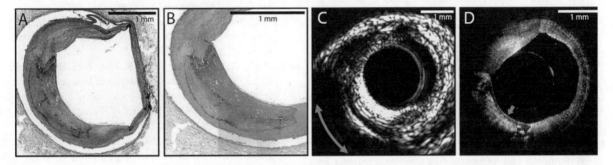

FIGURE 4.14 *In vitro* application of a hybrid IVUS and OCT catheter. Histological cross sections obtained from a human coronary artery stained with A Movat's pentachrome and B hematoxylin and eosin. A calcified superficial plaque span from 6 to 9 o'clock. C, D Corresponding intravascular ultrasound (IVUS) and optical coherence tomographic (OCT) images obtained during examination with a 4-F hybrid IVUS-OCT system that allows simultaneous image acquisition. (Figure adapted with permission from Feldchtein et al. 1996.)

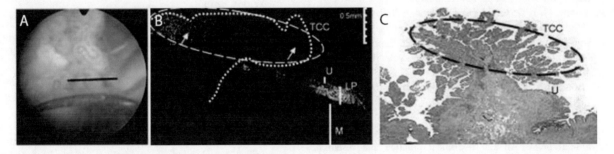

FIGURE 4.15 Cystoscopic OCT image of a human bladder cancer *in vivo*: A surface image, B cystoscopic OCT image, and C H&E stained histological section: U, urothelium; LP, lamina propria; M, muscularis. Image sizes for Panels A and B are 12 mm and 2 mm axially and 4.2 mm laterally, respectively. The black line in Panel A indicates the OCT scanning position. The cystoscopic OCT image clearly provides the margin between the tumor and the adjacent normal bladder wall, as indicated by the dashed curve. The lesion shown is papillary TCC low grade I/III, stage T2. (Figure adapted from Tate et al. 2014.)

ultrasound (IVUS). Similarly, fibrous cap thickness can only be assessed *in vivo* by OCT. The disadvantage of the current OCT systems lies in the optical penetration depth, which is typically less than 2 mm in tissue. Therefore, a multimodality IVUS/OCT catheter has been invented to exploit the synergistic advantages of IVUS and OCT [Bourantas et al., 2013; Feldchtein et al., 1998). The prototypes for such a hybrid catheter have already been tested *in vitro* (see Figure 4.14). However, technical limitations must be addressed before it can be used in clinical practice, including size, coregistration of the IVUS and OCT images, image quality, and the image acquisition rate.

4.3.5 Other Applications

Except for the aforementioned applications, OCT endoscopy has shown its potential role in many other medical fields such as urology (Wang et al., 2007; Tate et al., 2014), gynecology (Oltmanns et al., 2016), and otolaryngology. The schematics of the OCT endoscopes are specifically tailored to meet the requirements imposed by respective applications. For example, for bladder imaging, a forward-viewing OCT endoscope was developed and integrated with a conventional cystoscope (see Figure 4.15) (Tate et al., 2014)). The preliminary *in vivo* human study showed promising results with 91% sensitivity and 80% specificity for delineating transitional cell carcinomas based on the structural information.

For nasal cavity imaging, a side-viewing OCT probe was employed for quantitative *in vivo* assessment of structural abnormalities of the airways in patients with cystic fibrosis using the nasal mucosa as a model (Huang et al., 1991). It was implemented in the form of a hand probe with an outer diameter of 2.5 mm, and the imaging speed of 10 fps. The results showed that the thickness of the nasal mucosal layer in patients with cystic fibrosis was significantly different from that of normal subjects because of chronic inflammation.

REFERENCES

D. C. Adams et al., "Birefringence microscopy platform for assessing airway smooth muscle structure and function in vivo," *Science Translational Medicine*, vol. 8, no. 359, p. 359ra131, 2016.

D. C. Adler et al., "Three-dimensional optical coherence tomography of Barrett's esophagus and buried glands beneath neosquamous epithelium following radiofrequency ablation," (in English), *Endoscopy*, vol. 41, no. 9, pp. 773–776, 2009.

M. Arvanitakis et al., "Intraductal optical coherence tomography during endoscopic retrograde cholangiopancreatography for investigation of biliary strictures," (in English), *Endoscopy*, vol. 41, no. 8, pp. 696–701, 2009.

B. E. Bouma and G. J. Tearney, "Power-efficient nonreciprocal interferometer and linear-scanning fiber-optic catheter for optical coherence tomography," *Optics Letters*, vol. 24, no. 8, pp. 531–533, 1999.

B. E. Bouma et al., "Evaluation of intracoronary stenting by intravascular optical coherence tomography," (in English), *Heart*, vol. 89, no. 3, pp. 317–320, 2003.

B. E. Bouma, G. J. Tearney, C. C. Compton, and N. S. Nishioka, "High-resolution imaging of the human esophagus and stomach in vivo using optical coherence tomography," (in English), *Gastrointestinal Endoscopy*, vol. 51, no. 4, pp. 467–474, 2000.

C. V. Bourantas et al., "Hybrid Intravascular Imaging," (in English), *Journal of the American College of Cardiology*, vol. 61, no. 13, pp. 1369–1378, 2013.

T. G. Cao and H. L. Tey, "High-definition optical coherence tomography - An aid to clinical practice and research in dermatology," (in English), *Journal Der Deutschen Dermatologischen Gesellschaft*, vol. 13, no. 9, pp. 886–890, 2015.

M. Castro, and M. Kraft, *Clinical Asthma E-Book*. Elsevier Health Sciences, 2008.

Y. Chen et al., "Ultrahigh resolution optical coherence tomography of Barrett's esophagus: Preliminary descriptive clinical study correlating images with histology," (in English), *Endoscopy*, vol. 39, no. 7, pp. 599–605, 2007.

M. J. Cobb, X. M. Liu, and X. D. Li, "Continuous focus tracking for real-time optical coherence tomography," (in English), *Optics Letters*, vol. 30, no. 13, pp. 1680–1682, 2005.

P. Consolo, G. Strangio, C. Luigiano, G. Giacobbe, S. Pallio, and L. Familiari, "Optical coherence tomography in inflammatory bowel disease: Prospective evaluation of 35 patients," (in English), *Diseases of the Colon & Rectum*, vol. 51, no. 9, pp. 1374–1380, 2008.

A. Das et al., "High-resolution endoscopic imaging of the GI tract: A comparative study of optical coherence tomography versus high-frequency catheter probe EUS," (in English), *Gastrointestinal Endoscopy*, vol. 54, no. 2, pp. 219–224, 2001.

M. Ding et al., "Measuring airway remodeling in patients with different COPD staging using endobronchial optical coherence tomography," (in English), *Chest*, vol. 150, no. 6, pp. 1281–1290, 2016.

P. V. Draganov et al., "Diagnostic accuracy of conventional and cholangioscopy-guided sampling of indeterminate biliary lesions at the time of ERCP: A prospective, long-term follow-up study," (in English), *Gastrointestinal Endoscopy*, vol. 75, no. 2, pp. 347–353, 2012.

J. A. Evans and N. S. Nishioka, "The use of optical coherence tomography in screening and surveillance of Barrett's esophagus," *Clinical Gastroenterology and Hepatology*, vol. 3, no. 7 Supplement 1, pp. S8–S11, 2005.

F. I. Feldchtein et al., "Endoscopic applications of optical coherence tomography," (in English), *Optics Express*, vol. 3, no. 6, pp. 257–270, 1998.

H. L. Fu, Y. X. Leng, M. J. Cobb, K. Hsu, J. H. Hwang, and X. D. Li, "Flexible miniature compound lens design for high-resolution optical coherence tomography balloon imaging catheter," (in English), *Journal of Biomedical Optics*, vol. 13, no. 6, 060502, 2008.

J. G. Fujimoto, S. A. Boppart, G. J. Tearney, B. E. Bouma, C. Pitris, and M. E. Brezinski, "High resolution in vivo intra-arterial imaging with optical coherence tomography," *Heart*, vol. 82, no. 2, pp. 128–133, 1999.

M. J. Gora et al., "Imaging the upper gastrointestinal tract in unsedated patients using tethered capsule endomicroscopy," (in English), *Gastroenterology*, vol. 145, no. 4, pp. 723–725, 2013.

M. J. Gora et al., "Tethered capsule endomicroscopy enables less invasive imaging of gastrointestinal tract microstructure," (in English), *Nature Medicine*, vol. 19, no. 2, pp. 238–240, 2013.

M. J. Gora et al., "Tethered capsule endomicroscopy: From bench to bedside at a primary care practice," (in English), *Journal of Biomedical Optics*, vol. 21, no. 10, 104001, 2016.

M. J. Gora, M. J. Suter, G. J. Tearney, and X. D. Li, "Endoscopic optical coherence tomography: Technologies and clinical applications [Invited]," (in English), *Biomedical Optics Express*, vol. 8, no. 5, pp. 2405–2444, 2017.

R. C. Haggitt, "Barrett's esophagus, dysplasia, and adenocarcinoma," *Human Pathology*, vol. 25, no. 10, pp. 982–993, 1994.

L. P. Hariri et al., "Endobronchial Optical Coherence Tomography for Low-Risk Microscopic Assessment and Diagnosis of Idiopathic Pulmonary Fibrosis In Vivo," *American Journal of Respiratory and Critical Care Medicine*, vol. 197, no. 7, pp. 949–952, 2018.

W. Hatta et al., "A prospective comparative study of optical coherence tomography and EUS for tumor staging of superficial esophageal squamous cell carcinoma," *Gastrointestinal Endoscopy*, vol. 76, no. 3, pp. 548–555, 2012.

W. Hatta et al., "Optical coherence tomography for the staging of tumor infiltration in superficial esophageal squamous cell carcinoma," (in English), *Gastrointestinal Endoscopy*, vol. 71, no. 6, pp. 899–906, 2010.

M. R. Hee et al., "Optical coherence tomography of the human retina," (in English), *Archives of Ophthalmology*, vol. 113, no. 3, pp. 325–332, 1995.

P. R. Herz et al., "Micromotor endoscope catheter for in vivo, ultrahigh-resolution optical coherence tomography," (in English), *Optics Letters*, vol. 29, no. 19, pp. 2261–2263, 2004.

J. C. Hogg, P. T. Macklem, and W. M. Thurlbeck, "Site and nature of airway obstruction in chronic obstructive lung disease," *The New England Journal of Medicine*, vol. 278, no. 25, pp. 1355–1360, 1968.

A. D. Hopper, S. S. Cross, and D. S. Sanders, "Patchy villous atrophy in adult patients with suspected gluten-sensitive enteropathy: Is a multiple duodenal biopsy strategy appropriate?," (in English), *Endoscopy*, vol. 40, no. 3, pp. 219–224, 2008.

D. Huang et al., "Optical coherence tomography," (in English), *Science*, vol. 254, no. 5035, pp. 1178–1181, 1991.

S. Jackle et al., "In vivo endoscopic optical coherence tomography of esophagitis, Barrett's esophagus, and adenocarcinoma of the esophagus," (in English), *Endoscopy*, vol. 32, no. 10, pp. 750–755, 2000.

I. K. Jang et al., "Visualization of coronary atherosclerotic plaques in patients using optical coherence tomography: Comparison with intravascular ultrasound," (in English), *Journal of the American College of Cardiology*, vol. 39, no. 4, pp. 604–609, 2002.

W. Jung, D. T. McCormick, J. Zhang, L. Wang, N. C. Tien, and Z. P. Chen, "Three-dimensional endoscopic optical coherence tomography by use of a two-axis microelectromechanical scanning mirror," (in English), *Applied Physics Letters*, vol. 88, no. 16, pp. 163901, 2006.

K. H. Kim, J. A. Burns, J. J. Bernstein, G. N. Maguluri, B. H. Park, and J. F. de Boer, "In vivo 3D human vocal fold imaging with polarization sensitive optical coherence tomography and a MEMS scanning catheter," (in English), *Optics Express*, vol. 18, no. 14, pp. 14644–14653, 2010.

S. Lam et al., "In vivo optical coherence tomography imaging of preinvasive bronchial lesions," *Clinical Cancer Research*, vol. 14, no. 7, pp. 2006–2011, 2008.

H. C. Lee et al., "Circumferential optical coherence tomography angiography imaging of the swine esophagus using a micromotor balloon catheter," (in English), *Biomedical Optics Express*, vol. 7, no. 8, pp. 2927–2942, 2016.

H. C. Lee et al., "Endoscopic optical coherence tomography angiography microvascular features associated with dysplasia in Barrett's esophagus," (in English), *Gastrointestinal Endoscopy*, vol. 86, no. 3, pp. 476–484, 2017.

J. A. Li et al., "High speed miniature motorized endoscopic probe for optical frequency domain imaging," (in English), *Optics Express*, vol. 20, no. 22, pp. 24132–24138, 2012.

B. H. Li et al., "Hybrid intravascular ultrasound and optical coherence tomography catheter for imaging of coronary atherosclerosis," (in English), *Catheterization and Cardiovascular Interventions*, vol. 81, no. 3, pp. 494–507, 2013.

X. D. Li et al., "Optical coherence tomography: Advanced technology for the endoscopic imaging of Barrett's esophagus," (in English), *Endoscopy*, vol. 32, no. 12, pp. 921–930, 2000.

D. Li et al., "Parallel deep neural networks for endoscopic OCT image segmentation," *Biomedical Optics Express*, vol. 10, no. 3, pp. 1126–1135, 2019.

K. Y. Li et al., "Super-achromatic Optical Coherence Tomography Capsule for Ultrahigh-resolution Imaging of Esophagus," *Journal of Biophotonics*, Vol. 12, no. 3, e201800205, 2019.

X. Li, C. Chudoba, T. Ko, C. Pitris, and J. G. Fujimoto, "Imaging needle for optical coherence tomography," *Optics Letters*, vol. 25, no. 20, pp. 1520–1522, 2000.

K. Li, J. Mavadia-Shukla, W. Liang, and X. Li, "Ultrahigh-resolution tethered OCT endoscopic capsule at 800 nm (Conference Presentation)," In *Optical Coherence Tomography and Coherence Domain Optical Methods in Biomedicine XXI*, vol. 10053, p. 1005308. International Society for Optics and Photonics, J. G. Fujimoto, J. A. Izatt, and V. V. Tuchin Ed, 2017.

K. C. Liang et al., "Ultrahigh speed en face OCT capsule for endoscopic imaging," (in English), *Biomedical Optics Express*, vol. 6, no. 4, pp. 1146–1163, 2015.

K. C. Liang et al., "Volumetric Mapping of Barrett's Esophagus and Dysplasia With en face Optical Coherence Tomography Tethered Capsule," (in English), *American Journal of Gastroenterology*, vol. 111, no. 11, pp. 1664–1666, 2016.

L. B. Liu et al., "Imaging the subcellular structure of human coronary atherosclerosis using micro-optical coherence tomography," (in English), *Nature Medicine*, vol. 17, no. 8, pp. 1010–1132, 2011.

X. M. Liu, M. J. Cobb, Y. C. Chen, M. B. Kimmey, and X. D. Li, "Rapid-scanning forward-imaging miniature endoscope for real-time optical coherence tomography," (in English), *Optics Letters*, vol. 29, no. 15, pp. 1763–1765, 2004.

D. Lorenser, X. Yang, R. W. Kirk, B. C. Quirk, R. A. McLaughlin, and D. D. Sampson, "Ultrathin side-viewing needle probe for optical coherence tomography," *Optics Letters*, vol. 36, no. 19, pp. 3894–3896, 2011.

E. Masci et al., "Optical coherence tomography in pediatric patients: A feasible technique for diagnosing celiac disease in children with villous atrophy," (in English), *Digestive and Liver Disease*, vol. 41, no. 9, pp. 639–643, 2009.

J. E. McDonough et al., "Small-airway obstruction and emphysema in chronic obstructive pulmonary disease," *The New England Journal of Medicine*, vol. 365, no. 17, pp. 1567–1575, 2011.

S. McGuire, "World Cancer Report 2014. Geneva, Switzerland: World Health Organization, International Agency for Research on Cancer, WHO Press, 2015, " Advances in Nutrition, vol. 7, no. 2, pp. 418–419, 2016.

E. A. Mehanna, G. F. Attizzani, H. Kyono, M. Hake, and H. G. Bezerra, "Assessment of coronary stent by optical coherence tomography, methodology and definitions," (in English), *International Journal of Cardiovascular Imaging*, vol. 27, no. 2, pp. 259–269, 2011.

W. Mitzner, "Emphysema--a disease of small airways or lung parenchyma?," *The New England Journal of Medicine*, vol. 365, no. 17, pp. 1637–1639, 2011.

E. Montgomery et al., "Reproducibility of the diagnosis of dysplasia in Barrett esophagus: A reaffirmation," (in English), *Human Pathology*, vol. 32, no. 4, pp. 368–378, 2001.

S. Moon, S. W. Lee, M. Rubinstein, B. J. F. Wong, and Z. P. Chen, "Semi-resonant operation of a fiber-cantilever piezotube scanner for stable optical coherence tomography endoscope imaging," (in English), *Optics Express*, vol. 18, no. 20, pp. 21183–21197, 2010.

S. K. Nadkarni et al., "Measurement of collagen and smooth muscle cell content in atherosclerotic plaques using polarization-sensitive optical coherence tomography," (in English), *Journal of the American College of Cardiology*, vol. 49, no. 13, pp. 1474–1481, 2007.

U. Oltmanns et al., "Optical coherence tomography detects structural abnormalities of the nasal mucosa in patients with cystic fibrosis," (in English), *Journal of Cystic Fibrosis*, vol. 15, no. 2, pp. 216–222, 2016.

U. Oltmanns et al., "Optical coherence tomography detects structural abnormalitiesof the nasal mucosa in patients with cystic fibrosis," *Journal of Cystic Fibrosis*, vol. 15, no. 2, pp. 216–222, 2016.

Y. T. Pan, H. K. Xie, and G. K. Fedder, "Endoscopic optical coherence tomography based on a microelectromechanical mirror," (in English), *Optics Letters*, vol. 26, no. 24, pp. 1966–1968, 2001.

Y. Pan, T. Xie, C. Du, S. Bastacky, S. Meyers, and M. Zeidel, "Enhancing early bladder cancer detection with fluorescence-guided endoscopic optical coherent tomography," *Optics Letters*, vol. 28, no. 24, pp. 2485–2487, 2003.

J. M. Poneros et al., "Optical coherence tomography of the biliary tree during ERCP," (in English), *Gastrointestinal Endoscopy*, vol. 55, no. 1, pp. 84–88, 2002.

J. M. Poneros, "Diagnosis of Barrett's esophagus using optical coherence tomography," *Gastrointestinal Endoscopy Clinics of North America*, vol. 14, no. 3, pp. 573–588, 2004.

J. M. Poneros, S. Brand, B. E. Bouma, G. J. Tearney, C. C. Compton, and N. S. Nishioka, "Diagnosis of specialized intestinal metaplasia by optical coherence tomography," *Gastroenterology*, vol. 120, no. 1, pp. 7–12, 2001.

A. M. Rollins et al., "Real-time in vivo imaging of human gastrointestinal ultrastructure by use of endoscopic optical coherence tomography with a novel efficient interferometer design," (in English), *Optics Letters*, vol. 24, no. 19, pp. 1358–1360, 1999.

H. Schulz-Hildebrandt et al., "High-speed fiber scanning endoscope for volumetric multi-megahertz optical coherence tomography," (in English), *Optics Letters*, vol. 43, no. 18, pp. 4386–4389, 2018.

L. Scolaro, D. Lorenser, R. A. McLaughlin, B. C. Quirk, R. W. Kirk, and D. D. Sampson, "High-sensitivity anastigmatic imaging needle for optical coherence tomography," *Optics Letters*, vol. 37, no. 24, pp. 5247–5249, 2012.

U. Seitz et al., "First in vivo optical coherence tomography in the human bile duct," (in English), *Endoscopy*, vol. 33, no. 12, pp. 1018–1021, 2001.

A. Sergeev et al., "In vivo endoscopic OCT imaging of precancer and cancer states of human mucosa," *Optics Express*, vol. 1, no. 13, pp. 432–440, 1997.

B. Shen et al., "In vivo colonoscopic optical coherence tomography for transmural inflammation in inflammatory bowel disease," *Clin Gastroenterol Hepatol*, vol. 2, no. 12, pp. 1080–1087, 2004.

M. V. Sivak et al., "High-resolution endoscopic imaging of the GI tract using optical coherence tomography," (in English), *Gastrointestinal Endoscopy*, vol. 51, no. 4, pp. 474–479, 2000.

R. F. Souza and S. J. Spechler, "Concepts in the prevention of adenocarcinoma of the distal esophagus and proximal stomach," (in English), *Ca-a Cancer Journal for Clinicians*, vol. 55, no. 6, pp. 334–351, 2005.

M. Strathman et al., "MEMS scanning micromirror for optical coherence tomography," (in English), *Biomedical Optics Express*, vol. 6, no. 1, pp. 211–224, 2015.

M. Strathman, Y. B. Liu, X. D. Li, and L. Y. Lin, "Dynamic focus-tracking MEMS scanning micromirror with low actuation voltages for endoscopic imaging," (in English), *Optics Express*, vol. 21, no. 20, pp. 23934–23941, 2013.

J. J. Sun et al., "3D In Vivo optical coherence tomography based on a low-voltage, large-scan-range 2D MEMS mirror," (in English), *Optics Express*, vol. 18, no. 12, pp. 12065–12075, 2010.

M. J. Suter et al., "Comprehensive microscopy of the esophagus in human patients with optical frequency domain imaging," (in English), *Gastrointestinal Endoscopy*, vol. 68, no. 4, pp. 745–753, 2008.

M. J. Suter et al., "Esophageal-guided biopsy with volumetric laser endomicroscopy and laser cautery marking: A pilot clinical study," *Gastrointestinal Endoscopy*, vol. 79, no. 6, pp. 886–896, 2014.

K. M. Tan, M. Shishkov, A. Chee, M. B. Applegate, B. E. Bouma, and M. J. Suter, "Flexible transbronchial optical frequency domain imaging smart needle for biopsy guidance," *Biomedical Optics Express*, vol. 3, no. 8, pp. 1947–1954, 2012.

T. Tate, M. Keenan, E. Swan, J. Black, U. Utzinger, and J. Barton, "Optical design of an optical coherence tomography and multispectral fluorescence imaging endoscope to detect early stage ovarian cancer," In *International Optical Design Conference*, p. IM3B. 2. Optical Society of America, 2014.

G. J. Tearney et al., "In vivo endoscopic optical biopsy with optical coherence tomography," (in English), *Science*, vol. 276, no. 5321, pp. 2037–2039, 1997.

G. J. Tearney et al., "Scanning single-mode fiber optic catheter-endoscope for optical coherence tomography," (in English), *Optics Letters*, vol. 21, no. 7, pp. 543–545, 1996.

G. J. Tearney, I. K. Jang, and B. E. Bouma, "Optical coherence tomography for imaging the vulnerable plaque," *Journal of Biomedical Optics*, vol. 11, no. 2, p. 021002, 2006.

P. A. Testoni, A. Mariani, B. Mangiavillano, P. G. Arcidiacono, S. Di Pietro, and E. Masci, "Intraductal optical coherence tomography for investigating main pancreatic duct strictures," *The American Journal of Gastroenterology*, vol. 102, no. 2, pp. 269–274, 2007.

J. Thiboutot et al., "Current advances in COPD imaging," *Academic Radiology*, Vol. 26, no. 3, pp. 335–343, 2018.

P. H. Tran, D. S. Mukai, M. Brenner, and Z. P. Chen, "In vivo endoscopic optical coherence tomography by use of a rotational microelectromechanical system probe," (in English), *Optics Letters*, vol. 29, no. 11, pp. 1236–1238, 2004.

T. H. Tsai et al., "Endoscopic optical coherence angiography enables 3-dimensional visualization of subsurface microvasculature," (in English), *Gastroenterology*, vol. 147, no. 6, pp. 1219–1221, 2014.

T. H. Tsai et al., "Structural markers observed with endoscopic 3-dimensional optical coherence tomography correlating with Barrett's esophagus radiofrequency ablation treatment response," (in English), *Gastrointestinal Endoscopy*, vol. 76, no. 6, pp. 1104–1112, 2012.

T. H. Tsai et al., "Ultrahigh speed endoscopic optical coherence tomography using micromotor imaging catheter and VCSEL technology," (in English), *Biomedical Optics Express*, vol. 4, no. 7, pp. 1119–1132, 2013.

M. Tsuboi et al., "Optical coherence tomography in the diagnosis of bronchial lesions," *Lung Cancer*, vol. 49, no. 3, pp. 387–394, 2005.

M. Ulrich et al., "Dynamic optical coherence tomography in dermatology," (in English), *Dermatology*, vol. 232, no. 3, pp. 298–311, 2016.

B. J. Vakoc et al., "Comprehensive esophageal microscopy by using optical frequency-domain imaging (with video)," (in English), *Gastrointestinal Endoscopy*, vol. 65, no. 6, pp. 898–905, 2007.

J. N. van der Sijde, A. Karanasos, M. Villiger, B. E. Bouma, and E. Regar, "First-in-man assessment of plaque rupture by polarization-sensitive optical frequency domain imaging in vivo," (in English), *European Heart Journal*, vol. 37, no. 24, pp. 1932, 2016.

T. F. Vandamme, "Use of rodents as models of human diseases," *Journal of Pharmacy & Bioallied Sciences*, vol. 6, no. 1, p. 2, 2014.

M. Villiger et al., "Deep tissue volume imaging of birefringence through fibre-optic needle probes for the delineation of breast tumour," (in English), *Scientific Reports*, vol. 6, 28771, 2016.

L. Di Vito et al., "Identification and quantification of macrophage presence in coronary atherosclerotic plaques by optical coherence tomography," (in English), *European Heart Journal-Cardiovascular Imaging*, vol. 16, no. 7, pp. 807–813, 2015.

T. S. Wang et al., "Heartbeat OCT: In vivo intravascular megahertz-optical coherence tomography," (in English), *Biomedical Optics Express*, vol. 6, no. 12, pp. 5021–5032, 2015.

Z. Wang et al., "In vivo bladder imaging with microelectromechanical-systems-based endoscopic spectral domain optical coherence tomography," *Journal of Biomedical Optics*, vol. 12, no. 3, p. 034009, 2007.

T. S. Wang et al., "Intravascular optical coherence tomography imaging at 3200 frames per second," (in English), *Optics Letters*, vol. 38, no. 10, pp. 1715–1717, 2013.

Z. Wang, D. Durand, M. Schoenberg, and Y. Pan, "Fluorescence guided optical coherence tomography for the diagnosis of early bladder cancer in a rat model," *The Journal of Urology*, vol. 174, no. 6, pp. 2376–2381, 2005.

Y. C. Wu et al., "Robust high-resolution fine OCT needle for side-viewing interstitial tissue imaging," (in English), *IEEE Journal of Selected Topics in Quantum Electronics*, vol. 16, no. 4, pp. 863–869, 2010.

J. G. Wu, M. Conry, C. H. Gu, F. Wang, Z. Yaqoob, and C. H. Yang, "Paired-angle-rotation scanning optical coherence tomography forward-imaging probe," (in English), *Optics Letters*, vol. 31, no. 9, pp. 1265–1267, 2006.

J. F. Xi et al., "Diffractive catheter for ultrahigh-resolution spectral-domain volumetric OCT imaging," (in English), *Optics Letters*, vol. 39, no. 7, pp. 2016–2019, 2014.

J. Xi, L. Huo, Y. Wu, M. J. Cobb, J. H. Hwang, and X. Li, "High-resolution OCT balloon imaging catheter with astigmatism correction," *Optics Letters*, vol. 34, no. 13, pp. 1943–1945, 2009.

H. Yabushita et al., "Characterization of human atherosclerosis by optical coherence tomography," *Circulation*, vol. 106, no. 13, pp. 1640–1645, 2002.

V. X. D. Yang et al., "Endoscopic Doppler optical coherence tomography in the human GI tract: Initial experience," (in English), *Gastrointestinal Endoscopy*, vol. 61, no. 7, pp. 879–890, 2005.

V. X. Yang et al., "High speed, wide velocity dynamic range Doppler optical coherence tomography (Part III): In vivo endoscopic imaging of blood flow in the rat and human gastrointestinal tracts," *Optics Express*, vol. 11, no. 19, pp. 2416–2424, 2003.

X. J. Yu et al., "Toward High-Speed Imaging of Cellular Structures in Rat Colon Using Micro-optical Coherence Tomography," (in English), *IEEE Photonics Journal*, vol. 8, no. 4, 3900308, 2016.

W. Yuan, R. Brown, W. Mitzner, L. Yarmus, and X. D. Li, "Super-achromatic monolithic microprobe for ultrahigh-resolution endoscopic optical coherence tomography at 800 nm," (in English), *Nature Communications*, vol. 8, no.1, 1531, 2017.

S. H. Yun et al., "Comprehensive volumetric optical microscopy in vivo," (in English), *Nature Medicine*, vol. 12, no. 12, pp. 1429–1433, 2006.

E. Zagaynova, N. Gladkova, N. Shakhova, G. Gelikonov, and V. Gelikonov, "Endoscopic OCT with forward-looking probe: Clinical studies in urology and gastroenterology," (in English), *Journal of Biophotonics*, vol. 1, no. 2, pp. 114–128, 2008.

J. L. Zhang et al., "Automatic and robust segmentation of endoscopic OCT images and optical staining," (in English), *Biomedical Optics Express*, vol. 8, no. 5, pp. 2697–2708, 2017.

J. Zhang, Z. Chen, and G. Isenberg, "Gastrointestinal optical coherence tomography: Clinical applications, limitations, and research priorities," *Gastrointestinal Endoscopy Clinics of North America*, vol. 19, no. 2, pp. 243–259, 2009.

G. Zuccaro et al., "Optical coherence tomography of the esophagus and proximal stomach in health and disease," (in English), *American Journal of Gastroenterology*, vol. 96, no. 9, pp. 2633–2639, 2001.

5

Bioluminescence

Michael Conway, Tingting Xu, Amelia Brumbaugh, Anna Young, Dan Close, and Steven Ripp

CONTENTS

5.1 Introduction

A diverse set of organisms naturally emit light using endogenous chemical pathways. Modern biotechnology has exploited this luminescent biochemistry to extend our understanding of the natural world, dissect disease processes, and generate novel medications. This visually captivating luminescent biochemistry is driven by luciferases, which are a class of enzymes capable of oxidizing a chemical substrate to generate light. Luciferase's potential as a biological reporter was recognized with its initial cloning and ectopic expression (de Wet et al., 1985; Tatsumi et al., 1989) where, unlike fluorescent reporters, it has almost no subject background; an example of which is shown in Figure 5.1 in comparison to fluorescent-based imaging. This has made luciferase a widely deployed research tool and one that has undergone significant genetic manipulation to extend and improve its application. This chapter focuses on bioluminescent imaging (BLI) *in vivo* and how the development of modern luciferase systems enable this goal. Table 5.1 summarizes some of the most widely used beetle, marine, and bacterial luciferase technologies and notes examples of *in vivo* use.

5.2 The Chemistry of Bioluminescence

The appearance of bioluminescent creatures varies, as does the purpose for their emitted light, but the general chemistry of biological light production is remarkably similar. Insects, marine invertebrates, and bacteria all produce light by reacting a chemical substrate, generically termed a luciferin, with oxygen in a process catalyzed by a species-specific luciferase. During the reaction, the substrate reaches an electronically excited state that then decays concurrently with the emission of light.

In bacteria, the chemistry of bioluminescence is encoded in the *lux* operon, which contains genes for a luciferase (*luxAB*) and substrate synthesis (*luxCDE*) (Close et al., 2012). The *luxCDE* genes form a multienzyme fatty-acid reductase complex that produces a long chain aldehyde luciferin from endogenous metabolites, and the *luxAB* genes encode a heterodimeric luciferase. In some species, an additional flavin reductase, encoded by the *luxG/frp* gene, also serves to recycle FMN generated in the bioluminescent reaction (Tinikul and Chaiyen, 2016). The reaction between substrate, ATP, molecular oxygen, NADPH/H$_2$, and FMNH$_2$ catalyzed by the *luxAB* luciferase results in a bioluminescent signal near 490 nm (Tinikul and Chaiyen, 2016). Bacteria utilizing this mechanism include those in the *Aliivibrio, Vibrio,* and *Photorhabdus* genera.

The sea pansy *Renilla reniformis,* copepods *Gaussia princeps* and *Metridia longa,* and the jellyfish *Aequorea victoria* exemplify marine bioluminescence. A variety of marine luciferins exist, but these four species use coelenterazine in

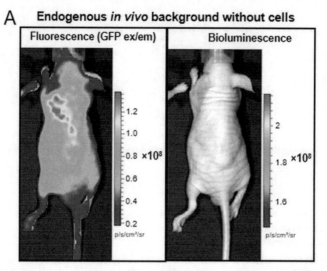

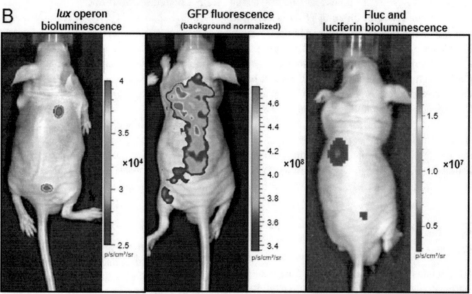

FIGURE 5.1   Comparison of bioluminescent and fluorescent imaging technologies for *in vivo* imaging. A. Nude mice with no exogenous cells or imaging agents were imaged under actual acquisition parameters in a PerkinElmer IVIS instrument. Excitation and emission filters used to collect GFP signal reveal significant endogenous background fluorescence emission (left) that must be processed to resolve a signal if one is present. In comparison, the nude mouse imaged for bioluminescence (right) exhibits very little to no endogenous bioluminescent signal. B. Nude mice with subcutaneous cell injections expressing *lux*, GFP, or Fluc illustrate the comparative signal intensity and resolution of each *in vivo* imaging technology. p/s/cm²/sr; photons/second/centimeter²/steradian

a uniquely ATP-independent, molecular-oxygen-dependent luciferase-catalyzed reaction to produce light in the 455–510 nm range (Ohmiya and Hirano, 1996; Wilson and Hastings, 1998; Markova and Vysotski, 2015). Despite sharing a common substrate, the marine luciferase reactions can be as idiosyncratic as the organisms themselves. For example, the *R. reniformis* luciferase system stores a 3-enol sulfate derivative of coelenterazine as a "pre-coelenterazine" that has a sulfate group removed prior to luciferase catalysis and light generation (Markova and Vysotski, 2015). *G. princeps* does not store coelenterazine like *R. reniformis* and is, instead, unique for secreting its luciferase (Verhaegent and Christopoulos, 2002). *A. victoria* and *R. reniformis* can also produce bioluminescence in response to calcium binding, a process that couples bioluminescence to their nervous systems, enabling a response

to external stimuli such as touch or fluctuation in electrical fields (Anderson et al., 1974; Lorenz et al., 1991). Finally, these species perform a natural version of resonance energy transfer, in which bioluminescence excites green fluorescent protein (GFP), a process detailed later in this chapter, to further modulate their light output (Anderson et al., 1974; Wilson and Hastings, 1998).

The third type of bioluminescent chemistry is endogenous to beetles, including fireflies. The beetle luciferin is called D-luciferin, a benzothiazoyl-thiazole, which is structurally distinct from coelenterazine. The monomeric firefly (*Photinus pyralis*) luciferase (Fluc) requires ATP in the presence of Mg^{2+} and oxygen to complete oxidation of D-luciferin and results in a peak emission of 560 nm (de Wet et al., 1987). However, peak emission wavelength is species-specific and can be above

TABLE 5.1

Summary of Beetle, Marine, and Bacterial Luciferase Systems and Their Use *In Vivo*

	Species and common name, luciferase name, and notes	Luciferin and Emission λ	Light reaction	Example of *in vivo* use
Beetle luciferase systems	*Photinus pyralis*, firefly, luciferase: Fluc (Green and McElroy, 1956; de Wet et al., 1985; de Wet et al., 1987). Related genetic derivatives: *luc2* (Wood, 1994), *effLuc* (Rabinovich et al., 2008) and red-shifted *PRE9* (Branchini et al., 2010).	D-luciferin, 560–620 nm	D-luciferin + O_2 + ATP + Mg^{2+} ↓ Light CO_2 + AMP + oxyluciferin	• Monitoring of T-cell homing with *effluc* at tumor site (Rabinovich et al., 2008). • Fluc in cl62 melanoma and M109 lung cancer cells to monitor organ metastasis (Zhang et al., 1994). • *luc2* based mammary gland tumor development monitoring in mouse (Kim et al., 2010). • Fluc-expressing glioblastoma cells to evaluate therapeutic effect of aldoxorubicin (Marrero et al., 2014). • Fluc-expressing stem-cell implant tracked in mice over 8 months (Vilalta et al., 2008).
	Pyrophorus plagiophthalamus, click beetle, luciferase: click beetle green (CBG), click beetle red (CBR) (Wood et al., 1989).	D-luciferin CBG 540 nm CBR 615 nm		
	Phrixothrix hirtus, railroad worm, luciferase red-shifted PhRE. *Phrixothrix vivianii*, railroad worm, luciferase: green PvGR (Viviani et al., 1999).	D-luciferin PvGR 542 nm PhRE 630 nm		
Marine luciferase systems	*Renilla reniformis*, sea pansy, luciferase: Rluc (Wilson and Hastings, 1998). Related genetic variants: red-shifted RLuc8 (Loening et al., 2006), RLuc8.6–535 (Loening et al., 2007).	Coelenterazine 480–535 nm	Coelenterazine + O_2 ↓ Light + CO_2 + coelenteramide	• Tail vein injected MTLn3 cells labeled with Rluc8 were used to image lung accumulation and monitor xenograft tumor growth (Rumyantsev et al., 2016). • Gluc mutant B16 melanoma cells used to reveal metastasis (Kim et al., 2011).
	Gaussia princeps, copepod, luciferase: Gluc. Related genetic variants: humanized hGluc (Tannous et al., 2005), I90L, 8990, Monsta (Kim et al., 2011), and M431, M43L, (Welsh et al., 2009).	Coelenterazine 480–500 nm		• Aequorin based Ca^{2+} reporter used to monitor slow muscle cell differentiation in zebrafish embryos (Cheung et al., 2011).
	Aequorea victoria, jelly fish, aequorin (Webb and Miller, 2012).	Coelenterazine 470 nm		• Rluc8, and Rluc8 based BRET used to image HT1080 fibrosarcoma cell metastasis in mice (De et al., 2009).
	Oplophorus gracilirostris, Shrimp, NanoLuc (Hall et al., 2012).	Furimazine 460 nm	Furimazine + O_2 ↓ Light + CO_2	
Bacterial luciferase systems	*Aliivibrio fischeri*, *Vibrio harveyi*, and *Photorhabdus luminescens*. Bacterial expression of the *lux* operon includes a dimeric luciferase, *luxA* and *luxB* (Engebrecht et al., 1983; Wilson and Hastings, 1998).	Endogenously synthesized long chain fatty aldehyde 490 nm	Long chain fatty aldehyde + $FMNH_2$ + O_2 ↓ Long chain fatty carboxylate + FMN + H_2O + light	• Tracking bacterial intestinal colonization in mice with the *lux* system (Foucault et al., 2010). • Monitoring mammalian host metabolism with synthetic *lux* during O157:H7 infection (Xu et al., 2015).
	A codon optimized version of the *lux* operon was engineered and expressed in mammalian cells, termed synthetic *lux* (Close et al., 2010; Xu et al., 2014a).			• Substrate free human cell imaging in mice (Close et al., 2011). • Mammalian expressed *lux* used to track bladder tumor development in mice (John et al., 2017).

600 nm as in the case of *Pyrophorus plagiophthalamus* (click beetle red) and *Phrixothrix hirtus* (railroad worm) (Xu et al., 2016). Because a common substrate is used, these emission differences are the result of luciferase structure (Wilson and Hastings, 1998). Luciferase amino acid changes have a strong influence on the emitted wavelength because the changes distinctly manipulate the geometries of the excited reaction intermediates, which in turn influences relaxation kinetics and emission wavelengths (Wilson and Hastings, 1998). This concept can be applied to substrate structure, where specific moieties also influence reaction intermediates. Indeed, synthetic luciferases and luciferin analogs have been zealously developed to improve wavelength emission spectra, among other properties (Adams and Miller, 2014; Evans et al., 2014).

5.3 A Diversity of Luciferases for Bioluminescent Imaging

The firefly luciferase was initially characterized in the mid-twentieth century (Green and McElroy, 1956) and first cloned in the 1980s (de Wet et al., 1985, 1987; Tatsumi et al., 1989). Shortly thereafter, coenzyme A was added to Fluc *in vitro* assays to improve stability and performance (Fraga et al., 2005). Luc2 was then introduced by Promega Corporation, featuring an optimized codon sequence, deleted peroxisome targeting and transregulatory sequences, and an improved light output (Wood, 1994). Presently, Fluc and its relatives are employed in a range of applications, the most basic of which is reporting cell viability. Because ATP is a requisite cofactor for Fluc and its proprietary variants, changes in the available ATP level resulting from changes in cell number or viability manifest as proportional bioluminescence output changes in the presence of Fluc and D-luciferin (Kangas et al., 1984; Crouch et al., 1993). A similar assay principle is used to track a range of cellular signaling events and biomarkers. For example, caspase-activated luciferin analogs can be employed that become accessible to luciferase upon cleavage to detect increasing caspase levels (Caspase-Glo 2 Assay, Promega). Fluc is also often cloned downstream of constitutive, inducible/repressible, or cell-specific promoters to study patterns of expression or fused to proteins to monitor expression and subcellular location. These techniques have been employed, for example, to study circadian rhythms (Yu and Hardin, 2007) and to map stem cell regulatory networks (Rodda et al., 2005; Suzuki et al., 2006; Chan et al., 2009).

In addition to Fluc, other beetle luciferases with red-shifted outputs, such as those derived from *P. plagiophthalamus* and *P. hirtus*, were developed for commercial and research applications after the adoption of Fluc. A common application for these red-shifted beetle luciferases is coexpression with Fluc. This allows the two populations to be monitored simultaneously (Chang et al., 2014), or permits an investigator to normalize the expression of one wavelength to the other to control for differences in transfection efficiency (Thorne et al., 2010). More recently, however, these red-shifted variants are increasingly utilized for *in vivo* imaging because their >600 nm output signals have significantly better deep tissue penetration than Fluc.

One such red-shifted luciferase is that from *R. reniformis* (Rluc). While Rluc was initially expressed in its native form (Lorenz et al., 1991), it has since been modified to reduce serum inactivation (Liu and Escher, 1999) and subjected to rational and evolutionary engineering to yield a synthetic version exhibiting improved light output (>500 nm) and stability (Lehmann et al., 2002; Loening et al., 2006). An advantage of Rluc is its comparatively small size. At 30 kDa, it is approximately half the size of Fluc (61 kDa). This makes it an attractive option for protein tagging and polycistronic expression, as well as for use in split luciferase applications (Paulmurugan and Gambhir, 2005). Like other coelenterazine-utilizing luciferases, Rluc is not ATP-dependent and, therefore, is useful in applications in which ATP concentrations are variable or low. However, Rluc signal strength is significantly reduced relative to Fluc, and therefore, it is more often used as a coreporter for signal normalization in multiplexed assays (Wood, 1994) or as a red-shifted tool for *in vivo* imaging (Loening et al., 2007).

Unlike other luciferases, the Gluc protein derived from *G. princeps* is naturally secreted and acts extracellularly. Gluc is also about one-third the size of Fluc, smaller even than Rluc, and thus highly amenable to coexpression or protein fusion. Together, its small size and secretion abilities have enabled researchers to use Gluc to record bioluminescent measurements without requiring concurrent cell lysis (Tannous, 2009) and provided a means for examining the cellular microenvironment by monitoring protein secretion and paracrine signaling (Stacer et al., 2013). However, its secretory nature also complicates its use as a cell viability and population size reporter, which has resulted in Fluc and Rluc being more commonly employed for these specific purposes.

NanoLuc (Nluc) and its synthetic substrate, furimazine, represent a new class of luciferases that are not refined natural products, but instead are carefully engineered enzymes (Hall et al., 2012; Chou and Moyle, 2014; Kim and Izumi, 2014). Nluc is based on a 19 kb subunit of the shrimp *Oplophorus gracilirostris* multiprotein luciferase complex. This subunit was disassembled *in silico* and subjected to computational protein structure prediction and modeling, followed by synthesis, screening, and refinement against a library of synthetic coelenterazine analogs to result in a system with significantly increased light output. Nluc is used interchangeably with established luciferase assays and has been employed to study gene expression, cell–cell signaling, and protein stability (England et al., 2016). One limitation of Nluc, however, is its 460 nm emission wavelength, which limits its use for *in vivo* imaging. Nonetheless, an emergent role for Nluc, especially given its size to signal ratio compared to traditional luciferases, has been as the activating luciferase partner in bioluminescence resonance energy transfer (BRET) based reporters (England et al., 2016).

Biomedical research rapidly adopted beetle and marine luciferases for their ease of use in the laboratory. Each is a single gene conducive to mammalian expression that catalyzes light emission in the presence of a small molecule. In contrast, the bacterial *lux* system is less amenable to ectopic *in vitro* molecular detection or mammalian expression, in part because of its multicistronic operon and $FMNH_2$ requirement. Despite these technical challenges, the bacterial *lux* system has become a valuable tool for studies that require continuous or triggered bioluminescent signal without exogenous substrate application. Self-supporting bacterial *lux* operons have been used to track bacterial infection in animals (Francis et al., 2000), act as bioreporters to indicate environmental contamination (Xu et al., 2014b), measure estrogen levels in environmental samples (Sanseverino et al., 2005; Iwanowicz et al., 2016), or signal the progress of remediation at contaminated sites, including active monitoring of the remediating organism (Sayler and Ripp, 2000). Recently, a modified bacterial *lux* reporter system capable of generating bioluminescence in response to estrogen exposure was demonstrated in human cells (Xu et al., 2014a) and, in a different study, used to monitor tumor development *in vivo* (John et al., 2017). The autonomously bioluminescent nature of this construct offers unique solutions to the problems associated with traditional luciferases when used in high throughput screening and animal studies in which substrate distribution kinetics and compound interactions can be a problem.

5.4 Applications of Bioluminescence for *In Vivo* Imaging

The following sections introduce how various bioluminescent technologies are leveraged *in vivo* for biological insight. Studying physiological and biochemical processes *in vivo* is essential for gaining a more complete understanding of animal development, cancer, and interventional medicine, among numerous other phenomena. BLI *in vivo* also presents challenges not typically found in two-dimensional, *in vitro* cell culture. The most consequential of which is the target's depth within an animal. On one hand, bioluminescent technologies are well suited for deep tissue imaging because their emitted signal is initiated by a chemical substrate delivered via endogenous circulation and diffusion, whereas fluorescent markers require excitation light, which is not easily deliverable in this case. Furthermore, the surrounding tissues emit essentially no bioluminescence, unlike fluorescence, which produces high signal-to-noise ratios (SNR) for BLI (Figure 5.1). But on the other hand, the majority of bioluminescent modalities at present have weaker overall signals compared to fluorescence and have emission wavelengths readily absorbed by surrounding tissues. However, red-shifted bioluminescent technologies are reducing this problem. With these advantages and limitations in mind, the remainder of this chapter will introduce traditional and commonly utilized bioluminescent technologies for *in vivo* imaging, as well as introduce recent and innovative technologies that improve deep tissue imaging or facilitate the study of molecular interactions.

5.4.1 Traditional *In Vivo* Bioluminescent Imaging Using Firefly Luciferase and D-Luciferin

Firefly luciferase is used extensively for *in vivo* BLI because it is genetically tractable and produces readily detectable light. BLI *in vivo* with Fluc is an experimental paradigm and, while newer, more specialized bioluminescent modalities have emerged. Fluc remains a mainstay for *in vivo* BLI research because of its established simplicity, reliability and economy.

Fluc BLI is frequently used for noninvasive monitoring of tumor progression, metastasis, and response to therapeutic intervention. In a typical experiment, the Fluc bearing tumor or potentially tumorous cells are introduced via injection or xenograft (usually within a mouse or small animal) and tracked by whole-body imaging for quantitation.

One of the first demonstrations of this paradigm occurred in 1994 when Zhang et al. (1994) transfected the firefly luciferase gene into murine cl62 melanoma and M109 lung cancer cell lines and, after establishing a correlation between cancer cell number and bioluminescent output *in vitro*, injected the cells intravenously into mice to track tumor formation and metastasis. The authors detected bioluminescence in the lungs but not in other organs. The Fluc-bearing cells were then injected intramuscularly and again found to produce tumors detectable by bioluminescence. In a final and convincing demonstration of Fluc's utility for *in vivo* tumor analysis, the authors found reduced luciferase activity in the lungs of mice receiving T-cell therapy compared to those without, indicating a decrease in tumor burden but also, importantly, the utility of Fluc to quantify such a decrease noninvasively.

The ultimate conclusion by Zhang et al. (1994), that a difference in tumor burden was detectable with bioluminescence, proved to be a pioneering idea that would be modeled over and over with Fluc in cancer research. It should be emphasized that, because Fluc signal is dependent on intracellular ATP levels, the resultant light signal can be used as an indicator for cell viability, proliferation, and tumor size. Because of this, BLI allows noninvasive measurement of tumor burden *in vivo* without the need to excise the tumor or sacrifice the animals. It also offers a significantly more quantitative measurement of the tumor compared to calipers, and can do so before the tumor is palpable. Longitudinal studies of tumor progression and response to drug treatment also benefit from BLI's ability to repetitively image animals over time.

A good example of the sensitivity of Fluc in *in vivo* BLI was shown by Kim et al. (2010) who labeled mouse mammary gland tumor 4T1 cells with an enhanced firefly luciferase gene (*luc2*) and demonstrated *in vivo* bioluminescent detection from as few as five 4T1-luc2 cells immediately following injection into a nude mouse model. The bioluminescent signal from the implanted cells was continually detectable from the time of implantation, allowing visualization and quantitative measurement of tumor burden as the tumor developed. In contrast, the same tumor could not be measured using the standard caliper method until Day 27 post-implantation. BLI also facilitated the early detection of metastasis of implanted 4T1-luc2 cells into the lung in less than 4 weeks postimplantation without sacrificing the animal.

Noninvasive Fluc *in vivo* BLI also enables imaging and quantification in tumor models that are intrinsically difficult to access, such as those occurring in the brain or skeletal system, and where caliper measurements or light excitation are not feasible. Marrero et al. (2014) developed an intracranial xenograft mouse model of human glioblastoma with Fluc-expressing human glioblastoma U87 cells to evaluate the antiglioblastoma therapeutic effect of aldoxorubicin. Bioluminescent signal intensity was used to track tumor burden kinetics over the course of treatment. After 10 days, the treatment group had significantly lower Fluc signal than the untreated control group, suggesting aldoxorubicin was capable of inhibiting intracranial glioblastoma growth. This BLI result was also validated by an improved median survival time of the aldoxorubicin treated group.

Thus far, examples of Fluc BLI *in vivo* have concentrated on tracking cancer, which is often introduced exogenously. When stem-cell culture became possible, there was, and continues to be, a need to test the efficacy and safety of stem cell-based regenerative and tissue engineering therapies. In response, many of the Fluc *in vivo* BLI experimental modalities developed for cancer have been adapted to study stem-cell-based therapies, as the two share the practical need to implant exogenous cells and track behavior *in vivo*. Indeed, there are numerous examples of *in vivo* BLI that extend beyond labeling cancer cells, and some examples are listed in Table 5.2.

With stem cells, Fluc BLI allows noninvasive and longitudinal tracking of the survival, proliferation, migration, engraftment, and differentiation of implanted cells in live animals. Vilalta et al. (2008) utilized BLI to follow the survival and biodistribution of Fluc-expressing human adipose-derived mesenchymal stem cells (hADMSCs) implanted into nude mice over a period of 8 months. BLI results indicated that, although 75% of hADMSCs injected intramuscularly died within one week of implantation, the remaining cells survived during the 8-month experimental period. Bioluminescent images also showed that a small portion of the intramuscularly injected cells migrated away from the injection site and accumulated in the liver.

Developmentally or tissue-specific events can be monitored with Fluc *in vivo* BLI because the *Fluc* gene can be cloned under the control of a specific promoter capable of responding to a biological event of interest. In this way, Fluc BLI enables real-time tracking of stem-cell differentiation *in vivo*. For example, dual labeled hADMSCs with constitutively expressed Rluc and tissue-specific promoter-regulated Fluc were used to simultaneously assess the viability and differentiation of hADMSCs in biomaterials implanted *in vivo* for tissue engineering scaffold development (Bagó et al., 2013). The constitutively expressed Rluc allowed the noninvasive measurement of viability and localization of the implanted cells, while the firefly luciferase was regulated by the platelet endothelia cell adhesion molecule-1 (PECAM-1) promoter or osteoblast-specific OC (osteocalcin) promoter to monitor hADMSC differentiation to endothelial and osteoblast lineages, respectively. Using this approach, Bago et al. (2013) demonstrated a 125-fold and 15-fold increase in PECAM-1 and OC promoter activities during an 8-week period postimplantation, indicating endothelial and osteogenic differentiation of the hADMSCs on the implanted scaffold *in vivo*.

5.4.2 *In Vivo* Bioluminescent Imaging Without Substrate

In contrast to the Fluc system discussed above, the bacterial *lux* system enables an alternative imaging approach, whereby the pathway producing the luciferin is genetically encoded and delivered into the target cells alongside the luciferase itself. These cells, therefore, generate an autonomous, "autobioluminescent" signal without requiring external substrate application. Furthermore, because the system is all genetically encoded, its expression can be subjected to the same types of transcriptional controls as can luciferase-only reporters. This allows the autobioluminescent signal to be generated continuously (controlled by a constitutive promoter), to self-initiate or self-terminate in response to natural gene regulation cycles (control by an endogenous promoter), or to cycle on and off in response to investigator-defined conditions (controlled by a synthetic regulatory gene circuit).

The genes (*luxA* and *luxB*) encoding the bacterial luciferase heterodimer were first cloned and expressed in *Escherichia coli* in 1982 (Belas et al., 1982), and the full cassette was

TABLE 5.2

Examples of Noncancer Focused *In Vivo* Bioluminescent Imaging

---

Interaction between bioluminescent bacteria and animal cells
    Tracking bacteria in animals (Francis et al., 2000; Francis et al., 2001; Avci et al., 2018; Lamb et al., 2018)

Examples in multipotent stem cells
    Long term mesenchymal stem cell (MSC) transplant tracking (Vilalta et al., 2008)
    Tissue-specific promoter-driven somatic cell differentiation tracking (Vilalta et al., 2009; Bagó et al., 2013)
    Monitor bone repair (Vila et al., 2014)
    Response to MSC infusion following surgically induced myocardial infarction (van der Bogt et al., 2008)
    MSC to wound-site homing and engraftment (Kidd et al., 2009)

Cellular and molecular activity monitoring
    Protein–protein interactions (Paulmurugan et al., 2002)
    Apoptosis (and caspase activity) (Coppola et al., 2008; Scabini et al., 2011)
    Cell migration (Shah et al., 2005; Ma et al., 2009)
    Cell–cell interaction (Sellmyer et al., 2013)
    Calcium transients (Bakayan et al., 2011)

Noncancer *in vivo* uses
    Kinetic, whole body T-cell imaging (Chewning et al., 2009)
    GFAP activation in brain tissue (Cho et al., 2009; Keller et al., 2009)
    Biodistribution monitoring. See Berger et al. (2008) for substrate biodistribution; see Sensebe and Fleury-Cappellesso (2013) for comprehensive review on injected cell biodistribution.
    Copper deficiency (Heffern et al., 2016)
    Long-term eye inflammation tracking (Gutowski et al., 2017)

cloned and expressed the next year (Engebrecht et al., 1983). During this initial development, it was believed that the system's dependence on FMNH$_2$ and operon-based expression architecture would prohibit it from functioning in eukaryotic organisms (Sambrook et al., 1990). However, using modern genetic tools, the system was reengineered to overcome these initial hurdles (Close et al., 2010). It was redesigned based on the *lux* genes of a terrestrial bioluminescent bacterium *Photorhabdus luminescens*, whose protein homologs displayed greater thermotolerance relative to its marine cousins. These genes were then codon optimized for the human genome, assembled as a pseudo-operon using viral 2A linker elements, and supplemented with coexpression of an oxidoreductase gene that shifted the cytosolic FMN/FMNH$_2$ balance to a more reduced state (Xu et al., 2014a). This enabled the humanized *lux* cassette to be expressed in mammalian cells and to be manipulated as a single open reading frame using the same molecular biology strategies as any traditional, single gene luciferase system. Under this deployment strategy, its expression has not been shown to significantly affect the basal metabolic activity level of its host cell (Close et al., 2010) and its autobioluminescent signal correlates strongly with overall metabolic activity level of the host (Xu et al., 2014a). Interestingly, the cassette can also be intentionally fragmented into two or more 2A-linked humanized operons to facilitate spatiotemporal imaging strategies.

Therefore, the humanized *lux* bioreporter is well suited for *in vivo* BLI. The autobioluminescent signal generated by the *lux* system shows significantly less variability than does a Fluc bioluminescent signal, which is subject to fluctuating luciferin availability. This was demonstrated *in vivo* by monitoring subcutaneous tumor models where all data within a 60-minute observational window were within the error of one another when using a *lux*-endowed cell, which was not observed with the Fluc system (Xu et al., 2014a). Continuous light generation also improves animal husbandry by reducing substrate injections and handling times, including time under anesthesia waiting for substrate kinetics to stabilize.

One recent example specifically highlights how the autobioluminescent *lux* system can be used to track the survival and tumorigenicity of implanted cells at a range of initial doses. John et al. (2017) used a *lux* expressing human bladder cancer-cell line implanted into mice and showed that tumors were detectable with bioluminescence prior to when detectable by physical means (palpation or caliper). This study suggested that the *in vivo* autobioluminescent cancer model is an improved tool for preclinical-level therapeutic compound dosing experimentation. This line of research is also especially promising, given that researchers have shown bioluminescent output of the *lux* system very closely parallels those of many *in vitro* ADME/Tox assays relevant to new compound discovery (Xu et al., 2014a).

In addition to the increased ease of use that comes from forgoing substrate addition in these types of traditional *in vivo* cancer biology imaging applications, the *lux* system could also be used when the delivery of an activating substrate may be difficult or impossible. One such example is bioluminescent imaging in the brain. Accumulation of D-luciferin in brain tissue is lower than in other organs (Berger et al., 2008), can be

influenced by the mode of anesthesia (Keyaerts et al., 2012; Aswendt et al., 2013), and is actively transported from the CNS along with coelenterazine (Pichler et al., 2004; Yasuda et al., 2015). Taken together, a significant amount of experimental coordination is required to ensure comparable or time stable data using traditional bioluminescent substrates. If used to image within the brain, the self-activated signal of the *lux* system could alleviate the hurdles involved, both physical and chemical, in supplying the target cells with an activation signal, while still allowing the system to be placed under the control of a specific promoter of interest. These unique capabilities, combined with its straightforward inject-and-monitor imaging format, positions the *lux* system as a flexible reporter option for *in vivo* imaging applications.

5.4.3 *In Vivo* Bioluminescent Imaging at Longer Wavelengths

BLI is well suited for *in vitro* assays and *in vivo* investigation of subcutaneous targets such as implants or tumors. However, as the bioluminescent target is increasingly covered by living tissue, more of its emitted light is absorbed, and the overall observed signal decreases. Hemoglobin is primarily responsible for this effect because of its strong absorption of light <600 nm (Xu et al., 2014a; Class et al., 2015), which eclipses the 450–600 nm range at which traditional luciferases emit (Xu et al., 2016). Thus, while deep tissue *in vivo* imaging is possible with current bioluminescent tools, they remain best suited for use in near-surface structures and from within small animals such as mice.

One strategy to improve deep tissue BLI has been to shift the bioluminescent emission range toward the red and infrared wavelengths (>600 nm) to circumvent absorption. These "next-generation" luciferase systems depart substantially from their native forms because significant genetic engineering has been applied to red-shift emission. Red-shifted bioluminescence was achieved by iteratively mutating and selecting Rluc variants to achieve a peak output of 535 nm, with approximately 20% of emitted light occurring at a wavelength greater than 600 nm (Loening et al., 2007). In a similar strategy, Fluc red-shifted variants were initially identified by random and rational mutagenesis (Branchini et al., 2005, 2007), then subjected to additional genetic engineering and codon optimization to yield variants emitting light at > 600 nm (Branchini et al., 2010). Red-shifting can also be achieved by manipulating the luciferin component of the system. By modifying the aromatic core of D-luciferin and improving its water solubility, Kuchimaru et al. (2016) were able to achieve a peak light output at 677 nm using an unmodified luciferase enzyme. Coelenterazine has also frequently been modified by substituent manipulation on its imidazopyrazinone ring system, which, in some cases, has delivered red-shifted emission (Coutant and Janin, 2015; Nishihara et al., 2015).

These red-shifted variants have demonstrated improved performance relative to their natural counterparts when used in *in vivo* imaging applications. Mezzanotte et al. (2010) compared a red-shifted Fluc variant with a peak emission at 612 nm with its wild type counterpart to image HepG2-derived subcutaneous tumors in white-coated mice. The red-shifted variant, despite having an overall lower photon flux *in vitro* than wild type Fluc, produced a net *in vivo* signal almost 60% higher, a difference attributed to loss of Fluc signal in the skin (Mezzanotte et al., 2010). In a similar study, a red-shifted Rluc showed an approximately 2-fold increase in signal emanating from intravenously administered bioluminescent cells accumulating in the lungs of mice relative to the non-red-shifted version (Loening et al., 2007).

5.4.4 Multiplexed *In Vivo* Bioluminescent Imaging

BLI *in vivo* has a high SNR and enables monitoring of genetic or biomolecular events, but it does not provide precise anatomic information. However, BLI can be multiplexed with positron emission tomography (PET), computed tomography (CT), or single photon emission computed tomography (SPECT) to permit simultaneous imaging of biological targets and achieve a higher anatomic resolution for diagnostic purposes. For example, Deroose et al. (2007) combined BLI, PET, and CT to detect melanoma metastasis in a mouse model. This work showed that while BLI was capable of detecting metastatic lesions at an early stage, even before lesions were detectable via PET/CT, the anatomic location of later stage lesions could be more accurately mapped with three-dimensional PET/CT images. In a second example, Hong et al. (2012) monitored tumor growth and lung metastasis following Fluc labeled murine breast cancer 4T1 cell injection. Once lung metastasis was detected by BLI, a CD105-specific imaging agent for both PET and near-infrared fluorescence (NIRF) imaging of tumor angiogenesis was applied to prove its efficacy in identifying tumor burden and demonstrate its potential as a clinical screening tool for early detection.

In addition to being multiplexed with PET, CT, and SPECT, BLI is frequently multiplexed with other bioluminescent or fluorescent modalities to increase the depth of biological characterization. As an example, one reporter's wavelength is used to monitor the event of interest (e.g. gene expression or ATP content), and the other reporter's wavelength is used to normalize between samples or provide spatiotemporal tracking. Alternatively, the multiple wavelengths can be used to collect different events in parallel or to track sequential and cyclical events. Beetle luciferase (e.g. Fluc, click beetle red, CBR) and marine luciferase (Rluc, Gluc) are frequently combined for these purposes because they demonstrate sufficient emission wavelength separation and luciferin specificity (D-luciferin vs. coelenterazine). In these pairings, the marine luciferase is often used as the internal control because of its weaker signal output, although many factors determine assignment. Promoter activity can be monitored with a dual luciferase approach (Alcaraz-Perez et al., 2008) as has apoptosis (Shah et al., 2003), cell migration (Shah et al., 2005), and stem-cell differentiation (Vilalta et al., 2009), and three targets were monitored in a mouse brain tumor system with Fluc utilizing D-luciferin, Gluc utilizing coelenterazine, and *Vargula hilgendorfii* luciferase (Vluc) utilizing the luciferin vargulin (Maguire et al., 2013).

5.4.5 *In Vivo* Bioluminescent Imaging Using a Caged Luciferin

The luciferase reactions discussed so far have been controlled genetically, in which the presence of the luciferase is a result of promoter activation or defined by the start of luciferin administration. However, the development of caged luciferins has allowed researchers to coordinate the onset of light production with a biochemical event. Caged luciferins are modified substrates that prevent luciferase interaction. When used, a second molecular event must take place for the caging moiety to be removed and make the luciferin available to the luciferase before light is generated. This is typically accomplished by expressing an enzyme to dissociate the caging moiety or engineering the modification to be targeted by an endogenous function. The Lugal system is one such example, whereby a galactose molecule is appended at the 6′ position of the luciferin to block luciferase activity until it is cleaved by β-galactosidase (β-gal) (Li et al., 2013; Porterfield et al., 2015). In the presence of the nonreactive luciferin-galactose substrate, Fluc expressing cells cannot produce light unless close to β-gal expressing cells, which create a local depot of cleaved luciferin available to produce light (Porterfield et al., 2015). Caged luciferins are similar to multiplexed BLI in that multiple biological events can be monitored (e.g. gene-regulated luciferin expression and dissociating enzyme activity); however, their most common application is for the study of biological proximity in which one cell expresses the luciferase and the other, the luciferin activating mechanism (Sellmyer et al., 2013).

5.4.6 Split Luciferases for *In Vivo* Bioluminescent Imaging

Caged luciferase systems provide insight into proximity, but they are not well suited to report physical molecular interactions. Much like the original two-hybrid system (Fields and Song, 1989), split luciferase systems are frequently used to study protein–protein and protein-nucleic acid interactions. They function by isolating luciferase protein domains onto separate target molecules, typically peptides or whole proteins, and thus are not capable of producing light until reunited. If the target molecules achieve sufficient proximity, the separated luciferase domains interact to recapitulate a functional enzyme (Azad et al., 2014). This technique was demonstrated when the N-terminal domain of Fluc was fused to MyoD, a myogenic regulatory protein, and the C-terminal Fluc domain was fused to Id, a negative regulator of myogenic differentiation (Paulmurugan et al., 2002). MyoD and Id interact when Id inhibits MyoD DNA binding (Jen et al., 1992). Split luciferases have elucidated membrane protein–protein signaling (Varnum et al., 2017), apoptotic events (Coppola et al., 2008), receptor ligand interactions (Kim et al., 2007), and even nonprotein small-molecule binding (e.g. Ca^{2+}) (Kaihara et al., 2008).

The Peroxy Caged Luciferin-2 (PCL-2) system developed by Van de Bittner et al. (2013) is an interesting amalgam of the caged luciferin and split luciferase concepts. The system essentially splits the Fluc D-luciferin substrate into two protoluciferin halves that require both hydrogen peroxide and caspase 8 to "uncage" each before the luciferin substrate can be recapitulated and enable BLI. The approach was used to image oxidative stress and inflammation *in vivo* (Van de Bittner et al., 2013) and was recently adapted to monitor copper status using bioluminescence (Heffern et al., 2016).

5.4.7 Bioluminescence Resonance Energy Transfer (BRET)

Bioluminescence resonance energy transfer (BRET) is another luciferase-based technique that is used to observe molecular interactions, typically between proteins, whose principle is akin to that exploited by split luciferases (Dragulescu-Andrasi et al., 2011), but the essential advantage of BRET is signal amplification. In BRET, one target is labeled with a luciferase while the other target is labeled with a fluorescent molecule. When the targets are sufficiently close, bioluminescent light excites the fluorophore, which registers as a change in both the wavelength and overall output of the measured signal (Sun et al., 2016). BRET is particularly useful for *in vivo* deep tissue imaging because it provides site-specific bioluminescent expression to excite an otherwise unexcitable fluorophore. The bioluminescent component alone is not sufficient for imaging because most of the < 600 nm light is absorbed by the surrounding tissue, and, likewise, the fluorescent component is too deep within an animal to be excited by externally supplied light. Therefore, the two imaging modalities work together to create an amplified signal that can penetrate the tissues for external detection (Dragulescu-Andrasi et al., 2011; Sun et al., 2016). This technique has enabled the imaging of deep tissue *in vivo* calcium transients in ectopically implanted cells in a mouse model (Bakayan et al., 2011), the study of G-protein coupled receptors (Pfleger and Eidne, 2005), and regulators of nuclear receptors (Germain-Desprez et al., 2003).

5.4.8 Fluorescence by Unbound Excitation from Luminescence (FUEL)

Fluorescence by unbound excitation from luminescence (FUEL) is similar to BRET in that bioluminescent light from one site excites a fluorophore at another site. However, BRET requires that both luminescent components interact within close proximity, typically through a molecular binding event. In contrast, FUEL reports interactions at greater distances without physical interaction. While the difference is subtle, FUEL is useful when the molecule(s) of interest have no binding interaction or when the molecular interactions are unknown. *A. victoria* and *R. reniformis* perform a natural version of this process, in which both excite a green fluorescent protein variant with endogenous bioluminescence as a way to modify light-output wavelengths (Wilson and Hastings, 1998). In another example of this technique, the lymphatic system of a mouse was mapped, including lymph nodes, to ultimately identify cancer metastasis (Kosaka et al., 2011; Xiong et al., 2012). Briefly, luciferase was fused to a near-infrared quantum dot and exposed to luciferin. The emitted bioluminescent light stimulated the quantum dot, which amplified the bioluminescent signal without a specific binding event and thus enabled the imaging of a structure that would otherwise obscure bioluminescent light and preclude excitation illumination. Adapting this approach to map other structures should be possible, provided the target quantum dots are freely available either in media or within an animal and are stimulated only when near a spatially restricted (e.g. tissue specific) source of bioluminescence.

5.4.9 Emerging Animal Models That Permit Endogenous *In Vivo* Bioluminescent Imaging

The bioluminescent modalities covered so far enable *in vivo* BLI in different ways but share one common theme: Each must be transferred into a cell of interest already in an animal or xenografted *in vivo* after bioluminescent endowment. The discovery of genome modifying technologies, notably CRISPR/Cas9, and stem and reproductive cell biology has enabled a new *in vivo* BLI modality to emerge in which the animal itself is endogenously bioluminescent. Such animals typically have the genetic constructs described above inserted into their genomes and introduced at the germline level. The resulting transgenic animals express the bioluminescent constructs ubiquitously or compartmentally, depending on the regulatory genetic architecture (Close et al., 2011). The primary advantage of endogenously bioluminescent animals is greater genetic penetration and access to finer time resolution. When bioluminescent operons must be delivered by xenograft, injection, or viral infection, there is significant dilution and time lag of the resulting signal. If a tissue or cell type of interest is potentially bioluminescent from fertilization, researchers are able to follow elements of development, cancer, and metabolism in real time.

One such example of an endogenously bioluminescent animal was described by Zagozdzon et al. (2012) who encoded Luc2 under the control of a mammary tissue-specific promoter and were able to identify early mammary tumor formation as well as track tumor progression. In a similar example, proliferative cells forming pancreatic ductal adenocarcinoma were visualized *in vivo* from a mouse with a genetically endogenous cell-cycle regulated Fluc (de Latouliere et al., 2016). In an interesting amalgamation of recent bioluminescent technologies, Matsushita et al. (2017) developed a transgenic mouse in which endogenously encoded BRET, consisting of an engineered Rluc and Venus, a YFP variant, was used to visualize and track blood vessel formation during normal development and tumor formation. Immune cell behaviors and fates are also readily amenable to study with BLI *in vivo*. Chewning et al. (2009) developed a transgenic mouse model, dubbed "T-Lux," in which T-cells express a commercially modified Fluc and enabled whole-body imaging of T-cell populations as well as kinetic tracking of T-lux CD3+ cells following adoptive transfer. Endogenously bioluminescent transgenic animals are also of significant value for brain imaging. Cho et al. (2009) created a transgenic mouse model expressing GFAP promoter-controlled Fluc to study GFAP gene expression in astrocytes in response to brain injury *in vivo*. Localization of the bioluminescent signal revealed that the GFAP promoter was highly active in the brain, expressed a very low level of activity in the heart, and showed no detectable activity in other tissues. When seizure was induced in the transgenic animals, significant activation of the GFAP promoter was detected within 6 hours, as indicated by the increase in bioluminescent signal. Similarly, Keller et al. (2009) identified that GFAP activation in Schwann cells was a marker for the onset of amyotrophic lateral sclerosis using BLI in a GFAP-Fluc transgenic mouse model.

5.5 Instrumentation for Bioluminescent Imaging

Instruments used to visualize bioluminescence *in vivo* all follow the same general format of an ultrasensitive camera that collects photons emitted from an animal housed in an attached light-tight chamber. Typically, a photographic image is initially taken, followed by a bioluminescent image. These images are then superimposed to display a pseudoimage of bioluminescent source emissions mapped to the subject. Integrated software then assists in determining bioluminescent intensities. With bioluminescence often being weakly emitted, cameras are capable of light collection under long integration times on the scale of tens of minutes. Fluorescence imaging capabilities are often built into these systems as well. Xenogen, Inc. was one of the pioneers of BLI instrumentation with their IVIS systems, which are now marketed by PerkinElmer (http://www.perk inelmer.com/catalog/category/id/in%20vivo%20imaging). Their IVIS Lumina and Spectrum series of imagers apply ultracooled charge-coupled devices (CCD) or ultrafast electron multiplying CCD (EMCCD) cameras for photon capture, along with accessories such as integrated anesthesia systems, heated stages, and ECG monitors. X-ray and microcomputed tomography (microCT) have also been paired with select systems to enable coregistration of bioluminescence against other imaging modalities for expanded informational content. Spectral Instruments Imaging (SI Imaging; http://www.spec-img.com/) offers their Lago and Ami lines of *in vivo* imagers with ultrasensitive CCD cameras and X-ray options. Bruker's line of In-Vivo Xtreme and PRO imaging systems offer CCD technologies along with similar X-ray options (https://ww w.bruker.com/products/preclinical-imaging/opticalx-ray-imaging.html). Biospace Lab markets their PhotonIMAGER system as modular in nature, with multiple options such as X-ray and 3D/4D kinetic rendering available as upgrades to a base unit (http://www.biospacelab.com/m-31-optical-imagi ng.html). LI-COR's Pearl Trilogy system is one of the most affordable *in vivo* imagers available for users that require sensitive BLI without the additional options (http://www .licor.com/bio/products/imaging_systems/pearl/). Berthold Technologies offers its NightOWL instrument for CCD-based *in vivo* animal imaging (https://www.berthold.com/en/bio/in% 20vivo%20imaging%20instruments). MI Labs is focused on multimode imaging systems that go beyond conventional bioluminescent/fluorescent imaging to additionally include platforms like SPECT and PET (http://www.milabs.com/preclini cal-imaging-systems/).

5.6 The Advantages and Disadvantages of Bioluminescent Imaging

The primary advantage of BLI is its high SNR. In addition to animals and animal husbandry supplies, cell cultureware, media, reagents, and cells all have low to undetectable native bioluminescence, whereas most of these items exhibit autofluorescence under common excitation wavelengths (Troy et al.,

2004) that causes the target fluorophore to experience photobleaching and imposes cellular phototoxicity (Daddysman et al., 2014). When cytomegalovirus (CMV) promoter-driven GFP, Fluc, or the bacterial *lux* operon were compared, Fluc had the highest average maximum radiance and required significantly fewer cells for detection (Close et al., 2011). *In vivo*, BLI has an additional advantage beyond high SNR that relates to how its light is "turned on." Because BLI is chemiluminescent, as long as the substrate reaches the site of luciferase expression, light will be emitted. In contrast, a fluorescent marker will emit light only if stimulated with excitatory light, which is challenging in a mouse, for example, either because light cannot be delivered or the autofluorescence would be overwhelming. In an *in vivo* example, cells expressing a commercial Fluc variant from the lungs of a mouse had a 35-fold SNR, whereas the otherwise identical fluorescently labeled cells were indistinguishable from background noise (Troy et al., 2004).

Another advantage of BLI is that luciferase-based reactions have a relatively rapid turnover rate. While the discussion above regarding genetic engineering of luciferases focused on improving wavelength and output, efforts have also been made to improve reaction kinetics to optimize "glow." In these efforts, the balance of luciferase degradation and reaction kinetics is optimized to produce a stable signal that decays conveniently for the purposes of the experiment, often to facilitate a monitoring time on the scale of the biological phenomena under study. Fluorescence, in contrast, can persist even after cells die and, therefore, may fail to provide sufficiently time-resolved information.

The limitations of BLI *in vitro* stem from the chemistry necessary to produce light, a fundamentally enzymatic process subject to modulation. This is especially pertinent to high-throughput screening, in which a subset of chemicals within a library will potentially interact with the reporting luciferase or substrate (Close et al., 2011). Fortunately, this effect can be partially controlled with counterscreening and combinatorial approaches.

The fundamental enzymatic nature of a luciferase reporter can also pose a challenge to *in vivo* imaging if uniform substrate distribution to the site of interest is delayed or impossible. Luciferins are typically injected intravenously or intraperitoneally, where they ultimately reach the target site by circulation and diffusion. The local concentration of luciferin is therefore transient, and only offers the researcher a limited window for reproducible experimentation (Close et al., 2011). Calibration can control this, but the task remains variable by animal and therefore is difficult to scale. The bioluminescent reaction may also be affected by local conditions. For example, tumors are characterized by leaky vasculature and often have regions of hypoxia, necrosis, and active apoptosis (Mehta et al., 2012). The amount of luciferin and oxygen reaching the tumor site may be heterogeneous depending on vasculature (or lack thereof) or could be adversely affected by local pH and enzyme profiles. Likewise, coelenterazine and D-luciferin can be actively expelled by P-glycoprotein and ABCG2 pumps, respectively, which may also bias bioluminescent output (Pichler et al., 2004; Bakhsheshian et al., 2013). Finally, spatial resolution can be a limitation for *in vivo* bioluminescence imaging due to absorbance of the bioluminescent signal by tissue.

The bacterial luciferase system may be better suited for *in vivo* imaging because it does not require externally supplied substrate and thus avoids many of these common hurdles. Cells expressing the *lux* operon autonomously produce bioluminescence from endogenous metabolites and available cofactors and, therefore, more reliably report on their true state. However, the primary drawbacks of the bacterial *lux* operon are that its emitted light signal is significantly less strong than that produced by Fluc when expressed in mammalian cells and that it is currently limited to wavelengths readily absorbed by hemoglobin (Close et al., 2011).

5.7 Conclusion and the Future of *In Vivo* Bioluminescent Imaging

In vivo imaging with bioluminescence is a product of the early study and cloning of naturally occurring enzymes found in a group of light emitting but otherwise morphologically disparate organisms. Modern biotechnology has diversified the bioluminescent tool set to answer basic questions about physiology as well as drive clinical innovation. This has led to the development of vastly different forms of bioluminescence from the simple expression of a single luciferase gene paired with its chemical substrate to luciferase derivatives with red-shifted emissions, to self-contained genetic operons that endow autonomous bioluminescence requiring no additional interaction, to fragmented luciferases that telegraph molecular interactions, to transgenic bioluminescent animals. Each of these bioluminescent formats has enabled a specific *in vivo* question to be addressed, and this, in turn, demands novel bioluminescent technologies to solve newly appreciated biological puzzles.

The future of BLI *in vivo* will continue to focus on red-shifted emissions and increased signal outputs. At present, these are the two greatest limitations of *in vivo* BLI. Of particular interest is the synthetic luciferases and luciferins (e.g., Nluc) (England et al., 2016). Red-shifting in synthetic luciferases is likely to proceed with the use of rational design, but semi- to fully random mutagenesis screening will factor in heavily as the price of gene synthesis continues to decrease and the cost of library screening, therefore, reduces. Genome modification should facilitate more precise spatiotemporal expression of bioluminescent operons, particularly when combined with ligand-induced regulatory circuits (Zabala et al., 2009). Also in this vein, real-time bioluminescent bioreporters should continue to emerge in which, like the caged and split luciferases, biomarkers are monitored with bioluminescence in addition to gene or protein expression. In the longer term, clinical therapies would benefit from the adoption of *in vivo* BLI in larger animals that better recapitulate human maladies, which should be, in part, facilitated by improved bioluminescent signal characteristics.

REFERENCES

Adams ST, Jr., Miller SC (2014) Beyond D-luciferin: Expanding the scope of bioluminescence imaging in vivo. *Curr Opin Chem Biol* 21:112–120.

Alcaraz-Perez F, Mulero V, Cayuela ML (2008) Application of the dual-luciferase reporter assay to the analysis of promoter activity in Zebrafish embryos. *BMC Biotechnol* 8:81.

Anderson JM, Charbonneau H, Cormier MJ (1974) Mechanism of calcium induction of *Renilla* bioluminescence. Involvement of a calcium-triggered luciferin binding protein. *Biochemistry* 13:1195–1200.

Aswendt M, Adamczak J, Couillard-Despres S, Hoehn M (2013) Boosting bioluminescence neuroimaging: An optimized protocol for brain studies. *PLoS One* 8:e55662.

Avci P, Karimi M, Sadasivam M, Antunes-Melo WC, Carrasco E, Hamblin MR (2018) In-vivo monitoring of infectious diseases in living animals using bioluminescence imaging. *Virulence* 9:28–63.

Azad T, Tashakor A, Hosseinkhani S (2014) Split-luciferase complementary assay: Applications, recent developments, and future perspectives. *Anal Bioanal Chem* 406:5541–5560.

Bagó JR, Aguilar E, Alieva M, Soler-Botija C, Vila OF, Claros S, Andrades JA, Becerra J, Rubio N, Blanco J (2013) In vivo bioluminescence imaging of cell differentiation in biomaterials: A platform for scaffold development. *Tissue Eng Pt A* 19:593–603.

Bakayan A, Vaquero CF, Picazo F, Llopis J (2011) Red fluorescent protein-aequorin fusions as improved bioluminescent Ca^{2+} reporters in single cells and mice. *PLoS One* 6:e19520.

Bakhsheshian J, Wei BR, Chang KE, Shukla S, Ambudkar SV, Simpson RM, Gottesman MM, Hall MD (2013) Bioluminescent imaging of drug efflux at the blood-brain barrier mediated by the transporter ABCG2. *Proc Natl Acad Sci USA* 110:20801–20806.

Belas R, Mileham A, Cohn D, Hilmen M, Simon M, Silverman M (1982) Bacterial bioluminescence - isolation and expression of the luciferase genes from *Vibrio harveyi*. *Science* 218:791–793.

Berger F, Paulmurugan R, Bhaumik S, Gambhir SS (2008) Uptake kinetics and biodistribution of 14C-D-luciferin--a radiolabeled substrate for the firefly luciferase catalyzed bioluminescence reaction: Impact on bioluminescence based reporter gene imaging. *Eur J Nucl Med Mol* I 35:2275–2285.

Branchini BR, Southworth TL, Khattak NF, Michelini E, Roda A (2005) Red- and green-emitting firefly luciferase mutants for bioluminescent reporter applications. *Anal Biochem* 345:140–148.

Branchini BR, Ablamsky DM, Murtiashaw MH, Uzasci L, Fraga H, Southworth TL (2007) Thermostable red and green light-producing firefly luciferase mutants for bioluminescent reporter applications. *Anal Biochem* 361:253–262.

Branchini BR, Ablamsky DM, Davis AL, Southworth TL, Butler B, Fan F, Jathoul AP, Pule MA (2010) Red-emitting luciferases for bioluminescence reporter and imaging applications. *Anal Biochem* 396:290–297.

Chan KK, Zhang J, Chia NY, Chan YS, Sim HS, Tan KS, Oh SK, Ng HH, Choo AB (2009) KLF4 and PBX1 directly regulate NANOG expression in human embryonic stem cells. *Stem Cells* 27:2114–2125.

Chang M, Anttonen KP, Cirillo SL, Francis KP, Cirillo JD (2014) Real-time bioluminescence imaging of mixed mycobacterial infections. *PLoS One* 9:e108341.

Cheung CY, Webb SE, Love DR, Miller AL (2011) Visualization, characterization and modulation of calcium signaling during the development of slow muscle cells in intact zebrafish embryos. *Int J Dev Biol* 55:153–174.

Chewning JH, Dugger KJ, Chaudhuri TR, Zinn KR, Weaver CT (2009) Bioluminescence-based visualization of CD4 T cell dynamics using a T lineage-specific luciferase transgenic model. *BMC Immunol* 10:44.

Cho W, Hagemann TL, Johnson DA, Johnson JA, Messing A (2009) Dual transgenic reporter mice as a tool for monitoring expression of glial fibrillary acidic protein. *J Neurochem* 110:343–351.

Chou TC, Moyle RL (2014) Synthetic versions of firefly luciferase and *Renilla* luciferase reporter genes that resist transgene silencing in sugarcane. *BMC Plant Bio* 14:92.

Class B, Thorne N, Aguisanda F, Southall N, Mckew J, Zheng W (2015) High-throughput viability assay using an autonomously bioluminescent cell line with a bacterial *lux* reporter. *J Lab Autom* 20:164–174.

Close D, Xu T, Smartt A, Rogers A, Crossley R, Price S, Ripp S, Sayler G (2012) The evolution of the bacterial luciferase gene cassette (*lux*) as a real-time bioreporter. *Sensors* 12:732–752.

Close DM, Patterson SS, Ripp S, Baek SJ, Sanseverino J, Sayler GS (2010) Autonomous bioluminescent expression of the bacterial luciferase gene cassette (*lux*) in a mammalian cell line. *PLoS One* 5:e12441.

Close DM, Hahn RE, Patterson SS, Baek SJ, Ripp SA, Sayler GS (2011) Comparison of human optimized bacterial luciferase, firefly luciferase, and green fluorescent protein for continuous imaging of cell culture and animal models. *J Biom Opt* 16:047003.

Coppola JM, Ross BD, Rehemtulla A (2008) Noninvasive imaging of apoptosis and its application in cancer therapeutics. *Clin Cancer Res* 14:2492–2501.

Coutant EP, Janin YL (2015) Synthetic routes to coelenterazine and other imidazo[1,2-a]pyrazin-3-one luciferins: Essential tools for bioluminescence-based investigations. *Chemistry* 21:17158–17171.

Crouch SP, Kozlowski R, Slater KJ, Fletcher J (1993) The use of ATP bioluminescence as a measure of cell proliferation and cytotoxicity. *J Immunol Methods* 160:81–88.

Daddysman MK, Tycon MA, Fecko CJ (2014) Photoinduced damage resulting from fluorescence imaging of live cells. *Methods Mol Biol* 1148:1–17.

De A, Ray P, Loening AM, Gambhir SS (2009) BRET3: A red-shifted bioluminescence resonance energy transfer (BRET)-based integrated platform for imaging protein-protein interactions from single live cells and living animals. *FASEB J* 23:2702–2709.

de Latouliere L, Manni I, Iacobini C, Pugliese G, Grazi GL, Perri P, Cappello P, Novelli F, Menini S, Piaggio G (2016) A bioluminescent mouse model of proliferation to highlight early stages of pancreatic cancer: A suitable tool for preclinical studies. *Ann Anat* 207:2–8.

de Wet JR, Wood KV, Helinski DR, DeLuca M (1985) Cloning of firefly luciferase cDNA and the expression of active luciferase in *Escherichia coli*. *Proc Natl Acad Sci USA* 82:7870–7873.

de Wet JR, Wood KV, DeLuca M, Helinski DR, Subramani S (1987) Firefly luciferase gene: Structure and expression in mammalian cells. *Mol Cell Biol* 7:725–737.

Deroose CM, De A, Loening AM, Chow PL, Ray P, Chatziioannou AF, Gambhir SS (2007) Multimodality imaging of tumor xenografts and metastases in mice with combined small-animal PET, small-animal CT, and bioluminescence imaging. *J Nuclear Med* 48:295–303.

Dragulescu-Andrasi A, Chan CT, De A, Massoud TF, Gambhir SS (2011) Bioluminescence resonance energy transfer (BRET) imaging of protein-protein interactions within deep tissues of living subjects. *Proc Natl Acad Sci USA* 108:12060–12065.

Engebrecht J, Nealson K, Silverman M (1983) Bacterial bioluminescence: Isolation and genetic analysis of functions from *Vibrio fischeri*. *Cell* 32:773–781.

England CG, Ehlerding EB, Cai W (2016) NanoLuc: A small luciferase is brightening up the field of bioluminescence. *Bioconjug Chem* 27:1175–1187.

Evans MS, Chaurette JP, Adams ST, Jr., Reddy GR, Paley MA, Aronin N, Prescher JA, Miller SC (2014) A synthetic luciferin improves bioluminescence imaging in live mice. *Nat Methods* 11:393–395.

Fields S, Song O (1989) A novel genetic system to detect protein-protein interactions. *Nature* 340:245–246.

Foucault ML, Thomas L, Goussard S, Branchini BR, Grillot-Courvalin C (2010) In vivo bioluminescence imaging for the study of intestinal colonization by *Escherichia coli* in mice. *Appl Environ Microbiol* 76:264–274.

Fraga H, Fernandes D, Fontes R, Esteves da Silva JC (2005) Coenzyme A affects firefly luciferase luminescence because it acts as a substrate and not as an allosteric effector. *FEBS J* 272:5206–5216.

Francis KP, Joh D, Bellinger-Kawahara C, Hawkinson MJ, Purchio TF, Contag PR (2000) Monitoring bioluminescent *Staphylococcus aureus* infections in living mice using a novel *luxABCDE* construct. *Infect Immun* 68:3594–3600.

Francis KP, Yu J, Bellinger-Kawahara C, Joh D, Hawkinson MJ, Xiao G, Purchio TF, Caparon MG, Lipsitch M, Contag PR (2001) Visualizing pneumococcal infections in the lungs of live mice using bioluminescent *Streptococcus pneumoniae* transformed with a novel gram-positive *lux* transposon. *Infect Immun* 69:3350–3358.

Germain-Desprez D, Bazinet M, Bouvier M, Aubry M (2003) Oligomerization of transcriptional intermediary factor 1 regulators and interaction with ZNF74 nuclear matrix protein revealed by bioluminescence resonance energy transfer in living cells. *J Biol Chem* 278:22367–22373.

Green AA, McElroy WD (1956) Crystalline firefly luciferase. *Biochim Biophys Acta* 20:170–176.

Gutowski MB, Wilson L, Van Gelder RN, Pepple KL (2017) In vivo bioluminescence imaging for longitudinal monitoring of inflammation in animal models of uveitis. *Invest Ophthalmol Vis Sci* 58:1521–1528.

Hall MP, Unch J, Binkowski BF, Valley MP, Butler BL, Wood MG, Otto P, Zimmerman K, Vidugiris G, Machleidt T, Robers MB, Benink HA, Eggers CT, Slater MR, Meisenheimer PL, Klaubert DH, Fan F, Encell LP, Wood KV (2012) Engineered luciferase reporter from a deep sea shrimp utilizing a novel imidazopyrazinone substrate. *ACS Chem Bio* 7:1848–1857.

Heffern MC, Park HM, Au-Yeung HY, Van de Bittner GC, Ackerman CM, Stahl A, Chang CJ (2016) In vivo bioluminescence imaging reveals copper deficiency in a murine model of nonalcoholic fatty liver disease. *Proc Natl Acad Sci USA* 113:14219–14224.

Hong H, Zhang Y, Severin GW, Yang Y, Engle JW, Niu G, Nickles RJ, Chen X, Leigh BR, Barnhart TE, Cai W (2012) Multimodality imaging of breast cancer experimental lung metastasis with bioluminescence and a monoclonal antibody dual-labeled with 89Zr and IRDye 800CW. *Mol Pharm* 9:2339–2349.

Iwanowicz LR, Blazer VS, Pinkney AE, Guy CP, Major AM, Munney K, Mierzykowski S, Lingenfelser S, Secord A, Patnode K, Kubiak TJ, Stern C, Hahn CM, Iwanowicz DD, Walsh HL, Sperry A (2016) Evidence of estrogenic endocrine disruption in smallmouth and largemouth bass inhabiting Northeast U.S. national wildlife refuge waters: A reconnaissance study. *Ecotoxicol Environ Saf* 124:50–59.

Jen Y, Weintraub H, Benezra R (1992) Overexpression of Id protein inhibits the muscle differentiation program: In vivo association of Id with E2A proteins. *Genes Dev* 6:1466–1479.

John BA, Xu T, Ripp S, Wang HR (2017) A real-time non-invasive auto-bioluminescent urinary bladder cancer xenograft model. *Mol Imaging Biol* 19:10–14.

Kaihara A, Umezawa Y, Furukawa T (2008) Bioluminescent indicators for Ca^{2+} based on split *Renilla* luciferase complementation in living cells. *Anal Sci* 24:1405–1408.

Kangas L, Gronroos M, Nieminen AL (1984) Bioluminescence of cellular ATP: A new method for evaluating cytotoxic agents in vitro. *Med Biol* 62:338–343.

Keller AF, Gravel M, Kriz J (2009) Live imaging of amyotrophic lateral sclerosis pathogenesis: Disease onset is characterized by marked induction of GFAP in schwann cells. *Glia* 57:1130–1142.

Keyaerts M, Remory I, Caveliers V, Breckpot K, Bos TJ, Poelaert J, Bossuyt A, Lahoutte T (2012) Inhibition of firefly luciferase by general anesthetics: Effect on in vitro and in vivo bioluminescence imaging. *PLoS One* 7:e30061.

Kidd S, Spaeth E, Dembinski JL, Dietrich M, Watson K, Klopp A, Battula VL, Weil M, Andreeff M, Marini FC (2009) Direct evidence of mesenchymal stem cell tropism for tumor and wounding microenvironments using in vivo bioluminescent imaging. *Stem Cells* 27:2614–2623.

Kim JB, Urban K, Cochran E, Lee S, Ang A, Rice B, Bata A, Campbell K, Coffee R, Gorodinsky A, Lu Z, Zhou H, Kishimoto TK, Lassota P (2010) Non-invasive detection of a small number of bioluminescent cancer cells in vivo. *PLoS One* 5:e9364.

Kim SB, Izumi H (2014) Functional artificial luciferases as an optical readout for bioassays. *Biochem Biophys Res Commun* 448:418–423.

Kim SB, Otani Y, Umezawa Y, Tao H (2007) Bioluminescent indicator for determining protein-protein interactions using intramolecular complementation of split click beetle luciferase. *Anal Chem* 79:4820–4826.

Kim SB, Suzuki H, Sato M, Tao H (2011) Superluminescent variants of marine luciferases for bioassays. *Anal Chem* 83:8732–8740.

Kosaka N, Mitsunaga M, Bhattacharyya S, Miller SC, Choyke PL, Kobayashi H (2011) Self-illuminating in vivo lymphatic imaging using a bioluminescence resonance energy transfer quantum dot nano-particle. *Contrast Media Mol Imaging* 6:55–59.

Kuchimaru T, Iwano S, Kiyama M, Mitsumata S, Kadonosono T, Niwa H, Maki S, Kizaka-Kondoh S (2016) A luciferin analogue generating near-infrared bioluminescence achieves highly sensitive deep-tissue imaging. *Nat Commun* 7:11856.

Lamb LE, Zhi X, Alam F, Pyzio M, Scudamore CL, Wiles S, Sriskandan S (2018) Modelling invasive group A streptococcal disease using bioluminescence. *BMC Microbiol* 18:60.

Lehmann M, Loch C, Middendorf A, Studer D, Lassen SF, Pasamontes L, van Loon AP, Wyss M (2002) The consensus concept for thermostability engineering of proteins: Further proof of concept. *Protein Eng* 15:403–411.

Li J, Chen L, Du L, Li M (2013) Cage the firefly luciferin! – a strategy for developing bioluminescent probes. *Chem Soc Rev* 42:662–676.

Liu J, Escher A (1999) Improved assay sensitivity of an engineered secreted *Renilla* luciferase. *Gene* 237:153–159.

Loening AM, Wu AM, Gambhir SS (2007) Red-shifted *Renilla reniformis* luciferase variants for imaging in living subjects. *Nat Methods* 4:641–643.

Loening AM, Fenn TD, Wu AM, Gambhir SS (2006) Consensus guided mutagenesis of *Renilla* luciferase yields enhanced stability and light output. *Protein Eng Des Sel* 19:391–400.

Lorenz WW, McCann RO, Longiaru M, Cormier MJ (1991) Isolation and expression of a cDNA encoding *Renilla reniformis* luciferase. *Proc Natl Acad Sci USA* 88:4438–4442.

Ma MS, Van Dam G, Meek M, Boddeke E, Copray S (2009) In vivo bioluminescent imaging of Schwann cells in a poly(DL-lactide-epsilon-caprolactone) nerve guide. *Muscle Nerve* 40:867–871.

Maguire CA, Bovenberg MS, Crommentuijn MH, Niers JM, Kerami M, Teng J, Sena-Esteves M, Badr CE, Tannous BA (2013) Triple bioluminescence imaging for in vivo monitoring of cellular processes. *Mol Ther Nucleic Acids* 2:e99.

Markova SV, Vysotski ES (2015) Coelenterazine-dependent luciferases. *Biochem Mosc* 80:714–732.

Marrero L, Wyczechowska D, Musto AE, Wilk A, Vashistha H, Zapata A, Walker C, Velasco-Gonzalez C, Parsons C, Wieland S, Levitt D, Reiss K, Prakash O (2014) Therapeutic efficacy of aldoxorubicin in an intracranial xenograft mouse model of human glioblastoma. *Neoplasia* 16:874–882.

Matsushita J, Inagaki S, Nishie T, Sakasai T, Tanaka J, Watanabe C, Mizutani KI, Miwa Y, Matsumoto K, Takara K, Naito H, Kidoya H, Takakura N, Nagai T, Takahashi S, Ema M (2017) Fluorescence and bioluminescence imaging of angiogenesis in Flk1-nano-lantern transgenic mice. *Sci Rep* 7:46597.

Mehta G, Hsiao AY, Ingram M, Luker GD, Takayama S (2012) Opportunities and challenges for use of tumor spheroids as models to test drug delivery and efficacy. *J Control Release* 164:192–204.

Mezzanotte L, Fazzina R, Michelini E, Tonelli R, Pession A, Branchini B, Roda A (2010) In vivo bioluminescence imaging of murine xenograft cancer models with a red-shifted thermostable luciferase. *Mol Imaging Biol* 12:406–414.

Nishihara R, Suzuki H, Hoshino E, Suganuma S, Sato M, Saitoh T, Nishiyama S, Iwasawa N, Citterio D, Suzuki K (2015) Bioluminescent coelenterazine derivatives with imidazopyrazinone C-6 extended substitution. *J Chem Soc Chem Commun* 51:391–394.

Ohmiya Y, Hirano T (1996) Shining the light: The mechanism of the bioluminescence reaction of calcium-binding photoproteins. *Chem Biol* 3:337–347.

Paulmurugan R, Gambhir SS (2005) Novel fusion protein approach for efficient high-throughput screening of small molecule-mediating protein-protein interactions in cells and living animals. *Cancer Res* 65:7413–7420.

Paulmurugan R, Umezawa Y, Gambhir SS (2002) Noninvasive imaging of protein-protein interactions in living subjects by using reporter protein complementation and reconstitution strategies. *Proc Natl Acad Sci USA* 99:15608–15613.

Pfleger KD, Eidne KA (2005) Monitoring the formation of dynamic G-protein-coupled receptor-protein complexes in living cells. *Biochemical J* 385:625–637.

Pichler A, Prior JL, Piwnica-Worms D (2004) Imaging reversal of multidrug resistance in living mice with bioluminescence: MDR1 P-glycoprotein transports coelenterazine. *Proc Natl Acad Sci USA* 101:1702–1707.

Porterfield WB, Jones KA, McCutcheon DC, Prescher JA (2015) A "caged" luciferin for imaging cell-cell contacts. *J Am Chem Soc* 137:8656–8659.

Rabinovich BA, Ye Y, Etto T, Chen JQ, Levitsky HI, Overwijk WW, Cooper LJ, Gelovani J, Hwu P (2008) Visualizing fewer than 10 mouse T cells with an enhanced firefly luciferase in immunocompetent mouse models of cancer. *Proc Natl Acad Sci USA* 105:14342–14346.

Rodda DJ, Chew JL, Lim LH, Loh YH, Wang B, Ng HH, Robson P (2005) Transcriptional regulation of nanog by OCT4 and SOX2. *J Biol Chem* 280:24731–24737.

Rumyantsev KA, Turoverov KK, Verkhusha VV (2016) Near-infrared bioluminescent proteins for two-color multimodal imaging. *Sci Rep* 6:36588.

Sambrook J, Fritsch E, Maniatis T (1990) *Molecular Cloning: A Laboratory Manual*, 2nd Edition. Cold Spring Harbor Laboratory Press, Cold Spring Harbor, New York

Sanseverino J, Gupta RK, Layton AC, Patterson SS, Ripp SA, Saidak L, Simpson ML, Schultz TW, Sayler GS (2005) Use of *Saccharomyces cerevisiae* BLYES expressing bacterial bioluminescence for rapid, sensitive detection of estrogenic compounds. *Appl Environ Microbiol* 71:4455–4460.

Sayler GS, Ripp S (2000) Field applications of genetically engineered microorganisms for bioremediation processes. *Curr Opin Biotechnol* 11:286–289.

Scabini M, Stellari F, Cappella P, Rizzitano S, Texido G, Pesenti E (2011) In vivo imaging of early stage apoptosis by measuring real-time caspase-3/7 activation. *Apoptosis* 16:198–207.

Sellmyer MA, Bronsart L, Imoto H, Contag CH, Wandless TJ, Prescher JA (2013) Visualizing cellular interactions with a generalized proximity reporter. *Proc Natl Acad Sci USA* 110:8567–8572.

Sensebe L, Fleury-Cappellesso S (2013) Biodistribution of mesenchymal stem/stromal cells in a preclinical setting. *Stem Cells Int* 2013:678063.

Shah K, Tang Y, Breakefield X, Weissleder R (2003) Real-time imaging of TRAIL-induced apoptosis of glioma tumors in vivo. *Oncogene* 22:6865–6872.

Shah K, Bureau E, Kim DE, Yang K, Tang Y, Weissleder R, Breakefield XO (2005) Glioma therapy and real-time imaging of neural precursor cell migration and tumor regression. *Ann Neurol* 57:34–41.

Stacer AC, Nyati S, Moudgil P, Iyengar R, Luker KE, Rehemtulla A, Luker GD (2013) NanoLuc reporter for dual luciferase imaging in living animals. *Mol Imaging* 12:1–13.

Sun S, Yang X, Wang Y, Shen X (2016) In vivo analysis of protein-protein interactions with bioluminescence resonance energy transfer (BRET): Progress and prospects. *Int J Mol Sci* 17: 1704.

Suzuki A, Raya A, Kawakami Y, Morita M, Matsui T, Nakashima K, Gage FH, Rodriguez-Esteban C, Izpisua Belmonte JC (2006) Nanog binds to Smad1 and blocks bone morphogenetic protein-induced differentiation of embryonic stem cells. *Proc Natl Acad Sci USA* 103:10294–10299.

Tannous BA (2009) *Gaussia* luciferase reporter assay for monitoring biological processes in culture and in vivo. *Nat Protoc* 4:582–591.

Tannous BA, Kim DE, Fernandez JL, Weissleder R, Breakefield XO (2005) Codon-optimized *Gaussia* luciferase cDNA for mammalian gene expression in culture and in vivo. *Mol Ther* 11:435–443.

Tatsumi H, Masuda T, Kajiyama N, Nakano E (1989) Luciferase cDNA from Japanese firefly, *Luciola cruciata*: Cloning, structure and expression in *Escherichia coli*. *J Biolum Chemilum* 3:75–78.

Thorne N, Inglese J, Auld DS (2010) Illuminating insights into firefly luciferase and other bioluminescent reporters used in chemical biology. *Chem Biol* 17:646–657.

Tinikul R, Chaiyen P (2016) Structure, mechanism, and mutation of bacterial luciferase. *Adv Biochem Eng Biotechnol* 154:47–74.

Troy T, Jekic-McMullen D, Sambucetti L, Rice B (2004) Quantitative comparison of the sensitivity of detection of fluorescent and bioluminescent reporters in animal models. *Mol Imaging* 3:9–23.

Van de Bittner GC, Bertozzi CR, Chang CJ (2013) Strategy for dual-analyte luciferin imaging: In vivo bioluminescence detection of hydrogen peroxide and caspase activity in a murine model of acute inflammation. *J Am Chem Soc* 135:1783–1795.

van der Bogt KE, Sheikh AY, Schrepfer S, Hoyt G, Cao F, Ransohoff KJ, Swijnenburg RJ, Pearl J, Lee A, Fischbein M, Contag CH, Robbins RC, Wu JC (2008) Comparison of different adult stem cell types for treatment of myocardial ischemia. *Circulation* 118 Supplement:S121–S129.

Varnum MM, Clayton KA, Yoshii-Kitahara A, Yonemoto G, Koro L, Ikezu S, Ikezu T (2017) A split-luciferase complementation, real-time reporting assay enables monitoring of the disease-associated transmembrane protein TREM2 in live cells. *J Biol Chem* 292:10651–10663.

Verhaegent M, Christopoulos TK (2002) Recombinant *Gaussia* luciferase. Overexpression, purification, and analytical application of a bioluminescent reporter for DNA hybridization. *Anal Chem* 74:4378–4385.

Vila OF, Martino MM, Nebuloni L, Kuhn G, Perez-Amodio S, Muller R, Hubbell JA, Rubio N, Blanco J (2014) Bioluminescent and micro-computed tomography imaging of bone repair induced by fibrin-binding growth factors. *Acta Biomater* 10:4377–4389.

Vilalta M, Jorgensen C, Degano IR, Chernajovsky Y, Gould D, Noel D, Andrades JA, Becerra J, Rubio N, Blanco J (2009) Dual luciferase labelling for non-invasive bioluminescence imaging of mesenchymal stromal cell chondrogenic differentiation in demineralized bone matrix scaffolds. *Biomaterials* 30:4986–4995.

Vilalta M, Degano IR, Bago J, Gould D, Santos M, Garcia-Arranz M, Ayats R, Fuster C, Chernajovsky Y, Garcia-Olmo D, Rubio N, Blanco J (2008) Biodistribution, long-term survival, and safety of human adipose tissue-derived mesenchymal stem cells transplanted in nude mice by high sensitivity non-invasive bioluminescence imaging. *Stem Cells Dev* 17:993–1003.

Viviani VR, Bechara EJ, Ohmiya Y (1999) Cloning, sequence analysis, and expression of active *Phrixothrix* railroad-worms luciferases: Relationship between bioluminescence spectra and primary structures. *Biochemistry* 38:8271–8279.

Webb SE, Miller AL (2012) Aequorin-based genetic approaches to visualize Ca²⁺ signaling in developing animal systems. *Biochim Biophys Acta* 1820:1160–1168.

Welsh JP, Patel KG, Manthiram K, Swartz JR (2009) Multiply mutated *Gaussia* luciferases provide prolonged and intense bioluminescence. *Biochem Biophys Res Commun* 389:563–568.

Wilson T, Hastings JW (1998) Bioluminescence. *Annu Rev Cell Dev Biol* 14:197–230.

Wood KV (1994) Firefly luciferase engineered for improved genetic reporting. *Promega Notes* 49:14–21.

Wood KV, Lam YA, Seliger HH, McElroy WD (1989) Complementary DNA coding click beetle luciferases can elicit bioluminescence of different colors. *Science* 244:700–702.

Xiong L, Shuhendler AJ, Rao J (2012) Self-luminescing BRET-FRET near-infrared dots for in vivo lymph-node mapping and tumour imaging. *Nat Commun* 3:1193.

Xu T, Ripp S, Sayler GS, Close DM (2014a) Expression of a humanized viral 2A-mediated *lux* operon efficiently generates autonomous bioluminescence in human cells. *PLoS One* 9:e96347.

Xu T, Close D, Smartt A, Ripp S, Sayler G (2014b) Detection of organic compounds with whole-cell bioluminescent bioassays. *Adv Biochem Eng Biotechnol* 144:111–151.

Xu T, Marr E, Lam H, Ripp S, Sayler G, Close D (2015) Real-time toxicity and metabolic activity tracking of human cells exposed to *Escherichia coli* O157:H7 in a mixed consortia. *Ecotoxicol* 24:2133–2140.

Xu T, Close D, Handagama W, Marr E, Sayler G, Ripp S (2016) The expanding toolbox of in vivo bioluminescent imaging. *Front Oncol* 6:150.

Yasuda K, Cline C, Lin YS, Scheib R, Ganguly S, Thirumaran RK, Chaudhry A, Kim RB, Schuetz EG (2015) In vivo imaging of human MDR1 transcription in the brain and spine of MDR1-luciferase reporter mice. *Drug Metab Dispos* 43:1646–1654.

Yu W, Hardin PE (2007) Use of firefly luciferase activity assays to monitor circadian molecular rhythms in vivo and in vitro. *Methods Mol Biol* 362:465–480.

Zabala M, Alzuguren P, Benavides C, Crettaz J, Gonzalez-Aseguinolaza G, Ortiz de Solorzano C, Gonzalez-Aparicio M, Kramer MG, Prieto J, Hernandez-Alcoceba R (2009) Evaluation of bioluminescent imaging for non-invasive monitoring of colorectal cancer progression in the liver and its response to immunogene therapy. *Mol Cancer* 8:2.

Zagozdzon AM, O'Leary P, Callanan JJ, Crown J, Gallagher WM, Zagozdzon R (2012) Generation of a new bioluminescent model for visualisation of mammary tumour development in transgenic mice. *BMC Cancer* 12:209.

Zhang L, Hellstrom KE, Chen L (1994) Luciferase activity as a marker of tumor burden and as an indicator of tumor response to antineoplastic therapy in vivo. *Clin Exp Metastasis* 12:87–92.

6

Macroscopic Fluorescence Imaging

Anand T. N. Kumar

CONTENTS

6.1 Introduction

Fluorescence imaging has played an important role in biological and clinical studies (Lakowicz, 1999), from microscopic to macroscopic length scales. This chapter will discuss fluorescence imaging in whole living organisms, or macroscopic fluorescence imaging (MFI). MFI involves light propagation in biological tissue that is several millimeters thick and is ideally performed using near infrared light (NIR)($\sim 650-850$ nm) because of the lower absorption and scattering of tissue in this spectral region. An important requirement for the successful biological application of MFI is the availability of intrinsic or extrinsic fluorescent markers that can specifically label a particular disease or biological component of interest. MFI can be performed using three types of detection techniques: steady state or continuous wave (CW) (Graves et al., 2003), frequency domain (FD) (Godavarty et al., 2005), or time domain (TD) (Kumar et al., 2008). Each of these methods has unique advantages and limitations, and the choice of which technique to use depends on the particular application of interest and the resources available to the researcher. By far, the most commonly used technique is CW fluorescence imaging, as it offers the cheapest and fastest approach to measure fluorescence in a whole subject. A major drawback of CW, however, is the inability to detect fluorescence lifetimes, which can be an important functional indicator (Lakowicz, 1999) as well as aid in multiplexing. Both FD and TD methods can detect lifetimes but are more expensive and complex than CW systems. Irrespective of the temporal detection technique used, MFI measurements can be either performed in planar reflectance or tomographic detection mode. Another aspect that can be used irrespective of the detection technique is multispectral detection (Xu and Rice, 2009; Pu et al., 2014), which allows for multiplexing using several fluorophores. Multiplexing is an important and unique advantage of optical methods and is primarily performed though either fluorescence spectrum or lifetime. We will introduce a general mathematical framework that allows direct comparison of spectral and lifetime multiplexing. The outline of the chapter is as follows: We first present an overview of experimental techniques used in MFI. Next, we detail the theoretical foundations of tomographic MFI, including the forward and inverse problems. The chapter will conclude with a discussion of some *in vivo* applications of MFI.

6.2 Experimental Techniques

We first present an overview of the various experimental schemes used for macroscopic fluorescence imaging. The schemes can be broadly classified based on the type of spatial and temporal measurements used for illumination (also called "excitation") of the sample and detection of the light emerging from the sample. Figure 6.1 shows a schematic of a typical experimental arrangement for MFI, illustrating the various possibilities for excitation and detection. Excitation can be either based on point-scanning (using for e.g. a microscopic objective or a collimation package), which is most ideally suited for tomographic acquisition, or it can employ spatially patterned (Cuccia et al., 2005), wide-field illumination (e.g. using digital micrometer devices), which is suitable for both planar imaging and depth-resolved three-dimensional (3D) tomography (Ducros et al., 2013). There are many advantages of a wide-field setup, including faster imaging speed and improved dynamic range performance (Venugopal et al., 2010; Venugopal and Intes, 2013). Wide-field structured illumination also allows the ability to selectively eliminate low spatial frequencies that are scattered more than high spatial frequencies, thereby allowing detection of short fluorescence lifetimes in a turbid medium (Kumar, 2013; Kumar et al., 2018).

Measurements can be made in the CW, FD, or TD modes. CW imaging uses light emitting diodes (LEDs), diode lasers, or simple white light sources for excitation. Detection is performed using multiple fiber-based detectors (such as photodiodes) in contact with the subject or using charge-coupled device (CCD) cameras that allow noncontact detection and arbitrary assignment of detectors as pixels on the CCD image. FD imaging uses MHz modulated excitation sources and detectors (Godavarty et al., 2005). TD imaging, which offers the most comprehensive information of all three methods, utilizes pulsed laser sources such as picosecond laser diodes, Ti:Sapphire lasers, or supercontinuum lasers (such as the SuperK laser series from NKT photonics). TD detection is performed using either time-correlated single-photon counting (TCSPC) method (Becker et al., 2006; Brambilla et al., 2008), which is suited for point measurements, or using widefield cameras such as gated image intensified CCDs (Turner et al., 2005; Kumar et al., 2008) (ICCD, e.g. PicoStar HRI, LaVision). Recently, an imaging platform based on time-resolved structured light and hyperspectral, single-pixel detection was demonstrated for macroscopic fluorescence lifetime imaging over a large field of view and multiple wavelengths (Pian et al. 2017). The use of single-pixel detection for TD imaging avoids the use of expensive intensified CCD (ICCD) cameras, and may be essential in spectral regions, such as the short wave IR (~ 1000 nm–1700 nm), where wide-field time-gated cameras are currently not available.

Irrespective of the type of temporal and spatial excitation or detection employed, MFI can be carried out either in the reflectance or in the transmission mode (Figure 6.1). Reflectance imaging, also called planar imaging, lacks depth information and can provide information to within only a few mm from the surface of the subject. However, reflectance imaging can by itself sometimes provide useful diagnostic information (Goergen et al., 2012; Klohs et al., 2009; Abulrob et al., 2007) although this information is usually neither quantitative nor depth-resolved. The low cost and ease of use of planar imaging systems makes them an attractive option for high-throughput imaging applications. Transmission imaging is essential to collect a tomographic

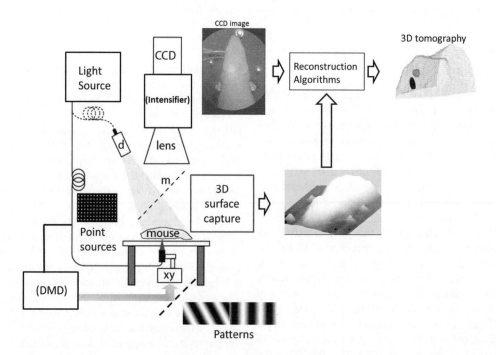

FIGURE 6.1   Schematic of a typical experimental setup for macroscopic fluorescence imaging. The source of excitation light is either a steady state or pulsed light source and is delivered to the sample either using a fiber (for point excitation) or a digital micromirror device (DMD) for patterned excitation. Detection is performed either using a CCD camera alone (CW imaging) or a CCD in conjunction with an image intensifier (TD imaging). For reflectance measurements, light is delivered either via the DMD or a diffuser (d). The 3D volume of the sample is acquired using a 3D surface capture device and is used in reconstruction algorithms to obtain tomographic images of fluorescence within the subject.

set of data points that will provide maximal spatial information regarding the imaging subject. Although complete angle systems would be ideal and have been developed previously, there is limited improvement in using complete angle coverage over fixed-angle tomography (Raymond et al. 2009). This can be attributed to the fact that NIR photons are highly scattered in tissue and do not travel straight so that a single detection angle with a sufficient number of excitation sources can detect photons that sample the entire 3D volume. An important requirement for tomography is the postprocessing and image reconstruction, which involves the use of mathematical algorithms to "invert" the raw fluorescence measurements to obtain the internal distribution of fluorophore(s) in the medium. The sections below will present further details behind the forward and inverse problems for fluorescence tomography.

An interesting question is the spatial resolution capability of tomographic fluorescence imaging. Although no systematic studies have been reported to date characterizing the resolution of fluorescence tomography across various length scales, submillimeter resolution has been reported using CW imaging in small animal-size phantoms (1.5 cm thickness) (Graves et al., 2003). However, this does not reflect true spatial resolution because the study used the edge-to-edge separation between 3 mm inner diameter tubes as a measure of resolution. A more correct measure of spatial resolution would be the center-to-center separation with much smaller targets (< 1 mm diameter). (Note that the center-to-center separation of the tubes was about 4.7 mm in Graves et al. (2003).) The use of smaller inclusions would provide a better approximation to the true point-spread function (Bertero and Boccacci, 1998). The actual spatial resolution for a medium thickness of 1.5 cm, and using NIR CW fluorescence tomography is about 4–5 mm and reflects the intrinsic resolution of diffuse optics in the NIR. When targets exhibit distinct fluorescence lifetimes, TD fluorescence tomography can dramatically improve on the intrinsic resolution limitation of diffuse optics, allowing the separation of fluorescent targets as close as 1.4 mm (Rice et al., 2013).

6.3 Theoretical Methods

The theoretical modeling of fluorescence in tissue is primarily of concern for tomographic imaging, which requires a detailed mathematical treatment of light propagation through the medium of interest (Arridge, 1999). There are two steps to performing fluorescence tomography reconstructions. The first is a physical model to describe light propagation through tissue (the "forward problem"). The second is the inversion step (the "inverse problem"), consisting of mathematical steps to invert the measured tomographic fluorescence data to obtain the 3D distributions of fluorophore properties in the medium. We will detail these individual steps further below.

6.3.1 The Forward Problem

6.3.1.1 The General TD Forward Problem

Because TD methods offer the most comprehensive optical information regarding a turbid medium, we will present the theoretical results below from the point of view of TD

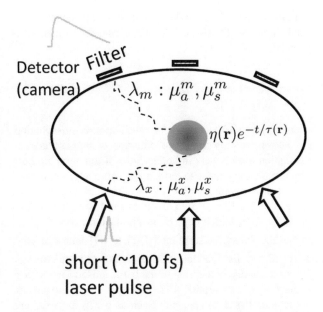

FIGURE 6.2 Schematic illustrating the steps involved in modeling fluorescence propagation in a macroscopic biological medium. The three stages of fluorescence signal generation are, propagation of excitation light from sources on the boundary (arrows) at wavelength λ_x, emission of fluorescence by a fluorophore with yield distribution of $\eta(\mathbf{r})$ and lifetime distribution of $\tau(\mathbf{r})$, where $\mathbf{r}$ the spatial coordinate, and propagation of the emitted fluorescence to the detectors on the surface, with emission filters at wavelength λ_m (rectangles). Note that the reemission of the fluorescence is ignored, but reabsorption of fluorescence by the fluorophores can be accounted for through the Green's functions. (μ_a^x, μ_s^x) and (μ_a^m, μ_s^m) denote the optical absorption and scattering distributions of the medium (which can include the fluorophore absorption) at the excitation and emission wavelengths.

measurements, and point to the CW and FD expressions along the way. The most general description of light transport in tissue is using the equation of radiative transfer (Chandrasekhar, 2013) and its various approximations (Arridge, 1999). A typical tomography measurement involves optical sources and detectors placed on the boundary of the imaging specimen (Figure 6.2). The detected fluorescence can be described as a sequential propagation of the excitation light from the source(s) to the fluorophore, fluorophore emission, and propagation of the emission field from the fluorophore to the detector. This is described using coupled equations for light transport at the excitation and emission wavelengths. Let the source and detector locations be $\mathbf{r}_s$ and $\mathbf{r}_d$. Let $\eta(\mathbf{r})$ represent the yield distribution (product of the quantum yield Q, concentration, and extinction coefficient) of the fluorophore, with $\mathbf{r}$ denoting the location of a point within the medium ("called a medium 'voxel'"). The expression for the detected fluorescence in the time domain (TD) can then be written as a double convolution of the excitation ($G^x(\mathbf{r}, \mathbf{r}_s, t)$) and emission ($G^m(\mathbf{r}_d, \mathbf{r}, t)$) Green's functions with the fluorescence decay term ($e^{-t/\tau(\mathbf{r})}$):

$$U(\mathbf{r}_s, \mathbf{r}_d, t) = \Theta \int_\Omega d^3r \, W(\mathbf{r}_s, \mathbf{r}_d, \mathbf{r}, t)\eta(\mathbf{r}), \qquad (6.1)$$

where Θ is a scaling factor that incorporates experimental constants such as source and detector coefficients, geometrical factors, and fluorescence filter attenuation. For simplicity, we will assume that $\Theta = 1$ for the remainder of this chapter. The weight function W (also called the "sensitivity" function) is given by:

$$W(\mathbf{r}_s, \mathbf{r}_d, \mathbf{r}, t) = \int_0^t dt' \int_0^{t'} dt'' G^m(\mathbf{r}_d, \mathbf{r}, t-t') e^{-\Gamma(\mathbf{r})(t'-t'')} G^x(\mathbf{r}, \mathbf{r}_s, t''), \quad (6.2)$$

where $\tau(\mathbf{r}) = 1/\Gamma(\mathbf{r})$ is the fluorescence lifetime distribution. The above equation neglects reemission of fluorescence, an assumption used widely in applications of tomographic fluorescence imaging and also termed the "Born approximation." Besides this approximation, the accuracy of Equation 6.2 depends on the level of rigor used for evaluating the Green's functions, $G^{x,m}$, which depend on the intrinsic tissue optical properties, namely absorption $\left(\mu_a^x(\mathbf{r}), \mu_a^m(\mathbf{r})\right)$ and scattering $\left(\mu_s^x(\mathbf{r}), \mu_s^m(\mathbf{r})\right)$ distributions at the excitation (λ_x) and emission (λ_m) wavelengths, in addition to the tissue anisotropy factor g. In general, the absorption and scattering are heterogeneous and include tissue components (such as water, melanin, and blood) and the absorption of the fluorophore (at both λ_x and λ_m). Generally, all the parameters, $\eta(\mathbf{r})$, $\tau(\mathbf{r})$, $\mu_a^{(x,m)}(\mathbf{r})$, $\mu_s^{(x,m)}(\mathbf{r})$ are unknown. A common starting point is the homogeneous approximation where the optical properties are assumed uniform throughout and fluorophore absorption is ignored for evaluating the Green's functions. In this case the Green's functions in Equation 6.2 are the solutions to the homogeneous diffusion or transport equations. Note that ($\mu_a^{(x,m)}$ and $\mu_s^{(x,m)}$) can, in practice, be determined independently using two separate "excitation" measurements at wavelengths λ_x and λ_m. In this case, Equations 6.1–6.2 can provide a highly accurate description of time resolved fluorescence in turbid media. With $\mu_a^{(x,m)}$ and $\mu_s^{(x,m)}$ known, the GFs can be calculated either using the diffusion approximation or the radiative transport equation.

6.3.1.2 Frequency Domain

Evaluation of the full TD sensitivity function in Equation 6.2 can become computationally intractable for a typical scenario with multiple source-detector (S-D) pairs ($\sim 10^2$–10^3), medium voxels ($\sim 10^4$), and time points ($\sim 10^3$). One simplifying approach is to solve the problem in frequency (Fourier) domain (FD), as the double convolution then simplifies to a product for each frequency so that the TD fluorescence in Equation 6.1 can be expressed as a Fourier integral:

$$U(\mathbf{r}_s, \mathbf{r}_d, t) = \int_\Omega d^3r \int_{-\infty}^{\infty} d\omega e^{-i\omega t} \widetilde{W}^B(\mathbf{r}_s, \mathbf{r}_d, \mathbf{r}, \omega) \underbrace{\left[\sum_n \frac{\tau_n \eta_n(\mathbf{r})}{(1-i\omega\tau_n)}\right]}_{F(\mathbf{r},\omega)}, \quad (6.3)$$

where

$$\widetilde{W}^B(\mathbf{r}_s, \mathbf{r}_d, \mathbf{r}, \omega) = \widetilde{G}^x(\mathbf{r}_s, \mathbf{r}, \omega)\widetilde{G}^m(\mathbf{r}_d, \mathbf{r}, \omega) \quad (6.4)$$

is the FD sensitivity obtained as a Fourier transform of the TD sensitivity (Equation 6.2), with ω as the modulation frequency. The standard approach in FD fluorescence tomography (Oleary et al., 1996) is to reconstruct the spatial

distribution of $F(\mathbf{r}, \omega)$ (quantity within the square brackets in Equation 6.3) from the FD measurements at a given frequency ω. The lifetime and yield distributions are obtained as the phase and the real part of $F(\mathbf{r}, \omega)$, respectively. Although this offers simplification of the problem (Soloviev et al., 2004, 2007, Nothdurft et al., 2009), FD modeling of TD data has the limitation that the Fourier transform is nontrivial for multiexponential decays and multiple frequencies are required to reliably extract multiple lifetimes or complex decay profiles. As clear from Equation 6.3, the TD data inherently contains all frequencies. However, handling multiple frequencies with the above forward problem becomes complicated because $F(\mathbf{r}, \omega)$ inseparably involves both a measurement parameter ω and the unknown lifetime $\tau(\mathbf{r})$. This necessitates a nonlinear approach as described in Milstein et al. (2003). However, it has also been shown that multiple frequencies do not necessarily improve the quality of the reconstruction (Milstein et al. 2004), with the number of useful frequencies restricted to the first three or four frequency components from zero. Other approaches to directly analyze TD data use overly simplistic models (such as assuming point fluorophores in an infinite homogeneous medium) (Hall et al., 2004), while the general formalism is quite intractable (Arridge and Schotland, 2009).

6.3.1.3 The Asymptotic Limit

An elegant simplification of the TD fluorescence problem can be achieved based on experimental observation that typical fluorescence lifetimes (τ) of fluorophores from the visible to NIR wavelengths are longer than the timescales for intrinsic timescales for diffuse light propagation in small volumes, τ_D (which is shorter than the absorption time scale $\tau_a = (v\mu_a)^{-1}$) (Haselgrove et al., 1992). Under the approximation that $\tau > \tau_D$, Equations 6.1–6.2 can be cast into an elegant and rigorous form that expresses the TD fluorescence (Kumar et al., 2005) as a multiexponential model. This tomographic fluorescence lifetime imaging (FLIM) approach can be derived in both the frequency domain, using complex integration, and directly in time domain from Equations 6.1–6.2. Before proceeding, it is convenient to recast Equations 6.1–6.2 in the following way, using the commutativity of the convolution:

$$U(\mathbf{r}_s, \mathbf{r}_d, t) = \int_\Omega d^3r \int_0^t dt' W^B(\mathbf{r}_s, \mathbf{r}_d, \mathbf{r}, t') \left[\sum_n e^{-\Gamma_n(t-t')}\eta_n(\mathbf{r})\right], \quad (6.5)$$

where we have defined a "background" weight function as (Fourier transform of $\widetilde{W}^B$ in Equation 6.4):

$$W^B(\mathbf{r}_s, \mathbf{r}_d, \mathbf{r}, t') = \int_0^{t'} dt'' G^x(\mathbf{r}_s, \mathbf{r}, t'-t'') G^m(\mathbf{r}, \mathbf{r}_d, t''), \quad (6.6)$$

and the $\Gamma_n = 1/\tau_n$ are discretized values of the *in vivo* lifetime distribution with corresponding yield distributions $\eta_n(\mathbf{r})$. In other words, each lifetime has a distinct yield distribution. Using the above equations, it can be shown that the TD forward problem in Equations 6.1–6.2 reduce to an elegant multiexponential form:

$$U(\mathbf{r}_s, \mathbf{r}_d, t) = \sum_n A_n(\mathbf{r}_s, \mathbf{r}_d, t) e^{-\Gamma_n t} \qquad (6.7)$$

with time dependent decay amplitudes A_n, given by:

$$A_n(\mathbf{r}_s, \mathbf{r}_d, t) = \int d^3 r \left[\int_0^t dt' W_n^B(\mathbf{r}_s, \mathbf{r}_d, \mathbf{r}, t') \right] \eta_n(\mathbf{r}). \qquad (6.8)$$

Equation 6.8 is a generalized multiexponential forward problem for tomographic FLIM that includes both early and late arriving photons and is rigorous within the radiative transport model of photon propagation in turbid media. The time dependence of the decay amplitudes A_n reflect the evolution of the background diffusive response. As shown previously (Kumar et al., 2006) (see Figure 6.3), $A(t)$ rapidly reaches a constant value, beyond which the temporal evolution of $U(\mathbf{r}_s, \mathbf{r}_d, t)$ is purely exponential. For long times such that $t \gg \tau_D$, the weight function $\widetilde{W}_n^B$ is just the CW (or time integrated) version of W_n^B so that $A_n(\mathbf{r}_s, \mathbf{r}_d, t) \to a_n(\mathbf{r}_s, \mathbf{r}_d)$ in the asymptotic limit. The time constant for the rise of $A(t)$ toward a_n (ref) will depend on the intrinsic tissue absorption, scattering, and size of the imaging volume. We define the "asymptotic regime" for times when the time-average of W_n^B (integrand of $A(t)$) in Equation 6.8 will become nearly time independent and approach the CW sensitivity function, which we denote by $\overline{W}_n^B \left(= \widetilde{W}_n^B(-i\Gamma_n) \right)$. The TD fluorescence signal in the asymptotic limit, therefore, becomes:

$$U(\mathbf{r}_s, \mathbf{r}_d, t) \xrightarrow{t > \tau_D} \sum_n e^{-\Gamma_n t} a_n(\mathbf{r}_s, \mathbf{r}_d) \qquad (6.9)$$

where the time-independent decay amplitudes a_n are given by:

$$a_n(\mathbf{r}_s, \mathbf{r}_d) = \int d_3 r \overline{W}_n^B(\mathbf{r}_s, \mathbf{r}_d, \mathbf{r}) \eta_n(\mathbf{r}). \qquad (6.10)$$

We will see below that the above form of TD fluorescence forward problem offers a powerful way for quantitative tomographic multiplexing and provides superior quantitation compared to both a direct inversion of TD data and spectral multiplexing.

6.3.1.4 Spatial Frequency Domain

All the results presented above can be expressed in the spatial frequency domain (SFD). Particularly, the TD fluorescence signal in the asymptotic region is factorized into functions of spatial-frequency and time (analogous to Equation 6.9) as:

$$U_F(\mathbf{k}_s, \mathbf{k}_d, t) \xrightarrow{t > \tau_a} \sum_{n=1}^N a_n(\mathbf{k}_s, \mathbf{k}_d) e^{-\Gamma_n t}, \qquad (6.11)$$

where $\mathbf{k}_s$, $\mathbf{k}_d$ are the source and detector spatial frequencies (SF) on the boundary of the medium and a_n is time-independent decay amplitudes in SFD, which are related to corresponding yield distributions η_n, analogous to Equation 6.10:

$$a_n(\mathbf{k}_s, \mathbf{k}_d) = \int d^3 r \overline{W}_n^B(\mathbf{k}_s, \mathbf{k}_d, \mathbf{r}) \eta_n(\mathbf{r}), \qquad (6.12)$$

where $\overline{W}_n^B = G_n^x(\mathbf{k}_s, \mathbf{r}) G_n^m(\mathbf{r}, \mathbf{k}_d)$ is a continuous wave (CW) fluorescence weight matrix evaluated as the product of transport Green's functions in SFD, $G_n^x(\mathbf{k}_s, \mathbf{r})$ and $G_n^m(\mathbf{r}, \mathbf{k}_d)$, computed at a reduced absorption $\mu_a^{x,m}(\mathbf{r}) - \Gamma_n/v$. An important advantage of SFD for TD measurements is that high spatial frequencies effectively attenuate highly scattered, late arriving

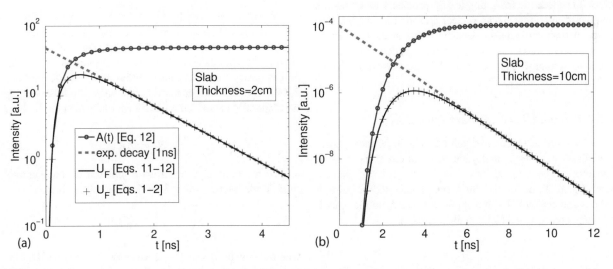

FIGURE 6.3 Simulations to elucidate the diffuse and pure fluorescent decay components as revealed by the time domain fluorescence model presented in Equations 6.1 and 6.5. The medium was an infinite slab of thickness 2 cm (left panel) and 10 cm (right panel), with optical properties $\mu_s^x = \mu_s^m = 10/\text{cm}$, $\mu_a^x = \mu_a^m = 0.1/\text{cm}$. The fluorescence signal was calculated for a single source detector pair with a small fluorescent inclusion at the center. The signal calculated using the full TD forward problem in Equation 6.1 (+ symbol) is compared with that calculated using an effective absorption-based model, viz. Equations 6.5 and 6.8 (solid black line). The time dependent decay amplitude, $A(t)$ (dotted line), and pure fluorescence decay (dashed line) are also delineated for both cases.

photons. This effectively increases the absorption (Bassi et al., 2008), thereby lowering τ_d, allowing the detection of shorter lifetimes than possible with the use of point illumination (Kumar, 2013). The exclusion of low spatial frequencies from the image reconstruction has been shown to improve quantitation and localization accuracy when using fluorophores with subnanosecond lifetimes (Kumar et al., 2018), which is typical for most fluorophores in the NIR spectral range.

6.3.2 The Inverse Problem

The general inverse problem in fluorescence tomography consists of determining the fluorescence yield and lifetime distributions (and the background tissue optical properties, if they are not known) from measured fluorescence and excitation data. In the previous section, we presented the TD fluorescence forward problem both in its complete form (Equations 6.1–6.2), as well as the multiexponential form in the asymptotic limit in spatial domain (Equations 6.9–6.10) and spatial frequency domain (Equations 6.11–6.12). This suggests two ways to perform inversion of TD data in the asymptotic region. The first way is to invert the full TD data to recover the lifetime ($\tau(\mathbf{r})$) and yield ($\eta(\mathbf{r})$) distributions using Equations 6.1–6.2. The second way involves a two-step approach that first recovers the decay amplitudes, a_n, for each discretized lifetime component, τ_n, using multiexponential fits to the asymptotic portion of the TD fluorescence. In the second step, the yield distribution for each lifetime is recovered using Equation 6.10 (or Equation 6.12 in SFD). The natural question is how these two methods for inversion compare in performance for reconstructing the 3D distributions of single or multiple fluorophores present within a turbid medium. In order to address this question, it is useful to cast the mathematical formulation of the forward and inverse problems in a matrix form. We will see below that the asymptotic method offers significantly improved quantitation and localization performance over the direct TD methods for multiplexing problems and that the matrix formulation provides a clear and rigorous explanation for the distinct performance between the methods. For further background related to the material in this section, we refer the reader to standard texts on general linear inverse problems (Bertero and Boccacci, 1998).

6.3.2.1 Direct Time Domain Inversion

Consider a turbid medium of volume Ω with N fluorophores of distinct lifetimes τ_n and yield distributions $\eta_n(\mathbf{r})$, $\mathbf{r} \in \Omega$. Discretizing the medium into V voxels, the full TD fluorescence signal (Equation 6.1) for L time points and M pairs of sources and detectors located at points $\mathbf{r}_s$ and $\mathbf{r}_d$ on the boundary can be represented by the matrix equation:

$$y = W\eta, \qquad (6.13)$$

where y is a ($ML \times 1$) measurement vector, $W = [W_1, \ldots, W_N]$ is the ($ML \times NV$) TD weight matrix (Equation 6.2) and $\eta = [\eta_1, \ldots, \eta_N]^T$ is a ($NV \times 1$) vector of unknown fluorescence yields of all lifetimes. The central quantities of interest are the fluorescence yield distributions, $\eta_n(\mathbf{r})$, of each fluorophore with

the assumption that the discrete lifetimes τ_n are either known or are independently retrieved through global analysis methods (Raymond et al., 2010). A straight-forward solution to the above linear problem, which we refer to as the direct TD (DTD) approach, uses Tikhonov inversion of Equation 6.13, resulting in a reconstructed yield distribution:

$$\hat{\eta}_{DTD} = \widehat{W}_{DTD} y, \qquad (6.14)$$

where the inverse operator $\widehat{W}_{DTD}$ is given by the standard Tikhonov expression with regularization parameter λ:

$$\widehat{W}_{DTD} = W^T(WW^T + \lambda I)^{-1}. \qquad (6.15)$$

The DTD inversion can be interpreted statistically by using the well-known connection between Tikhonov regularization and Bayesian inversion (Bertero and Boccacci, 1998). Given common assumptions that the measurement noise, n and unknown yield, η can be modeled as white Gaussian random vectors, η_{DTD} is equivalent to the minimum mean square error (MMSE) solution if the regularization parameter is chosen as $\lambda = (\sigma_n/\sigma_\eta)^2$, where σ_n and σ_η are the variances of the measurement noise and yield, respectively. Although providing a MMSE solution, DTD does not explicitly restrict the cross talk between the η_ns and can lead to severe cross talk between yield distributions of distinct lifetimes. The cross talk translates into poor spatial localization and quantitation (Hou et al., 2016b). The DTD approach is thus not the method of choice for multiplexing applications in which the relative amounts of multiple overlapping parameters (which, in the present case, are the yield distributions) are of interest. Therefore, the DTD approach is ideal if only a single fluorophore of interest is present in the medium.

6.3.2.2 Asymptotic Time Domain Inversion

An alternate approach to TD fluorescence tomography is to consider the TD forward problem in the asymptotic limit, i.e. Equation 6.10, which is valid for times longer than the intrinsic diffuse timescale τ_D of the medium (Kumar et al., 2006) and assuming the widely held condition, $\tau_n > \tau_D$ (Kumar et al., 2005). The TD weight matrix now takes the following spatiotemporally factorized matrix form in the asymptotic limit:

$$W \overset{t \gg \tau_D}{=} A\overline{W} \qquad (6.16)$$

and correspondingly, the forward problem in the asymptotic limit, Equation 6.9, takes the following matrix form:

$$y \overset{t \gg \tau_D}{=} A\overline{W}\eta, \qquad (6.17)$$

where $A = [\exp(-t/\tau_1) \times I, \ldots, \exp(-t/\tau_N) \times I]$ is a ($ML \times MN$) basis matrix of exponential decays, I is a ($M \times M$) identity matrix, and $\overline{W} = \mathrm{diag}(\overline{W}_n)$ is a ($MN \times V\,N$) block diagonal matrix containing CW weight matrices, $\overline{W}_n^B$, which are evaluated with the background absorption reduced by $1/v\tau_n$ (Kumar et al., 2006).

It is now easy to see that a simple manipulation of Equation 6.17 can allow a fundamentally different approach to solving the TD inverse problem. Since A is a well-conditioned matrix, it can be first inverted without regularization by multiplication with its Moore–Penrose pseudoinverse $A^\dagger$. The left hand side of Equation 6.17, after premultiplication with $A^\dagger$, is then simply a linear least squares solution for a multiexponential analysis of U. In other words, $A^\dagger y = a$, where $a = [a_1, \ldots, a_N]^T$ is a $(MN \times 1)$ vector of decay amplitudes for all source/detector pairs and lifetimes. In the next step, we apply Tikhonov regularization to invert the CW sensitivity matrix $\overline{W}$:

$$\eta^{ATD} = \overline{W}^T (\overline{W}\,\overline{W}^T + \lambda I)^{-1} A^\dagger y. \qquad (6.18)$$

This two-step inversion is a matrix representation of the previously derived asymptotic TD (ATD) approach (Kumar et al., 2005, 2006, 2008). Analogous to the DTD inverse operator, Equation 6.15, we can write the ATD inverse operator as:

$$W^{ATD} = \overline{W}^T (\overline{W}\,\overline{W}^T + \lambda I)^{-1} A^\dagger. \qquad (6.19)$$

It should be noted that although the forward problems of DTD (Equations 6.1–6.2) and ATD (Equations 6.9–6.10) are equivalent in the asymptotic regime, the corresponding inverse problems for direct TD (Equation 6.26) and ATD (Equation 6.19) are distinct and will produce different reconstructions even when applied to the same measurement y. The key aspect of the ATD approach is that the basis matrix A is removed from the regularization step. This ensures that the measurements in y are directly separated using the exponential basis function of each fluorophore. Additionally, given the block diagonal nature of $\overline{W}$, Equation 6.18 essentially reduces to completely separate inverse problems for each yield distribution η_n. This should be contrasted with the DTD inversion in Equation 6.14, in which the inverse problem for each η_n is not separable. Consequently, the ATD approach results in significantly lower cross talk between the yields of multiple lifetimes than the DTD approach.

6.3.2.3 Optimal Estimator

While it is clear that the ATD approach provides zero cross talk between multiple lifetime components, the DTD approach provides the least error solution. A natural question is whether there exist linear estimators that can provide zero cross talk while also providing the least error. We call this the "optimal estimator" because it combines a minimum error solution with optimal cross-talk performance. Such an approach was presented by Hou et al. (2016b). We will briefly summarize the key steps behind the derivation of the optimal estimator.

As a first step, we provide a rigorous definition of interparameter cross talk using model resolution matrix, which takes the following form for a general linear imaging system with forward operator W and its inverse operator (estimator) $\widehat{W}$:

$$R = \widehat{W} W. \qquad (6.20)$$

For single parameter problems, R can be interpreted in the standard way (Bertero and Boccacci, 1998), i.e. the c'th column of

R represents the point spread function for the c'th voxel, and the j'th row represents the contribution at the j'th voxel due to all the other voxels in the medium. To see the usefulness of the resolution matrix for the multiparameter case, we consider the case of two lifetimes, τ_1 and τ_2 with corresponding yield distributions η_1 and η_2, although the results can be readily generalized to any number of parameters. Both W and $\widehat{W}$ can be split into two separate submatrices for each FL:

$$W = \begin{bmatrix} W_1 & W_2 \end{bmatrix}, \; \widehat{W} = \begin{bmatrix} \widehat{W}_1 \\ \widehat{W}_2 \end{bmatrix}, \qquad 6.(21)$$

where W_1 and W_2 are each of dimension $ML \times V$. The resolution matrix R is then a block matrix with four quadrants:

$$R = \begin{bmatrix} \widehat{W}_1 W_1 & \widehat{W}_1 W_2 \\ \widehat{W}_2 W_1 & \widehat{W}_2 W_2 \end{bmatrix} \equiv \begin{bmatrix} R_{11} & R_{12} \\ R_{21} & R_{22} \end{bmatrix}. \qquad (6.22)$$

The diagonal blocks, R_{11} and R_{22} can be interpreted similarly to the single fluorophore-resolution matrix as described above. The off-diagonal blocks have special significance for multiplexing. The columns of R_{12} can be interpreted as the cross talk into the η_1 distribution from a point inclusion with FL τ_2 (and vice versa for R_{21}). Because R_{12} and R_{21} provide a complete and quantitative measure of the intuitive notion of cross talk, we directly incorporate these terms into the optimization problem.

The optimal estimator, also called the cross-talk constrained TD (CCTD) estimator is obtained as an MMSE solution with an imposed zero-cross-talk constraint on R_{12} and R_{21}. Let the first and second order moments for noise, n and η be given by $E[n]=0$, $E[\eta]=0$, $cov(n)=C_n$, $cov(\eta)=C_\eta$. The optimization problem takes the form:

$$\widehat{W}_{CCTD} = \arg\min_{\widehat{W}} E \left[\| \eta - \hat{\eta} \|^2 \right] \qquad (6.23)$$

with the constraints:

$$R_{12} = 0 \text{ and } R_{21} = 0. \qquad (6.24)$$

The above optimization problem can be categorized as a quadratic programming problem with linear equality constraints and can be solved analytically. Assuming the factorization in Equation 6.16 and model and data covariance matrices, $C_\eta = \sigma_\eta^2 I$ and $C_n = \sigma_n^2 I$, the CCTD estimator for the TD fluorescence problem in the asymptotic region takes the final form:

$$\widehat{W}_{CCTD} = \overline{W}^T \left(\overline{W}\,\overline{W}^T + \lambda \frac{\text{diag}(\mathbf{C}_a)}{C_n} \right)^\dagger A^\dagger, \qquad (6.25)$$

Where $\mathbf{C}_a = (A^T A / \sigma_n^2)^{-1}$, $A^\dagger$ is the Moore–Penrose pseudoinverse (Press et al., 1996) of the well-conditioned matrix A, $diag(X)$ sets all off-diagonal blocks of a matrix X to zero, and $\lambda = (\sigma_n/\sigma_\eta)^2$. Equation 6.25 offers a compact expression for a novel estimator for tomographic FL multiplexing that achieves zero cross talk between multiple lifetimes while also minimizing MSE. The matrix $\mathbf{C}_a$ is immediately recognized from linear regression theory (Press et al., 1996) as the covariance matrix for the decay amplitudes obtained using linear fitting with basis functions A. The diagonal terms of $\mathbf{C}_a$ are the uncertainties of each amplitude, while the off-diagonal terms correspond to the covariances between the amplitudes for distinct lifetimes. Equation 6.25, therefore, has the remarkable interpretation that the optimal estimator achieves zero cross talk by simply setting the off-diagonal elements of the decay amplitude covariance, $\mathbf{C}_a$, to zero. To further appreciate the significance of this result, we write the DTD inverse operator (Equation 6.15) in the asymptotic region, by applying the asymptotic factorization in Equation 6.16:

$$\widehat{W}_{\mathrm{DTD}} = \overline{W}^T \left(\overline{WW}^T + \lambda \frac{(\mathbf{C}_a)}{C_n} \right)^{-1} A^\dagger. \qquad (6.26)$$

Direct comparison of Equations 6.25 and 6.26 shows that the only difference between the optimal estimator and the DTD approach is that the optimal estimator sets the off-diagonal blocks of the matrix of the $\mathbf{C}_a$ to zero, while the DTD retains the full covariance of the decay amplitudes, thereby resulting in higher cross talk between multiple lifetimes. It is now interesting to compare the resolution matrices for ATD, DTD, and CCTD estimators in the asymptotic limit. Using the definition of the resolution matrix (Equation 6.20) for each method, $R_{\mathrm{method}} = \widehat{W}_{\mathrm{method}} W$, and using Equations 6.26, 6.19, and 6.25, we have:

$$R_{\mathrm{DTD}} = \overline{W}^T \left(\overline{WW}^T + \lambda \mathbf{C}_a / C_n \right)^{-1} \overline{W}, \qquad (6.27)$$

$$R_{\mathrm{ATD}} = \overline{W}^T \left(\overline{WW}^T + \lambda I \right)^{-1} \overline{W}. \qquad (6.28)$$

$$R_{\mathrm{CCTD}} = \overline{W}^T \left(\overline{WW}^T + \lambda \mathrm{diag}(\mathbf{C}_a) / C_n \right)^{-1} \overline{W}. \qquad (6.29)$$

Here, we have used the identity $A^\dagger A = I$, given A is full column rank provided $N \ll L$, i.e. when only a few discrete lifetimes are considered compared to the number of time gates. From the above equations, the key difference between the above three estimators is clear. The resolution matrices for the CCTD and ATD estimators are fully block-diagonal (note that W is block diagonal), therefore, providing zero cross talk for an arbitrary distribution of fluorophores. However, the DTD estimator has off-diagonal terms, resulting in significant cross talk and poor localization (Hou et al., 2016b). It is also clear that the ATD approach is identical to the CCTD estimator to within a quantitative correction because of the amplitude uncertainties (diagonal elements on $\mathbf{C}_n$). The incorporation of the amplitude uncertainties in the regularization reduces the mean square error compared to ATD.

6.3.2.4 Spectral vs. Lifetime Multiplexing

Analogous to lifetime multiplexing, spectral multiplexing employs fluorophores with distinct spectral shapes (absorption, emission, or both) to simultaneously label multiple components of disease. In many respects, spectral multiplexing is similar to lifetime multiplexing. Both methods employ basis functions that are highly overlapping: In the case of TD imaging, the basis functions are the temporal decay curves of distinct lifetimes, and in the case of spectral imaging, the basis is comprised of the excitation or emission spectra. Both techniques provide comparable performance for imaging *in vitro* samples, i.e. in the absence of tissue scattering. However, for tomographic imaging in thick tissue, multispectral fluorescence tomography (MSFT) and tomographic lifetime multiplexing (TFLM) provide entirely different performance. This is most generally understood using the same mathematical formalism as presented above.

Consider a turbid medium containing N fluorophores with distinct excitation or emission spectra, $b_n(\lambda)$, and lifetimes, τ_n, $n = 1 \ldots N$. The forward problem for MSFT takes the following matrix form for V medium voxels, M measurement pairs (number of sources $\times$ detectors) and K wavelengths:

$$y = W_S c, \qquad (6.30)$$

where $W_S = \left[W_{S_1}, \ldots, W_{S_N} \right]$ is the $(KM \times NV)$ spectral weight matrix, y is a measurement vector with dimensions $(KM \times 1)$ and $c = [c_1, \ldots, c_N]^T$ is a $(NV \times 1)$ parameter vector containing the unknown concentrations for each fluorophore. Analogous to the factorization of TD fluorescence weight matrix in the asymptotic limit (Equation 6.16), the spectral weight matrix, W_S, can also be factorized into a product of a basis matrix containing the spectral functions alone and a spatially varying matrix representing the diffuse propagation in the medium. For MSFT, the factorized form is:

$$W_S = \overline{W}_S A_S, \qquad (6.31)$$

where $A_S = [b_1(\lambda) \otimes I, \ldots, b_N(\lambda) \otimes I]$ is a $(KV \times NV)$ spectral basis matrix containing Kronecker products ($\otimes$) of the excitation or emission spectra and the $(V \times V)$ identity matrix, I, and $\overline{W}_S = \mathrm{diag}\left[\overline{W}_{S_1}, \ldots, \overline{W}_{S_K} \right]$ is a $(KM \times KV)$ block diagonal matrix containing CW weight matrices for each wavelength. The fundamental difference between the spectral and TD forward problems is immediately clear from Equations 6.31 and 6.16. In spectral multiplexing, the "mixing" of the unknown fluorophore concentrations occurs at the location of the individual voxels (through A_S). The mixed concentrations are then propagated through the medium by wavelength-dependent CW weight matrices ($\overline{W}_S$). For lifetime multiplexing, the individual concentrations are first propagated through the medium by reduced absorption CW matrices ($\overline{W}$) followed by mixing with temporal basis functions (A). This flip in order of the mixing and diffuse propagation steps (illustrated schematically in Figure 6.4) results in a significant difference between the capability of each method to quantitatively recover multiple fluorophore distributions (Hou et al., 2016a).

To further appreciate the difference between MSFT and TFLM, we compare the inverse problems for the two methods. The TFLM inverse problem was presented above. Considering the spectral inverse problem, it is first possible to show that there are two ways to perform spectral inversion, analogous to the DTD and ATD for the TD problem. The first approach is

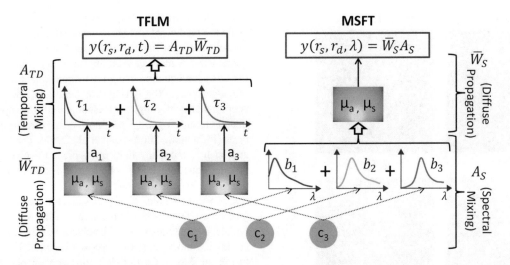

FIGURE 6.4 Schematic showing the fundamental difference in order of the mixing and diffuse propagation steps in TFLM and MSFT forward problems for three fluorophores with concentrations, c_1, c_2, c_3, lifetimes τ_1, τ_2, τ_3, and spectral basis functions b_1, b_2, b_3.

an inversion of the full spectral weight matrix, W_S, in Equation 6.30, called the direct spectral (DS) method (Li et al., 2004). Using Tikhonov regularization (Bertero and Boccacci, 1998) for inverting W_S and substituting Equation 6.31, we get the following resolution matrix for the direct spectral case:

$$R_{DS} = \widehat{W}_S W_S = A_S^T \overline{W}_S^T \left(\overline{W}_S A_S A_S^T \overline{W}_S^T + \lambda I \right)^{-1} \left[\overline{W}_S A_S \right], \quad (6.32)$$

where λ is the Tikhonov regularization parameter. Alternately, the factorized form in Equation 6.31 can be used to invert the spectral data in two stages, similar to the ATD approach. First, $\overline{W}_S$ is inverted using its Tikhonov-regularized inverse matrix, $\widehat{W}_S$. Next, because of the well-conditioned nature of A_S, it is inverted without regularization by multiplication with its Moore–Penrose pseudoinverse, $A_S^\dagger$. The latter step is equivalent to performing a linear fit to the spectral basis functions, $b_n(\lambda)$ at each voxel. The resolution matrix for this indirect spectral (IS) method is given by:

$$R_{IS} = A_S^\dagger \overline{W}_S W_S = A_S^\dagger \left[\overline{W}_S^T \left(\overline{W}_S \overline{W}_S^T + \lambda I \right)^{-1} \right] \overline{W}_S A_S. \quad (6.33)$$

where we have again used Equation 6.31 for W_S. The nondiagonal spectral basis matrices, A_S and $A_S^\dagger$, in both Equations 6.30 and 6.31 implies that, in general, both R_{DS} and R_{IS} are not block diagonal. However, inspection of Equations 6.30 and 6.31 reveals that R_{DS} and R_{IS} are block diagonal under certain conditions. Considering the DS case, Equation 6.32, first and switching to the equivalent overdetermined form of the inverse matrix, we can rewrite the resolution matrix as (using Equation 6.31):

$$R_{DS} = \left(A_S^T \overline{W}_S^T \overline{W}_S A_S + \lambda I \right)^{-1} \left(A_S^T \overline{W}_S^T \overline{W}_S A_S \right). \quad (6.34)$$

When the basis functions $b_j(\lambda_j)$ are nonoverlapping (i.e. $b_l(\lambda) b_m(\lambda_j) = 0 \; \forall j$ for all fluorophore pairs (l, m)), it is clear that the off-diagonal blocks $\sum_j b_l(\lambda_j) b_m(\lambda_j) = \overline{W}_{S_j}^T \overline{W}_{S_j}$ of the

$A_S^T \overline{W}_S^T \overline{W}_S A_S$ are zero. Hence R_{DS} also becomes block diagonal. For the IS case, we can rewrite Equation 6.33 as

$$R_{IS} = \left(A_S^T A_S \right)^{-1} A_S^T \widetilde{W} A_S \quad (6.35)$$

where $\widetilde{W} = \left[\overline{W}_S^T \left(\overline{W}_S \overline{W}_S^T + \lambda I \right)^{-1} \right] \overline{W}_S = \mathrm{diag}\left(\alpha_1, \alpha_2, ..., \alpha_K \right)$ is a block diagonal matrix with the $V \times V$ matrices α_j along the diagonal blocks, and we have applied the definition of the Moore–Penrose pseudoinverse for full column rank matrices, $A_S^\dagger = \left(A_S^T A_S \right)^{-1} A_S^T$. When the basis functions are nonoverlapping, $\left(A_S^T A_S \right)^{-1}$ becomes block diagonal and $A_S^T \widetilde{W} A_S$, whose off-diagonal blocks are equal to $\sum_j b_l(\lambda_j) b_m(\lambda_j) \alpha_j$, is also block diagonal. Hence, R_{IS} also becomes block diagonal. Additionally, when the optical properties are wavelength independent, we have $\overline{W}_{S_1} = \overline{W}_{S_2} ... = \overline{W}_{S_K} (\alpha_1 = \alpha_2 = ... \alpha)$, and R_{IS} can be simplified to $R_{IS} = \left(A_S^T A_S \right)^{-1} A_S^T A_S \times \mathrm{diag}(\alpha, ..., \alpha) = \mathrm{diag}(\alpha, ..., \alpha)$, which is a $NV \times NV$ block diagonal matrix. Thus, we can summarize the conditions for zero cross talk in spectral multiplexing as follows:

1) R_{IS} becomes block diagonal when the blocks along the diagonal of $\overline{W}_S$ are equal. This occurs when tissue optical properties (and correspondingly, the $\overline{W}_{Sk}$) are wavelength independent.

2) Both R_{IS} and R_{DS} become block diagonal when the spectral basis functions are non overlapping. For biomedical applications, which involve heterogeneous tissue with strongly wavelength dependent absorption and scattering, the first condition is rarely satisfied. The second condition is also hard to satisfy for NIR fluorophores, which typically exhibit broad and overlapping spectra. Figure 6.5 numerically illustrates the form of resolution matrix for TD and spectral methods (Equations 6.27, 6.28, 6.32 and 6.33) for a pair of NIR fluorescent proteins, iRFP702

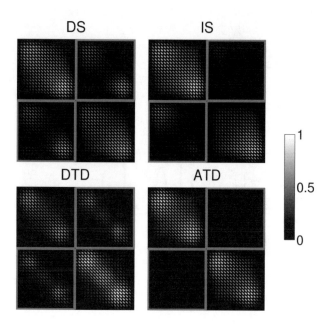

FIGURE 6.5 Resolution matrices for spectral and lifetime tomography, R_{DS}, R_{IS}, R_{DTD}, and R_{ATD} as given in Equations 6.32, 6.33, 6.27, and 6.28. Each resolution matrix was generated for a 2 cm thick diffuse medium for multiplexing of two NIR fluorophores (iRFP702 and iRFP720). Off-diagonal blocks of each resolution matrix represent cross talk between different fluorophores.

and iRFP720 (Rice et al., 2015). To aid in visualization, the rows and columns were binned by a factor of 80. Figure 6.5 shows that cross-talk terms for R_{DS} are positive and generally symmetric between the two fluorophores. On the other hand, R_{IS} shows more asymmetry between the two fluorophores with a major portion of the cross talk being negative. For the TD case, while R_{DTD} contains off-diagonal terms as expected, R_{ATD} is a block diagonal matrix with zero cross talk.

6.4 Computational Methods

The previous section dealt with the theoretical aspects of modeling fluorescent light transport and inversion techniques to recover fluorescence distributions within turbid media. Practical implementation of these models requires numerical computation of the diffusion or transport Green's functions involved in the theoretical expressions. Once the Green's functions are computed at the excitation and emission wavelengths, the fluorescence signal can be readily calculated for CW, FD, or TD forward problems. A wide range of sophisticated techniques have been developed to solve both the radiative transfer equation and the diffusion equation using various numerical techniques for arbitrary geometries and tissue optical properties. We briefly summarize these methods in this section and refer the reader to the cited references for more details (Arridge, 1999; Arridge and Schotland, 2009).

The simplest approach to modeling fluorescence is the use of analytical or semianalytical formulas for simple medium geometries, such as semi-infinite or infinite slab (Patterson

et al., 1989). In this case, boundary conditions are handled using the method of images to ensure that specified conditions on both the light fluence and its gradient are satisfied at the boundaries. A more general approach is to numerically solve the diffusion equation using the finite element method (FEM) (Arridge et al., 2000; Arridge and Schotland, 2009). The FEM is computationally highly efficient and allows rapid computation of Green's functions for fluorescence tomography applications (Joshi et al., 2006; Song et al., 2007). The more rigorous approach to modeling light transport is to solve the radiative transfer equation (RTE), which provides more accurate solutions even under conditions in which the diffusion approximation is valid (Kienle, 2007). In some simple geometries, it is possible to obtain analytical solution of the RTE (Liemert and Kienle, 2013). Numerical RTE solvers exist based on the FEM. By far the most common approach for modeling the RTE is to use the Monte Carlo (MC) approach for photon transport (Wang et al., 1995; Boas et al., 2002). The MC approach tracks the propagation of each photon as it absorbs and scatters in tissue, and provides a rigorous solution to the RTE in the limit of a large number ($> 10^8$) of photons. As such, the MC approach is computationally intensive. However, the use of graphics processing units (GPU) can dramatically accelerate MC computation (Fang and Boas, 2009), allowing the computation of the entire Green's function for a source within seconds. Using sophisticated techniques for mesh-based surface modeling (Fang, 2010; Chen et al., 2012), FEM methods can handle complex shaped objects such as small animals, the human brain, and breast. Recent implementations of GPU-MC also allow arbitrary input light patterns, such as sinusoidal waveforms, thereby making them powerful computational tools for fluorescence tomography. Given the ever increasing power of GPUs, it is likely that MC based implementations of light transport will become a staple in future fluorescence tomography systems both in preclinical and clinical settings.

6.5 Biological Applications

We conclude this chapter with a brief review of biological applications of MFI. The purpose of this section is not to provide a comprehensive review of all the latest developments in the field of molecular imaging, but to present select applications that illustrate the use of MFI for biological imaging. Because the key element in MFI is contrast, we will present applications of MFI from the point of view of the type of fluorescence labeling used, namely, intrinsic and targeted or activatable.

6.5.1 Intrinsic Fluorescence

A great challenge in MFI is the highly *in vivo* specific labeling of the target molecule of interest. While targeting specific disease signatures *in vivo* using extrinsic probes poses many challenges and is an extensive area of research, the use of intrinsic fluorescence contrast can alleviate the problem of specific labeling. The primary means of achieving intrinsic fluorescence contrast for *in vivo* imaging is using fluorescent protein (FPs) (Giepmans et al., 2006), which have revolutionized biological imaging by enabling highly specific labeling of gene expression events and tumor cell proliferation (Jain et al., 2002). The brightest FPs, such as the green fluorescent protein (GFP), excite and emit in the visible region and are thus not ideal for macroscopic, whole-body fluorescence imaging because of the high absorption of visible light by tissue chromophores. The use of far-red FPs, such as tdTomato and mCherry, with excitation wavelengths up to 600 nm somewhat alleviate the problem of poor light penetration, and tomographic imaging using these far-red FPs has been reported in mice (Deliolanis et al., 2008). More recently, NIR FPs (or iRFPs) were synthesized with excitation spectra extending beyond 700 nm and allow imaging of deep-seated tumors (Shcherbakova and Verkhusha, 2013; Shcherbakova et al., 2015). However, tissue autofluorescence (AF) still remains a major confound that limits sensitivity when using CW detection, which cannot distinguish FP fluorescence from tissue AF. Therefore, CW imaging iRFPs can be used only to detect subcutaneous or large tumors ($>10^6$ cells) and requires very high laser powers and long imaging times (Deliolanis et al., 2014; Jiguet-Jiglaire et al., 2014). Multispectral unmixing can be employed to improve the sensitivity of FP detection in deep organs by exploiting the distinct absorption spectra of FPs compared to the background, as in photoacoustic tomography (PAT) (Razansky et al., 2009). However, sensitivity will still be limited by the fact that fluorescence spectra are altered because of propagation, an effect discussed in Section 6.3.2.4. With the use of TD detection, the lifetime contrast between AF and several FPs can be exploited to improve detection sensitivity by more than 10-fold (Kumar et al., 2009; Rice and Kumar, 2014). Recently, the detection of 25K cells in deep organs such as the lungs and ~1.5K cells subcutaneously was demonstrated using iRFP720 labeled cancer cells (Rice et al., 2015). Figure 6.6 shows a table comparing the performance of CW and TD tomography for detecting FP labeled cancer cells in whole animals. The vast potential of FPs for whole-animal preclinical studies has yet to be fully exploited, particularly for multiplexing (Figure 6.7), given that FPs exhibit a wide range of fluorescence lifetimes and spectra.

6.5.2 Extrinsic Fluorescence

While intrinsic fluorescence labeling offers the great advantage of high specificity, the need for genetically encoding fluorescent reporters makes this approach clinically infeasible. The development of extrinsic, molecularly targeted NIR fluorescent probes is an extensive area of investigation (Zhang et al., 2017). A variety of molecules that target specific disease mechanisms and markers have been developed and used for *in vitro* and *in vivo* studies of molecular and cellular processes. The use of such disease-targeted fluorophores for clinical imaging has been limited and is only emerging. Specifically, the NIR fluorophores IRDye800CW and Chlorotoxin-Cy5.5 (tumor paint) (Veiseh et al., 2007; Butte et al., 2014) allow specific tumor targeting and are currently in clinical trials. Currently, only indocyanine green (ICG), methylene blue, and 5-aminolevulinic acid (5-ALA) are clinically approved. Of these, ICG offers greater brightness and longer excitation wavelength and has primarily been applied

	CW fluorescence	TD fluorescence
Sensitivity[a] (deep, e.g., lungs)	$>10^6$ cells [59,60]	$0.02 - 0.05 \times 10^6$ cells [42]
Sensitivity[a] (shallow, e.g., lymph)	5×10^4 cells [60]	0.15×10^4 cells [42]
Multiplexing	Spectral[b]	Spectral and lifetime
Depth	Whole mouse (20 mm)	Whole mouse (20 mm)
Resolution[c]	4 - 5 mm[d] [2]	$1 - 2$ mm[e]
Imaging time	5 min[f] (head) [60] - 10 min (torso)	15 min (head) – 30 min (torso)

FIGURE 6.6 Table 6.1 Comparison of CW and TD fluorescence imaging techniques for whole-animal imaging fluorescent proteins. (a) Minimum number of iRFP cells as reported in literature cited. (b) Limited sensitivity with spectral unmixing due to broad spectra and tissue red-shifting (Hou et al., 2016a). (c) At a depth of 7 mm in a 1.5 cm thick scattering medium. (d) Center-to-center distance in Graves et al. (2003). (e) Assuming targets have distinct lifetimes. (f) For large ($>10^6$) tumors. Imaging times can be longer (>60 min) to achieve sufficient contrast for smaller tumors.

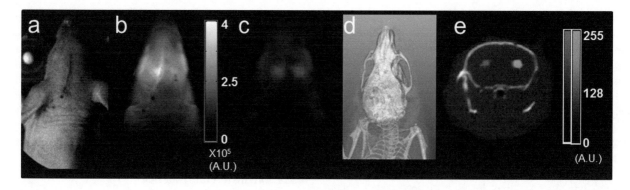

FIGURE 6.7 Tomographic lifetime multiplexing of iRFP tumors in the mouse brain. (a) White light image, (b) planar transmission CW fluorescence (integrated TD signal) and (c) fluorescence decay amplitude images for excitation source locations spanning the head of a sacrificed nude mouse following cerebral stereotactic injection with 5×10^5 MTLn3-iRFP720 (left) and 1.5×10^6 MTLn3-iRFP702 cells (right) in 3 μL each. (c) shows the decay amplitudes for tissue autofluorescence (blue), iRFP720 (red), and iRFP702 (green) computed from the full TD data using linear fit with basis functions (Rice and Kumar, 2014) for AF and the iRFP lifetimes. (d) Volumetric and (e) coronal slice from tomographic reconstruction of iRFP720 (red) and iRFP702 (green) fluorescence yield distributions coregistered with the head X-ray CT (grayscale).

for lymph node mapping and angiography during reconstructive surgery (Xiong et al., 2014; Sevick-Muraca et al., 2008; Liberale et al., 2015; Vahrmeijer et al., 2013; van der Vorst et al., 2013). For tumor imaging, the enhanced permeability and retention (EPR) of ICG has been exploited for imaging liver (Ishizawa et al., 2009, 2010; Kokudo and Ishizawa, 2012), ovarian (Horowitz et al., 2006), head and neck (Yokoyama et al., 2013, 2014), and breast (Poellinger et al., 2011; Hagen et al., 2009; Corlu et al., 2007) cancers. Despite EPR, a significant amount of ICG remains in tissue for several hours postinjection, thereby reducing contrast and sensitivity and specificity (Tummers et al., 2015). However, it has been shown that beyond 24 hours after injection, ICG is cleared from normal tissue and is solely confined within the tumor environment (Kumar et al., 2017). At these late times, further improvement in sensitivity and specificity can be achieved by using TD detection and exploiting the lifetime contrast between tumor bound-ICG and background tissue autofluorescence. In some scenarios, it is feasible to directly label the target molecule of interest using extrinsic fluorophores. One such study used NIR fluorescence labeled macrophages to detect inflammatory response in granuloma formation (Eisenblätter et al., 2009). This approach avoids the complications involved in designing specific targeting *in vivo*.

6.5.3 Activatable Fluorescence

An important type of targeted fluorophores is specifically designed to shift their fluorescence intensity, lifetime, or spectrum upon sensing disease specific signatures. These probes are usually based on one of two kinds of intramolecular energy transfer mechanisms, namely static and dynamic quenching (Lakowicz, 1999). Static quenching occurs when a pair of interacting molecules form a complex in the ground state, thereby significantly altering their absorption or emission spectra. In contrast, dynamic quenching occurs because of energy transfer when a molecule is in the excited state. While static quenching is usually accompanied by a change in fluorescence intensity alone, dynamic quenching results in a change in both intensity and lifetime. A powerful example is

the phenomenon of Forster resonance energy transfer (FRET) (Selvin 2000), which has found wide application in fluorescence spectroscopy as a way to measure distances between fluorophores on the nanometer scale. FRET occurs when an excited donor fluorophore transfers its energy to an adjacent ground-state acceptor fluorophore through dipole coupling. This results in both a decrease in donor fluorescence intensity and its lifetime. This process depends strongly on the distance between molecules in the 1–10 nm range, and, therefore, can be exploited as a spectroscopic ruler [2]. With the use of fluorescence proteins, FRET has found an ever increasing range of applications, such as tracking protein–protein interactions in cellular processes (Elangovan et al., 2002), probing DNA/RNA dynamics (Blanchard et al., 2004), and high throughput drug screening (Kumar et al., 2011). *In vivo* tomographic lifetime-FRET has also recently been demonstrated using NIR-labeled transferrins in breast cancer cells and tumors (Pian et al., 2017). This approach has allowed the ability to distinguish cellular uptake within tumors from mere accumulation within the tumor environment through the EPR effect. Other than FP-based FRET probes, NIR fluorescent oligodeoxyribonucleotide (ODN) reporters have been developed that can sense transcription factor NF-κB p50 protein binding (Zhang et al., 2008). These reporters are based on two closely spaced NIR fluorophores, Cy5.5 and Cy7, that were linked to two complementary ODNs that encoded p50 binding sites.

Besides FRET probes, another highly promising class of activatable probes is based on self-quenched macromolecular NIR sensors that undergo either a change in fluorescence intensity alone (static quenching) or intensity and lifetime changes (dynamic quenching) upon activation (Weissleder et al., 1999; Kobayashi and Choyke, 2010). Such probes have been primarily developed for *in vivo* imaging of proteolysis (Verdoes and Verhelst, 2016; Edgington et al., 2011). Despite promising initial results for detecting protease activity *in vivo* (Nahrendorf et al., 2007), a major drawback is the large nonspecific activation from other organs, such as the liver, that can dominate the overall fluorescence intensity when measuring nearby organs such as the heart (Goergen et al., 2012). Recently, it was shown using planar reflectance TD fluorescence imaging that the fluorescence lifetimes of an activated probe in the liver was significantly different from the heart under myocardial infraction. Capitalizing on this lifetime contrast, TD detection can provide a 10-fold increase in contrast to background compared to CW imaging (Goergen et al., 2012). The mechanism behind these activatable probes is not yet fully understood, although initial studies suggest that a combination of both static (changes in fluorescence intensity) and dynamic (fluorescence lifetime and intensity change) quenching could be involved (Kumar et al., 2016). While activatable probes hold great promise for molecular imaging, it would be important to fully understand the mechanisms behind their action so that they can be optimized for quantitative *in vivo* imaging.

6.6 Summary

We have presented an overview of experimental and theoretical methods for macroscopic fluorescence imaging with particular emphasis on *in vivo* tomographic imaging. Although whole-body fluorescence imaging is at least two decades old, the future is still bright for the field given the potential for multiplexing using spectral and fluorescence lifetime. Several commercial NIR fluorescent dyes are already available with capability for targeting specific disease pathologies in small animals. By a careful selection of dyes with lifetime contrast, the MFI with lifetime contrast will thus create new avenues to visualize multiple biological processes noninvasively, accelerating the drug discovery process in preclinical imaging and can potentially offer interesting applications in the clinical settings as well.

The central challenge for applying tomographic FLIM in the clinical setting (or whole- body human imaging) is the same as that for diffuse optical NIR spectroscopy, namely, the larger imaging volumes and the high absorption and scattering suffered by light photons in tissue. Although sophisticated algorithms exist for modeling light transport through complex tissue, the limited penetration of diffuse light implies poor depth penetration and spatial resolution. However, diffuse optical tomography has been applied for functional contrast in the human breast and brain (Fang et al., 2009). The same clinical applications can be envisaged with extrinsic contrast agents. As more disease-specific fluorochromes become available for human use, tomographic lifetime multiplexing is likely to play an important role in the clinical setting as well.

REFERENCES

Abulrob, A., Brunette, E., Slinn, J., Baumann, E., and Stanimirovic, D. (2007). In vivo time domain optical imaging of renal ischemia-reperfusion injury: Discrimination based on fluorescence lifetime. *Molecular Imaging*, 6(5):304–314.

Arridge, S. R. (1999). Optical tomography in medical imaging. *Inverse Problems*, 15(2):R41–R93.

Arridge, S. R., Dehghani, H., Schweiger, M., and Okada, E. (2000). The finite element model for the propagation of light in scattering media: A direct method for domains with nonscattering regions. *Medical Physics*, 27(1):252–264.

Arridge, S. R. and Schotland, J. C. (2009). Optical tomography: Forward and inverse problems. *Inverse Problems*, 25(12).

Bassi, A., D'Andrea, C., Valentini, G., Cubeddu, R., and Arridge, S. (2008). Temporal propagation of spatial information in turbid media. *Optics Letters*, 33(23):2836–2838.

Becker, W., Bergmann, A., Haustein, E., Petrasek, Z., Schwille, P., Biskup, C., Kelbauskas, L., Benndorf, K., Klocker, N., Anhut, T., Riemann, I., and Konig, K. (2006). Fluorescence lifetime images and correlation spectra obtained by multidimensional time-correlated single photon counting. *Microscopy Research and Technique*, 69(3):186–195.

Bertero, M. and Boccacci, P. (1998). *Introduction to Inverse Problems in Imaging*. CRC Press.

Blanchard, S. C., Kim, H. D., Gonzalez, R. L., Puglisi, J. D., and Chu, S. (2004). TRNA dynamics on the ribosome during translation. *Proceedings of the National Academy of Sciences of the United States of America*, 101(35):12893–12898.

Boas, D. A., Culver, J. P., Stott, J. J., and Dunn, A. K. (2002). Three dimensional Monte Carlo code for photon migration through complex heterogeneous media including the adult human head. *Optics Express*, 10(3):159–170.

Brambilla, M., Spinelli, L., Pifferi, A., Torricelli, A., and Cubeddu, R. (2008). Time-resolved scanning system for double reflectance and transmittance fluorescence imaging of diffusive media. *Review of Scientific Instruments*, 79(1).

Butte, P. V., Mamelak, A., Parrish-Novak, J., Drazin, D., Shweikeh, F., Gangalum, P. R., Chesnokova, A., Ljubimova, J. Y., and Black, K. (2014). Near-infrared imaging of brain tumors using the tumor paint blz-100 to achieve near-complete resection of brain tumors. *Neurosurgical Focus*, 36(2):E1.

Chandrasekhar, S. (2013). *Radiative Transfer*. Courier Corporation.

Chen, J., Fang, Q., and Intes, X. (2012). Mesh-based Monte Carlo method in time-domain widefield fluorescence molecular tomography. *Journal of Biomedical Optics*, 17(10):106009.

Corlu, A., Choe, R., Durduran, T., Rosen, M. A., Schweiger, M., Arridge, R., Schnall, M. D., and Yodh, A. G. (2007). Three-dimensional in vivo fluorescence diffuse optical tomography of breast cancer in humans. *Optics Express*, 15(11):6696–6716.

Cuccia, D. J., Bevilacqua, F., Durkin, A. J., and Tromberg, B. J. (2005). Modulated imaging: Quantitative analysis and tomography of turbid media in the spatial-frequency domain. *Optics Letters*, 30(11):1354–1356.

Deliolanis, N. C., Ale, A., Morscher, S., Burton, N., Schaefer, K., Radrich, K., Razansky, D., and Ntziachristos, V. (2014). Deep tissue reporter gene imaging with fluorescence and optoacoustic tomography: A performance overview. *Molecular Imaging and Biology*, 16:652–660.

Deliolanis, N. C., Kasmieh, R., Wurdinger, T., Tannous, B. A., Shah, K., and Ntziachristos, V. (2008). Performance of the red-shifted fluorescent proteins in deep-tissue molecular imaging applications. *Journal of Biomedical Optics*, 13(4).

Ducros, N., Bassi, A., Valentini, G., Canti, G., Arridge, S., and D'Andrea, C. (2013). Fluorescence molecular tomography of an animal model using structured light rotating view acquisition. *Journal of Biomedical Optics*, 18(2).

Edgington, L. E., Verdoes, M., and Bogyo, M. (2011). Functional imaging of proteases: Recent advances in the design and application of substrate-based and activity-based probes. *Current Opinion in Chemical Biology*, 15(6):798–805.

Eisenblätter, M., Ehrchen, J., Varga, G., Sunderkötter, C., Heindel, W., Roth, J., Bremer, C., and Wall, A. (2009). In vivo optical imaging of cellular inflammatory response in granuloma formation using fluorescence-labeled macrophages. *Journal of Nuclear Medicine*, 50(10):1676–1682.

Elangovan, M., Day, R. N., and Periasamy, A. (2002). Nanosecond fluorescence resonance energy transfer-fluorescence lifetime imaging microscopy to localize the protein interactions in a single living cell. *Journal of Microscopy (Oxford)*, 205:3–14.

Fang, Q. (2010). Mesh-based Monte Carlo method using fast ray-tracing in plücker coordinates. *Biomedical Optics Express*, 1(1):165–175.

Fang, Q. Q. and Boas, D. A. (2009). Monte Carlo simulation of photon migration in 3d turbid media accelerated by graphics processing units. *Optics Express*, 17(22):20178–20190.

Fang, Q. Q., Carp, S. A., Selb, J., Boverman, G., Zhang, Q., Kopans, D. B., Moore, R. H., Miller, E. L., Brooks, D. H., and Boas, D. A. (2009). Combined optical imaging and mammography of the healthy breast: Optical contrast derived from breast structure and compression. *ieee Transactions on Medical Imaging*, 28(1):30–42.

Giepmans, B. N. G., Adams, S. R., Ellisman, M. H., and Tsien, R. Y. (2006). Review - the fluorescent toolbox for assessing protein location and function. *Science*, 312(5771):217–224.

Godavarty, A., Sevick-Muraca, E. M., and Eppstein, M. J. (2005). Three-dimensional fluorescence lifetime tomography. *Medical Physics*, 32(4):992–1000.

Goergen, C. J., Chen, H. H., Bogdanov, A., Sosnovik, D. E., and Kumar, A. T. N. (2012). In vivo fluorescence lifetime detection of an activatable probe in infarcted myocardium. *Journal of Biomedical Optics*, 17(5).

Graves, E. E., Ripoll, J., Weissleder, R., and Ntziachristos, V. (2003). A submillimeter resolution fluorescence molecular imaging system for small animal imaging. *Medical Physics*, 30(5):901–911.

Hagen, A., Grosenick, D., Macdonald, R., Rinneberg, H., Burock, S., Warnick, P., Poellinger, A., and Schlag, P. M. (2009). Late-fluorescence mammography assesses tumor capillary permeability and differentiates malignant from benign lesions. *Optics Express*, 17(19):17016–17033.

Hall, D., Ma, G. B., Lesage, F., and Yong, W. (2004). Simple time-domain optical method for estimating the depth and concentration of a fluorescent inclusion in a turbid medium. *Optics Letters*, 29(19):2258–2260.

Haselgrove, J. C., Schotland, J. C., and Leigh, J. S. (1992). Long-time behavior of photon diffusion in an absorbing medium: Application to time-resolved spectroscopy. *Applied Optics*, 31(15):2678–2683.

Horowitz, N. S., Penson, R. T., Kassis, E. N., Foster, R., Seiden, M. V., Weissleder, R., and Fuller, A. F. (2006). Laparoscopy in the near infrared with ICG detects microscopic tumor in women with ovarian cancer: 0078. *International Journal of Gynecological Cancer*, 16:622.

Hou, S. S., Bacskai, B. J., and Kumar, A. T. (2016a). Comparison of tomographic fluorescence spectral and lifetime multiplexing. *Optics Letters*, 41(22):5337–5340.

Hou, S. S., Bacskai, B. J., and Kumar, A. T. N. (2016b). Optimal estimator for tomographic fluorescence lifetime multiplexing. *Optics Letters*, 41(7):1352–1355.

Ishizawa, T., Bandai, Y., Ijichi, M., Kaneko, J., Hasegawa, K., and Kokudo, N. (2010). Fluorescent cholangiography illuminating the biliary tree during laparoscopic cholecystectomy. *British Journal of Surgery*, 97(9):1369–1377.

Ishizawa, T., Fukushima, N., Shibahara, J., Masuda, K., Tamura, S., Aoki, T., Hasegawa, K., Beck, Y., Fukayama, M., and Kokudo, N. (2009). Real time identification of liver cancers by using indocyanine green fluorescent imaging. *Cancer*, 115(11):2491–2504.

Jain, R. K., Munn, L. L., and Fukumura, D. (2002). Dissecting tumour pathophysiology using intravital microscopy. *Nature Reviews Cancer*, 2(4):266–276.

Jiguet-Jiglaire, C., Cayol, M., Mathieu, S., Jeanneau, C., Bouvier-Labit, C., Ouafik, L., and El-Battari, A. (2014). Noninvasive near-infrared fluorescent protein-based imaging of tumor progression and metastases in deep organs and intraosseous tissues. *Journal of Biomedical Optics*, 19(1):6.

Joshi, A., Bangerth, W., Hwang, K., Rasmussen, J. C., and Sevick-Muraca, E. M. (2006). Fully adaptive FEM based fluorescence optical tomography from time-dependent measurements with area illumination and detection. *Medical Physics*, 33(5):1299–1310.

Kienle, A. (2007). Anisotropic light diffusion: An oxymoron? *Physical Review Letters*, 98(21):218104.

Klohs, J., Steinbrink, J., Bourayou, R., Mueller, S., Cordell, R., Licha, K., Schirner, M., Dirnagl, U., Lindauer, U., and Wunder, A. (2009). Near-infrared fluorescence imaging with fluorescently labeled albumin: A novel method for non-invasive optical imaging of blood–brain barrier impairment after focal cerebral ischemia in mice. *Journal of Neuroscience Methods*, 180(1):126–132.

Kobayashi, H. and Choyke, P. L. (2010). Target-cancer-cell-specific activatable fluorescence imaging probes: Rational design and in vivo applications. *Accounts of Chemical Research*, 44(2):83–90.

Kokudo, N. and Ishizawa, T. (2012). Clinical application of fluorescence imaging of liver cancer using indocyanine green. *Liver Cancer*, 1(1):15–21.

Kumar, A., Raymond, S., Bacskai, B., and Boas, D. A. (2008). A time domain fluorescence tomography system for small animal imaging. *IEEE Transactions on Medical Imaging*, 27(8):1152–1163.

Kumar, A. T., Hou, S. S., and Rice, W. L. (2018). Tomographic fluorescence lifetime multiplexing in the spatial frequency domain. *Optica*, 5(5):624–627.

Kumar, A. T., Raymond, S. B., Boverman, G., Boas, D. A., and Bacskai, J. (2006). Time resolved fluorescence tomography of turbid media based on lifetime contrast. *Optics Express*, 14(25):12255–12270.

Kumar, A. T., Rice, W. L., López, J. C., Gupta, S., Goergen, C. J., and Bogdanov Jr, A. A. (2016). Substrate-based near-infrared imaging sensors enable fluorescence lifetime contrast via built-in dynamic fluorescence quenching elements. *ACS Sensors*, 1(4):427–436.

Kumar, A. T. N. (2013). Fluorescence lifetime detection in turbid media using spatial frequency domain filtering of time domain measurements. *Optics Letters*, 38(9):1440–1442.

Kumar, A. T. N., Carp, S. A., Yang, J., Ross, A., Medarova, Z., and Ran, B. (2017). Fluorescence lifetime-based contrast enhancement of indocyanine green-labeled tumors. *Journal of Biomedical Optics*, 22(4):040501–040501.

Kumar, A. T. N., Chung, E., Raymond, S. B., van de Water, J., Shah, K., Fukumura, D., Jain, R. K., Bacskai, B. J., and Boas, D. A. (2009). Feasibility of in vivo imaging of fluorescent proteins using lifetime contrast. *Optics Letters*, 34(13):2066–2068.

Kumar, A. T. N., Skoch, J., Bacskai, B. J., Boas, D. A., and Dunn, A. K. (2005). Fluorescence-lifetime-based tomography for turbid media. *Optics Letters*, 30(24):3347–3349.

Kumar, S., Alibhai, D., Margineanu, A., Laine, R., Kennedy, G., McGinty, J., Warren, S., Kelly, D., Alexandrov, Y., Munro, I., et al. (2011). FLIM FRET technology for drug discovery: Automated multiwell-plate high-content analysis, multiplexed readouts and application in situ. *ChemPhysChem*, 12(3):609–626.

Lakowicz, J. R. (1999). *Principles of Fluorescence Spectroscopy*. Springer, 2nd edition.

Li, A., Zhang, Q., Culver, J. P., Miller, E. L., and Boas, D. A. (2004). Reconstructing chromosphere concentration images directly by continuous-wave diffuse optical tomography. *Optics Letters*, 29(3):256–258.

Liberale, G., Vankerckhove, S., Galdon, M. G., Donckier, V., Larsimont, D., and Bourgeois, P. (2015). Fluorescence imaging after intraoperative intravenous injection of indocyanine green for detection of lymph node metastases in colorectal cancer. *European Journal of Surgical Oncology (EJSO)*, 41(9):1256–1260.

Liemert, A. and Kienle, A. (2013). Exact and efficient solution of the radiative transport equation for the semi-infinite medium. *Scientific Reports*, 3:2018.

Milstein, A. B., Oh, S., Webb, K. J., Bouman, C. A., Zhang, Q., Boas, D. A., and Millane, R. P. (2003). Fluorescence optical diffusion tomography. *Applied Optics*, 42(16): 3081–3094.

Milstein, A. B., Stott, J. J., Oh, S., Boas, D. A., Millane, R. P., Bouman, C. A., and Webb, K. J. (2004). Fluorescence optical diffusion tomography using multiple- frequency data. *Journal of the Optical Society of America a-Optics Image Science and Vision*, 21(6):1035–1049.

Nahrendorf, M., Sosnovik, D. E., Waterman, P., Swirski, F. K., Pande, A. N., Aikawa, E., Figueiredo, J. L., Pittet, M. J., and Weissleder, R. (2007). Dual channel optical tomographic imaging of leukocyte recruitment and protease activity in the healing myocardial infarct. *Circulation Research*, 100(8):1218–1225.

Nothdurft, R. E., Patwardhan, S. V., Akers, W., Ye, Y. P., Achilefu, S., and Culver, J. P. (2009). In vivo fluorescence lifetime tomography. *Journal of Biomedical Optics*, 14(2):7.

Oleary, M. A., Boas, D. A., Li, X. D., Chance, B., and Yodh, A. G. (1996). Fluorescence lifetime imaging in turbid media. *Optics Letters*, 21(2):158–160.

Patterson, M. S., Chance, B., and Wilson, B. C. (1989). Time resolved reflectance and transmittance for the noninvasive measurement of tissue optical properties. *Applied Optics*, 28(12):2331–2336.

Pian, Q., Yao, R., Sinsuebphon, N., and Intes, X. (2017). Compressive hyperspectral time-resolved wide-field fluorescence lifetime imaging. *Nature Photonics*, 11(7):411.

Poellinger, A., Burock, S., Grosenick, D., Hagen, A., Ldemann, L., Diekmann, F., Engelken, F., Macdonald, R., Rinneberg, H., and Schlag, P.-M. (2011). Breast cancer: Early-and late-fluorescence near-infrared imaging with indocyanine green–a preliminary study. *Radiology*, 258(2):409–416.

Press, W. H., Teukolsky, S. A., Vetterling, W. T., and Flannery, B. P. (1996). *Numerical Recipes in C*, Vol. 2. Cambridge University Press.

Pu, H., Zhang, G., He, W., Liu, F., Guang, H., Zhang, Y., Bai, J., and Luo, J. (2014). Resolving fluorophores by unmixing multispectral fluorescence tomography with independent component analysis. *Physics in Medicine & Biology*, 59(17):5025.

Raymond, S. B., Boas, D. A., Bacskai, B. J., and Kumar, A. T. N. (2010). Lifetime-based tomographic multiplexing. *Journal of Biomedical Optics*, 15(4):046011.

Raymond, S. B., Kumar, A. T. N., Boas, D. A., and Bacskai, B. J. (2009). Optimal parameters for near infrared fluorescence imaging of amyloid plaques in Alzheimer's disease mouse models. *Physics in Medicine and Biology*, 54(20):6201–6216.

Razansky, D., Distel, M., Vinegoni, C., Ma, R., Perrimon, N., Köster, R. W., and Ntziachristos, V. (2009). Multispectral opto-acoustic tomography of deep- seated fluorescent proteins in vivo. *Nature Photonics*, 3(7):412.

Rice, W. and Kumar, A. T. N. (2014). Preclinical whole body time domain fluorescence lifetime multiplexing of fluorescent proteins. *Journal of Biomedical Optics*, 19(4):046004.

Rice, W. L., Hou, S., and Kumar, A. T. N. (2013). Resolution below the point spread function for diffuse optical imaging using fluorescence lifetime multiplexing. *Optics Letters*, 38(12):2038–2040.

Rice, W. L., Shcherbakova, D. M., Verkhusha, V. V., and Kumar, A. T. (2015). In vivo tomographic imaging of deep-seated cancer using fluorescence lifetime contrast. *Cancer Research*, 75(7):1236–1243.

Selvin, P. R. (2000). The renaissance of fluorescence resonance energy transfer. *Nature Structural Biology*, 7(9):730–734.

Sevick-Muraca, E. M., Sharma, R., Rasmussen, J. C., Marshall, M. V., Wendt, J. A., Pham, H. Q., Bonefas, E., Houston, J. P., Sampath, L., Adams, K. E., Blanchard, D. K., Fisher, R. E., Chiang, S. B., Elledge, R., and Mawad, M. E. (2008). Imaging of lymph flow in breast cancer patients after microdose administration of a near-infrared fluorophore: Feasibility study. *Radiology*, 246(3):734–741.

Shcherbakova, D. M., Baloban, M., and Verkhusha, V. V. (2015). Near-infrared fluorescent proteins engineered from bacterial phytochromes. *Current Opinion in Chemical Biology*, 27:52–63.

Shcherbakova, D. M. and Verkhusha, V. V. (2013). Near-infrared fluorescent proteins for multicolor in vivo imaging. *Nature Methods*, 10(8):751.

Soloviev, V. Y., McGinty, J., Tahir, K. B., Neil, M. A. A., Sardini, A., Hajnal, J. V., Arridge, S. R., and French, P. M. W. (2007). Fluorescence lifetime tomography of live cells expressing enhanced green fluorescent protein embedded in a scattering medium exhibiting background autofluorescence. *Optics Letters*, 32(14):2034–2036.

Soloviev, V. Y., Wilson, D. F., and Vinogradov, S. A. (2004). Phosphorescence lifetime imaging in turbid media: The inverse problem and experimental image reconstruction. *Applied Optics*, 43(3):564–574.

Song, X. L., Wang, D. F., Chen, N. G., Bai, J., and Wang, H. (2007). Reconstruction for free-space fluorescence tomography using a novel hybrid adaptive finite element algorithm. *Optics Express*, 15(26):18300–18317.

Tummers, Q. R. J. G., Hoogstins, C. E. S., Peters, A. A. W., de Kroon, C. D., Trimbos, J. B. M. Z., van de Velde, C. J. H., Frangioni, J. V., Vahrmeijer, A. L., and Gaarenstroom, K. N. (2015). The value of intraoperative near-infrared fluorescence imaging based on enhanced permeability and retention of indocyanine green: Feasibility and false-positives in ovarian cancer. *PloS One*, 10(6):e0129766.

Turner, G. M., Zacharakis, G., Soubret, A., Ripoll, J., and Ntziachristos, V. (2005). Complete-angle projection diffuse optical tomography by use of early photons. *Optics Letters*, 30(4):409–411.

Vahrmeijer, A. L., Hutteman, M., van der Vorst, J. R., van de Velde, C. J. H., and Frangioni, J. V. (2013). Image-guided cancer surgery using near-infrared fluorescence. *Nature Reviews Clinical Oncology*, 10(9):507–518.

van der Vorst, J. R., Schaafsma, B. E., Verbeek, F. P. R., Keereweer, S., Jansen, J. C., van der Velden, L.-A., Langeveld, A. P. M., Hutteman, M., Lwik, C. W. G. M., and van de Velde, C. J. H. (2013). Near-infrared fluorescence sentinel lymph node mapping of the oral cavity in head and neck cancer patients. *Oral Oncology*, 49(1):15–19.

Veiseh, M., Gabikian, P., Bahrami, S. B., Veiseh, O., Zhang, M., Hackman, R. C., Ravanpay, A. C., Stroud, M. R., Kusuma, Y., and Hansen, S. J. (2007). Tumor paint: A chlorotoxin: Cy5. 5 bioconjugate for intraoperative visualization of cancer foci. *Cancer Research*, 67(14):6882–6888.

Venugopal, V., Chen, J., Lesage, F., and Intes, X. (2010). Full-field time-resolved fluorescence tomography of small animals. *Optics Letters*, 35(19):3189–3191.

Venugopal, V. and Intes, X. (2013). Adaptive wide-field optical tomography. *Journal of Biomedical Optics*, 18(3):036006.

Verdoes, M. and Verhelst, S. H. (2016). Detection of protease activity in cells and animals. *Biochimica et Biophysica Acta (BBA) - Proteins and Proteomics*, 1864(1):130–142.

Wang, L. H., Jacques, S. L., and Zheng, L. Q. (1995). Mcml - Monte-Carlo modeling of light transport in multilayered tissues. *Computer Methods and Programs in Biomedicine*, 47(2):131–146.

Weissleder, R., Tung, C. H., Mahmood, U., and Bogdanov, A. (1999). In vivo imaging of tumors with protease-activated near-infrared fluorescent probes. *Nature Biotechnology*, 17(4):375–378.

Xiong, L., Gazyakan, E., Yang, W., Engel, H., Hanerbein, M., Kneser, U., and Hirche, C. (2014). Indocyanine green fluorescence-guided sentinel node biopsy: A meta-analysis on detection rate and diagnostic performance. *European Journal of Surgical Oncology (EJSO)*, 40(7):843–849.

Xu, H. and Rice, B. W. (2009). In-vivo fluorescence imaging with a multivariate curve resolution spectral unmixing technique. *Journal of Biomedical Optics*, 14(6):064011.

Yokoyama, J., Fujimaki, M., Ohba, S., Anzai, T., Yoshii, R., Ito, S., Kojima, M., and Ikeda, K. (2013). A feasibility study of nir fluorescent image-guided surgery in head and neck cancer based on the assessment of optimum surgical time as revealed through dynamic imaging. *OncoTargets and Therapy*, 6:325–330.

Yokoyama, J., Ooba, S., Fujimaki, M., Anzai, T., Yoshii, R., Kojima, M., and Ikeda, K. (2014). Impact of indocyanine green fluorescent image-guided surgery for parapharyngeal space tumours. *Journal of Cranio-Maxillofacial Surgery*, 42(6):835–838.

Zhang, R. R., Schroeder, A. B., Grudzinski, J. J., Rosenthal, E. L., Warram, J. M., Pinchuk, A. N., Eliceiri, K. W., Kuo, J. S., and Weichert, J. P. (2017). Beyond the margins: Real-time detection of cancer using targeted fluorophores. *Nature Reviews Clinical Oncology*.

Zhang, S., Metelev, V., Tabatadze, D., Zamecnik, P. C., and Bogdanov, A. (2008). Fluorescence resonance energy transfer in near-infrared fluorescent oligonucleotide probes for detecting protein-DNA interactions. *Proceedings of the National Academy of Sciences of the United States of America*, 105(11):4156–4161.

7

Optical Coherence Tomography

Roshan Dsouza and Stephen A. Boppart

CONTENTS

7.1 Introduction

Optical coherence tomography (OCT) is an established biomedical imaging technology that has been successfully translated from the laboratory and implemented in clinical practice. OCT provides real-time three-dimensional (3D) images of the tissue architecture depicting details that can be correlated to histological findings, without any necessity of tissue excision. OCT is analogous to ultrasound imaging except light waves are used instead of sound waves. OCT acquires cross-sectional images by measuring the magnitude and time delay of backscattered light (axial scans or A-scan) and by scanning the incident beam in the transverse direction (B-scan). Though penetration depth of OCT is limited to ~2 mm in most tissue

(Schmitt, 1999), depending on the optical scattering properties of the tissue, integrating OCT in fiber optic endoscopic or needle probes has enabled imaging within the body.

OCT can be compared to other imaging modalities in terms of imaging resolution and imaging depth of penetration. Figure 7.1 shows the resolution and imaging depth of different imaging modalities, highlighting that higher resolution is customarily associated with reduced penetration depths. OCT occupies and bridges the gap between optical microscopy and ultrasound imaging. The axial resolution in OCT is determined by the spectral bandwidth of the light source. OCT systems commonly have axial resolutions ranging from 1–10 μm, which is 10–100 times better than ultrasound imaging. The diagnostic potential of OCT is enhanced with the use of various endogenous contrast mechanisms, such as tissue birefringence, spectroscopic properties, blood flow, and oxygenation or with the use of exogenous contrast agents or in combination with other imaging modalities.

OCT has found widespread clinical applications (Drexler and Fujimoto, 2015; Lumbroso et al., 2015; Wang et al., 2017), mainly in ophthalmology because of the transparent tissue of the eye and is now a standard imaging modality for retinal imaging. The resolving capability of OCT enables visualization of intricate retinal structures and markers of disease (Wojtkowski et al., 2005). In cardiology, OCT has been applied to detect vulnerable plaques (Bouma et al., 2003) and to guide intravascular procedures such as stent placement. OCT has also been applied for diagnosis of various cancers as well as for visualizing airway profiles (McLaughlin et al., 2014). Other major applications are in gastroenterology, dermatology, and otolaryngology.

The scope of this chapter is to present the background of the principles of OCT. The chapter begins with an overview of the fundamental aspects of OCT, system performance, and various contrast techniques. This chapter also provides an overview of biomedical applications of OCT and concludes by summarizing the translational application of OCT.

7.2 Principles of OCT

7.2.1 Low-Coherence Interferometry

The physical principle of OCT relies on a classical optical measurement technique called low-coherence interferometry (LCI), or white-light interferometry, which was first reported by Sir Isaac Newton. Interferometry is a powerful technique that detects the echo time delay and magnitude of backscattered light with high sensitivity. LCI was initially used to detect flaws in optical fibers and waveguide devices in telecommunication networks (Takada et al., 1987). In 1988, Fercher et al. reported the first biological application by measuring the axial length of the eye using LCI, which was soon followed by other schemes that were developed for biological applications (Schmitt et al., 1993; Tanno and Ichimura, 1994).

Interferometry uses a correlation method by interfering the backscattered light from the sample with light that has traveled through a reference path with known optical delay. Figure 7.2 shows a schematic of a Michelson-type interferometer. The incident light source is divided into two paths: a sample beam and a reference beam. Light reflected back from the sample and the reference arm are combined to form an interference signal that is detected by a photodetector. Constructive interference generates a fringe pattern with good fringe visibility within the coherence length of the light source. The electric field at the output of the interferometer is a sum of the sample and reference fields. The detector then

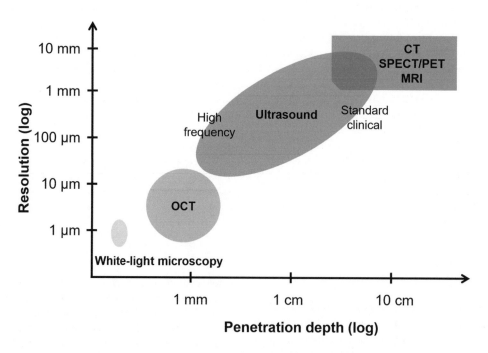

FIGURE 7.1 Comparison of OCT with standard clinical imaging modalities. An observable trade-off and trend is that resolution generally improves at the sacrifice of imaging depth.

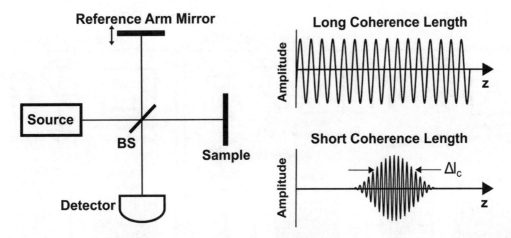

FIGURE 7.2 Schematic shows the concept of low coherence interferometry. Using a short coherence length light source, interference is observed only when the path lengths of the two interferometer arms are matched to within the coherence length (Δl_c) of the light source. BS: Beam splitter.

measures the intensity, which is proportional to the square of the electromagnetic field:

$$I_0 \sim |E_r|^2 + |E_s|^2 + 2 \cdot E_r \cdot E_s \cos(2 \cdot k \cdot \Delta L), \quad (7.1)$$

where ΔL is the path length difference between the sample and reference arms and k is the wavenumber. If the reference path is scanned with a constant velocity, then the interference signal is generated as a function of time, producing a Doppler shift of the reference field. If the probing beam is from a highly coherent source rather than a low-coherence source, then the interference signal will be produced over a wide range of relative path length delays of the sample and reference arms with a uniform maximum intensity. However, in LCI, a source with a broad bandwidth is used to more narrowly localize and then detect the optical interference. Because a low-coherence light source (phase discontinuities over a distance) is used, the interference of the light occurs only when the optical path lengths of the reference and sample arms are matched. The axial depth scan of a sample is measured by varying the reference arm pathlength, while the detected signal is bandpass filtered and demodulated to acquire the envelope of the signal.

The addition of lateral scanning of the sample arm beam led to the breakthrough paper published in *Science*, calling this technique OCT (Huang et al., 1991). OCT enabled cross-sectional, high-resolution imaging of a biological sample, based solely on the spatial optical scattering properties within the tissue. The position of the lateral beam can be scanned in either or both transverse directions to generate two-, three-, and four- (volumes over time) dimensional images, which will be discussed in a later section. The intensity of the back-reflected signal can be displayed in false-color or gray-scale as an OCT image. OCT can be implemented in either a time domain OCT (TD-OCT) or Fourier domain OCT (FD-OCT) system.

7.2.2 OCT System Variants

7.2.2.1 Time-Domain Optical Coherence Tomography

TD-OCT (Figure 7.3) operates as described above in the LCI section, in which a reference mirror is scanned at a constant velocity. In TD-OCT, the backscattered signal is detected using a single photodiode or multiple detectors, depending on the detection configuration, such as in point-scanning (Boppart et al., 1997a), line-field (Koch et al., 2004), or full-field TD-OCT (Beaurepaire et al., 1998). Here, the broadband light is split into the reference and sample arms and the backscattered light interference is then detected by a photodiode. The signal is demodulated and then processed on a computer and displayed as an image. Point-scanning OCT is a relatively simple setup, but depth information is limited to the time required to translate the reference mirror. A major limitation of this approach is the maximum rate that can be achieved with mechanical reference-arm scanning, which limits the image acquisition rate and gives rise to motion artifacts in the images. Alternatively, to improve acquisition speed and eliminate mechanical scanning, configurations such as line-field OCT and full-field OCT have been developed (Beaurepaire et al., 1998; Koch et al., 2004). Nonetheless, implementation and clinical use of TD-OCT have been reduced because of the significant sensitivity improvement offered by FD-OCT systems.

7.2.2.2 Frequency-Domain Optical Coherence Tomography

The conventional TD-OCT system provides limited sensitivity (90–95 dB) and imaging speed (typically <4 kHz A-line rate), which limits the use of the system in many *in vivo* imaging applications. Because the imaging speed has a relationship to the electronic detection bandwidth, increasing the A-line rate requires a concomitant increase in the electronic detection bandwidth, and hence the sensitivity drops. Many biological applications require a sensitivity over 100 dB for sufficient imaging depth penetration and cannot afford to compromise sensitivity with an increased A-line rate. This limitation has led to the development of frequency domain OCT, which is now the most dominant implementation of OCT. The concept of spectral domain detection was first reported in 1995 (Fercher et al., 1995), followed by reports on its sensitivity improvements (Ha Usler and Lindner, 1998). In 2003, the OCT community fully realized and utilized the advantages of

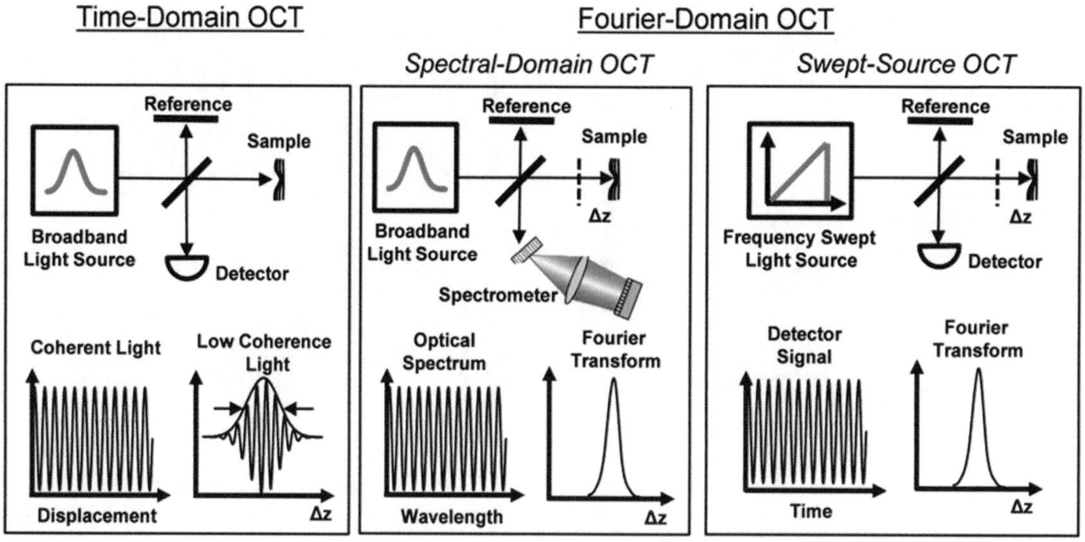

FIGURE 7.3 Schematic of different OCT modalities. OCT systems can be classified as either time-domain (TD) or Fourier-domain (FD) systems. FD-OCT systems can be further divided into spectrometer-based spectral-domain (SD) and swept-source (SS)-based systems. Reproduced with permission from Swanson and Fujimoto (2017).

FD-OCT over TD-OCT (Choma et al., 2003; de Boer et al., 2003; Leitgeb et al., 2003). FD-OCT has the advantage of obtaining full depth-resolved information without moving the reference mirror. FD-OCT systems are subdivided into spectral-domain OCT (SD-OCT) or swept-source OCT (SS-OCT) configurations.

In a SD-OCT system (Figure 7.3), the light source remains the same as in TD-OCT, but the reference arm is kept stationary. The spectral interference signal from all depths in the sample is dispersed by a spectrometer and collected simultaneously on an array detector such as a charge-coupled device (CCD) or complementary metal-oxide semiconductor (CMOS) camera. The depth information is then obtained by taking the Fourier transform of the spectrally resolved interference signal. However, the spectrometer generates a spectrum that is a function of wavelength, and hence the spectral signals obtained with the spectrometer are not evenly spaced in k-space. To obtain a proper depth signal, all SD-OCT processing employs a background subtraction and then the spectral signal must be numerically resampled from λ-space to k-space using various calibration methods (Wojtkowski et al., 2002).

The maximum imaging depth range is dependent on the spectral resolution and the detector size of the linear detector. The maximum range is defined as:

$$I_{max} = \frac{1}{4n} \frac{\lambda_0^2}{\Delta\lambda_s} N, \qquad (7.2)$$

where n is the refractive index of the sample, λ_0 is the center wavelength of the source, $\Delta\lambda_s$ is the spectrometer bandwidth, and N is the number of array detectors.

SD-OCT systems enable OCT imaging at rates more than 100 times faster compared to first-generation OCT systems (Wojtkowski, 2010). It must be noted that SD-OCT must be operated at higher speeds because motion artifacts will induce averaging of interference fringes if the acquisition speed is too slow. To date, SD-OCT has had a dramatic impact in ophthalmology because of its improved imaging capabilities (Fujimoto and Swanson, 2016). High-speed imaging has a major advantage as it provides a greater number of A-scans per second, and hence either a larger number of A-scans per B-mode image or a greater number of transverse B-mode scans (Potsaid et al., 2008), allowing larger fields-of-view or the capture of more dynamic biological motion or processes. Although SD-OCT has better speed and sensitivity, it also has artifacts that are not present in a TD-OCT system, such a "mirror" artifact and a detection sensitivity that degrades as a function of depth. The Fourier transform of the spectral signal cannot distinguish between positive and negative delays, and hence forms an OCT image from both positive and negative delays. Therefore, if the sample is at zero delay, the OCT image produces an overlapping or folded signal and produces a "mirror" artifact. This will result in an axial scan of $N/2$ pixels; therefore, a larger number of N pixels is required to create a larger depth scan with high pixel resolution (refer to above equation). Alternatively, full-range complex imaging methods have been developed to overcome the "mirror" artifact (Yasuno et al., 2006; Baumann et al., 2007). In addition, SD-OCT has a sensitivity roll-off over depth. SD-OCT is more sensitive to the reflections at zero delay, and the sensitivity decreases away from this zero delay. This is due to the limited spectral resolution of the spectrometer line array camera, as the longer path length (away from the zero-delay position) leads to modulations of the spectrum with higher frequencies.

Swept-source OCT (SS-OCT) (Figure 7.3), also called optical frequency domain imaging (OFDI), is the second type of Fourier domain detection used in OCT systems. SS-OCT uses an interferometer, a narrow band, tunable light source, and a single-element detector to measure the interference signal as a function of time. OCT using swept-source detection was first demonstrated in 1997, but the sensitivity and speed advantages were not recognized in this initial study (Chinn et al., 1997;

Golubovic et al., 1997). In 2003, researchers demonstrated OCT imaging with swept-source technology and realized the sensitivity advantages over TD-OCT (Choma et al., 2003; Yun et al., 2003b).

The output of a swept source is split into sample and reference arms, as before. Light in the sample is backscattered from the inherent refractive index variations of the sample as the reference beam is reflected from a stationary mirror at a fixed delay. The sample and reference beams have a time offset that is determined by the path length difference. Because the frequency (wavelength) of the light is swept as a function of time, the light backscattered from different depths in the sample will have a frequency offset from the reference beam and will subsequently produce a modulation of the interference signal. The depth-resolved signal is then obtained by taking the inverse Fourier transform of the interference signal. The axial resolution in SS-OCT is determined by the sweep range of the laser. Because the frequency sweeps are not linear in time, it is necessary to rescale the digitized signal linearly in k-space. If the calibration of the swept source is available, then the correction can be done computationally. To avoid the computational time, modern swept sources are equipped with a k-clock that can rescale the spectrum during the acquisition process.

SS-OCT has additional advantages over SD-OCT systems. SS-OCT can be implemented using a balance-mode detection scheme to cancel the strong background reflection in the interferometer. SD-OCT typically uses silicon detectors for wavelengths in the 800–1060 nm range. More expensive and slower line-array cameras using indium gallium arsenide (InGaAs) detectors are required for wavelengths around 1300 nm. SS-OCT, however, can more readily operate at longer wavelengths and can often operate at much faster acquisition rates than SD-OCT systems. Additionally, signal roll-off over depth in SS-OCT has been significantly reduced, mainly because of improvements in the coherence properties of the laser source and the highly efficient broad bandwidth detectors. The conventional tunable laser uses bulk or fiber components that result in a relatively longer resonator cavity length and limit high sweep rates because of the long round trip time (Yun et al., 2003a; Choma et al., 2005; Yasuno et al., 2005; Okabe et al., 2012). To overcome this issue, two novel source technologies have been developed: a Fourier-domain mode locking (FDML) laser and a vertical cavity surface-emitting laser (VCSEL). Both sources have significantly improved sweep rates (in kHz–MHz) with an increased depth range from tens of millimeters to a meter (Jayaraman et al., 2012; Wieser et al., 2014; Wang et al., 2016b).

7.2.3 OCT System Performance

7.2.3.1 Axial Resolution

Important parameters for characterizing OCT systems include the imaging resolution (both axial and transverse), the signal-to-noise ratio, the phase stability, and the imaging speed. Unlike conventional microscopy, one advantage of OCT is that the axial resolution and transverse resolution are independent. The axial resolution of an OCT system is governed by the

Spatial Resolution

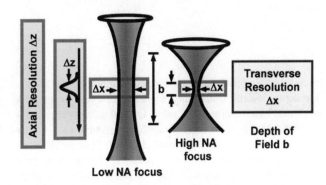

FIGURE 7.4 Resolution limits of OCT. OCT can achieve high-axial resolution independent of the numerical aperture. Using low-coherence interferometry, the axial resolution is inversely proportional to the bandwidth of the light source, while the transverse resolution is given by the focal spot size. The depth-of-field is determined by the confocal parameter of the focused beam. Reproduced with permission from Swanson and Fujimoto (2017).

coherence length of the light source and is inversely proportional to the source bandwidth (Schmitt, 1999):

$$\Delta z = \frac{2 \ln 2}{\pi} \cdot \frac{\lambda^2}{\Delta \lambda}, \tag{7.3}$$

where λ is the center wavelength of the source and $\Delta \lambda$ is the bandwidth of the source. To achieve a higher axial resolution, a source with a larger spectral bandwidth is desired. Axial resolution can also be increased by decreasing the center wavelength of the source. However, shorter wavelengths are more highly scattered and will result in lower penetration in a biological sample. Higher axial resolution is particularly important for detecting early structural changes in tissue from various diseases occurring at the cellular level. In OCT, wavelengths typically around 800 nm, 1060 nm, and 1300 nm are used for imaging because these wavelengths fall within the "optical window" in tissue, where light attenuation is more dependent on scattering than on absorption.

7.2.3.2 Transverse Resolution

The transverse resolution Δx is governed by the numerical aperture of the focusing optics and is given as the focused spot size. The transverse resolution is given by (Saleh and Teich, 1991):

$$\Delta x = \frac{4\lambda}{\pi} \cdot \frac{f}{d}, \tag{7.4}$$

where d is the diameter of the beam incident on the focusing lens and f is its focal length. Figure 7.4 shows the relation between the focused spot size and the depth-of-field for high and low numerical apertures. Higher transverse resolution is achieved by focusing the beam to a small spot size by using an objective with a higher numerical aperture. However, there is a trade-off between transverse resolution and the depth-of-field.

Depth-of-field is defined by the confocal parameter $2z_R$, which is two times the Rayleigh range and is given by (Saleh and Teich, 1991):

$$2z_R = \frac{\pi\Delta x^2}{2\lambda}. \tag{7.5}$$

Increasing (improving) the transverse resolution will reduce the depth-of-field. Commonly, higher transverse resolution is preferred based on the application. Optical coherence microscopy (OCM) provides higher transverse resolution but with a limited depth-of-field. Depth-of-field limitations in OCM or OCT can be overcome in time domain detection by using a dynamic focusing method (Lexer et al., 1999). In spectral-domain detection methods, with their superior phase stability, computational techniques such as interferometric synthetic aperture microscopy (ISAM) (Ralston et al., 2007) or computational adaptive optics (CAO) (Adie et al., 2012) can be utilized to overcome such depth-of-field limitations.

7.2.3.3 System Sensitivity

The sensitivity of an OCT system is defined as the ratio of the signal obtained from a perfect reflector placed at the focus to the noise of the system. The sensitivity of the system is also proportional to the optical power incident on the sample and inversely proportional to the A-scan rate and is given by (Fercher et al., 2003):

$$S = \frac{1}{4} \cdot \frac{\eta\lambda}{hc} \cdot \frac{P_s}{B}, \tag{7.6}$$

where η is the quantum efficiency of the detector, h is Plank's constant, c is the speed of the light, P_s is the optical power incident on the sample, and B is the electronic detection bandwidth. FD-OCT systems provide a significant improvement in detection sensitivity and acquisition speed compared to TD-OCT systems. The sensitivity advantage of a SD-OCT system over a TD-OCT system is mainly due to the significant reduction of shot noise by replacing the single-element detector with multielement arrays of detectors (Nassif et al., 2004). In TD-OCT systems, the shot noise generated at one particular wavelength is present at all frequencies, hence affecting the SNR at all wavelengths. In SD-OCT, the spectrum is dispersed to separate detectors and therefore significantly reduces the shot noise with this configuration.

7.2.3.4 Phase Stability

Phase stability of an OCT system is of utmost importance for a wide range of phase-sensitive processing techniques, including computational optical interferometric techniques such as CAO and ISAM and others described in sections "Phase-Signal Based OCTA Techniques" and "Complex-Signal Based OCTA Techniques." At optical wavelengths, the phase of scattered light is sensitive to motion. This may be due to fluctuations in the reference arm, transverse and axial disturbances, electric noise, and low SNR (Ralston et al., 2006; Vakoc et al., 2009; Shemonski et al., 2014a).

Phase changes of an OCT system are measured by calculating the phase difference between consecutive A-lines at a depth position Δz using a mirror as a sample (Shemonski et al., 2014b):

$$\Delta\varphi = 2n_s k_0 \Delta z, \tag{7.7}$$

where n_s is the overall refractive index of the material, $k_0 = 2\pi / \lambda_0$ is the wave number in air, and λ_0 is the central wavelength of the source. The factor of 2 is due to the double-pass configuration in OCT.

Comparing TD-, SD-, and SS-OCT systems, SD-OCT is known to have the best phase stability because of the absence of moving parts (i.e. absence of a scanning reference mirror or a scanning light source). The second most stable system is SS-OCT; however, this type of system is less stable compared to SD-OCT systems because of the scanning light source (Choma et al., 2006; Sarunic et al., 2006). Recent developments in new swept sources that do not include any mechanical movements tend to have better phase stability (Bonesi et al., 2014; Sasaki et al., 2014). Finally, TD-OCT is the least stable OCT modality because of the slow imaging speed, low SNR, and the moving reference arm components used to vary the optical path length.

7.2.3.5 Acquisition Rates

Imaging speed is one of the important parameters used to evaluate an OCT system. First-generation TD-OCT system speed achieved a maximum speed of 1 kHz and was limited by the speed of the scanning mirror in the reference arm (Huang et al., 1991). With further improvement in axial scan speed, fast dynamic events, such as the beating heart of a tadpole, were able to be captured (Boppart et al., 1997b; Rollins et al., 1998). The development of FD-OCT enabled a breakthrough in OCT technology with improved sensitivity and speed (Choma et al., 2003; de Boer et al., 2003; Leitgeb et al., 2003). Because the FD-OCT technology no longer needed a scanning mirror, the imaging speed was limited only by the CCD line scan camera rate or the sweeping rate of the optical source. Modern OCT systems can achieve an axial scan rate of 1–10 MHz and enable real-time volumetric OCT (4D) imaging (Wieser et al., 2014).

7.3 Contrast Techniques

In conventional OCT, the mechanism of contrast relies on the variations in optical refractive index, and hence the backscattering of light, from the sample. However, the spectrum of the backscattered light contains additional contrast that has significantly improved the diagnostic capability of OCT. A number of intrinsic properties of tissues have been utilized by developing functional imaging techniques such as refractive index, polarization, spectroscopic, elastography, and blood flow imaging techniques. Additionally, these techniques have enabled the imaging of tissues to understand their normal physiology as well as the pathology of tissues. This section briefly discusses the application of these techniques for improved diagnostics.

7.3.1 Refractive Index

Variations in refractive index distribution within the sample are the primary source of the OCT signal. The refractive index of tissue has been found to be altered by the hydration state (Kim et al., 2004), calcification (Fujimoto et al., 2000), and tumor malignancy (Das et al., 1997; Zysk and Boppart, 2006; Zysk et al., 2009), among others. At the cellular level, studies have shown that refractive index is sensitive to nuclear changes resulting from mitosis (Boppart et al., 1998) and dysplasia (Gurjar et al., 2001). The group refractive index and the thickness of the sample is measured by obtaining the optical path length and optical shift of the samples. Based on this principle, several techniques for measuring the refractive index of the sample have been developed. This includes a focus-tracking method (Tearney et al., 1995), projected index computed tomography (Zysk et al., 2003), bifocal optical coherence reflectometry (Zvyagin et al., 2003), a needle-based sensing device (Zysk et al., 2007a), a Fresnel reflection coefficients method (Zysk et al., 2007b), and a linear regression analysis method (de Freitas et al., 2013).

7.3.2 Polarization-Sensitive OCT

Polarization-sensitive OCT (PS-OCT) provides an additional contrast mechanism that utilizes changes in the depth-dependent polarization states of the detected light. Several materials and tissues change the polarization state of the incident light due to properties such as birefringence, dichroism, and optic axis orientation. The first reported study on the use of polarization-sensitive detection for one-dimensional measurements was in 1992 (Hee et al., 1992). Since then, PS-OCT has been widely used for volumetric imaging at high speed and sensitivity and applied in various research investigations. A PS-OCT system is similar to a conventional TD-OCT or FD-OCT system except for the addition of a polarizer, polarization beam splitter, quarter wave plate, and an additional detector. The extraction of the complete polarization properties is directly determined by studying Stokes parameters (Boer et al., 1997; Hitzenberger et al., 2001) or by determining Jones or Mueller matrices (Yao and Wang, 1999; Park et al., 2004). PS-OCT has been used in a wide range of applications such as to determine burn depth in skin with decreased in birefringence (Park et al., 2001), to measure the retinal nerve fiber layer in ophthalmology (Duan et al., 2012), for brain imaging (Wang et al., 2016a), for changes in collagen density in oncology (South et al., 2014), and for assessing airway smooth muscle (Adams et al., 2016).

7.3.3 Spectroscopic OCT

Spectroscopic OCT (SOCT) (Leitgeb et al., 2000; Morgner et al., 2000) is an extension of OCT that not only provides structural information but also rich spectroscopy information of the sample. SOCT data is retrieved by utilizing the spectral bandwidth of a light source. SOCT can be implemented either in hardware (Kim et al., 2014; Tanaka et al., 2015), which uses two or more light sources with different bands of wavelength, or by using software approaches and the time-frequency transformation (Graf and Wax, 2007; Oldenburg et al., 2007; Robles et al., 2009). SOCT provides information on the wavelength-dependent scattering or absorption of the sample and can be used to detect endogenous molecules or exogenous agents. There is an inherent trade-off between the spectral and spatial resolutions, and this can be mitigated in OCM, which uses only a narrow depth-of-field (Xu et al., 2006). Performing SOCT with visible light sources is currently gaining more interest because of the fact that hemoglobin absorption is more than an order-of-magnitude greater at visible wavelengths compared to near-infrared wavelengths (Faber et al., 2005). The spectral analysis is usually mapped in a false color scale for the NIR region and in true color that matches our visual perception for the visible region (Robles et al., 2011; Li et al., 2012b). SOCT has been widely used for measuring the concentrations of hemoglobin and blood oxygen saturation levels (Oldenburg et al., 2009; Boustany et al., 2010; Chong et al., 2015). These parameters have also been useful for detecting lipid in atherosclerotic plaques (Fleming et al., 2013; Nam et al., 2016) and evaluating burn injuries (Maher et al., 2014; Zhao et al., 2015).

7.3.4 Optical Coherence Elastography

The mechanical properties of tissue are important parameters to assess for physiological and pathological changes. Quantification of such parameters may provide early detection of structural changes in the tissue. Optical coherence elastography (OCE) is a method that maps the mechanical properties of the tissue from displacement measurements. OCE was first proposed and demonstrated in 1998 (Schmitt, 1998). Following this, several techniques to measure displacement have been developed. The quality of the elastogram is mainly dependent on the accuracy and dynamic range of the displacement measured. The displacement of the tissue in OCE is measured using several different techniques: speckle tracking (Schmitt, 1998; Chan et al., 2004), phase-sensitive detection (Wang et al., 2006, 2007b), and the use of the Doppler spectrum (Szkulmowski et al., 2008; Adie et al., 2009; Kennedy et al., 2012b). A wide variety of OCE techniques have been developed, largely based on the method by which a mechanical perturbation is induced in the tissue. These include compression (Kennedy et al., 2011, 2012a), surface acoustic waves (Li et al., 2012a; Manapuram et al., 2012), acoustic radiation forces (Soest et al., 2007; Qi et al., 2012), magnetomotive forces (Oldenburg and Boppart, 2010), and mechanical spectroscopy methods (Adie et al., 2010). The application of OCE has been used for the generation of strain maps in engineered tissues *in vitro* and in the *Xenopus laevis* tadpole *in vivo* (Ko et al., 2006), in arterial plaques (Rogowska et al., 2006), for intraoperative breast tumor margin identification (Kennedy et al., 2015), and for monitoring therapeutic procedures on the cornea (Singh et al., 2016).

7.3.5 Optical Coherence Tomography Angiography

An OCT-based angiography (OCTA) method provides the blood flow information in tissue by using either phase or speckle information. This section is divided into three categories: phase-signal based OCTA, intensity-signal based OCTA, and complex-signal based OCTA.

7.3.5.1 Phase-Signal Based OCTA Techniques

Doppler OCT combines the Doppler principle with OCT to obtain high-resolution images of static and moving scatterers simultaneously in highly scattering biological samples. Doppler OCT signals can be retrieved by measuring the fringe carrier frequency (Izatt et al., 1997) or by processing sequential phase-resolved A-scans (Zhao et al., 2000). Doppler OCT has been used to measure the blood flow in the retina in various ocular diseases (Blatter et al., 2013) and in brain research (Srinivasan et al., 2010). Doppler OCT has also been used to monitor *in vivo* blood flow during photodynamic therapy (Standish et al., 2008). In the future, Doppler OCT could advance high-resolution imaging of blood flow in living systems for fundamental and translational research.

Phase variance OCT (PV-OCT) is an alternative angiography technique that calculates the phase variance of the motion-corrected phase changes acquired between adjacent B-scans (Fingler et al., 2008). PV-OCT is capable of imaging changes across a wide range of velocities and orientations, in contrast to Doppler OCT (Fingler et al., 2009). Implementation of PV-OCT in a SS-OCT system showed higher sensitivity and reduced fringe wash-out for high blood flow velocity compared to a SD-OCT system (Poddar et al., 2014). The angiograms from PV-OCT of patients with retinal disease showed similar vascular information as provided by fluorescein angiography (Schwartz et al., 2014). This indicates the significant potential of PV-OCT in ophthalmology. In a recent study, PV-OCT was applied to track the regeneration of the microvascular network in a diabetic wound-healing study (Li et al., 2017a).

7.3.5.2 Intensity-Signal Based OCTA Techniques

Speckles in OCT play a dual role, both as a source of noise and as a carrier of information, such as in the case of blood flow (Schmitt et al., 1999). The adaptation of speckle for blood flow imaging in OCT was developed as an alternative technique to Doppler OCT (Barton and Stromski, 2005). Intensity amplitude-based angiography maps are popular, as these techniques are not dependent on the phase of the signal. The intensity amplitude-based methods include speckle variance (Mariampillai et al., 2010), correlation mapping (Jonathan et al., 2011), split-spectrum amplitude decorrelation angiography (SSADA) (Jia et al., 2012), and correlation masked speckle variance (Choi et al., 2014). These techniques take advantage of the speckle effect in the vascular (flow) regions compared to the bulk (nonflow) regions. All the above-mentioned techniques have demonstrated flow networks in brain, retina, and skin. In particular, the SSADA algorithm can detect the retinal vasculature in the human retina and is able to differentiate between normal and diseased eyes (Spaide et al., 2015). This technique has already been implemented in commercial clinical grade OCT systems (AngioVueHD, Optovue) for imaging the retinal vasculature along with the standard structural imaging.

7.3.5.3 Complex-Signal Based OCTA Techniques

The optical microangiography (OMAG) technique provides 3D images of the blood flow by analyzing the spatial frequency of the time-varying spectral interference signal. This analysis separates and differentiates the signal that is backscattered from static and moving particles. The OMAG technique can be implemented by modulating the reference arm mirror (Wang et al., 2007a) or by using the modified Hilbert transform (Wang et al., 2010). The OMAG technique provides high-contrast 3D images of the microvasculature of the human retina for diagnosing various retinal diseases (Zhang et al., 2016) and has recently been adapted in commercial OCT systems (ZEISS AngioPlex™). The OMAG technique has also been investigated for use in skin diseases (Choi and Wang, 2014), in the brain (Jia and Wang, 2010), and in cochlear research (Subhash et al., 2011).

7.3.6 Exogenous Contrast Agents for OCT

OCT generates morphological images based on the spatially varying scattering properties or refractive indices of tissue and is not directly sensitive to molecular composition. Exogenous contrast agents, which have been developed and used in every other biomedical imaging modality, can be functionalized with antibodies or molecules to enhance the diagnostic capabilities by imaging specific tissue or cell types. Several groups have explored and developed novel contrast agents and methods such as magnetomotive OCT (Crecea et al., 2009), pump-probe OCT (Rao et al., 2003), and photothermal OCT (Skala et al., 2008) for OCT imaging. Newly engineered scattering-based contrast agents, such as protein-shell oil-core microspheres, have been developed and characterized for OCT imaging and were found to produce strong contrast enhanced OCT imaging in the mouse liver (Lee et al., 2003). *In vivo* measurements have demonstrated that the scattering variations were strong in the vascular regions of exposed mouse intestinal wall and with minimal changes observed in an avascular region (Boppart et al., 2005). Novel magnetic nanoparticles have been developed and were functionalized to target the human epidermal growth factor receptor 2 (HER2 neu) protein. Magnetomotive OCT provides contrast by temporally modulating an external magnetic field gradient. Using magnetic nanoparticles, *in vivo* magnetomotive OCT in a preclinical mammary tumor model was demonstrated (John et al., 2010). The feasibility of photothermal OCT using endogenous (melanin) and exogenous (gold nanorods) contrast in the eye has been reported (Lapierre-Landry et al., 2017). The pump-probe OCT technique provides molecular contrast by spatially resolving the pump-probe interaction. *In vivo* imaging using methylene blue dye showed that the pump-probe method could clearly identify *Xenopus laevis* vasculature (Carrasco-Zevallos et al., 2015).

7.3.7 Multimodal OCT

A limitation of OCT is its inability to acquire molecular information from a sample, which can make it difficult to distinguish between pathological and normal tissue. Multimodal imaging, using OCT in combination with other optical imaging methods, offers the potential to improve the accuracy and efficacy of imaging for diagnostics and provides a more comprehensive characterization of the tissue response.

Combining laser-induced fluorescence (LIF) spectroscopy with OCT provides both biochemical composition and structural information of the tissue. The combination of OCT with LIF spectroscopy was used to image tumor boundaries in the cervix and showed fewer false positive results than either single modality alone (Kuranov et al., 2002). Several other studies have showed improved efficiency and sensitivity for early diagnosis of bladder cancer (Pan et al., 2003), colon cancer (Tumlinson et al., 2004; Hariri et al., 2006), and diagnosis of macular degeneration (Rosen et al., 2009).

Furthermore, the combination of Raman spectroscopy (RS) with OCT also provides molecular information of the specimen. The Raman signal is derived from the rotational and/or vibration states probed in the sample. This combination has led to the identification of early dental caries (Ko et al., 2005), differentiation between normal and cancerous skin (Patil et al., 2008), and to discriminate colon tissue (Ashok et al., 2013). In all the aforementioned studies, the obtained RS signal was volume integrated from different depths, resulting in an average RS signal. Depth-resolved RS, enabled by implementing a confocal Raman configuration, eliminates the out-of-focus signal (Khan et al., 2014).

Another study reported using the coherent properties of nonlinear processes, such as coherent anti-Stokes Raman scattering (CARS) and second harmonic generation (SHG), to enhance OCT contrast and hence named this as molecularly sensitive OCT (Vinegoni et al., 2004; Bredfeldt et al., 2005). Enhancement in contrast was also obtained with a method called nonlinear interferometric vibrational imaging (NIVI) (Marks and Boppart, 2004). Molecularly sensitive OCT imaging of a biological specimen was demonstrated for a thin, lipid-dense layer of beef tissue sandwiched between two glass slides and showed improved OCT contrast (Bredfeldt et al., 2005).

The combination of multiphoton microscopy (MPM) and OCT provides enhanced 3D optical sectioning from thick scattering samples (Vinegoni et al., 2006). MPM is comprised of a set of nonlinear processes such as two-photon excited fluorescence (TPF) and SHG. Multiphoton microscopy and OCT found wide-spread biomedical applications in tissue engineering and cell biology (Vinegoni et al., 2006), cellular imaging (Chong et al., 2013), dermatology (König et al., 2009; Alex et al., 2013), and oncology (Jo et al., 2010). Compact all-fiber-optic scanning multimodal endomicroscopes have also been developed for simultaneous OCT and TPF imaging (Xi et al., 2012).

The above-mentioned multimodal techniques provide additional contrast for differentiating specimens but cannot enable deeper imaging depths. To address this issue, several groups have reported dual-modality OCT and photoacoustic imaging (PI). PI uses photoacoustic effects to detect laser pulse-induced photoacoustic waves, and a 3D chromophore has been reconstructed with this method (Zhang et al., 2006). PI detects broad bandwidths from different imaging depths, which can be up to a centimeter deep in a specimen (Laufer et al., 2012). The combination of OCT with high-resolution PI has been utilized in *in vivo* studies such as in determining the metabolic rate of oxygen in the mouse ear (Liu et al., 2011), retinal imaging in rats (Jiao et al., 2010), neovascularization in a porous scaffold (Cai et al., 2013), and in epilepsy mapping (Tsytsarev

et al., 2013). With the increased depth configuration, OCT/PI is being used for high-resolution 3D imaging of vascular structures in murine and human skin (Zhang et al., 2011). The combination of OCT with other modalities has shown higher sensitivity and higher specificity and, therefore, enhances the diagnostic capabilities of OCT.

7.4 OCT Applications: From Cellular to Animal Imaging

7.4.1 Cellular Imaging

The application of OCT in developmental biology and cellular biology research has grown rapidly because of the improvement in axial resolution and imaging speed (Drexler et al., 2001a). Excisional biopsy, followed by histopathological assessment, is still considered the gold-standard for assessment of a disease condition at the cellular and tissue levels. However, histopathological reports can be subjected to false negative reports due to sampling error. Optical imaging techniques, with resolutions at or near that of histopathology, could improve the capability of identifying early-stage diseases that occur at a cellular level.

An *in vitro* study from a carcinogen-induced rat mammary tumor model enabled visualization of anatomical features and good correlation with histology was obtained (Luo et al., 2005). This study has led OCT to be used in intraoperative imaging for assessing human lymph nodes for metastatic cancer (Nguyen et al., 2009; Nolan et al., 2016). Several other studies have used OCM for imaging malignant pathologies in multiple tissues and. in particular, to discriminate cancerous tissue from normal tissue (Clark et al., 2004; Zhou et al., 2010; Lee et al., 2012). With improved axial and transverse resolution and dynamic contrast, detection of most of the retinal cell types and their dynamics were revealed in macaque retinal explants (Thouvenin et al., 2017). As shown in Figure 7.5, dynamic cells with stationary (fibrous – red), slow moving (cells with large nuclei – green), and fast moving (smaller structures – blue) organelles are revealed.

7.4.2 Tissue Engineering

Over the past decades, tissue engineering has emerged as a promising therapeutic solution in regenerative medicine. Tissue engineering comprises methods in which specific cells are engineered or processed to grow into the required tissue architecture *in vitro*. Cells, scaffolds, and culture environments are the three crucial elements in monitoring the cell and engineered tissue growth profile. The potential of using OCT and multimodal imaging for tissue engineering has emerged with extended functions and new contrast abilities.

7.4.2.1 Imaging of Scaffolds

Scaffolds used in tissue engineering act as a template for tissue regeneration and the parameters, such as material, pore size, and porosity, affects the cell activity and distribution within the scaffold. OCT, along with a novel processing algorithm,

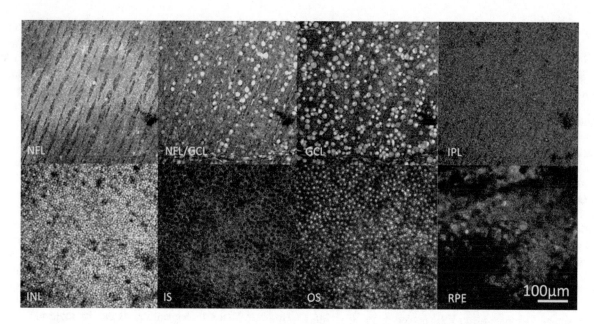

FIGURE 7.5 Color coding the dynamic signal for rate of signal decorrelation time allows for the discrimination of different structures in retinal layers. Shown are color-coded, three-channel stacks of dynamic FF-OCT images of the macaque retina at 2 hours postmortem; color coding for speed of movement/signal decorrelation: the red channel shows low temporal frequencies (< 0.5 Hz); green shows the intermediate frequencies (between 0.5 and 5 Hz); and blue shows the fastest pixels (> 5 Hz). Fibrous structures are slowest (red); cells are intermediate (green); and subcellular details are fastest (blue). NFL: Nerve fiber layer; GCL: Ganglion cell layer; IPL: Inner plexiform layer; INL: inner nuclear layer; OS: Outer segments; RPE: Retinal pigment epithelium. Reproduced with permission from Thouvenin et al. (2017).

was used to investigate parameters such as a porous scaffold, pore interconnectivity, and size (Chen et al., 2011). OCT images of porous hydrogel scaffolds were able to resolve micropores and hydrogel in both cross-sectional and *en face* images (Chen et al., 2011). Integration of perfusion methods such as Doppler OCT has led to identifying heterogeneous distributions of flow velocity in the porous structures (Jia et al., 2009). The combination of Doppler OMAG enabled the mapping of slow flow rates more precisely, which was difficult to map with Doppler OCT alone (Wang and An, 2009). In addition to above-mentioned properties, the mechanical properties of the scaffold are important and affect the regeneration of the tissue (Hunter et al., 2002). OCE has the potential to detect the extracellular matrix that is altered mainly when the scaffold is seeded with cells. Figure 7.6 shows an array of structural and OCE images acquired from a 3D scaffold. In this investigation, half of the scaffold was seeded with cells and the other half served as a cell-free control. The samples were imaged for 10 days, and the OCE maps showed increased stiffness over the culture period. These results demonstrated that OCE could be used to measure the stiffness and biomechanical properties in a nondestructive manner (Ko et al., 2006).

7.4.2.2 Cell Dynamics

In addition to scaffolds, cellular activities in these environments determine the fate of tissue regeneration. OCT has been utilized to investigate cellular dynamics in engineered tissue because of the refractive index difference between cells and scaffolds (Yang et al., 2006; Liang et al., 2009; Levitz et al., 2010). A longitudinal study revealed that OCT could accurately measure the cell growth and dynamics in a chitosan

scaffold (Tan et al., 2004). The quantitative analysis of cell proliferation on a poly (lactide) (PLA) scaffold reported a statistically significant difference compared to a blank scaffold (Yang et al., 2005). In addition to cell proliferation, OCT has also been utilized to study cell migration. Repeated OCT scans for several hours showed the migration of macrophage cells that moved through a 3D agarose scaffold in response to a chemoattractant gradient (Tan et al., 2006) and with lung cancer cells seeded on a collagen scaffold (Yang et al., 2004).

Integrating OCM with other optical methods (MPM and fluorescence lifetime imaging microscopy [FLIM]) offers unique advantages, as it provides new contrast capabilities and can probe cell dynamics in better ways (Zhao et al., 2012). A study showed that FLIM images were clearly able to distinguish hydrogel from cells based on different lifetimes, which were not well separated in the MPM images (Zhao et al., 2012). Post-transplant dynamics of the engineered tissue have also been investigated to evaluate the efficiency of treatment. OCT has been used to monitor wound healing processing after the implantation of a hydrogel scaffold (Yuan et al., 2010). This study revealed the structures of wound beds in normal and diabetic after engraftment of hydrogels and was confirmed with histological images in parallel. A recent investigation integrated OCM with a multimodal imaging system to enable imaging of the scaffold after removal from the wound bed of mice (Bower et al., 2017). Multimodal images of the scaffold revealed the absence of DiI-labeled fibroblasts with vertically oriented microchannels, as shown in Figure 7.7c. However, with randomly oriented microchannels, some patches of clustered DiI-labeled fibroblasts were observed (Figure 7.7d). This study concluded that cells were more likely to be delivered from the scaffolds into the wound bed with vertically

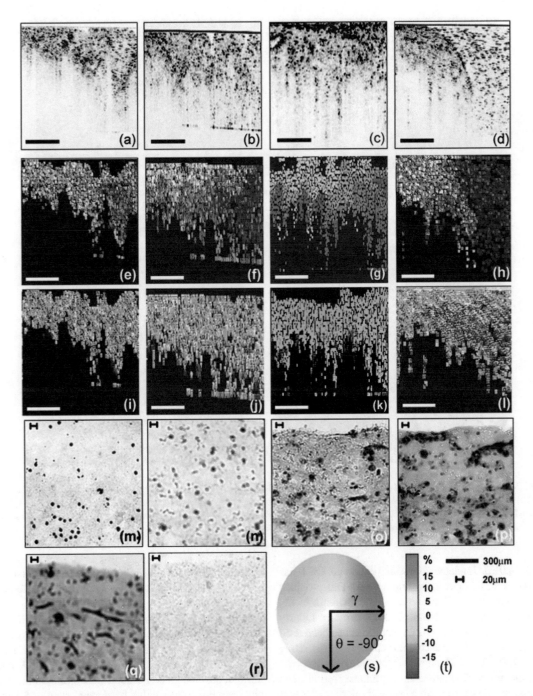

FIGURE 7.6 Images of cell-seeded engineered tissue scaffolds. (a–d) Structural OCT images on Days 0, 3, 7, and 10, respectively, of the boundary between the cell-seeded region (left) and the cell-free region (right). (e–h) Displacement maps on Days 0, 3, 7, and 10, respectively, color coded using the scale in s. (i–l) Strain maps on Days 0, 3, 7, and 10, respectively, using the color scale in (t). (m–p) Corresponding histology from the cell-seeded tissue regions. (q) Histological image of cells after 10 days of incubation without embedded microspheres. (r) Histological image of a cell-free scaffold and microspheres. Scale bar = 300 μm in a–l; 20 μm in m–r. Reproduced with permission from Ko et al. (2006).

orientated microchannels compared to those with randomly orientated microchannels.

7.4.3 Small Animal Imaging

7.4.3.1 Ophthalmology

OCT has become the standard-of-care imaging technique in the field of ophthalmology. Retinal OCT provides high-resolution images of retinal features that were previously not accessible with fundus photography, fluorescein angiography, and ophthalmic ultrasound. In animal models, OCT has been used to understand retinal disease progression and investigate new retinal therapies (Srinivasan et al., 2006; Doukas et al., 2008; Bai et al., 2010). The development of computational, FF-OCT, and adaptive optics methods has led to the visualization of individual retinal photoreceptors at high resolution (Shemonski et al., 2015; Hillmann et al., 2016;

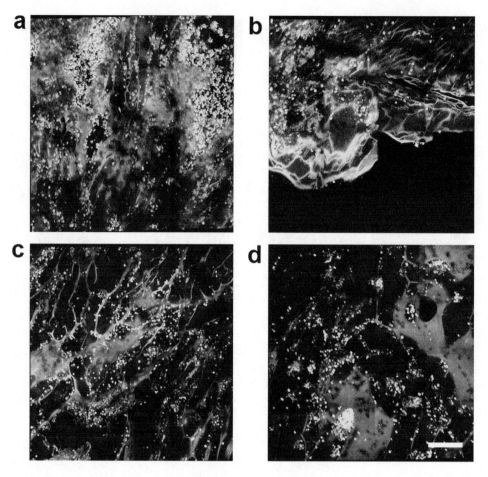

FIGURE 7.7   Multimodal optical images of microchanneled alginate hydrogel scaffolds after removal from the wound 2 days following treatment with (a) vertical microchannels cultured without fibroblasts, (b) random microchannels without fibroblasts, (c) vertical microchannels with fibroblasts, and (d) random microchannels with fibroblasts. Color key: Yellow, TPEF at 800 nm; Green, GFP fluorescence; Cyan, OCM; Red, DiI fluorescence from labeled fibroblasts. Scale bar = 250 μm. Reproduced with permission from Bower et al. (2017).

Jonnal et al., 2016). Implementation of functional OCT has increased the contrast of certain retinal layers and provided new disease biomarkers (Kemp et al., 2005; Makita et al., 2006; Wang et al., 2007c; Pircher et al., 2011; Lapierre-Landry et al., 2017). For example, the OCTA technique has been used to detect and resolve microvasculature changes in the retina (Zhi et al., 2014). Figure 7.8 shows wide-field depth-resolved images of the microvasculature in the mouse retina. The microvascular network forms three layers of microvessels that were clearly resolved. This study concluded that OMAG could detect early microvasculature changes in diabetic retinopathy.

7.4.3.2 Cardiology

Over the past decade, OCT has been extensively applied in the field of cardiology to examine the vasculature, specifically the coronary arteries, and analyze the cellular and molecular features. OCT has been demonstrated as a tool for the *in vivo* identification of plaque types, the content in a lesion, and to characterize attenuation and layer thickness changes such as lipid-rich regions and calcifications (MacNeill et al., 2005; van der Meer et al., 2005; Tearney et al., 2006). In addition to the presence of macrophages, OCT has been used to investigate

clinically relevant composition in the detected plaques, which have associated risk factors (van der Meer et al., 2005; Stamper et al., 2006).

Intravascular OCT in combination with near-infrared fluorescence (NIRF) has been successfully used for the visualization of cardiac interventions such as stent placement (Hara et al., 2017). Figure 7.9 shows the NIRF-OCT imaging of fibrin deposition on and around stent struts. In this study, rabbits were implanted with a bare metal stent of 3.5 mm diameter in the aorta. NIRF-OCT images were validated with fluorescence microscopy and confirmed the presence of a NIRF-fibrin signal within the stent. This study confirmed that NIRF-OCT fibrin molecular imaging can accurately assess fibrin deposition in stents *in vivo*. Other functional modalities such as magnetomotive OCT and fluorescence imaging have also been investigated to evaluate the properties of the arterial walls (Kim et al., 2016; Li et al., 2017b).

7.4.3.3 Musculoskeletal Imaging

Musculoskeletal disorders can affect muscles, nerves, tendons, and bones, and these represent some of the leading causes of disability. The application of OCT in musculoskeletal imaging

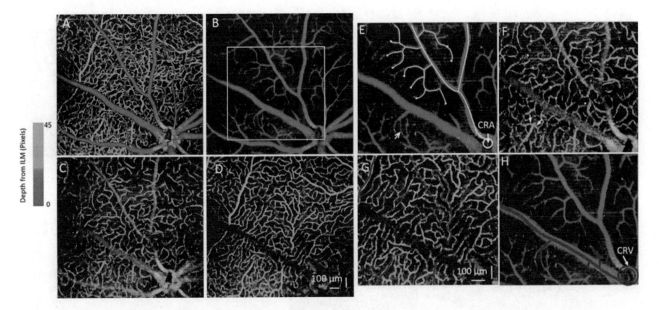

FIGURE 7.8 Depth-resolved retinal microvascular network and 3D tracking of the microvessels are shown. (a) Depth-coded color angiography of the whole mouse retina is shown. (b–d) Microvascular networks are shown within three different layers: (b) ganglion cell layer (GCL), (c) inner plexiform layer (IPL), and (d) outer plexiform layer (OPL). The microvascular organization within the three layers has distinct patterns. (e–h) Tracking of the microvessel paths of an arteriole and venule is demarcated in the white rectangular region in (b). (e) In the superficial microvessels of the GCL, the blood flows through arterioles branched from the central retinal artery (CRA), shown as the yellow lines. The yellow dots denote where the vessels dive to a separate layer. The arrow points to a direct arteriovenular connection in the superficial GCL layer. (f) An intermediate capillary layer in the IPL is shown. Yellow dots indicate the vessels that are diving to this layer from the superficial layer in (e). These capillaries flow as pink lines and then dive into a deeper layer at the pink dots. Arrows point to capillaries joining the postcapillary venules. (g) Deep-layer microvessel in OPL is shown, in which the pink dots from IPL indicate diving vessels and blue lines indicate flow. All of the blood flow will then be collected by the postcapillary venules and returned to the superficial layer and to the major venules (h), which will join the central retinal vein (CRV). Reproduced with permission from Zhi et al. (2014).

has been reported for osteoarthritis monitoring, mainly imaging the bone-cartilage interface and to evaluate collagen organization. Most biological tissues exhibit birefringence, and detecting those changes reflect the tissue organization. Several groups have shown structural musculoskeletal changes between normal and pathological conditions using OCT, PS-OCT, and OCE (Drexler et al., 2001b; Han et al., 2003; Patel et al., 2005; Beaudette et al., 2012; Chin et al., 2014). PS-OCT has also been used to detect exercise-induced structural changes in normal and genetically altered (*mdx*) murine skeletal muscle as shown in Figure 7.10 (Pasquesi et al., 2006). Muscle from *mdx* mice exhibited a marked decrease in birefringence after exercise, whereas normal mice showed highly birefringent muscle tissue both before and after exercise. The authors concluded that PS-OCT can detect birefringence changes occurring because of ultrastructural changes in skeletal muscle due to genetic mutations and damage following exercise.

7.4.3.4 Dermatology

OCT in dermatology has demonstrated the ability to identify dermal epidermal layers and observe conditions including inflammation, blistering, and various other skin conditions (Schmitt et al., 1995; Gladkova et al., 2000; Pierce et al., 2004). Use of endogenous contrast mechanisms such as with PS-OCT has led to the identification of decreases in birefringence due to burn damage (Park et al., 2001; Kim et al., 2012). The dynamics of wound-healing processes in the skin of bone-marrow

(BM)-transplanted GFP mice has also been studied with multimodal imaging techniques (Graf et al., 2013). The reported study was able to track the dynamic changes during skin regeneration over long time periods (Figure 7.11). The combination of multiple modalities provided a wide range of complementary information and holds promise as a tool for assessing cell dynamics *in vivo*. OCT has the potential to detect a wide range of structural and functional changes in skin, and the implementation of various contrast-enhancing techniques will yield a noninvasive method for diagnosing skin-related disease.

7.5 Conclusion

Over the past 25 years, OCT has emerged as a powerful imaging technology that provides unique capabilities for biomedical research. Its wide range of applications, ranging from biological to medical to industrial, indicates the breadth of impact and the success of the OCT technology. The feasibility of utilizing OCT for the evaluation and characterization of the salient features of a normal and a diseased condition can be extensively explored using suitable animal models. This allows for the use of techniques and simulations that are not directly feasible for human subjects. Preclinical studies, therefore, enable an investigator to follow a process to its natural endpoint with frequent sampling throughout the course of investigation, beyond *in vitro* studies. Animal models play a critically important role and enable the application

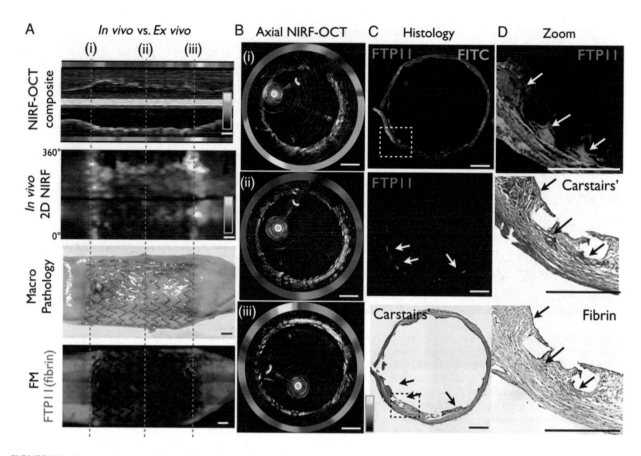

FIGURE 7.9 Intravascular near-infrared fluorescence-optical coherence tomography (NIRF-OCT) imaging of fibrin deposition on stents. Rabbits underwent implantation of a 3.5 mm diameter bare metal stent in the aorta. At Day 7, the fibrin molecular imaging agent FTP11-CyAm7 was intra-veneously injected, followed by near-infrared fluorescence-optical coherence tomography. (a) *In vivo* 2D near-infrared fluorescence map of stent (upper two rows) and corresponding *ex vivo* imaging after longitudinally opened (lower two rows). (b) Representative cross-sectional near-infrared fluorescence-optical coherence tomography images at the stent distal edge (i), middle (ii), and proximal edge (iii). (c and d) Histological sections of the proximal stent edge (dotted line in a (i) are shown.) Near-infrared fluorescence signal of FTP11-CyAm7 (red) colocalized with fibrin (bright red in Carstairs' staining) and fibrin immunostaining. All scale bars = 1 mm. Reproduced with permission from Hara et al. (2017).

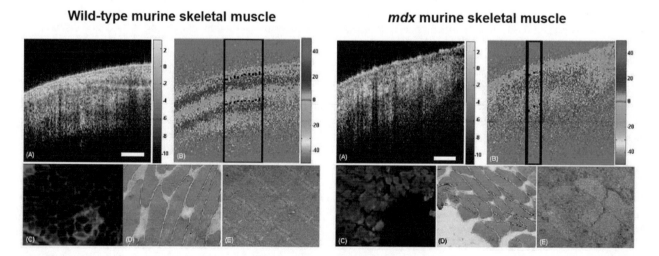

FIGURE 7.10 PS-OCT of exercised skeletal muscle of wild-type (left) and *mdx* (right) murine. (a) Structural OCT image. (b) False color birefrin-gence image. The black box outlines the region over which the mean zero-crossing distance was calculated, and the black points represent the zero-crossing points identified by the algorithm. Histological images are shown using (c) Evans blue fluorescence at 20x, (d) H&E at 20x, and (e) TEM at 5000x magnifications. Scale bar = 500 μm. Reproduced and modified with permission from Pasquesi et al. (2006).

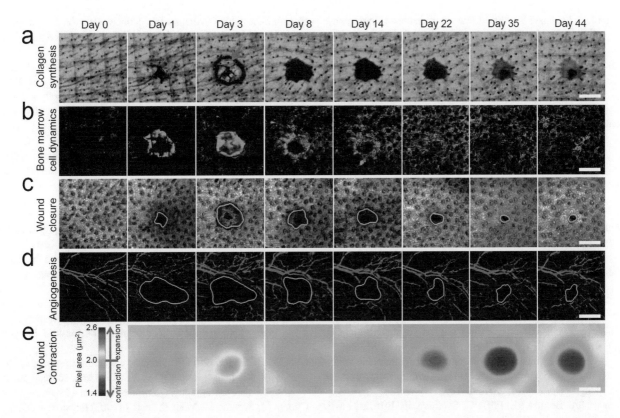

FIGURE 7.11 Multimodal, time-lapse sequence of cutaneous wound-healing processes. (a) Axial projections of the SHG volumes showing scab formation (red arrow) and synthesis of new collagen. (b) Axial projections of the TPEF volumes showing the dynamics of the BM-derived, GFP-expressing cell populations. (c) *En face* sections from the dermis layer of the structural OCT volume, showing closure of the wound region (area within the green outline). (d) Axial projections of the microvascular OCT volume showing the growth of the microvascular network. Area within the green outline indicates region lacking blood vessels. (e) Wound contraction maps calculated relative to the initial time point showing that wound contraction stabilizes by 4 weeks. Scale bars = 500 μm. Reproduced with permission from Graf et al. (2013).

of imaging techniques such as OCT to be translated into clinical human studies. The versatility of OCT has enabled its use for imaging across many different size scales, from cells, to tissues, to small animals, and to humans. OCT will continue to facilitate both biological discovery as well as clinical application.

7.6 Acknowledgments

This chapter was prepared with support from grants from the National Institutes of Health (R01 CA213149, R01 EB023232, and R01 EB013723). We wish to thank all of our colleagues who have helped drive the advancement and application of the OCT technology since its inception. While we have attempted to capture the incredible breadth of the field, we regret that not all the impactful studies could be highlighted. We thank the Association for Research in Vision and Ophthalmology (ARVO) for granting copyright permission.

Additional information can be found at: http://biophotonics.illinois.edu.

REFERENCES

Adams DC, Hariri LP, Miller AJ, Wang Y, Cho JL, Villiger M, Holz JA, Szabari MV, Hamilos DL, Harris RS, Griffith JW, Bouma BE, Luster AD, Medoff BD, Suter MJ (2016) Birefringence microscopy platform for assessing airway smooth muscle structure and function *in vivo*. *Sci Transl Med* 8:359ra131–359ra131.

Adie SG, Graf BW, Ahmad A, Carney PS, Boppart SA (2012) Computational adaptive optics for broadband optical interferometric tomography of biological tissue. *Proc Natl Acad Sci* 109:7175–7180.

Adie SG, Kennedy BF, Armstrong JJ, Alexandrov SA, Sampson DD (2009) Audio frequency *in vivo* optical coherence elastography. *Phys Med Biol* 54:3129–3139.

Adie SG, Liang X, Kennedy BF, John R, Sampson DD, Boppart SA (2010) Spectroscopic optical coherence elastography. *Opt Express* 18:25519–25534.

Alex A, Weingast J, Weinigel M, Kellner-Höfer M, Nemecek R, Binder M, Pehamberger H, König K, Drexler W (2013) Three-dimensional multiphoton/optical coherence tomography for diagnostic applications in dermatology. *J Biophotonics* 6:352–362.

Ashok PC, Praveen BB, Bellini N, Riches A, Dholakia K, Herrington CS (2013) Multi-modal approach using Raman spectroscopy and optical coherence tomography for the discrimination of colonic adenocarcinoma from normal colon. *Biomed Opt Express* 4:2179–2186.

Bai Y, Xu J, Brahimi F, Zhuo Y, Sarunic MV, Saragovi HU (2010) An agonist TrkB mAb causes sustained TrkB activation, delays RGC death, and protects the retinal structure in optic nerve axotomy and in glaucoma. *Invest Ophthalmol Vis Sci* 51:4722–4731.

Barton JK, Stromski S (2005) Flow measurement without phase information in optical coherence tomography images. *Opt Express* 13:5234–5239.

Baumann B, Pircher M, Götzinger E, Hitzenberger CK (2007) Full range complex spectral domain optical coherence tomography without additional phase shifters. *Opt Express* 15:13375–13387.

Beaudette K, Strupler M, Benboujja F, Parent S, Aubin C-E, Boudoux C (2012) Optical coherence tomography for the identification of musculoskeletal structures of the spine: A pilot study. *Biomed Opt Express* 3:533–542.

Beaurepaire E, Boccara AC, Lebec M, Blanchot L, Saint-Jalmes H (1998) Full-field optical coherence microscopy. *Opt Lett* 23:244–246.

Blatter C, Grajciar B, Schmetterer L, Leitgeb RA (2013) Angle independent flow assessment with bidirectional Doppler optical coherence tomography. *Opt Lett* 38:4433–4436.

Boer JF de, Milner TE, Gemert MJC van, Nelson JS (1997) Two-dimensional birefringence imaging in biological tissue by polarization-sensitive optical coherence tomography. *Opt Lett* 22:934–936.

Bonesi M, Minneman MP, Ensher J, Zabihian B, Sattmann H, Boschert P, Hoover E, Leitgeb RA, Crawford M, Drexler W (2014) Akinetic all-semiconductor programmable swept-source at 1550 nm and 1310 nm with centimeters coherence length. *Opt Express* 22:2632–2655.

Boppart SA, Bouma BE, Pitris C, Southern JF, Brezinski ME, Fujimoto JG (1998) *In vivo* cellular optical coherence tomography imaging. *Nat Med* 4:861–865.

Boppart SA, Bouma BE, Pitris C, Tearney GJ, Fujimoto JG, Brezinski ME (1997a) Forward-imaging instruments for optical coherence tomography. *Opt Lett* 22:1618–1620.

Boppart SA, Oldenburg AL, Xu C, Marks DL (2005) Optical probes and techniques for molecular contrast enhancement in coherence imaging. *J Biomed Opt* 10:41208.

Boppart SA, Tearney GJ, Bouma BE, Southern JF, Brezinski ME, Fujimoto JG (1997b) Noninvasive assessment of the developing *Xenopus* cardiovascular system using optical coherence tomography. *Proc Natl Acad Sci U S A* 94:4256–4261.

Bouma BE, Tearney GJ, Yabushita H, Shishkov M, Kauffman CR, DeJoseph Gauthier D, MacNeill BD, Houser SL, Aretz HT, Halpern EF, Jang I-K (2003) Evaluation of intracoronary stenting by intravascular optical coherence tomography. *Heart Br Card Soc* 89:317–320.

Boustany NN, Boppart SA, Backman V (2010) Microscopic imaging and spectroscopy with scattered light. *Annu Rev Biomed Eng* 12:285–314.

Bower AJ, Mahmassani Z, Zhao Y, Chaney EJ, Marjanovic M, Lee MK, Graf BW, De Lisio M, Kong H, Boppart MD, Boppart SA (2017) *In vivo* assessment of engineered skin cell delivery with multimodal optical microscopy. *Tissue Eng Part C Methods* 23:434–442.

Bredfeldt JS, Vinegoni C, Marks DL, Boppart SA (2005) Molecularly sensitive optical coherence tomography. *Opt Lett* 30:495–497.

Cai X, Zhang Y, Li L, Choi S-W, MacEwan MR, Yao J, Kim C, Xia Y, Wang LV (2013) Investigation of neovascularization in three-dimensional porous scaffolds *in vivo* by a combination of multiscale photoacoustic microscopy and optical coherence tomography. *Tissue Eng Part C Methods* 19:196–204.

Carrasco-Zevallos O, Shelton RL, Kim W, Pearson J, Applegate BE (2015) *In vivo* pump-probe optical coherence tomography imaging in *Xenopus laevis*. *J Biophotonics* 8:25–35.

Chan RC, Chau AH, Karl WC, Nadkarni S, Khalil AS, Iftimia N, Shishkov M, Tearney GJ, Kaazempur-Mofrad MR, Bouma BE (2004) OCT-based arterial elastography: Robust estimation exploiting tissue biomechanics. *Opt Express* 12:4558–4572.

Chen C-W, Betz MW, Fisher JP, Paek A, Chen Y (2011) Macroporous hydrogel scaffolds and their characterization by optical coherence tomography. *Tissue Eng Part C Methods* 17:101–112.

Chin L, Kennedy BF, Kennedy KM, Wijesinghe P, Pinniger GJ, Terrill JR, McLaughlin RA, Sampson DD (2014) Three-dimensional optical coherence micro-elastography of skeletal muscle tissue. *Biomed Opt Express* 5:3090–3102.

Chinn SR, Swanson EA, Fujimoto JG (1997) Optical coherence tomography using a frequency-tunable optical source. *Opt Lett* 22:340–342.

Choi WJ, Reif R, Yousefi S, Wang RK (2014) Improved microcirculation imaging of human skin *in vivo* using optical micro-angiography with a correlation mapping mask. *J Biomed Opt* 19:36010.

Choi WJ, Wang RK (2014) Volumetric cutaneous microangiography of human skin *in vivo* by VCSEL swept-source optical coherence tomography. *Quantum Electron* 44:740.

Choma M, Sarunic M, Yang C, Izatt J (2003) Sensitivity advantage of swept source and Fourier domain optical coherence tomography. *Opt Express* 11:2183–2189.

Choma MA, Ellerbee AK, Yazdanfar S, Izatt JA (2006) Doppler flow imaging of cytoplasmic streaming using spectral domain phase microscopy. *J Biomed Opt* 11:024014.

Choma MA, Hsu K, Izatt JA (2005) Swept source optical coherence tomography using an all-fiber 1300-nm ring laser source. *J Biomed Opt* 10:44009.

Chong SP, Lai T, Zhou Y, Tang S (2013) Tri-modal microscopy with multiphoton and optical coherence microscopy/tomography for multi-scale and multi-contrast imaging. *Biomed Opt Express* 4:1584–1594.

Chong SP, Merkle CW, Leahy C, Radhakrishnan H, Srinivasan VJ (2015) Quantitative microvascular hemoglobin mapping using visible light spectroscopic optical coherence tomography. *Biomed Opt Express* 6:1429–1450.

Clark AL, Gillenwater A, Alizadeh-Naderi R, El-Naggar AK, Richards-Kortum R (2004) Detection and diagnosis of oral neoplasia with an optical coherence microscope. *J Biomed Opt* 9:1271–1280.

Crecea V, Oldenburg AL, Liang X, Ralston TS, Boppart SA (2009) Magnetomotive nanoparticle transducers for optical rheology of viscoelastic materials. *Opt Express* 17:23114–23122.

Das BB, Liu F, Alfano RR (1997) Time-resolved fluorescence and photon migration studies in biomedical and model random media. *Rep Prog Phys* 60:227.

de Boer JF, Cense B, Park BH, Pierce MC, Tearney GJ, Bouma BE (2003) Improved signal-to-noise ratio in spectral-domain compared with time-domain optical coherence tomography. *Opt Lett* 28:2067–2069.

de Freitas C, Ruggeri M, Manns F, Ho A, Parel J-M (2013) *In vivo* measurement of the average refractive index of the human crystalline lens using optical coherence tomography. *Opt Lett* 38:85–87.

Doukas J, Mahesh S, Umeda N, Kachi S, Akiyama H, Yokoi K, Cao J, Chen Z, Dellamary L, Tam B, Racanelli-Layton A, Hood J, Martin M, Noronha G, Soll R, Campochiaro PA (2008) Topical administration of a multi-targeted kinase inhibitor suppresses choroidal neovascularization and retinal edema. *J Cell Physiol* 216:29–37.

Drexler W, Fujimoto JG eds. (2015) *Optical Coherence Tomography: Technology and Applications*, 2nd ed. Springer International Publishing.

Drexler W, Morgner U, Ghanta RK, Kärtner FX, Schuman JS, Fujimoto JG (2001a) Ultrahigh-resolution ophthalmic optical coherence tomography. *Nat Med* 7:502–507.

Drexler W, Stamper D, Jesser C, Li X, Pitris C, Saunders K, Martin S, Lodge MB, Fujimoto JG, Brezinski ME (2001b) Correlation of collagen organization with polarization sensitive imaging of *in vitro* cartilage: Implications for osteoarthritis. *J Rheumatol* 28:1311–1318.

Duan L, Yamanari M, Yasuno Y (2012) Automated phase retardation oriented segmentation of chorio-scleral interface by polarization sensitive optical coherence tomography. *Opt Express* 20:3353–3366.

Faber DJ, Mik EG, Aalders MCG, Leeuwen TG van (2005) Toward assessment of blood oxygen saturation by spectroscopic optical coherence tomography. *Opt Lett* 30:1015–1017.

Fercher AF, Drexler W, Hitzenberger CK, Lasser T (2003) Optical coherence tomography - principles and applications. *Rep Prog Phys* 66:239.

Fercher AF, Hitzenberger CK, Kamp G, El-Zaiat SY (1995) Measurement of intraocular distances by backscattering spectral interferometry. *Opt Commun* 117:43–48.

Fercher AF, Mengedoht K, Werner W (1988) Eye-length measurement by interferometry with partially coherent light. *Opt Lett* 13:186–188.

Fingler J, Readhead C, Schwartz DM, Fraser SE (2008) Phase-contrast OCT imaging of transverse flows in the mouse retina and choroid. *Invest Ophthalmol Vis Sci* 49:5055–5059.

Fingler J, Zawadzki RJ, Werner JS, Schwartz D, Fraser SE (2009) Volumetric microvascular imaging of human retina using optical coherence tomography with a novel motion contrast technique. *Opt Express* 17:22190–22200.

Fleming CP, Eckert J, Halpern EF, Gardecki JA, Tearney GJ (2013) Depth resolved detection of lipid using spectroscopic optical coherence tomography. *Biomed Opt Express* 4:1269–1284.

Fujimoto J, Swanson E (2016) The development, commercialization, and impact of optical coherence tomography. *Invest Ophthalmol Vis Sci* 57: OCT1–OCT13.

Fujimoto JG, Pitris C, Boppart SA, Brezinski ME (2000) Optical coherence tomography: An emerging technology for biomedical imaging and optical biopsy. *Neoplasia N Y N* 2:9–25.

Gladkova ND, Petrova GA, Nikulin NK, Radenska-Lopovok SG, Snopova LB, Chumakov YP, Nasonova VA, Gelikonov VM, Gelikonov GV, Kuranov RV, Sergeev AM, Feldchtein FI (2000) *In vivo* optical coherence tomography imaging of human skin: Norm and pathology. *Skin Res Technol* 6:6–16.

Golubovic B, Bouma BE, Tearney GJ, Fujimoto JG (1997) Optical frequency-domain reflectometry using rapid wavelength tuning of a Cr^{4+}: Forsterite laser. *Opt Lett* 22:1704–1706.

Graf BW, Chaney EJ, Marjanovic M, Adie SG, De Lisio M, Valero MC, Boppart MD, Boppart SA (2013) Long-term time-lapse multimodal intravital imaging of regeneration and bone-marrow-derived cell dynamics in skin. *Technol Elmsford N* 1:8–19.

Graf RN, Wax A (2007) Temporal coherence and time-frequency distributions in spectroscopic optical coherence tomography. *JOSA* A 24:2186–2195.

Gurjar RS, Backman V, Perelman LT, Georgakoudi I, Badizadegan K, Itzkan I, Dasari RR, Feld MS (2001) Imaging human epithelial properties with polarized light-scattering spectroscopy. *Nat Med* 7:1245–1248.

Ha Usler G, Lindner MW (1998) "Coherence radar" and "spectral radar"-new tools for dermatological diagnosis. *J Biomed Opt* 3:21–31.

Han CW, Chu CR, Adachi N, Usas A, Fu FH, Huard J, Pan Y (2003) Analysis of rabbit articular cartilage repair after chondrocyte implantation using optical coherence tomography. *Osteoarthritis Cartilage* 11:111–121.

Hara T, Ughi GJ, McCarthy JR, Erdem SS, Mauskapf A, Lyon SC, Fard AM, Edelman ER, Tearney GJ, Jaffer FA (2017) Intravascular fibrin molecular imaging improves the detection of unhealed stents assessed by optical coherence tomography *in vivo*. *Eur Heart J* 38:447–455.

Hariri LP, Tumlinson AR, Besselsen DG, Utzinger U, Gerner EW, Barton JK (2006) Endoscopic optical coherence tomography and laser-induced fluorescence spectroscopy in a murine colon cancer model. *Lasers Surg Med* 38:305–313.

Hee MR, Huang D, Swanson EA, Fujimoto JG (1992) Polarization-sensitive low-coherence reflectometer for birefringence characterization and ranging. *J Opt Soc Am B* 9:903–908.

Hillmann D, Spahr H, Pfäffle C, Sudkamp H, Franke G, Hüttmann G (2016) *In vivo* optical imaging of physiological responses to photostimulation in human photoreceptors. *Proc Natl Acad Sci U S A* 113:13138–13143.

Hitzenberger CK, Götzinger E, Sticker M, Pircher M, Fercher AF (2001) Measurement and imaging of birefringence and optic axis orientation by phase resolved polarization sensitive optical coherence tomography. *Opt Express* 9:780–790.

Huang D, Swanson EA, Lin CP, Schuman JS, Stinson WG, Chang W, Hee MR, Flotte T, Gregory K, Puliafito CA, Fujimoto JG (1991) Optical coherence tomography. *Science* 254:1178–1181.

Hunter CJ, Imler SM, Malaviya P, Nerem RM, Levenston ME (2002) Mechanical compression alters gene expression and extracellular matrix synthesis by chondrocytes cultured in collagen I gels. *Biomaterials* 23:1249–1259.

Izatt JA, Kulkarni MD, Yazdanfar S, Barton JK, Welch AJ (1997) *In vivo* bidirectional color Doppler flow imaging of picoliter blood volumes using optical coherence tomography. *Opt Lett* 22:1439–1441.

Jayaraman V, Cole GD, Robertson M, Burgner C, John D, Uddin A, Cable A (2012) Rapidly swept, ultra-widely-tunable 1060 nm MEMS-VCSELs. *Electron Lett* 48:1331–1333.

Jia Y, Bagnaninchi PO, Yang Y, Haj AE, Hinds MT, Kirkpatrick SJ, Wang RK (2009) Doppler optical coherence tomography imaging of local fluid flow and shear stress within microporous scaffolds. *J Biomed Opt* 14:034014.

Jia Y, Tan O, Tokayer J, Potsaid B, Wang Y, Liu JJ, Kraus MF, Subhash H, Fujimoto JG, Hornegger J, Huang D (2012) Split-spectrum amplitude-decorrelation angiography with optical coherence tomography. *Opt Express* 20:4710.

Jia Y, Wang RK (2010) Label-free *in vivo* optical imaging of functional microcirculations within meninges and cortex in mice. *J Neurosci Methods* 194:108–115.

Jiao S, Jiang M, Hu J, Fawzi A, Zhou Q, Shung KK, Puliafito CA, Zhang HF (2010) Photoacoustic ophthalmoscopy for *in vivo* retinal imaging. *Opt Express* 18:3967–3972.

Jo JA, Applegate BE, Park J, Shrestha S, Pande P, Gimenez-Conti IB, Brandon JL (2010) *In vivo* simultaneous morphological and biochemical optical imaging of oral epithelial cancer. *IEEE Trans Biomed Eng* 57:2596–2599.

John R, Rezaeipoor R, Adie SG, Chaney EJ, Oldenburg AL, Marjanovic M, Haldar JP, Sutton BP, Boppart SA (2010) *In vivo* magnetomotive optical molecular imaging using targeted magnetic nanoprobes. *Proc Natl Acad Sci U S A* 107:8085–8090.

Jonathan E, Enfield J, Leahy MJ (2011) Correlation mapping method for generating microcirculation morphology from optical coherence tomography (OCT) intensity images. *J Biophotonics* 4:583–587.

Jonnal RS, Kocaoglu OP, Zawadzki RJ, Liu Z, Miller DT, Werner JS (2016) A review of adaptive optics optical coherence tomography: Technical advances, scientific applications, and the future. *Invest Ophthalmol Vis Sci* 57:OCT51–OCT68.

Kemp NJ, Park J, Zaatari HN, Rylander HG, Milner TE (2005) High-sensitivity determination of birefringence in turbid media with enhanced polarization-sensitive optical coherence tomography. *JOSA* A 22:552–560.

Kennedy BF, Koh SH, McLaughlin RA, Kennedy KM, Munro PRT, Sampson DD (2012a) Strain estimation in phase-sensitive optical coherence elastography. *Biomed Opt Express* 3:1865–1879.

Kennedy BF, Liang X, Adie SG, Gerstmann DK, Quirk BC, Boppart SA, Sampson DD (2011) *In vivo* three-dimensional optical coherence elastography. *Opt Express* 19:6623–6634.

Kennedy BF, McLaughlin RA, Kennedy KM, Chin L, Wijesinghe P, Curatolo A, Tien A, Ronald M, Latham B, Saunders CM, Sampson DD (2015) Investigation of optical coherence microelastography as a method to visualize cancers in human breast tissue. *Cancer Res* 75:3236–3245.

Kennedy BF, Wojtkowski M, Szkulmowski M, Kennedy KM, Karnowski K, Sampson DD (2012b) Improved measurement of vibration amplitude in dynamic optical coherence elastography. *Biomed Opt Express* 3:3138–3152.

Khan KM, Krishna H, Majumder SK, Rao KD, Gupta PK (2014) Depth-sensitive Raman spectroscopy combined with optical coherence tomography for layered tissue analysis. *J Biophotonics* 7:77–85.

Kim J, Ahmad A, Li J, Marjanovic M, Chaney EJ, Suslick KS, Boppart SA (2016) Intravascular magnetomotive optical coherence tomography of targeted early-stage atherosclerotic changes in *ex vivo* hyperlipidemic rabbit aortas. *J Biophotonics* 9:109–116.

Kim KH, Pierce MC, Maguluri G, Park BH, Yoon SJ, Lydon M, Sheridan R, de Boer JF (2012) *In vivo* imaging of human burn injuries with polarization-sensitive optical coherence tomography. *J Biomed Opt* 17:066012.

Kim TS, Jang S-J, Oh N, Kim Y, Park T, Park J, Oh W-Y (2014) Dual-wavelength band spectroscopic optical frequency domain imaging using plasmon-resonant scattering in metallic nanoparticles. *Opt Lett* 39:3082–3085.

Kim YL, Jr JTW, Goldstickk TK, Glucksberg MR (2004) Variation of corneal refractive index with hydration. *Phys Med Biol* 49:859.

Ko AC-T, Choo-Smith L-P 'ing, Hewko M, Leonardi L, Sowa MG, Dong CCS, Williams P, Cleghorn B (2005) *Ex vivo* detection and characterization of early dental caries by optical coherence tomography and Raman spectroscopy. *J Biomed Opt* 10:031118.

Ko H-J, Tan W, Stack R, Boppart SA (2006) Optical coherence elastography of engineered and developing tissue. *Tissue Eng* 12:63–73.

Koch P, Hüttmann G, Schleiermacher H, Eichholz J, Koch E (2004) Linear optical coherence tomography system with a downconverted fringe pattern. *Opt Lett* 29:1644–1646.

König K, Speicher M, Bückle R, Reckfort J, McKenzie G, Welzel J, Koehler MJ, Elsner P, Kaatz M (2009) Clinical optical coherence tomography combined with multiphoton tomography of patients with skin diseases. *J Biophotonics* 2:389–397.

Kuranov RV, Sapozhnikova VV, Shakhova NM, Gelikonov VM, Zagainova EV, Petrova SA (2002) Combined application of optical methods to increase the information content of optical coherent tomography in diagnostics of neoplastic processes. *Quantum Electron* 32:993.

Lapierre-Landry M, Gordon AY, Penn JS, Skala MC (2017) *In vivo* photothermal optical coherence tomography of endogenous and exogenous contrast agents in the eye. *Sci Rep* 7:9228.

Laufer J, Johnson P, Zhang E, Treeby B, Cox B, Pedley B, Beard P (2012) *In vivo* preclinical photoacoustic imaging of tumor vasculature development and therapy. *J Biomed Opt* 17:056016.

Lee H-C, Zhou C, Cohen DW, Mondelblatt AE, Wang Y, Aguirre AD, Shen D, Sheikine Y, Fujimoto JG, Connolly JL (2012) Integrated optical coherence tomography and optical coherence microscopy imaging of *ex vivo* human renal tissues. *J Urol* 187:691–699.

Lee TM, Oldenburg AL, Sitafalwalla S, Marks DL, Luo W, Toublan FJ-J, Suslick KS, Boppart SA (2003) Engineered microsphere contrast agents for optical coherence tomography. *Opt Lett* 28:1546–1548.

Leitgeb R, Hitzenberger C, Fercher A (2003) Performance of Fourier domain vs. time domain optical coherence tomography. *Opt Express* 11:889–894.

Leitgeb R, Wojtkowski M, Kowalczyk A, Hitzenberger CK, Sticker M, Fercher AF (2000) Spectral measurement of absorption by spectroscopic frequency-domain optical coherence tomography. *Opt Lett* 25:820–822.

Levitz D, Hinds MT, Choudhury N, Tran NT, Hanson SR, Jacques SL (2010) Quantitative characterization of developing collagen gels using optical coherence tomography. *J Biomed Opt* 15:026019.

Lexer F, Hitzenberger CK, Drexler W, Molebny S, Sattmann H, Sticker M, Fercher AF (1999) Dynamic coherent focus OCT with depth-independent transversal resolution. *J Mod Opt* 46:541–553.

Li C, Guan G, Reif R, Huang Z, Wang RK (2012a) Determining elastic properties of skin by measuring surface waves from an impulse mechanical stimulus using phase-sensitive optical coherence tomography. *J R Soc Interface* 9:831–841.

Li J, Bower AJ, Arp Z, Olson EJ, Holland C, Chaney EJ, Marjanovic M, Pande P, Alex A, Boppart SA (2017a) Investigating the healing mechanisms of an angiogenesis-promoting topical treatment for diabetic wounds using multimodal microscopy. *J Biophotonics*: e201700195.

Li Y, Jing J, Qu Y, Miao Y, Zhang B, Ma T, Yu M, Zhou Q, Chen Z (2017b) Fully integrated optical coherence tomography, ultrasound, and indocyanine green-based fluorescence tri-modality system for intravascular imaging. *Biomed Opt Express* 8:1036–1044.

Li YL, Seekell K, Yuan H, Robles FE, Wax A (2012b) Multispectral nanoparticle contrast agents for true-color spectroscopic optical coherence tomography. *Biomed Opt Express* 3:1914–1923.

Liang X, Graf BW, Boppart SA (2009) Imaging engineered tissues using structural and functional optical coherence tomography. *J Biophotonics* 2:643–655.

Liu T, Wei Q, Wang J, Jiao S, Zhang HF (2011) Combined photoacoustic microscopy and optical coherence tomography can measure metabolic rate of oxygen. *Biomed Opt Express* 2:1359–1365.

Lumbroso B MD, Huang D MD,Ph D, Jia Y Ph D, Fujimoto JG Ph D, Rispoli M MD (2015) *Clinical Guide to Angio-OCT: Non Invasive, Dyeless OCT Angiography*, 1st ed. Jaypee Brothers Medical Pub.

Luo W, Nguyen FT, Zysk AM, Ralston TS, Brockenbrough J, Marks DL, Oldenburg AL, Boppart SA (2005) Optical biopsy of lymph node morphology using optical coherence tomography. *Technol Cancer Res Treat* 4:539–548.

MacNeill BD, Bouma BE, Yabushita H, Jang I-K, Tearney GJ (2005) Intravascular optical coherence tomography: Cellular imaging. *J Nucl Cardiol Off Publ Am Soc Nucl Cardiol* 12:460–465.

Maher JR, Jaedicke V, Medina M, Levinson H, Selim MA, Brown WJ, Wax A (2014) *In vivo* analysis of burns in a mouse model using spectroscopic optical coherence tomography. *Opt Lett* 39:5594–5597.

Makita S, Hong Y, Yamanari M, Yatagai T, Yasuno Y (2006) Optical coherence angiography. *Opt Express* 14:7821–7840.

Manapuram RK, Aglyamov SR, Monediado FM, Mashiatulla M, Li J, Emelianov SY, Larin KV (2012) *In vivo* estimation of elastic wave parameters using phase-stabilized swept source optical coherence elastography. *J Biomed Opt* 17:100501.

Mariampillai A, Leung MKK, Jarvi M, Standish BA, Lee K, Wilson BC, Vitkin A, Yang VXD (2010) Optimized speckle variance OCT imaging of microvasculature. *Opt Lett* 35:1257–1259.

Marks DL, Boppart SA (2004) Nonlinear interferometric vibrational imaging. *Phys Rev Lett* 92:123905.

McLaughlin RA, Noble PB, Sampson DD (2014) Optical coherence tomography in respiratory science and medicine: From airways to alveoli. *Physiology* 29:369–380.

Morgner U, Drexler W, Kärtner FX, Li XD, Pitris C, Ippen EP, Fujimoto JG (2000) Spectroscopic optical coherence tomography. *Opt Lett* 25:111–113.

Nam HS, Song JW, Jang S-J, Lee JJ, Oh W-Y, Kim JW, Yoo H (2016) Characterization of lipid-rich plaques using spectroscopic optical coherence tomography. *J Biomed Opt* 21:075004.

Nassif N, Cense B, Park BH, Yun SH, Chen TC, Bouma BE, Tearney GJ, Boer JF de (2004) *In vivo* human retinal imaging by ultrahigh-speed spectral domain optical coherence tomography. *Opt Lett* 29:480–482.

Nguyen FT, Zysk AM, Chaney EJ, Kotynek JG, Oliphant UJ, Bellafiore FJ, Rowland KM, Johnson PA, Boppart SA (2009) Intraoperative evaluation of breast tumor margins with optical coherence tomography. *Cancer Res* 69:8790–8796.

Nolan RM, Adie SG, Marjanovic M, Chaney EJ, South FA, Monroy GL, Shemonski ND, Erickson-Bhatt SJ, Shelton RL, Bower AJ, Simpson DG, Cradock KA, Liu ZG, Ray PS, Boppart SA (2016) Intraoperative optical coherence tomography for assessing human lymph nodes for metastatic cancer. *BMC Cancer* 16:144.

Okabe Y, Sasaki Y, Ueno M, Sakamoto T, Toyoda S, Yagi S, Naganuma K, Fujiura K, Sakai Y, Kobayashi J, Omiya K, Ohmi M, Haruna M (2012) 200 kHz swept light source equipped with KTN deflector for optical coherence tomography. *Electron Lett* 48:201–202.

Oldenburg AL, Boppart SA (2010) Resonant acoustic spectroscopy of soft tissues using embedded magnetomotive nanotransducers and optical coherence tomography. *Phys Med Biol* 55:1189–1201.

Oldenburg AL, Hansen MN, Ralston TS, Wei A, Boppart SA (2009) Imaging gold nanorods in excised human breast carcinoma by spectroscopic optical coherence tomography. *J Mater Chem* 19:6407.

Oldenburg AL, Xu C, Boppart SA (2007) Spectroscopic optical coherence tomography and microscopy. *IEEE J Sel Top Quantum Electron* 13:1629–1640.

Pan YT, Xie TQ, Du CW, Bastacky S, Meyers S, Zeidel ML (2003) Enhancing early bladder cancer detection with fluorescence-guided endoscopic optical coherence tomography. *Opt Lett* 28:2485–2487.

Park BH, Pierce MC, Cense B, Boer JF de (2004) Jones matrix analysis for a polarization-sensitive optical coherence tomography system using fiber-optic components. *Opt Lett* 29:2512–2514.

Park BH, Saxer CE, Srinivas SM, Nelson JS, Boer JF de (2001) *In vivo* burn depth determination by high-speed fiber-based polarization sensitive optical coherence tomography. *J Biomed Opt* 6:474–480.

Pasquesi JJ, Schlachter SC, Boppart MD, Chaney E, Kaufman SJ, Boppart SA (2006) *In vivo* detection of exercise-induced ultrastructural changes in genetically-altered murine skeletal muscle using polarization-sensitive optical coherence tomography. *Opt Express* 14:1547–1556.

Patel NA, Zoeller J, Stamper DL, Fujimoto JG, Brezinski ME (2005) Monitoring osteoarthritis in the rat model using optical coherence tomography. *IEEE Trans Med Imaging* 24:155–159.

Patil CA, Bosschaart N, Keller MD, Leeuwen TG van, Mahadevan-Jansen A (2008) Combined Raman spectroscopy and optical coherence tomography device for tissue characterization. *Opt Lett* 33:1135–1137.

Pierce MC, Strasswimmer J, Hyle Park B, Cense B, de Boer JF (2004) Birefringence measurements in human skin using polarization-sensitive optical coherence tomography. *J Biomed Opt* 9:287–291.

Pircher M, Hitzenberger CK, Schmidt-Erfurth U (2011) Polarization sensitive optical coherence tomography in the human eye. *Prog Retin Eye Res* 30:431–451.

Poddar R, Kim DY, Werner JS, Zawadzki RJ (2014) *In vivo* imaging of human vasculature in the chorioretinal complex using phase-variance contrast method with phase-stabilized 1-μm swept-source optical coherence tomography. *J Biomed Opt* 19: 126010.

Potsaid B, Gorczynska I, Srinivasan VJ, Chen Y, Jiang J, Cable A, Fujimoto JG (2008) Ultrahigh speed Spectral / Fourier domain OCT ophthalmic imaging at 70,000 to 312,500 axial scans per second. *Opt Express* 16:15149–15169.

Qi W, Chen R, Chou L, Liu G, Zhang J, Zhou Q, Chen Z (2012) Phase-resolved acoustic radiation force optical coherence elastography. *J Biomed Opt* 17:110505.

Ralston TS, Marks DL, Carney PS, Boppart SA (2006) Phase stability technique for inverse scattering in optical coherence tomography. In *3rd IEEE International Symposium on Biomedical Imaging: Nano to Macro*. pp. 578–581.

Ralston TS, Marks DL, Scott Carney P, Boppart SA (2007) Interferometric synthetic aperture microscopy. *Nat Phys* 3:129–134.

Rao KD, Choma MA, Yazdanfar S, Rollins AM, Izatt JA (2003) Molecular contrast in optical coherence tomography by use of a pump-probe technique. *Opt Lett* 28:340–342.

Robles F, Graf RN, Wax A (2009) Dual window method for processing spectroscopic optical coherence tomography signals with simultaneously high spectral and temporal resolution. *Opt Express* 17:6799–6812.

Robles FE, Wilson C, Grant G, Wax A (2011) Molecular imaging true-colour spectroscopic optical coherence tomography. *Nat Photonics* 5:744–747.

Rogowska J, Patel N, Plummer S, Brezinski ME (2006) Quantitative optical coherence tomographic elastography: Method for assessing arterial mechanical properties. *Br J Radiol* 79:707–711.

Rollins A, Yazdanfar S, Kulkarni M, Ung-Arunyawee R, Izatt J (1998) *In vivo* video rate optical coherence tomography. *Opt Express* 3:219–229.

Rosen RB, Hathaway M, Rogers J, Pedro J, Garcia P, Dobre GM, Podoleanu AG (2009) Simultaneous OCT/SLO/ICG imaging. *Invest Ophthalmol Vis Sci* 50:851–860.

Saleh BEA, Teich MC (1991) *Fundamentals of Photonics*. Wiley.

Sarunic MV, Weinberg S, Izatt JA (2006) Full-field swept-source phase microscopy. *Opt Lett* 31:1462–1464.

Sasaki Y, Fujimoto M, Yagi S, Yamagishi S, Toyoda S, Kobayashi J (2014) Ultrahigh-phase-stable swept source based on KTN electro-optic deflector towards Doppler OCT and polarization-sensitive OCT. *Proc. Soc. Photo. Opt. Instrum. Engs* 89342Y:1–6.

Schmitt JM (1998) OCT elastography: Imaging microscopic deformation and strain of tissue. *Opt Express* 3:199–211.

Schmitt JM (1999) Optical coherence tomography (OCT): A review. *IEEE J Sel Top Quantum Electron* 5:1205–1215.

Schmitt JM, Knüttel A, Bonner RF (1993) Measurement of optical properties of biological tissues by low-coherence reflectometry. *Appl Opt* 32:6032–6042.

Schmitt JM, Xiang SH, Yung KM (1999) Speckle in optical coherence tomography. *J Biomed Opt* 4:95–106.

Schmitt JM, Yadlowsky MJ, Bonner RF (1995) Subsurface imaging of living skin with optical coherence microscopy. *Dermatol Basel Switz* 191:93–98.

Schwartz DM, Fingler J, Kim DY, Zawadzki RJ, Morse LS, Park SS, Fraser SE, Werner JS (2014) Phase-variance optical coherence tomography: A technique for noninvasive angiography. *Ophthalmology* 121:180–187.

Shemonski ND, Adie SG, Liu Y-Z, South FA, Carney PS, Boppart SA (2014a) Stability in computed optical interferometric tomography (Part I): Stability requirements. *Opt Express* 22:19183–19197.

Shemonski ND, Ahmad A, Adie SG, Liu Y-Z, South FA, Carney PS, Boppart SA (2014b) Stability in computed optical interferometric tomography (Part II): *In vivo* stability assessment. *Opt Express* 22:19314–19326.

Shemonski ND, South FA, Liu Y-Z, Adie SG, Carney PS, Boppart SA (2015) Computational high-resolution optical imaging of the living human retina. *Nat Photonics* 9:440–443.

Singh M, Li J, Vantipalli S, Wang S, Han Z, Nair A, Aglyamov SR, Twa MD, Larin KV (2016) Noncontact elastic wave imaging optical coherence elastography for evaluating changes in corneal elasticity due to crosslinking. *IEEE J Sel Top Quantum Electron* 22:266–276.

Skala MC, Crow MJ, Wax A, Izatt JA (2008) Photothermal optical coherence tomography of epidermal growth factor receptor in live cells using immunotargeted gold nanospheres. *Nano Lett* 8:3461–3467.

Soest G van, Bouchard RR, Mastik F, Jong N de, Steen AFW van der (2007) *Proc. Soc. Photo. Opt. Instrum. Engs* 6627:1–10.

South FA, Chaney EJ, Marjanovic M, Adie SG, Boppart SA (2014) Differentiation of *ex vivo* human breast tissue using polarization-sensitive optical coherence tomography. *Biomed Opt Express* 5:3417–3426.

Spaide RF, Klancnik JM, Cooney MJ, Yannuzzi LA, Balaratnasingam C, Dansingani KK, Suzuki M (2015) Volume-rendering optical coherence tomography angiography of macular telangiectasia Type 2. *Ophthalmology* 122:2261–2269.

Srinivasan VJ, Ko TH, Wojtkowski M, Carvalho M, Clermont A, Bursell S-E, Song QH, Lem J, Duker JS, Schuman JS, Fujimoto JG (2006) Noninvasive volumetric imaging and morphometry of the rodent retina with high-speed, ultrahigh-resolution optical coherence tomography. *Invest Ophthalmol Vis Sci* 47:5522–5528.

Srinivasan VJ, Sakadžić S, Gorczynska I, Ruvinskaya S, Wu W, Fujimoto JG, Boas DA (2010) Quantitative cerebral blood flow with optical coherence tomography. *Opt Express* 18:2477–2494.

Stamper D, Weissman NJ, Brezinski M (2006) Plaque characterization with optical coherence tomography. *J Am Coll Cardiol* 47:C69–C79.

Standish BA, Lee KKC, Jin X, Mariampillai A, Munce NR, Wood MFG, Wilson BC, Vitkin IA, Yang VXD (2008) Interstitial Doppler optical coherence tomography as a local tumor necrosis predictor in photodynamic therapy of prostatic carcinoma: An *in vivo* study. *Cancer Res* 68:9987–9995.

Subhash HM, Davila V, Sun H, Nguyen-Huynh AT, Shi X, Nuttall AL, Wang RK (2011) Volumetric *in vivo* imaging of microvascular perfusion within the intact cochlea in mice using ultra-high sensitive optical microangiography. *IEEE Trans Med Imaging* 30:224–230.

Swanson EA, Fujimoto JG (2017) The ecosystem that powered the translation of OCT from fundamental research to clinical and commercial impact. *Biomed Opt Express* 8:1638–1664.

Szkulmowski M, Szkulmowska A, Bajraszewski T, Kowalczyk A, Wojtkowski M (2008) Flow velocity estimation using joint spectral and time domain optical coherence tomography. *Opt Express* 16:6008–6025.

Takada K, Yokohama I, Chida K, Noda J (1987) New measurement system for fault location in optical waveguide devices based on an interferometric technique. *Appl Opt* 26:1603–1606.

Tan W, Oldenburg AL, Norman JJ, Desai TA, Boppart SA (2006) Optical coherence tomography of cell dynamics in three-dimensional tissue models. *Opt Express* 14:7159–7171.

Tan W, Sendemir-Urkmez A, Fahrner LJ, Jamison R, Leckband D, Boppart SA (2004) Structural and functional optical imaging of three-dimensional engineered tissue development. *Tissue Eng* 10:1747–1756.

Tanaka M, Hirano M, Murashima K, Obi H, Yamaguchi R, Hasegawa T (2015) 1.7-µm spectroscopic spectral-domain optical coherence tomography for imaging lipid distribution within blood vessel. *Opt Express* 23:6645–6655.

Tanno N, Ichimura T (1994) Reproduction of optical reflection-intensity-distribution using multimode laser coherence. *Electron Commun Jpn Part II Electron* 77:10–19.

Tearney GJ, Brezinski ME, Southern JF, Bouma BE, Hee MR, Fujimoto JG (1995) Determination of the refractive index of highly scattering human tissue by optical coherence tomography. *Opt Lett* 20:2258.

Tearney GJ, Jang I-K, Bouma BE (2006) Optical coherence tomography for imaging the vulnerable plaque. *J Biomed Opt* 11:021002.

Thouvenin O, Boccara C, Fink M, Sahel J, Pâques M, Grieve K (2017) Cell motility as contrast agent in retinal explant imaging with full-field optical coherence tomography. *Invest Ophthalmol Vis Sci* 58:4605–4615.

Tsytsarev V, Rao B, Maslov KI, Li L, Wang LV (2013) Photoacoustic and optical coherence tomography of epilepsy with high temporal and spatial resolution and dual optical contrasts. *J Neurosci Methods* 216:142–145.

Tumlinson AR, Hariri LP, Utzinger U, Barton JK (2004) Miniature endoscope for simultaneous optical coherence tomography and laser-induced fluorescence measurement. *Appl Opt* 43:113–121.

Vakoc BJ, Tearney GJ, Bouma BE (2009) Statistical properties of phase-decorrelation in phase-resolved Doppler optical coherence tomography. *IEEE Trans Med Imaging* 28:814–821.

van der Meer FJ, Faber DJ, Perrée J, Pasterkamp G, Baraznji Sassoon D, van Leeuwen TG (2005) Quantitative optical coherence tomography of arterial wall components. *Lasers Med Sci* 20:45–51.

Vinegoni C, Bredfeldt J, Marks D, Boppart S (2004) Nonlinear optical contrast enhancement for optical coherence tomography. *Opt Express* 12:331–341.

Vinegoni C, Ralston T, Tan W, Luo W, Marks DL, Boppart SA (2006) Integrated structural and functional optical imaging combining spectral-domain optical coherence and multiphoton microscopy. *Appl Phys Lett* 88:053901.

Wang H, Akkin T, Magnain C, Wang R, Dubb J, Kostis WJ, Yaseen MA, Cramer A, Sakadžić S, Boas D (2016a) Polarization sensitive optical coherence microscopy for brain imaging. *Opt Lett* 41:2213–2216.

Wang J, Xu Y, Boppart SA (2017) Review of optical coherence tomography in oncology. *J Biomed Opt* 22:121711.

Wang RK, An L (2009) Doppler optical micro-angiography for volumetric imaging of vascular perfusion *in vivo*. *Opt Express* 17:8926–8940.

Wang RK, An L, Saunders S, Wilson DJ (2010) Optical micro-angiography provides depth-resolved images of directional ocular blood perfusion in posterior eye segment. *J Biomed Opt* 15:020502.

Wang RK, Jacques SL, Ma Z, Hurst S, Hanson SR, Gruber A (2007a) Three dimensional optical angiography. *Opt Express* 15:4083–4097.

Wang RK, Kirkpatrick S, Hinds M (2007b) Phase-sensitive optical coherence elastography for mapping tissue microstrains in real time. *Appl Phys Lett* 90:164105.

Wang RK, Ma Z, Kirkpatrick SJ (2006) Tissue Doppler optical coherence elastography for real time strain rate and strain mapping of soft tissue. *Appl Phys Lett* 89:144103.

Wang Y, Bower BA, Izatt JA, Tan O, Huang D (2007c) *In vivo* total retinal blood flow measurement by Fourier domain Doppler optical coherence tomography. *J Biomed Opt* 12:041215.

Wang Z, Potsaid B, Chen L, Doerr C, Lee H-C, Nielson T, Jayaraman V, Cable AE, Swanson E, Fujimoto JG (2016b) Cubic meter volume optical coherence tomography. *Optica* 3:1496–1503.

Wieser W, Draxinger W, Klein T, Karpf S, Pfeiffer T, Huber R (2014) High definition live 3D-OCT *in vivo*: Design and evaluation of a 4D OCT engine with 1 GVoxel/s. *Biomed Opt Express* 5:2963–2977.

Wojtkowski M (2010) High-speed optical coherence tomography: Basics and applications. *Appl Opt* 49:D30–D61.

Wojtkowski M, Leitgeb R, Kowalczyk A, Bajraszewski T, Fercher AF (2002) *In vivo* human retinal imaging by Fourier domain optical coherence tomography. *J Biomed Opt* 7:457–464.

Wojtkowski M, Srinivasan V, Fujimoto JG, Ko T, Schuman JS, Kowalczyk A, Duker JS (2005) Three-dimensional retinal imaging with high-speed ultrahigh-resolution optical coherence tomography. *Ophthalmology* 112:1734–1746.

Xi J, Chen Y, Zhang Y, Murari K, Li M-J, Li X (2012) Integrated multimodal endomicroscopy platform for simultaneous *en face* optical coherence and two-photon fluorescence imaging. *Opt Lett* 37:362–364.

Xu C, Vinegoni C, Ralston TS, Luo W, Tan W, Boppart SA (2006) Spectroscopic spectral-domain optical coherence microscopy. *Opt Lett* 31:1079–1081.

Yang Y, Bagnaninchi PO, Wood MA, Haj AJE, Guyot E, Dubois A, Wang RK (2005) Monitoring cell profile in tissue engineered constructs by OCT. *Proc. Soc. Photo. Opt. Instrum. Engs* 5695:51–57.

Yang Y, Dubois A, Qin X, Li J, El Haj A, Wang RK (2006) Investigation of optical coherence tomography as an imaging modality in tissue engineering. *Phys Med Biol* 51:1649–1659.

Yang Y, Sulé-Suso J, El Haj AJ, Hoban PR, Wang R (2004) Monitoring of lung tumour cell growth in artificial membranes. *Biosens Bioelectron* 20:442–447.

Yao G, Wang LV (1999) Two-dimensional depth-resolved Mueller matrix characterization of biological tissue by optical coherence tomography. *Opt Lett* 24:537–539.

Yasuno Y, Madjarova VD, Makita S, Akiba M, Morosawa A, Chong C, Sakai T, Chan K-P, Itoh M, Yatagai T (2005) Three-dimensional and high-speed swept-source optical coherence tomography for *in vivo* investigation of human anterior eye segments. *Opt Express* 13:10652–10664.

Yasuno Y, Makita S, Endo T, Aoki G, Itoh M, Yatagai T (2006) Simultaneous B-M-mode scanning method for real-time full-range Fourier domain optical coherence tomography. *Appl Opt* 45:1861–1865.

Yuan Z, Zakhaleva J, Ren H, Liu J, Chen W, Pan Y (2010) Noninvasive and high-resolution optical monitoring of healing of diabetic dermal excisional wounds implanted with biodegradable *in situ* gelable hydrogels. *Tissue Eng Part C Methods* 16:237–247.

Yun SH, Boudoux C, Tearney GJ, Bouma BE (2003a) High-speed wavelength-swept semiconductor laser with a polygon-scanner-based wavelength filter. *Opt Lett* 28:1981–1983.

Yun SH, Tearney GJ, de Boer JF, Iftimia N, Bouma BE (2003b) High-speed optical frequency-domain imaging. *Opt Express* 11:2953–2963.

Zhang EZ, Povazay B, Laufer J, Alex A, Hofer B, Pedley B, Glittenberg C, Treeby B, Cox B, Beard P, Drexler W (2011) Multimodal photoacoustic and optical coherence tomography scanner using an all optical detection scheme for 3D morphological skin imaging. *Biomed Opt Express* 2:2202–2215.

Zhang HF, Maslov K, Stoica G, Wang LV (2006) Functional photoacoustic microscopy for high-resolution and noninvasive *in vivo* imaging. *Nat Biotechnol* 24:848–851.

Zhang Q, Lee CS, Chao J, Chen C-L, Zhang T, Sharma U, Zhang A, Liu J, Rezaei K, Pepple KL, Munsen R, Kinyoun J, Johnstone M, Gelder RNV, Wang RK (2016) Wide-field optical coherence tomography based microangiography for retinal imaging. *Sci Rep* 6:srep22017.

Zhao Y, Chen Z, Saxer C, Xiang S, de Boer JF, Nelson JS (2000) Phase-resolved optical coherence tomography and optical Doppler tomography for imaging blood flow in human skin with fast scanning speed and high velocity sensitivity. *Opt Lett* 25:114.

Zhao Y, Graf BW, Chaney EJ, Mahmassani Z, Antoniadou E, Devolder R, Kong H, Boppart MD, Boppart SA (2012) Integrated multimodal optical microscopy for structural and functional imaging of engineered and natural skin. *J Biophotonics* 5:437–448.

Zhao Y, Maher JR, Kim J, Selim MA, Levinson H, Wax A (2015) Evaluation of burn severity *in vivo* in a mouse model using spectroscopic optical coherence tomography. *Biomed Opt Express* 6:3339–3345.

Zhi Z, Chao JR, Wietecha T, Hudkins KL, Alpers CE, Wang RK (2014) Noninvasive imaging of retinal morphology and microvasculature in obese mice using optical coherence tomography and optical microangiography. *Invest Ophthalmol Vis Sci* 55:1024–1030.

Zhou C, Cohen DW, Wang Y, Lee H-C, Mondelblatt AE, Tsai T-H, Aguirre AD, Fujimoto JG, Connolly JL (2010) Integrated optical coherence tomography and microscopy for *ex vivo* multiscale evaluation of human breast tissues. *Cancer Res* 70:10071–10079.

Zvyagin A, Silva KKMB, Alexandrov S, Hillman T, Armstrong J, Tsuzuki T, Sampson D (2003) Refractive index tomography of turbid media by bifocal optical coherence refractometry. *Opt Express* 11:3503–3517.

Zysk AM, Adie SG, Armstrong JJ, Leigh MS, Paduch A, Sampson DD, Nguyen FT, Boppart SA (2007a) Needle-based refractive index measurement using low-coherence interferometry. *Opt Lett* 32:385–387.

Zysk AM, Boppart SA (2006) Computational methods for analysis of human breast tumor tissue in optical coherence tomography images. *J Biomed Opt* 11:054015.

Zysk AM, Marks DL, Liu DY, Boppart SA (2007b) Needle-based reflection refractometry of scattering samples using coherence-gated detection. *Opt Express* 15:4787–4794.

Zysk AM, Nguyen FT, Chaney EJ, Kotynek JG, Oliphant UJ, Bellafiore FJ, Johnson PA, Rowland KM, Boppart SA (2009) Clinical feasibility of microscopically-guided breast needle biopsy using a fiber-optic probe with computer-aided detection. *Technol Cancer Res Treat* 8:315–321.

Zysk AM, Reynolds JJ, Marks DL, Carney PS, Boppart SA (2003) Projected index computed tomography. *Opt Lett* 28:701–703.

8

Multiscale Photoacoustic Imaging

Tianxiong Wang and Song Hu

CONTENTS

8.1 Introduction

In photoacoustic imaging (PAI), biological tissues are typically excited by short-pulsed laser light. As shown in Figure 8.1, the light energy is absorbed by optically absorbing molecules in the tissue and partially converted into heat. The transient heating induces thermoelectric expansion and an initial pressure rise, which then propagates as an acoustic wave and is finally captured by a single-element or array-based ultrasonic detector to form an image(Wang and Wu, 2012). Presenting an elegant marriage of optics and ultrasound, PAI is uniquely capable of organelle-cell-tissue imaging in the optical ballistic and quasiballistic regimes (within 1 mm from the surface of the tissue) and tissue-organ imaging in the quasidiffusive and diffusive regimes (up to cm-depth), bridging the gaps between pure optical microscopy and tomography. Moreover, PAI is exquisitely sensitive to optical absorption, highly complementary to the contrasts provided by optical coherence tomography and fluorescence imaging. In the past couple of decades, PAI has drawn increasing attention from various research communities, including imaging, chemistry, and biomedicine.

In this chapter, we introduce *in vivo* PAI at different spatial scales, spanning from individual organelles to the whole body of small animals. Detailed methodologies are discussed, and promising applications are reviewed. The frontiers of this emerging technology are also highlighted.

8.2 Spatial Scalability of PAI

As shown in Figure 8.2, there are two major implementations of PAI with complementary spatial resolution and tissue penetration: optical-resolution PAI and acoustic-resolution PAI.

In optical-resolution PAI (often termed optical-resolution photoacoustic microscopy or OR-PAM), focused light excitation is employed to spatially confine the generation of acoustic waves. Thus, the lateral resolution of OR-PAM is determined by the focal spot size of the beam. For example, focusing a 532-nm beam with a 0.1-numerical aperture (NA) condenser lens leads to a 2.6-μm lateral resolution, which is sufficient to resolve single red blood cells (RBCs) *in vivo* (Hu et al., 2011). By contrast, the axial resolution of OR-PAM is determined by the bandwidth of the acoustic detector and can be estimated as $R_a = 0.88c/\Delta f$, where c is the speed of sound and Δf is the detection bandwidth (Zhang et al., 2012).

For effective optical focusing, the penetration of OR-PAM is typically within the optical diffusion limit (~1 mm). To acquire high-resolution photoacoustic images beyond this limit, where acoustic generation can be hardly confined, a focused acoustic detector is required to spatially confine the detection of acoustic waves. Thus, both the lateral and axial resolutions of the acoustic-resolution PAI are determined by ultrasound. Using weakly focused or unfocused light and single-element focused transducers of different center frequencies, acoustic-resolution photoacoustic microscopy (AR-PAM) (Zhang et al., 2006) and macroscopy (AR-PAMac) (Song and Wang, 2012) have achieved scalable tissue penetration (from 3 mm to 4 cm). Similar to OR-PAM, AR-PAM or AR-PAMac requires two-dimensional (2D) raster scan to obtain a volumetric image, which severely limits the imaging speed. Using a transducer array instead, photoacoustic computed tomography (PACT) has enabled high-speed imaging at depth. In PACT, each array element has a relatively large acceptance angle within the field of view (FOV). Thus, a tomographic image can be instantly obtained by integrating data from all elements without raster

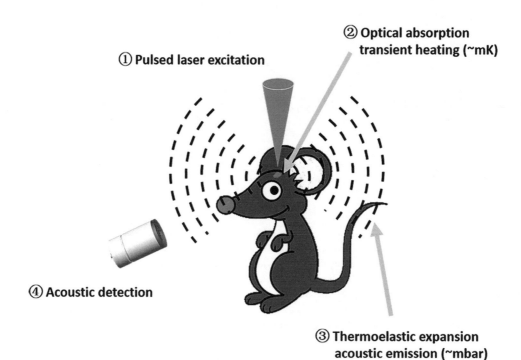

FIGURE 8.1 Principle of photoacoustic imaging.

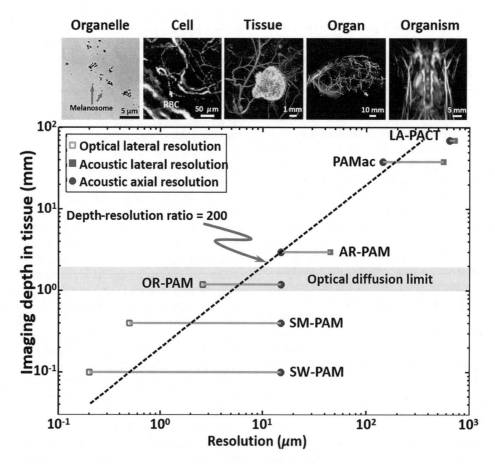

FIGURE 8.2 Tissue penetration versus spatial resolution in multiscale photoacoustic imaging.

scan. The lateral resolution of PACT is codetermined by the acceptance angle and center frequency of individual transducer elements, and the axial resolution is determined by the detection bandwidth of the elements.

8.3 Photoacoustic Imaging at the Organelle Level

By tightly focusing the excitation light, OR-PAM can achieve submicron lateral resolution for organelle imaging *in vivo*, which is highly desired for biological and pathological studies. For example, the nuclei of cancer cells have specific morphological features, including relatively large size and irregular shape. With tightly focused ultraviolet light excitation, OR-PAM is able to image single cell nucleus *in vivo* (Yao et al, 2010; Yao et al, 2012), thereby holding the potential for high-resolution detection of the tumor margin.

As shown in Figure 8.3a, the ultraviolet ns-pulsed light is focused by a condenser lens, purified by a pinhole, and then focused into the tissue through a 0.1-NA objective lens. A ring-shaped ultrasonic transducer is coaxially and confocally aligned with the optical excitation for maximal sensitivity. A water tank filled with distilled water is used to couple the excited photoacoustic signal to the transducer. The water tank and the animal are mounted on a two-axis (x–y plane) scanning stage to realize raster scan. Given that both nucleic acids (e.g. thymus DNA) and cytoplasmic proteins (e.g. glutamate dehydrogenase) are optically absorptive in the ultraviolet region, the excitation wavelength needs to be optimized for maximal contrast-to-noise ratio (CNR). By examining the CNR of the OR-PAM images acquired in the range of

245–275 nm with the same laser pulse energy and a 5-nm spectral interval, the optimal wavelength for imaging cell nuclei is determined to be 250 nm. With a pulse energy of 20 nJ, individual cell nuclei in the live mouse skin is clearly visualized by OR-PAM (Figure 8.3b), based on the average nuclear diameter (Figure 8.3c), and the internuclear distance of keratinocytes can be quantified (Figure 8.3d).

By using a water-immersion microscope objective with 1.23 NA, Zhang et al. (2010) has pushed the lateral resolution of OR-PAM to subwavelength, achieving 220 nm with 532-nm excitation. With the subwavelength OR-PAM, single melanosomes in the live ear of a pigmented mouse can be clearly visualized.

8.4 Photoacoustic Imaging at the Cellular Level

Slightly relaxing the light focus of OR-PAM allows single-cell imaging *in vivo* with extended depth of focus. Rich in hemoglobin, RBCs are the primary oxygen carrier in the circulation of mammals. Dynamic imaging of oxygen transport at the single RBC level based on the endogenous contrast of hemoglobin may shed new light on the tissue function and metabolism. To this end, Wang et al. (2013) has developed an OR-PAM system that allows real-time functional imaging of single RBCs *in vivo*.

As shown in Figure 8.4a, the light emitted from two synchronized lasers with wavelengths of 532 and 560 nm are combined and then focused by a condenser lens, spatially filtered by a pinhole, and coupled into a single-mode fiber (SMF). A beam sampler is employed to partially reflect the light into a high-speed photodiode to monitor the fluctuation in laser pulse

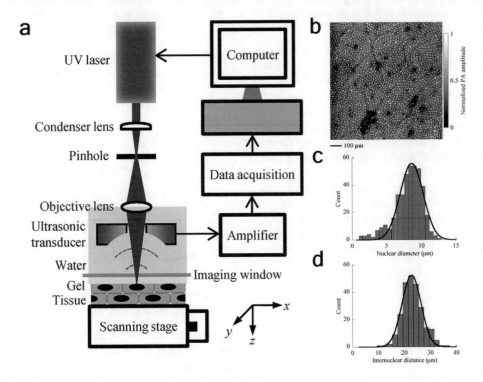

FIGURE 8.3 Photoacoustic imaging at the organelle level. (a) Schematic of the ultraviolet OR-PAM. (b) OR-PAM of cell nuclei in the live mouse skin. (c, d) Histograms of the nuclear diameter and the internuclear distance of keratinocytes in the mouse skin.

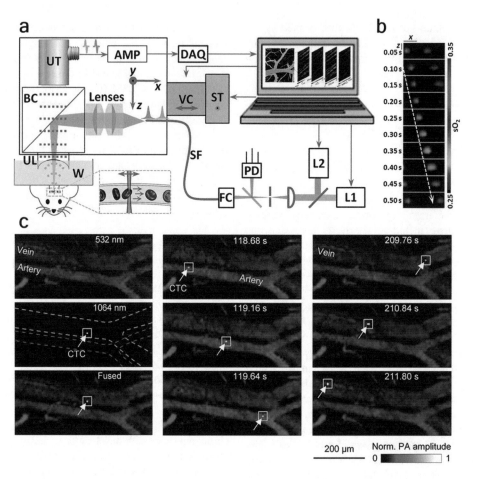

FIGURE 8.4 Photoacoustic imaging at the cellular level. (a) Schematic of single-cell OR-PAM. AMP, signal amplifiers and filters; BC, acoustic optical beam combiner; DAQ, digitizer; FC, fiber coupler; L1, pulsed laser 1, 560 nm; L2, pulsed laser 2, 532 nm; Lenses, two achromatic doublets; PD, photodiode; SF, single-mode optical fiber; ST, linear stage (y-axis); UL, ultrasonic lens; UT, ultrasonic transducer; VC, voice-coil scanner (x-axis); W, water. (b) Sequential snapshots of single RBCs releasing oxygen in the mouse brain. Scale bars: x = 10 μm and z = 30 μm. Blood flows from left to right. The dashed arrow follows the trajectory of a single flowing RBC. (c) Snapshots showing single circulating melanoma cells traveling in an artery-vein pair.

energy. The output of the SMF is mapped into the object to be imaged by a pair of achromatic doublets. A self-developed acoustic-optical beam combiner is employed to combine the incident light beam and the generated ultrasonic wave. A flat transducer and an acoustic lens are glued to the top and bottom of the beam combiner, respectively, for focused ultrasonic detection. The detected signal is amplified by an amplifier and then acquired by a digitizer. A voice-coil scanner and a linear stage are combined for high-speed 2D raster scan.

To measure the hemoglobin concentration and oxygen saturation of individual RBCs (C_{Hb} and sO_2, respectively), the dual-wavelength laser pulses sequentially excite nearly the same region of a RBC to acquire two depth-resolved photoacoustic signals (i.e. A-lines), whose amplitudes are (Wang and Wu, 2012).

$$PA_{532} = K\Gamma\eta_{th}F_{532}\left(\sigma_{HbR,532}N_{HbR} + \sigma_{HbO_2,532}N_{HbO_2}\right) \quad (8.1)$$

$$PA_{560} = K\Gamma\eta_{th}F_{560}\left(\sigma_{HbR,560}N_{HbR} + \sigma_{HbO_2,560}N_{HbO_2}\right) \quad (8.2)$$

where K is a constant determined by the transducer response and the gain of the electrical amplifier, Γ is the Grüneisen coefficient, η_{th} is the heat conversion efficiency, F_{532} and F_{560}

are the optical fluence at 532 and 560 nm, $\sigma_{HbR,532}$, $\sigma_{HbO_2,532}$, $\sigma_{HbR,560}$ and $\sigma_{HbO_2,560}$ are the absorption cross sections of deoxy- and oxyhemoglobin (HbR and HbO$_2$, respectively) at 532 and 560 nm, and N_{HbR} and N_{HbO_2} are the number density of HbR and HbO$_2$.

According to Equation (8.1) and (8.2), the sO_2 can be derived as:

$$sO_2 = \frac{N_{HbO_2}}{N_{HbO_2} + N_{HbR}}$$

$$= \frac{PA_{532}F_{560}\sigma_{HbR,560} - PA_{560}F_{532}\sigma_{HbR,532}}{PA_{560}F_{532}\left(\sigma_{HbO_2,532} - \sigma_{HbR,532}\right) + PA_{532}F_{560}\left(\sigma_{HbR,560} - \sigma_{HbO_2,560}\right)} \quad (8.3)$$

The C_{Hb} can be estimated as:

$$C_{Hb} = \frac{PA_{532}}{K\Gamma\eta_{th}F_{532}\left(\sigma_{HbR,532}\left(1 - sO_2\right) + \sigma_{HbO_2,532} \cdot sO_2\right)} \quad (8.4)$$

Because the B-scan rate is fast enough to capture single RBCs flowing through the FOV, the RBC flow speed can be estimated from the cross correlation between two consecutive B-scan images acquired at 532 nm. With μm-level resolution, >100-Hz 2D imaging rate, and 20-μs oxygenation detection time, this OR-PAM system enables simultaneous imaging of C_{Hb}, sO_2, flow speed, and oxygen release rate of single RBCs in real time. As shown in Figure 8.4b, to record the dynamic oxygen release of individual RBCs, a set of time-lapse B-scan images of the live mouse brain are acquired, during which a single RBC flows from the left to the right side of the FOV. The oxygen release of this RBC is clearly captured, showing 3% decrease in sO_2 over the 32-μm distance. This technique holds the potential to study neurovascular coupling at the fundamental level.

Recently, He et al. (2016 have expanded the scope of the single-cell OR-PAM to detect circulating tumor cells (CTCs) *in vivo*. By using a water-immersible microelectromechanical system (MEMS) mirror, both the light excitation and ultrasound detection can be steered simultaneously for high-speed scan along the fast axis. The object to be imaged is mounted on a linear stage for orthogonal scan along the slow axis. Using a high-repetition-rate pulsed laser, this OR-PAM system is capable of volumetric imaging at 10 Hz over a FOV of 3×0.25 mm^2. The lateral resolution at 1064 nm is 7 μm, which is sufficient to capture single circulating melanoma cells. As shown in Figure 8.4c, the vascular structure is imaged with high contrast at 532 nm because of the strong optical absorption of hemoglobin. However, circulating melanoma cells can hardly be distinguished from RBCs in these images because of the strong absorption of both melanosome and hemoglobin at this wavelength. At 1064 nm, in contrast, melanoma cells produce 5-fold stronger photoacoustic signals than that of the tissue background, while RBC signals are at the noise level. Thus, with the dual-wavelength excitation, the flowing of single CTCs in the blood vessels can be clearly visualized.

8.5 Photoacoustic Imaging at the Tissue Level

Further relaxing the light focus permits high-resolution imaging at the tissue level in the quasiballistic and quasidiffusive regimes using OR-PAM (Maslov et al., 2008) and AR-PAM (Zhang et al., 2007), respectively. Capitalizing on the optical absorption of hemoglobin, PAM is, heretofore, the only technology that allows label-free comprehensive characterization of the microvascular structure and function, from which the oxygen metabolism of live tissues can be estimated.

Recently, we have developed the first-of-a-kind multiparametric PAM for simultaneous imaging of blood perfusion, oxygenation, and flow *in vivo* (Ning et al., 2015a; Ning et at., 2015b). With the aid of vessel segmentation, the regional metabolic rate of oxygen (MRO_2) can be derived using Fick's law (Cao et al., 2017).

As shown in Figure 8.5a, the state-of-the-art multiparametric PAM (Wang et al., 2016) employs a 532-nm ns-pulsed laser with a repetition rate of 300 kHz. The laser beam passes through an electro-optical modulator (EOM) and a half-wave plate (HWP), before being expanded by a pair of condenser lenses. The EOM and HWP together act as a high-speed optical switch. When a high voltage is applied to the EOM, the beam polarization is rotated by 90° and becomes vertical. Thus, it is reflected by the polarization beam splitter (PBS) and coupled into a 5-meter-long polarization-maintaining

single-mode fiber (PM-SMF) through an optical objective. When the 532-nm beam propagates inside the PM-SMF, its wavelength will be red-shifted due to the Raman scattering. The output of the PM-SMF is collimated by a collimator and purified by an optical band-pass filter to isolate the 558-nm component. When a low voltage is applied, the 532-nm beam remains horizontally polarized and passes through the PBS without wavelength conversion. Thus, alternating the EOM voltage allows pulse-by-pulse switching of the laser wavelength. The 532 and 558-nm beam are combined by a dichroic mirror (DM) and coupled into a 2-meter-long regular SMF. A beam sampler is placed before the SMF to tap off a small fraction (~5%) of the light into a high-speed photodiode for pulse energy monitoring. According to our test, regular SMFs with such a short length do not generate noticeable Raman shift to the optical wavelength.

In the imaging head of the multiparametric PAM, the dual-wavelength beam coming from the SMF is collimated by an achromatic doublet, reflected by a two-axis galvanometer scanner, and focused into the object to be imaged by a second doublet. A correction lens is placed beneath the doublet to compensate for the optical aberration caused by the air-water interface. A ring-shaped transducer is placed beneath the correction lens and immersed in a water tank to detect the generated photoacoustic signal. The imaging head is mounted on two motorized linear stages for 2D raster scan.

This system combines one-dimensional optical scan and 2D mechanical scan for high-speed multiparametric image acquisition. Specifically, the galvanometer scanner steers the laser beam along the y-axis within the 50-μm focus of the ring transducer at a round-trip rate of 2.1 kHz, while the linear stage mechanically translates the optical-acoustic dual foci along the x-axis at a constant speed of 0.88 mm/s. During the optical-mechanical hybrid scanning process, the laser output is switched between 532 and 558 nm at a 3.3-μs interval to produce dual-wavelength A-line pairs. The hybrid scan allows 20 B-scans to be simultaneously acquired, leading to a 20-fold increase in imaging speed over our first-generation multiparametric PAM that is based on pure mechanical scan (Ning et al., 2015b).

Simultaneous PAM of blood flow, C_{Hb} and sO_2 at the microscopic level is realized by correlation, statistical and spectral analyses of individual B-scans. Specifically, PAM is insensitive to sO_2 at 532 nm, a near-isosbestic point of hemoglobin. Fluctuation in the photoacoustic signal acquired at this wavelength encodes both the flow and Brownian motion of RBCs. The blood flow speed is quantified by the decorrelation rate of successively acquired A-lines. Theoretically, the correlation coefficient between two adjacent A-lines depends on their time interval. The dependence follows a Gaussian decay, of which the decay constant is linearly proportional to the flow speed. By fitting the experimentally measured decorrelation curve with the theoretical model, the flow speed can be quantified. Given the 2.1-kHz round-trip rate of the optical scan and the 0.88-mm/s speed of the mechanical scan, 49 A-lines can be acquired when the linear stage travels 10 μm along the B-scan direction. The decorrelation curve is obtained by calculating the correlation coefficients between the central A-line and each of the 48 preceding and subsequent A-lines, as well as

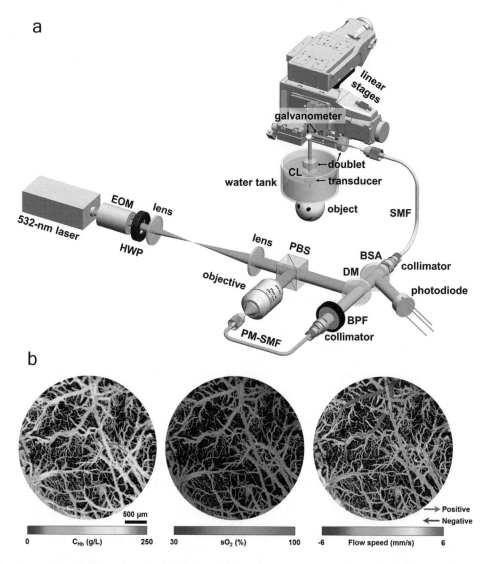

FIGURE 8.5 Photoacoustic imaging at the tissue level. (a) Schematic of the high-speed multiparametric OR-PAM. EOM, electro-optical modulator; HWP, half-wave plate; PBS, polarizing beam splitter; PM-SMF, polarization-maintaining single-mode fiber; BPF, band-pass filter; DM, dichroic mirror; BSA, beam sampler; SMF, regular single-mode fiber; CL, correction lens. (b) Simultaneous high-speed OR-PAM of (left) C_{Hb}, (middle) sO_2, and (right) CBF (both speed and direction) in the mouse brain.

the corresponding time delays. This correlation analysis allows blood flow quantification at the spatial resolution comparable to the average diameter of capillaries. In parallel, C_{Hb} can be derived in absolute values by analyzing the Brownian motion-induced statistical fluctuation in the amplitudes of the same 49 A-lines, which is known to depend on the number of RBCs within the detection volume of PAM but not on the flow speed (Ning et al., 2015a; Zhou et al, 2014). The higher the C_{Hb}, the larger the A-line amplitude and the higher the fluctuation. By comparing the readouts at both wavelengths (532 and 558 nm), sO_2 can be simultaneously quantified. As shown in Figure 8.5b, C_{Hb}, sO_2, and blood flow speed of the microvasculature in the mouse brain can be acquired by a single hybrid scan.

When imaging tissues with uneven surfaces, the out-of-focus issue may lead to reduced accuracy in the quantitative measurements. To maintain high resolution for accurate quantification of C_{Hb}, sO_2, and blood flow, we have developed ultrasound-aided multiparametric PAM (Ning et al., 2015b).

By operating the ring transducer in the pulse-echo mode, we can perform scanning acoustic microscopy (SAM) concurrently with multiparametric PAM. The larger depth of focus of SAM, which spans several hundred microns, and the high ultrasound contrast of structural heterogeneity together allows more reliable detection of the contour of the tissue surface. A three-step procedure has been developed to extract the surface contour. The procedure begins with a rapid SAM of the region of interest. Then, the depth of the maximum signal in each A-line is identified. Integrating the depth information extracted from individual A-lines leads to a three-dimensional (3D) map of the surface contour. Finally, the ultrasonically extracted contour map is interpolated (down to the same step sizes planned for the PAM scan) and smoothed (with a span of 15% of the B-scan length) for the contour-guided multiparametric PAM.

Besides traditional PAM, which are based on piezoelectric transducers, optical-based ultrasound detectors have also been utilized for PAI at the tissue level. For example, Jathoul et al. (2015) have developed a Fabry–Pérot (FP) sensor-based PAI. The FP sensor consists of an etalon overlaying a 10-mm-thick wedged polymethyl methacrylate backing substrate. The etalon contains a pair of DMs and a 40-μm-thick Parylene C polymer spacer in between. To transmit the photoacoustic excitation light and reflect the FP probe light, the DMs are designed to be highly reflective in the range of 1500–1600 nm and transparent at shorter wavelengths (600–1200 nm). Following photoacoustic excitation, the generated acoustic waves travel through the tissue and hit the etalon. The pressure-induced modulation of the FP cavity is then detected by laser interferometry at 1500–1600 nm. To maximize the detection sensitivity, the probe laser wavelength is tuned to the peak derivative of the etalon's reflectance transfer function. Under this condition, subtle changes in the etalon cavity are linearly converted to temporal modulations of the light interference. To map the spatial distribution of the photoacoustic waves, the probe beam is optically scanned over the etalon surface, and the time-resolved photoacoustic signals are recorded at each scan point. To compensate for variations in the optical thickness of the polymer spacer, the probe wavelength is optimized for each scan point. To form a 3D image, a reconstruction algorithm based on time reversal is used (Treeby et al., 2010; Treeby and Cox., 2010). The all-optical PAI has demonstrated high-resolution imaging of live tissues at depths approaching 1 cm with superb image quality (Jathoul et al., 2015).

8.6 Photoacoustic Imaging at the Organ Level

Internal organ imaging is typically carried out by acoustic-resolution PAI. With weakly focused or unfocused light excitation, photoacoustic signals can be generated in the diffusive regime and detected by either a single-element focused transducer or a transducer array. The relatively low acoustic attenuation of biological tissues enables acoustic-resolution PAI to achieve deep penetration for organ imaging. Besides traditional AR-PAM and PACT, photoacoustic endoscopy allows the imaging device to physically access the internal organ, thereby reducing the penetration burden.

Yang et al. (2012) have developed simultaneous photoacoustic and ultrasound dual-mode endoscopy. As shown in Figure 8.6a, a multimode optical fiber is used to deliver the dual-wavelength laser beam to the internal organ for photoacoustic excitation. The output end of the fiber is inserted into a ring-shaped transducer for coaxial alignment of the optical excitation and ultrasonic detection. The excitation light beam and the generated acoustic wave are both reflected by a scanning mirror. A built-in actuator, consisting of a micromotor and other mechanical components for water sealing, is employed to rotate the mirror for circumferential sector scan. A signal wire is used to control the transducer to fire ultrasound pulses and detect both the echo signal and the photoacoustic signal. The entire endoscope probe has an outer diameter of 3.8 mm and rigid distal length of 38 mm, which enables simultaneous

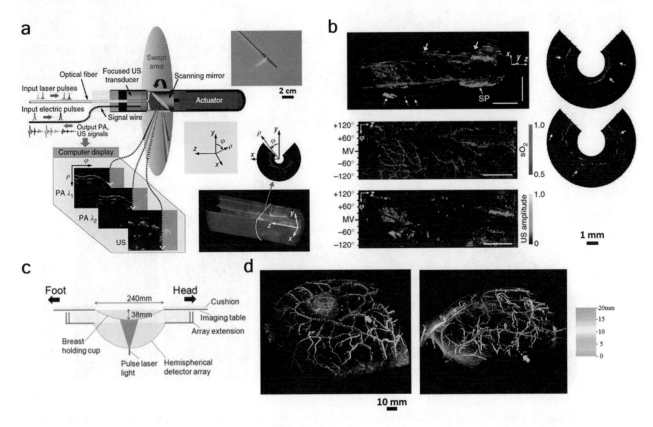

FIGURE 8.6 Photoacoustic imaging at the organ level. (a) Illustration of simultaneous ultrasonic and multiwavelength photoacoustic endoscopy. (b) Simultaneous, coregistered, ultrasonic and photoacoustic images of a rat colon *in vivo*. (c) Schematic of the PACT for breast imaging. (d) Depth-encoded projection images of the human breast.

photoacoustic and ultrasonic imaging of the rat colon *in vivo* (Figure 8.6b).

Toi et al. (2017) have developed a PACT system for noninvasive imaging of the human breast. As shown in Figure 8.6c, the system consists of a breast holding cup, a hemispherical transducer array, and a wavelength-tunable laser excitation source. An unfocused dual-wavelength (755 and 795 nm) laser beam is used to measure both the vessel structure and sO_2. The relative long excitation wavelength ensures deep tissue penetration. The image reconstruction is carried out using a universal back-projection algorithm, in which both the average sound speed of the patient breast and that of water are considered for impedance matching to reduce the degradation of the image resolution *in vivo* (Xu and Wang, 2005). This PACT system allows high-resolution imaging of the vascular network in the human breast with 20-mm penetration depth (Figure 8.6d), holding great potential for breast cancer diagnosis.

8.7 Photoacoustic Imaging of the Small-Animal Whole Body

Recent advances in high-speed PACT also enable whole-body imaging of small animals. Li et al. have developed a single-impulse panoramic PACT (SIP-PACT) that is capable of anatomical and functional imaging of the whole body of the adult mouse with 125-μm in-plane resolution, 50-μs/frame data acquisition, 50-Hz frame rate and full-view fidelity.

To achieve whole-body penetration, the excitation wavelength is selected within the optical window (650–1350 nm), where mammalian tissues least attenuate light. As shown in Figure 8.7a, the SIP-PACT employs a 1064-nm ns-pulsed laser, an OPO laser, and a Ti:sapphire laser. The laser beams at different wavelengths are combined via a beam combiner and then homogenized and expanded by a light diffuser, before being delivered into the mouse. Two different configurations are utilized to image the brain and trunk, respectively. For brain imaging, top illumination and side detection is used. The excitation beam is uniformly shined on the cortex after passing through the diffuser. For trunk imaging, in contrast, side illumination and side detection are used. A conical lens is employed to convert the light to a ring-shaped pattern, which is then refocused by a condenser lens. The illuminated area is located within the elevational focal zone of the transducer. To achieve 2D panoramic acoustic detection, a 512-element full-ring transducer array is used, in which each element is cylindrically focused. A 512-channel preamplifier is directly connected to the housing of the array to minimize the length of the connection cable for reduced noise. The amplified signals are then digitized by a 512-channel data acquisition

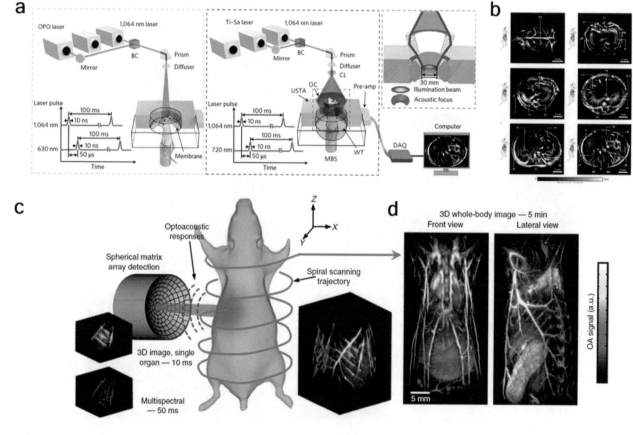

FIGURE 8.7 Photoacoustic imaging of the whole body of small animal. (a) Schematic of the SIP-PACT system for mouse brain and trunk imaging. BC, beam combiner; CL, conical lens; DAQ, data acquisition system; MBS, magnetic base scanner; OC, optical condenser; OPO, optical parametric oscillator; USTA, (full-ring) ultrasonic transducer array; WT, water tank. (b) Label-free SIP-PACT of whole-body anatomy of the mouse, from the brain to the trunk. (c) Schematic of the SVOT system. (d) SVOT of the whole body of the mouse.

system and fed into a half-time dual-speed-of-sound (in tissue and water) universal back-projection algorithm for image reconstruction (Xu and Wang, 2005; Anastasio et al., 2005). As shown in Figure 8.7b, SIP-PACT can generate high-resolution tomographic views of the whole body of the mouse in real time, with suborgan anatomical and functional details.

Also, Deán-Ben et al. have developed spiral volumetric optoacoustic tomography (SVOT), which offers the unique capability to efficiently bridge the visualization of dynamic processes at multiple scales (Deán-Ben et al., 2017). In SVOT, unfocused near-infrared laser light is used to excite photoacoustic signals (Figure 8.7c). The beam is guided through a custom-made fiber bundle inserted through the central opening of a 256-element spherical transducer array. The generated acoustic signal is collected at multiple locations around the imaging volume, allowing for on-the-fly image reconstruction and real-time (100 frames/second) volumetric rendering over a 1-cm³ FOV following each laser pulse excitation. A near-isotropic resolution of ~200 μm is measured around the geometrical center of the FOV, which gradually degrades to ~600 μm at the periphery. In addition, multispectral information from the imaging volume can be readily collected by fast sweeping of the laser wavelength, which can be tuned to any desired value within the near-infrared spectral window on a single-pulse basis.

Small-animal whole-body imaging by SVOT is realized by fast motion of the spherical array along a spiral trajectory and real-time acquisition of multiple volumetric datasets. Image reconstruction of the individual volumetric frames is performed by a back-projection algorithm, and the time-resolved signals are deconvolved with the frequency response of the transducer array and band-pass filtered. Image rendering

of the whole body is performed by stitching the volumetric images acquired at each position (Figure 8.7d).

8.8 Frontiers in Photoacoustic Imaging

A major advantage of acoustic-resolution PACT-based molecular imaging over the fluorescence tomography-based technique is the high-spatial resolution at depth. However, the sensitivity and specificity of molecular PACT is limited by the strong background from endogenous absorbers (e.g. hemoglobin). To address this challenge, Yao et al. (2015) have developed a reversibly switchable photochromic probe for molecular PACT.

BphP1, a phytochrome from the bacterium *Rhodopseudomonas palustris*, has a natural photochromic behavior. It undergoes Pfr→Pr photoconversion upon light illumination at 730–790 nm and reverse photoconversion upon illumination at 630–690 nm. For simplicity, Pr and Pfr are referred to as the off and on state, respectively. The relaxation time from off to on is ~210 s. Capitalizing on the reversible photoconversion of BphP1, differential detection can be performed for molecular PACT with high sensitivity and specificity. To this end, 16 s of laser illumination at 780 nm is applied to switch BphP1 from on to off, during which photoacoustic images are acquired at the beginning and end of the illumination. Then, 630-nm laser is utilized to switch it back on. BphP1 has very different optical absorption in the two states, in striking contrast to the unchanged blood absorption. Thus, pixel-wise subtraction of the off image from the on image results in a differential image with all blood background removed.

The switchable PACT has enabled high-resolution tumor detection at depth. As shown in Figure 8.8a, 1 week after the

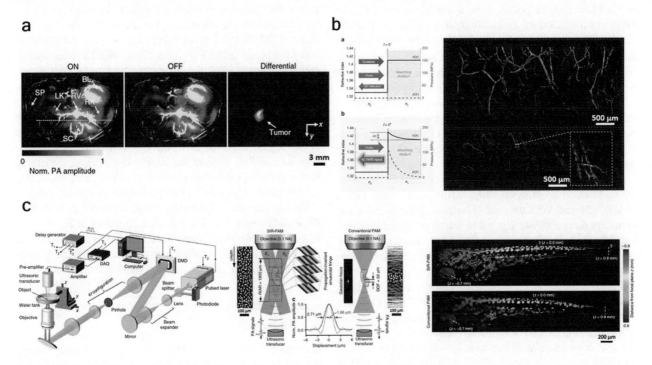

FIGURE 8.8 Frontiers in photoacoustic imaging. (a) Photo-switchable PACT for high-resolution molecular imaging of BphP1-labeled tumor in the mouse kidney *in vivo*. (b) Diagram of the PARS mechanism and its application for noncontact imaging of the mouse ear *in vivo*. (c) Principle and schematic of SIR-PAM and its application for zebrafish embryo imaging *in vivo* with extended depth of focus.

injection of BphP1-expressing U87 cells into the mouse kidney, switchable PACT of the kidney is acquired. Both the on and off images are overwhelmed by the excessive blood background. Excitingly, subtraction of the off image from the on image results in a high CNR image of the BphP1-expressing tumor.

Compared with other light microscopy techniques, a major disadvantage of OR-PAM is the contact detection of ultrasound. To address this challenge, Hajireza et al. (2017) have developed a noninterferometric photoacoustic remote sensing (PARS) microscopy.

As shown in Figure 8.8b, the light-excited photoacoustic wave can modulate the local refractive index of the tissue $n(x,t)$ because of the elasto-optical effect:

$$n(x,t) = n(x) + \delta n(x,t) = n_0(x)\left(1 + \frac{\eta n_0(x)^2 p(x,t)}{2\rho c^2}\right) \quad (8.5)$$

where $n(x)$ and $\delta n(x,t)$ are the intrinsic refractive index of the tissue and the photoacoustic wave-induced change, respectively, η is the elasto-optic coefficient, $p(x,t)$ is the pressure field, ρ is the mass density, and c is the speed of sound in the tissue. Thus, the local optical absorption of the tissue can be estimated by measuring δn through the ultrasound-modulated optical reflectivity:

$$\Delta R = \left|\frac{n_1 + \delta n - n_2}{n_1 + \delta n + n_2}\right|^2 - R \quad (8.6)$$

where $R = \left|\dfrac{n_1 - n_2}{n_1 + n_2}\right|^2$ is the intrinsic reflectivity of the tissue without the photoacoustic perturbation, n_1 and n_2 are the refractive indices of the absorbing and surrounding media, respectively, and δn is the modulated refractive index of absorbing medium. Simplifying Equation (8.4) leads to $\Delta R \propto \delta n(n_1 - n_2)$. Because δn is proportional to local optical absorption, measuring ΔR provides an optical means to measure the absorption contrast.

In PARS microscopy, a continuous-wave probe laser is utilized in combination with a ns-pulsed excitation laser. To measure the photoacoustic modulation of the optical reflectivity, a photodetector is used to monitor the intensity of the back-reflected probe beam. This system is sensitive to reflectivity modulations over the entire depth of focus of the probe beam. Thus, to avoid signal mixture, the probe beam should be low coherent. With the remote sensing approach, no ultrasound coupling medium (e.g. gel or water) is required. *In vivo* imaging of the mouse ear is realized by simultaneous optical scanning of both the excitation beam and the probe beam (Figure 8.8b).

Similar to other light microscopy techniques, OR-PAM suffers from the limited depth of focus. In OR-PAM, Gaussian beam is typically employed for excitation, leading to a trade-off between the lateral resolution and depth of focus. Although μm-level lateral resolution can be achieved, the depth of focus is limited to less than 100 μm. To address this challenge, Yang et al. (2017) have recently developed a motionless volumetric OR-PAM with spatially invariant resolution (SIR-PAM).

As shown in Figure 8.8c, a small fraction of the ns-pulsed laser output is tapped off by a beam splitter and monitored by a high-speed photodiode to compensate for the pulse energy fluctuation. The residual beam is expanded by a beam expander, reflected by a mirror, modulated by a digital micromirror device (DMD), focused by a lens, spatially filtered by a pinhole, collimated by a second lens, converged by a relay lens, and finally focused into the object to be imaged by a microscope objective. The excited photoacoustic signal is detected by a focused transducer, amplified by an amplifier, and digitized by a digitizer. A delay generator is used to synchronize the electronics.

To achieve the spatially invariant resolution, SIR-PAM employs a series of propagation-invariant sinusoidal fringes (PISFs) generated by the DMD with different spatial frequencies and phases. The normalized 3D intensity profile of a PISF can be expressed as

$$I_\varphi(x,y,z) = \frac{1}{2}\left[\cos(2\pi f_x x + 2\pi f_y y + \varphi) + 1\right] \quad (8.7)$$

where f_x and f_y are spatial frequencies, and φ is a shifting phase. Thus, the excited photoacoustic signal can be expressed as

$$s_\varphi(f_x, f_y, z) = \iint\limits_{R(z)} \mu_a(x,y,z) \times I_\varphi(x,y,z)\, dx\, dy$$

$$= \frac{1}{2}\iint\limits_{R(z)} \mu_a(x,y,z) \times \left[\cos(2\pi f_x x + 2\pi f_y y + \varphi) + 1\right] dx\, dy \quad (8.8)$$

where $R(z)$ is the illuminated area with intensity modulation at depth z, $\mu_a(x,y,z)$ is the optical absorption distribution in the object. By analyzing the time-of-flight information of photoacoustic signals, the signal $s_\varphi(f_x, f_y, z)$ from each depth z can be resolved. Then, the Fourier coefficient at z can be extracted using the phase-shifting method. Thus, with a certain illumination frequency PISF, the corresponding Fourier coefficients of all cross sections can be derived from the received photoacoustic signals. Combining the PISFs with all frequencies (f_x, f_y) in the Fourier domain, the Fourier coefficients of each cross section can be accurately retrieved. Finally, the 3D image of the object is reconstructed layer by layer using the inverse Fourier transformation of the Fourier coefficients. Benefiting from the PISFs, modulation transfer functions at all depths are the same, thereby enabling the spatially invariant resolution.

For comparison, a zebrafish embryo is imaged by both SIR-PAM and conventional PAM *in vivo*. As shown in Figure 8.8c, SIR-PAM provides uniform lateral resolution, regardless of the depth, while the resolution of conventional PAM quickly degrades when the embryo is out of the depth of focus.

8.9 Conclusion

Over the past couple of decades, PAI has been rapidly emerging and evolving toward finer spatiotemporal resolution, higher detection sensitivity and larger penetration depth. The application of PAI in biomedicine has also been greatly expanded across all five systems levels, spanning from single organelles to the small-animal whole body. Various PAI schemes have been developed and demonstrated superior performance.

Although still facing challenges, PAI has identified its unique niche in functional and metabolic imaging and has found broad applications in brain, cardiovascular, and cancer research, among many others. With the highly desired optical absorption contrast and spatial scalability across the optical and ultrasonic dimensions, PAI is expected to find more high-impact applications in both basic research and clinical practice.

Acknowledgments

This work is supported in part by the National Institutes of Health (NS099261, AG052062 and EB020843) and American Heart Association (15SDG25960005) to SH.

REFERENCES

Anastasio, M. A., et al. (2005). Half-time image reconstruction in thermoacoustic tomography. *IEEE Trans. Med. Imaging*, *24*(2): 199–210.

Cao, R., et al. (2017). Functional and oxygen-metabolic photoacoustic microscopy of the awake mouse brain. *Neuroimage*, *150*: 77–87.

Deán-Ben, X. L., Fehm, T. F., Ford, S. J., Gottschalk, S., and Razansky, D. (2017). Spiral volumetric optoacoustic tomography visualizes multi-scale dynamics in mice. *Light Sci. Appl.*, *6*(4).

Hajireza, P., Shi, W., Bell, K., Paproski, R. J., and Zemp, R. J. (2017). Non-interferometric photoacoustic remote sensing microscopy. *Light Sci. Appl.*, *6*(6).

He, Y., et al. (2016). In vivo label-free photoacoustic flow cytography and on-the-spot laser killing of single circulating melanoma cells. *Sci. Rep.*, *6*: 1–8.

Hu, S., Maslov, K., and Wang, L. V. (2011). Second-generation optical-resolution photoacoustic microscopy with improved sensitivity and speed. *Opt. Lett.*, *36*(7): 1134–1136.

Jathoul, A. P., et al. (2015). Deep in vivo photoacoustic imaging of mammalian tissues using a tyrosinase-based genetic reporter. *Nat. Photonics*, *9*(4): 239–246.

Maslov, K., Zhang, H. F., Hu, S., and Wang, L. V. (2008). Optical-resolution photoacoustic microscopy for in vivo imaging of single capillaries. *Opt. Lett.*, *33*(9): 929.

Ning, B., et al. (2015a). Simultaneous photoacoustic microscopy of microvascular anatomy, oxygen saturation, and blood flow. *Opt. Lett.*, *40*(6): 910.

Ning, B., et al. (2015b). Ultrasound-aided multi-parametric photoacoustic microscopy of the mouse brain. *Sci. Rep.*, *5*: 18775.

Song, K. H., and Wang, L. V. (2012). Deep reflection-mode photoacoustic imaging of biological tissue. *J. Biomed. Opt.*, *12*: 60503.

Toi, M., et al. (2017). Visualization of tumor-related blood vessels in human breast by photoacoustic imaging system with a hemispherical detector array. *Sci. Rep.*, *7*.

Treeby, B. E., and Cox, B. T. (2010). k-Wave: MATLAB toolbox for the simulation and reconstruction of photoacoustic wave fields. *J. Biomed. Opt.*, *15*(2): 21314.

Treeby, B. E., Zhang, E. Z., and Cox, B. T. (2010). Photoacoustic tomography in absorbing acoustic media using time reversal. *Inverse Probl.*, *26*(11).

Wang, L. V., and Wu, H. I. (2012). Biomedical optics: Principles and imaging. *Biomedical Optics: Principles and Imaging*. doi:10.1002/9780470177013

Wang, L., Maslov, K., and Wang, L. V. (2013). Single-cell label-free photoacoustic flowoxigraphy in vivo. *Proc. Natl. Acad. Sci. U.S.A.*, *110*(15): 5759–5764.

Wang, T., et al. (2016). Multiparametric photoacoustic microscopy of the mouse brain with 300-kHz A-line rate. *Neurophotonics*, *3*(4): 45006.

Xu, M., and Wang, L. V. (2005). Universal back-projection algorithm for photoacoustic computed tomography. *Phys. Rev. E Stat. Nonlinear, Soft Matter Phys.*, *71*(1 Pt 2).

Yang, J., et al. (2017). Motionless volumetric photoacoustic microscopy with spatially invariant resolution. *Nat. Commun.*, *8*(1).

Yang, J. M., et al. (2012). Simultaneous functional photoacoustic and ultrasonic endoscopy of internal organs in vivo. *Nat. Med.*, *18*(8): 1297–1302.

Yao, D.-K., Chen, R., Maslov, K. I., Zhou, Q., and Wang, L. V. (2012). Optimal ultraviolet wavelength for in vivo photoacoustic imaging of cell nuclei. *J. Biomed. Opt.*, *17*(5): 56004.

Yao, D.-K., Maslov, K., Shung, K. K., Zhou, Q., and Wang, L. V. (2010). In vivo label-free photoacoustic microscopy of cell nuclei by excitation of DNA and RNA. *Opt. Lett.*, *35*(24): 4139–4141.

Yao, J., et al. (2015). Multiscale photoacoustic tomography using reversibly switchable bacterial phytochrome as a near-infrared photochromic probe. *Nat. Methods*, *13*(1): 67–73.

Zhang, C., Maslov, K., and Wang, L. V. (2010). Subwavelength-resolution label-free photoacoustic microscopy of optical absorption in vivo. *Opt. Lett.*, *35*(19): 3195–3197.

Zhang, C., Maslov, K., Yao, J., and Wang, L. V. (2012). In vivo photoacoustic microscopy with 7.6-μm axial resolution using a commercial 125-MHz ultrasonic transducer. *J. Biomed. Opt.*, *17*(11): 116016.

Zhang, H. F., Maslov, K., Sivaramakrishnan, M., Stoica, G., and Wang, L. V. (2007). Imaging of hemoglobin oxygen saturation variations in single vessels in vivo using photoacoustic microscopy. *Appl. Phys. Lett.*, *90*.

Zhang, H. F., Maslov, K., Stoica, G., and Wang, L. V. (2006). Functional photoacoustic microscopy for high-resolution and noninvasive in vivo imaging. *Nat. Biotechnol.*, *24*(7): 848–851.

Zhou, Y., Yao, J., Maslov, K. I., and Wang, L. V. (2014). Calibration-free absolute quantification of particle concentration by statistical analyses of photoacoustic signals in vivo. *J. Biomed. Opt.*, *19*(3): 37001.

9

Fluorescence Lifetime: Techniques, Analysis, and Applications in the Life Sciences

Jenu Varghese Chacko, Md Abdul Kader Sagar, and Kevin W. Eliceiri

CONTENTS

9.1 Introduction

The scientific advances in optical microscopy have greatly accelerated in the last 20 years with the advent of new and improved imaging techniques, objectively to obtain the most descriptive information possible from cell and animal models. The essential feature of most of these methods is to tag a biological target in its spatial dimensions with attributes of interest, such as the number of fluorescent photons, the color of the photon, and so forth. Microscopy based on *luminescence attributes* of a molecule provide specificity in imaging and allows us to understand biomolecules and their interaction. Fluorescence microscopy played a substantial role in this progress as a tool for identifying biomolecules. The application of fluorescence microscopy spans from studying small molecules inside a cell to clinically relevant fluorescence aided surgical tools. Fluorescence microscopy yields many specific physical attributes from an object that can be used to tag a molecule and to identify them. This made fluorescence microscopy a very powerful tool in the past century for looking at spatial and temporal changes at cellular and subcellular features. Quantification has become a key aspect of modern microscopy; quantification makes the data more functional and the attributes more fathomable in their respective spatial and temporal dimensions. Quantification in fluorescence microscopy measures the physical parameters from a fluorophore, such as its quantum yield (the ratio of emission photons to excitation photons), polarization state of the molecules (the ratio of molecules in parallel to perpendicular orientations of electric field with respect the excitation), emission spectrum of the molecule (energy level distribution), and fluorescence lifetime decay of the fluorophore. This aim to advance microscopy to record not only the structural and temporal dynamics but to quantify that information and achieve a deeper understanding of the sample and its microenvironment, makes quantitative fluorescence microscopy the tool of choice for many types of experiments.

Fluorescence microscopy comprises of multiple non-invasive imaging methods, such as fluorescent correlation spectroscopy methods to study protein dynamics, fluorescent lifetime distributions to investigate biomolecular interactions and transient states, hyperspectral imaging, superresolution techniques to identify molecular interactions at the nanoscale, and so on. These techniques can model biochemical systems down to their atomic levels and expand the knowledge base of biology and medicine. One such interest is the study of cellular microenvironments, which landed in the spotlight as early as 1989 (Kallinowski et al., 1989), after the identification of its role in cancer tumor progression and the significance of heterogeneous distribution of cells and their impact and synergy in disease progression (Warburg, 1956; Heiden, Cantley, and Thompson, 2009; Hanahan and Weinberg, 2011; Jemal et al., 2011). This microenvironment study can be modeled in small animals and can be investigated down to cellular-scale biochemistry. The interest in the biochemical mapping of cells, tissues, and small animals has advanced with the discovery of fluorescent markers to decipher these complex biological systems. The fluorescent tags, which are either endogenous or exogenous are targeted due to its fluorescence properties (Dean and Palmer,

2014). Fluorescence microscopy has helped to tackle diverse problems in cell microenvironment and has recently developed into a guidance tool for tumor surgery. Among the fluorescence applications currently employed, cells and cellular interactions are predominant mainly because deciphering cellular level responses plays a direct role in current pharmacology and medicine. Methods based on endogenous fluorescence and fluorescent markers can be scaled up to clinical application and are being used in medicine and surgery. The techniques used to identify these markers collectively, called *fluorescence microscopy*, use intensity, lifetime, spectrum, and polarization as the chief physical investigation parameters.

Fluorescence lifetime is the average time spent by an excited molecule in its excited state before returning to the ground state by the emission of a fluorescent photon. Fluorescence lifetime techniques are methods that can measure the decay rate of the excited molecule population. This decay rate can provide information about the fluorescent molecule and its immediate environment. Fluorescence lifetime techniques can be of macroscopic nature, measured in bulk materials and cuvettes or by microscope, known as fluorescence lifetime imaging microscopy (FLIM). The quantitation aspect of fluorescence lifetime measurements is distinct from other fluorescence methods because the fluorescent lifetime of the molecule is independent of the absolute number of photons emitted by the molecule and is based on the rate of relative change in the number of photons per excitation cycle. The fluorescence lifetime is impacted by changes in the microenvironment such as pH, binding, and the presence of multiple fluorophores. Fluorescence lifetime techniques originated as a transient state chemical characterization method and have emerged from the physical chemistry realm into the multidisciplinary approach we know today. With the emergence of modern computers and electronics, it became feasible to map fluorescence lifetime onto an image known as FLIM. This method generated a wide interest in biomedical and biochemical applications to study microscopic spatial heterogeneities. This essential unification of chemical information with spatial localization is realized by modern FLIM. Fluorescence lifetime can constructively offer additional quantitative biochemical information to any measurement sized down to molecular levels. Because of the clinical relevance of noninvasive imaging, there is great interest in applying FLIM techniques in medicine and many applications of using FLIM in animal models and humans.

As an imaging technique, FLIM has a unique advantage: The lifetime is independent of the other physical parameters of fluorescence, such as intensity or spectrum. This puts FLIM as an often desired quantitative target among the fluorescence attributes in imaging. The FLIM based biochemical sensors currently span over many targeted proteins, their activation, binding, and aggregation states. The power of FLIM to interpret oxygen levels in the system drove the development of phosphorescence lifetime imaging microscopy (PLIM) based oxygen sensors. The sensitivity of FLIM to binding states allowed FLIM-based metabolic imaging methods. The sensitivity to dipole-dipole interactions resulted in us Forster resonance energy transfer (FRET) methods that are sensitive to nanometer resolution. A priori information of the quantum yield of the fluorophore measured in a FLIM image can calculate the concentration of molecules. Rapid FLIM imaging methods have allowed real-time measurements that are translational in a multitude of biomedical applications. These applications are further explained in Section 9.4. These key aspects make FLIM and time-resolved applications unique in their ability to study protein dynamics with noninvasive autofluorescence imaging (metabolic imaging), and FRET. In the last decade, superresolution techniques and flow cytometry methods have started using fluorescence lifetime as a cogent marker. This chapter explains the different types of FLIM instrumentation with an emphasis on analysis techniques, applications, and their challenges.

In this chapter, we will introduce the basic concepts of fluorescence lifetime and advance to the different instrumentation schemes used to realize FLIM in the biological research laboratory, common methods to analyze FLIM data, and representative FLIM applications. The chapter is organized as follows: Section 9.1 introduces the fundamentals of fluorescence and FLIM. Section 9.2 reviews various optical and electronic instrumentation schemes used for FLIM implementation and explains the current challenges. Section 9.3 discusses FLIM analysis techniques and their applicability and presents a few

representative FLIM analysis software packages. In Section 9.4, we discuss various biological applications of FLIM including cellular and multiscale imaging examples.

9.1.1 Fluorescence

Fluorescence is the emission of lower frequency light (photon) from a material after the absorption of light of higher energy (or frequency). The molecule that absorbs and goes to a higher energy level (a fluorescent molecule) relaxes back to the ground energy levels after a short duration, simultaneously emitting a photon. The consequent relaxation accompanied radiation after the absorption of light from the excited state to the ground state can take from hundreds of picoseconds (ps) to microseconds (µs). The average time spent by the molecule in the excited state is a characteristic of that molecule known as its *fluorescence lifetime* (τ). If the photon emission delays take longer than a few hundred nanoseconds (ns), terms such as *delayed fluorescence* and *phosphorescence* are used to describe the relaxation processes (schematic shown in Figure 9.1A). The figure shows

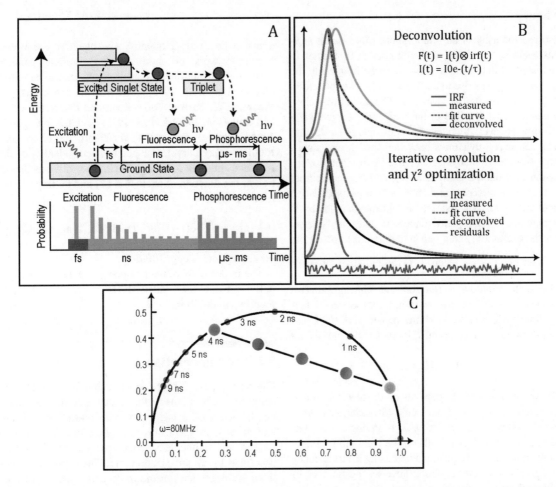

FIGURE 9.1 Fluorescence lifetime. A. Schematic of fluorescence and phosphorescence and their respective probability in time. B. TCSPC photon counting histogram. The panel shows a simulated time correlated single photon counting histogram. The histogram is fitted to an exponential curve (dashed-red line). The signal is convolved by the instrument response function (IRF) (magenta line) and presents as the measurement (blue line). This can be removed using deconvolution methods, back to the original decay curve (cyan line) with the knowledge of the IRF. C. Phasor representation of lifetime: Phasor plots representation of lifetime on an 80MHz (12.5 ns) universal circle. The lifetime values from 0 ns to 9 ns are shown in red circles. A binomial curve is represented in a line connecting two single exponential points (cyan, pink), with each circle showing the varying fractional contribution of each (fraction of cyan/pink).

the excitation–relaxation events in the axis of time. If the information content is recorded in a time-resolved fashion, in the order of nanoseconds, to study the decay kinetics of a system, the method serves to be a fluorescence lifetime technique. There are many factors that affect the decay rate ($k = 1/\tau$) of a fluorescent molecule, such as *dynamic quenching* that nonradiatively removes the excited state energy through collision; intersystem crossing into metastable levels, including long-lived phosphorescence levels; dipole-dipole coupling energy transfer through Forster resonance energy transfer (FRET), and others. These processes can be investigated by time-resolved spectroscopy or fluorescence lifetime measurements. Excited molecules can form complexes or dimers with excited or nonexcited molecules, termed *excimers*, with a characteristic biexponential decay in FLIM. The counterpart to dynamic quenching, *static quenching*, is a lifetime independent quenching mechanism that forms a dark complex with the ground state fluorophore. This can be studied by observing the contrast in lifetime and quantum yield together (Lakowicz, 1983).

9.1.2 Theory

Time-resolved intensity of any fluorophore (*fluorescent molecule*) ensemble is a function of the number of molecules in the excited state at that instant. This first-order relation to the excited population gives rise to a kinetic model of exponential decay of fluorescence intensity over time. The excitation–relaxation dynamics are a function of the molecule's ground state energy absorption levels, associated excited energy states, and subsequent relaxation dynamics back into the ground states. The energy from an excited state can also be dissipated without emission of a photon (*nonradiative*), and hence the final relaxation rate of a molecule is a combination of its radiative and nonradiative rates. The theoretical lifetime (denoted by τ) is the average time a fluorophore remains in its excited energy state after the excitation, and it can be defined in terms of the decay rates as $\tau = \dfrac{1}{k_r + k_{nr}}$, where k_r is the radiative rate constant and k_{nr} is the nonradiative rate constant. The nonradiative decay (k_{nr}) is the sum of the rate constant for vibrational relaxation (k_{ic}), intersystem crossing to the triplet state (k_{isc}), and quenching rates due to interaction with other molecules (K_q).

$$k_{nr} = k_{ic} + k_{isc} + K_q.$$

The radiative rate constant, k_r, depends on the absorption and emission spectra of the probe molecule (Strickler and Berg, 1962) and the refractive index of its surroundings. The nonradiative rate constant, k_{nr}, depends on the local environment of the probe molecule and can be studied using lifetime decay curves. The following rate equation governs the number of excited fluorophores as a function of time

$$\left(N(t)\right): dN = (k_r + k_{nr})N(t)dt,$$

where dN is the change in the number of excited molecules in a short time interval, dt. Integrating this equation for a time

period (t) derives the number of photons measurable as a result of the decay of fluorescence or intensity,

$$I(t) = I_0\, e^{-t/\tau},$$

where I_0 represents the fluorescence intensity at $t = 0$. The decay constant parameter for this decay (τ) is the same as fluorescence lifetime. This is an excitation kinetics parameter of a system and is independent of factors such as excitation intensity, fluorophore concentration (below quenching concentration levels), detector efficiency, optical pathlength or scattering (Becker, 2015). This aspect of FLIM that keeps the physical parameter of decay independent from measurement parameters (O'Leary et al., 1996) makes it uniquely quantitative and valuable to study different quenching processes.

Note that K_q is a concentration dependent term that increases with the concentration of quenching molecules [Q] as ($K_q = k_q[Q]$). The static and dynamic quenching processes can be characterized using the modified Stern–Volmer equation (Weber, 1948; Lakowicz, 1983; Gratton et al., 1984):

$$F_0 / F = \left(1 + \tau_0\, k_{dynamic}\left[Q\right]\right)\left(1 + k_{static}\left[Q\right]\right)$$

and can be used to quantify the quenching concentration. Extensive work on modified *Stern–Volmer equations* to accommodate nonlinear quenching effects can be found in literature (Lakowicz, 1983; Gratton et al., 1984; Lakowicz and Weber, 1973; Parson, 2007; Zeng and Durocher, 1995; Boaz and Rollefson, 1950; Keizer, 1983; Seidel, Schulz, and Sauer, 1996). (Refer to Figure 9.4b for simulation of quenching on lifetime decay curve). For practical considerations, the fluorescence emission intensity (F) can be written as the convolution of the function of excitation (E) and an impulse response function (δ) as

$$F(t) = \int_0^t E(t')\delta(t - t')dt'. \text{ (Eichorst, Teng, and Clegg, 2014;}$$

Periasamy and Clegg, 2009)

The impulse/instrument response function (IRF) is a result of the spread of the width of the short excitation pulse, jitter in precise photon timing, etc.

9.2 Instrumentation

Fluorescence is a multiparameter, signal-by-signal processing definition because every fluorescent photon distribution can be characterized using its intensity, wavelength of emission and absorption spectrum, relaxation time characterized by fluorescent lifetime decay curve, and electric field orientation described by its polarization, etc. These parameter distributions are dependent on the molecular structure of the molecule, interaction with the physical microenvironments of the molecule, and changes in the structure with effect to the solvent. Correlative methods used in multiscale imaging schemes use one or more of these parameters to investigate features of interest such as spectrum and lifetime, polarization and lifetime, polarization and spectrum, and so on.

9.2.1 Fluorometry and FLIM

Fluorimeters (*steady-state*) and fluorometers (*time resolved*) are the two common classes of fluorescence spectrometers that are currently used in laboratories. The former carries out measurements such as absorbance and transmission spectrum that evaluate equilibrium conditions, while the latter class measures properties that are a function of time using fast detection methods. These two classes are analogous to ground-state absorption studies vs. transient state absorption (pump-probe spectroscopy) studies. Following fluorescence emission kinetics using time-resolved techniques have the potential to unravel chemical reaction intermediates in a complex formation or disintegration and other rate changing steps in fluorescence emission kinetics. Fluorometry mapped onto spatial coordinates x, y, and z generate images of characteristic lifetime or decay amplitude changes in space. A considerable complexity was introduced when fluorometry was mapped in these spatial coordinates in order to create an image in the context of understanding a matrix of exponential curves and their representations in the image (Figure. 9.2a). Hence, by definition, an image with each pixel representing a lifetime curve is a FLIM image.

9.2.2 Spectral and Polarization Resolved FLIM

Every photon collected from a fluorescent sample has measurable physical parameters, including color (a function of its energy), polarization (direction of electric field oscillation), and time or phase of the origin of the photon with respect to the excitation event. A collective measure of these parameters yields the *fluorescence emission spectrum, fluorescence*

anisotropy, and *fluorescence lifetime.* Polarization and spectrum measured in a time-resolved manner can produce additional quantitative correlative information to better understand the fluorophore and microenvironment interaction (Beule et al., 2007).

For a polarized excitation, the time-dependent intensity collected in two perpendicular detector channels (to the light path) are

$$I_\parallel = \left(1 + 2r(t)\right) I(t) / 3 \text{ and } I_\perp = \left(1 - r(t)\right) I(t) / 3,$$

where the function *r(t)* is the *emission anisotropy decay*

$$r(t) = \sum_{i=1}^{N} r_{0i} e^{-t/\tau_{ri}} \text{ and } I(t) = F.e^{(-t/\tau)} \text{ (Vogel et al., 2009)}.$$

It can be shown that the difference in anisotropy decay (lifetime and amplitude) is a function of the limiting anisotropy (r_0), rotational correlation time (τ_{ri}), radiative decay rate (τ), and intensity scaling factor for the two detectors (F). Conventional isotropic fluorescence lifetime measurements are also modulated by depolarization effects during the collection time and can be solved by placing a polarizer at 54.7 degrees to match both polarization components of the dipole (Becker, 2005, 2015; Vogel et al., 2015). The time-dependent depolarization (anisotropy) is a signature for homo-FRET (donor and acceptor with the same energy levels) measurements (Vogel et al., 2015). Anisotropy studies usually plot the two perpendicular lifetime curves from each pixel in either linear scale or log scale as shown in Figure 9.3a.

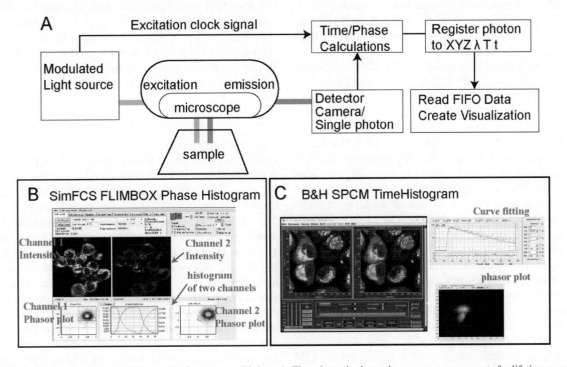

FIGURE 9.2 Instrumentation schematic used in fluorescence lifetime. A. The schematic shows the necessary components for lifetime acquisition. A pulsed excitation source that can be correlated with the detector signal for single photon arrival time is the basis for all fluorescence-based timing devices. The arrival times of photons are further registered with spatial or other dimensions to generate useful information. B. and C. Screenshots of two commercial scan-based lifetime measurement software: digital frequency domain (SimFCS) and time domain (SPCM).

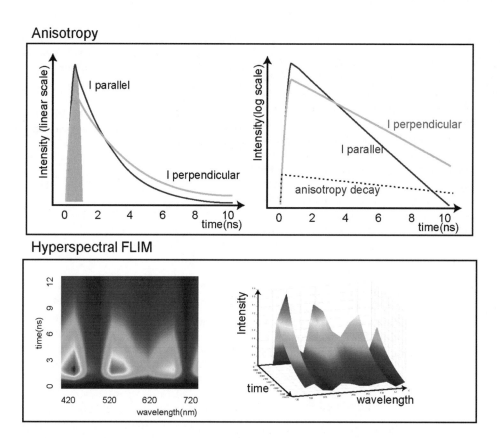

FIGURE 9.3 Lifetime representations used in literature: (a) Anisotropy decay from two perpendicular channels are compared together. Presented in both logscale and linear scale of frequency. (b) Hyperspectral lifetime decay curves are normally plotted as 2D matrices or surface or contour plots as shown here. This representation identifies the peak of the (λ-τ) matrix.

Hyperspectral FLIM methods are typically implemented using multiple detectors with different spectral bands of detection with an implementation of dispersive element (Favreau et al., 2013) imaged using single-photon avalanche diode (SPAD) arrays (Popleteeva et al., 2015), cameras (Beule et al., 2007), multichannel photomultiplier tube (PMTs) (Becker, Bergmann, and Biskup, 2007), quadrant anodes (Vitali et al., 2011), etc. Hyperspectral implementation has shown to be useful in a number of fluorescence studies, such as autofluorescence analysis in cardiac cells under respiration modulators (Chorvatova, Mateasik, and Chorvat Jr, 2013; Datta et al., 2016) and specifically for evaluation of an unknown autofluorescent species (Marcu et al., 2004) and tissue diagnosis (Sun et al., 2008). These techniques traditionally suffer from lower signal-to-noise ratio because the use of multiple detectors limits the photon counts per detector. The evolution of new noise-reduction methods has improved hyperspectral FLIM in the last decade. For example, the spectral FLIM separation using frequency domain (FD) Hadamard transform based multiplexing has shown high sensitivity in the quantification of (λ wavelength-τ lifetime) matrix (Pian et al., 2017). Another separation method of interest is the multidimensional phasor analysis (Malacrida, Jameson, and Gratton, 2017). An entire hyperspectral FLIM data is plotted in Leray et al. (2009). Figure 9.3b shows a representative plot of hyperspectral FLIM data from one single pixel of an image.

Time-resolved spectral and anisotropy techniques increase the precision in separating and identifying molecular fingerprints. This is more prevalent when mathematical fitting of exponential curves reaches a limit in separating different species from FLIM data or if the species are inseparable. The spatial association of pixels in an image to these biochemical states in FLIM can be used for real-time species separation and to plot dynamic information in video microscopy or endoscopy (Elson, Jo, and Marcu, 2007).

9.2.3 Laser Scanning Microscopes and FLIM

There are many methods for measuring fluorescence lifetime in confocal (Amos and White, 2003) or multiphoton (Denk et al., 1990) laser scanning microscopes. For the majority of the current methods, fluorescence lifetime is measured from the decay in the intensity, ($I(t)$), after an excitation with a short pulse of light, in which the duration of the excitation pulse is ideally much shorter than the lifetime to be measured. A pulsed excitation produces a fluorescence emission, which decreases the intensity exponentially with the time that can be read with the photodetector and time-resolving electronics. For a sample containing a single species, the fluorescence lifetime is given by the time over which the intensity of the fluorescent drops to about $1/e$, or 37%, of its initial value. Two major instrumentation schemes exist for the calculation of the exponential decay from *pulsed excitation*: 1) *time correlated time domain photon histogram* (example: used in TCSPC (O'Connor, 2012) electronics, Figure 9.3b), and 2) *phase correlated FD histogram* (example: used in FLIM box (Colyer, Lee, and Gratton, 2008) electronics, Figure 9.3b) (Gratton, 2016). Both schemes use fast, programmable electronics, such as field programmable gate array (FPGA) logic, to tabulate a photon timing histogram with high time resolution and calculates the representative lifetime values. Another widely used approach based on CW *(continuous wave) excitation* avoids the use of the more expensive pulsed laser setup. This scheme measures the fluorescence intensity from a phase-locked modulated excitation-modulated detector gain (sensitivity) and uses signal processing tools to demodulate and estimate lifetime or phase. This mode of measurement of lifetime uses a modulated CW excitation pulse and measures the emission wave and estimates the lifetime from the demodulation parameters (phase and amplitude) of the emission signal. This FD approach can be implemented as either a homodyned scheme or a heterodyned demodulation scheme, either in digital or analog mode (Chen and Gratton, 2013; Periasamy and Clegg, 2009).

For pulsed systems, there are three instrumentation architectures to determine the decay histogram. The time-digital conversion (TDC) architecture, time amplitude converter–analog digital converter (TAC-ADC) architecture, and the phase cross-correlation (pCC) measurement architecture (Spencer and Weber, 1969; Becker and Bergmann, 2003; Tamborini et al., 2013; Wahl, 2014; Kalisz, 2004; Wahl et al., 2013; Gratton and Barbieri, 1993). The FD collection has the advantage of working without a deconvolution routine that provides faster component separation with a lower cost of calculation. For example, fluorescence anisotropy in the FD method calculates the phase difference between two polarized channels for quantification, while time domain (TD) requires the fitting and estimation of the difference in the decay rates (Beechem et al., 2002). This rapid lifetime estimation can be realized in real-time FLIM by methods like *rapid FLIM* (Koenig et al., 2017; Orthaus-Mueller et al., n.d.) and *fast FLIM* (Li, Yu, and Chen, 2015; Colyer, Lee, and Gratton, 2008) approaches. This faster species separation with a lower cost of calculation is one of the main advantages of FD FLIM (Jameson, 2016; Baeyens, 1990). On the other hand, time-domain measurements have the advantage of straightforward fitting analysis routines done at a high temporal resolution data (Becker et al., 2003). An extensive comparison of these two collection domains be found in literature (Gratton et al., 2003; Becker, 2015), as well as comparisons of the same lifetime standards collected in both domains (Schlachter et al., 2009; Elder, Schlachter, and Kaminski, 2008; Elder et al., 2006; Elder, Kaminski, and Frank, 2009).

The majority of time-resolved imaging methods used now in laser scanning microscopy are based on a single photon counting detector and a pulsed laser as an excitation source. The detector can be a photon counting PMT, such as the H7422P from Hamamatsu, or an analog detector aided with a discriminator unit to detect single photons. Many custom-made acquisition circuits may need an extra digital integrator to convert photon counting pulses into digital pulses (such as transistor-transistor logic (TTL)). The common detector units, such as APDs (avalanche photo diodes), SPADs, and hybrid PMTs, enable single photon counting with high sensitivity. The instrumentation schemes used for FLIM are improving to support fast and video-rate imaging. Detectors used in

time-resolved imaging are characterized for their quantum efficiency, spectral range, signal-background ratio, transient time spread (TTS), dead time, detection area, etc. The main timing aspects in selecting detectors are timing resolution and deadtime. The selection of one particular detector is difficult when the cost, efficiency, and dynamic applications are considered. For example, conventional, fast digital PMTs like the H7422 (effective time resolution of ~500ps) can be tuned up to ~230ps time resolution (Becker, 2015), while microchannel plate (MCP) detectors can have a resolution below 50ps, but are more expensive (Becker, 2005). Emerging alternatives like hybrid PMTs (Becker et al., 2011) can generate cleaner decay curves after removing after-pulsing noise.

9.2.4 Camera-Based FLIM Microscopes

The fluorescence lifetime imaging of large areas is achieved using widefield detectors like MCPs (both in TD (Becker et al., 2016; Michalet et al., 2006) and FD (Colyer et al., 2012)), gated cameras (Elson et al., 2006; Cole et al., 2001; Colyer, Lee, and Gratton, 2008) and streak cameras (Krishnan et al., 2003). It is difficult to obtain high time resolution on camera-based systems because of current electronic limitations. But this can be circumvented using high-sensitivity detector arrays, time-gating, and other methods. This can enable the development of camera based FLIM methods, such as structured illumination FLIM (SI-FLIM) (Hinsdale et al., 2017; Cole et al., 2001), that can image thicker samples. Hirvonen and Suhling (2017), Gerritsen et al. (2002), Michalet et al. (2013) have excellent reviews comparing widefield FLIM techniques based on Microchannel plate (MCP), SPAD arrays, and superconducting detector arrays (Becker, 2015). The current instrumentation schemes used for small-animal imaging FLIM are vast in range, and we list some of them here with literature support on those schemes. The detectors used vary from intensified charged-couple device (ICCD) modules (Straub and Hell, 1998; Zhao et al., 2013) gated optical imager (GOI) (Rinnenthal et al., 2013), complementary metal-oxide semiconductor (CMOS) based-FLIM (Chen, Holst, and Gratton, 2015; Hirvonen, Festy, and Suhling, 2014; Buntz et al., 2016), analog ICCD (Chen and Gratton, 2013; Elder et al., 2006; Sun et al., 2013; Papour et al., 2015), and so on.

However, a cost-effective solution to such expensive camera systems is to rely on commercial gating devices (Dowling et al., 1998; Gioux et al., 2010) that can convert a standard electron multiplying CCD (EMCCD) or scientific CMOS (sCMOS) camera used widely for widefield fluorescence intensity measurements into FLIM imaging cameras (Straub and Hell, 1998). The typical approach of gated FLIM is as follows: Lifetime information in the image can be sectioned in time in different ranges to create time-gated images. For example, a 9ns lifetime curve can be divided into three gates: A) 0–1ns, B) 1–4 ns, C) 4–9 ns (Figure 9.4a). The histogram can be split among three gates, and each one treated like a spectral channel. This approach in FLIM is useful and effective in generating a fit-free representation of lifetime data (even for scanning systems). Conversely, if image acquisition can be hardware-gated in time, using a GOI, high-rate imager (HRI) (Webb et al., 2002; Beule et al., 2007) or other means, we can

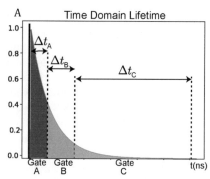
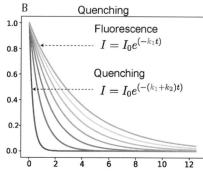
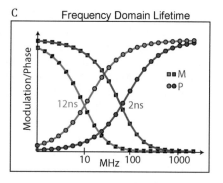

FIGURE 9.4 Interpreting frequency domain lifetime and time domain lifetime. (a) Lifetime gating techniques: The exponential decay curve is temporally divided into different gates of widths ($\Delta\tau_A$, $\Delta\tau_B$, $\Delta\tau_C$) and summed to give three representative values instead of the all the measurable time windows of the histogram. These three values can be directly used to color the pixels and represent the fraction of photons in the representative gate. (b) Effect of quenching. The change in the lifetime of one species with the effect of multiple quenching processes. The fluorescence lifetime curve (red) with the decay rate ($k1$) with the presence of one quenching rate ($k2$) or two quenching rates ($k2$ and $k3$). (c) Frequency domain plots: Simulated representation of Modulation and Phase values for different fluorescent lifetimes. i) A multifrequency plot for increasing lifetime values, ii) phase decay curves for different TP-phase lifetime, iii) modulation decay curve TM-modulation lifetime.

generate the lifetime histogram using sequential gating to build the histogram. This can be achieved using either analog (Gadella, Jovin, and Clegg, 1993) or digital (Chen and Gratton, 2013) FD methods (Periasamy et al., 1996). This class of imaging devices belongs to a broader category of phase sensitive imaging tools (time of flight (TOF) imaging) that can support ultrafast imaging. These gated-imaging approaches are designed for camera-based FLIM, providing higher imaging frame rates and larger fields of view when compared to raster scanning techniques. A comparison of currently used image intensifiers used for FLIM can be found in Sparks et al. (2017).

9.2.5 Challenges in FLIM Instrumentation for Live-Animal Imaging

Two common FLIM instrumentation types used in live-animal (intravital) imaging are: 1) with a *multiphoton imaging scheme* to offer higher penetration depth and 2) *widefield FLIM imaging* of the near infrared (NIR) probes. Two-photon fluorescence signal from the sample can be simplified as $F = \tau_{fl}\left[C\right]\psi$, where $[C]$ is the concentration and ψ is the system parameter. ψ is a function of excitation emission wavelengths, average laser intensity, optical resolution, absorption cross section at that excitation wavelength, the repetition rate of the laser, pulse duration, and numerical aperture (NA) of the objective lens, etc. Deep imaging brings attenuation with penetration depth and scattering effects that distort the focusing excitation wavefront. Current solutions to these challenges in animal imaging are to use an adaptive optics correction for the excitation wave front as a depth calibrated function (Feeks and Hunter, 2017) or to increase penetration depth using a longer wavelength of excitation (two-photon or three-photon) or use NIR that emits at longer wavelengths (Horton et al., 2013).

For intravital imaging, physical constraints, such as accessibility, movement artifacts, tracking labels, and intrinsic fluorescence, present imaging challenges. Proper imaging schemes, improved contrast agents, and animal restraints can solve many of these imaging problems. However, FLIM imaging in animals is also severely restricted by the speed

of imaging when compared to analog detection schemes. For scanning FLIM systems, lifetime estimation needs minimum collection time to obtain >100 photons per exponential decay curve and thus limits a final scan rate to >1 kHz pixel. Gated-detection schemes, spinning disk scanning, new timing electronics, and new estimating technologies such as rapid FLIM and faster scanning methods achieve frame rates below a second (pixel rate < 1kHz) (Akers et al., 2008; Orthaus-Mueller et al., n.d.). The development of new NIR probes (Rao, Dragulescu-Andrasi, and Yao, 2007) (>700nm) has also rendered tumor targeting and whole-animal FLIM imaging possible (Bloch et al., 2005). Nanoparticle encapsulated fluorophores have also been demonstrated to help in fluorescence lifetime multiplexing (Hoffmann et al., 2014).

9.2.6 Key Points

- FLIM instrumentation is often customized for multiparametric acquisition that includes fluorescence polarization, fluorescence dynamics of the molecule, and fluorescence spectrum of the molecule.
- Widefield FLIM and point scanning FLIM have their merits and demerits. Research on FLIM cameras is an ongoing field of research.
- Live-animal imaging FLIM with a focus toward human clinical FLIM imaging implementation is a focus of FLIM development over the next decade.

9.3 Analysis

The conventional lifetime decay analysis includes the fitting of a fluorescence decay curve to a single or multiexponential curve to identify the characteristic lifetime and amplitude. Another widely used method is phasor analysis, which is used for both TD and FD data analysis SimFCS (ISS Instruments/LFD) (Clayton, Hanley, and Verveer, 2004; Redford and Clegg, 2005; Digman et al., 2008; Chen et al., 2017). The fitting method is limited by the number of photons used in the fit and other goodness of fit parameters. Commercial software such as SPCImage (Becker and Hickl), SymPhoTime (PicoQuant), and DAS6 (Horiba) use exponential curve fitting of TCSPC data. Advanced lifetime determination methods have their own advantages, including 1) *Bayesian methods* that could work better even with a lower number of photons, 2) *stretched exponential analysis* that can probe local heterogeneity, 3) *moment analysis* that can estimate lifetime faster, etc. High content screening for FLIM imaging has also been achieved using high-speed screening FLIM with minimal analysis (Talbot et al., 2008). All these analysis methods are detailed briefly in this section.

9.3.1 Exponential Decay Curve Fitting

Exponential curve fitting fits the photon arrival histogram into an exponential decay function and estimates the lifetime (at $t =$ mean lifetime, the intensity $I = I_0/e = 36\%$ of peak intensity). This lifetime value can be evaluated as the average time spent by a fluorescent molecule in its excited electronic state

$$\int_0^\infty t\, I(t)\,dt \,/ \int_0^\infty I(t)\,dt = \tau, \ for\, I(t) = e^{-t/\tau}.$$

Conventional lifetime measuring systems use a light source and detectors that convolve the fluorescence decay with a characteristic timing jitter in tagging a time coordinate to single photon pulse, known as the instrument response time (IRF (t)). TCSPC implementation ensures that the transit time spread (TTS) of photon detection pulses from the detector is shorter than the detector response minimizing the IRF.

Essentially, the fitting function for N component fit can be written as

$$I(t) = \sum_{i=1}^{N} a_i\, e^{\frac{t-shift}{\tau_i}} + s.R(t) + offset,$$

where a_i is the amplitude of the ith exponential component with lifetime τ_i. The two commonly used estimated parameters from a multiexponential fit are 1) mean lifetime

$$\tau_{mean} = \sum_{i=1}^{N} a_i \tau_i \,/ \sum_{i=1}^{N} a_i,$$

which is weighted by the amplitude coefficients and 2) apparent lifetime or intensity weighted lifetime

$$\tau_{apparent} = \sum_{i=1}^{N} a_i \tau_i^2 \,/ \sum_{i=1}^{N} a_i \tau_i,$$

which is weighted by the integral intensity or photons in each component. The apparent lifetime is the correct representation of the average lifetime (Fišerová and Kubala, 2012). We can validate the definition function for τ,

$$\int_0^\infty t\, I(t)\,dt \,/ \int_0^\infty I(t)\,dt,$$

for two components: intensity

$$I(t) = a_1 e^{-t/\tau_1} + a_2 e^{-t/\tau_2} \ as \ \sum_1^2 a_i \tau_i^2 \,/ \sum_1^2 a_i \tau_i$$

The numerical fitting of these exponential curves is evaluated by residual analysis or chi-square values. For a single exponential curve, experimental data involves dependencies to other parameters, such as amplitude, shift (with respect to IRF), offset, scattering photon fraction from Raman scattering, or second harmonic generation (SHG) as additional parameters; this renders fitting a deft task. There are many fine elements in fitting that can make a FLIM user's day difficult, such as incomplete decay, photon pile-up, presence of a lifetime component shorter than IRF, or inhomogeneous background.

Lifetime data can be used to deduce the quantum efficiency and fractional contribution of two species. If f_1 and f_2 are the fractional contribution of each species in the lifetime curve, the concentration ($C_i=f_i/q_i$) contribution of one species to the total fluorescence is intrinsically connected to its quantum yield as

$$C_1 = f_1 \,/ (f_2 Q_r + f_1),$$

where $Q_r = Q_1 / Q_2$, the relative quantum efficiency of the first species with respect to the second and $f_2 = (1 - f_1)$.

9.3.2 Fitting Routines

Most of the fitting routines use the minimization of fitting residual using a quality parameter such as chi-square (Beechem et al., 2002).

$$\chi^2 = (1/T - p) \sum_{t=1}^{T} (d(t) - f(t))^2 \,/ d(t),$$

where T is the number of points in the timing histogram, $d(t)$ the data points, and $f(t)$ the fitted points. This reduced (normalized) version of chi-square is normally reported in FLIM analysis instead of conventional Neyman's or Pearson's chi-square.

One of the misconceptions associated with exponential fits minimization can be seen in the log-scale fitting. The equation of $y = Ae^{-kx}$ is analyzed by taking the log of both sides and performing a least square regression to estimate the least square error as:

$$f(A,k) = \sum_{i=1}^{N} (y_i - (Ae^{-kxi}))^2 .$$

instead the linear case will calculate the values of a and b that minimize a different function

$$g(A,k) = \sum_{i=1}^{N} (\ln(y_i) - \ln(A) - kxi))^2 .$$

9.3.2.1 Maximum Likelihood Estimation (MLE) – Nonlinear Least-Square (NLS) Fitting

The least square method is the most used analysis of lifetime histograms. Nevertheless, there are many debates on the nature of the underlying distribution being Poisson or even polynomial nature instead of Gaussian (Maus et al., 2001). Maximum likelihood estimation (MLE) uses a calculated probability distribution (P) and maximizes the total probability with respect to one parameter of interest $dP/dt = 0$ (Bialkowski, 1989; Turton, Reid, and Beddard, 2003; Santra et al., 2016). This approach gains more power with prior knowledge of the system and yields accurate estimates of the parameter. Application of the likelihood function can replace the conventional chi-square estimate as can derive fits and quality parameter $2I*$,

$$\left[2I^* = \sum_i^k \left(n_i - g_i \right)^2 / n_i \right],$$

where n is the number of photon counts and k is the number of channels of decay histogram (Maus et al., 2001). This process has a higher cost of computation compared to iterative nonlinear least square (NLS) fitting, but MLE can be modeled to the noise characteristics and other known parameters. The main difference between MLE and NLS is that MLE depends only on τ, while NLS depends on multiple parameters: τ, amplitude, and others.

9.3.2.2 Rapid Lifetime Determination – Levenberg Marquardt Algorithm

Rapid lifetime determination (RLD) is a lifetime gated-image calculation that allows for a rapid estimate of amplitude and rate of decay. This can be used in a nonlinear regression analysis, such as a least square approach, to initialize the parameters for fitting. Rather than recording and analyzing the complete decay curve, RLD looks at areas under different regions of the decay to estimate the lifetime. Sharman et al. (1999) extended the RLD technique for single exponential to double-exponential decay. Most of the lifetime fitting routines employ a nonlinear least square fitting based on the Levenberg–Marquardt algorithm (LMA) or damped least-squares (DLS), and RLD can speed up and guide the LMA by proper initialization. The simplest two-gate algorithms, such as RLD, can be improvised into generalized and correlated uncorrelated RLD algorithms (Li et al., 2012; Sharman et al., 1999; Ballew and Demas, 1989).

9.3.2.3 Global Analysis

Global analysis methods obtain quantitative results by using the spatial (or other available) dimension information in a low-signal image to improve the fitting precision (Verveer and Bastiaens, 2003; Verveer, Squire, and Bastiaens, 2000). Global approaches gain better estimates of lifetime values by borrowing photons from the nearby pixels for building the photon histogram (Beechem et al., 2002). This spatial invariance assumption can accurately model fitting parameters better than single pixel-by-pixel analysis. This was successfully demonstrated in lifetime data in cuvette based samples (Knutson, Beechem, and Brand, 1983), FD FLIM (Verveer, Squire, and Bastiaens, 2000), anisotropy (Beechem et al., 2002), spectrally resolved data (Previte et al., 2008), and segmented global analysis (Knutson, Beechem, and Brand, 1983; Beechem et al., 2002; Pelet et al., 2004).

9.3.2.4 Moment Analysis

Moment analysis is a fast parameter estimate method to calculate lifetime and lifetime error in an exponential curve (Isenberg and Dyson, 1969). This approach calculates the first moment of the distribution without fitting and estimates a qualitative lifetime value from the curve (Libertini and Small, 1989). The moment by definition is

$$moment_{first} = \sum_{i=1}^{Nchannels} \left(N_{photons\ in\ channel\ i}\ t_i \right) / N_{total\ photons}.$$

The difference between the moments of any decay and IRF will give the lifetime (Esposito, Gerritsen, and Wouters, 2005). The standard deviation of this lifetime can be written as

$$\sigma_\tau = \sqrt{2}\sigma_{IRF} / \sqrt{N}.$$

Practically, this timing accuracy can be validated over multiple lifetime measurements of the same standard. Similar to RLD methods, the results from moment analysis can be used in the fitting to initialize the parameters.

9.3.2.5 Stretched Exponentials

The use of the stretched-exponential functions for analyzing fluorescence decay profiles has been shown to provide excellent tissue discrimination and spatial image quality in whole-field FLIM maps (Lee et al., 2001). Even more importantly, the stretched-exponential function not only describes the decay profiles almost exactly, but also derives from the more realistic decay model of continuous lifetime distributions in biological tissue, rather than from an arbitrary assumption of single or multiple discrete exponential decay components. Such continuous distributions have been reported for tryptophan (Alcala, Gratton, and Prendergast, 1987) in tissue.

This new approach yields, besides the mean lifetime $\langle\tau\rangle$ of the distribution, an additional parameter of interest (h), which is related to the width of the lifetime distribution, that is a direct measure for the local heterogeneity of the sample. This heterogeneity parameter should permit the study of mechanisms that broaden the lifetime distributions in complex samples, and it also provides an additional means to contrast different biological components. Fluorophores that have continuous distribution have been shown to be better represented by the stretched exponential. The time behavior of the emission is described by a continuous distribution of exponential decays. The stretched exponential is represented by this equation $F(t) = A \exp\left[-\left(t/t_k \right)^{\frac{1}{h}} \right]$, where t_k is a characteristics time constant and h is the heterogeneity parameter. For many biological samples, stretched exponential is likely to provide a truer representation of underlying fluorescence dynamics for which there is no prior knowledge of multiple discrete decay profiles. But stretched exponential requires more precision in timing, and it is more susceptible to noise from additional fit parameters and requires a significantly longer acquisition and computation time (Galletly et al., 2008).

9.3.2.6 Laguerre Expansion: Model Free Deconvolution

Laguerre deconvolution techniques are used in time-resolved data analysis to generate projections in the orthonormal axis of Laguerre coefficients. The discrete Laguerre expansion (LE),

$$h(r,n) = \sum_{j=0}^{L-1} c_j(r) b_j^{\alpha}(n) \quad \text{(Jo et al. 2004)},$$

for L number of orthonormal functions (b) of jth order and their unknown Laguere expansion (LE) coefficients (c) at pixel r. LE is applied instead of exponential deconvolution used in traditional lifetime measurements to interpret multiexponential data (Pande and Jo, 2011). Currently this nonparametric approach has been demonstrated with a hyperspectral collection modality in *in vivo* samples and clinical applications (Jo et al., 2006).

9.3.2.7 Blind Deconvolution

Blind deconvolution algorithms are formulated as an alternative to conventional fitting using IRF (Campos-Delgado et al., 2015). The nonlinear estimation process can cause errors in fitting in fitting if there are variations in the IRF position, a parameter commonly termed as IRF shift. This blind deconvolution approach is one of the more convenient ways out of this shift variance. There are only very few implementations of this approach mainly because of the cost of computation associated with correction of shift (Campos-Delgado et al., 2015).

9.3.2.8 Maximum Entropy Method

Maximum entropy method (MEM) (Esposito, Gerritsen, and Wouters, 2007; Swaminathan and Periasamy, 1996; Vecer and Herman, 2011; Brochon et al., 1990) is currently one of the best multiexponential decay analysis methods and has similarities to stretched-exponential fitting. MEM employs a different minimization scheme than that of chi-square. The chi-square-based parametric minimization procedure normally optimizes the curve fitting in three or fewer components. The optimum distribution in MEM is defined as the estimator that maximized the value of Shannon-Jaynes Entropy function $S = -\sum p_i \log(p_i)$, where $p_i = \alpha_i / \sum \alpha_i$ in addition to keeping the minimal chi-square value. This is referred to as the classical maximum entropy theory (Brochon, 1994), and this can be implemented in both phasor plots and histogram analysis.

9.3.2.9 Bayesian Analysis

In cases of low photon numbers for FLIM acquisition, Bayesian analysis provides a better decay time estimation than a traditional curve fitting algorithm (Barber et al., 2010). High numbers of photon counts require longer acquisition time and higher intensity that might damage the sample. However, for the cases when photon count is low for either low fluorophore concentration or dynamic processes, difficulties increase to determine the correct lifetime.

The Bayesian approach allows a decay time estimation with a much better confidence limit than Levenberg–Marquardt or maximum likelihood estimator fitting when the data has a low number of photons (Rowley et al., 2011a). If the photon number is higher, this approach is equivalent to traditional fitting approaches. Rowley et al. (2011a) assumed a monoexponential fluorescence decay model of the form,

$$p(\Delta t|\omega_0, \omega_1) = \left(\frac{\omega_0}{T}\right) + (1 - \omega_0)\bar{F}(\Delta t|\omega_1),$$

where ω_0 is the fraction of the detected photon due to a uniform background. ω_1 is the relaxation time of the photon generating emission process, and $\bar{F}(\Delta t|\omega_1)$ describes the likelihood of a photon due to the emission process being detected at time, Δt. The maximum likelihood of a photon being detected at a time, Δt, can be simplified to $F(\Delta t|\omega_1) = \left(1 + 1/e^{\frac{T_m}{\omega_1}} - 1\right)e^{\frac{\Delta t}{\omega_1}}$.

After defining the fluorescence decay model, Rowley et al. (Barber et al., 2010; Rowley et al., 2011b) applied Bayesian formalism and determined expressions to extract statistics of the model parameters. They showed that the accuracy of 10% was obtained with a photon count of mere 200 photons. For the same photon count of 200, MLE and phasor methods produced an accuracy of 10%, and the least square method produced uncertainty of 20%.

9.3.3 Phasor Analysis

Phasor analysis is a fitting-free Fourier transform method that helps us generate a fast representation of the FLIM data. This is achieved by reducing the temporal dimensions to two or four Fourier coefficients. The phasor coordinate system is derived from fundamental FD representation of waves in a unit circle. Although phasor plots are universal for both time and FD approaches, they prove to be indispensable for FD lifetime imaging. For FD lifetime measurements, the sample is excited with a modulated light (even pulsed light) of angular frequency, ω, and the fluorescence photons are collected and demodulated to find the phase lag (φ) and modulation ratio (M), which are functions of the lifetime (τ) as:

$$\varphi(\tau) = \tan^{-1}(\omega\tau) \text{ and } M(\tau) = 1/\sqrt{1 + \omega^2\tau^2},$$

where ω is the angular frequency of the excitation light. These equations are nonlinear, and phasor/polar plot schemes are commonly employed to interpret these values. $M(\tau)$ and $\varphi(\tau)$ in a multicomponent system are decomposed into two orthogonal axes, g $(M\cos(\varphi))$ and s $(M\sin(\varphi))$ for a time period $(T = 1/\text{repetition frequency})$. g and s are defined as

$$g = \int_{t=i}^{T} I(t) M(t)\cos(\varphi(t)) \text{ and } s = \int_{t=i}^{T} I(t) M(t)\sin(\varphi(t)), \text{ that}$$

equates to $g = \sum_{i=1}^{N} [(\alpha_i\tau_i)/(1 + \omega^2\tau_i^2)] / \sum_{i=1}^{N} \alpha_i\omega_i$ and $s = \sum_{i=1}^{N} [(\alpha_i$

$$\omega\tau_i^2)/(1 + \omega^2\tau_i^2)] / \sum_{i=1}^{N} \alpha_i\omega_i.$$

In this coordinate system, for N multicomponents, the final position is a linear vector sum of individual components and their fractional contribution:

$$(g, s) = \left(\sum_{i=1}^{N} f_i\, g_i,\ \sum_{i=1}^{N} f_i\, s_i, \right),\ \text{where } f = \alpha_i \tau_i / \sum_{i=1}^{N} \alpha_i \tau_i \text{ is the}$$

same fractional intensity component used in the apparent lifetime estimation. Practically, for any N component, exponential decay, (g, s) of nth harmonic can be estimated from the from the nth harmonic Fourier coefficients of the decay curve over a period (T) as

$$g(n) = \sum_{t=0}^{T} \cos(2\pi nt / T).\, I(t) \text{ and } s(n) = \sum_{t=0}^{T} \sin(2\pi nt / T).$$

$I(t)$ and total intensity, $I_{dc} = \sum_{t=0}^{T} I(t)$. M and φ can be calculated from nth harmonic $(g_n,\, s_n)$ as $\varphi_n = \tan^{-1}(s_n / g_n)$, and $M_n = \left(\sqrt{g_n^2 + s_n^2} \right)$. This is convenient using discrete Fourier transforms (DFT).

Both M and φ can be substituted in their definitions to find two representative lifetime values as modulation lifetime τ_{M} and phase lifetime τ_{p} that correspond to that harmonic and frequency (van Munster and Gadella, 2005). Full rendering of a lifetime is possible only through a multifrequency plot to identify where τ_{p} is equal to τ_{m} (Figure 9.3) (Yahav et al., 2017; Baeyens, 1990). A simulated representation of modulation and phase values for different fluorescent lifetimes can be seen in Figure 9.4c. This figure shows a multi-frequency plot, where the phase and modulation values are plotted across varying modulation frequencies. A set of identical phase (TP) and modulation (TM) lifetimes are shown. The MEM approaches can be applied in multifrequency lifetime methods for component separation without phasor plots (Brochon et al., 1990). Phasor coordinate systems can calculate the chi-square using the equation (Verveer and Hanley, 2009)

$$\chi^2(x,y) = \sum_i (s_i - x - y\, g_i)^2 / (\sigma_{s,i}^2 + \sigma_{g,i}^2).$$

The phasor cluster from a FLIM image is advantageous in identifying FLIM signatures in a pattern recognition context and its reversible mapping to spatial coordinates to identify the pixels of interest (Gregor and Patting, 2014). The summary of the above-mentioned instrumentation and analysis scheme is summarized in the schematic shown in Figure 9.5.

9.3.4 Software for FLIM Curve Fitting

9.3.4.1 Commercial Options

Some of the most popular commercial FLIM analysis software packages that are used primarily with laser scanning microscopy are SPCImage (B&H), Symphotime (Picoquant), and Vistavision (ISS, analogous to SimFCS from Globals/LFD). Symphotime has the advantage of scripting features and built-in time-tagged time-resolved (TTTR) analysis for fluorescence lifetime correlation spectroscopy (FLCS) and real-time FLIM mapping. SimFCS can handle different file formats from both FD collection and TD collection and perform phasor-based analysis. SPCM has large hardware support and works with high time-resolution data. SimFCS supports different lifetime analyses and offers multiple image correlation analysis and supports a multitude of microscope hardware. Most labs use the software provided by their instrumentation vendors, such as Becker and Hickl, Picoquant, Horiba, pCO, and so on, however, we found multiple home-built field programmable gate array–time-digital conversion (FPGA-TDC) implementations in the literature that work for individual labs with their custom electronics (mentioned earlier in the instrumentation section). The screenshots of two commercial acquisition software packages are shown in Figure 9.2b and c. One of them works with a phase histogram and the other with time.

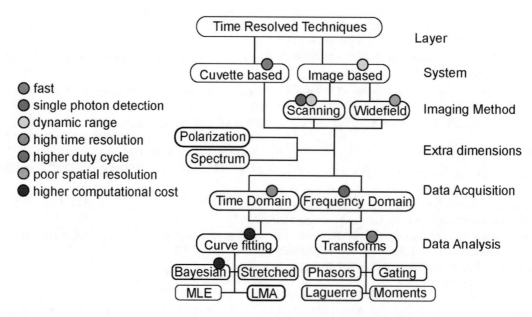

FIGURE 9.5 Flowchart of the relevant techniques and analysis.

9.3.4.2 *FLIMFit*

FLIMFit (Warren et al., 2013) is an open source MATLAB-based client for FLIM data analysis using the OMERO platform. The FLIMFit software tool was developed at the Imperial College London photonics group. FLIMFit can be used to analyze most types of FLIM data using the Bio-Formats file-format library (Linkert et al., 2010), including *.sdt files, *.ptu files, and *.msr files. FLIMFit can also be used to analyze time-resolved fluorescence anisotropy data and multichannel data.

9.3.4.3 *SLIM Curve Library*

SLIM Curve is an exponential curve-fitting library developed for analyzing FLIM and spectral fluorescence lifetime imaging microscope (SLIM) data. It was developed in a collaboration between Paul Barber of the Advanced Technology Group at the Cancer Research UK and the Laboratory for Optical and Computational Instrumentation at the University of Wisconsin-Madison. SLIM Curve is an open-source project licensed under the GNU General Public License v3 or later. Because of the open-source nature of the curve fitting library, it can be used in a wide variety of third-party software.

The library offers nonlinear least-squares fitting methods, such as LMA, advanced algorithms, such as maximum likelihood and global analysis optimized for FLIM, (Barber et al., 2009), and computationally simpler methods, such as RLD and phasor domain analysis. The library is written in ANSI C and can be run as a native executable accessible by MATLAB, C++, or Python. Time Resolved Imaging (TRI2), a graphical software package developed by the Barber-led advanced technology group, was released in 2004 and uses SLIM Curve as a core library to analyze FLIM images. TRI2 can be downloaded from the University of Oxford website (http://users.ox.ac.uk/~atdgroup/software). One of the several benefits of being an open-source library is the flexibility that is inherent to open-source projects. It can be integrated to independently develop curve fitting software complemented by other image processing plugins like region of interest (ROI) based segmentation.

9.3.5 Standards

Many of the current lifetime standards for different imaging techniques can be found in Boens et al. (2007) 2P (Two-Photon) lifetime standards (Kristoffersen et al., 2014), and standards for dyes are listed within (http://www.iss.com/resources/reference/data_tables/FL_LifetimeStandards.html).

9.3.6 Key points

- FLIM analysis is tied to the instrumentation used and its temporal resolution features.
- Fitting and nonfitting routines offer their own merits and demerits.
- Newer GPU-based and open-source fitting routines released in the past decade have created a push toward systematic perspective in the unification of FLIM analysis.

9.4 Applications

FLIM, as an imaging method, has had a wide impact on label-free imaging and fluorescence imaging (Berezin and Achilefu, 2010). Applications of FLIM include segregating different lifetimes of different fluorophores or identifying the change in the same fluorophore as a state of its interaction. These applications span from quantifying binding kinetics, oxygen concentration, pH, ion concentration, aggregation, and viscosity in cell culture, animal models, and human clinical trials (Becker, 2012; Berezin and Achilefu, 2010; Roberts et al., 2011).

9.4.1 FRET

FRET is a molecular-level resonance energy transfer between a donor molecule and acceptor molecule with overlapping spectrums and preferable spatial orientation and short intermolecular distance between them (Periasamy and Day, 2011; Clegg, 2009; Joo et al., 2008). The FRET energy transfer can be measured using an intensity-based approach or a lifetime-based approach. The rate constant for FRET is additional decay for the donor and can be measured by the change in the donor lifetime. The rate

$$k = 1 / \tau_D \left[R_0 / R \right]^6,$$

where R is the distance between donor and acceptor and

$$R_0^6 = \left(8.785 \times 10^{-5} \right) \left(\kappa^2 \varnothing_D J \right) / n^4,$$

is the Förster distance. The Förster distance is a function of the orientation factor κ^2, quantum yield ($\varnothing_D$) of the donor, spectral overlap (J), and the refractive index of the medium or solvent (n). FRET uses a nanoscopic ruler because of its sensitivity toward the sixth power of the distance, R (<10nm). Abe et al. (2013) reported that the NIR fluorescence lifetime FRET technique is capable of noninvasively detecting bound and internalized forms of transferrin in cancer cells and tumors within a live small-animal model, and their results are quantitatively consistent when compared to well-established intensity-based FRET microscopy methods used in *in vitro* experiments. Intravital FRET has been used for a wide variety of applications such as small-animal biosensors (RhoA (Nobis et al., 2017)-Rac (Johnsson et al., 2014)), neuronal dysfunction (Radbruch et al., 2015; Rinnenthal et al., 2013), treatment response (Nobis et al., 2013; Janssen et al., 2013), cancer applications (Rajoria et al., 2014), etc.

9.4.2 Autofluorescence and Clinical Applications

FLIM has been demonstrated to be an advantageous tool in the discrimination between free and bound forms of the intrinsically fluorescence metabolite reduced nicotinamide adenine dinucleotide (NADH), providing both lifetimes with their relative concentrations (Scott et al., 1970; Lakowicz et al., 1992). There is extensive published literature on NADH and flavin adenine dinucleotide (FAD) based FLIM studies for metabolic imaging and tumor detection. A combination of fluorescence lifetime of NADH with fluorescence intensity relative to the emission of tryptophan (an endogenous fluorescent amino acid) as an internal standard can differentiate normal from cervical cancer cells (Galletly et al., 2008; Periasamy et al., 2017).

FLIM has clinical implications in skin cancer detection (Dancik et al., 2013). The skin contains fluorophores including elastin, keratin, collagen, FAD, and NADH. These endogenous contrast agents can be used as a marker of homeostasis in *in vivo* skin imaging when compared against tumor or wound-healing models (Krasieva et al., 2012). For example, distinguishing basal cell carcinomas presents a contrast with respect to the surrounding, uninvolved skin in early skin cancer (Galletly et al., 2008). Collagen fluorescence lifetime can also yield contrast for a cancer diagnosis; the disruption of collagen by tumor associated matrix metalloproteinases from dermal basal cell carcinoma decreases the lifetime of the autofluorescence component of collagen. Cell types present differences in their mean lifetime; melanocytes and keratinocytes present a clear difference between malignant and benign cells (Dimitrow et al., 2009). Hyperspectral imaging of skin cancer can yield results at the level of individual cells (Patalay et al., 2011)

FLIM applied to intrinsic tissue autofluorescence generates detectable levels of disparity in lifetime and intensity in a range of surface tissue tumors, such as colon cancer. With the help of compact, clinically deployable flexible endoscopic instrumentation, FLIM helps to distinguish between cancerous and healthy colon tissue (McGinty et al., 2010). FLIM has been used to image oral carcinoma (Sun et al., 2009). These endoscopic implementations collect time-resolved images, minimally affected by tissue morphology, endogenous absorbers, and illumination. These results demonstrate the potential of FLIM as an intraoperative diagnostic technique. FLIM techniques for brain tissue characterization can generate optical contrast in underlying tissue structures of normal and malignant brain tissue (Papour et al., 2013) and reveal cerebral metabolism in live cells (Yaseen et al., 2013; Trinh et al., 2017). Genetically encoded NADH-NAD+ fluorescence lifetime sensors like Peredox (Tejwani et al., 2017) have also been used in brain tissue redox imaging.

9.4.2.1 Development of FLIM signatures

Endogenous fluorescent biomarkers can identify various cell types based on specific metabolic markers in live tissue or fresh biopsies, helping researchers to monitor their activities unadulterated by external reagents or antibodies. Other traditional options, such as antibody labeling, exogenous dyes, or genetic manipulation, can directly perturb cellular processes. Multiphoton FLIM can be used to exploit the endogenous fluorescence of NADH versus FAD already present in tissue to identify changes in metabolism (Chance, 1976). Szulczewski et al. (2016) and Garcia et al. (2016; Alfonso-García et al. 2016) used such distinct metabolic signatures to identify macrophages and separate them from the intrinsic fluorescence signature of tumor cells in live tissues, suggesting a role of metabolic reprogramming in the regulation of the innate inflammatory response (Orihuela, McPherson, and Harry, 2016). Stringari et al. (2012) and Squirrell et al. (2012)

separately identified cellular differentiation states in live cultured cells, as well as small animals, using NADH metabolic signatures (Chacko and Eliceiri, 2019). Several notable works on cancer cell metabolism by Skala (Walsh et al, 2013), Gratton et al (Pate et al., 2014), and others in the past decade have provided the basis for identification of metastatic states associated with cellular proliferation and differentiation states (Blacker et al., 2014; Hou et al., 2016). There are many phasors fingerprinting-based work identifying differences in metabolism and environment such as Long Lifetime Species (LLS) for oxidized lipids (Datta, 2016; Datta et al., 2016), bacterial species (Bhattacharjee et al., 2017), etc. These trends have also been extended to tissues, including cell differentiation in kidney tissue (Hato et al., 2017), fat content in liver tissue (Ranjit et al., 2017), and renal fibrosis quantification (Ranjit et al., 2016). FLIM can quantify fluorescent markers on apparatus including lab-on-a-chip, bio-reactors, microfluidic devices and others. Such diagnostic framework can facilitate large-scale oncogene studies and quantitative live-cell studies (Datta et al., 2016; Li et al., 2010).

9.4.2.2 FLIM Endoscopy

FLIM based contrast seen in local microenvironments can be applied into early cancer detection using an endoscope implementation. Current implementations of FLIM endoscopes (Mizeret et al., 1999) have demonstrated the FLIM contrast in excised human urinary bladder and oral mucosa using an FD FLIM setup (Mizeret et al., 1999) and larynx and neoplastic mouse bowel using TD FLIM (Sparks et al., 2015). e-FLIM (Fruhwirth et al., 2010) implementation shows miniaturized FLIM endoscopes that can measure FRET in live cells.

9.4.3 Fluorescence Lifetime Correlation Spectroscopy (FLCS)

FLIM can be combined with fluorescence correlation spectroscopy (FCS) techniques to derive dynamic information of complex molecules. Two-dimensional (2D) FLCS is one of the leading techniques in this field, which can generate TCSPC FCS data and paint 2D emission delay correlation maps that can help decipher dynamics of single molecules along with excited state decay kinetics (Ishii, Otosu, and Tahara, 2014; Kapusta et al., 2012; Eggeling et al., 2006). FLCS methods can be applied with superresolution schemes, such as stimulated emission depletion microscopy (STED), to decode single molecule dynamics as shown in separation of photons by lifetime tuning (SPLIT-FLCS) technique of enhanced green fluorescence protein (EGFP) in live cells (Lanzanò et al., 2017). The use of burst integrated fluorescence lifetime (BIFL) (Fries et al., 1998; Weidtkamp-Peters et al., 2009; Widengren et al., 2006) is a single molecule identification time-resolved technique capable of determining concentrations of multiple species in a pixel.

9.4.4 Flow Cytometry (FCM) Aided with Fluorescence Lifetime

Fluorescence lifetime can be used with high throughput flow cytometry methods as a marker to resolve or sort cells, as well

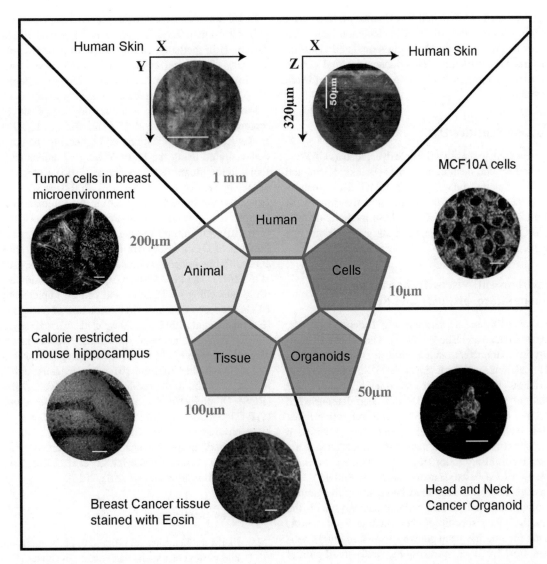

FIGURE 9.6 Multiscale representation of lifetime imaging: The image shows the applications of endogenous FLIM imaging done in an increasing size scale (clockwise) from cells, groups of cells, tissues, animals, to humans. The applications showed in the figure are 1) MCF-10A cells used to study the effect of respiration inhibition and confluency (Bird et al., 2005; Chacko and Eliceiri, 2018) 2) Head and neck cancer cell spheroids analyzed to distinguish the metabolic signatures (Shah and Skala, 2016; Skala et al., 2005). 3) Tissue characterization for endogenous and eosin labeled tissue sections (Martin et al., 2016; Conklin et al., 2009). 4) *In vivo* imaging of tumor and ECM in mice (Szulczewski et al., 2016) 5) Human skin imaging that reaches a penetration <300um and covers a cross section in millimeters (Roberts et al., 2011; Patalay et al., 2011; Koenig, 2012).

as to study the interaction of fluorophores with the surroundings. FLIM can function better than intensity-based sorting because of its convenient concentration independent quantification scheme and improved background separation caused by unbound extrinsic dye and scattering (Steinkamp and Crissman, 1993; Deka and Steinkamp, 1999). Cao et al. (2016) has used FD cytometry (known as *phase sensitive FCM or ps-FCM)* to estimate the population in yeast cells and mammalian cells. A phase filtered cell sorting system was demonstrated with FCM on mammalian cells with ethidium bromide/propidium iodide staining pair (Cao, Pankayatselvan, and Houston, 2013). The ps-FCM can separate cellular DNA and RNA and even binding status with the help of lifetime information (Cui et al., 2003). Unsupervised machine learning screening using FLIM has been demonstrated for high-content, high-throughput image cytometry (Esposito et al., 2007).

9.4.5 Superresolution

Superresolution imaging techniques (Chacko, Zanacchi, and Diaspro, 2013; Diaspro and van Zandvoort, 2016) can enable imaging beyond the diffraction limit. FLIM imaging with superresolution techniques such as STED microscopy enables exploiting FRET to study protein interaction localized well below the traditional limit. A STED FLIM technique has been demonstrated that uses a supercontinuum source and can switch between confocal and STED (Auksorius et al., 2008) mode. Time-resolved FLIM imaging is employed in STED, as gated-STED offers better image resolution (Vicidomini et al., 2014) and innovative approaches such as SPLIT (Lanzanò et al. 2015) and three-dimensional (3D) FCS (Lanzanò et al., 2017). These advanced gating techniques help to increase the signal to background ratio. This can significantly improve resolution when used in CW-gated STED (Moffitt, Osseforth, and Michaelis, 2011), which uses a CW-depletion beam that brings atoms to their ground state using stimulated emission with no characteristic lifetime. This depletion signal can be separated using background subtraction schemes or hardware gating techniques.

9.4.6 Voltage Sensitive Dyes

Emerging new voltage sensitive lifetime dyes, such as CAESR, are well suited for cell membrane studies (Brinks, Klein, and Cohen, 2015). Calcium imaging using FRET dyes, such as cerulean-citrine (Heim et al. 2007), are also a trending FLIM application (Rinnenthal et al., 2013). These methods demonstrate TCSPC FLIM implementations imaging at a frame rate above 10 fps.

9.4.7 Environment: Viscosity, Refractive Index, Temperature, pH, pO_2, and ROS

The fluorescence lifetime of a sample is highly dependent on its environment (Chacko and Eliceiri, 2018). The solvent dynamics and physical parameters, such as viscosity (Kuimova et al., 2008), pH (Lakowicz and Szmacinski, 1993), and temperature (Ni and Melton, 1996) have their own roles in changing the excited state of a fluorophore. FLIM has been demonstrated to be effective as a probe to measure local refractive index – a valuable technique to investigate local heterogeneities in the cellular structure such as the membrane (van Manen et al., 2008; Jones and Suhling, 2006; Tregidgo, Levitt, and Suhling, 2008). Many FLIM-based sensors, such as BODIPY-C12 (Kuimova et al., 2008), can detect local viscosity of the sample.

FRET biosensors can predict cellular pathways with the help of small GTPase specific markers such as Rac1, RhoA, Cdc42. This method has been shown to work in single cell-level membrane ruffling, protrusions, and small-animal models to identify the activation pathway. These biosensors can be used for visualizing different protein activities and monitoring cellular behavior, such as changes in morphology and tumorigenesis steps, including metastasis and complex responses to treatment (Machacek et al., 2009; Nobis et al., 2017). FRET can also be used for the maximum likelihood estimation (MLE) of the rotational mobility of a protein complex and measure subresolution distances. This nanoscale information can interpret the DNA hybridization mechanism (Paul et al., 2016), salt bridge formation in an amyloid chain (Nag et al., 2013), and so on.

pH dynamics is one of the most prominent features in living organisms affecting many physiological processes. Different dyes are used to probe pH in different cellular machinery. Seminaphthofluorescein dyes c-SNAFL-1, SNAFL-2, and c-SNAFL-2 are lifetime probes for cytosolic pH recordings in different cell lines, whereas lysosomal pH measurements could be successfully performed with the acidic lifetime probes DM-NERF dextrans, OG-514 carboxylic acid dextrans and LysoSensor DND-160 (Lin, Herman, and Lakowicz, 2003; Andersson et al., 2000; Carlsson et al., 2000). Hille et al. (2008) reviewed the performance of these dyes as a pH sensing tool using FLIM.

FLIM has been used with special temperature sensitive polymers to map the temperature in living cells. FLIM of rhodamine B in methanol can measure temperature with ± 3 °C accuracy (Benninger et al., 2006), while novel temperature sensitive polymers have been designed for physiological range (37 °C) with higher accuracy and stability (Okabe et al., 2012).

Most oxygen sensors are currently based on phosphorescence and quenching dyes (Sakadžić et al., 2010, 2011). However, FLIM applications of O_2 sensing dyes have been demonstrated using the Stern–Volmer equations can also be found in the literature (Yaseen et al., 2015). Ruthenium based dyes (Zhong, Urayama, and Mycek, 2003) are widely used for oxygen concentration-based lifetime measurements in biology. Oxygen is a quencher for fluorescence and phosphorescence in a diffusion limited regime. The partial pressure for pO_2 (amount of O_2 in tissue) can be estimated using phosphorescence lifetime imaging microscopy (PLIM) pO_2 sensors (Rück et al., 2017). The PLIM-based oxygen sensors can be modeled using quenching coefficients and can be implemented using LEDs because of the low time resolution of phosphorescence (Hartmann et al., 1997; Kurokawa et al., 2015; Hosny, Lee, and Knight, 2012). It has been shown that hypoxia presents as a distinct effect on FLIM metabolic imaging (Datta et al., 2015) and shows higher lifetime structures in cells and holds an intricate relation to the production of reactive oxygen species (ROS). Other FLIM based O_2 monitoring uses either FRET (PtP-C343 modified with PEG-amines (Sakadžić et al., 2010)) or porphyrin compounds. Combined imaging scheme for fluorescence and phosphorescence are implemented for a correlative imaging purpose (Shcheslavskiy et al., 2016; Yaseen et al., 2015; Jahn, Buschmann, and Hille, 2015).

9.4.8 Key points

- FLIM applications are tailored to probe micromolecular parameters such as molecular distances (FRET), quenching of fluorescence, rotational dynamics, and others.

- FLIM-based molecular signatures are developed and used in the identification of many disease markers, notably related to metabolism and tumor-metabolic markers.

- Current FLIM applications in literature can be found with superresolution imaging, environmental sensing, and molecular identification and targeting.

- Correlative approaches using FLIM have been on the rise in the past decade, offering reinforced validation of markers using multiple methods of probing.

9.5 Discussion

FLIM serves as one of the powerful quantitative imaging techniques to probe diverse biological interactions including protein binding, lipid binding, DNA binding, oxygen concentration through quenching kinetics, ion concentration, pH values, and aggregation through changes in k_{nr} and k_r, and transient local environments relative to a fluorophore. Current

FLIM implementations are broadly classified into a flowchart (Figure 9.5). This list is definitely not exclusive, but rather a representation of different schemes.

FD lifetime techniques and TD lifetime techniques exist in parallel (Gratton et al., 2003), working in similar fashions to generate time-resolved information (Gerritsen et al., 2009; Verveer and Hanley, 2009). FLIM methods require careful analysis to avoid misinterpretation of the photon statistics and especially in exponential curve fitting routines. The current list of published exponential decay analysis techniques is overwhelming. The advantage of pattern recognition analysis is that, like phasors and PCA plots (Le Marois et al., 2016) based reduction of lifetime features, it allows for faster recognition of fluorescent lifetime fingerprints. FLIM presents other sets of challenges and limitations in the collection, analysis, and visualization schemes. The time required to extract single photon statistics is another limitation, and with the current fast detectors, electronics and algorithms, many of these limitations are being overcome. Another noticeable problem in FLIM is the lack of effective standards to validate and estimate the current FLIM techniques. This leads to consequent variability in published results in FLIM and lifetime values because of incomplete representation of results or flaws in methods used. The lack of real-time plotting tools for FLIM are another limitation, especially for validating a long measurement, such as a time lapse or 3D video FLIM microscopy. A different challenging aspect of FLIM is the cost of implementation, especially with mode-locked lasers and timing electronics.

FLIM users, especially those interested in metabolic process measurements, should keep in mind that the variation in signal over the collection period of lifetime data records an average waveform for the entire collection period. A roll-over timing gate over the signal can decode this dynamic information instead of the conventional fitting. Possible pile-up artifacts in single-photon statistics in highly fluorescent samples can limit the count rates below 40MHz and even 10 kHz in some cases. A conventional rule of thumb to estimate the quality of a lifetime imaging technique is to calculate the normalized relative root mean square noise or F (figure of merit) value (Philip and Carlsson, 2003; Gadella, 2009; Gerritsen et al., 2009) $\sqrt{N}\sigma_\tau / \tau$ for standard deviation (σ) of measured lifetime (τ). Multiexponential decays need to be analyzed to identify whether they present incomplete exponential curves, which can change the fitting parameters.

Self-quenching by high fluorophore concentration and color quenching by an external absorber are other measurement artifacts arising from the sample. One must pay careful attention to physical parameters, such as the fluorescence cross section, IRF of the instrument, and deconvolved parameter interpretations, to prevent this widescale confusion seen in the literature. We also recommend using intensity weighted lifetime parameters for multiexponential analysis techniques and faster plots, such as universal phasor plots, which is the first harmonic plot to 80MHz (the laser fundamental frequency of most of the titanium-sapphire lasers). When it comes to techniques, different schools of techniques often follow their own standards and modifications in the measurements, and this situation results in the need for user-friendly, interoperable platforms.

9.6 Conclusion

FLIM is a quantitative and versatile method to measure molecular function in a wide range of biological applications from oxygen sensing to metabolic imaging (Kalinina et al., 2015; Becker, 2015; Suhling, French, and Phillips, 2005; Berezin and Achilefu, 2010). A combination of techniques to provide correlative microscopy investigations can localize events spatially as well as unveiling kinetics and signaling processes. Fluorescence lifetime is an extra physical parameter that can be extracted from your current fluorescence microscope with the careful addition of FLIM electronics and analysis methods. The pixel-based information in FLIM constitutes a large statistical data set that can be rigorously processed for quantitative analysis. Scientific advances in the fields of computation and microscopy have allowed FLIM to be developed for animal imaging and human applications with real-time FLIM visualization emerging for some implementations. FLIM is also beginning to be adapted for future clinical applications with promising candidate clinical studies for FLIM-based contrast for tumor margin detection during surgery and to aid in the diagnosis of infection in the respiratory tract. As FLIM improves in its sensitivity, speed and robustness, biomedical and clinical applications will only increase.

9.7 Glossary

FLIM	Fluorescence lifetime imaging microscopy
FRET	Forster/Fluorescence resonance energy transfer
PLIM	Phosphorescence lifetime imaging microscopy
Anisotropy	Time dependent depolarization
Homo-FRET	FRET between donor and acceptor with identical energy levels
SHG	Second harmonic generation
TTTR	Time-tagged time-resolved collection of photons
χ^2	Greek letter chi squared; term used to denote statistical goodness of fit
$\sigma\tau$	Sigma, term used to denote standard deviation of a distribution
NIR	Near infrared
CW	Continuous wave
MHz	Megahertz, unit for 1 million Hertz
kHz	Kilohertz, unit for 1 thousand Hertz
Hz	Hertz, unit for frequency or number of cycles (seconds-1)
ns	Nanoseconds (10(-9) seconds)
ps	Picoseconds (10(-12) seconds)
μs	Microseconds (10(-6) seconds)
ms	Milliseconds (10(-3) seconds)
k	Decay rate (in unit of seconds -1)
τ	Decay lifetime/fluorescence lifetime (in units of seconds)
τfl	Fluorescence lifetime (in units of seconds)
kr	Radiative decay rate

knr	Nonradiative decay rate
kic	Internal conversion decay rate/vibrational decay rate
kisc	Intersystem-crossing decay rate/triplet-state decay rate
Kq	Quenching rate
k-dynamic	Quenching rate from dynamic quenching events
k-static	Quenching rate from static quenching events
N(t)	Number of molecules as a function of time (t)
I(t)	Intensity as a function of time (t)
F(t)	Fluorescence emission intensity as a function of time (t)
[Q]	Concentration of quenching molecules
ψ	System parameter of fluorescence emission
[C]	Concentration
Qr	Quantum (radiative) efficiency
τp	Phase lifetime
τm	Modulation lifetime
MLE	Maximum likelihood estimation
RLD	Rapid lifetime determination
DLS	Damped least-squares
LMA	Levenberg–Marquardt algorithm
LE	Laguerre expansion
M	Modulation ratio
(φ)	Phase lag
(ω)	Angular frequency
φD	Quantum yield of donor in FRET
R	Distance between donor and acceptor (in FRET)
(g,s)	Coordinate system used in frequency domain plots
(x,y,z)	Coordinate system used to denote 3D space
TCSPC	Time correlated single-photon counting module
IRF	Instrument/impulse response function
SPCM	Commercial software for FLIM analysis by Becker& Hickl
SimFCS	Commercial software for FLIM analysis by Global Software
SymPhoTime	Commercial software for FLIM analysis by Picoquant
ISS/LFD	Commercial FLIM vendor ISS Inc/Lab. for Fluorescence Dynamics
SPAD	Single-photon avalanche diode
PMT	Photomultiplier tube
APDs	Avalanche photo diodes
MCP	Microchannel plate
ICCD	Intensified charged-couple device
CMOS	Complementary metal-oxide semiconductor
CCD	Charged-coupled device
sCMOS	Scientific CMOS
EMCCD	Electron multiplying CCD
HRI	High-rate imager
TTS	Transit time spread
GOI	Gated optical imager
TIF	Time of flight
FPGA	Field programmable gate array
Homodyned scheme	Identical modulation and demodulation frequency
Heterodyned scheme	Demodulation frequency is shifted from modulation frequency
TDC	Time-digital conversion
TAC-ADC	Time amplitude converter–analog digital converter
pCC	Phase cross correlation
TTL	Transistor-transistor logic
NADH	Reduced nicotinamide adenine dinucleotide
NAD+	Nicotinamide adenine dinucleotide (oxidized form)
FAD	Flavin adenine dinucleotide
FADH2	Flavin adenine dinucleotide (reduced form)
LLS	Long-lifetime species
FD	Frequency domain
TD	Time domain
STED	Stimulated emission depletion microscopy
FLCS	Fluorescence lifetime correlation spectroscopy
EGFP	Enhanced green fluorescence protein
BIFL	Burst integrated fluorescence lifetime
ps-FCM	Phase sensitive FCM
FCM	Flow cytometry
FCS	Fluorescence correlation spectrometry
SPLIT	Separation of photons by lifetime tuning
NLS	Nonlinear least-squares method

Acknowledgments

We thank Dr. Ellen Arena for useful edits and comments. We acknowledge funding from NIH R01 CA185251 (KWE).

REFERENCES

Abe, Ken, Lingling Zhao, Ammasi Periasamy, Xavier Intes, and Margarida Barroso. 2013. "Non-Invasive In Vivo Imaging of Near Infrared-Labeled Transferrin in Breast Cancer Cells and Tumors Using Fluorescence Lifetime FRET." *PLOS ONE* 8 (11): e80269. doi: 10.1371/journal.pone.0080269

Akers, Walter J., Mikhail Y. Berezin, Hyeran Lee, and Samuel Achilefu. 2008. "Predicting In Vivo Fluorescence Lifetime Behavior of NIR Fluorescent Contrast Agents Using In Vitro Measurements." *Journal of Biomedical Optics* 13 (5): 054042. doi: 10.1117/1.2982535

Alcala, Jose Ricardo, E. Gratton, and F. G. Prendergast. 1987. "Fluorescence Lifetime Distributions in Proteins." *Biophysical Journal* 51 (4): 597–604. doi: 10.1016/S0006-3495(87)83384-2

Alfonso-García, Alba, Tim D. Smith, Rupsa Datta, Thuy U. Luu, Enrico Gratton, Eric O. Potma, and Wendy F. Liu. 2016. "Label-Free Identification of Macrophage Phenotype by

Fluorescence Lifetime Imaging Microscopy." *Journal of Biomedical Optics* 21 (4): 046005. doi: 10.1117/1. JBO.21.4.046005

Amos, W. B., and J. G. White. 2003. "How the Confocal Laser Scanning Microscope Entered Biological Research." *Biology of the Cell* 95 (6): 335–42. doi: 10.1016/ S0248-4900(03)00078-9

Andersson, Ronnie M, Kjell Carlsson, Anders Liljeborg, and Hjalmar Brismar. 2000. "Characterization of Probe Binding and Comparison of Its Influence on Fluorescence Lifetime of Two PH-Sensitive Benzo[c]Xanthene Dyes Using Intensity-Modulated Multiple-Wavelength Scanning Technique." *Analytical Biochemistry* 283 (1): 104–10. doi: 10.1006/abio.2000.4652

Auksorius, Egidijus, Bosanta R. Boruah, Christopher Dunsby, Peter M. P. Lanigan, Gordon Kennedy, Mark A. A. Neil, and Paul M. W. French. 2008. "Stimulated Emission Depletion Microscopy with a Supercontinuum Source and Fluorescence Lifetime Imaging." *Optics Letters* 33 (2): 113–15. doi: 10.1364/OL.33.000113

Baeyens, Willy R. G. 1990. *Luminescence Techniques in Chemical and Biochemical Analysis*. CRC Press.

Ballew, Richard M., and J. N. Demas. 1989. "An Error Analysis of the Rapid Lifetime Determination Method for the Evaluation of Single Exponential Decays." *Analytical Chemistry* 61 (1): 30–3. doi: 10.1021/ac00176a007

Barber, Paul R., S. M. Ameer-Beg, J. Gilbey, L. M. Carlin, M. Keppler, T. C. Ng, and B. Vojnovic. 2009. "Multiphoton Time-Domain Fluorescence Lifetime Imaging Microscopy: Practical Application to Protein–Protein Interactions Using Global Analysis." *Journal of The Royal Society Interface* 6 (Suppl 1): S93–105. doi: 10.1098/rsif.2008.0451.focus

Barber, Paul R., S. M. Ameer-Beg, S. Pathmananthan, M. Rowley, and A. C. C. Coolen. 2010. "A Bayesian Method for Single Molecule, Fluorescence Burst Analysis." *Biomedical Optics Express* 1 (4): 1148–58. doi: 10.1364/ BOE.1.001148

Becker, Wolfgang. 2005. *Advanced Time-Correlated Single Photon Counting Techniques*. Springer Science & Business Media.

Becker, Wolfgang. 2015. *Advanced Time-Correlated Single Photon Counting Applications*. Vol. 111. Springer Series in Chemical Physics. Springer International Publishing.

Becker, Wolfgang. 2012. "Fluorescence Lifetime Imaging - Techniques and Applications: Fluorescence Lifetime Imaging." *Journal of Microscopy* 247 (2): 119–36. doi: 10.1111/j.1365-2818.2012.03618.x

Becker, Wolfgang, B. Su, O. Holub, and K. Weisshart. 2011. "FLIM and FCS Detection in Laser-Scanning Microscopes: Increased Efficiency by GaAsP Hybrid Detectors." *Microscopy Research and Technique* 74 (9): 804–11. doi: 10.1002/jemt.20959

Becker, Wolfgang, and Axel Bergmann. 2003. "Lifetime Imaging Techniques for Optical Microscopy." http://www.boselec. com/products/documents/LifetimeImagingTechniquesfor Microscopy-41p.pdf

Becker, Wolfgang, Axel Bergmann, and Christoph Biskup. 2007. "Multispectral Fluorescence Lifetime Imaging by TCSPC." *Microscopy Research and Technique* 70 (5): 403–9. doi: 10.1002/jemt.20432

Becker, Wolfgang, Axel Bergmann, Christoph Biskup, Laimonas Kelbauskas, Thomas Zimmer, Nikolaj Klocker, and Klaus Benndorf. 2003. "High Resolution TCSPC Lifetime Imaging." In *Biomedical Optics 2003*, 175–184. International Society for Optics and Photonics.

Becker, Wolfgang, Liisa M. Hirvonen, James Milnes, Thomas Conneely, Ottmar Jagutzki, Holger Netz, Stefan Smietana, and Klaus Suhling. 2016. "A Wide-Field TCSPC FLIM System Based on an MCP PMT with a Delay-Line Anode." *Review of Scientific Instruments* 87 (9): 093710. doi: 10.1063/1.4962864

Beechem, Joseph M., Enrico Gratton, Marcel Ameloot, Jay R. Knutson, and Ludwig Brand. 2002. "The Global Analysis of Fluorescence Intensity and Anisotropy Decay Data: Second-Generation Theory and Programs." In *Topics in Fluorescence Spectroscopy*, edited by Joseph R. Lakowicz, 241–305. Springer US.

Benninger, Richard K. P., Yasemin Koç, Oliver Hofmann, Jose Requejo-Isidro, Mark A. A. Neil, Paul M. W. French, and Andrew J. deMello. 2006. "Quantitative 3D Mapping of Fluidic Temperatures within Microchannel Networks Using Fluorescence Lifetime Imaging." *Analytical Chemistry* 78 (7): 2272–78. doi: 10.1021/ac051990f

Berezin, Mikhail Y., and Samuel Achilefu. 2010. "Fluorescence Lifetime Measurements and Biological Imaging." *Chemical Reviews* 110 (5): 2641–84. doi: 10.1021/cr900343z

Beule, Pieter De, Dylan M. Owen, Hugh B. Manning, Clifford B. Talbot, Jose Requejo-Isidro, Christopher Dunsby, James Mcginty, et al. 2007. "Rapid Hyperspectral Fluorescence Lifetime Imaging." *Microscopy Research and Technique* 70 (5): 481–84. doi: 10.1002/jemt.20434

Bhattacharjee, Arunima, Rupsa Datta, Enrico Gratton, and Allon I. Hochbaum. 2017. "Metabolic Fingerprinting of Bacteria by Fluorescence Lifetime Imaging Microscopy." *Scientific Reports* 7 (1): 1–10. doi: 10.1038/s41598-017-04032-w

Bialkowski, Stephen E. 1989. "Data Analysis in the Shot Noise Limit. 1. Single Parameter Estimation with Poisson and Normal Probability Density Functions." *Analytical Chemistry* 61 (22): 2479–83. doi: 10.1021/ac00197a006

Bird, Damian K., Long Yan, Kristin M. Vrotsos, Kevin W. Eliceiri, Emily M. Vaughan, Patricia J. Keely, John G. White, and Nirmala Ramanujam. 2005. "Metabolic Mapping of MCF10A Human Breast Cells via Multiphoton Fluorescence Lifetime Imaging of the Coenzyme NADH." *Cancer Research* 65 (19): 8766–73. doi: 10.1158/0008-5472. CAN-04-3922

Blacker, Thomas S., Zoe F. Mann, Jonathan E. Gale, Mathias Ziegler, Angus J. Bain, Gyorgy Szabadkai, and Michael R. Duchen. 2014. "Separating NADH and NADPH Fluorescence in Live Cells and Tissues Using FLIM." *Nature Communications* 5 (May): 3936. doi: 10.1038/ ncomms4936

Bloch, Sharon R., Frédéric Lesage, Laura McIntosh, Amir H. Gandjbakhche, Kexiang Liang, and Samuel Achilefu. 2005. "Whole-Body Fluorescence Lifetime Imaging of a Tumor-Targeted near-Infrared Molecular Probe in Mice." *Journal of Biomedical Optics* 10 (5): 054003. doi: 10.1117/1.2070148

Boaz, Harold, and G. K. Rollefson. 1950. "The Quenching of Fluorescence. Deviations from the Stern-Volmer Law." *Journal of the American Chemical Society* 72 (8): 3435–43.

Boens, Noël, Wenwu Qin, Nikola Basarić, Johan Hofkens, Marcel Ameloot, Jacques Pouget, Jean-Pierre Lefèvre, et al. 2007. "Fluorescence Lifetime Standards for Time and Frequency Domain Fluorescence Spectroscopy." *Analytical Chemistry* 79 (5): 2137–49. doi: 10.1021/ac062160k

Brinks, Daan, Aaron J. Klein, and Adam E. Cohen. 2015. "Two-Photon Lifetime Imaging of Voltage Indicating Proteins as a Probe of Absolute Membrane Voltage." *Biophysical Journal* 109 (5): 914–21. doi: 10.1016/j.bpj.2015.07.038

Brochon, Jean-Claude. 1994. "Maximum Entropy Method of Data Analysis in Time-Resolved Spectroscopy." In *Methods in Enzymology*, Vol. 240, 262–311. Part B: Numerical Computer Methods. Academic Press.

Brochon, Jean-Claude, Alastair K. Livesey, Jacques Pouget, and Bernard Valeur. 1990. "Data Analysis in Frequency-Domain Fluorometry by the Maximum Entropy Method — Recovery of Fluorescence Lifetime Distributions." *Chemical Physics Letters* 174 (5): 517–22. doi: 10.1016/S0009-2614(90)87189-X

Buntz, Annette, Sarah Wallrodt, Eva Gwosch, Michael Schmalz, Sascha Beneke, Elisa Ferrando-May, Andreas Marx, and Andreas Zumbusch. 2016. "Real-Time Cellular Imaging of Protein Poly(ADP-Ribos)Ylation." *Angewandte Chemie International Edition* 55 (37): 11256–60. doi: 10.1002/anie.201605282

Campos-Delgado, Daniel U., Omar Gutierrez-Navarro, Edgar R. Arce-Santana, Melissa C. Skala, Alex J. Walsh, and Javier A. Jo. 2015. "Blind Deconvolution Estimation of Fluorescence Measurements through Quadratic Programming." *Journal of Biomedical Optics* 20 (7): 075010.

Cao, Ruofan, Patrick Jenkins, William Peria, Bryan Sands, Mark Naivar, Roger Brent, and Jessica P. Houston. 2016. "Phasor Plotting with Frequency-Domain Flow Cytometry." *Optics Express* 24 (13): 14596–607. doi: 10.1364/OE.24.014596

Cao, Ruofan, Varayini Pankayatselvan, and Jessica P. Houston. 2013. "Cytometric Sorting Based on the Fluorescence Lifetime of Spectrally Overlapping Signals." *Optics Express* 21 (12): 14816–31. doi: 10.1364/OE.21.014816

Carlsson, Kjell, A. Liljeborg, R. M. Andersson, and Hjalmar Brismar. 2000. "Confocal PH Imaging of Microscopic Specimens Using Fluorescence Lifetimes and Phase Fluorometry: Influence of Parameter Choice on System Performance." *Journal of Microscopy* 199 (2): 106–14. doi: 10.1046/j.1365-2818.2000.00722.x

Chacko, Jenu V., and Kevin W. Eliceiri. 2018. "Autofluorescence Lifetime Imaging of Cellular Metabolism: Sensitivity toward Cell Density, PH, Intracellular, and Intercellular Heterogeneity: FLIM Sensitivity towards PH, Heterogeneity and Confluency." *Cytometry Part A*, October. doi: 10.1002/cyto.a.23603

Chacko, Jenu Varghese, and Kevin W. Eliceiri. 2019. "NAD(P)H Fluorescence Lifetime Measurements in Fixed Biological Tissues." *Methods and Applications in Fluorescence* 7 (4): 044005. doi: 10.1088/2050-6120/ab47e5

Chacko, Jenu Varghese, Francesca Cella Zanacchi, and Alberto Diaspro. 2013. "Probing Cytoskeletal Structures by Coupling Optical Superresolution and AFM Techniques for a Correlative Approach." *Cytoskeleton* 70 (11): 729–40.

Chance, Britton. 1976. "Pyridine Nucleotide as an Indicator of the Oxygen Requirements for Energy-Linked Functions of Mitochondria." *Circulation Research* 38 (5 Suppl 1): I31–38.

Chen, Hongtao, and Enrico Gratton. 2013. "A Practical Implementation of Multifrequency Widefield Frequency-Domain Fluorescence Lifetime Imaging Microscopy: Multifrequency Widefield Fd-Flim." *Microscopy Research and Technique* 76 (3): 282–89. doi: 10.1002/jemt.22165

Chen, Hongtao, Gerhard Holst, and Enrico Gratton. 2015. "Modulated CMOS Camera for Fluorescence Lifetime Microscopy." *Microscopy Research and Technique* 78 (12): 1075–81. doi: 10.1002/jemt.22587

Chen, Sez-Jade, Nattawut Sinsuebphon, Margarida Barroso, Xavier Intes, and Xavier Michalet. 2017. "AlliGator: A Phasor Computational Platform for Fast *in Vivo* Lifetime Analysis." In *Optics in the Life Sciences Congress (2017), Paper OmTu2D.2*, OmTu2D.2. Optical Society of America.

Chorvatova, Alzbeta, Anton Mateasik, and Dusan Chorvat Jr. 2013. "Spectral Decomposition of NAD(P)H Fluorescence Components Recorded by Multi-Wavelength Fluorescence Lifetime Spectroscopy in Living Cardiac Cells." *Laser Physics Letters* 10 (12): 125703. doi: 10.1088/1612-2011/10/12/125703

Clayton, A. H. A., Q. S. Hanley, and P. J. Verveer. 2004. "Graphical Representation and Multicomponent Analysis of Single-Frequency Fluorescence Lifetime Imaging Microscopy Data." *Journal of Microscopy* 213 (1): 1–5. doi: 10.1111/j.1365-2818.2004.01265.x

Clegg, Robert M. 2009. "Förster Resonance Energy Transfer— FRET What Is It, Why Do It, and How It's Done." *Laboratory Techniques in Biochemistry and Molecular Biology* 33: 1–57.

Cole, M. J., J. Siegel, S. E. D. Webb, R. Jones, K. Dowling, M. J. Dayel, D. Parsons-Karavassilis, et al. 2001. "Time-Domain Whole-Field Fluorescence Lifetime Imaging with Optical Sectioning." *Journal of Microscopy* 203 (3): 246–57. doi: 10.1046/j.1365-2818.2001.00894.x

Colyer, Ryan A., Claudia Lee, and Enrico Gratton. 2008. "A Novel Fluorescence Lifetime Imaging System That Optimizes Photon Efficiency." *Microscopy Research and Technique* 71 (3): 201–13. doi: 10.1002/jemt.20540

Colyer, Ryan A., Oswald H. W. Siegmund, Anton S. Tremsin, John V. Vallerga, Shimon Weiss, and Xavier Michalet. 2012. "Phasor Imaging with a Widefield Photon-Counting Detector." *Journal of Biomedical Optics* 17 (1). doi: 10.1117/1.JBO.17.1.016008

Conklin, Matthew W., Paolo P. Provenzano, Kevin W. Eliceiri, Ruth Sullivan, and Patricia J. Keely. 2009. "Fluorescence Lifetime Imaging of Endogenous Fluorophores in Histopathology Sections Reveals Differences Between Normal and Tumor Epithelium in Carcinoma In Situ of the Breast." *Cell Biochemistry and Biophysics* 53 (3): 145–57. doi: 10.1007/s12013-009-9046-7

Cui, H. Helen, Joseph G. Valdez, John A. Steinkamp, and Harry A. Crissman. 2003. "Fluorescence Lifetime-Based Discrimination and Quantification of Cellular DNA and RNA with Phase-Sensitive Flow Cytometry." *Cytometry* 52A (1): 46–55. doi: 10.1002/cyto.a.10022

Dancik, Yuri, Amandine Favre, Chong Jin Loy, Andrei V. Zvyagin, and Michael S. Roberts. 2013. "Use of Multiphoton Tomography and Fluorescence Lifetime Imaging to Investigate Skin Pigmentation in Vivo." *Journal of Biomedical Optics* 18 (2): 026022. doi: 10.1117/1.JBO.18.2.026022

Datta, Rupsa. 2016. "Label-Free Fluorescence Lifetime Imaging Microscopy (FLIM) to Study Metabolism and Oxidative Stress in Biological Systems." Ph.D., United States, California: University of California, Irvine.

Datta, Rupsa, Alba Alfonso-García, Rachel Cinco, and Enrico Gratton. 2015. "Fluorescence Lifetime Imaging of Endogenous Biomarker of Oxidative Stress." *Scientific Reports* 5 (May): 9848. doi: 10.1038/srep09848

Datta, Rupsa, Christopher Heylman, Steven C. George, and Enrico Gratton. 2016. "Label-Free Imaging of Metabolism and Oxidative Stress in Human Induced Pluripotent Stem Cell-Derived Cardiomyocytes." *Biomedical Optics Express* 7 (5): 1690–701. doi: 10.1364/BOE.7.001690

Datta, Rupsa, Agua Sobrino, Christopher Hughes, and Enrico Gratton. 2016. "Fluorescence Lifetime Imaging Microscopy to Study Metabolism in a Microfluidic Device Based Tumor Microenvironment." In *Optical Tomography and Spectroscopy*, JW4A–2. Optical Society of America.

Dean, Kevin M., and Amy E. Palmer. 2014. "Advances in Fluorescence Labeling Strategies for Dynamic Cellular Imaging." *Nature Chemical Biology* 10 (7): 512–23. doi: 10.1038/nchembio.1556

Deka, Chiranjit, and John A. Steinkamp. 1999. Time-resolved fluorescence decay measurements for flowing particles. US5909278 A, filed July 29, 1997, and issued June 1, 1999.

Denk, Winfried, James H. Strickler, Watt W. Webb, and others. 1990. "Two-Photon Laser Scanning Fluorescence Microscopy." *Science* 248 (4951): 73–6.

Diaspro, Alberto, and Marc AMJ van Zandvoort. 2016. *Super-Resolution Imaging in Biomedicine*. CRC Press.

Digman, Michelle A., Valeria R. Caiolfa, Moreno Zamai, and Enrico Gratton. 2008. "The Phasor Approach to Fluorescence Lifetime Imaging Analysis." *Biophysical Journal* 94 (2): L14–16. doi: 10.1529/biophysj.107.120154

Dimitrow, Enrico, Iris Riemann, Alexander Ehlers, Martin Johannes Koehler, Johannes Norgauer, Peter Elsner, Karsten König, and Martin Kaatz. 2009. "Spectral Fluorescence Lifetime Detection and Selective Melanin Imaging by Multiphoton Laser Tomography for Melanoma Diagnosis." *Experimental Dermatology* 18 (6): 509–15. doi: 10.1111/j.1600-0625.2008.00815.x

Dowling, K., M. J. Dayel, M. J. Lever, P. M. W. French, J. D. Hares, and A. K. L. Dymoke-Bradshaw. 1998. "Fluorescence Lifetime Imaging with Picosecond Resolution for Biomedical Applications." *Optics Letters* 23 (10): 810–12. doi: 10.1364/OL.23.000810

Eggeling, Christian, Jerker Widengren, Leif Brand, Jörg Schaffer, Suren Felekyan, and Claus A. M. Seidel. 2006. "Analysis of Photobleaching in Single-Molecule Multicolor Excitation and Förster Resonance Energy Transfer Measurements†." *The Journal of Physical Chemistry A* 110 (9): 2979–95. doi: 10.1021/jp054581w

Eichorst, John Paul, Kaiwen Teng, and Robert M. Clegg. 2014. "Fluorescence Lifetime Imaging Techniques: Frequency-Domain FLIM." In *Fluorescence Lifetime Spectroscopy and Imaging*, 165–86. CRC Press.

Elder, Alan D., J. H. Frank, J. Swartling, X. Dai, and C. F. Kaminski. 2006. "Calibration of a Wide-Field Frequency-Domain Fluorescence Lifetime Microscopy System Using Light Emitting Diodes as Light Sources." *Journal of Microscopy* 224 (2): 166–80. doi: 10.1111/j.1365-2818.2006.01689.x

Elder, Alan D., Clemens F. Kaminski, and Jonathan H. Frank. 2009. "φ²FLIM: A Technique for Alias-Free Frequency Domain Fluorescence Lifetime Imaging." *Optics Express* 17 (25): 23181–203. doi: 10.1364/OE.17.023181

Elder, Alan, Simon Schlachter, and Clemens F. Kaminski. 2008. "Theoretical Investigation of the Photon Efficiency in Frequency-Domain Fluorescence Lifetime Imaging Microscopy." *JOSA* A 25 (2): 452–62. doi: 10.1364/JOSAA.25.000452

Elson, Daniel S., J. A. Jo, and Laura Marcu. 2007. "Miniaturized Side-Viewing Imaging Probe for Fluorescence Lifetime Imaging (FLIM): Validation with Fluorescence Dyes, Tissue Structural Proteins and Tissue Specimens." *New Journal of Physics* 9 (5): 127. doi: 10.1088/1367-2630/9/5/127

Elson, Daniel S., Neil Galletly, Clifford Talbot, Jose Requejo-Isidro, James McGinty, Christopher Dunsby, Peter M. P. Lanigan, et al. 2006. "Multidimensional Fluorescence Imaging Applied to Biological Tissue." In *Reviews in Fluorescence*, 477–524. Springer.

Esposito, Alessandro, Christoph P. Dohm, Matthias Bähr, and Fred S. Wouters. 2007. "Unsupervised Fluorescence Lifetime Imaging Microscopy for High Content and High Throughput Screening." *Molecular & Cellular Proteomics* 6 (8): 1446–54. doi: 10.1074/mcp.T700006-MCP200

Esposito, Alessandro, Hans C. Gerritsen, and Fred S. Wouters. 2005. "Fluorescence Lifetime Heterogeneity Resolution in the Frequency Domain by Lifetime Moments Analysis." *Biophysical Journal* 89 (6): 4286–99. doi: 10.1529/biophysj.104.053397

Esposito, Alessandro, Hans C. Gerritsen, and Fred S. Wouters. 2007. "Optimizing Frequency-Domain Fluorescence Lifetime Sensing for High-Throughput Applications: Photon Economy and Acquisition Speed." *JOSA* A 24 (10): 3261–73. doi: 10.1364/JOSAA.24.003261

Favreau, Peter F., Clarissa Hernandez, Ashley Stringfellow Lindsey, Diego F. Alvarez, Thomas C. Rich, Prashant Prabhat, and Silas J. Leavesley. 2013. "Thin-Film Tunable Filters for Hyperspectral Fluorescence Microscopy." *Journal of Biomedical Optics* 19 (1): 011017. doi: 10.1117/1.JBO.19.1.011017

Feeks, James A., and Jennifer J. Hunter. 2017. "Adaptive Optics Two-Photon Excited Fluorescence Lifetime Imaging Ophthalmoscopy of Exogenous Fluorophores in Mice." *Biomedical Optics Express* 8 (5): 2483–95. doi: 10.1364/BOE.8.002483

Fišerová, Eva, and Martin Kubala. 2012. "Mean Fluorescence Lifetime and Its Error." *Journal of Luminescence* 132 (8): 2059–64. doi: 10.1016/j.jlumin.2012.03.038

"Frequency-Domain Fluorometry." 1990. In *Luminescence Techniques in Chemical and Biochemical Analysis (Luminescence Spectroscopy, Vol. 12)*.

Fries, Joachim R., Leif Brand, Christian Eggeling, Malte Köllner, and Claus A. M. Seidel. 1998. "Quantitative Identification of Different Single Molecules by Selective Time-Resolved Confocal Fluorescence Spectroscopy." *The Journal of Physical Chemistry A* 102 (33): 6601–13. doi: 10.1021/jp980965t

Fruhwirth, Gilbert O., Simon Ameer-Beg, Richard Cook, Timothy Watson, Tony Ng, and Frederic Festy. 2010. "Fluorescence Lifetime Endoscopy Using TCSPC for the Measurement of FRET in Live Cells." *Optics Express* 18 (11): 11148–58. doi: 10.1364/OE.18.011148

Gadella, Theodorus W. J., ed. 2009. *Laboratory Techniques in Biochemistry and Molecular Biology.* 1st ed., vol. 33. Elsevier.

Gadella, Theodorus W. J., Thomas M. Jovin, and Robert M. Clegg. 1993. "Fluorescence Lifetime Imaging Microscopy (FLIM): Spatial Resolution of Microstructures on the Nanosecond Time Scale." *Biophysical Chemistry* 48 (2): 221–39. doi: 10.1016/0301-4622(93)85012-7

Galletly, Neil P., J. McGinty, C. Dunsby, F. Teixeira, J. Requejo-Isidro, I. Munro, D. S. Elson, et al. 2008. "Fluorescence Lifetime Imaging Distinguishes Basal Cell Carcinoma from Surrounding Uninvolved Skin." *The British Journal of Dermatology* 159 (1): 152–61. doi: 10.1111/j.1365-2133.2008.08577.x

Garcia, Alba Alfonso, Tim Smith, Rupsa Datta, Enrico Gratton, Eric O. Potma, and Wendy Liu. 2016. "Visualizing Cellular Metabolic Processes With Combined Nonlinear Optical Microscopy." In *Biomedical Optics 2016 (2016), Paper OTh4C.7*, OTh4C.7. Optical Society of America.

Gerritsen, Hans C., A. V. Agronskaia, A. N. Bader, and A. Esposito. 2009. "Time Domain FLIM: Theory, Instrumentation, and Data Analysis." *Laboratory Techniques in Biochemistry and Molecular Biology* 33: 95–132.

Gerritsen, Hans C., M. A. H. Asselbergs, A. V. Agronskaia, and WGJHM Van Sark. 2002. "Fluorescence Lifetime Imaging in Scanning Microscopes: Acquisition Speed, Photon Economy and Lifetime Resolution." *Journal of Microscopy* 206 (3): 218–24.

Gioux, Sylvain, Stephen J. Lomnes, Hak Soo Choi, and John V. Frangioni. 2010. "Low-Frequency Wide-Field Fluorescence Lifetime Imaging Using a High-Power near-Infrared Light-Emitting Diode Light Source." *Journal of Biomedical Optics* 15 (2). doi: 10.1117/1.3368997

Gratton, Enrico, D. M. Jameson, G. Weber, and B. Alpert. 1984. "A Model of Dynamic Quenching of Fluorescence in Globular Proteins." *Biophysical Journal* 45 (4): 789–94. doi: 10.1016/S0006-3495(84)84223-X

Gratton, Enrico. 2016. "Measurements of Fluorescence Decay Time by the Frequency Domain Method." In *Perspectives on Fluorescence*, edited by David M. Jameson, Vol. 17, 67–80. Springer International Publishing.

Gratton, Enrico, and Beniamino Barbieri. 1993. High speed cross-correlation frequency domain fluorometry-phosphorimetry. US5212386 A, filed December 13, 1991, and issued May 18, 1993.

Gratton, Enrico, Sophie Breusegem, Jason Sutin, Qiaoqiao Ruan, and Nicholas Barry. 2003. "Fluorescence Lifetime Imaging for the Two-Photon Microscope: Time-Domain and Frequency-Domain Methods." *Journal of Biomedical Optics* 8 (3): 381–90. doi: 10.1117/1.1586704

Gregor, Ingo, and Matthias Patting. 2014. "Pattern-Based Linear Unmixing for Efficient and Reliable Analysis of Multicomponent TCSPC Data." In *Advanced Photon Counting*, edited by Peter Kapusta, Michael Wahl, and Rainer Erdmann, 241–63. Springer Series on Fluorescence 15. Springer International Publishing.

Hanahan, Douglas, and Robert A. Weinberg. 2011. "Hallmarks of Cancer: The Next Generation." *Cell* 144 (5): 646–74. doi: 10.1016/j.cell.2011.02.013

Hartmann, Paul, Werner Ziegler, Gerhard Holst, and Dietrich W. Lübbers. 1997. "Oxygen Flux Fluorescence Lifetime Imaging." *Sensors and Actuators B: Chemical* 38 (1): 110–15. doi: 10.1016/S0925-4005(97)80179-7

Hato, Takashi, Seth Winfree, Richard Day, Ruben M. Sandoval, Bruce A. Molitoris, Mervin C. Yoder, Roger C. Wiggins, et al. 2017. "Two-Photon Intravital Fluorescence Lifetime Imaging of the Kidney Reveals Cell-Type Specific Metabolic Signatures." *Journal of the American Society of Nephrology* 28 (8): 2420–30. doi: 10.1681/ASN.2016101153

Heiden, Matthew G. Vander, Lewis C. Cantley, and Craig B. Thompson. 2009. "Understanding the Warburg Effect: The Metabolic Requirements of Cell Proliferation." *Science* 324 (5930): 1029–33. doi: 10.1126/science.1160809

Heim, Nicola, Olga Garaschuk, Michael W. Friedrich, Marco Mank, Ruxandra I. Milos, Yury Kovalchuk, Arthur Konnerth, and Oliver Griesbeck. 2007. "Improved Calcium Imaging in Transgenic Mice Expressing a Troponin C–Based Biosensor." *Nature Methods* 4 (2): 127–29. doi: 10.1038/nmeth1009

Hille, Carsten, Maik Berg, Lena Bressel, Dorit Munzke, Philipp Primus, Hans-Gerd Löhmannsröben, and Carsten Dosche. 2008. "Time-Domain Fluorescence Lifetime Imaging for Intracellular PH Sensing in Living Tissues." *Analytical and Bioanalytical Chemistry* 391 (5): 1871–79. doi: 10.1007/s00216-008-2147-0

Hinsdale, Taylor, Cory Olsovsky, Jose J. Rico-Jimenez, Kristen C. Maitland, Javier A. Jo, and Bilal H. Malik. 2017. "Optically Sectioned Wide-Field Fluorescence Lifetime Imaging Microscopy Enabled by Structured Illumination." *Biomedical Optics Express* 8 (3): 1455–65. doi: 10.1364/BOE.8.001455

Hirvonen, Liisa M., Frederic Festy, and Klaus Suhling. 2014. "Wide-Field Time-Correlated Single-Photon Counting (TCSPC) Lifetime Microscopy with Microsecond Time Resolution." *Optics Letters* 39 (19): 5602–5. doi: 10.1364/OL.39.005602

Hirvonen, Liisa M., and Klaus Suhling. 2017. "Wide-Field TCSPC: Methods and Applications." *Measurement Science and Technology* 28 (1): 012003. doi: 10.1088/1361-6501/28/1/012003

Hoffmann, Katrin, Thomas Behnke, Markus Grabolle, and Ute Resch-Genger. 2014. "Nanoparticle-Encapsulated Vis- and NIR-Emissive Fluorophores with Different Fluorescence Decay Kinetics for Lifetime Multiplexing." *Analytical and Bioanalytical Chemistry* 406 (14): 3315–22. doi: 10.1007/s00216-013-7597-3

Horton, Nicholas G., Ke Wang, Demirhan Kobat, Catharine G. Clark, Frank W. Wise, Chris B. Schaffer, and Chris Xu. 2013. "In Vivo Three-Photon Microscopy of Subcortical Structures within an Intact Mouse Brain." *Nature Photonics* 7 (3): 205–9. doi: 10.1038/nphoton.2012.336

Hosny, Neveen A., David A. Lee, and Martin M. Knight. 2012. "Single Photon Counting Fluorescence Lifetime Detection of Pericellular Oxygen Concentrations." *Journal of Biomedical Optics* 17 (1): 016007. doi: 10.1117/1.JBO.17.1.016007

Hou, Jue, Heather J. Wright, Nicole Chan, Richard Tran, Olga V. Razorenova, Eric O. Potma, and Bruce J. Tromberg. 2016. "Correlating Two-Photon Excited Fluorescence Imaging of Breast Cancer Cellular Redox State with Seahorse Flux Analysis of Normalized Cellular Oxygen Consumption." *Journal of Biomedical Optics* 21 (6): 060503. doi: 10.1117/1.JBO.21.6.060503

Isenberg, Irvin, and Robert D. Dyson. 1969. "The Analysis of Fluorescence Decay by a Method of Moments." *Biophysical Journal* 9 (11): 1337–50.

Ishii, Kunihiko, Takuhiro Otosu, and Tahei Tahara. 2014. "Lifetime-Weighted FCS and 2D FLCS: Advanced Application of Time-Tagged TCSPC." In *Advanced Photon Counting*, edited by Peter Kapusta, Michael Wahl, and Rainer Erdmann, 111–28. Springer Series on Fluorescence 15. Springer International Publishing.

Jahn, Karolina, Volker Buschmann, and Carsten Hille. 2015. "Simultaneous Fluorescence and Phosphorescence Lifetime Imaging Microscopy in Living Cells." *Scientific Reports* 5 (September): 14334. doi: 10.1038/srep14334

Jameson, David M. 2016. "A Fluorescent Lifetime: Reminiscing About Gregorio Weber." In *Perspectives on Fluorescence*, edited by David M. Jameson, 1–16. Springer Series on Fluorescence 17. Springer International Publishing.

Janssen, Aniek, Evelyne Beerling, René Medema, and Jacco van Rheenen. 2013. "Intravital FRET Imaging of Tumor Cell Viability and Mitosis during Chemotherapy." *PLOS ONE* 8 (5): e64029. doi: 10.1371/journal.pone.0064029

Jemal, Ahmedin, Freddie Bray, Melissa M. Center, Jacques Ferlay, Elizabeth Ward, and David Forman. 2011. "Global Cancer Statistics." *CA: A Cancer Journal for Clinicians* 61 (2): 69–90. doi: 10.3322/caac.20107

Jo, Javier A., Qiyin Fang, Thanassis Papaioannou, J. Dennis Baker, Amir Dorafshar, Todd Reil, Jianhua Qiao, Michael C. Fishbein, Julie A. Freischlag, and Laura Marcu. 2006. "Laguerre-Based Method for Analysis of Time-Resolved Fluorescence Data: Application to *in-Vivo* Characterization and Diagnosis of Atherosclerotic Lesions." *Journal of Biomedical Optics* 11 (2): 021004. doi: 10.1117/1.2186045

Jo, Javier A., Qiyin Fang, Thanassis Papaioannou, and Laura Marcu. 2004. "Fast Model-Free Deconvolution of Fluorescence Decay for Analysis of Biological Systems." *Journal of Biomedical Optics* 9 (4): 743–52. doi: 10.1117/1.1752919

Johnsson, Anna-Karin E., Yanfeng Dai, Max Nobis, Martin J. Baker, Ewan J. McGhee, Simon Walker, Juliane P. Schwarz, et al. 2014. "The Rac-FRET Mouse Reveals Tight Spatiotemporal Control of Rac Activity in Primary Cells and Tissues." *Cell Reports* 6 (6): 1153–64. doi: 10.1016/j.celrep.2014.02.024

Jones, Carolyn, and Klaus Suhling. 2006. "Refractive Index Sensing Using Fluorescence Lifetime Imaging (FLIM)." *Journal of Physics: Conference Series* 45 (1): 223. doi: 10.1088/1742-6596/45/1/031

Joo, Chirlmin, Hamza Balci, Yuji Ishitsuka, Chittanon Buranachai, and Taekjip Ha. 2008. "Advances in Single-Molecule Fluorescence Methods for Molecular Biology." *Annual Review of Biochemistry* 77 (1): 51–76. doi: 10.1146/annurev.biochem.77.070606.101543

Kalinina, Sviatlana, Dominik Bisinger, Jasmin Breymayer, and A. Ruck. 2015. "Cell Metabolism, FLIM and PLIM and Applications." In *Multiphoton Microscopy in the Biomedical Sciences XV*, Vol. 9329, 93290C. International Society for Optics and Photonics.

Kalisz, Józef. 2004. "Review of Methods for Time Interval Measurements with Picosecond Resolution." *Metrologia* 41 (1): 17. doi: 10.1088/0026-1394/41/1/004

Kallinowski, F., K. H. Schlenger, S. Runkel, M. Kloes, M. Stohrer, P. Okunieff, and P. Vaupel. 1989. "Blood Flow, Metabolism, Cellular Microenvironment, and Growth Rate of Human Tumor Xenografts." *Cancer Research* 49 (14): 3759–64.

Kapusta, Peter, Radek Macháň, Aleš Benda, and Martin Hof. 2012. "Fluorescence Lifetime Correlation Spectroscopy (FLCS): Concepts, Applications and Outlook." *International Journal of Molecular Sciences* 13 (10): 12890–910. doi: 10.3390/ijms131012890

Keizer, Joel. 1983. "Nonlinear Fluorescence Quenching and the Origin of Positive Curvature in Stern-Volmer Plots." *Journal of the American Chemical Society* 105 (6): 1494–98.

Knutson, Jay R., Joseph M. Beechem, and Ludwig Brand. 1983. "Simultaneous Analysis of Multiple Fluorescence Decay Curves: A Global Approach." *Chemical Physics Letters* 102 (6): 501–7. doi: 10.1016/0009-2614(83)87454-5

Koenig, Karsten. 2012. "Hybrid Multiphoton Multimodal Tomography of in Vivo Human Skin." *IntraVital* 1 (1): 11–26. doi: 10.4161/intv.21938

Koenig, Marcelle, Sandra Orthaus-Mueller, Rhys Dowler, Benedikt Kraemer, Astrid Tannert, Olaf Schulz, Tino Roehlicke, et al. 2017. "Rapid Flim: The New and Innovative Method for Ultra-Fast Imaging of Biological Processes." *Biophysical Journal* 112 (3): 298a. doi: 10.1016/j.bpj.2016.11.1614

Krasieva, Tatiana B., Chiara Stringari, Feng Liu, Chung-Ho Sun, Yu Kong, Mihaela Balu, Frank L. Meyskens, Enrico Gratton, and Bruce J. Tromberg. 2012. "Two-Photon Excited Fluorescence Lifetime Imaging and Spectroscopy of Melanins in Vitro and in Vivo." *Journal of Biomedical Optics* 18 (3): 031107. doi: 10.1117/1.JBO.18.3.031107

Krishnan, R. V., H. Saitoh, H. Terada, V. E. Centonze, and B. Herman. 2003. "Development of a Multiphoton Fluorescence Lifetime Imaging Microscopy System Using a Streak Camera." *Review of Scientific Instruments* 74 (5): 2714–21. doi: 10.1063/1.1569410

Kristoffersen, Arne S., Svein R. Erga, Børge Hamre, and Øyvind Frette. 2014. "Testing Fluorescence Lifetime Standards Using Two-Photon Excitation and Time-Domain Instrumentation: Rhodamine B, Coumarin 6 and Lucifer Yellow." *Journal of Fluorescence* 24 (4): 1015–24. doi: 10.1007/s10895-014-1368-1

Kuimova, Marina K., Gokhan Yahioglu, James A. Levitt, and Klaus Suhling. 2008. "Molecular Rotor Measures Viscosity of Live Cells via Fluorescence Lifetime Imaging." *Journal of the American Chemical Society* 130 (21): 6672–73. doi: 10.1021/ja800570d

Kurokawa, Hiromi, Hidehiro Ito, Mai Inoue, Kenji Tabata, Yoshifumi Sato, Kazuya Yamagata, Shinae Kizaka-Kondoh, et al. 2015. "High Resolution Imaging of Intracellular Oxygen Concentration by Phosphorescence Lifetime." *Scientific Reports* 5 (June): 10657. doi: 10.1038/srep10657

Lakowicz, Joseph R. 1983. "Quenching of Fluorescence." In *Principles of Fluorescence Spectroscopy*, 257–301. Springer.

Lakowicz, Joseph R., and Henryk Szmacinski. 1993. "Fluorescence Lifetime-Based Sensing of PH, Ca2+, K+ and Glucose." *Sensors and Actuators B: Chemical* 11 (1): 133–43. doi: 10.1016/0925-4005(93)85248-9

Lakowicz, Joseph R., Henryk Szmacinski, Kazimierz Nowaczyk, and Michael L. Johnson. 1992. "Fluorescence Lifetime Imaging of Free and Protein-Bound NADH." *Proceedings of the National Academy of Sciences of the United States of America* 89 (4): 1271–75.

Lakowicz, Joseph R., and Gregorio Weber. 1973. "Quenching of Protein Fluorescence by Oxygen. Detection of Structural Fluctuations in Proteins on the Nanosecond Time Scale." *Biochemistry* 12 (21): 4171–79. doi: 10.1021/bi00745a021

Lanzanò, Luca, Iván Coto Hernández, Marco Castello, Enrico Gratton, Alberto Diaspro, and Giuseppe Vicidomini. 2015. "Encoding and Decoding Spatio-Temporal Information for Super-Resolution Microscopy." *Nature Communications* 6 (April): 6701. doi: 10.1038/ncomms7701

Lanzanò, Luca, Lorenzo Scipioni, Melody Di Bona, Paolo Bianchini, Ranieri Bizzarri, Francesco Cardarelli, Alberto Diaspro, and Giuseppe Vicidomini. 2017. "Measurement of Nanoscale Three-Dimensional Diffusion in the Interior of Living Cells by STED-FCS." *Nature Communications* 8 (1): 65. doi: 10.1038/s41467-017-00117-2

Le Marois, Alix, Simon Labouesse, Klaus Suhling, and Rainer Heintzmann. 2016. "Noise-Corrected Principal Component Analysis of Fluorescence Lifetime Imaging Data." *Journal of Biophotonics*, December. doi: 10.1002/jbio.201600160

Lee, K. C., J. Siegel, S. E. Webb, S. Lévêque-Fort, M. J. Cole, R. Jones, K. Dowling, M. J. Lever, and P. M. French. 2001. "Application of the Stretched Exponential Function to Fluorescence Lifetime Imaging." *Biophysical Journal* 81 (3): 1265–74. doi: 10.1016/S0006-3495(01)75784-0

Leray, A., C. Spriet, D. Trinel, and Laurent Héliot. 2009. "Three-Dimensional Polar Representation for Multispectral Fluorescence Lifetime Imaging Microscopy." *Cytometry Part A* 75A (12): 1007–14. doi: 10.1002/cyto.a.20802

Li, David Day-Uei, Simon Ameer-Beg, Jochen Arlt, David Tyndall, Richard Walker, Daniel R. Matthews, Viput Visitkul, Justin Richardson, and Robert K. Henderson. 2012. "Time-Domain Fluorescence Lifetime Imaging Techniques Suitable for Solid-State Imaging Sensor Arrays." *Sensors* 12 (12): 5650–69. doi: 10.3390/s120505650

Li, David Day-Uei, Hongqi Yu, and Yu Chen. 2015. "Fast Bi-Exponential Fluorescence Lifetime Imaging Analysis Methods." *Optics Letters* 40 (3): 336–39. doi: 10.1364/OL.40.000336

Li, Day-Uei, Jochen Arlt, Justin Richardson, Richard Walker, Alex Buts, David Stoppa, Edoardo Charbon, and Robert Henderson. 2010. "Real-Time Fluorescence Lifetime Imaging System with a 32 × 32 013μm CMOS Low Dark-Count Single-Photon Avalanche Diode Array." *Optics Express* 18 (10): 10257. doi: 10.1364/OE.18.010257

Libertini, Louis J., and Enoch W. Small. 1989. "Application of Method of Moments Analysis to Fluorescence Decay Lifetime Distributions." *Biophysical Chemistry* 34 (3): 269–82.

Lin, Hai-Jui, Petr Herman, and Joseph R. Lakowicz. 2003. "Fluorescence Lifetime-Resolved PH Imaging of Living Cells." *Cytometry Part A* 52A (2): 77–89. doi: 10.1002/cyto.a.10028

Linkert, Melissa, Curtis T. Rueden, Chris Allan, Jean-Marie Burel, Will Moore, Andrew Patterson, Brian Loranger, et al. 2010. "Metadata Matters: Access to Image Data in the Real World." *The Journal of Cell Biology* 189 (5): 777–82. doi: 10.1083/jcb.201004104

Machacek, Matthias, Louis Hodgson, Christopher Welch, Hunter Elliott, Olivier Pertz, Perihan Nalbant, Amy Abell, Gary L. Johnson, Klaus M. Hahn, and Gaudenz Danuser. 2009. "Coordination of Rho GTPase Activities during Cell Protrusion." *Nature* 461 (7260): 99–103. doi: 10.1038/nature08242

Malacrida, Leonel, David M. Jameson, and Enrico Gratton. 2017. "A Multidimensional Phasor Approach Reveals LAURDAN Photophysics in NIH-3T3 Cell Membranes." *Scientific Reports* 7 (1): 9215. doi: 10.1038/s41598-017-08564-z

Manen, Henk-Jan van, Paul Verkuijlen, Paul Wittendorp, Vinod Subramaniam, Timo K. van den Berg, Dirk Roos, and Cees Otto. 2008. "Refractive Index Sensing of Green Fluorescent Proteins in Living Cells Using Fluorescence Lifetime Imaging Microscopy." *Biophysical Journal* 94 (8): L67–69. doi: 10.1529/biophysj.107.127837

Marcu, Laura, Javier A. Jo, Pramod V. Butte, William H. Yong, Brian K. Pikul, Keith L. Black, and Reid C. Thompson. 2004. "Fluorescence Lifetime Spectroscopy of Glioblastoma Multiforme." *Photochemistry and Photobiology* 80 (1): 98–103. doi: 10.1111/j.1751-1097.2004.tb00055.x

Martin, Stephen A., Tyler M. DeMuth, Karl N. Miller, Thomas D. Pugh, Michael A. Polewski, Ricki J. Colman, Kevin W. Eliceiri, et al. 2016. "Regional Metabolic Heterogeneity of the Hippocampus Is Nonuniformly Impacted by Age and Caloric Restriction." *Aging Cell* 15 (1): 100–10. doi: 10.1111/acel.12418

Maus, Michael, Mircea Cotlet, Johan Hofkens, Thomas Gensch, Frans C. De Schryver, J. Schaffer, and C. A. M. Seidel. 2001. "An Experimental Comparison of the Maximum Likelihood Estimation and Nonlinear Least-Squares Fluorescence Lifetime Analysis of Single Molecules." *Analytical Chemistry* 73 (9): 2078–86. doi: 10.1021/ac000877g

McGinty, James, Neil P. Galletly, Chris Dunsby, Ian Munro, Daniel S. Elson, Jose Requejo-Isidro, Patrizia Cohen, et al. 2010. "Wide-Field Fluorescence Lifetime Imaging of Cancer." *Biomedical Optics Express* 1 (2): 627–40. doi: 10.1364/BOE.1.000627

Michalet, X., R. A. Colyer, G. Scalia, A. Ingargiola, R. Lin, J. E. Millaud, S. Weiss, et al. 2013. "Development of New Photon-Counting Detectors for Single-Molecule Fluorescence Microscopy." *Philosophical Transactions of the Royal Society B: Biological Sciences* 368 (1611): 20120035. doi: 10.1098/rstb.2012.0035

Michalet, X., O. H. W. Siegmund, J. V. Vallerga, P. Jelinsky, J. E. Millaud, and S. Weiss. 2006. "Photon-Counting H33D Detector for Biological Fluorescence Imaging." *Nuclear Instruments and Methods in Physics Research Section A: Accelerators, Spectrometers, Detectors and Associated Equipment* 567 (1): 133–36. doi: 10.1016/j.nima.2006.05.155

Mizeret, Jérôme, Thomas Stepinac, Marc Hansroul, André Studzinski, Hubert van den Bergh, and Georges Wagnières. 1999. "Instrumentation for Real-Time Fluorescence Lifetime Imaging in Endoscopy." *Review of Scientific Instruments* 70 (12): 4689–701. doi: 10.1063/1.1150132

Moffitt, Jeffrey R., Christian Osseforth, and Jens Michaelis. 2011. "Time-Gating Improves the Spatial Resolution of STED Microscopy." *Optics Express* 19 (5): 4242–54. doi: 10.1364/OE.19.004242

Munster, Erik B. van, and Theodorus W. J. Gadella. 2005. "Fluorescence Lifetime Imaging Microscopy (FLIM)." In *Microscopy Techniques*, edited by Jens Rietdorf, Vol. 95, 143–75. Springer Berlin Heidelberg.

Nag, Suman, Bidyut Sarkar, Muralidharan Chandrakesan, Rajiv Abhyanakar, Debanjan Bhowmik, Mamata Kombrabail, Sucheta Dandekar, Eitan Lerner, Elisha Haas, and Sudipta Maiti. 2013. "A Folding Transition Underlies the Emergence of Membrane Affinity in Amyloid-β." *Physical Chemistry Chemical Physics* 15 (44): 19129–33. doi: 10.1039/C3CP52732H

Ni, Tuqiang, and Lynn A. Melton. 1996. "Two-Dimensional Gas-Phase Temperature Measurements Using Fluorescence Lifetime Imaging." *Applied Spectroscopy* 50 (9): 1112–16.

Nobis, Max, David Herrmann, Sean C. Warren, Shereen Kadir, Wilfred Leung, Monica Killen, Astrid Magenau, et al. 2017. "A RhoA-FRET Biosensor Mouse for Intravital Imaging in Normal Tissue Homeostasis and Disease Contexts." *Cell Reports* 21 (1): 274–88. doi: 10.1016/j.celrep.2017.09.022

Nobis, Max, Ewan J. McGhee, Jennifer P. Morton, Juliane P. Schwarz, Saadia A. Karim, Jean Quinn, Mike Edward, et al. 2013. "Intravital FLIM-FRET Imaging Reveals Dasatinib-Induced Spatial Control of Src in Pancreatic Cancer." *Cancer Research* 73 (15): 4674–86. doi: 10.1158/0008-5472.CAN-12-4545

O'Connor, Desmond. 2012. *Time-Correlated Single Photon Counting.* Elsevier Science.

Okabe, Kohki, Noriko Inada, Chie Gota, Yoshie Harada, Takashi Funatsu, and Seiichi Uchiyama. 2012. "Intracellular Temperature Mapping with a Fluorescent Polymeric Thermometer and Fluorescence Lifetime Imaging Microscopy." *Nature Communications* 3 (February): 705. doi: 10.1038/ncomms1714

O'Leary, M. A., D. A. Boas, X. D. Li, B. Chance, and A. G. Yodh. 1996. "Fluorescence Lifetime Imaging in Turbid Media." *Optics Letters* 21 (2): 158–60. doi: 10.1364/OL.21.000158

Orihuela, Ruben, Christopher A McPherson, and Gaylia Jean Harry. 2016. "Microglial M1/M2 Polarization and Metabolic States." *British Journal of Pharmacology* 173 (4): 649–65. doi: 10.1111/bph.13139

Orthaus-Mueller, Sandra, Ben Kraemer, Rhys Dowler, André Devaux, Astrid Tannert, Tino Roehlicke, Michael Wahl, Hans-Juergen Rahn, and Rainer Erdmann. n.d. "RapidFLIM: The New and Innovative Method for Ultra Fast FLIM Imaging."

Pande, Paritosh, and Javier A. Jo. 2011. "Automated Analysis of Fluorescence Lifetime Imaging Microscopy (FLIM) Data Based on the Laguerre Deconvolution Method." *IEEE Transactions on Biomedical Engineering* 58 (1): 172–81. doi: 10.1109/TBME.2010.2084086

Papour, Asael, Zachary D. Taylor, Adria J. Sherman, Desiree Sanchez, Gregory Lucey, Linda Liau, Oscar M. Stafsudd, William H. Yong, and Warren S. Grundfest. 2013. "Optical Imaging for Brain Tissue Characterization Using Relative Fluorescence Lifetime Imaging." *Journal of Biomedical Optics* 18 (6): 060504. doi: 10.1117/1.JBO.18.6.060504

Papour, Asael, Zachary Taylor, Oscar Stafsudd, Irena Tsui, and Warren S. Grundfest. 2015. "Imaging Autofluorescence Temporal Signatures of the Human Ocular Fundus *in Vivo*." *Journal of Biomedical Optics* 20 (11): 110505. doi: 10.1117/1.JBO.20.11.110505

Parson, William W. 2007. *Modern Optical Spectroscopy: With Examples from Biophysics and Biochemistry.* Springer.

Patalay, Rakesh, Clifford Talbot, Yuriy Alexandrov, Ian Munro, Hans Georg Breunig, Karsten König, Sean Warren, et al. 2011. "Non-Invasive Imaging of Skin Cancer with Fluorescence Lifetime Imaging Using Two Photon Tomography." In *European Conference on Biomedical Optics*, Vol. 8087, 808718-808718-8. Optical Society of America.

Pate, Kira T., Chiara Stringari, Stephanie Sprowl-Tanio, Kehui Wang, Tara TeSlaa, Nate P. Hoverter, Miriam M. McQuade, et al. 2014. "Wnt Signaling Directs a Metabolic Program of Glycolysis and Angiogenesis in Colon Cancer." *The EMBO Journal* 33 (13): 1454–73.

Paul, Tapas, Subhas Chandra Bera, Nidhi Agnihotri, and Padmaja P. Mishra. 2016. "Single-Molecule FRET Studies of the Hybridization Mechanism during Noncovalent Adsorption and Desorption of DNA on Graphene Oxide." *The Journal of Physical Chemistry B* 120 (45): 11628–36. doi: 10.1021/acs.jpcb.6b06017

Pelet, S., M.J.R. Previte, L.H. Laiho, and P.T. C. So. 2004. "A Fast Global Fitting Algorithm for Fluorescence Lifetime Imaging Microscopy Based on Image Segmentation." *Biophysical Journal* 87 (4): 2807–17. doi: 10.1529/biophysj.104.045492

Periasamy, Ammasi, Shagufta R. Alam, Zdenek Svindrych, and Horst Wallrabe. 2017. "FLIM-FRET Image Analysis of Tryptophan in Prostate Cancer Cells." In *European Conference on Biomedical Optics*, Vol. 10414, 1041402-1041402-5. Optical Society of America.

Periasamy, Ammasi, and Robert M. Clegg, eds. 2009. *FLIM Microscopy in Biology and Medicine.* 1 edition. Chapman and Hall/CRC.

Periasamy, Ammasi, and Richard Day. 2011. *Molecular Imaging: FRET Microscopy and Spectroscopy.* Elsevier.

Periasamy, Ammasi, Pawel Wodnicki, Xue F. Wang, Seongwook Kwon, Gerald W. Gordon, and Brian Herman. 1996. "Time-resolved Fluorescence Lifetime Imaging Microscopy Using a Picosecond Pulsed Tunable Dye Laser System." *Review of Scientific Instruments* 67 (10): 3722–31. doi: 10.1063/1.1147139

Philip, Johan, and Kjell Carlsson. 2003. "Theoretical Investigation of the Signal-to-Noise Ratio in Fluorescence Lifetime Imaging." *JOSA* A 20 (2): 368–79. doi: 10.1364/JOSAA.20.000368

Pian, Qi, Ruoyang Yao, Nattawut Sinsuebphon, and Xavier Intes. 2017. "Compressive Hyperspectral Time-Resolved Wide-Field Fluorescence Lifetime Imaging." *Nature Photonics* 11 (7): 411–14. doi: 10.1038/nphoton.2017.82

Popleteeva, Marina, Kalina T. Haas, David Stoppa, Lucio Pancheri, Leonardo Gasparini, Clemens F. Kaminski, Liam D. Cassidy, et al. 2015. "Fast and Simple Spectral FLIM for Biochemical and Medical Imaging." *Optics Express* 23 (18): 23511–25. doi: 10.1364/OE.23.023511

Previte, Michael J. R., Serge Pelet, Ki Hean Kim, Christoph Buehler, and Peter T. C. So. 2008. "Spectrally Resolved Fluorescence Correlation Spectroscopy Based on Global Analysis." *Analytical Chemistry* 80 (9): 3277–84. doi: 10.1021/ac702474u

Radbruch, Helena, Daniel Bremer, Ronja Mothes, Robert Günther, Jan Leo Rinnenthal, Julian Pohlan, Carolin Ulbricht et al. 2015. "Intravital FRET: Probing Cellular and Tissue Function in Vivo." *International Journal of Molecular Sciences* 16 (5): 11713–27. doi: 10.3390/ijms160511713

Rajoria, Shilpi, Lingling Zhao, Xavier Intes, and Margarida Barroso. 2014. "FLIM-FRET for Cancer Applications." *Current Molecular Imaging* 3 (2): 144–61.

Ranjit, Suman, Evgenia Dobrinskikh, John Montford, Alexander Dvornikov, Allison Lehman, David J. Orlicky, Raphael Nemenoff, et al. 2016. "Label-Free Fluorescence Lifetime and Second Harmonic Generation Imaging Microscopy Improves Quantification of Experimental Renal Fibrosis." *Kidney International* 90 (5): 1123–28. doi: 10.1016/j.kint.2016.06.030

Ranjit, Suman, Alexander Dvornikov, Evgenia Dobrinskikh, Xiaoxin Wang, Yuhuan Luo, Moshe Levi, and Enrico Gratton. 2017. "Measuring the Effect of Western Diet on Liver Tissue Architecture by FLIM Autofluorescence and Harmonic Generation Microscopy: Erratum." *Biomedical Optics Express* 8 (7): 3501. doi: 10.1364/BOE.8.003501

Rao, Jianghong, Anca Dragulescu-Andrasi, and Hequan Yao. 2007. "Fluorescence Imaging in Vivo: Recent Advances." *Current Opinion in Biotechnology, Analytical Biotechnology* 18 (1): 17–25. doi: 10.1016/j.copbio.2007.01.003

Redford, Glen I., and Robert M. Clegg. 2005. "Polar Plot Representation for Frequency-Domain Analysis of Fluorescence Lifetimes." *Journal of Fluorescence* 15 (5): 805–15. doi: 10.1007/s10895-005-2990-8

Rinnenthal, Jan Leo, Christian Börnchen, Helena Radbruch, Volker Andresen, Agata Mossakowski, Volker Siffrin, Thomas Seelemann, et al. 2013. "Parallelized TCSPC for Dynamic Intravital Fluorescence Lifetime Imaging: Quantifying Neuronal Dysfunction in Neuroinflammation." *PLOS ONE* 8 (4): e60100. doi: 10.1371/journal.pone.0060100

Roberts, Michael S., Yuri Dancik, Tarl W. Prow, Camilla A. Thorling, Lynlee L. Lin, Jeffrey E. Grice, Thomas A. Robertson, Karsten König, and Wolfgang Becker. 2011. "Non-Invasive Imaging of Skin Physiology and Percutaneous Penetration Using Fluorescence Spectral and Lifetime Imaging with Multiphoton and Confocal Microscopy." *European Journal of Pharmaceutics and Biopharmaceutics* 77 (3): 469–88. doi: 10.1016/j.ejpb.2010.12.023

Rowley, Mark I., Paul R. Barber, Anthony C. C. Coolen, and Borivoj Vojnovic. 2011a. "Bayesian Analysis of Fluorescence Lifetime Imaging Data." In *Multiphoton Microscopy in the Biomedical Sciences XI*, Vol. 7903, 790325-790325-12. International Society for Optics and Photonics.

Rück, Angelika C., S. Kalinina, P. Schäfer, B. von Einem, and C. von Arnim. 2017. "Correlated Oxygen-Sensing PLIM, Cell Metabolism FLIM and Applications." In *Multiphoton Microscopy in the Biomedical Sciences XVII*, edited by Ammasi Periasamy, Peter T. C. So, Karsten König, and Xiaoliang S. Xie, 100691I. International Society for Optics and Photonics.

Sakadžić, Sava, Emmanuel Roussakis, Mohammad A. Yaseen, Emiri T. Mandeville, Vivek J. Srinivasan, Ken Arai, Svetlana Ruvinskaya, et al. 2010. "Two-Photon High-Resolution Measurement of Partial Pressure of Oxygen in Cerebral Vasculature and Tissue." *Nature Methods* 7 (9): 755–59. doi: 10.1038/nmeth.1490

Sakadzic, Sava. 2011. "Cerebral Blood Oxygenation Measurement Based on Oxygen-Dependent Quenching of Phosphorescence." *Journal of Visualized Experiments: JoVE*, no. 51 (May). doi: 10.3791/1694

Santra, Kalyan, Jinchun Zhan, Xueyu Song, Emily A. Smith, Namrata Vaswani, and Jacob W. Petrich. 2016. "What Is the Best Method to Fit Time-Resolved Data? A Comparison of the Residual Minimization and the Maximum Likelihood Techniques As Applied to Experimental Time-Correlated, Single-Photon Counting Data." *The Journal of Physical Chemistry B* 120 (9): 2484–90. doi: 10.1021/acs.jpcb.6b00154

Schlachter, S., A. D. Elder, A. Esposito, G. S. Kaminski, J. H. Frank, L. K. van Geest, and C. F. Kaminski. 2009. "MhFLIM: Resolution of Heterogeneous Fluorescence Decays in Widefield Lifetime Microscopy." *Optics Express* 17 (3): 1557–70. doi: 10.1364/OE.17.001557

Scott, T. Gordon, Richard D. Spencer, Nelson J. Leonard, and Gregorio Weber. 1970. "Emission Properties of NADH. Studies of Fluorescence Lifetimes and Quantum Efficiencies of NADH, AcPyADH, [Reduced Acetylpyridineadenine Dinucleotide] and Simplified Synthetic Models." *Journal of the American Chemical Society* 92 (3): 687–95. doi: 10.1021/ja00706a043

Seidel, Claus A. M., Andreas Schulz, and Markus H. M. Sauer. 1996. "Nucleobase-Specific Quenching of Fluorescent Dyes. 1. Nucleobase One-Electron Redox Potentials and Their Correlation with Static and Dynamic Quenching Efficiencies." *The Journal of Physical Chemistry* 100 (13): 5541–53. doi: 10.1021/jp951507c

Shah, Amy T., and Melissa C. Skala. 2016. "Metabolic Microscopy of Head and Neck Cancer Organoids." In *Multiphoton Microscopy in the Biomedical Sciences XVI SPIE BiOS*, 97120V–97120V. International Society for Optics and Photonics.

Sharman, Kristin K., Ammasi Periasamy, Harry Ashworth, and J. N. Demas. 1999. "Error Analysis of the Rapid Lifetime Determination Method for Double-Exponential Decays and New Windowing Schemes." *Analytical Chemistry* 71 (5): 947–52. doi: 10.1021/ac981050d

Shcheslavskiy, V. I., A. Neubauer, R. Bukowiecki, F. Dinter, and Wolfgang Becker. 2016. "Combined Fluorescence and Phosphorescence Lifetime Imaging." *Applied Physics Letters* 108 (9): 091111. doi: 10.1063/1.4943265

Skala, Melissa C., Jayne M. Squirrell, Kristin M. Vrotsos, Jens C. Eickhoff, Annette Gendron-Fitzpatrick, Kevin W. Eliceiri, and Nirmala Ramanujam. 2005. "Multiphoton Microscopy of Endogenous Fluorescence Differentiates Normal, Precancerous, and Cancerous Squamous Epithelial Tissues." *Cancer Research* 65 (4): 1180–86. doi: 10.1158/0008-5472.CAN-04-3031

Sparks, Hugh, F. Görlitz, D. J. Kelly, S. C. Warren, P. A. Kellett, E. Garcia, A. K. L. Dymoke-Bradshaw, et al. 2017. "Characterisation of New Gated Optical Image Intensifiers for Fluorescence Lifetime Imaging." *Review of Scientific Instruments* 88 (1): 013707. doi: 10.1063/1.4973917

Sparks, Hugh, Sean Warren, Joana Guedes, Nagisa Yoshida, Tze Choong Charn, Nadia Guerra, Taranjit Tatla, Christopher Dunsby, and Paul French. 2015. "A Flexible Wide-Field FLIM Endoscope Utilising Blue Excitation Light for Label-Free Contrast of Tissue." *Journal of Biophotonics* 8 (1–2): 168–78. doi: 10.1002/jbio.201300203

Spencer, Richard D., and Gregorio Weber. 1969. "Measurements of Subnanosecond Fluorescence Lifetimes with a Cross-Correlation Phase Fluorometer*." *Annals of the New York*

Academy of Sciences 158 (1): 361–76. doi: 10.1111/j.1749-6632.1969.tb56231.x

Squirrell, Jayne M., Jimmy J. Fong, Carlos A. Ariza, Amber Mael, Kassondra Meyer, Nirupama K. Shevde, Avtar Roopra, et al. 2012. "Endogenous Fluorescence Signatures in Living Pluripotent Stem Cells Change with Loss of Potency." *PloS One* 7 (8): e43708.

Steinkamp, John A., and Harry A. Crissman. 1993. "Resolution of Fluorescence Signals from Cells Labeled with Fluorochromes Having Different Lifetimes by Phase-Sensitive Flow Cytometry." *Cytometry* 14 (2): 210–16. doi: 10.1002/cyto.990140214

Straub, M., and S. W. Hell. 1998. "Fluorescence Lifetime Three-Dimensional Microscopy with Picosecond Precision Using a Multifocal Multiphoton Microscope." *Applied Physics Letters* 73 (13): 1769–71. doi: 10.1063/1.122276

Strickler, S. J., and Robert A. Berg. 1962. "Relationship between Absorption Intensity and Fluorescence Lifetime of Molecules." *The Journal of Chemical Physics* 37 (4): 814–22. doi: 10.1063/1.1733166

Stringari, Chiara, Robert A. Edwards, Kira T. Pate, Marian L. Waterman, Peter J. Donovan, and Enrico Gratton. 2012. "Metabolic Trajectory of Cellular Differentiation in Small Intestine by Phasor Fluorescence Lifetime Microscopy of NADH." *Scientific Reports* 2 (August): srep00568. doi: 10.1038/srep00568

Suhling, Klaus, Paul M. W. French, and David Phillips. 2005. "Time-Resolved Fluorescence Microscopy." *Photochemical & Photobiological Sciences* 4 (1): 13–22. doi: 10.1039/B412924P

Sun, Yinghua, Rui Liu, Daniel S. Elson, Christopher W. Hollars, Javier A. Jo, Jesung Park, Yang Sun, and Laura Marcu. 2008. "Simultaneous Time- and Wavelength-Resolved Fluorescence Spectroscopy for near Real-Time Tissue Diagnosis." *Optics Letters* 33 (6): 630–32. doi: 10.1364/OL.33.000630

Sun, Yinghua, Jennifer Phipps, Daniel S. Elson, Heather Stoy, Steven Tinling, Jeremy Meier, Brian Poirier, Frank S. Chuang, D. Gregory Farwell, and Laura Marcu. 2009. "Fluorescence Lifetime Imaging Microscopy: In Vivo Application to Diagnosis of Oral Carcinoma." *Optics Letters* 34 (13): 2081–83.

Sun, Yinghua, Jennifer Phipps, Jeremy Meier, Nisa Hatami, Brian Poirier, Daniel S Elson, D Gregory Farwell, and Laura Marcu. 2013. "Endoscopic Fluorescence Lifetime Imaging for In Vivo Intraoperative Diagnosis of Oral Carcinoma." *Microscopy and Microanalysis: The Official Journal of Microscopy Society of America, Microbeam Analysis Society, Microscopical Society of Canada* 19 (May): 1–8. doi: 10.1017/S1431927613001530

Swaminathan, R., and N. Periasamy. 1996. "Analysis of Fluorescence Decay by the Maximum Entropy Method: Influence of Noise and Analysis Parameters on the Width of the Distribution of Lifetimes." *Proceedings of the Indian Academy of Sciences - Chemical Sciences* 108 (1): 39. doi: 10.1007/BF02872511

Szulczewski, Joseph M., David R. Inman, David Entenberg, Suzanne M. Ponik, Julio Aguirre-Ghiso, James Castracane, John Condeelis, Kevin W. Eliceiri, and Patricia J. Keely. 2016. "*In Vivo* Visualization of Stromal Macrophages via Label-Free FLIM-Based Metabolite Imaging." *Scientific Reports* 6 (May): srep25086. doi: 10.1038/srep25086

Talbot, Clifford B., James McGinty, David M. Grant, Ewan J. McGhee, Dylan M. Owen, Wei Zhang, Tom D. Bunney, et al. 2008. "High Speed Unsupervised Fluorescence Lifetime Imaging Confocal Multiwell Plate Reader for High Content Analysis." *Journal of Biophotonics* 1 (6): 514–21. doi: 10.1002/jbio.200810054

Tamborini, Davide, Bojan Markovic, Simone Tisa, Federica Alberta Villa, and Alberto Tosi. 2013. "TDC with 1.5% DNL Based on a Single-Stage Vernier Delay-Loop Fine Interpolation." In *2013 IEEE Nordic-Mediterranean Workshop on Time-to-Digital Converters (NoMe TDC)*, 1–6. IEEE.

Tejwani, Vijay, Franz-Josef Schmitt, Svea Wilkening, Ingo Zebger, Marius Horch, Oliver Lenz, and Thomas Friedrich. 2017. "Investigation of the NADH/NAD+ Ratio in Ralstonia Eutropha Using the Fluorescence Reporter Protein Peredox." *Biochimica et Biophysica Acta (BBA) - Bioenergetics* 1858 (1): 86–94. doi: 10.1016/j.bbabio.2016.11.001

Tregidgo, Carolyn, James A. Levitt, and Klaus Suhling. 2008. "Effect of Refractive Index on the Fluorescence Lifetime of Green Fluorescent Protein." *Journal of Biomedical Optics* 13 (3): 031218. doi: 10.1117/1.2937212

Trinh, Andrew L., Hongtao Chen, Yumay Chen, Yuanjie Hu, Zhenzhi Li, Eric R. Siegel, Mark E. Linskey, Ping H. Wang, Michelle A. Digman, and Yi-Hong Zhou. 2017. "Tracking Functional Tumor Cell Subpopulations of Malignant Glioma by Phasor Fluorescence Lifetime Imaging Microscopy of NADH." *Cancers* 9 (12): 168. doi: 10.3390/cancers9120168

Turton, David A., Gavin D. Reid, and Godfrey S. Beddard. 2003. "Accurate Analysis of Fluorescence Decays from Single Molecules in Photon Counting Experiments." *Analytical Chemistry* 75 (16): 4182–87. doi: 10.1021/ac034325k

Vecer, Jaroslav, and Petr Herman. 2011. "Maximum Entropy Analysis of Analytically Simulated Complex Fluorescence Decays." *Journal of Fluorescence* 21 (3): 873–81. doi: 10.1007/s10895-009-0589-1

Verveer, P. J., and P. I. H. Bastiaens. 2003. "Evaluation of Global Analysis Algorithms for Single Frequency Fluorescence Lifetime Imaging Microscopy Data." *Journal of Microscopy* 209 (1): 1–7. doi: 10.1046/j.1365-2818.2003.01093.x

Verveer, Peter J., and Quentin S. Hanley. 2009. "Chapter 2 Frequency Domain FLIM Theory, Instrumentation, and Data Analysis." In *Laboratory Techniques in Biochemistry and Molecular Biology*, Vol. 33, 59–94. Fret and Flim Techniques. Elsevier.

Verveer, Peter J., Anthony Squire, and Philippe IH Bastiaens. 2000. "Global Analysis of Fluorescence Lifetime Imaging Microscopy Data." *Biophysical Journal* 78 (4): 2127–37.

Vicidomini, Giuseppe, Ivan Coto Hernández, Marta d'Amora, Francesca Cella Zanacchi, Paolo Bianchini, and Alberto Diaspro. 2014. "Gated CW-STED Microscopy: A Versatile Tool for Biological Nanometer Scale Investigation." *Methods, Advanced Light Microscopy* 66 (2): 124–30. doi: 10.1016/j.ymeth.2013.06.029

Vitali, Marco, Fernando Picazo, Yury Prokazov, Alessandro Duci, Evgeny Turbin, Christian Götze, Juan Llopis, et al. 2011. "Wide-Field Multi-Parameter FLIM: Long-Term Minimal Invasive Observation of Proteins in Living Cells." *PLOS ONE* 6 (2): e15820. doi: 10.1371/journal.pone.0015820

Vogel, Steven S., Tuan A. Nguyen, Paul S. Blank, and B. Wieb van der Meer. 2015. "An Introduction to Interpreting Time Resolved Fluorescence Anisotropy Curves." In *Advanced*

Time-Correlated Single Photon Counting Applications, 385–406. Springer Series in Chemical Physics. Springer.

Vogel, Steven S., Christopher Thaler, Paul S. Blank, and Srinagesh V. Koushik. 2009. "Time Resolved Fluorescence Anisotropy." *FLIM Microscopy in Biology and Medicine* 1: 245–88.

Wahl, Michael. 2014. "Modern TCSPC Electronics: Principles and Acquisition Modes." In *Advanced Photon Counting*, edited by Peter Kapusta, Michael Wahl, and Rainer Erdmann, 1–21. Springer Series on Fluorescence 15. Springer International Publishing.

Wahl, Michael, Tino Röhlicke, Hans-Jürgen Rahn, Rainer Erdmann, Gerald Kell, Andreas Ahlrichs, Martin Kernbach, Andreas W. Schell, and Oliver Benson. 2013. "Integrated Multichannel Photon Timing Instrument with Very Short Dead Time and High Throughput." *The Review of Scientific Instruments* 84 (4): 043102. doi: 10.1063/1.4795828

Walsh, Alex J., Cook, Rebecca S., Charles Manning H., Hicks, Donna J., Lafontant A., Arteaga Carlos L., Skala, Melissa C. 2013. "Optical Metabolic Imaging Identifies Glycolytic Levels, Subtypes, and Early-Treatment Response in Breast Cancer." *Cancer Res* 73 (20): 6164–6174. doi: 10.1158/0008-5472.CAN-13-0527

Warburg, Otto. 1956. "On the Origin of Cancer Cells." *Science* 123 (3191): 309–14.

Warren, Sean C., Anca Margineanu, Dominic Alibhai, Douglas J. Kelly, Clifford Talbot, Yuriy Alexandrov, Ian Munro, et al. 2013. "Rapid Global Fitting of Large Fluorescence Lifetime Imaging Microscopy Datasets." *PloS One* 8 (8): e70687. doi: 10.1371/journal.pone.0070687

Webb, S. E. D., Y. Gu, S. Lévêque-Fort, J. Siegel, M. J. Cole, K. Dowling, R. Jones, et al. 2002. "A Wide-Field Time-Domain Fluorescence Lifetime Imaging Microscope with Optical Sectioning." *Review of Scientific Instruments* 73 (4): 1898–907. doi: 10.1063/1.1458061

Weber, G. 1948. "The Quenching of Fluorescence in Liquids by Complex Formation. Determination of the Mean Life of the Complex" 44: 185–89. doi: 10.1039/TF9484400185

Weidtkamp-Peters, Stefanie, Suren Felekyan, Andrea Bleckmann, Rüdiger Simon, Wolfgang Becker, Ralf Kühnemuth, and Claus A. M. Seidel. 2009. "Multiparameter Fluorescence Image Spectroscopy to Study Molecular Interactions." *Photochemical & Photobiological Sciences* 8 (4): 470–80. doi: 10.1039/B903245M

Widengren, Jerker, Volodymyr Kudryavtsev, Matthew Antonik, Sylvia Berger, Margarita Gerken, and Claus A. M. Seidel. 2006. "Single-Molecule Detection and Identification of Multiple Species by Multiparameter Fluorescence Detection." *Analytical Chemistry* 78 (6): 2039–50. doi: 10.1021/ac0522759

Yahav, Gilad, Eran Barnoy, Nir Roth, Lior Turgeman, and Dror Fixler. 2017. "Reference-Independent Wide Field Fluorescence Lifetime Measurements Using Frequency-Domain (FD) Technique Based on Phase and Amplitude Crossing Point." *Journal of Biophotonics* 10 (9): 1198–207. doi: 10.1002/jbio.201600220

Yaseen, Mohammad A., Sava Sakadžić, Weicheng Wu, Wolfgang Becker, Karl A. Kasischke, and David A. Boas. 2013. "In Vivo Imaging of Cerebral Energy Metabolism with Two-Photon Fluorescence Lifetime Microscopy of NADH." *Biomedical Optics Express* 4 (2): 307. doi: 10.1364/BOE.4.000307

Yaseen, Mohammad A., Vivek J. Srinivasan, Iwona Gorczynska, James G. Fujimoto, David A. Boas, and Sava Sakadžić. 2015. "Multimodal Optical Imaging System for in Vivo Investigation of Cerebral Oxygen Delivery and Energy Metabolism." *Biomedical Optics Express* 6 (12): 4994–5007. doi: 10.1364/BOE.6.004994

Zeng, Hualing, and Gilles Durocher. 1995. "Analysis of Fluorescence Quenching in Some Antioxidants from Non-Linear Stern—Volmer Plots." *Journal of Luminescence* 63 (1): 75–84. doi: 10.1016/0022-2313(94)00045-E

Zhao, Lingling, Ken Abe, Margarida Barroso, and Xavier Intes. 2013. "Near Infrared FRET Using Wide-Field Fluorescence Lifetime Imaging in Live Animals." In *Novel Biophotonic Techniques and Applications II (2013), Paper 88010A*, 88010A. Optical Society of America.

Zhong, Wei, Paul Urayama, and Mary-Ann Mycek. 2003. "Imaging Fluorescence Lifetime Modulation of a Ruthenium-Based Dye in Living Cells: The Potential for Oxygen Sensing." *Journal of Physics D: Applied Physics* 36 (14): 1689–95. doi: 10.1088/0022-3727/36/14/306

Section II

Imaging Cellular Behavior

10

Imaging Cell Metabolism: Analyzing Cellular Metabolism by NAD(P)H and FAD

Ruofan Cao, Horst Wallrabe, Karsten H. Siller, and Ammasi Periasamy

CONTENTS

10.1 Introduction

Evolution has provided life on earth with several methods of generating cellular energy. None is more complex than the metabolism of mammalian cells. Carbohydrate and fatty acid starting materials are converted by cascades of biochemical reactions to produce substrates for glycolysis, citric acid (Krebs) cycle, and oxidative phosphorylation (OXPHOS) in the electron transport chain (ETC) of the mitochondria, resulting in the energy molecule adenosine triphosphate (ATP) (Zheng, 2012). Critical effectors are the coenzymes NAD(P)H (reduced nicotinamide adenosine dinucleotide (phosphate)[H]) and FAD (oxidized flavin adenine dinucleotide), the former being present in cytosolic glycolysis and mitochondrial citric acid and ETC/OXPHOS, the latter mainly in the mitochondria. Both coenzymes are autofluorescent in their reduced (NAD(P)H) and oxidized (FAD) forms and, therefore, suitable targets for light microscopy exploration (Chance, 1996). Disturbed and/or remodeled cellular metabolism is present in a number of human pathologies, e.g. cancer, neurodegenerative diseases, diabetes, and inflammatory processes, to name a few, and monitoring these changes provides critical information for diagnosis and disease progression (Koppenol et al., 2011; Lin and Beal, 2006; King and Blom, 2017; Eguchi et al., 2012). One of the markers of metabolic health is the cellular reduction/oxidation (redox) state and quantification of the interrelationship between glycolysis and OXPHOS (Gibson, 2005).

In this chapter, we will discuss and provide details on how best to apply fluorescence lifetime microscopy (FLIM) to those autofluorescent cellular components and analyze the data by individual, segmented cells using HeLa cervical cancer cells as a model. Best-practice imaging and lifetime fitting processes using SPCImage software and challenges are described first, followed by image processing using NAD(P)H and FAD as examples; more detailed biological background is provided later, in which quantitative analysis elucidates the effects of anticancer treatment, cell-by-cell response, and OXPHOS vs. glycolysis/mitochondrial vs. cytosolic data, as per Sections 10.6–10.9.

10.2 Multiphoton Fluorescence Lifetime Imaging Microscopy (MP-FLIM), Time-Correlated Single Photon Counting (TCSPC)

Applications of FLIM have grown exponentially in a broad range of life sciences and industrial fields, a reflection of specific advantages over intensity-based microscopy (Periasamy and Clegg, 2009; Ghukasyan and Heikal, 2014; Becker, 2015a; Marcu, French, and Elson, 2014; Meleshina et al., 2016). Importantly, fluorescence lifetime is independent of fluorophore concentration, which makes it a valuable tool for quantitative studies in scattering and absorbing samples. One- and two-photon excitation microscopy configuration has been widely used for FLIM imaging (Abe et al., 2013). Multiphoton FLIM is an added advantage over one-photon FLIM, as MP-FLIM provides less autofluorescence from cells and tissues. Multiphoton excitation conveniently excites molecules that would otherwise require excitation in the ultraviolet (UV) region, generally injurious to live cells at longer exposure. Both, frequency-domain and time-domain FLIM methods have been applied (Lakowicz, 1988; Stringari et al., 2011; Ruck et al., 2014; Cao et al. 2013). The technical aspects have been described with further detail in the previous chapter. Here, we use time domain, also called time-correlated single photon counting (TCSPC) (Becker et al., 2014). A Zeiss LSM-780 NLO confocal/multiphoton microscopy system consists of an inverted Axio Observer (Zeiss) microscope, motorized stage for automated scanning, Chameleon Vision-II (Coherent Inc.) ultrafast Ti:sapphire laser with dispersion compensation to deliver shorter pulses at the specimen plane (690–1060 nm, 80 MHz, 150 fs) for multiphoton excitation, and a standard set of dry and immersion objectives. Two HPM-100-40 hybrid GaAsP detectors (Becker and Hickl) are connected to the nondescanned (NDD) port of the microscope using two T-adapters (Zeiss) with proper dichroics and band pass filters to collect as much fluorescence as possible in the spectral ranges Ch1: 460–500 nm (NAD(P)H) and Ch2: 520–560 nm (FAD). NADH and FAD channels also contain 690 nm short pass filter (Zeiss) in the beam path. Two SPC-150 cards (Becker and Hickl) synchronized with the pulsed laser and the Zeiss LSM-780 scan head signals collect the time-resolved fluorescence in TCSPC mode using SPCM (9.74) acquisition software.

TABLE 10.1

Excitation and Emission Settings for NAD(P)H and FAD Channels

Channels	Excitation Wavelength	Emission Filters
NAD(P)H	740 nm	480/40 nm
FAD	890 nm	540/40 nm

10.3 Optimizing Autofluorescent NAD(P)H and FAD FLIM Acquisition

As has been pointed out here and in the preceding chapter, FLIM is a powerful tool for the life sciences, but its application requires careful optimization of many variables, which we will address in this section.

10.3.1 Excitation Wavelengths, Filters, Laser Power vs. Acquisition Times

Optimization of all variables, not just imaging, is the key factor for successful experimentation. As shown in Table 10.1, the two autofluorescent cellular targets of interest (NAD(P)H, FAD) all absorb light in the UV spectrum, which is not desirable for live specimens; therefore, 2-photon FLIM is a better option than 1-photon intensity acquisition. With a 2-detector TCSPC system, the autofluorescent signals of NAD(P)H can be captured at 740 nm excitation, and, for FAD, the 2-photon excitation wavelength is switched to 890 nm for optimal FLIM acquisition. Because there is potential emission overlap, which would contaminate (bleed-through) each other's signal, correct emission filters are chosen to minimize this effect while not greatly affecting the signal, which is achieved with the NAD(P)H band-pass (BP) filter at 480/40 nm and FAD at 540/40 nm, respectively (Table 10.1).

To optimize photon counts, necessary for correct off-line fitting processing (although fitting on-the-fly is possible), a trade-off between acquisition time and laser power level is unavoidable. For live specimens, the lowest laser power is desirable to avoid photodamage. To collect enough photons for NAD(P), we have found the acquisition time at 60 seconds at 2% (7 mW) power of 740 nm to be optimal. FAD signals are also acquired at 60 seconds and ~7 mW average laser power of 890 nm. There is an ongoing debate about what is a sufficient photon count; we will return to this subject in Section 10.4.2. A step-wise image acquisition sequence is described in the supplement of reference (Wallrabe et al., 2018a).

10.3.2 Testing Possibility of Photodamage

Long acquisition time and high excitation intensity can result in photobleaching (Schleusener et al., 2017), photo-damage (Melis, 1999) and other cellular defects (Rounds, 1965). This can be tested by simultaneously taking bright-field images on the confocal instrument during FLIM acquisition to look for any "blebbing" or other morphological changes in cells. This is equally relevant to the previous point when optimizing acquisition time vs. laser power. We tested the effects of extended acquisition times at constant laser power (7 mW for NAD(P)H and 7 mW for FAD) on lifetime parameters (results not shown); in short, lifetimes did not change, even after 5 minutes of acquisition, showing that any lifetime parameter changes – e.g. after treatment under our imaging conditions – are driven by biological events in the specimen. Signs of photodamage become apparent when reduction in photon counts (intensity) occur and/or morphology changes (blebbing) live cells. To avoid this kind of issue, we suggest decreasing the laser power and increasing the acquisition time to protect the live specimens.

10.3.3 Pretesting Fixed Cells vs. Live

The effects on lifetime of fixing specimens with formaldehyde or methanol have been published with mixed results for fluorescently labeled proteins, depending on types of buffers and/or mounting media (Joosen et al., 2014). Autofluorescent lifetimes were compared in formalin-fixed and fresh ovarian cancer tissue sections, showing some increase in fluorescence but generally little impact on FAD lifetimes (Adur et al., 2012). Another publication found that the fundamental properties of NADH/FAD endogenous fluorescence were preserved during formalin fixation, paraffin-embedding, and histology slide preparation (Conklin et al., 2009). The previous references are based on tissue sections; we instead have compared live and fixed HeLa cells and acquired FLIM images for NAD(P)H and FAD at conditions described in this chapter. Our results showed a noticeable change in lifetime after fixing, accompanied by an intensity change of NADH and FAD (Figure 10.1). Clearly, autofluorescent lifetime signals can be obtained successfully in fixed samples but, depending on the scientific objectives, need to be compared with other standards (e.g. histology) or live controls. As a principle, live specimens should be used for FLIM imaging.

10.3.4 Exploring the Application of "Stitching" and "Z-Stacks"

For large tissue sections, FLIM images can be recorded by a series of image frames that are then "stitched" into one image; this is a function of the mosaic Becker and Hickl (B&H) FLIM software combined with the Zeiss 780 Zen system. From an image-acquisition point of view, this "stitching" FLIM facility is very useful for tissue samples, as tissue samples usually require larger area analysis.

Another attractive application of mosaic FLIM is to generate a z-stack of several focal planes. For this application, the transition from one mosaic element to the next is associated with the confocal z-scanning software. Z-stack FLIM has versatile uses. For example, cell layers rarely come in identical surface heights, and the focal plane of any cell in a field of view (FOV) will likely be positioned at slightly different height. While this may not matter much statistically, taking focal sections (z-stacks) provides a method to test this hypothesis. Having evaluated the potential risk of photodamage (Section 10.3.2), a limited number of focal planes above and below the brightest one can be taken and matched by photon counts and quantitatively analyzed. An additional benefit arises when several FOV are to be reimaged after treatment (see Section 10.3.5); here, treatment focal planes can be chosen to best match the morphology of the control. Likewise, time-lapse images can be mapped into element sequences through mosaic FLIM. Thus, a FLIM video can be created. Moreover, by combining z-stack and time-lapse mosaic FLIM, a 3D FLIM video is possible.

10.3.5 The Choice of Imaging Same FOVs for Controls and Treatment

If experimental objectives allow, and it is part of the design, multiple control FOVs can be taken and reimaged at identical x-y stage coordinates and at different points in time, such as taking control images followed by treatment on a microscope stage. Achieving the exact focal plane (z) of the control can be a minor challenge. With practice and putting the control image on the screen to compare, this becomes manageable. An additional step is to take a small z-stack to match focal planes at the analysis stage based on morphology (see Section 10.3.4). The major benefit of generating identical FOVs lies in the ability to segment individual cells, follow them through the treatment phases, and quantitate their heterogeneous (if any) response to interventions (see Figure 10.2).

In addition to a treatment time course experiment, a control group that is imaged under identical conditions but receives no treatment is recommended. This control is designed to highlight any effects of confounding variables such as time

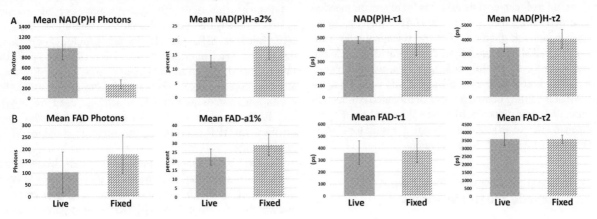

FIGURE 10.1 Live-cell control vs. formaldehyde fixing on microscopy stage, comparing identical FOVs. (a) NAD(P)H and (b) FAD show lifetime variable parameter changes (Cao et al., 2020).

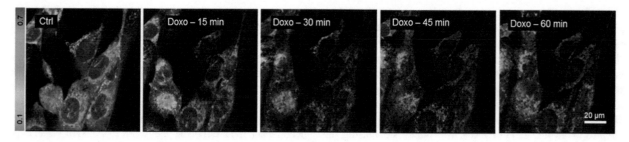

FIGURE 10.2 An example of following cells within the same FOV over time after treatment. Cell segmentation provides additional information about potential heterogeneity of response to treatment (Wallrabe et al., 2018).

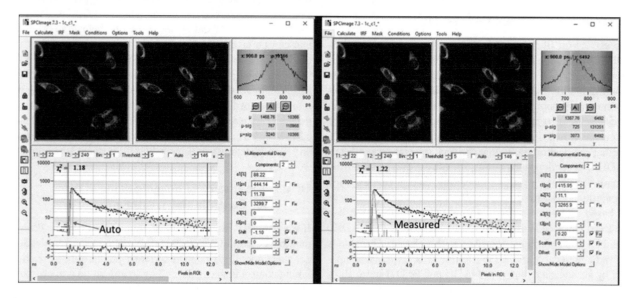

FIGURE 10.3 An example of auto IRF (left) vs. measured IRF (right). The lifetime distribution, lifetime images, and even lifetime parameters in the two cases are very close (Cao et al., 2020).

or environmental factors; thus, it becomes a tool to optimize experimental conditions because temperature control and gas mixes are required in live-cell imaging.

10.4 FLIM-Fitting Process in B&H SPCImage Software

The B&H handbook is the ultimate source of information and a treasure trove of technical details. It is highly recommended for those dedicating their careers to TCSPC microscopy to study this handbook diligently. For life scientists who want to apply this FLIM technique, this section provides hands-on, practical steps for successful implementation, without claiming to provide a comprehensive picture of this versatile technique. All of the fitting parameters (see highlighted parameters in Figure 10.4) can influence lifetime values and need to be optimized to achieve an acceptable best fit χ^2 result; for clarity, we have dealt with each of them separately.

10.4.1 Instrument Response Function, Automatic and/or Measured?

The fluorescence lifetime parameters are derived from a model containing lifetime components and fractions used

to fit the measured data. The measured decay data are fitted with the convolution of the model function and the instrument response function (IRF). Therefore, an IRF is required to get the fluorescence lifetime parameters from the model function. Usually a deconvolution step is involved. IRF can be measured and, in certain situations, should be measured; alternatively, the automatic software function can be used. The recommendation made in this section for automatic IRF follows that of the B&H handbook.

For measuring IRF in 2p systems, urea crystals are suggested to measure second harmonic generation (SHG) as IRF. Because SHG signals do not provide fluorescence decay, the SHG approach is ideal for IRF for 2p-FLIM. For 1p-FLIM, different light scatter must be used to avoid any fluorescence decays in that IRF (Sun et al., 2009). The excitation sources and emission settings used for NADH and FAD channels are listed in Table 10.1. Under these settings, the IRF is recorded the same as standard FLIM experiments.

The IRF is the function, the FLIM system would record when an excitation pulse is measured directly. This is determined by the quality of excitation source, detector(s) response, and instrument parameters. The IRF in modern systems has a much smaller width and clear shape compared to fluorescence lifetime and can be viewed as an impulse function in

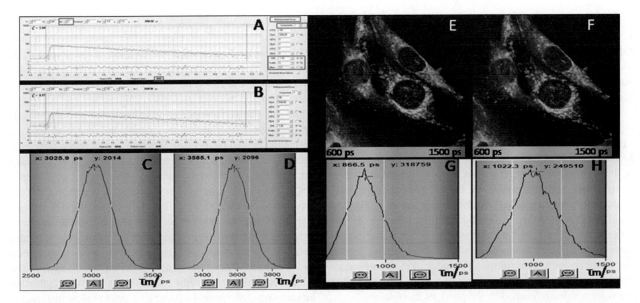

FIGURE 10.4 FLIM fitting model and parameter settings using ATTO 425 solution as a FLIM standard. A. Multiexponential model and fitting parameters. B. Incomplete model and fitting parameters. C. Multiexponential model results. D. Incomplete model results. E. NAD(P)H FLIM image of HeLa cells using multiexponential model. F. NAD(P)H FLIM image of HeLa cells using incomplete model. G. HeLa NAD(P)H multiexponential model tm histogram. H. HeLa NAD(P)H incomplete model tm histogram (Cao et al., 2020).

most cases. The B&H FLIM software can generate an IRF (auto IRF) from the data themselves and uses this IRF for fitting process. A comparison of auto IRF and measured IRF is given in Figure 10.3, showing that results are comparable; the green trace (IRF) below the peak decay curve should be coincidental with this peak and ideally have a narrow base. The left vertical bar/marker should be moved to the base of the rising decay curve, if necessary. The lifetime distribution, lifetime images, and even lifetime parameters in each pixel are identical. For the most accurate analysis, it is highly recommended that FLIM users apply the measured IRF procedure. This is important as different FLIM setups may have different after-pulsing tails in the measured IRF (unlike the computer-generated one). A different FLIM setup may also suffer from internal reflection of the excitation pulses that may skew the rise time of the decay. However, when the auto IRF does not show much difference, as demonstrated in our results, an auto IRF can be used, accompanied by taking periodic IRF measurements with urea crystals.

10.4.2 Fitting Models and Parameters

In FLIM image analyses, the interconnected parameters for optimal fitting have to be established (Figure 10.4) (a) The model – multiexponential vs. incomplete multiexponential; (b) Offset – float vs. fixed; (c) Scatter - float vs. fixed; (d) Bin – as low as possible, adjusted for sufficient photon count; (e) Shift – float first for optimal χ^2, then fixed at the level of optimal χ^2.

Model: Multiexponential or Incomplete Multiexponential? The standard multiexponential decay model will fit the data with a single exponential decay function or the sum of two or three exponential functions. The incomplete-decay model will do the same, but it includes the fluorescence remaining from the decay of previous laser pulse in the model

function. Using the incomplete model is recommended for high repetition lasers (Ti:sapphire) and low offset detectors (hybrid detectors) (Becker, 2015b). Of course, ambient light must be avoided in all cases. For the short fluorescence lifetime (when there are negligible signals on the left side of the decay rising edge), both incomplete and multiexponential models work. For the long fluorescence lifetime, the incomplete multiexponential model is suggested. To simplify, we suggest using the incomplete model for general experiments when there is negligible background noise (ambient light, electronic noise, etc.) contribution.

A comparison of the multiexponential decay model and the incomplete decay model is shown in Figure 10.4. The fluorophore ATTO 425 in solution is used as a FLIM standard, where the fluorescence lifetime is known to be 3.6 ns. Figure 10.4a shows the decay trace and fitting parameters using the multiexponential decay model. Figure 10.4b shows the decay trace and fitting parameters using the incomplete decay model. In the decay trace illustrated by Figure 10.4a and 10.4b, there is a noticeable contribution from fluorescence excited by the previous laser pulses on the left side of the rising edge of the fluorescence pulse in both cases. The multiexponential decay model interprets this signal as an offset. As a result, too short a fluorescence lifetime is calculated because a fraction of the photons in the tail of the decay curve is interpreted as background photons, and an incorrect result is generated. This flexible offset also results in a broader distribution, which is shown in Figure 10.4c; the peak lifetime distribution is incorrectly displayed at ~3 ns. The incomplete decay model interprets the signal correctly at 3.6 ns, the known lifetime of the ATTO 425 FLIM standard, and the fixed offset generates a more robust distribution shown in Figure 10.4d. Figure 10.4e–h shows the HeLa FLIM results. Figure 10.4e–g shows the multiexponential FLIM tm (amplitude weighted average lifetime = $(\tau1*a1)+(\tau2*a1)/(a1+a2))$ image and histogram.

Figure 10.4f and 10.4h are the incomplete FLIM tm image and histogram. Comparing Figure 10.4h with 10.4g, we see the different results using the two models. The multiexponential model generates 800 ps τ_m, and the incomplete model generates 1000 ps τ_m. Based on standard ATTO 425 experiment, the incomplete model should generate the right results.

Components. It is mostly known from the literature whether a particular cellular target or fluorophore has a single or multiple decay character. This can be tested by changing the component level. In a one-component decay, one amplitude (a1) will dominate when fitted with a two-component decay, or alternatively, both lifetimes ($\tau1$ & $\tau2$) will be relatively close.

Offset ('baseline offset of the decay curve'). For the multiexponential decay model, the baseline offset should be unfixed, i.e. allowed to float; the baseline offset in the incomplete multiexponential model should be fixed to zero so that the signal on the left side of the rising edge of the fluorescence decay is interpreted as the contribution from the unfinished fluorescence decay, excited by the previous laser pulse.

Scatter. "Scatter" is the amount of scattered excitation light detected or the amount of other unspecified emission, such as second harmonic generation in tissues. Scatter is used to fit a SHG or scattering component in multiphoton FLIM data. For one- or multiphoton FLIM without SHG or scattering, it should be fixed to zero

Binning/Photon Counts. Binning of photon counts is the combination of adjacent pixels into one for better signal to noise ratio (S/N). Binning is used to increase the number of photons by combing the adjacent pixels (e.g. Bin1=9 pixels, Bin2=25 pixels), reaching an acceptable fitting level, but reducing the spatial resolution. Generally, binning for NAD(P)H and FAD is set to Bin1 or 2 to achieve a photon count/pixel of ~400 in the time channel of the maximum of fluorescence decay, fitted at two components. For specimens at one component, fitting a level of 100 or just below is usually sufficient.

Shift and χ^2. These are relative positions between IRF and the fluorescence decay curve. After all the aforementioned fitting parameters are determined, the "shift" value is then fixed to meet the best χ^2 by testing several locations on cell morphologies. A good fitting is indicated by the χ^2 number – the closer the χ^2 is to 1, the better the fitting (we suggest a range of 0.7–2 for good fitting).

Lifetime results of the first image are calculated, followed by batch calculation of the remaining images. If batch processing is used, we recommend checking individual processed images to make sure all the images produce acceptable χ^2. Unacceptable χ^2 images must be reprocessed individually. The FLIM data are then exported for further analysis.

10.4.3 Phasor Plots

While phasor plots are usually associated with frequency domain FLIM (Cao et al., 2016), TCSPC data can also be plotted as phasor plots through Fourier transform. Phasor plots are a tool used to visualize the fluorescence lifetime distribution "on-the-fly." As discussed in Chapter 9, within a phasor plot, a semicircle with a radius 0.5 and centered at [0.5, 0] is plotted as a reference. Monoexponential fluorescence lifetime pixels appear directly on the semicircle while multiexponential

fluorescence lifetime pixels appear inside the semicircle of the phasor plot (Redford and Clegg, 2005). In the case of a two-components lifetime system, the location of the pixel will be on the line between the locations of the two single components. In a nutshell, the FLIM parameters' (lifetimes and fractions) changes can be mapped into different locations, and correlated populations can be mapped in one phasor plot by different colors. Biological autofluorescence usually has multiple lifetimes, corresponding to bound and unbound components. Using the phasor plot does not require a fitting process; it allows for quick interpretation of the data showing the changes of lifetimes and fractions. For example, in Figure 10.5, the red population represents the NAD(P)H FLIM data in HeLa cells without treatment, and the green population after 60 min doxorubicin treatment. The line intersecting with the semicircle generates two points with lifetimes 3.9 ns and 0.4 ns, which corresponds to the bound ($\tau2$) and unbound NAD(P)H ($\tau1$) lifetimes. The data population moves along the line to the left ($\tau2$) indicating an increase of the bound NAD(P)H fractions after treatment.

10.4.4 Data Export Choices

Once the fitting process is completed, data for single FOVs or batch export is chosen, the latter is most convenient. At any time later, fitted img files can be reloaded, and data for any parameter can be exported; we routinely export everything to spot-check or investigate parameters that were not originally thought to be a target of the analysis rather than going back to export left-out data categories later.

10.5 Processing SPCImage: Exported Data in ImageJ/Fiji Software

In this section several connected, but partially manual and separate steps are described: (1) assembly of NAD(P)H intensity images into .tiff stacks for batch regions of interest (ROI) selection (Figure 10.6a); (2) manually zeroed or thresholded nuclear signals for each image in the stack; generating a set of ROIs manually in Fiji with ROI manager by segmenting each cell in each FOV (Figure 10.6b); (3) normalization of photon images in the stack to equalize varying intensity levels so that later the more prominent mitochondrial signals can be separated from lower intensity cytosolic signals (Figure 10.6c); (4) creation of two sets (mitochondrial and cytosolic) of 1 or 4 pixel ROIs by thresholding with custom plugin for each FOV in the stack; (5) allocating ROIs generated in (4) above to each segmented cell ROI; (6) finally, all FLIM data in the selected ROIs are exported.

10.5.1 Preparing Photon Images for ROI Selection

NAD(P)H intensity ("photon") images are extracted from the exported data files, copied, and saved in a new folder; all individual intensity images are converted into a .tiff stack for batch-processing to avoid having to deal with each individual image later. Next, nuclear signals are removed manually (Fiji – freehand selection tool, Edit > Clear) to exclude nonmetabolic signals. Figure 10.6 shows the next steps. To

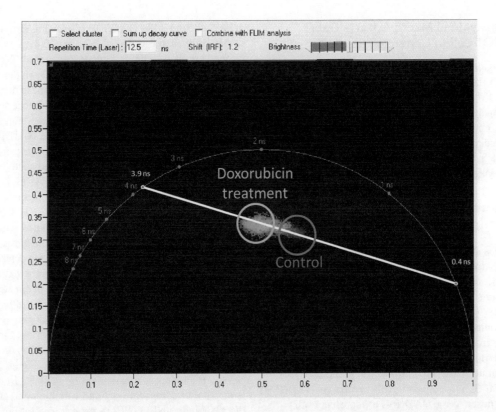

FIGURE 10.5 An example of NAD(P)H fluorescence lifetime phasor plot. The location [1, 0] represents 0 ns, and the location of origin [0, 0] represents an infinite lifetime. Fluorescence lifetimes along the semicircle increase counterclockwise to the left. In this coordinate system, the measured value is the intensity-weighted average of the lifetime components. The red population represents the NAD(P)H FLIM data in HeLa cells without treatment, and the green population after doxorubicin (Cao et al., 2020) treatment.

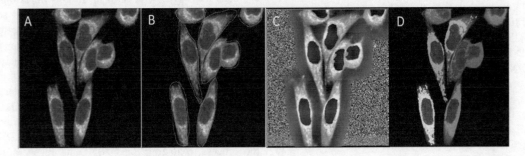

FIGURE 10.6 Process of ROI selections. A. ROI selections start with exported NAD(P)H intensity images. B. Segmented cell ROIs are generated. C. To overcome the inhomogeneous gray levels in different cells, the images are normalized. D. 1–4 pixels ROIs are generated by a Fiji plugin and thresholding to capture a particular morphology by FOV, in this case, the mitochondria and cytosol based on normalized gray levels. Both ROI sets are allocated in two steps to the segmented cell ROIs (Cao et al., 2019).

overcome the inevitable intensity differences between images in the stack – particularly in a time series – and still be able to pick-up all morphological features with specific threshold ranges, the stack intensities are "normalized" (Fiji: Plugins > Integral Image Filters > Normalize Local Contrast). We are now ready to select ROIs.

10.5.2 Selecting Three Sets of ROIs: Mitochondrial, Cytosolic, and Segmented Cells

All three ROI sets are based on the same normalized image stack described in the previous section. While some Fiji segmentation protocols might work, we have chosen to segment individual cell boundaries manually with the ROI Manager that is saved in a separate folder. Mitochondrial and cytosolic ROI sets are generated separately in two steps for each complete FOV in the stack. NAD(P)H signals occupy the brightest image locations and are easily separated from cytosolic data by setting the low threshold matching mitochondrial morphology. For the cytosolic signals, the upper threshold is a couple of gray-level values below the low threshold of the mitochondrial range, while the lower threshold is set well above background. ROI sets are saved in separate folders. Finally, mitochondrial and cytosolic ROIs are allocated and merged into the segmented cell ROIs; a small number may fall outside the segmented boundaries and are not analyzed. These two final ROI sets by individual cell – mitochondrial and cytosolic – become the basis for all data analysis headings.

10.5.3 An ImageJ/Fiji Macro Extracts Data by ROI/Cell to Generate "Results"

A custom ImageJ/Fiji plugin applies the ROI selections to the exported FLIM data and creates data sets by ROI within FOVs for some 50 FLIM parameters in the form of .txt "Results." These results can be analyzed for each compartment (whole-cell without nuclear region, mitochondria, cytosol) for any chosen subpopulation with any suitable quantitation software (EXCEL, MatLab, "R", Python, etc.). We have written a custom Python code for the final analysis, which is applied to the Sections 10.7–10.9.

10.6 NAD(P)H and FAD Biological Background

The coenzymes NADH and FAD are involved in catabolic reactions of amino acid and fatty acid oxidation, glycolysis, citric acid cycle, and in electron transport chain (ETC), which ultimately results in energy generation of ATP by oxidative phosphorylation (OXPHOS). NADPH is mainly involved in anabolic reactions, which use energy for biosynthesis; other NADPH roles are in excessive reactive oxygen species (ROS) defense (Forstermann, 2008). As NADH and NADPH cannot be readily separated based on their spectral properties (Blacker and Buchen, 2016), it is conventionally written as NAD(P)H; we have also chosen to use NAD(P)H to represent both.

Mitochondrial OXPHOS activity consumes NADH (increased NADH-enzyme-bound fraction) and produces FAD (diminished FAD enzyme-bound, quenched fraction). Both the coenzymes in their reduced (NAD(P)H and $FADH_2$) and oxidized (NAD(P)$^+$ and FAD) forms participate in the cellular oxidation-reduction reactions critical for cell physiology (Skala et al., 2007; Alam et al., 2017). Tracking the autofluorescent signals of the coenzymes NAD(P)H and FAD in combination with an intensity-based FAD/NAD(P)H redox ratio has been well established by Chance et al. (1979) as a basis to measure the overall redox state in cells, as the FAD/NAD(P)H are near oxidation-reduction equilibrium, the ratio of the two fluorescence intensities, suitably normalized, approximates the *in vivo* oxidation-reduction, which offers a foundation for the resolution of the redox states in 2- and 3-dimensions. Unfortunately, light scattering and absorption – especially in tissue specimens – makes intensity-based methods problematic or unusable. We expanded the common FLIM assay parameters by introducing a novel fluorescence lifetime redox ratio (FLIRR) measurement, NAD(P)H-a2%/ FAD-a1% (Wallrabe et al., 2018b). The most prominent NAD(P) H signal occurs in the mitochondrial compartment, which, as we have shown previously, matches mitotracker labeling well (Alam et al., 2017), allowing us to differentiate mitochondrial OXPHOS from cytosolic glycolysis, as described in Section 10.5.2.

10.7 FLIM-Based Redox Ratio (FLIRR) Assay

Here, we will address why a fluorescence lifetime-based measurement for redox is preferable to an intensity-based metric. Lifetime measurements of NAD(P)H and FAD generate a short and long lifetime in nanosecond ranges and their respective fractions – by pixel. Enzyme-bound fractions of bound

(a2%)/unbound (a1%) NAD(P)H display higher/lower lifetimes, respectively in the order of >2.5 ns/~0.4 ns. FAD enzyme-bound/quenched (a1%)/unbound/unquenched (a2%) fractions display lower/higher lifetimes, respectively in the order of ~0.4 ns/>3.0 ns (Lakowicz, 1988). Changes in these important bound fractions are used in the FLIRR measurement as markers for redox states. The rationale for FLIRR, therefore, is based on the following: during the OXPHOS cycle, NADH is converted to NAD$^+$ by the enzyme NADH dehydrogenase; for this to happen NADH has to be in an enzyme bound state (more NADH-enzyme-bound fraction-a2%); whereas, $FADH_2$ is converted to nonenzyme bound FAD by the enzyme succinate dehydrogenase (less FAD enzyme-bound fraction-a1%). The a1 and a2 are the preexponential parameters associated with the shorter (τ_1) and longer (τ_2) lifetime components of a biexponential fluorescence decay model. These parameters are determined by fitting the model to the measured fluorescence decay data on a per pixel basis. a1% and a2% are normalized parameters according to a1% = a1 / (a1 + a2) and a2% = a2 / (a1 + a2), thus avoiding the effect of the overall fluorescence intensity on the parameters. We validated FLIRR against the intensity ratio for consistency, which is possible in monolayer cell cultures, as proof-of-principle under the three different interventions (Figure 10.7).

Why track cancer metabolism by FLIRR? In cancer, there is metabolic reprogramming and variable interaction between the glycolytic and OXPHOS energy generation. Unlike normal cells, cancer cells often produce energy via glycolysis followed by the production of lactate, even in presence of oxygen (Warburg Effect) (Warburg, 1956). Usually, cancer cells have glycolytic rates up to 200 times higher when compared to their respective normal tissue and some have defective OXPHOS activity as a strategy to interfere in the apoptotic pathways (Alam et al., 2017). A higher glycolytic rate in cancer is a less efficient way of producing energy (2Pyruvate+2ATP+2NADH), than the low glycolytic rate and mitochondrial oxidation of pyruvate (36 ATP) seen in normal cells (Pecqueur et al., 2013). However, cancer cells shift their metabolism to the production of lactate from pyruvate in the cytosol by the enzyme lactate dehydrogenase (LDH) and, in the process, oxidizing the NADH and regenerating NAD+ required for ATP production through glycolysis.

Tracking the autofluorescent signals of the coenzymes NAD(P)H and FAD in combination with an intensity-based FAD/NAD(P)H redox ratio has been well established in this field (Chance et al., 1979). Genetically encoded fluorescent redox sensors (Lukyanov and Belousov, 2014) offer alternative approaches to investigate cellular metabolic states in a variety of specimen types, particularly in cancer applications. As mentioned, light scattering and absorption – especially in tissue specimens – make intensity-based methods less suitable than FLIM approaches described here.

10.8 Quantitative Analysis, Charting, and Statistics

As mentioned above, users may analyze the data with their favorite software, write their own script, macros, etc. to create additional ratios, correlations to analyze subpopulations,

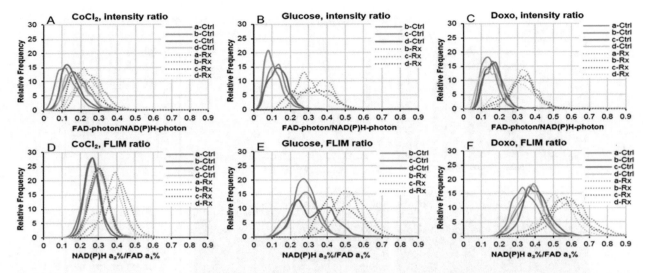

FIGURE 10.7 Comparison between intensity ratios and FLIRR for redox states for each of the three experimental groups. (a–c) Charts display the common intensity-based redox ratios FAD/NAD(P)H for the three experimental conditions explored in this manuscript. (d–f) Charts display the suggested alternative FLIRR – NAD(P)H-a2%/FAD-a1% – the ratios of the enzyme-bound fractions of each coenzyme. See text for further details (Wallrabe et al., 2018).

generate charts and statistics, design models, and the like. In this chapter, we have used a custom Python code and produced charts with Excel software. Depending on the complexity of experimental objectives, users may further automate certain steps in this analysis sequence.

The basic data unit is a ROI, cross-referenced to an individual cell and FOV – for any of the exported FLIM parameters. The custom Python script affords any analysis combination of segmented cells by "compartment" (i.e. whole-cell without nuclear region, mitochondria, cytosol) or FOV, either merged or for individual cells. Medians, means with standard deviations, quartiles, frequency distribution histograms/data, correlations, and more can be added to this open-source software. This section will first look at the total merged data for a treatment time course, then take a reductionist approach and examine individual segmental cell data, although, here, this is mainly the control and last treatment time point of 60 minutes, which mostly shows the greatest treatment effect. Finally, to reduce the single-cell data complexity, cells are allocated to three treatment response categories based on their median percent rise of FLIRR to explore the effect of changing redox states on other parameters. One of the objectives here is to examine the redox state based on mitochondrial OXPHOS vs. cytosolic glycolysis. Of course, depending on different analysis objectives, a different categorization method will serve just as well.

10.8.1 Separate Analysis of Whole-Cell, Mitochondrial, and Cytosolic Data

We start with an overview of the merged cell data comparing the redox marker FLIRR for whole-cell, mitochondria, and cytosol (without nuclear signals) – Figure 10.8. Any other FLIM parameters can be explored in this manner for a quick assessment of specific trends.

FLIRR is driven by the increasing NAD(P)H–a2% bound fraction and decreasing FAD-a1% bound/quenched fraction,

while the commonly applied redox intensity ratio is dominated by increasing FAD photon rise. Having made the case in Section 10.7 for FLIRR, we shall not return to the subject of intensity-based ratio in this chapter.

10.8.2 Segmented Cell Data and Categorization Based on FLIRR Response to Treatment

Individually segmented cell ROI data points displayed as frequency histograms or scatter charts will detect subtle differences in any of the chosen parameters, which may otherwise not be detected using just cell averages. Without losing this detailed information, additional analysis strategies must be introduced to reduce the complexity of large data sets. Using a meaningful parameter, cells and their ROI data can be categorized into specific ranges and intra- and interrange differences can be visualized by cell.

By way of example, in Figure 10.9, individual FLIRR cell median data and frequency distribution are shown by the compartment (mitochondria and cytosol) category. The histogram data highlights differences between OXPHOS and glycolysis and the heterogeneity of cells data.

10.9 Application of FLIRR Assay to a Tumor Xenograft Mouse Model

The motivation to develop an alternative FLIM-based redox measurement arose from our work with live murine xenografts, for which intensity-based measurements are unsuitable because of wavelength- and depth-dependent light scattering and absorption (Wallrabe, 2018b). In a mouse cancer xenograft model, the tumor was imaged at different depths collecting NADH and FAD images at 740 nm and 890 nm illumination, respectively (Figure 10.10). FLIM redox ratio values are consistent across all depths within the tumor, even though the excitation light intensity was

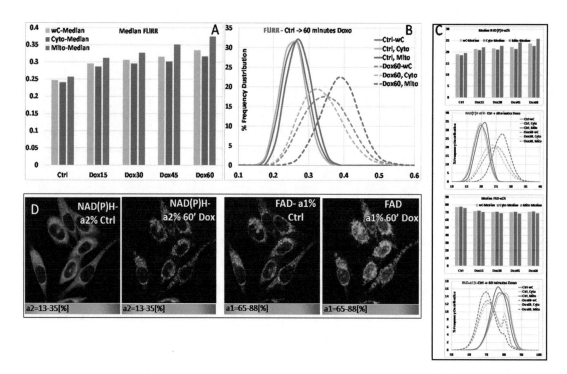

FIGURE 10.8 Comparing whole-cell, mitochondrial, and cytosolic compartments. (a) FLIRR median levels by compartment, showing changes at 15, 30, 45, and 60-minute time points after treatment with doxorubicin. (b) Percent frequency distribution by compartment control vs. 60 minutes of treatment. (c) Details of the two enzyme-bound fractions driving FLIRR – NAD(P)H-a2% and FAD-a1%. (d) Color-coded images of the enzyme bound fractions of NAD(P)H and FAD at control and after 60 minutes treatment (Cao et al., 2019).

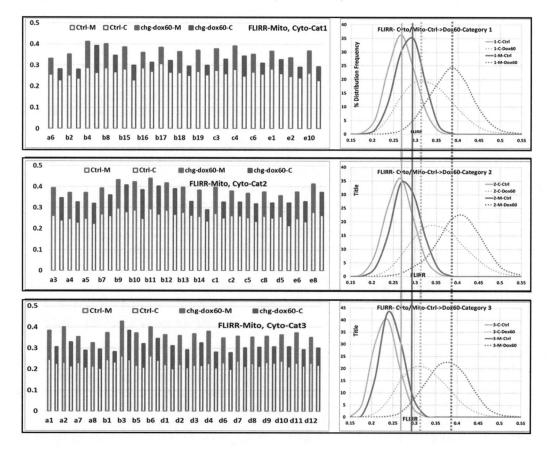

FIGURE 10.9 Categorization 1–3 for individual, segmented cell data. Mitochondrial FLIRR, median response to 60-minute treatment of doxorubicin and frequency distribution by merge category. Bar charts show control levels in open bars and the incremental rise of FLIRR median after 60-minute treatment. (red-Mito, blue-Cyto). A is the lowest response-level category. B is the middle response-level category. C is the highest response-level category (Cao et al., 2019).

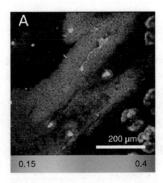
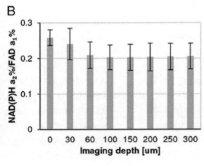
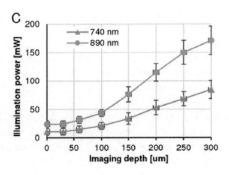

FIGURE 10.10 Application of FLIRR *in vivo* in a tumor xenograft mouse model. A. Image of FLIRR, 60 um below skin surface (25 × 0.8NA immersion lens). B. Mean values of FLIRR at different imaging depths, based on the same ROI assay applied in the HeLa cell monolayer experiments. ROI selections are made based on the intensity of NAD(P)H and FAD fluorescence, avoiding bright features (e.g. sebaceous glands). At depths of 0 and 30 μm, the fluorescence originates predominantly from the epidermis and FLIRR deviates from that of the tumor. C. The excitation power levels were increased with increasing imaging depth to account for light scattering and absorption. The overall fluorescence intensity was kept close to 105 photons/s to avoid saturation of the FLIM detection system (Wallrabe et al., 2018.

increased significantly with increasing imaging depth. The combined effects of light scattering, absorption, and optical aberrations of the excitation laser light cause rapid decrease of the peak illumination intensity in the focus with increasing imaging depth within an inhomogeneous tissue. While these effects can be partially counteracted by increasing the excitation power, it is impossible to maintain constant two-photon excitation intensity at different depths and two different wavelengths, which is a requirement for measuring intensity-based redox ratio deep in tissue, making the application of a two-photon intensity-based redox measurement unachievable.

10.10 Summary

The application of FLIM and the redox assay presented in this chapter offers exciting opportunities to explore metabolic changes based on the autofluorescent coenzymes NAD(P)H and FAD. This imaging technique has been used in a variety of research areas and, with advances in instrument development, will further expand the field. Yet, executing FLIM assays successfully requires careful optimization in all areas covered in this chapter: imaging conditions, lifetime fitting procedures, and detailed analysis tasks. In the end, it is a rewarding application for exploring cellular processes, either enhancing intensity-based microscopy or generating results, that are not suitable for intensity-based exploration.

Acknowledgment

We are grateful for financial support from University of Virginia, and NIH/Office of the Director for funds to purchase a Zeiss 780 multiphoton/FLIM (OD016446 and NIH RO1-AG067048 to AP) and ISS ALBA FLIM FD microscopes (RR027409 to AP) that were used throughout these studies.

REFERENCES

Abe, K., et al. (2013). Non-invasive in vivo imaging of breast cancer cell internalization of transferrin by near infrared FRET. *PLoS One, 8*(11): e80269.

Adur, J., et al. (2012). Optical biomarkers of serous and mucinous human ovarian tumor assessed with nonlinear optics microscopies. *PLoS One, 7*(10): e47007.

Alam, S. R., et al. (2017). Investigation of mitochondrial metabolic response to doxorubicin in prostate cancer cells: An NADH, FAD and tryptophan FLIM assay. *Scientific Reports, 7*(1): 10451.

Blacker, T. S., and Duchen, M. R. (2016). Investigating mitochondrial redox state using NADH and NADPH autofluorescence. *Free Radical Biology and Medicine, 100*: 53–65.

Becker, W. (2015a). *Advanced time-correlated single photon counting applications*. Vol. 111. Springer.

Becker, W. (2015b). *The BH TCSPC handbook*. Becker & Hickl GmbH.

Becker, W., et al. (2004). Fluorescence lifetime imaging by time-correlated single-photon counting. *Microscopy Research and Technique, 63*(1): 58–66.

Cao, R., et al. (2016). Phasor plotting with frequency-domain flow cytometry. *Optics Express, 24*(13): 14596–14607.

Cao, R., et al. (2019). Single-cell redox states analyzed by fluorescence lifetime metrics and tryptophan FRET interaction with NAD (P) H. *Cytometry Part A, 95*(1): 110–121.

Cao, R., et al. (2020). Optimization of FLIM imaging, fitting and analysis for auto-fluorescent NAD (P) H and FAD in cells and tissues. *Methods and Applications in Fluorescence, 8*(2): 024001.

Cao, R., Pankayatselvan, V., and Houston, J. (2013). Cytometric sorting based on the fluorescence lifetime of spectrally overlapping signals. *Optics Express, 21*(12): 14816–14831.

Chance, B. (1996). The use of intrinsic fluorescent signals for characterizing tissue metabolic states in health and disease. *Advances in Laser and Light Spectroscopy to Diagnose Cancer and Other Diseases Iii: Optical Biopsy, Proceedings, 2679*: 2–7.

Chance, B., et al. (1979). Oxidation-reduction ratio studies of mitochondria in freeze-trapped samples. NADH and flavoprotein fluorescence signals. *Journal of Biological Chemistry, 254*(11): 4764–4771.

Conklin, M. W., et al. (2009). Fluorescence lifetime imaging of endogenous fluorophores in histopathology sections reveals differences between normal and tumor epithelium in carcinoma in situ of the breast. *Cell Biochem Biophys, 53*(3): 145–157.

Eguchi, K., Manabe, I., and Nagai, R. (2012). Inflammation induced by free-fatty acid-tlr4/myd88 signaling in non-immune cells play a pivotal role in both cardiovascular and metabolic diseases. *Circulation, 126*(21).

Forstermann, U. (2008). Oxidative stress in vascular disease: Causes, defense mechanisms and potential therapies. *Nature Clinical Practice Cardiovascular Medicine, 5*(6): 338–349.

Ghukasyan, V. V., and Heikal, A. A. (2014). *Natural biomarkers for cellular metabolism: Biology, techniques, and applications.* CRC Press.

Gibson, B. W. (2005). The human mitochondrial proteome: Oxidative stress, protein modifications and oxidative phosphorylation. *International Journal of Biochemistry & Cell Biology, 37*(5): 927–934.

Islam, M. S., et al. (2013). pH dependence of the fluorescence lifetime of FAD in solution and in cells. *International Journal of Molecular Sciences, 14*(1): 1952–1963.

Joosen, L., et al. (2014). Effect of fixation procedures on the fluorescence lifetimes of Aequorea victoria derived fluorescent proteins. *Journal of Microscopy, 256*(3): 166–176.

King, B. C., and Blom, A. M. (2017). Non-traditional roles of complement in type 2 diabetes: Metabolism, insulin secretion and homeostasis. *Molecular Immunology, 84*: 34–42.

Koppenol, W. H., Bounds, P. L., and Dang, C. V. (2011). Otto Warburg's contributions to current concepts of cancer metabolism. *Nature Reviews Cancer, 11*(5): 325–337.

Lakowicz, J. R. (1988). Principles of frequency-domain fluorescence spectroscopy and applications to cell membranes. *Subcell Biochem, 13*: 89–126.

Lin, M. T., and Beal M. F. (2006). Mitochondrial dysfunction and oxidative stress in neurodegenerative diseases. *Nature, 443*(7113): 787–795.

Lukyanov, K. A., and Belousov, V. V. (2014). Genetically encoded fluorescent redox sensors. *Biochim Biophys Acta, 1840*(2): 745–756.

Marcu, L., French, P. M., and Elson, D. S. (2014). *Fluorescence lifetime spectroscopy and imaging: Principles and applications in biomedical diagnostics.* CRC Press.

Meleshina, A. V., et al. Probing metabolic states of differentiating stem cells using two-photon FLIM. *Scientific Report*, 2016. 6: 21853.

Melis, A. (1999). Photosystem-II damage and repair cycle in chloroplasts: What modulates the rate of photodamage in vivo? *Trends in Plant Science, 4*(4): 130–135.

Pecqueur, C., et al. (2013). Targeting metabolism to induce cell death in cancer cells and cancer stem cells. *International Journal of Cell Biology, 2013*: 805975.

Periasamy, A., and Clegg, R. M. (2009). *FLIM microscopy in biology and medicine.* Chapman and Hall/CRC.

Redford, G. I., and Clegg, R. M. (2005). Polar plot representation for frequency-domain analysis of fluorescence lifetimes. *Journal of Fluorescence, 15*(5): 805–815.

Rounds, D. (1965). *Effects of laser radiation on cell cultures.* Pasadena Foundation for Medical Research.

Ruck, A., et al. (2014). Spectrally resolved fluorescence lifetime imaging to investigate cell metabolism in malignant and nonmalignant oral mucosa cells. *Journal of Biomedical Optics, 19*(9): 96005.

Schleusener, J., Lademann, J., and Darvin, M. E. (2017). Depth-dependent autofluorescence photobleaching using 325, 473, 633, and 785 nm of porcine ear skin ex vivo. *Journal of Biomedical Optics, 22*(9): 091503.

Skala, M. C., et al. (2007). In vivo multiphoton microscopy of NADH and FAD redox states, fluorescence lifetimes, and cellular morphology in precancerous epithelia. *Proceedings of the National Academy of Sciences, 104*(49): 19494–19499.

Stringari, C., et al. (2011). Phasor approach to fluorescence lifetime microscopy distinguishes different metabolic states of germ cells in a live tissue. *Proceedings of the National Academy of Sciences of the United States of America, 108*(33): 13582–13587.

Sun, Y., et al. (2009). Characterization of an orange acceptor fluorescent protein for sensitized spectral fluorescence resonance energy transfer microscopy using a white-light laser. *Journal of Biomedical Optics, 14*(5): 054009.

Wallrabe, H., et al. (2018a). Segmented cell analyses to measure redox states of autofluorescent NAD (P) H, FAD & Trp in cancer cells by FLIM. *Scientific Reports, 8*(1): 1–11.

Wallrabe, H., et al. (2018b). Segmented cell analyses to measure redox states of autofluorescent NAD (P) H, FAD & Trp in cancer cells by FLIM. *Scientific Reports, 8*(1): 79.

Warburg, O. (1956). On respiratory impairment in cancer cells. *Science, 124*(3215): 269–270.

Zheng, J. (2012). Energy metabolism of cancer: Glycolysis versus oxidative phosphorylation. *Oncology Letters, 4*(6): 1151–1157.

11

Intravital Imaging of Cancer Cell Migration In Vivo

David Entenberg, Maja H. Oktay, and John Condeelis

CONTENTS

11.1 Introduction

Cancer is a collection of diseases, all characterized by the uncontrolled proliferation of cells. In addition, most cancers are also characterized by the ability of cancer cells to invade local structures and disseminate to distant sites. Cancer is typically categorized by the organ of origin, and the type of tissues affected by the neoplastic process. Cancers derived from epithelial cells, connective tissues (bone, muscles, tendons, nerves, etc.), blood cells in circulation, blood cells in the lymphatic system, and those resembling embryonic tissues are characterized as carcinomas, sarcomas, leukemias, lymphomas, and blastomas, respectively. When occurring in solid tissues, uncontrolled cellular proliferation results in the formation of local swelling of tissue called a tumor. Cancers of the circulating white blood cells (e.g. lymphocytes), meanwhile, cause significant congestion of the circulatory system and have come be known as "liquid tumors." Solid tumors are by far the most common type of cancer, with carcinomas, i.e. malignant tumors of the epithelial origin, making up 80–90% of all solid tumor cases.

Despite being the second leading cause of death worldwide, cancer mortality is not predominantly caused by the primary tumor. Instead, it has been estimated that approximately 90% of cancer-associated mortality is due to the spread of tumor cells from their location of origin, or primary tumor, to other parts of the body, a process known as metastasis (Seyfried and Huysentruyt, 2013).

This is reflected in the current 10-year survival rates for breast cancer, the most common type of cancer in women. When diagnosed early in the disease progression, before cancer cells spread beyond the basement membrane of the ducts (*in situ* disease), the 10-year survival rate is ~93%. If the cancer cells spread beyond the basement membrane boundary (invasive disease), the average survival rate drops to ~73%. However, the outlook for patients whose cancer has spread to distant sites (e.g. lungs, bones) is decidedly more dismal with a 10-year survival rate of only 17% (metastatic disease) (Narod et al., 2015).

While the progression of the tumor from the *in situ*, through invasive, and on to metastatic categories has traditionally been considered a function of time-elapsed since initiation of tumor growth, new evidence is emerging that tumor cells may disseminate to distant locations extremely early in the disease progression (Harper et al., 2016; Hosseini et al., 2016), indicating that tumor dissemination is independent of tumor growth. This can place signification limitations upon the efficacy of traditional cancer therapeutics, which focus on limiting the primary tumor growth either by inhibiting tumor cell division, or preventing the establishment of primary tumor support mechanisms such as the generation of new blood vessels, a process known as angiogenesis. Indeed, in both cases, clinical trials have demonstrated that while the evolution of the primary tumor may be held in check (as measured by the metric called progression-free survival), overall survival is most often not significantly impacted (Sachdev and Jahanzeb, 2016; Aalders et al., 2017). Thus, while great improvements in survival have been observed over the past several decades, this has been achieved by many modest reductions rather than one or even a few significant advances in treatment (Joy, 2008). Attention is now turning toward understanding the process of metastasis from both primary and secondary sites as the potential source of that significant advancement in treatment.

Central to this investigation is understanding how tumor cells move from the primary tumor and arrive at secondary sites. Motion of cells in these processes spans a wide range of spatial and temporal scales. Temporally, cell-migration speeds span several orders of magnitude, from fractions, to hundreds of microns per hour. Spatially, motility is directed through a combination of chemotactic, haptotactic, and paracrine signaling with neighboring cells (Haessler et al., 2011; Roussos et al., 2011b; Weber et al., 2013; Rosen and Roarty, 2014) to allow travel over a wide range of distances (Teddy and Kulesa, 2004). Motility *in vivo* can then be classified into different motility phenotypes including cells migrating singly (e.g. single-cell intravasation), as linear streams of unattached cells, as coordinated groups of attached cells, and even as entire sheets (e.g. collective migration of cells at the invasive front) (Roussos et al., 2011a).

This chapter will describe the latest developments in techniques to analyze and measure the motility of cancer cells over all of the wide spatial and temporal scales, and in the most physiologically relevant manner possible, through multiphoton microscopy of living animals, also known as multiphoton microscopy based intravital imaging. Though applicable to many fields, many of these techniques were developed in cancer research in an effort to visualize and understand the impact cellular motility has in the process of metastasis. As such, we also summarize the current understanding and primary evidence for the role of the different motility phenotypes listed above in metastasis in mammals.

11.2 Traditional Tools for Studying Cancer Progression and Dissemination Are Limited

Traditionally, cancer progression and metastatic dissemination have been studied with two main tools: histopathologic analysis of fixed tissues and *in vitro* assays containing isolated tumor cells (Nobre et al., 2018). These tools have proven through the years to be extremely powerful and useful techniques, as fixed tissues, particularly when taken from patients, accurately reflect the disease as it occurs at the moment of fixation in the clinic (van den Tweel and Taylor, 2010; Kumar et al., 2017), while *in vitro* assays are able to dissect and reveal the molecular biological mechanisms underlying different cancer cell phenotypes as they occur *in vitro* (Repesh, 1989; Liang et al., 2007). However, despite the advantages, these techniques also have significant limitations, which have reduced their applicability to the study of dynamic processes, such as metastasis. Excision, chemical fixation, and mechanical sectioning, steps necessary for the preparation of fixed tissues for histopathologic analysis, unavoidably stops the biological process at the point of analysis and prevents the observation of any further progression. This means that only a single snapshot view of a sample can be acquired from each patient, with the rest of the disease progression, before and after excision, completely hidden. This forces speculations about the process to be drawn by comparing many samples from many different patients, prevents the definitive determination of cause-and-effect relationships, and results instead in correlations and inferred mechanisms.

Similarly, *in vitro* assays are limited by their lack of connection to the rest of the biological systems found in the body.

This is particularly significant, as it is becoming recognized more and more that, while genetic mutations initiate tumor growth, the ultimate metastatic fate of these cells depends heavily upon epigenetic changes controlled by the tumor microenvironment (DeClerck et al., 2017), which consists of nontumor cells types (e.g. immune, stromal, endothelial), acellular structures (e.g. extracellular matrix), and microenvironmental conditions (e.g. oxygen levels, growth factor concentrations) surrounding the tumor cells. In addition, tumor and nontumor cells within tumor microenvironments secrete a plethora of chemokines, cytokines, and growth factors that induce systemic effects on other organs. In particular, the bone marrow and immune system respond to tumor secreted mediators by releasing various soluble factors and cellular progenitors which in turn affect tumor progression (Shaked and Voest, 2009; Roodhart et al., 2013). Moreover, new evidence indicates that the tumor microenvironment is implicated in a form of drug resistance, called environment-mediated drug resistance, also known as the host response (Cukierman and Bassi, 2012; Borriello and DeClerck, 2014; Shaked, 2016; Karagiannis et al., 2018). Further, tumor produced chemokines, growth factors, and signaling molecules can influence distant sites in a variety of ways. For example, these molecules can attract immune cells from the blood and stimulate the growth and function of stromal cells (Muller et al., 2001; Palomino and Marti, 2015). All these stimuli are lacking in *in vitro* systems, and as such, phenotypes observed in these assays are not necessarily representative of the tumor phenotypes observed *in vivo*. Taken together, these limitations mean that the conclusions drawn from fixed tissues or *in vitro* assays must ultimately be validated by observing tumor cell phenotypes, and their interactions, *in vivo*.

11.3 Intravital Imaging

Ideally, what would be required to best study dynamic biological processes such as metastasis would be the ability to perform mechanistic *in vivo* studies longitudinally during cancer progression to define and test hypotheses in real-time and at single-cell resolution. To this end, many groups have turned to imaging live animals (intravital imaging) with multiphoton microscopy as the tool of choice (Chambers et al., 1995; Dunn et al., 2007; Masedunskas et al., 2013; Zomer et al., 2013; Entenberg et al., 2017). Multiphoton microscopy relies upon the ability of fluorescent molecules to simultaneously absorb multiple photons when their density becomes sufficiently high (Zipfel et al., 2003). Sufficient photon densities can be attained by compressing the light output from a laser in all three spatial dimensions (Zipfel et al., 2003). Compression along the beam's axis of propagation is attained by the use of ultrafast pulsed lasers (~100 fs), while compression lateral to the beam is accomplished by the use of lenses with high convergence angles (high numerical apertures). Because the simultaneous absorption of multiple photons is an unlikely event outside of this zone of high density, the technique has the ability to illuminate isolated slices of tissue deep within the whole sample (a process known as optical sectioning). This generates images strikingly similar to those obtained in histopathology (Figure 11.1a–b) in a nondestructive manner that allows a direct view of the living

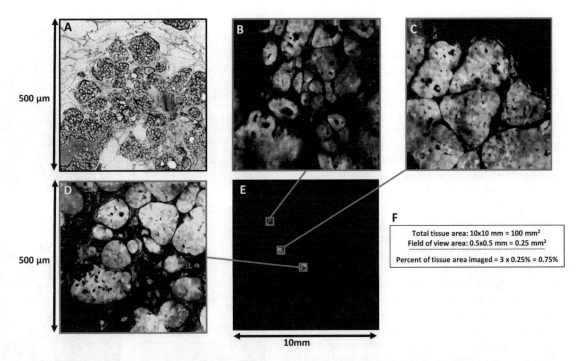

FIGURE 11.1 Traditional approaches to intravital imaging produce images similar to those used in histopathology but leave much of the tissue unimaged. (A) Histopathological image of breast carcinoma in a PyMT mouse model stained for tumor cells (brown = cytokeratin) and blood vessels (blue = CD31). (B–D) Single, high resolution fields of view of breast carcinoma in a PyMT mouse model in which the tumor cells express GFP; the vasculature has been labeled with a high molecular weight fluorescent dextran, and the macrophages express CFP (green = GFP, red = fluorescent dextran, blue = CFP). (E) Traditionally, intravital multiphoton imaging focuses on a few individual high magnification fields of view randomly positioned across the tissue (B–D). Placing these fields of view within the context of the full 1cm tissue (black area) demonstrates how little of the full tissue is imaged. (F) Comparison of the areas of the fields of view with the full tissue areas shows that three 500µm fields of view cover less than one percent of the tissue.

cells composing the tumor, in their native environs, and over extended periods of time (Squirrell et al., 1999).

In our laboratory, we have developed our own custom-built multiphoton microscope (Entenberg et al., 2011) which utilizes several laser sources and multiple detectors. This enables the detection of many signals simultaneously to capture as many of the cell–cell and cell-stromal interactions as possible (Entenberg et al., 2013).

11.4 Surgical Engineering Expands the Capabilities of Intravital Imaging

While intravital imaging is a powerful technique, it is one that is generally considered to be technically difficult, and its adoption and utilization has been limited (Cahalan et al., 2003). In recent years, however, several new protocols have been published detailing instrumentation and methods that reduce this difficulty and make intravital imaging accessible even to laboratories that do not specialize in the technique. These methods cover imaging in diverse locations such as mammary tumors (Wyckoff et al., 2011; Entenberg et al., 2013; Harney et al., 2016) and glands (Harper et al., 2016; Entenberg et al., 2017); the heart (Vinegoni et al., 2012); salivary glands (Masedunskas et al., 2013); lymph nodes (Jeong et al., 2015; Entenberg et al., 2017); visceral organs, such as the liver (Marques et al., 2015); spleen (Ferrer et al., 2012); kidney (Ritsma et al., 2012); and even the lungs (Looney et al., 2011; Entenberg et al., 2018).

These methods are direct applications of the field of surgical engineering, which is one that has traditionally focused on utilizing engineering skills to develop new devices, materials, and technologies for use in the operating room. Efforts in the field have produced advances such as: minimally invasive surgical protocols (Gholami et al., 2017; Lianos et al., 2017); novel, biocompatible materials, including surgical sealants (Peng and Shek, 2010) and implants (Velnar et al., 2016); tools and robotics for assisting surgeons (Ibrahim et al., 2017); and engineered tissues for use in surgery (Lalan et al., 2001). For intravital imaging, however, this traditional model is turned around, and instead, the techniques and skills of the surgeon are brought into the optics laboratory. This results in methods and protocols that simplify intravital imaging, make the technique accessible to more labs, expand the utility of mouse models, and allow imaging of more tissue, over longer periods of time, and in locations that have previously been considered extremely challenging to image (Harper et al., 2016; Entenberg et al., 2018).

11.5 Large-Volume High-Resolution Intravital Imaging

Successful application of these new techniques allows intravital imaging to be utilized in a manner that is somewhat different from traditional intravital imaging and closer to how standard histopathologic analysis is performed. In histopathology, sectioned and stained tissues are first examined (with bright-field microscopy) at low magnification in order to understand the overall structure and architecture of the tissue. Interesting regions are next selected for analysis at higher magnification, and then these regions are examined at single cell-resolution in order to identify individual cells based on their staining and morphology (Kothari et al., 2013; Heindl et al., 2015; Lloyd et al., 2015).

In traditional intravital imaging (because of the optical limitations in field diameter associated with high numerical aperture objective lenses), a sample is placed on the microscope, a few fields of view to be investigated identified, and time-lapse microscopy is used to follow them over time (Figure 11.1b–d). However, as can be seen in Figure 11.1e, the total amount of tissue covered by these fields of view spans an exceedingly small portion of the tumor tissue (sometimes less than 1%, see Figure 11.1f). Given the heterogeneity of tumors (Meacham and Morrison, 2013), this adds considerable uncertainty to the significance of the imaging results.

Unfortunately, simply increasing the number of fields of view (FOV) acquired is not enough to rectify this limitation. Even complete acquisition of all FOVs, while giving much more information about the underlying subject, still fails to present a coherent picture. The concept is demonstrated in Figure 11.2a where it can be seen that the acquisition of just a few individual, high-magnification frames from a well-known image fail to give an understanding of the subject. While complete coverage of the subject (Figure 11.2b) allows significantly more information to be gleaned about the subject (e.g. the subject is a person, the person is Caucasian.), other information is difficult or impossible to discern (e.g. subject's sex, subject's identity).

Combining multiscale imaging with intravital imaging would thus allow intravital imaging to be utilized in the same way as histopathology. One simple solution to this undersampling can be garnered from the fields of satellite imagery and digital pathology in which large samples are digitally recorded by the serial acquisition of many individual high-resolution FOVs which are then mosaicked and stitched to form a large, low-magnification image. This process, called large-volume high-resolution (LVHR) imaging, is utilized in many fields including endoscopy, oceanography, and X-ray microscopy (Loo et al., 2000; Seshamani et al., 2006; De Zanet et al., 2016; Kwasnitschka et al., 2016), maintains the spatial relationships between the individually acquired fields, and allows the overall context of the imaged subject to be captured. Figure 11.2c demonstrates how the images are serially acquired and stitched to form the complete image (Figure 11.2d).

The ability to capture large volumes of tissue at high resolution is particularly important for biological samples. As an example, Deryugina et al. (2017) recently used LVHR imaging to look at fixed tumors and evaluated the quantity and location of intravasation sites across the entire tissue. The spatial context provided by LVHR enabled the researchers to determine that >98% of intravasated cells are found at the tumor interior. This is an interesting result, as invasiveness of tumor cells into adjacent stroma is considered to be a sign of increased malignancy of tumors (Hanahan and Weinberg, 2011). However, our research using intravital imaging and other techniques has shown that cancer cell intravasation occurs within the tumor core and not at the tumor margin (Robinson et al., 2009;

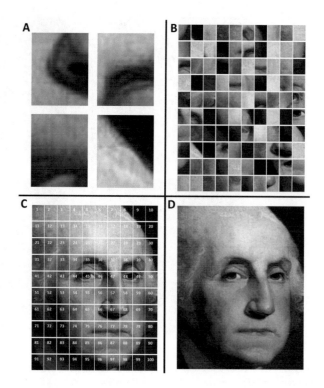

FIGURE 11.2 Demonstration of how multiscale imaging provides detail and context by acquiring many high-magnification images sequentially and stitching them together to form a low magnification overview image. (A) The acquisition of just a few individual fields of view does not provide enough context to understand fully the imaged subject. (B) Information about the subject is improved by increasing the number of acquisitions; however, without maintaining the spatial relationship between the individual images, the overall context remains obscured, even with full coverage of the sample. (C) Acquisition and arrangement of the images in a specific order (numbers) maintains the spatial relationship of the images. (D) Final stitching of the image reveals the overall context while preserving the original resolution of the underlying images.

Roussos et al., 2010, 2011b; Pignatelli et al., 2014; Rohan et al., 2014; Harney et al., 2015).

Combining LVHR imaging with multiphoton intravital imaging, a process called large-volume high-resolution intravital imaging (LVHR-IVI), combines the benefits of the two techniques and generates multiscale images that can be used in the same way that pathologists examine static histological sections (Sipkins et al., 2005; Pineda et al., 2015). However, unlike in pathology, which produces static images, LVHR-IVI captures the underlying dynamics of the living tissue, longitudinally in time (Entenberg et al., 2017).

Using surgical engineering, a series of mechanical designs for fixturing plates (used to immobilize tissues relative to the imaging axis) and imaging windows, combined with surgical protocols for their implantation, completely immobilizes tissues over the extended periods of time required for time-lapse mosaicked acquisition (Entenberg et al., 2017). These protocols can be used for both chronically implantable imaging windows (Kedrin et al., 2008; Ritsma et al., 2012) and skin-flap surgeries (Harney et al., 2016) over many tissues, including lymph nodes, mammary glands, and mammary tumors of all stages, as well as internal organs, such as the liver, spleen, and lungs.

Thus, LVHR-IVI has several advantages over traditional intravital imaging and provides: 1) a better understanding of the context and architecture of the tissue, 2) a guide to identify which areas are most suitable for further analysis (e.g. with high speed time-lapse imaging), 3) the ability to capture events that are rare in time or in space, and 4) large volumes of data for machine learning algorithms, such as deep learning (Litjens et al., 2017) or support vector machine analysis (Gligorijevic et al., 2014).

Support vector machine analysis is particularly well suited for discerning complex relationships between many variables. As an example, we were able to analyze large numbers of high-resolution intravital images and observed that the motility of tumor cells in each FOV demonstrated either a fast (>1μm/min) or a slow migratory phenotype with the latter having average speeds 10–100 times slower than the former (Gligorijevic et al., 2014). Very few FOVs were observed to contain tumor cells in both categories. Through the application of a series of image filters, microenvironmental information (e.g. density of collagen fibers, number of macrophages, diameter of blood vessels, etc.) could be extracted and quantified for each of the FOVs. These microenvironmental parameters were then used as inputs to a nonlinear support vector machine algorithm whose classifier outputs were either the fast or slow migratory phenotype. Varying degrees of predictive ability were attained when using three, four, or five microenvironmental parameters, with a maximum accuracy of 92% when all five parameters were used. In this way, the composition of the microenvironments associated with a cell phenotype *in vivo*, such as either fast or slow tumor cell migration, was revealed and has led to the identification of the microenvironmental components, such as various stromal cell types, responsible for these tumor cell phenotypes (Gligorijevic et al., 2014).

Thus, large-volume imaging proves to be a powerful technique to accurately measure tumor phenotypes over a wide range of spatial and temporal scales and to determine the microenvironments responsible for them *in vivo*.

11.6 Types of Motility: Collective, Single Cell, and Single Cell Streaming

The migration of cells, cancerous or healthy, may take on dramatically distinct patterns depending upon the number of cells, their degree of coordination, and their connectedness (Figure 11.3). At one extreme, single cells migrate individually, independent of surrounding neighbors. This type of migration has been extensively studied and is exemplified by the motility observed in leukocytes (Imhof and Dunon, 1997; Weninger et al., 2014) and metastasizing cancer cells (Condeelis et al., 2005; Roussos et al., 2011a; Patsialou et al., 2013; Harney et al., 2015).

When numerous, independently migrating cells are directed by a chemotactic signal, large cohorts of cells can respond simultaneously. In invasive carcinoma, tumor cells that do not make continuous contact are frequently observed to form a linear pattern of movement called streaming.

Streaming migration has been observed in different types of cancer (Roussos et al., 2011b; Beerling et al., 2016). In this phenotype, tumor cells and sometimes tumor cells and immune cells, align in single-file lines and comigrate together along extracellular matrix (ECM) fibers (Wyckoff et al., 2004; Patsialou et al., 2009; Leung et al., 2017). This migratory pattern is dependent upon a combination of chemotactic, haptotactic, autocrine, and paracrine signals that direct the tumor cells and tumor/immune cells toward blood vessels (Wyckoff et al., 2004; Leung et al., 2017). Intravital imaging has proven to be crucial to demonstrating that these motile cells are responsible for hematogenous dissemination and metastasis by directly visualizing their intravasation *in vivo* (Wyckoff et al., 2007; Entenberg et al., 2011; Harney et al., 2015; Harper et al., 2016).

Another category of motion is the collective migration of groups of cells that maintain intact cell–cell junctions. This migratory phenotype is best characterized by the extension and invasion of the ductal tree through the mammary fat pad during puberty, but it has also been observed in cancer *in vivo* when the cells are confined to preexisting tracts and channels, such as between muscle fibers (Alexander et al., 2008). Morphologically, collectively migrating cancer cells can take on a number of different phenotypes, ranging from wide sheets of cells to strands one or two cells in diameter.

In cancer, motility of tumor cells has a significant impact on clinical outcome, as tumor cell migration is required for metastatic dissemination. Each of the types of motility listed above has been directly observed to occur in various carcinomas, and thus each has been speculated to contribute to the metastatic dissemination of tumor cells. However, the primary evidence for each motility phenotype's contribution to hematogenous dissemination is unequal, with the contribution of streaming migration having been extensively elucidated, while that of collective migration remains unclear. At the time of this writing, only three studies in the literature are cited for the direct, *in vivo* observation of collective migration of orthotopically injected tumor cells (Giampieri et al., 2009; Weigelin et al., 2012; Ilina et al., 2018), and none have demonstrated that

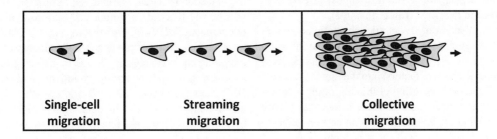

| Single-cell migration | Streaming migration | Collective migration |

FIGURE 11.3 Migrating cells may take on different motility phenotypes depending on the number of cells and their connectedness.

collective migration is involved in the invasion of tumor cells into the blood vasculature. Indeed, in their study, Giampieri et al. (2009) concluded that, "Single cell motility is essential for blood-borne metastasis, whereas cohesive invasion [into stroma] is capable of lymphatic spread."

Despite this paucity of direct evidence, the number of studies looking *in vitro* at the molecular mechanisms of collective migration and invasion is rapidly growing. As of this writing, a Pubmed search for the keywords "collective invasion AND (tumor OR cancer)," and that excludes the word "development," results in 223 papers published in the past 5 years. In fact, only three published studies have looked at collective migration of tumor cells using intravital imaging. This includes the above mentioned studies by Giampieri et al. (2009) and Weigelin et al. (2012) and an additional study by Beerling et al. (2016) in which they "visualized the migratory behavior of primary tumor cells in a genetically engineered pancreatic cancer mouse model and found that pancreatic tumor cells migrate with a mesenchymal morphology as single individual cells or as a stream of non-cohesive single motile cells." Thus, while there is a great deal of *in vitro* work on collective migration, there is a great paucity of *in vivo* work on collective migration, and it remains unclear if collective migration is involved in tumor cell dissemination *in vivo*.

11.7 Single Cell Streaming Motility Results in Hematogenous Dissemination

In the case of single-cell and streaming migration, direct observation of the intravasation event has been accomplished on numerous occasions (Wyckoff et al., 2000b, 2007; Kedrin et al., 2008; Entenberg et al., 2013; Harney et al., 2015), and the molecular mechanisms underlying the process have been well elucidated.

In this process, a small fraction of individual cancer cells within the tumor comigrate with macrophages at high velocity (>3μm/min) in single-file streams along ECM fibers but without cell–cell junctions (Farina et al., 1998). Cancer cell-macrophage pair streaming motility is supported by relay chemotaxis, consisting of a paracrine loop involving macrophage-derived epidermal growth factor (EGF) and carcinoma cell-derived colony stimulating factor-1 (CSF-1) (Wyckoff et al., 2004). The unidirectional motility of these cancer cell-macrophage pairs toward blood vessels is mediated by endothelial cell-secreted hepatocyte growth factor (HGF) (Leung et al., 2016). Importantly, streaming migration has been observed in both xenograft as well as transgenic mouse models of breast cancer, the latter of which recapitulate the histological disease progression observed in the clinic (Lin et al., 2003).

Isolation and collection of the motile, chemotaxis-competent fraction of the tumor was made possible by the insertion of "artificial blood vessels" (microneedles filled with perivascular chemoattractant growth factors) into the tumor (Wyckoff et al., 2000a). Expression profiling of this migrating subpopulation of tumor cells identified the protein Mena as highly upregulated (Wang et al., 2004). Mena is an actin regulatory protein from the Ena/Vasp family that is highly conserved across species (Krause et al., 2003). Deletion of the Mena gene from animal models of cancer inhibits intravasation, greatly reduces the frequency of metastatic dissemination to the lung, and eliminates mortality and morbidity, all without affecting primary tumor growth, implicating the involvement of Mena in the process of cancer cell dissemination (Roussos et al., 2010; Karagiannis et al., 2017).

Mena additionally has several splice-variant isoforms (Goswami et al., 2009; Shapiro et al., 2011) including Mena11a, an isoform whose expression in breast cancer cells causes the formation of poorly metastatic tumors with a highly epithelial architecture not capable of responding to EGF chemotactic cues *in vivo* (Roussos et al., 2011b). In a tumor, the fraction of Mena lacking the metastasis suppressor 11a isoform is measured with a multiplexed quantitative immunofluorescence test called MenaCalc. High MenaCalc scores reflect the presence of metastatic competent Mena11a-low tumor cells and correlate with increased metastatic risk in patient cohorts (Agarwal et al., 2012; Forse et al., 2015).

Intravital imaging of metastatic tumors showed that successful tumor cell intravasation events overwhelmingly occur in association with macrophages (Wyckoff et al., 2007). This confluence of a Mena overexpressing tumor cell, a macrophage, and an endothelial cell thus came to be termed the tumor microenvironment of metastasis (TMEM) (Robinson et al., 2009). TMEM represents a doorway for tumor cell intravasation (Harney et al., 2015), and TMEM count in patient samples is prognostic for metastatic outcome in several clinical cohorts (Robinson et al., 2009; Rohan et al., 2014; Sparano et al., 2017). High-resolution multiphoton intravital imaging of TMEM doorways has revealed that these tripartite structures can be found predominantly at vascular branch points and are stationary doorways of cancer cell intravasation that persist for extended periods of time (Harney et al., 2015). The macrophage-tumor cell contact that occurs in TMEM also stimulates the formation of extracellular matrix degrading structures (invadopodia) in the tumor cell (Gligorijevic et al., 2014; Roh-Johnson et al., 2014; Eddy et al., 2017) which then combine with the vascular endothelial growth factor (VEGF) production and release by the TMEM-Tie2[Hi] macrophage to open a gap in the vascular wall, resulting in transient leakage of blood serum into the interstitium. Concurrent with, and spatially adjacent to this transient vascular leakage, a high rate of tumor cell intravasation is also observed. Ablating macrophages or deleting the VEGF gene from macrophages dramatically reduces the transient vasculature leakage and the intravasation of tumor cells (Harney et al., 2015).

Further, rebastinib, a small molecule inhibitor of Tie2, was found to reduce tumor growth and metastasis in orthotopic mouse models of metastatic mammary carcinoma through the reduction of Tie2+ myeloid cell infiltration, antiangiogenic effects, and blockade of tumor cell intravasation mediated by perivascular Tie2[Hi]/VEGFA[Hi] macrophages in TMEM. The antitumor effects of rebastinib enhance the efficacy of microtubule inhibiting chemotherapeutic agents by reducing tumor volume and metastasis and improving overall survival (Harney et al., 2017; Karagiannis et al., 2017). The doorway function of TMEM and inhibition of TMEM function by rebastinib were revealed by multiphoton intravital imaging, illustrating how imaging is essential for establishing cause and effect relationships *in vivo*.

In summary, the role of streaming migration in hematogenous dissemination of cells is a well-investigated process, the

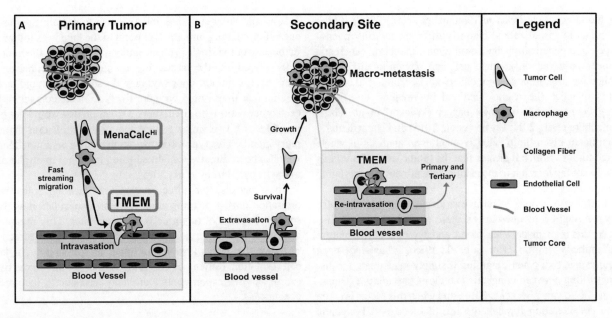

FIGURE 11.4 Mechanism of migration and dissemination for which there is direct evidence from IVI: The role of streaming migration, MenaCalc, and TMEM in the primary and secondary sites. (A) In the primary tumor, MenaCalc[Hi] tumor cells pair with macrophages within the tumor core and comigrate along collagen fibers toward blood vessels. Upon arriving at a blood vessel, streaming cells either form a new, or interact with an existing, TMEM doorway. TMEM doorways are capable of opening the vasculature to allow other streaming cells to intravasate and hematogenously disseminate. (B) In the secondary site, circulating tumor cells physically lodge within the vasculature where they reside until they are able to extravasate, aided by macrophages. Survival in the parenchyma of the secondary site eventually leads to the growth of a metastasis. Blood vessels within the metastasis can then support assembly of secondary-site TMEM doorways, which allow reintravasation and dissemination to locations in the secondary site, back to the primary site, or onward to tertiary sites.

steps and mechanisms of which are well known. The entire process is summarized by the model depicted in Figure 11.4a in which a minority population of tumor cells within the bulk tumor undergo an altered gene expression pattern to become MenaCalc[Hi], pair with macrophages, and comigrate at high velocities along collagen fibers toward blood vessels. Upon reaching the vasculature, the macrophages and tumor cells stop their migration and either become new stationary TMEM doorways, or the tumor cells intravasate at existing TMEM doorways and disseminate hematogenously to secondary sites such as the lung. New preliminary evidence (Entenberg et al., 2018) indicates that this process is then recapitulated in metastatic lesions where secondary TMEM doorways form and allow tumor cells to redisseminate to new locations in the secondary organ, back to the primary site, or on to tertiary locations (Figure 11.4b).

11.8 Collective Vascular Invasion

A proposed mechanism for tumor cell dissemination to distant sites is the collective migration of groups of cells that maintain adherens junctions while migrating into the peritumoral stroma. This speculated migration mode is proposed to play a role based upon the observation of finger-like protrusions and clusters of tumor cells within stromal tissue seen in fixed histological sections (Leighton et al., 1960; Einenkel et al., 2007; Bronsert et al., 2014; Khalil et al., 2017) or within the extracellular matrix surrounding *in vitro* cultures (Cheung et al., 2013) and cultured explants (Friedl et al., 1995). These histological protrusions have been correlated with a worse prognosis for metastatic-free survival (Khalil et al., 2017).

The mechanisms underlying this correlation, and just how collective migration of cells may lead to metastatic dissemination, has been unclear however. The prevailing theory is that tumor cells may grow into the vasculature and break off as clusters of cells in a process known as "collective vascular invasion." Using analysis of fixed tissue sections of xenograft tumors (MCH66, a murine breast carcinoma cell line), Sugino et al. (2002) observed that as tumors grew, their angiogenic blood vessels developed into fused, dilated sinuses that enveloped entire tumor nests, such that the tumor islands appeared to be entirely surrounded with blood. This observation led the group to speculate that these tumor islands could then be carried in the tumor vasculature to distant sites (Sugino et al., 2002).

However, this speculation has not been proven because all the primary evidence gathered thus far is indirect. The speculation is based on the observation of: 1) "finger-like" protrusions of tumor cells into the peritumoral stromal regions in fixed tissues (Leighton et al., 1960; Einenkel et al., 2007; Bronsert et al., 2014; Khalil et al., 2017), 2) intravascular cell clusters, again in fixed tissues (de Mascarel et al., 1998; Kato et al., 2003), and 3) tumor cell clusters in murine and patient blood samples (Liotta et al., 1976; Kats-Ugurlu et al., 2009; Duda et al., 2010; Aceto et al., 2014; Cheung et al., 2016). Direct observation, by intravital imaging of live tissues, of collective migration in a genetically engineered mouse model of cancer has never been reported nor has direct observation of collective vascular invasion in any model.

Further, the direct observation of collective migration has only been made with xenografts of cancer cell lines (Giampieri et al., 2009; Weigelin et al., 2012; Ilina et al., 2018) with one of the studies concluding that collective migration is not involved in blood vascular invasion (Giampieri et al., 2009). Other attempts to visualize and quantify different motility phenotypes in a genetically engineered mouse model of pancreatic cancer observed only single cell and streaming migration; collective migration was not observed (Beerling et al., 2016). This matches the results obtained by imaging genetically engineered models of breast cancer (Wyckoff et al., 2007; Entenberg et al., 2011; Harney et al., 2015) and is particularly significant given the previously mentioned analysis of whole fixed tumors, which indicates that the tumor-stromal border is likely not the location of metastatic dissemination (Deryugina and Kiosses, 2017).

Early observations of intravascular cell clusters have also been brought into question as these studies did not definitively distinguish between blood and lymphatic vasculature by antibody staining (Krasna et al., 1988). It has since been determined that generic staining is simply inadequate for distinguishing between lymphatic and blood vasculature (Gujam et al., 2014; van Wyk et al., 2014), and when this factor is taken into consideration through the use of blood and lymphatic vessel specific antibody staining, >97% of the cases of intravascular tumor cell clusters observed in fixed tissues actually involve lymphatic not blood vasculature (Mohammed et al., 2007, 2011).

Studies that detected cell clusters exiting from tumors required experimental procedures that are not physiological, and as such, their observations may have been the result of experimentally induced tissue damage. Both Liotta et al. (1976) and Kats-Ugurlu et al. (2009) used manual massage of the tumor tissue to induce cell cluster generation, while the study by Duda et al. (2010) involved cannulating, and mechanically pumping fluid through the tumor's vasculature, possibly inducing vessel swelling and breakage. While the methods used in these studies are not physiological, the question of tumor cell shedding during surgery is one that has long caused concern (Tyzzer, 1913; Foss et al., 1966; Hansen et al., 1995; Katharina, 2011) without a way to definitively determine the impact these cells have. The study by Aceto et al. (2014) attempted to address this with a sophisticated experimental setup that utilized tumor cells expressing either green or red fluorescent proteins injected orthotopically into the mouse as either multicolored mixtures in a single mammary fat pad, or as single colors in contralateral fat pads. The authors isolated circulating tumor cells (CTC) (after growth of the primary tumor) from the blood of the mouse and then compared them with the metastases that eventually developed in the lung. They found that when there was one primary tumor consisting of mixed colors, >90% of the CTC clusters were multicolored, however, when two separate primary tumors each of a single color were present, only about 4% of the clusters were multicolored. They take this to mean that the tumor cells entered into the vasculature already clustered. Given the relative rarity of circulating tumor cell clusters (<3% of total circulating cells), and the assumption that the multicolored lung metastases derive directly from them, they calculate that clusters are approximately 50 times more potent at generating metastases as compared to single cells. This speculation underscores the importance of understanding tumor cell dissemination and the role of single cells vs. clusters.

However, an equally possible scenario to explain the presence of circulating tumor cell clusters in the lung is as follows: in the case of the single primary multicolored tumor, the tumor cells entered into the vasculature as single cells, but, because they use a common doorway, they do so in such rapid succession that they would be more likely to make contact with each other within the circulation, a situation that would not be possible with two single color tumors in contralateral mammary glands. This type of intravasation using a common doorway has been directly visualized using intravital multiphoton microscopy (Harney et al., 2015).

Thus, because the direct evidence is lacking, collective migration during dissemination remains a speculation as the source of spread of tumor cells to distant sites. This situation illustrates the importance of intravital imaging to directly visualize the events involved in tumor progression at single-cell resolution. Relying on correlations inferred from fixed tissue or *in vitro* experiments alone may lead one to incorrect conclusions.

11.9 Surgical Engineering Enables Direct Visualization of the Lung Vasculature, Serially

A major tool to permit the direct observation of the mechanism of tumor cell dissemination is the ability to image, at single cell resolution, both the primary tumor and secondary sites, serially, during dissemination. Mammary imaging windows (Figure 11.5a) for the observation of primary mammary tumors have been available for more than a decade (Shan et al., 2003; Kedrin et al., 2008) and a panoply of other windows have also been developed for imaging a variety of metastatic sites, including the liver (Figure 11.5b) (Ritsma et al., 2012), the lung (Figure 11.5c) (Entenberg et al., 2018), the brain (Kienast et al., 2010), the bone marrow of the calvaria (Lo Celso et al., 2011) (Figure 11.5d), the lymph node (Figure 11.5e) (Ito et al., 2012), and the femoral bone marrow (Figure 11.5f) (Reismann et al., 2017; Lo Celso et al., 2011). Finally, the dorsal skin fold chamber (Papenfuss et al., 1979; Lehr et al., 1993) has proven useful for imaging ectopic tumors (Figure 11.5g). These tissues can be accessed on either inverted (Figure 11.5h) microscopes, where the imaging window can be captured and stabilized within the x–y stage (Entenberg et al., 2017), or on upright microscopes (Figure 11.5i), where tissue stabilization often requires employing a stabilizing ring and/or bar to remove the motion artifacts due to breathing and muscle contractions. Several excellent reviews outline how these windows have allowed intravital imaging-based investigations into the behavior and dynamics of tumor cells in these sites and the biological discoveries that have been made as a result (Gligorijevic and Condeelis, 2009; Hak et al., 2010; Alieva et al., 2014; Kitamura et al., 2017).

The lung, the most common metastatic site in mice with primary tumors, is a particularly challenging tissue to image due to its delicate nature and constant motion. *In vivo* imaging of the lung has garnered interest for nearly a century with work dating back to that of Wearn and German in 1926 (Wearn et al.,

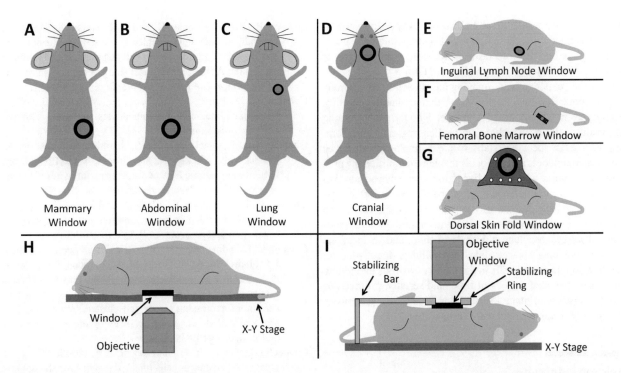

FIGURE 11.5 Overview of implantable imaging windows for intravital microscopy of cancer. (A–F) Implantable imaging windows give access to a variety of tissues for serial imaging. These include (A) the mammary gland and mammary tumors, (B) the visceral organs within the abdomen (e.g. liver, pancreas, spleen, intestine, etc.), (C) the lungs, (D) the brain or the marrow of the calvaria, (E) the inguinal lymph node, (F) the femoral bone marrow. (G) Dorsal skin fold windows also give access to the deep dermal layers and have been used for ectopic transplantation of a number of cancer cell lines. (H) Each of these windows can be used on an inverted microscope. Rigid attachment of the window directly to the x–y stage plate stabilizes residual motion of the window and the underlying tissue. (I) Alternatively, each of these windows can be used on an upright microscope. Stabilization of the window and the underlying tissue is accomplished through the use of a stabilizing ring attached to a stabilizing bar connected to the x–y stage.

1926). Since that time, many techniques have been developed to gain access to the lung, mostly focusing on large animals such as dogs, cats, rabbits, and rats (Terry, 1939; De Alva and Rainer, 1963; Wagner and Filley, 1965; Wagner, 1969; Groh et al., 1992; Fingar et al., 1994; Lamm et al., 2005), but more recently, the technique has been miniaturized for use in mice (Tabuchi et al., 2008) with the latest works utilizing vacuum stabilization of the lung tissue. While this does enable single-cell resolution microscopy in the lung (Looney et al., 2011; Presson et al., 2011; Entenberg et al., 2015; Rodriguez-Tirado et al., 2016), the invasiveness of the surgery and vacuum stabilization prevent long-term imaging (max of ~12 hrs), potentially introduce artifacts, and limit data acquisition to a single imaging session.

To remove these technical limitations, our group has again turned to the field of surgical engineering, which we used to develop a permanent Window for High-Resolution Imaging the murine Lung (WHRIL) (Entenberg et al., 2018). The surgical protocol for this window involves intubation of the mouse (to maintain ventilation during the time the thoracic cavity is breached), resection of a small section (~5 mm) of ribs 6 and 7, and implantation of the window within this opening. Using a combination of sutures and adhesive, the thoracic cavity is then resealed allowing the mouse to be extubated, awaken from anesthesia, and resume normal daily activity. The permanence of this window makes it possible to repeatedly image the lung tissue under the window serially over a period of days to weeks. This window has allowed the first direct observation, with subcellular

resolution, of the arrival, extravasation, and growth of disseminated cells into lung metastases (Entenberg et al., 2018).

Thus, the new development of optical imaging techniques for directly viewing the lung vasculature, particularly serially, as is now possible with the WHRIL, make addressing the question of what role tumor cell clusters play in metastatic dissemination simple and definitive. The technique also holds the promise to elucidate many other biological processes in the lung that have previously required indirect or correlative approaches.

11.10 Conclusion

Cellular motility is a crucial biological process in both health and disease. In particular, different cellular motility phenotypes, such as single-cell and streaming, have been implicated in metastatic spread. While the connection between streaming migration and hematogenous dissemination of tumor cells has been well established, more work still needs to be done to definitively establish whether collective migration is involved in metastatic spread. In particular, it is still to be determined whether the speculated process of collective vascular invasion occurs *in vivo*. The current evidence obtained by intravital imaging points to a model of metastasis, summarized in Figure 11.4a, in which a minority of tumor cells within the bulk tumor are able to dissociate from their neighboring cells, migrate in close association with macrophages toward blood vessels, and then enter the blood vasculature at TMEM doorways. Upon arrival to the secondary site, these tumor cells extravasate and subsequently grow into macrometastases, whereupon the process of dissemination may begin again (Figure 11.4b)

Over the years, intravital imaging has proven to be a powerful technique that can directly visualize and quantify motility phenotypes in a nondestructive manner. Though the technique has traditionally been technically challenging, and, therefore, limited to a few specialized labs, recent technological advances have generated tools and protocols that have simplified the process to the point where it is becoming accessible to any lab with access to a multiphoton microscope. The latest technological advances have dramatically expanded the utility of mouse models by imaging, with subcellular resolution, greater quantities of tissue, over longer periods of time, and in locations that were previously inaccessible. This ability to directly locate, identify, and quantitate cells in their native environs and follow them over extended periods of time, and serially, promises to provide answers to many longstanding biological questions in the fields of developmental and cancer biology.

Acknowledgments

This work was was supported by Einstein's Gruss-Lipper Biophotonics Center and its associated Integrated Imaging Program and by Jane A. and Myles P. Dempsey.

REFERENCES

Aalders, K. C., Tryfonidis, K., Senkus, E., and Cardoso, F. (2017). Anti-angiogenic treatment in breast cancer: Facts, successes, failures and future perspectives. *Cancer Treatment Reviews, 53*: 98–110.

Aceto, N., Bardia, A., Miyamoto, D. T., Donaldson, M. C., Wittner, B. S., Spencer, J. A., Yu, M., Pely, A., Engstrom, A., Zhu, H., Brannigan, B. W., Kapur, R., Stott, S. L., Shioda, T., Ramaswamy, S., Ting, D. T., Lin, C. P., Toner, M., Haber, D. A., and Maheswaran, S. (2014). Circulating tumor cell clusters are oligoclonal precursors of breast cancer metastasis. *Cell, 158*: 1110–1122.

Agarwal, S., Gertler, F. B., Balsamo, M., Condeelis, J. S., Camp, R. L., Xue, X., Lin, J., Rohan, T. E., and Rimm, D. L. (2012). Quantitative assessment of invasive mena isoforms (menacalc) as an independent prognostic marker in breast cancer. *Breast Cancer Research, 14*: R124.

Alexander, S., Koehl, G. E., Hirschberg, M., Geissler, E. K., and Friedl, P. (2008). Dynamic imaging of cancer growth and invasion: A modified skin-fold chamber model. *Histochemistry and Cell Biology, 130*: 1147–1154.

Alieva, M., Ritsma, L., Giedt, R. J., Weissleder, R., and van Rheenen, J. (2014). Imaging windows for long-term intravital imaging: General overview and technical insights. *Intravital, 3*: e29917.

Beerling, E., Oosterom, I., Voest, E., Lolkema, M., and van Rheenen, J. (2016). Intravital characterization of tumor cell migration in pancreatic cancer. *Intravital, 5*: e1261773.

Borriello, L., and DeClerck, Y. A. (2014). Tumor microenvironment and therapeutic resistance process. *Médecine Sciences (Paris), 30*: 445–451.

Bronsert, P., Enderle-Ammour, K., Bader, M., Timme, S., Kuehs, M., Csanadi, A., Kayser, G., Kohler, I., Bausch, D., Hoeppner, J., and Hopt, U. T. (2014). Cancer cell invasion and EMT marker expression: A three-dimensional study of the human cancer-host interface. *The Journal of Pathology, 234*: 410–422.

Cahalan, M. D., Parker, I., Wei, S. H., and Miller, M. J. (2003). Real-time imaging of lymphocytes in vivo. *Current Opinion in Immunology, 15*: 372–377.

Chambers, A. F., MacDonald, I. C., Schmidt, E. E., Koop, S., Morris, V. L., Khokha, R., and Groom, A. C. (1995). Steps in tumor metastasis: New concepts from intravital videomicroscopy. *Cancer Metastasis Reviews, 14*: 279–301.

Cheung, K. J., Gabrielson, E., Werb, Z., and Ewald, A. J. (2013). Collective invasion in breast cancer requires a conserved basal epithelial program. *Cell, 155*: 1639–1651.

Cheung, K. J., Padmanaban, V., Silvestri, V., Schipper, K., Cohen, J. D., Fairchild, A. N., Gorin, M. A., Verdone, J. E., Pienta, K. J., Bader, J. S., and Ewald, A. J. (2016). Polyclonal breast cancer metastases arise from collective dissemination of keratin 14-expressing tumor cell clusters. *Proceedings of the National Academy of Sciences of the United States of America, 113*: E854–E863.

Condeelis, J., Singer, R. H., and Segall, J. E. (2005). The great escape: When cancer cells hijack the genes for chemotaxis and motility. *Annual Review of Cell and Developmental Biology, 21*: 695–718.

Cukierman, E., and Bassi, D. E. (2012). The mesenchymal tumor microenvironment: A drug-resistant niche. *Cell Adhesion and Migration, 6*: 285–296.

De Alva, W. E., and Rainer, W. G. (1963). A method of high speed in vivo pulmonary microcinematography under physiologic conditions. *Angiology, 14*: 160–164.

de Mascarel, I., Bonichon, F., Durand, M., Mauriac, L., MacGrogan, G., Soubeyran, I., Picot, V., Avril, A., Coindre, J. M., and Trojani, M. (1998). Obvious peritumoral emboli: An elusive prognostic factor reappraised. Multivariate analysis of 1320 node-negative breast cancers. *European Journal of Cancer, 34*: 58–65.

De Zanet, S., Rudolph, T., Richa, R., Tappeiner, C., and Sznitman, R. (2016). Retinal slit lamp video mosaicking. *International Journal of Computer Assisted Radiology and Surgery, 11*: 1035–1041.

DeClerck, Y. A., Pienta, K. J., Woodhouse, E. C., Singer, D. S., and Mohla, S. (2017). The tumor microenvironment at a turning point knowledge gained over the last decade, and challenges and opportunities ahead: A white paper from the NCI TME Network. *Cancer Research, 77*: 1051–1059.

Deryugina, E. I., and Kiosses, W. B. (2017). Intratumoral cancer cell intravasation can occur independent of invasion into the adjacent stroma. *Cell Reports, 19*: 601–616.

Duda, D. G., Duyverman, A. M., Kohno, M., Snuderl, M., Steller, E. J., Fukumura, D., and Jain, R. K. (2010). Malignant cells facilitate lung metastasis by bringing their own soil. *Proceedings of the National Academy of Sciences of the United States of America, 107*: 21677–21682.

Dunn, K. W., Sutton, T. A., and Sandoval, R. M. (2007). Live-animal imaging of renal function by multiphoton microscopy. *Current Protocols in Cytometry, 41*: 12–19.

Eddy, R. J., Weidmann, M. D., Sharma, V. P., and Condeelis, J. S. (2017). Tumor cell invadopodia: Invasive protrusions that orchestrate metastasis. *Trends Cell Biology, 27*: 595–607.

Einenkel, J., Braumann, U. D., Horn, L. C., Kuska, J. P., and Hockel, M. (2007). 3-D analysis of the invasion front in squamous cell carcinoma of the uterine cervix: Histopathologic evidence for collective invasion per continuitatem. *Analytical and Quantitative Cytology and Histology, 29*: 279–290.

Entenberg, D., Kedrin, D., Wyckoff, J., Sahai, E., Condeelis, J., and Segall, J. E. (2013). Imaging tumor cell movement in vivo. *Current Protocols in Cell Biology*, vol (58).

Entenberg, D., Rodriguez-Tirado, C., Kato, Y., Kitamura, T., Pollard, J. W., and Condeelis, J. (2015). In vivo subcellular resolution optical imaging in the lung reveals early metastatic proliferation and motility. *Intravital 4*: 1–11.

Entenberg, D., Wyckoff, J., Gligorijevic, B., Roussos, E. T., Verkhusha, V. V., Pollard, J. W., and Condeelis, J. (2011). Setup and use of a two-laser multiphoton microscope for multichannel intravital fluorescence imaging. *Nature Protocols, 6*: 1500–1520.

Entenberg, D., Pastoriza, J. M., Oktay, M. H., Voiculescu, S., Wang, Y., Sosa, M. S., Aguirre-Ghiso, J., and Condeelis, J. (2017). Time-lapsed, large-volume, high-resolution intravital imaging for tissue-wide analysis of single cell dynamics. *Methods, 128*: 65–77.

Entenberg, D., Voiculescu, S., Guo, P., Borriello, L., Wang, Y., Karagiannis, G. S., Jones, J., Baccay, F., Oktay, M., and Condeelis, J. (2018). A permanent window for the murine lung enables high-resolution imaging of cancer metastasis. *Nature Methods, 15*: 73–80.

Farina, K. L., Wyckoff, J. B., Rivera, J., Lee, H., Segall, J. E., Condeelis, J. S., and Jones, J. G. (1998). Cell motility of tumor cells visualized in living intact primary tumors using green fluorescent protein. *Cancer Research, 58*: 2528–2532.

Ferrer, M., Martin-Jaular, L., Calvo, M., and del Portillo, H. A. (2012). Intravital microscopy of the spleen: Quantitative analysis of parasite mobility and blood flow. *Journal of Visualized Experiments*, vol 59.

Fingar, V. H., Taber, S. W., and Wieman, T. J. (1994). A new model for the study of pulmonary microcirculation: Determination of pulmonary edema in rats. *Journal of Surgical Research, 57*: 385–393.

Forse, C. L., Agarwal, S., Pinnaduwage, D., Gertler, F., Condeelis, J. S., Lin, J., Xue, X., Johung, K., Mulligan, A. M., Rohan, T. E., Bull, S. B., and Andrulis, I. L. (2015). Menacalc, a quantitative method of metastasis assessment, as a prognostic marker for axillary node-negative breast cancer. *BMC Cancer, 15*: 483.

Foss, O. P., Brennhovd, I. O., Messelt, O. T., Efskind, J., and Liverud, K. (1966). Invasion of tumor cells into the bloodstream caused by palpation or biopsy of the tumor. *Surgery, 59*: 691–695.

Friedl, P., Noble, P. B., Walton, P. A., Laird, D. W., Chauvin, P. J., Tabah, R. J., Black, M., and Zanker, K. S. (1995). Migration of coordinated cell clusters in mesenchymal and epithelial cancer explants in vitro. *Cancer Research, 55*: 4557–4560.

Gholami, S., Cassidy, M. R., and Strong, V. E. (2017). Minimally invasive surgical approaches to gastric resection. *Surgical Clinics of North America. 97*: 249–264.

Giampieri, S., Manning, C., Hooper, S., Jones, L., Hill, C. S., and Sahai, E. (2009). Localized and reversible TGFbeta signalling switches breast cancer cells from cohesive to single cell motility. *Nature Cell Biology, 11*: 1287–1296.

Gligorijevic, B., and Condeelis, J. (2009). Stretching the timescale of intravital imaging in tumors. *Cell Adhesion and Migration, 3*: 313–315.

Gligorijevic B., Bergman, A., and Condeelis, J. (2014). Multiparametric classification links tumor microenvironments with tumor cell phenotype. *PLoS Biology, 12*: e1001995.

Goswami, S., Philippar, U., Sun, D., Patsialou, A., Avraham, J., Wang, W., Di Modugno, F., Nistico, P., Gertler, F. B., and Condeelis, J. S. (2009). Identification of invasion specific splice variants of the cytoskeletal protein Mena present in mammary tumor cells during invasion in vivo. *Clinical & Experimental Metastasis, 26*: 153–159.

Groh, J., Kuhnle, G. E., Kuebler, W. M., and Goetz, A. E. (1992). An experimental model for simultaneous quantitative analysis of pulmonary micro- and macrocirculation during unilateral hypoxia in vivo. *Research in Experimental Medicine 192*: 431–441.

Gujam, F. J., Going, J. J., Edwards, J., Mohammed, Z. M., and McMillan, D. C. (2014). The role of lymphatic and blood vessel invasion in predicting survival and methods of detection in patients with primary operable breast cancer. *Critical Reviews in Oncology/Hematology 89*: 231–241.

Haessler, U., Pisano, M., Wu, M., and Swartz, M. A. (2011). Dendritic cell chemotaxis in 3D under defined chemokine gradients reveals differential response to ligands CCL21 and CCL19. *Proceedings of the National Academy of Sciences of the United States of America 108*: 5614–5619.

Hak, S., Reitan, N. K., Haraldseth, O., and de Lange Davies, C. (2010). Intravital microscopy in window chambers: A unique tool to study tumor angiogenesis and delivery of nanoparticles. *Angiogenesis, 13*: 113–130.

Hanahan, D., and Weinberg, R. A. (2011). Hallmarks of cancer: The next generation. *Cell, 144*: 646–674.

Hansen, E., Wolff, N., Knuechel, R., Ruschoff, J., Hofstaedter, F., and Taeger, K. (1995). Tumor cells in blood shed from the surgical field. *Archives of Surgery, 130*: 387–393.

Harney, A. S., Wang, Y., Condeelis, J. S., and Entenberg, D. (2016). Extended time-lapse intravital imaging of real-time multicellular dynamics in the tumor microenvironment. *Journal of Visualized Experiments: JoVE*: vol 142, e54042.

Harney, A. S., Arwert, E. N., Entenberg, D., Wang, Y., Guo, P., Qian, B. Z., Oktay, M. H., Pollard, J. W., Jones, J. G., and Condeelis, J. S. (2015). Real-time imaging reveals local, transient vascular permeability, and tumor cell intravasation stimulated by TIE2hi macrophage-derived VEGFA. *Cancer Discovery, 5*: 932–943.

Harney, A. S., Karagiannis, G. S., Pignatelli, J., Smith, B. D., Kadioglu, E., Wise, S. C., Hood, M. M., Kaufman, M. D., Leary, C. B., Lu, W. P., Al-Ani, G., Chen, X., Entenberg, D., Oktay, M. H., Wang, Y., Chun, L., De Palma, M., Jones, J. G., Flynn, D. L., and Condeelis, J. S. (2017). The selective Tie2 inhibitor rebastinib blocks recruitment and function of Tie2(Hi) macrophages in breast cancer and pancreatic neuroendocrine tumors. *Molecular Cancer Therapeutics, 16*: 2486–2501.

Harper, K. L., Sosa, M. S., Entenberg, D., Hosseini, H., Cheung, J. F., Nobre, R., Avivar-Valderas, A., Nagi, C., Girnius, N., Davis, R. J., Farias, E. F., Condeelis, J., Klein, C. A., and Aguirre-Ghiso, J. A. (2016). Mechanism of early dissemination and metastasis in Her2(+) mammary cancer. *Nature, 540*: 589–612.

Heindl, A., Nawaz, S., and Yuan, Y. (2015). Mapping spatial heterogeneity in the tumor microenvironment: A new era for digital pathology. *Laboratory Investigation, 95*: 377–384.

Hosseini, H., Obradović, M. M., Hoffmann, M., Harper, K. L., Sosa, M. S., Werner-Klein, M., Nanduri, L. K., Werno, C., Ehrl, C., Maneck, M., and Patwary, N. (2016). Early dissemination seeds metastasis in breast cancer. *Nature, 540*: 552–558.

Ibrahim, A. E., Sarhane, K. A., and Selber, J. C. (2017). New frontiers in robotic-assisted microsurgical reconstruction. *Clinics in Plastic Surgery, 44*: 415–423.

Ilina, O., Campanello, L., Gritsenko, P. G., Vullings, M., Wang, C., Bult, P., Losert, W., and Friedl, P. (2018). Intravital microscopy of collective invasion plasticity in breast cancer. *Disease Models & Mechanisms, 11*: dmm034330.

Imhof, B. A., and Dunon, D. (1997). Basic mechanism of leukocyte migration. *Hormone and Metabolic Research, 29*: 614–621.

Ito, K., Smith, B. R., Parashurama, N., Yoon, J. K., Song, S. Y., Miething, C., Mallick, P., Lowe, S., and Gambhir, S. S. (2012). Unexpected dissemination patterns in lymphoma progression revealed by serial imaging within a murine lymph node. *Cancer Research, 72*: 6111–6118.

Jeong, H. S., Jones, D., Liao, S., Wattson, D. A., Cui, C. H., Duda, D. G., Willett, C. G., Jain, R. K., and Padera, T. P. (2015). Investigation of the lack of angiogenesis in the formation of lymph node metastases. *Journal of the National Cancer Institute, 107*: 1–11.

Joy, A. A. (2008). San Antonio Breast Cancer Symposium 2007 - Adjuvant endocrine therapy update: ATAC 100 highlights. *Current Oncology, 15*: 68–69.

Karagiannis, G. S., Condeelis, J. S., and Oktay, M. H. (2018). Chemotherapy-induced metastasis: Mechanisms and translational opportunities. *Clinical & Experimental Metastasis, 35*: 269–284.

Karagiannis, G. S., Pastoriza, J. M., Wang, Y., Harney, A. S., Entenberg, D., Pignatelli, J., Sharma, V. P., Xue, E. A., Cheng, E., D'Alfonso, T. M., Jones, J. G., Anampa, J., Rohan, T. E., Sparano, J. A., Condeelis, J. S., and Oktay, M. H. (2017). Neoadjuvant chemotherapy induces breast cancer metastasis through a TMEM-mediated mechanism. *Science Translational Medicine, 9*: 1–15.

Katharina, P. (2011). Tumor cell seeding during surgery-possible contribution to metastasis formations. *Cancers, 3*: 2540–2553.

Kato, T., Kameoka, S., Kimura, T., Nishikawa, T., and Kobayashi, M. (2003). The combination of angiogenesis and blood vessel invasion as a prognostic indicator in primary breast cancer. *British Journal of Cancer, 88*: 1900–1908.

Kats-Ugurlu, G., Roodink, I., de Weijert, M., Tiemessen, D., Maass, C., Verrijp, K., van der Laak, J., de Waal, R., Mulders, P., Oosterwijk, E., and Leenders, W. (2009). Circulating tumour tissue fragments in patients with pulmonary metastasis of clear cell renal cell carcinoma. *The Journal of Pathology, 219*: 287–293.

Kedrin, D., Gligorijevic, B., Wyckoff, J., Verkhusha, V. V., Condeelis, J., Segall, J. E., and van Rheenen, J. (2008). Intravital imaging of metastatic behavior through a mammary imaging window. *Nature Methods, 5*: 1019–1021.

Khalil, A. A., Ilina, O., Gritsenko, P. G., Bult, P., Span, P. N., and Friedl, P. (2017). Collective invasion in ductal and lobular breast cancer associates with distant metastasis. *Clinical & Experimental Metastasis. 34*: 421–429.

Kienast, Y., von Baumgarten, L., Fuhrmann, M., Klinkert, W. E., Goldbrunner, R., Herms, J., and Winkler, F. (2010). Real-time imaging reveals the single steps of brain metastasis formation. *Nature Methods, 16*: 116–122.

Kitamura, T., Pollard, J. W., and Vendrell, M. (2017). Optical windows for imaging the metastatic tumour microenvironment in vivo. *Trends in Biotechnology, 35*: 5–8.

Kothari, S., Phan, J. H., Stokes, T. H., and Wang, M. D. (2013). Pathology imaging informatics for quantitative analysis of whole-slide images. *Journal of the American Medical Informatics Association, 20*: 1099–1108.

Krasna, M. J., Flancbaum, L., Cody, R. P., Shneibaum, S., and Ben Ari, G. (1988). Vascular and neural invasion in colorectal carcinoma. Incidence and prognostic significance. *Cancer, 61*: 1018–1023.

Krause, M., Dent, E. W., Bear, J. E., Loureiro, J. J., and Gertler, F. B. (2003). Ena/VASP proteins: Regulators of the actin cytoskeleton and cell migration. *Annual Review of Cell and Developmental Biology, 19*: 541–564.

Kumar, V., Abbas, A. K., and Aster, J. C. (2017). *Robbins basic pathology e-book.* Elsevier Health Sciences.

Kwasnitschka T., Koser, K., Sticklus, J., Rothenbeck, M., Weiss, T., Wenzlaff, E., Schoening, T., Triebe, L., Steinfuhrer, A., Devey, C., and Greinert, J. (2016). DeepSurveyCam–A deep ocean optical mapping system. *Sensors, 16*: 164.

Lalan, S., Pomerantseva, I., and Vacanti, J. P. (2001). Tissue engineering and its potential impact on surgery. *World Journal of Surgery, 25*: 1458–1466.

Lamm, W. J., Bernard, S. L., Wagner, W. W., Jr., and Glenny, R. W. (2005). Intravital microscopic observations of 15-micron microspheres lodging in the pulmonary microcirculation. *Journal of Applied Physiology, 98*: 2242–2248.

Lehr, H. A., Leunig, M., Menger, M. D., Nolte, D., and Messmer, K. (1993). Dorsal skinfold chamber technique for intravital microscopy in nude mice. *The American Journal of Pathology, 143*: 1055–1062.

Leighton, J., Kalla, R. L., Turner, J. M., Jr., and Fennell, R. H., Jr. (1960). Pathogenesis of tumor invasion. II. Aggregate replication. *Cancer Research, 20*: 575–586.

Leung, E., Xue, A., Wang, Y., Rougerie, P., Sharma, V. P., Eddy, R., Cox, D., and Condeelis, J. (2016). Blood vessel endothelium-directed tumor cell streaming in breast tumors requires the HGF/C-Met signaling pathway. *Oncogene, 36*: 2680.

Leung, E., Xue, A., Wang, Y., Rougerie, P., Sharma, V. P., Eddy, R., Cox, D., and Condeelis, J. (2017). Blood vessel endothelium-directed tumor cell streaming in breast tumors requires the HGF/C-Met signaling pathway. *Oncogene, 36*: 2680–2692.

Liang, C. C., Park, A. Y., and Guan, J. L. (2007). In vitro scratch assay: A convenient and inexpensive method for analysis of cell migration in vitro. *Nature Protocols, 2*: 329–333.

Lianos, G. D., Christodoulou, D. K., Katsanos, K. H., Katsios, C., and Glantzounis, G. K. (2017). Minimally invasive surgical approaches for pancreatic adenocarcinoma: Recent trends. *Journal of Gastrointestinal Cancer, 48*: 129–134.

Lin, E. Y., Jones, J. G., Li, P., Zhu, L., Whitney, K. D., Muller, W. J., and Pollard, J. W. (2003). Progression to malignancy in the polyoma middle T oncoprotein mouse breast cancer model provides a reliable model for human diseases. *The American Journal of Pathology, 163*: 2113–2126.

Liotta, L. A., Saidel, M. G., and Kleinerman, J. (1976). The significance of hematogenous tumor cell clumps in the metastatic process. *Cancer Research, 36*: 889–894.

Litjens, G., Kooi, T., Bejnordi, B. E., Setio, A. A. A., Ciompi, F., Ghafoorian, M., van der Laak, J., van Ginneken, B., and Sanchez, C. I. (2017). A survey on deep learning in medical image analysis. *Medical Image Analysis, 42*: 60–88.

Lloyd, M. C., Rejniak, K. A., Brown, J. S., Gatenby, R. A., Minor, E. S., and Bui, M. M. (2015). Pathology to enhance precision medicine in oncology: Lessons from landscape ecology. *Advances in Anatomic Pathology, 22*: 267–272.

Lo Celso, C., Lin, C. P., and Scadden, D. T. (2011). In vivo imaging of transplanted hematopoietic stem and progenitor cells in mouse calvarium bone marrow. *Nature Protocols, 6*: 1–14.

Loo, B. W., Jr., Meyer-Ilse, W., and Rothman, S. S. (2000). Automatic image acquisition, calibration and montage assembly for biological X-ray microscopy. *Journal of Microscopy, 197*: 185–201.

Looney, M. R., Thornton, E. E., Sen, D., Lamm, W. J., Glenny, R. W., and Krummel, M. F. (2011). Stabilized imaging of immune surveillance in the mouse lung. *Nature Methods, 8*: 91–96.

Marques, P. E., Oliveira, A. G., Chang, L., Paula-Neto, H. A., and Menezes, G. B. (2015). Understanding liver immunology using intravital microscopy. *Journal of Hepatology, 63*: 733–742.

Masedunskas, A., Porat-Shliom, N., Tora, M., Milberg, O., and Weigert, R. (2013). Intravital microscopy for imaging subcellular structures in live mice expressing fluorescent proteins. *Journal of Visualized Experiments: JoVE, 79*: e50558.

Meacham, C. E., and Morrison, S. J. (2013). Tumour heterogeneity and cancer cell plasticity. *Nature, 501*: 328–337.

Mohammed, R. A., Martin, S. G., Gill, M. S., Green, A. R., Paish, E. C., and Ellis, I. O. (2007). Improved methods of detection of lymphovascular invasion demonstrate that it is the predominant method of vascular invasion in breast cancer and has important clinical consequences. *The American Journal of Surgical Pathology, 31*: 1825–1833.

Mohammed R. A., Martin, S. G., Mahmmod, A. M., Macmillan, R. D., Green, A. R., Paish, E. C., and Ellis, I. O. (2011). Objective assessment of lymphatic and blood vascular invasion in lymph node-negative breast carcinoma: Findings from a large case series with long-term follow-up. *The Journal of Pathology, 223*: 358–365.

Muller, A., Homey, B., Soto, H., Ge, N., Catron, D., Buchanan, M., McClanahan, T., Murphy, E., Yuan, W., Wagner, S., Barrera, J., Mohar, A., Verastegui, E., and Zlotnik, A. (2001). Involvement of chemokine receptors in breast cancer metastasis. *Nature, 410*: 50–56.

Narod, S. A., Iqbal, J., and Miller, A. B. (2015). Why have breast cancer mortality rates declined? *Journal of Cancer Policy, 5*: 8–17.

Nobre, A. R., Entenberg, D., Wang, Y., Condeelis, J., and Aguirre-Ghiso, J. A. (2018). The different routes to metastasis via hypoxia-regulated programs. *Trends Cell Biology, 28*: 941–956.

Palomino, D. C., and Marti, L. C. (2015). Chemokines and immunity. *Einstein, 13*: 469–473.

Papenfuss, H. D., Gross, J. F., Intaglietta, M., and Treese, F. A. (1979). A transparent access chamber for the rat dorsal skin fold. *Microvascular Research, 18*: 311–318.

Patsialou, A., Wyckoff, J., Wang, Y., Goswami, S., Stanley, E. R., and Condeelis, J. S. (2009). Invasion of human breast cancer cells in vivo requires both paracrine and autocrine loops involving the colony-stimulating factor-1 receptor. *Cancer Research, 69*: 9498–9506.

Patsialou, A., Bravo-Cordero, J. J., Wang, Y., Entenberg, D., Liu, H., Clarke, M., and Condeelis, J. S. (2013). Intravital multiphoton imaging reveals multicellular streaming as a crucial component of in vivo cell migration in human breast tumors. *Intravital, 2*: e25294.

Peng, H. T., and Shek, P. N. (2010). Novel wound sealants: Biomaterials and applications. *Expert Review of Medical Devices, 7*: 639–659.

Pignatelli, J., Goswami, S., Jones, J. G., Rohan, T. E., Pieri, E., Chen, X., Adler, E., Cox, D., Maleki, S., Bresnick, A., Gertler, F. B., Condeelis, J. S., and Oktay, M. H. (2014). Invasive breast carcinoma cells from patients exhibit MenaINV- and macrophage-dependent transendothelial migration. *Science Signaling, 7*: ra112.

Pineda, C. M., Park, S., Mesa, K. R., Wolfel, M., Gonzalez, D. G., Haberman, A. M., Rompolas, P., and Greco, V. (2015). Intravital imaging of hair follicle regeneration in the mouse. *Nature Protocols, 10*: 1116–1130.

Presson, R. G., Jr., Brown, M. B., Fisher, A. J., Sandoval, R. M., Dunn, K. W., Lorenz, K. S., Delp, E. J., Salama, P., Molitoris, B. A., and Petrache, I. (2011). Two-photon imaging within the murine thorax without respiratory and cardiac motion artifact. *The American Journal of Pathology, 179*: 75–82.

Reismann, D., Stefanowski, J., Gunther, R., Rakhymzhan, A., Matthys, R., Nutzi, R., Zehentmeier, S., Schmidt-Bleek, K., Petkau, G., Chang, H. D., Naundorf, S., Winter, Y., Melchers, F., Duda, G., Hauser, A. E., and Niesner, R. A. (2017). Longitudinal intravital imaging of the femoral bone marrow reveals plasticity within marrow vasculature. *Nature Communications, 8*: 2153.

Repesh, L. A. (1989). A new in vitro assay for quantitating tumor cell invasion. *Invasion Metastasis, 9*: 192–208.

Ritsma, L., Steller, E. J., Beerling, E., Loomans, C. J., Zomer, A., Gerlach, C., Vrisekoop, N., Seinstra, D., van Gurp, L., Schafer, R., Raats, D. A., de Graaff, A., Schumacher, T. N., de Koning, E. J., Rinkes, I. H., Kranenburg, O., and van Rheenen, J. (2012). Intravital microscopy through an abdominal imaging window reveals a pre-micrometastasis stage during liver metastasis. *Science Translational Medicine, 4*: 158ra145.

Robinson B. D., Sica, G. L., Liu, Y. F., Rohan, T. E., Gertler, F. B., Condeelis, J. S., and Jones, J. G. (2009). Tumor microenvironment of metastasis in human breast carcinoma: A potential prognostic marker linked to hematogenous dissemination. *Clinical Cancer Research, 15*: 2433–2441.

Rodriguez-Tirado, C., Kitamura, T., Kato, Y., Pollard, J. W., Condeelis, J. S., and Entenberg, D. (2016). Long-term high-resolution intravital microscopy in the lung with a vacuum stabilized imaging window. *Journal of Visualized Experiments: JoVE, 116*, e54603.

Roh-Johnson, M., Bravo-Cordero, J. J., Patsialou, A., Sharma, V. P., Guo, P., Liu, H., Hodgson, L., and Condeelis, J. (2014). Macrophage contact induces RhoA GTPase signaling to trigger tumor cell intravasation. *Oncogene, 33*: 4203–4212.

Rohan, T. E., Xue, X., Lin, H. M., D'Alfonso, T. M., Ginter, P. S., Oktay, M. H., Robinson, B. D., Ginsberg, M., Gertler, F. B., Glass, A. G., Sparano, J. A., Condeelis, J. S, and Jones, J. G. (2014). Tumor microenvironment of metastasis and risk of distant metastasis of breast cancer. *Journal of the National Cancer Institute, 106.*

Roodhart, J. M., He, H., Daenen, L. G., Monvoisin, A., Barber, C. L., van Amersfoort, M., Hofmann, J. J., Radtke, F., Lane, T. F., Voest, E. E., and Iruela-Arispe, M. L. (2013). Notch1 regulates angio-supportive bone marrow-derived cells in mice: Relevance to chemoresistance. *Blood, 122*: 143–153.

Rosen, J. M., and Roarty, K. (2014). Paracrine signaling in mammary gland development: What can we learn about intratumoral heterogeneity? *Breast Cancer Research: BCR, 16*: 202.

Roussos, E. T., Condeelis, J. S., and Patsialou, A. (2011a). Chemotaxis in cancer. *Nature Reviews Cancer, 11*: 573–587.

Roussos E.T., Wang, Y., Wyckoff, J. B., Sellers, R. S., Wang, W., Li, J., Pollard, J. W., Gertler, F. B., Condeelis, J. S. (2010). Mena deficiency delays tumor progression and decreases metastasis in polyoma middle-T transgenic mouse mammary tumors. *Breast Cancer Research: BCR, 12*: R101.

Roussos, E. T., Balsamo, M., Alford, S. K., Wyckoff, J. B., Gligorijevic, B., Wang, Y., Pozzuto, M., Stobezki, R., Goswami, S., Segall, J. E., Lauffenburger, D. A., Bresnick, A. R., Gertler, F. B., and Condeelis, J. S. (2011b). Mena invasive (MenaINV) promotes multicellular streaming motility and transendothelial migration in a mouse model of breast cancer. *Journal of Cell Science, 124*: 2120–2131.

Sachdev, J. C., and Jahanzeb, M. (2016). Use of cytotoxic chemotherapy in metastatic breast cancer: Putting taxanes in perspective. *Clin Breast Cancer, 16*: 73–81.

Seshamani, S., Lau, W., and Hager, G. (2006). Real-time endoscopic mosaicking. *Medical Image Computing and Computer-Assisted Intervention, 9*: 355–363.

Seyfried, T. N., and Huysentruyt, L. C. (2013). On the origin of cancer metastasis. *Critical Reviews in Oncogenesis, 18*: 43–73.

Shaked, Y. (2016). Balancing efficacy of and host immune responses to cancer therapy: The yin and yang effects. *Nature Reviews Clinical Oncology, 13*: 611–626.

Shaked, Y., and Voest, E. E. (2009). Bone marrow derived cells in tumor angiogenesis and growth: Are they the good, the bad or the evil? *Biochimica et Biophysica Acta, 1796*: 1–4.

Shan, S., Sorg, B., and Dewhirst, M. W. (2003). A novel rodent mammary window of orthotopic breast cancer for intravital microscopy. *Microvascular Research, 65*: 109–117.

Shapiro, I. M., Cheng, A. W., Flytzanis, N. C., Balsamo, M., Condeelis, J. S., Oktay, M. H., Burge, C. B., and Gertler, F. B. (2011). An EMT-driven alternative splicing program occurs in human breast cancer and modulates cellular phenotype. *PLoS Genetics, 7*: e1002218.

Sipkins, D. A., Wei, X., Wu, J. W., Runnels, J. M., Cote, D., Means, T. K., Luster, A. D., Scadden, D. T., and Lin, C. P. (2005). In vivo imaging of specialized bone marrow endothelial microdomains for tumour engraftment. *Nature, 435*: 969–973.

Sparano, J. A., Gray, R., Oktay, M. H., Entenberg, D., Rohan, T., Xue, X., Donovan, M., Peterson, M., Shuber, A., Hamilton, D. A., D'Alfonso, T., Goldstein, L. J., Gertler, F., Davidson, N. E., Condeelis, J., and Jones, J. (2017). A metastasis biomarker (MetaSite Breast Score) is associated with distant recurrence in hormone receptor-positive, HER2-negative early-stage breast cancer. *Nature PJ Breast Cancer, 3*: 42.

Squirrell, J. M., Wokosin, D. L., White, J. G., and Bavister, B. D. (1999). Long-term two-photon fluorescence imaging of mammalian embryos without compromising viability. *Nature Biotechnology, 17*: 763–767.

Sugino, T., Kusakabe, T., Hoshi, N., Yamaguchi, T., Kawaguchi, T., Goodison, S., Sekimata, M., Homma, Y., and Suzuki, T. (2002). An invasion-independent pathway of blood-borne metastasis: A new murine mammary tumor model. *The American Journal of Pathology, 160*: 1973–1980.

Tabuchi, A., Mertens, M., Kuppe, H., Pries, A. R., and Kuebler, W. M. (2008). Intravital microscopy of the murine pulmonary microcirculation. *Journal of Applied Physiology, 104*: 338–346.

Teddy, J. M., and Kulesa, P. M. (2004). In vivo evidence for short- and long-range cell communication in cranial neural crest cells. *Development, 131*: 6141–6151.

Terry, R. J. (1939). A thoracic window for observation of the lung in a living animal. *Science, 90*: 43–44.

Tyzzer, E. E. (1913). Factors in the production and growth of tumor metastases. *The Journal of Medical Research, 28*: 309–332.

van den Tweel, J. G., and Taylor, C. R. (2010). A brief history of pathology: Preface to a forthcoming series that highlights milestones in the evolution of pathology as a discipline. *Virchows Archiv, 457*: 3–10.

van Wyk, H. C., Roxburgh, C. S., Horgan, P. G., Foulis, A. F., and McMillan, D. C. (2014). The detection and role of lymphatic and blood vessel invasion in predicting survival in patients with node negative operable primary colorectal cancer. *Critical Reviews in Oncology/Hematology, 90*: 77–90.

Velnar, T., Bunc, G., Klobucar, R., Gradisnik, L. (2016). Biomaterials and host versus graft response: A short review. *Bosnian Journal of Basic Medical Sciences, 16*: 82–90.

Vinegoni, C., Lee, S., Gorbatov, R., and Weissleder, R. (2012). Motion compensation using a suctioning stabilizer for intravital microscopy. *Intravital, 1*: 115–121.

Wagner, W. W., Jr. (1969). Pulmonary microcirculatory observations in vivo under physiological conditions. *Journal of Applied Physiology, 26*: 375–377.

Wagner, W. W., Jr., and Filley, G. F. (1965). Microscopic observation of the lung in vivo. *Vascular Disease, 2*: 229–241.

Wang, W., Goswami, S., Lapidus, K., Wells, A. L., Wyckoff, J. B., Sahai, E., Singer, R. H., Segall, J. E., and Condeelis, J. S. (2004). Identification and testing of a gene expression signature of invasive carcinoma cells within primary mammary tumors. *Cancer Research, 64*: 8585–8594.

Wearn, J. T., Barr, J. S., and German, W. J. (1926). The Behavior of the arterioles and capillaries of the lung. *Experimental Biology and Medicine, 24*: 114–115.

Weber, M., Hauschild, R., Schwarz, J., Moussion, C., de Vries, I., Legler, D. F., Luther, S. A., Bollenbach, T., and Sixt, M. (2013). Interstitial dendritic cell guidance by haptotactic chemokine gradients. *Science, 339*: 328–332.

Weigelin, B., Bakker, G.-J., and Friedl, P. (2012). Intravital third harmonic generation microscopy of collective melanoma cell invasion: Principles of interface guidance and microvesicle dynamics. *IntraVital, 1*: 32–43.

Weninger, W., Biro, M., and Jain, R. (2014). Leukocyte migration in the interstitial space of non-lymphoid organs. *Nature Reviews Immunology, 14*:232–246.

Wyckoff, J., Gligorijevic, B., Entenberg, D., Segall, J., and Condeelis, J. (2011). High-resolution multiphoton imaging of tumors in vivo. *Cold Spring Harbor Protocols, 2011*: 1167–1184.

Wyckoff, J., Wang, W., Lin, E. Y., Wang, Y., Pixley, F., Stanley, E. R., Graf, T., Pollard, J. W., Segall, J., and Condeelis, J. (2004). A paracrine loop between tumor cells and macrophages is required for tumor cell migration in mammary tumors. *Cancer Research, 64*: 7022–7029.

Wyckoff, J. B., Segall, J. E., and Condeelis, J. S. (2000a). The collection of the motile population of cells from a living tumor. *Cancer Research, 60*: 5401–5404.

Wyckoff, J. B., Jones, J. G., Condeelis, J. S., and Segall, J. E. (2000b). A critical step in metastasis: In vivo analysis of intravasation at the primary tumor. *Cancer Research, 60*: 2504–2511.

Wyckoff, J. B., Wang, Y., Lin, E. Y., Li, J. F., Goswami, S., Stanley, E. R., Segall, J. E., Pollard, J. W., and Condeelis, J. (2007). Direct visualization of macrophage-assisted tumor cell intravasation in mammary tumors. *Cancer Research, 67*: 2649–2656.

Zipfel, W. R., Williams, R. M., and Webb, W. W. (2003). Nonlinear magic: Multiphoton microscopy in the biosciences. *Nature Biotechnology, 21*: 1369–1377.

Zomer, A., Ellenbroek, S. I., Ritsma, L., Beerling, E., Vrisekoop, N., and Van Rheenen, J. (2013). Intravital imaging of cancer stem cell plasticity in mammary tumors. *Stem Cells, 31*:602–606.

12

Imaging Cellular Signaling In Vivo Using Fluorescent Protein Biosensors

Christopher A. Reissaus, Richard N. Day, and Kenneth W. Dunn

CONTENTS

12.1 Introduction

Eukaryotic cells have spatial and temporal responses to extracellular changes that can be missed by many conventional endpoint assays, including immunostaining or western blotting. To monitor cell-signaling events in real-time, researchers developed genetically encoded fluorescent reporter probes herein referred to as biosensors. These biosensors typically utilize the fluorescent proteins (FPs), such as GFP or one of its variants, to provide a quantitative output as a cell responds to a stimulus. The collection of biosensors continues to grow annually, but the most common probes have focused on measuring protein activity (e.g. protein kinase A – AKAR) (Tao et al., 2015) and intracellular ions and metabolites (e.g. calcium – GCaMP's) (Chen et al., 2013). The characterization of biosensors *in vitro* using tissue culture models has opened the door to more complex studies in living organisms. The development of transgenic mouse models expressing the biosensors, or viral transduction of vectors encoding the biosensors, have been used successfully to monitor cell-signaling events in living animals (Tao et al., 2015; Warren et al., 2018).

This chapter discusses the application of two-photon laser-scanning microscopy (TPLSM) to monitor biosensor activity in living animals. This intravital microscopy (IVM) approach enables measurements of cellular and subcellular activities within an intact organ, providing important information about cellular physiology and pathophysiology in the context of the living organism. This is critical because pathologies originate in the organized cellular environment *in vivo* and can only be fully understood in this context (Follain et al., 2017).

However, translating FP imaging methods developed in cultured cells into methods that can be used to analyze cell biology *in vivo* are complicated by several unique challenges. First, IVM of genetically encoded FP-based biosensors requires developing methods to target the probe to the cells of interest at suitable levels, as well as validation of the behavior of that probe *in vivo*. In addition, high resolution imaging at depth in scattering tissue typically requires TPLSM, which imposes additional constraints on the choice of FPs and approach used to image them. Furthermore, quantitative imaging in the living animal requires methods that ensure the physiological state of the animal while on the microscope stage. Finally, the methods need to be comparatively insensitive to the loss of signal that occurs with imaging at depth, while minimizing the degradation of images caused by respiratory and cardiac motion artifacts in the living animal. Altogether, TPLSM is required for most IVM studies but should be approached with caution when FP biosensors are incorporated in the experimental scheme.

12.2 Fluorescent Protein Biosensors

The directed evolution of the FPs has produced an extensive palette of fluorescent proteins that emit across the visible spectrum (Follain et al., 2017). Many recently developed FPs have been engineered to overcome deficiencies associated with prior versions, including problems with their production in living cells, their intrinsic brightness, photostability, and sensitivity to different cellular environments. These improved FP variants are being incorporated into new genetically encoded biosensors. Generally, the biosensors utilize either a single FP that changes its spectral characteristics when bound by analyte or chemically modified or dual FP probes that exploit changes in FRET when the sensor domain undergoes conformational change (Oldach and Zhang, 2014). A brief list of biosensors that have been specifically utilized in the study of cellular signaling within visceral organs can be found in Table 12.1.

In single FP biosensors, a sensing peptide sequence, which can bind analytes or be chemically altered by enzymes, is attached to or inserted with the sequence for the FP (Oldach and Zhang, 2014; Miyawaki, 2011; DiPilato and Zhang, 2010; Zhang et al., 2002). Conformational changes in the probe lead to changes in fluorescence. Most single FP biosensors exploit changes in fluorescent intensity that occur with analyte binding or chemical modification. Some of these probes,

however, change the excitation or emission spectrum of the FP in response to cellular events (e.g. RO-GFP2 (Gutscher et al., 2008), allowing the biosensor activity to be measured ratiometrically. Ratiometric probes provide several advantages over "intensity" probes; the readout is independent of FP concentration, light path, and field illumination heterogeneity. Ratiometric values from different cells can be compared or even calibrated to reflect specific analyte concentrations.

Some biosensors are designed to exploit Förster resonance energy transfer (FRET) to directly measure changes in probe conformation. FRET is a nonradiative, quenching pathway that allows a fluorophore in the excited-state (the donor) to transition to the ground state by transferring energy to another nearby molecule (the acceptor). This process depletes the donor's excited-state energy, quenching its emission while causing sensitized emission from the acceptor. Because FRET occurs through near-field dipole interactions, the transfer of energy between donor and acceptor FPs is limited to separation distances less than about 80 Å (Periasamy and Day, 2005).

The transgene for FRET-based biosensors encodes a reporter module that consists of donor and acceptor FPs (Figure 12.1A). Like single FP biosensors, FRET probes are responsive to the activity of kinases or the binding of ions, small molecules, or protein ligands (Oldach and Zhang, 2014; Miyawaki, 2011; DiPilato and Zhang, 2010; Zhang et al., 2002). Changes in the biosensor conformation that accompany the cell-signaling event alter FRET between the FPs in the reporter module. Because FRET quenches the donor signal while causing increased emission from the acceptor (sensitized emission),

TABLE 12.1

List of Biosensors That Have Been Utilized to Visualize Intracellular Signaling within Abdominal Organs

Biosensor Analyte or Target	Biosensor Name	Organ
Protein Kinase A	AKAR (Tao et al., 2015)	Liver
RhoA	RhoA-FRET (Nobis et al., 2017)	Intestine, pancreas
Calcium	GCaMP (Burford et al., 2014)	Kidney
	Chameleon (Yoshikawa et al., 2016)	Spleen
Rac1	Raichu-Rac1 (Warren et al., 2018; Johnsson et al., 2014)	Intestine, pancreas
ERK	EKAREV (Sano et al., 2016)	Bladder

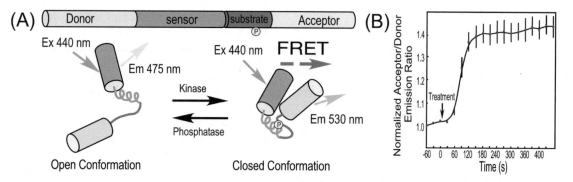

FIGURE 12.1 Schematic representation of a FRET-based biosensor for kinase activity. (A) The biosensor reporter module (e.g. AKAR) consists of donor and acceptor FPs that are linked by a sensing unit designed to detect phosphorylation by a specific kinase. The sensor peptide unit contains the kinase's recognition sequence, which becomes phosphorylated when the kinase is active. The phosphorylation event alters the biosensor tertiary structure, leading to altered energy transfer between the two fluorescent proteins. (B) FRET quenches the donor signal while causing increased emission from the acceptor. Ratiometric imaging of the acceptor to donor emission intensities allows measurement of dynamic changes in probe conformation.

the ratio of the acceptor and donor emission intensities provides a sensitive and dynamic measure of probe conformational changes (Figure 12.1B).

The ratiometric imaging of FRET-based biosensors enables a straightforward method to monitor changing biosensor activities (Warren et al., 2018; Follain et al., 2017; Day and Davidson, 2014; Oldach and Zhang, 2014). Because the donor and acceptor FPs are linked to one another, the biosensors can be internally calibrated. Variations in the ratio of the acceptor to the donor emission over time provide a quantitative measure of changing probe conformation that results from cell-signaling events. A distinct advantage of ratiometric imaging compared to other intensity or time-resolved techniques is the speed of acquisition. Because the donor and acceptor intensity images are acquired simultaneously, the approach is only limited by the integration time necessary to obtain sufficient signal for analysis (depending on the size of the region of interest analyzed, this can range from 10s of milliseconds to a few seconds for each time point). The advantage of the ratiometric method for IVM is that it requires only a single 2PE wavelength to monitor the activity of genetically encoded biosensors produced in the cells of living animals (see Section 12.3).

12.3 Challenges of IVM of Biosensors – Imaging Living Tissues

12.3.1 The Limitations of Multiphoton Microscopy

Wide-field microscopy has been used for intravital studies for nearly 200 years (Cohnheim, 1889; Wagner, 1839) and to collect fluorescence images of the internal organs of living animals for nearly 100 years (Ellinger and Hirt, 1929). However, the out-of-plane fluorescence of living tissues is such that high-resolution imaging typically requires some form of optical sectioning, such as selective plane illumination, confocal, or multiphoton microscopy. Selective plane illumination microscopy (SPIM) is a very powerful approach for imaging deep into biological tissues, up to a few millimeters, at high speed without sacrificing resolution. However, the orthogonal arrangement of the illumination and imaging planes limits SPIM to studies of small organisms that can be placed at the intersection of the illumination and imaging planes. Because our focus is on studies of mammalian systems, we refer the reader to recent review articles for more information about the use of SPIM in intravital microscopy (Power and Huisken, 2017; Strobl et al., 2017). Intravital microscopy of the organs in mice and rats can be accomplished with confocal microscopy, using either single-point scanning (Farina et al.,; Wyckoff et al., 2000) or spinning disk designs (Jenne et al., 2011; Falati et al., 2002; Mitsuoka et al., 1999). However, confocal imaging of living tissues is limited to depths of only the first 10–20 microns because of the effects of light scattering, which degrade the focus and deflect emissions away from the confocal pinhole. Consequently, most intravital studies are conducted using multiphoton microscopy. While some *in vivo* applications have been developed using 3-photon excitation (Rowlands et al., 2017), two-photon excitation (2PE) is by far the most common for IVM, herein referred to as TPLSM. The advantages of this approach are reduced scattering of the longer wavelength excitation light and that the emission signal is limited to just the focal volume. This enables TPLSM to be used to collect high-resolution images hundreds of microns into tissues (Centonze and White, 1998; Theer et al., 2003; Theer and Denk, 2006).

While multiphoton microscopy provides the opportunity to collect images at subcellular resolution deep into living tissues, it also introduces several challenges for quantitative imaging of fluorescent biosensors. First, the vast majority of fluorescent biosensors have been extensively utilized and validated in studies of cultured cells, using conventional single-photon fluorescence excitation. *In vivo* utilization of an FP biosensor may thus require first identifying the appropriate wavelength for excitation (because two photon cross sections are not predictable from single-photon excitation spectra). Only a subset of FP probes is compatible with the range of useable wavelengths provided by the titanium sapphire lasers most commonly used for multiphoton fluorescence excitation microscopy (~700–900 nm). FPs also need to be evaluated for untoward consequences of multiphoton excitation, including excessive photobleaching or photoswitching. A second issue with multiphoton fluorescence excitation is that the two-photon cross sections of FPs are so broad that it is difficult to selectively stimulate the fluorescence of a single probe (Zipfel et al., 2003). For FRET-based fluorescent biosensors, this has the effect of increasing direct excitation of the acceptor probe to the degree that it can be difficult to detect the sensitized emission because of FRET. Finally, because most multiphoton microscope systems are equipped with a single laser, they are poorly suited to excitation ratiometric measurements. Although titanium lasers can be tuned to different wavelengths, changing wavelengths can significantly shift the alignment of the illumination. This shift in alignment results in differences in the illumination fields at the two wavelengths, which can significantly affect excitation ratio measurements. These challenges highlight the rigor required to adapt FRET-based biosensors to IVM.

This problem is demonstrated in Figure 12.2, which shows TPLSM ratiometric measurements collected from a field of HEK-293 cells expressing the redox sensor, RO-GFP2. Exposure to peroxide increases the fluorescence of RO-GFP2 when excited at 800 nm and decreases the fluorescence when excited at 900 nm. Accordingly, the ratio of the two provides a sensitive readout of redox that is independent of the amount of biosensor expressed by each cell. One limitation of this approach is that tuning the laser for the different wavelengths is time-consuming (seconds to minutes), limiting the ability to capture dynamic events and producing artifacts in ratio measurements. For example, some of the cells in the ratio image shown in Figure 12.2B have very high ratios (red) at their edges, and this could reflect a localized accumulation of peroxide. Alternatively, these high ratios might reflect registration artifacts caused by cell shape changes that occurred during the time that lapsed between tuning of the laser for the collection of the 800 nm and 900 nm images. Moreover, we observed a surprising variability in the responses of different cells (Figure 12.2). Close examination of Figure 12.2B shows that there was a distinct spatial pattern to this variability, likely reflecting differences in the alignment of the laser when tuned to either 800 or 900 nm, and this is illustrated in Figure 12.2C.

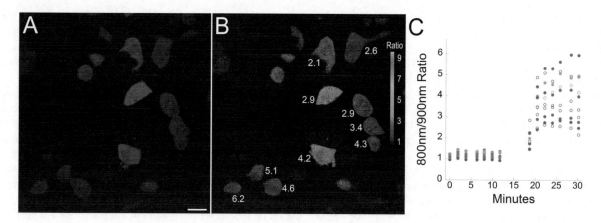

FIGURE 12.2 Multiphoton excitation ratiometric detection of redox changes in cultured cells. HEK-293 were transduced with adeno-RO-GFP2 and imaged before (A) and 10 minutes after addition of tert-butyl hydroperoxide (B). Images are displayed as the ratio of emissions excited at 800 nm to emissions excited at 900 nm. Regions of interest in each of 10 cells were quantified over time, and the resulting ratios are displayed in (C). While RO-GFP2 shows a robust change in fluorescence ratio following addition of tert-butyl hydroperoxide at 12 minutes, the responses of individual cells varied significantly, depending upon their position in the field. This spatial component (displayed in Panel B, where the final ratios are displayed beside each cell) reflects differences in the alignment of the infrared laser when tuned to 800 vs. 900 nm. Scale bar = 20 microns.

In addition to compromising ratio excitation measurements of probes such as RO-GFP2, these problems also confound measurements of sensitized emission of FRET-based probes, which typically require analysis of images collected at two different excitation wavelengths. These problems can be circumvented by equipping a multiphoton microscope with two carefully aligned lasers, but for most users, the ideal biosensors for IVM should need only a single 2PE wavelength to allow rapid acquisition of ratiometric measurements without retuning of the laser. In Section 12.5, we discuss the characterization of FRET-based biosensors optimized for IVM using a single 2PE wavelength.

12.3.2 Depth-Dependent Signal Attenuation

A second unique challenge of intravital microscopy of biosensors results from the attenuation of signals when imaging at increasing depth in biological tissues. While multiphoton microscopy extends the depth at which fluorescence images can be collected in biological tissues, signal levels, nonetheless, continuously decline with depth, particularly in highly scattering tissues, such as the kidney or liver. For this reason, it is very difficult to relate a probe concentration to fluorescence intensity. At a more practical level, depth-dependent signal attenuation complicates pooling of data collected from different fields, or from different animals.

One solution to this problem is to measure a property that is independent of fluorescence intensity, such as the velocities and directionality of migrating FP-expressing cells. Alternatively, imaging with a TPLSM system that is capable of time-resolved measurements of fluorescence lifetime changes for a biosensor can provide measurements that are independent of intensity (Farina et al., 1998). A distinct advantage of the ratiometric probes discussed above, however, is that the intensity ratio is internally calibrated and essentially corrects for the effects of depth. For example, the fluorescence intensity of the calcium-sensitive biosensor GCamp6 is dependent on changes in calcium binding, and this can be measured as

the ratio to the initial fluorescence. Ratios of this kind correct not only for the effects of depth, but also for field illumination heterogeneity and differences in protein expression. However, this approach is suitable only for evaluating acute effects. This is discussed in detail in Section 12.4.

The effects of depth can also be addressed through ratiometric imaging of fluorescent probes whose excitation or emission spectrum depends on analyte concentration. This approach is most commonly applied for measuring sensitized emission of FRET-based fluorescent biosensors, for which the ratio of the acceptor to donor fluorescence can be used to estimate FRET efficiency. Because these ratiometric probes are internally calibrated, they are less affected by depth-dependent signal attenuation than intensity-based imaging methods. While wavelength-dependent differences in scattering can, in principle, introduce depth-dependent effects on FRET ratios (Radbruch et al., 2015; Heim et al., 2007), the effects are minimized in most FRET probes in which the emission of donor and acceptor differs by only ~50 nm. In practice, we find that useful signals in both the donor and acceptor channels of a FRET probe diminish before there is an obvious depth-dependent effect on the emission ratio.

12.3.3 Tissue Motion

One of the most difficult challenges confronting an investigator in studies of living animals is the startling amount of tissue motion. Characterizing cell biology *in vivo* imposes draconian limits on tissue motion, requiring stabilization to submicron levels despite centimeter-scale respiratory motions. For studies of the skull or brain the effects of respiration can be easily addressed with stereotaxic devices, but studies of most internal organs require development of specialized techniques and devices. For our studies of the kidney, liver, and pancreas of mice and rats, we have developed an approach in which the surgically exposed organ is placed into the bottom of a coverslip-bottomed dish and imaged using an inverted microscope (Tao et al., 2015; Day et al., 2016; Dunn and Ryan, 2017;

Dun et al., 2002; Hato et al., 2017; Ryan et al., 2014; Tanner et al., 2004). In many cases, the weight of the animal is sufficient to drastically minimize respiratory motion, but it may also be necessary to use cyanoacrylate to directly secure the periphery of the organ to the dish. We have found that this approach is effective for short-term studies in which the animal is sacrificed following microscopy. However, a similar approach is also used in a "chronic" window system for longitudinal studies of various internal organs over periods of months (Ritsma et al., 2012; Ritsma et al., 2013). The solutions for presenting the internal organs of living animals for high-resolution microscopy are highly specific to the organ and application, and to our knowledge, there are no commercially available window systems.

The effects of tissue motion can also be reduced through "gated imaging" a procedure in which image collection is synchronized with intervals between respiration, an approach that can be applied for imaging highly mobile organs such as the heart (Lee et al., 2012) or lung (Presson et al., 2011; Kreiel et al., 2010). Motion artifacts can also be addressed after collection using methods of automated image processing that either discard distorted parts of images (Lee et al., 2014) or correct distortions in the affected images (Presson et al., 2011; Lorenz et al., 2012). When time-gated image collection is paired with automated image processing software, such as Galene, high-quality intravital data can be gathered from difficult-to-image tissues (Warren et al., 2018).

12.3.4 Physiological Maintenance of Small Animals

With the possible exception of studies of the skin, intravital microscopy of mammals is necessarily invasive to some extent, as it requires anesthesia and surgical exposure of the tissue of interest. While intravital microscopy provides the opportunity to evaluate cell biology in a "normal" physiological context, the relevance of these measurements is critically dependent on how accurately the physiology is represented in an anesthetized rodent whose internal organs have been surgically exposed on the stage of a microscope.

Minimizing the trauma of intravital microscopy depends largely upon the specialized skills of a small animal surgeon who must present the organ of interest for imaging with minimal collateral damage, while simultaneously monitoring temperature, blood pressure, heart, and respiratory rate. Although miniaturized microscope systems have been developed for intravital microscopy of nonanesthetized ("awake") mice (Helmchen et al., 2001; Lutcke et al., 2010; Sasportas and Gambhir, 2014), most IVM studies also require the use of anesthesia. Best practices dictate that investigators utilize the minimal level of anesthesia sufficient to induce a state of unconsciousness while still maintaining a stable physiological state. However, anesthetics themselves are known to alter physiology in ways that can complicate interpretations of intravital studies. For example, several anesthetics used in intravital microscopy have been observed to alter glucose tolerance and insulin secretion in mice (Windelov et al., 2016); therefore, this is a factor that must be considered in studies that focus on cellular events that are glucose or hormone dependent. As much as possible, the properties evaluated in intravital studies

should be validated by comparison with studies conducted in nonanesthetized animals.

Because anesthesia also compromises thermoregulation, additional sources of heat (heating pads, heating lamp) must be provided to maintain animal temperature during surgery and imaging, and animal temperature should be continuously monitored. Mice are particularly susceptible to hypothermia, owing to their small size. A less obvious consideration is the need for a heater for the objective lens. Because of its considerable mass relative to the tissue, the objective lens can act as a heat sink, particularly in a cold microscope room, depressing the local temperature of the adjacent tissue. Overall, intravital imaging introduces a high level of complexity that will test an investigator's commitment to developing and using it as a research tool. The laboratory will need to acquire the instrumentation necessary for imaging, develop the specialized ancillary devices necessary for tissue presentation, develop skills in small animal surgery sufficient for reproducible animal preparation, and develop a workflow compatible with an additional layer of institutional oversight that penalizes poor planning.

12.4 Challenges of IVM of Biosensors – *In Vivo* Expression of FP Biosensors

12.4.1 Transgenic FP Biosensor Mice

Transgenic expression of biosensors can be achieved *in vivo* in a variety of organisms, including *Drosophila*, *C. elegans*, zebrafish, and rodents (Johnsson et al., 2014; Chen et al., 2017). However, to obtain data that might be relevant and translatable to humans, rodent models are a must. Furthermore, these rodent models are a required facet of competitive research grants. To fill the wide array of research studies, transgenic mice that express a variety of different probes, including sensors of chloride (Berglund et al., 2006), calcium (Haustein et al., 2014; Madisen et al., 2015), and voltage (Akermann et al., 2012), are available commercially (The Jackson Laboratory, Bar Harbor, Maine) and might be useful for specific intravital imaging applications. Transgenic mice expressing fluorescent biosensors for PKA, Erk, Rac, and Ras are also available from the National Institutes of Biomedical Innovation, Health and Nutrition (Osaka, Japan).

Some transgenic biosensor mice have ubiquitous tissue distribution (Kamioka et al., 2012), while others have restricted tissue expression based on an established cre-lox system (Hara et al., 2004; Erami et al., 2016). Transgenic expression between tissues can be highly variable when the biosensor is driven by synthetic promoters such as CMV or CAG and even endogenous promoters such as ROSA26 (Irion et al., 2007; Zariwala et al., 2012). To overcome this variable and potentially low expression of certain promoters, a secondary element, such as the woodchuck hepatitis virus post-translation regulatory element (WPRE), can be incorporated into the biosensor transgene design to boost expression (Zariwala et al., 2012). Most successful transgenic mouse studies using biosensors have relied on the single-FP sensors, such as the GCaMPs. There has been more limited success with transgenic animals expressing the FRET-based sensors, especially when they are imaged early in development (Johnsson et al., 2014). The reason for the limited success is likely related to the observation that the FRET-based sensors undergo recombination and become inactivated during development (Komatsu et al., 2011).

12.4.2 *In Vivo* Gene Transfer via Viral Transduction

The use of viral transduction is an alternative approach that offers rapid probe delivery in living animals without the need for lengthy breeding strategies to achieve tissue expression. The injection of viral particles, including but not limited to adenovirus (Ad), lentivirus, and adenoviral-associated vectors (AAVs), can infect a variety of tissues (Bouard et al., 2009). The challenge for *in vivo* administration of viral vectors, however, is to obtain selective expression of the biosensor in the relevant cell types. While stereotactic injection of a virus into the brain can specifically target certain brain regions, systemic administration of a virus intravenously or intraperitoneally generally results in accumulation in the liver and spleen (Buchholz et al., 2015). This is an immense challenge when trying to deliver biosensors to other organs within the abdominal or thoracic cavities. Importantly, there are many serotypes of AAV that differ in their capsid protein structures, and this enables distinct tissue tropism for the different serotypes (Buchholz et al., 2015; Wang et al., 2006). For example, AAV4 has tropism for the lung, AAV8 has tropism for the pancreas, and AAV9 has tropism for the heart (Wang et al., 2006; Zincarelli et al., 2008). Additionally, custom or unique promoters can be incorporated into the construct design to drive expression in the cells of interest (Gray et al., 2011; Koh et al., 2017).

Adenovirus (Ad) transduction has broad tropism and provides robust, but transient, expression of the transgenes at comparatively low titers and costs. Ad, though, stimulates an inflammatory response that often limits its use in prolonged or immune system-related studies. Like Ad, lentivirus has broad tropism and yields robust expression, along with an interferon response within the liver (Brown et al., 2007). However, stable integration of the lentiviral-packaged construct can sustain expression for several weeks (Dittgen et al., 2004). AAV transduction provides robust, sustained expression without significant immune response (Wang et al., 2006), but the narrow tropism and costly production scheme for new viruses can be prohibitive to large-scale studies.

12.4.3 Transplantation of *Ex Vivo* Transduced Tissue

Because viral infection has limited targeting *in vivo*, a tissue of interest can be isolated, transduced *ex vivo*, and transplanted back into a recipient animal. This has been demonstrated with tumorigenic cells that were preloaded with a biosensor and monitored during cancer progression (Johnsson et al., 2014). The xenograph approach greatly increases the level of biosensor expression compared to transgenic mice. This also enhances the imaging success rates because the transplanted tissue is easily differentiated from the native tissue at the transplant site. While characterized extensively, engraftment is still a major consideration when using this approach.

In Figure 12.3, we show one novel use of this approach in which pancreatic islets were isolated from donor animals and infected with Ad-packaged GCaMP6s in culture prior to transplantation under the kidney capsule. This novel study was conducted to characterize calcium oscillations in islet beta cells *in vivo*. While the ideal experiment is to measure *in situ* beta cell calcium dynamics, this currently requires transgenic expression of a calcium biosensor. Furthermore, discovery of superficial endocrine tissue is difficult because islets only make up ~2–3% of the overall pancreas mass. As an alternative, we isolated wild-type mouse islets and infected them with Ad-packaged GCaMP6s, then transplanted them under the kidney capsule of recipient mice. While the kidney is not the native location for the islets, it does recapitulate many aspects of normal tissue physiology, including vascularization and perfusion. After 2 weeks to allow engraftment, we exposed the kidney and recorded GCaMP6s signal over time. The fluorescence intensity of GCaMP6s collected over 8 minutes is graphed in Figure 12.3 and demonstrates the generation of synchronized calcium oscillations from individual cells within the islets following glucose stimulation (2g/kg). Images from the same focal plane collected over an 8-minute time frame, as well as the normalized fluorescence intensity profile for the indicated cells, are shown in Figure 12.3B. The coupling of beta cells within the islet and calcium oscillation frequency observed here is required for normal cellular function and is similar to that reported for *ex vivo* experiments (Benninger et al., 2008).

12.5 Caveats and Considerations for IVM of FRET Biosensors

12.5.1 Ratiometric Measurements of FRET *In Vivo*

As discussed above, FRET can be measured using acceptor photobleaching, fluorescence lifetimes, or ratiometric measures of sensitized emission. Other methods for measuring changing FRET signals are difficult to implement in IVM. For example, acceptor photobleaching FRET measurements are slow and difficult to do using multiphoton excitation because photobleaching increases as the square (or cube) of illumination (Patterson and Piston, 2000). Further, this approach is an endpoint measurement, so it cannot be used to monitor changes. Alternatively, time-resolved fluorescence lifetime measurements can be used to monitor changes in donor lifetime because of FRET and have been used for IVM (Farina et al., 1998). However, the lifetime measurements are time-consuming, requiring integrations over seconds to minutes, making it difficult to monitor dynamics and would be more susceptible to motion artifacts than ratiometric methods.

Compared to the intensity or time-resolved techniques, ratiometric measurement of sensitized emissions has a significant advantage in speed of acquisition. Because the donor and acceptor intensity images are acquired simultaneously, the approach is limited only by the integration time necessary to obtain sufficient signal for analysis (depending on the size of the region of interest analyzed, this can range from 10s of milliseconds to a few seconds for each time point). Most

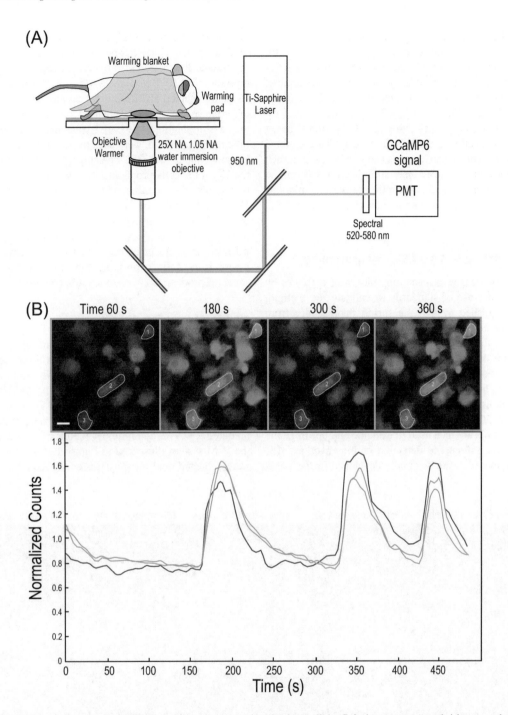

FIGURE 12.3 IVM measurements of GCaMP6s intensity responses to changing intracellular Ca2+ in mouse pancreatic islets transplanted under the kidney capsule. (A) Schematic diagram of the IVM system used for intensity measurements of GCaMP6s. (B) IVM measurements of the response of the GCaMP6s probe to glucose from pancreatic islet cells. The GCaMP6s biosensor was introduced to islets prior to transplantation. *Ex vivo* islets were transduced with an adenoviral vector encoding GCaMP6s, resulting in extensive expression in the transplanted tissue up to 14 days later. 3D image volumes (25 planes spanning 10 microns) were collected every 6 seconds following intravenous injection of glucose (2 mg/kg). Each volume was then summed and regions of interest for cells in the field and the full image intensities were plotted (Scale bar = 10 µm).

ratiometric measurements of FRET were developed for single-photon excitation, and may require the use of two separate excitation wavelengths to acquire images of the donor and the acceptor fluorophores. This approach would be difficult to utilize for IVM because, as described above, IVM systems are usually equipped with a single infrared laser, so switching to another wavelength requires retuning and realigning the laser between acquisitions of the two separate images, limiting temporal resolution and introducing artifacts. Therefore, the ideal approach should require excitation at only a single 2PE wavelength to allow rapid acquisition of ratiometric measurements.

Earlier, we evaluated the 2PE characteristics of range of potential FP pairs for FRET-based biosensor and found that mTurquoise and mVenus were well suited for measurements using IVM (Tao et al., 2015; Day et al., 2016). The newer mTurquoise FP is optimally excited near the power maximum of the titanium-sapphire lasers typically used for 2PE microscopy and has improved photophysical characteristics (Goedhart et al., 2012). In addition, mVenus is an optimal FRET acceptor for mTurquoise, but has minimal 2PE between 800 nm and 820 nm (Zipfel et al., 2003). This reduces the fluorescence crosstalk caused by the direct excitation of mVenus at the wavelengths needed to excite mTurquoise. The combined fluorescent properties of mTurquoise and mVenus are such that mTurquoise can be selectively stimulated at 810 nm so that FRET can be measured as the ratio of mVenus emissions (520–580 nm) to mTurquoise emissions (454–494 nm) (Tao et al., 2015).

12.5.2 Validating *In Vivo* FRET Measurements

The inherent challenges of measuring biosensor activities by IVM demand the use of standards to validate the method. This is accomplished by first measuring fluorescence from cells producing either the donor (mTurquoise) or the acceptor (mVenus) alone to determine the spectral bleed through (SBT) components resulting from single excitation wavelength of 810 nm. An estimation of the fractional excitation of the acceptor (mVenus) at 810 nm is obtained using cells that produce a mixture of mTurquoise and mVenus as described earlier (Tao et al., 2015Day et al., 2016).

FRET standards are then used to evaluate and optimize the IVM system for ratiometric detection of biosensor activity. The FRET standards were originally developed in the Vogel

laboratory (Thaler et al., 2005) and consist of donor and acceptor FPs separated by linker peptides of increasing length. When the standards are produced in cells in culture, they have very reproducible FRET efficiencies (E_{FRET}) (Koushik et al., 2006). Here, transfected HEK293 cells expressing a 1:1 mixture of mTurquoise and mVenus were used as a negative control, while the cells producing a low FRET standard, mTurquoise-TRAF-mVenus (E_{FRET} ~5%), were used to assess the sensitivity of the imaging system for FRET detection (Figure 12.4). By contrast, the E_{FRET} for cells producing the two high FRET standards, mTurquoise-5AA (amino acid)-Venus and mTurquoise-10AA-Venus were easily measured, and the two probes could be clearly distinguished from one-another. The E_{FRET} was determined from the cells expressing the highest E_{FRET} standard; mTurquoise-5AA-Venus was about 45%, whereas measurements from cells expressing the mTurquoise-10AA-Venus standard produce E_{FRET} of about 36% (Figure 12.4).

12.5.3 Measuring FRET *In Vivo*

The *in vitro* measurements of the standards and the biosensor provides the basis for moving forward with measurements of cellular function in the living animal. We used Ad transduction to achieve high-level expression of the AKAR4.1 biosensor in the liver of mice (Tao et al., 2015). Seven days after injection with Ad AKAR4.1, the mice were fasted for 3 h, prepared for IVM, and ratiometric images were acquired with the IVM system illustrated in Figure 12.5A. Baseline images were collected, and imaging was continued after IP injection

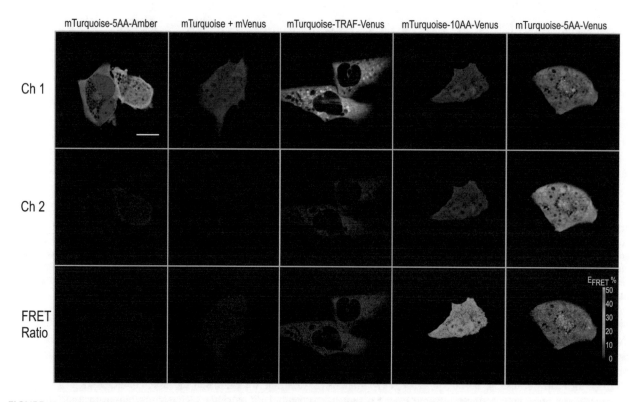

FIGURE 12.4 Validation of biosensor measurements with the FRET standards. Representative donor intensity (Ch. 1, 454–494 nm; scale bar = 10 μm), acceptor intensity (Ch. 2, 520–580 nm), and FRET efficiency (E_{FRET}) determined for HEK293 cells expressing the indicated FRET standard probes (scale bar = 10 μm, and the look-up table indicates higher E_{FRET} with warmer color). Reproduced from Tao et al. (2015).

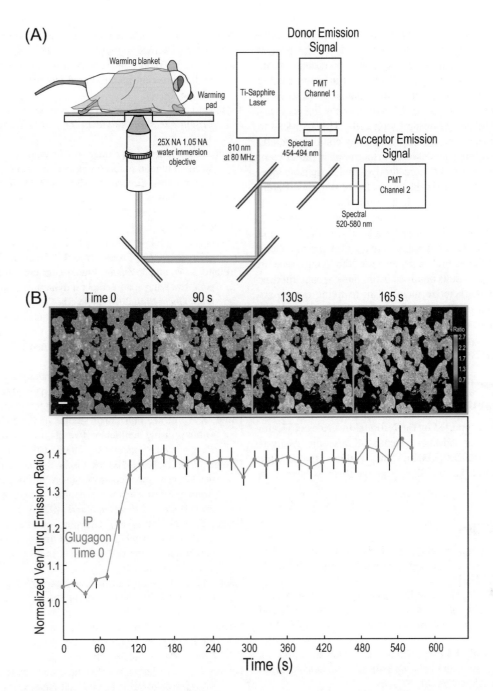

FIGURE 12.5 IVM ratiometric measurements of PKA activity in mouse liver. (A) Schematic diagram of the IVM system used for ratiometric measurements of biosensors. (B) IVM measurements of the response of the AKAR4.1 probe to glucagon in hepatocytes in the intact mouse liver. The AKAR4.1 biosensor was introduced into mice by tail vein injection of an adenoviral vector encoding AKAR4.1, resulting in extensive expression in the liver 7 days later. Mice were fasted for 3 h prior to imaging by IVM, and 3D image volumes (10 planes spanning 10 microns) were collected over time prior to and following IP injection of glucagon (200 μg/kg). Ratio images from a single image plane in mouse liver were acquired at the indicated time points (scale bar = 10 μm, and the look-up table indicates higher FRET ratio with warmer color). Each volume was then summed and background-corrected mVen/mTurq ratios were determined for regions of interest from cells in the field. The results for AKAR4.1 are from the 10 individual cells (± SE). Reproduced from Tao et al. (2015).

of glucagon (200 μg/kg), a treatment that will rapidly activate the PKA pathway in the hepatocytes of fasted mice (Miller et al., 2013). The IVM of the AKAR4.1 biosensor revealed a 1.4-fold change in the mVenus/mTurquoise emission ratio following the glucagon treatment, demonstrating a rapid and sustained activation of PKA, which is detected at the cellular level in the liver of the living mouse (Figure 12.5B).

12.6 IVM Future Directions

Genetically encoded FP biosensors have enabled new strategies to investigate the function of proteins and dynamics of signaling pathways in their natural environment. Measurements obtained from cells in the intact, living organism provide the most physiologically relevant information about cellular behavior currently available, particularly in the context of disease or injury. The studies described here demonstrate that IVM can be used to quantify changes in cell-signaling events in the intact tissues of living organisms. However, a general limitation to the use of FP biosensors is the necessity of producing the probes inside living cells. This can lead to artifacts associated with overexpression that could potentially interfere with cellular processes. Furthermore, a direct comparison between inorganic indicator dyes and biosensors measuring the same signal pathway demonstrated slightly different results, suggesting the probe might perturb the underlying cellular mechanisms that they are designed to detect (Lock et al., 2015). Therefore, there is a need for a label-free IVM method that can provide a fast, quantitative readout of the metabolic state of cells in the natural environment of the living tissue. In this regard, recent results obtained using fluorescence lifetime microscopy to characterize metabolism in living tissues are particularly exciting (Datta et al., 2016; Stringari et al., 2011). To the degree that IVM studies require the expression of exogenous proteins, careful control experiments must be conducted to verify protein function and behavior.

Acknowledgments

This research is supported by the National Institutes of Health O'Brien Center for Advanced Renal Microscopic Analysis (NIH-NIDDK P30DK079312). The authors thank Drs. Malgorzata Kamocka and Wen Tao for their assistance in microscopy.

REFERENCES

Akemann, W., et al., Imaging neural circuit dynamics with a voltage-sensitive fluorescent protein. *J Neurophysiol*, 2012. **108**(8): 2323–37.

Benninger, R.K., M. Hao, and D.W. Piston, Multi-photon excitation imaging of dynamic processes in living cells and tissues. *Rev Physiol Biochem Pharmacol*, 2008. **160**: 71–92.

Berglund, K., et al., Imaging synaptic inhibition in transgenic mice expressing the chloride indicator, clomeleon. *Brain Cell Biol*, 2006. **35**(4–6): 207–28.

Bouard, D., D. Alazard-Dany, and F.L. Cosset, Viral vectors: from virology to transgene expression. *Br J Pharmacol*, 2009. **157**(2): 153–65.

Brown, B.D., et al., In vivo administration of lentiviral vectors triggers a type I interferon response that restricts hepatocyte gene transfer and promotes vector clearance. *Blood*, 2007. **109**(7): 2797–805.

Buchholz, C.J., T. Friedel, and H. Buning, Surface-engineered viral vectors for selective and cell type-specific delivery. *Trends Biotechnol*, 2015. **33**(12): 777–90.

Burford, J.L., et al., Intravital imaging of podocyte calcium in glomerular injury and disease. *J Clin Invest*, 2014. **124**(5): 2050–8.

Centonze, V.E. and J.G. White, Multiphoton excitation provides optical sections from deeper within scattering specimens than confocal imaging. *Biophys J*, 1998. **75**(4): 2015–24.

Chen, T.W., et al., Ultrasensitive fluorescent proteins for imaging neuronal activity. *Nature*, 2013. **499**(7458): 295–300.

Chen, Z., T.M. Truong, and H.W. Ai, Illuminating brain activities with fluorescent protein-based biosensors. *Chemosensors*, 2017. **5**(4):1–36.

Cohnheim, J., *Lectures on General Pathology: A Handbook for Practitioners and Students*. The New Sydenham Society, 1889.

Datta, R., et al., Label-free imaging of metabolism and oxidative stress in human induced pluripotent stem cell-derived cardiomyocytes. *Biomed Opt Express*, 2016. **7**(5): 1690–701.

Day, R.N. and M.W. Davidson (Eds.). *The Fluorescent Protein Revolution*. CRC Press, 2014.

Day, R.N., W. Tao, and K.W. Dunn, A simple approach for measuring FRET in fluorescent biosensors using two-photon microscopy. *Nat Protoc*, 2016. **11**(11): 2066–80.

DiPilato, L.M. and J. Zhang, Fluorescent protein-based biosensors: resolving spatiotemporal dynamics of signaling. *Curr Opin Chem Biol*, 2010. **14**(1): 37–42.

Dittgen, T., et al., Lentivirus-based genetic manipulations of cortical neurons and their optical and electrophysiological monitoring in vivo. *Proc Natl Acad Sci U S A*, 2004. **101**(52): 18206–11.

Dunn, K.W. and J.C. Ryan, Using quantitative intravital multiphoton microscopy to dissect hepatic transport in rats. *Methods*, 2017. **128**: 40–51.

Dunn, K.W., et al., Functional studies of the kidney of living animals using multicolor two-photon microscopy. *Am J Physiol Cell Physiol*, 2002. **283**(3): C905–16.

Ellinger, P. and A. Hirt, Mikroskopische Beobachtungen an lebenden Organen mit Demonstrationen (Intravitalmikroskopie). *Naunyn-Schmiedebergs Archiv für experimentelle Pathologie und Pharmakologie*, 1929. **147**(1): 63.

Erami, Z., et al., Intravital FRAP imaging using an E-cadherin-GFP mouse reveals disease- and drug-dependent dynamic regulation of cell-cell junctions in live tissue. *Cell Rep*, 2016. **14**(1): 152–67.

Falati, S., et al., Real-time in vivo imaging of platelets, tissue factor and fibrin during arterial thrombus formation in the mouse. *Nat Med*, 2002. **8**(10): 1175–81.

Farina, K.L., et al., Cell motility of tumor cells visualized in living intact primary tumors using green fluorescent protein. *Cancer Res*, 1998. **58**(12): 2528–32.

Follain, G., et al., Seeing is believing - multi-scale spatio-temporal imaging towards in vivo cell biology. *J Cell Sci*, 2017. **130**(1): 23–38.

Goedhart, J., et al., Structure-guided evolution of cyan fluorescent proteins towards a quantum yield of 93%. *Nat Commun*, 2012. **3**: 751.

Gray, S.J., et al., Optimizing promoters for recombinant adeno-associated virus-mediated gene expression in the peripheral and central nervous system using self-complementary vectors. *Hum Gene Ther*, 2011. **22**(9): 1143–53.

Gutscher, M., et al., Real-time imaging of the intracellular glutathione redox potential. *Nat Methods*, 2008. **5**(6): 553–9.

Hara, M., et al., Imaging endoplasmic reticulum calcium with a fluorescent biosensor in transgenic mice. *Am J Physiol Cell Physiol*, 2004. **287**(4): C932–8.

Hato, T., et al., Two-photon intravital fluorescence lifetime imaging of the kidney reveals cell-type specific metabolic signatures. *J Am Soc Nephrol*, 2017. **28**(8): 2420–30.

Haustein, M.D., et al., Conditions and constraints for astrocyte calcium signaling in the hippocampal mossy fiber pathway. *Neuron*, 2014. **82**(2): 413–29.

Heim, N., et al., Improved calcium imaging in transgenic mice expressing a troponin C-based biosensor. *Nat Methods*, 2007. **4**(2): 127–9.

Helmchen, F., et al., A miniature head-mounted two-photon microscope. high-resolution brain imaging in freely moving animals. *Neuron*, 2001. **31**(6): 903–12.

Irion, S., et al., Identification and targeting of the ROSA26 locus in human embryonic stem cells. *Nat Biotechnol*, 2007. **25**(12): 1477–82.

Jenne, C.N., et al., The use of spinning-disk confocal microscopy for the intravital analysis of platelet dynamics in response to systemic and local inflammation. *PLoS One*, 2011. **6**(9): e25109.

Johnsson, A.E., et al., The Rac-FRET mouse reveals tight spatio-temporal control of Rac activity in primary cells and tissues. *Cell Rep*, 2014. **6**(6): 1153–64.

Kamioka, Y., et al., Live imaging of protein kinase activities in transgenic mice expressing FRET biosensors. *Cell Struct Funct*, 2012. **37**(1): 65–73.

Koh, W., et al., AAV-mediated astrocyte-specific gene expression under human ALDH1L1 promoter in mouse thalamus. *Exp Neurobiol*, 2017. **26**(6): 350–61.

Komatsu, N., et al., Development of an optimized backbone of FRET biosensors for kinases and GTPases. *Mol Biol Cell*, 2011. **22**(23): 4647–56.

Koushik, S.V., et al., Cerulean, Venus, and VenusY67C FRET reference standards. *Biophys J*, 2006. **91**(12): L99–101.

Kreisel, D., et al., In vivo two-photon imaging reveals monocyte-dependent neutrophil extravasation during pulmonary inflammation. *Proc Natl Acad Sci U S A*, 2010. **107**(42): 18073–8.

Lee, S., et al., Real-time in vivo imaging of the beating mouse heart at microscopic resolution. *Nat Commun*, 2012. **3**: 1054.

Lee, S., et al., Automated motion artifact removal for intravital microscopy, without a priori information. *Sci Rep*, 2014. **4**: 4507.

Lock, J.T., I. Parker, and I.F. Smith, A comparison of fluorescent Ca(2)(+) indicators for imaging local Ca(2)(+) signals in cultured cells. *Cell Calcium*, 2015. **58**(6): 638–48.

Lorenz, K.S., et al., Digital correction of motion artefacts in microscopy image sequences collected from living animals using rigid and nonrigid registration. *J Microsc*, 2012. **245**(2): 148–60.

Lutcke, H., et al., Optical recording of neuronal activity with a genetically-encoded calcium indicator in anesthetized and freely moving mice. *Front Neural Circuits*, 2010. **4**: 9.

Madisen, L., et al., Transgenic mice for intersectional targeting of neural sensors and effectors with high specificity and performance. *Neuron*, 2015. **85**(5): 942–58.

Miller, R.A., et al., Biguanides suppress hepatic glucagon signalling by decreasing production of cyclic AMP. *Nature*, 2013. **494**(7436): 256–60.

Mitsuoka, H., et al., Intravital laser confocal microscopy of pulmonary edema resulting from intestinal ischemia-reperfusion injury in the rat. *Crit Care Med*, 1999. **27**(9): 1862–68.

Miyawaki, A., Development of probes for cellular functions using fluorescent proteins and fluorescence resonance energy transfer. *Annu Rev Biochem*, 2011. **80**: 357–73.

Nobis, M., et al., A RhoA-FRET biosensor mouse for intravital imaging in normal tissue homeostasis and disease contexts. *Cell Rep*, 2017. **21**(1): 274–88.

Oldach, L. and J. Zhang, Genetically encoded fluorescent biosensors for live-cell visualization of protein phosphorylation. *Chem Biol*, 2014. **21**(2): 186–97.

Patterson, G.H. and D.W. Piston, Photobleaching in two-photon excitation microscopy. *Biophys J*, 2000. **78**(4): 2159–62.

Periasamy, A. and R.N. Day (Eds.). *Molecular Imaging: FRET Microscopy and Spectroscopy*. American Physiological Society, Oxford University Press, 2005.

Power, R.M. and J. Huisken, A guide to light-sheet fluorescence microscopy for multiscale imaging. *Nat Methods*, 2017. **14**(4): 360–73.

Presson, R.G., Jr., et al., Two-photon imaging within the murine thorax without respiratory and cardiac motion artifact. *Am J Pathol*, 2011. **179**(1): 75–82.

Radbruch, H., et al., Intravital FRET: Probing cellular and tissue function in vivo. *Int J Mol Sci*, 2015. **16**(5): 11713–27.

Ritsma, L., et al., Intravital microscopy through an abdominal imaging window reveals a pre-micrometastasis stage during liver metastasis. *Sci Transl Med*, 2012. **4**(158): 158ra145.

Ritsma, L., et al., Surgical implantation of an abdominal imaging window for intravital microscopy. *Nat Protoc*, 2013. **8**(3): 583–94.

Rowlands, C.J., et al., Wide-field three-photon excitation in biological samples. *Light Sci Appl*, 2017. **6**: e16255.

Ryan, J.C., K.W. Dunn, and B.S. Decker, Effects of chronic kidney disease on liver transport: quantitative intravital microscopy of fluorescein transport in the rat liver. *Am J Physiol Regul Integr Comp Physiol*, 2014. **307**(12): R1488–92.

Sano, T., et al., Intravital imaging of mouse urothelium reveals activation of extracellular signal-regulated kinase by stretch-induced intravesical release of AT. *Physiol Rep*, 2016. **4**(21).

Sasportas, L.S. and S.S. Gambhir, Imaging circulating tumor cells in freely moving awake small animals using a miniaturized intravital microscope. *PLoS One*, 2014. **9**(1): e86759.

Stringari, C., et al., Phasor approach to fluorescence lifetime microscopy distinguishes different metabolic states of germ cells in a live tissue. *Proc Natl Acad Sci U S A*, 2011. **108**(33): 13582–7.

Strobl, F., A. Schmitz, and E.H.K. Stelzer, Improving your four-dimensional image: traveling through a decade of light-sheet-based fluorescence microscopy research. *Nat Protoc*, 2017. **12**(6): 1103–9.

Tanner, G.A., R.M. Sandoval, and K.W. Dunn, Two-photon in vivo microscopy of sulfonefluorescein secretion in normal and cystic rat kidneys. *Am J Physiol Renal Physiol*, 2004. **286**(1): F152–60.

Tao, W., et al., A practical method for monitoring FRET-based biosensors in living animals using two-photon microscopy. *Am J Physiol Cell Physiol*, 2015. **309**(11): C724–35.

Thaler, C., et al., Quantitative multiphoton spectral imaging and its use for measuring resonance energy transfer. *Biophys J*, 2005. **89**(4): 2736–49.

Theer, P. and W. Denk, On the fundamental imaging-depth limit in two-photon microscopy. *J Opt Soc Am A Opt Image Sci Vis*, 2006. **23**(12): 3139–49.

Theer, P., M.T. Hasan, and W. Denk, Two-photon imaging to a depth of 1000 microm in living brains by use of a Ti:Al2O3 regenerative amplifier. *Opt Lett*, 2003. **28**(12): 1022–4.

Wagner, R., *Erlauterungstafeln zur Physiologie und Entwicklungsgeschichte.* Leopold Voss, 1839.

Wang, Z., et al., Widespread and stable pancreatic gene transfer by adeno-associated virus vectors via different routes. *Diabetes*, 2006. **55**(4): 875–84.

Warren, S.C., et al., Removing physiological motion from intravital and clinical functional imaging data. *Elife*, 2018. **7**, e35800.

Windelov, J.A., J. Pedersen, and J.J. Holst, Use of anesthesia dramatically alters the oral glucose tolerance and insulin secretion in C57Bl/6 mice. *Physiol Rep*, 2016. **4**(11).

Wyckoff, J.B., et al., A critical step in metastasis: In vivo analysis of intravasation at the primary tumor. *Cancer Res*, 2000. **60**(9): 2504–11.

Yoshikawa, S., et al., Intravital imaging of Ca(2+) signals in lymphocytes of Ca(2+) biosensor transgenic mice: indication of autoimmune diseases before the pathological onset. *Sci Rep*, 2016. **6**: 18738.

Zariwala, H.A., et al., A Cre-dependent GCaMP3 reporter mouse for neuronal imaging in vivo. *J Neurosci*, 2012. **32**(9): 3131–41.

Zhang, J., et al., Creating new fluorescent probes for cell biology. *Nat Rev Mol Cell Biol*, 2002. **3**(12): 906–18.

Zincarelli, C., et al., Analysis of AAV serotypes 1–9 mediated gene expression and tropism in mice after systemic injection. *Mol Ther*, 2008. **16**(6): 1073–80.

Zipfel, W.R., R.M. Williams, and W.W. Webb, Nonlinear magic: multiphoton microscopy in the biosciences. *Nat Biotechnol*, 2003. **21**(11): 1369–77.

13

Imaging Cell Adhesion and Migration

Chandrani Mondal, Julie Di Martino, and Jose Javier Bravo-Cordero

CONTENTS

13.1 Introduction

In the seventeenth century, Antonie van Leeuwenhoek, a microscopist and scientist, documented what is considered to be the first cell motility event in the history of cell biology (Dunn and Jones, 2004). van Leeuwenhoek constructed a simple microscope consisting of a single lens mounted between two thin plates, making up the body of the instrument. A specimen was placed on a sharp pin in front of the lens, and its distance from the lens was tunable with screws. This basic device allowed van Leeuwenhoek to visualize the movement of a bacterium, which is documented in his original drawings; thus, he established the notion that microscopy could capture the "dynamics of life" (Lane, 2015). From this first observation in 1683, the continuous development of the microscopy field has given birth to sophisticated technologies that have revealed the intricate machinery that regulates cellular movement. Techniques such as single molecule imaging and *in vivo* multiphoton microscopy have shown that migration is a tightly orchestrated process that requires regulation of multiple cellular components and signaling pathways in both space and time. In the following chapter, we will discuss the basics of cell motility and how imaging technologies have contributed to the understanding of this cellular process in both development and disease.

13.1.1 Cell Motility Cycle

Cell movement is a key process that is required during the development of multicellular organisms. Alterations of its basic functions can have consequences leading to pathologies, such as cancer metastasis (Franz et al., 2002). Studies of different migratory cells (fibroblasts, tumor cells) state that in order for a cell to move and translocate its cell body, a series of steps need to be accomplished. These steps are commonly known as the cell motility cycle and include: protrusion, extension/retraction, adhesion, contraction of the trailing edge, and degradation of extracellular matrix when cells migrate in complex environments (Bravo-Cordero et al., 2012) (Figure 13.1). These steps are orchestrated by cycles of actin polymerization and depolymerization, achieving a net movement of cells.

Studies of highly motile tumor cells have revealed that the formation of a leading-edge protrusion is the initial step of the motility cycle and is driven by actin polymerization. Different types of membrane protrusions at the leading edge have been described based on their shape (lamellipodia-flat shape, pseudopodia-round, filopodia-needle shape, lobopodia-cylindrical) (Taylor and Condeelis, 1979), and all of them contribute to the efficient movement of cells. The formation of these protrusions depends on the control of actin polymerization by actin regulatory molecules, such as cofilin, Arp2/3, and GTPases, such as RhoA, RhoC, and Rac1 (Bravo-Cordero et al., 2013a). Adhesion to the substrate during migration is mediated by focal adhesions, which are complex structures that anchor the cell to the extracellular matrix (Burridge, 2017). At these adhesion sites, clustering of integrin receptors transmits signals intracellularly to modulate cellular function. The first identification of these structures by electron microscopy comes from

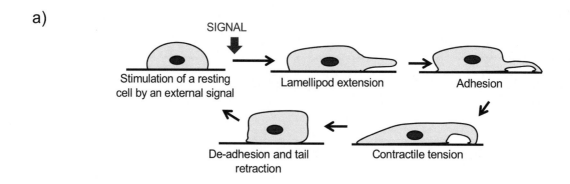

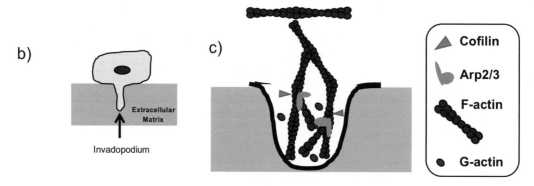

FIGURE 13.1   Formation of cellular protrusions. a) Schematic of the cell motility cycle of migratory cells. b) Drawing of a tumor cell forming an invadopodium protrusion. c) Cytoskeletal architecture at the invadopodium core.

the studies of Abercrombie et al. (1971) in fibroblasts. Finally, retraction at the back of the cell through contraction mediates the translocation of the cell body (Figure 13.1a).

In vivo, the migration of cells in tissues takes place in a three-dimensional (3D) context that requires the cell to degrade the extracellular matrix. When crossing extracellular matrix barriers, such as basement membranes, cells can rearrange the actin cytoskeleton into a membrane-degrading protrusion termed invadopodia (from "invasive [inva] feet [podia]") (Medwig and Matus, 2017) (Figure 13.1b, c). This protrusive structure is formed by tumor cells when crossing endothelial barriers (Roh-Johnson et al., 2014) and also in cells crossing basement membranes during development (Lohmer et al., 2014).

Another type of motility, referred to as *blebbing*, can be used by cells to move across extracellular matrices. In this type of contraction-based movement, the flow of cytosol pushes the membrane to form a bleb, which facilitates the movement of the cell (Paluch and Raz, 2013). For the purpose of this chapter, we will focus on actin-driven protrusions, as described above.

While microscopy has been used to visualize cell motility, the understanding of the signaling pathways regulating each of the individual steps contributing to motility is a more complex challenge that requires the development of new imaging tools (i.e. pathway biosensors to measure changes in activity of key motility regulatory molecules, such as GTPases). As we describe below, high-resolution imaging techniques can now

uncover how signaling pathways are spatiotemporally regulated during cell migration.

13.1.2 FRET Biosensors for Imaging Signaling Pathway Activation in Real Time

FRET (fluorescence resonance energy transfer) is a physical phenomenon that takes place when two fluorescent molecules with certain spectral properties (i.e. the emission of the donor fluorophore overlaps with the excitation of the acceptor fluorophore) are in close proximity (< than 10 nm). The use of FRET as a tool to study the activation of signaling pathways in the context of motility has revealed new signaling molecules regulating this process. FRET microscopy has also been used *in vitro* to study the direct interaction of molecules using antibody probes and fluorescent proteins (Shrestha et al., 2015). We will not discuss the details of FRET imaging in this chapter; rather, we will focus on how this technology has been applied to study the spatiotemporal regulation of signaling pathways in live specimens.

13.1.2.1 RhoGTPase Biosensors

RhoGTPase activation plays a major role in regulating the steps of the cell motility cycle. GTPases cycle between an inactive state bound to guanosine diphosphate (GDP) and an active state bound to guanosine triphosphate (GTP), which can bind to effectors and activate downstream signaling (Figure 13.2a). GTPase activation is driven by guanine nucleotide exchange factors (GEFs) and deactivation by GTPase activating proteins (GAPs). During the early 2000s, the development of biosensors to study the activation of Rho GTPase signaling pathways in living cells by using FRET constituted a major scientific development in the field. These new tools allowed for the study of GTPase signaling with high spatiotemporal resolution during cell migration. Several groups developed single-chain biosensors containing the following: two fluorescent molecules (e.g. a FRET pair, such as CFP and YFP, or any of their variants) separated by a linker, a C-terminal GTPase protein, and an N-terminal fragment of a GTPase effector that can only bind to the GTPase in its active state (GTP-bound) (Bravo-Cordero et al., 2013b) (Figure 13.2b). The design of these sensors has been optimized for different GTPases to improve FRET measurements (Aoki and Matsuda, 2009; Bravo-Cordero et al., 2011; Komatsu et al., 2011; Hanna et al., 2014; Moshfegh et al., 2014; Wu et al., 2015; Miskolci et al., 2016). When the GTPase is activated (by the exchange of GDP for GTP), it can bind the GTPase effector, resulting in a structural change that brings the two fluorescent molecules together, thus decreasing their distance and increasing the FRET signal. The sensitized FRET emission signal emitted by the acceptor when the donor is excited is used to calculate FRET (Spiering et al., 2013; Donnelly et al., 2014).

By using these tools, different groups have revealed new aspects of spatiotemporal GTPase signaling and have uncovered new functions of GTPases (Donnelly et al., 2014). GTPase biosensors have elucidated the regulation of GTPase activation by upstream regulators (i.e. GEFs, GAPs, and guanine nucleotide dissociation inhibitors), how multiple GTPases have coordinated activity during migration, and the activation of different GTPases at both lamellipodium and invadopodium protrusions, contributing to major breakthroughs in the cell biology field (Kurokawa and Matsuda, 2005; Pertz et al., 2006; Bravo-Cordero et al., 2013). As an example, in the context of tumor cell migration, by using biosensor imaging, a signaling node including p190RhoGEF/p190RhoGAP/RhoC was shown

to coordinate invadopodia formation for efficient matrix degradation and lamellipodia formation (Bravo-Cordero et al., 2011, 2013, 2014). The spatial activation of RhoC at invasive protrusions was characterized by using biosensor technology and the molecular pathways upstream and downstream of RhoC were described, revealing how subcellular activation patterns of RhoC (around the invadopodium core or behind the leading edge) mediate efficient invadopodia and lamellipodia formation.

13.2 Imaging Cell Motility

13.2.1 Reconstituting Cell Motility Behavior *In Vitro*

Mammalian cells sense, polarize, and migrate in response to chemotactic, haptotactic, and durotactic gradients during key physiological and pathological processes, including embryonic development, immune cell trafficking, wound healing, and tumor cell invasion (Kim and Peyton, 2012; Kim and Wu, 2012). In order to mimic these steps, many laboratories have developed *in vitro* assays in combination with microscopy techniques to model cell motility. Classical assays used to analyze migration toward or away from soluble cues, respectively, include chambers with a source and sink (Pujic et al., 2009), such as the Boyden (transwell) (1962), Zigmond (1977) and Dunn (Zicha et al., 1991) chambers, as well as the under-agarose (Nelson et al., 1975) and micropipette (Gerisch and Keller, 1981) assays. The Zigmond, Dunn, under-agarose and micropipette assays can be imaged with time-lapse microscopy; the Boyden chamber is an endpoint assay (Kim and Wu, 2012). A key benefit of these conventional assays is their ease of use; however, they are often limited to two-dimensional (2D) studies and lack the capability of producing a stable chemical gradient over time (Kim and Wu, 2012).

The development of microfluidic devices has provided the option of performing migration studies in a more physiologically relevant context with accurate chemical gradient formation and high-resolution imaging. Both convective-flow

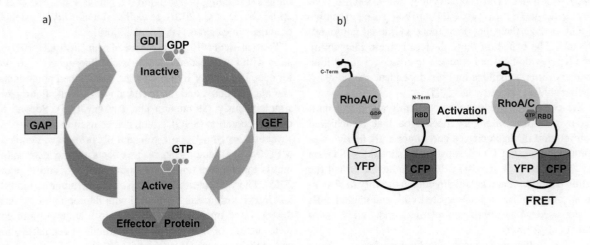

FIGURE 13.2 RhoGTPase biosensors. a) GTPase activity cycle. GEF = guanine nucleotide exchange factor, GAP = GTPase activating protein, GDI = guanine nucleotide dissociation inhibitor. b) Schematic of the RhoGTPase biosensor design for RhoA and RhoC.

and diffusion-based microfluidic devices can establish stable chemical gradients, but convective-flow devices have the disadvantage of fluid contact with cells, potentially altering cellular function through the introduction of shear stress and washing away autocrine and paracrine signaling factors, and diffusion-based devices are restricted in the time necessary to develop a stable gradient. In addition, both types of devices are largely limited to 2D studies (Kim and Wu, 2012).

To obviate these issues, recent platforms have utilized 3D hydrogel-based microfluidics (Cheng et al., 2007), which more readily mimic the microenvironment. Peptides, signaling molecules, growth factors, and nutrients are easily diffusible through the hydrogel matrix, which can consist of synthetic (e.g. polyethylene glycol, or PEG) or natural monomers (e.g. collagen, fibrin) (Lühmann et al., 2009). In addition, the hydrogel is both mechanically and chemically tunable (Lühmann et al., 2009). Hydrogel matrices can be used to generate stable gradients of soluble cues (Haessler et al., 2009; Yoon et al., 2016), as well as matrix-bound cues, which is how many guidance cues are presented *in vivo*. It has been well-established that cells migrate quite differently in 2D compared to 3D environments (Cukierman et al., 2001); substrate-bound protein gradients can be used to analyze haptotaxis (Kim and Peyton, 2012) and can be generated through microfluidic patterning (Allazzetta et al., 2011) and a myriad of other technologies (Kim and Peyton, 2012; Ricoult et al., 2015). Time-lapse confocal microscopy is used to image the formation of molecular gradients of both chemotaxis and haptotaxis assays using microfluidic devices (Keenan and Folch, 2008), and software such as Imaris (Bitplane, Zurich, Switzerland) and ImageJ (NIH, USA) can be used to analyze 3D imaging.

Micropatterning has been used to deconstruct the complexity of cell motility in the microenvironment into the analysis of single parameters; both micropatterned lines and microfabricated channels have been used to figure out key aspects of cell motility (Lautenschlager and Piel, 2013). For example, the use of adhesive micropatterned lines demonstrated that one-dimensional (1D) migration of fibroblasts more closely resembles 3D migration in fibrillar ECM (Doyle et al., 2009). Microchannels, which are made of polydimethylsiloxane (PDMS) with a restricted geometry (e.g. wavy, ratchet, constrictive patterns) (Heuze et al., 2011), are being used to confine cells and analyze their motion without additional confounding factors. The benefit of these devices is their adaptability to time-lapse wide-field and confocal microscopy; acquisition parameters (time, magnification) are dependent on the question being studied (Heuze et al., 2011).

In the context of breast cancer cell migration, the multicellular streaming migratory phenotype that is achieved by comigration of macrophages and tumor cells *in vivo* was recapitulated by using 1D micropatterned substrates (Sharma et al., 2012; Leung et al., 2017). These studies identified the signaling pathways that regulate directional motility *in vivo*. In addition, microfluidic channels coated with endothelial cells have been used to characterize circulating tumor cell behavior (Follain et al., 2018a).

The GTPase biosensor imaging technology has also been used in the context of microfluidic devices (Lin et al., 2012; Martin et al., 2016). Recent studies using a Rac1 biosensor

have shown the dynamics of Rac1 activity upon activation of an upstream GEF Tiam1 while migrating in microfluidic chambers (Lin et al., 2012). In addition, work from the Pertz lab used a microfluidic chip to precisely control the stimulation of fibroblasts and measure changes in GTPase activation during protrusion dynamics (Martin et al., 2016). The combination of *in vitro* systems with the biosensor technology is an interesting avenue to study the activation of GTPases in more physiological contexts.

13.2.2 Imaging Migration *In Vivo*

Intravital imaging is a recent technological advance that is used to examine cell motility *in vivo* with high spatial and temporal resolution. Because of their optical transparency and immature immune system at early developmental stages, zebrafish embryos have been used as a model system to study circulating tumor cell dynamics. Labeled tumor cells are injected into zebrafish embryos immobilized in a glass-bottom dish, followed by time-lapse imaging with a confocal microscope. Because of the wide variety of zebrafish transgenic lines, circulating tumor cell behavior and labeled vasculature can be imaged simultaneously (Follain et al., 2018b); recent work from Follain et al. utilized this model to demonstrate how blood flow affects circulating tumor cell arrest and extravasation (Follain et al., 2018a). To image with a quick acquisition speed and limited phototoxicity *in vivo*, lattice light-sheet microscopy has been developed to image ultrathin light sheets that can capture 3D dynamics, including the movement of circulating tumor cells in the zebrafish embryo (Liu et al., 2018), as well as cells in a developing Drosophila embryo for many hours (Chen et al., 2014).

Multiphoton laser scanning microscopy (MPLSM) and imaging windows have been used to image tumor cell motility in tumors developed from transgenic and xenograft mouse models (Ellenbroek and van Rheenen, 2014), tumor cell motility in the lung (Entenberg et al., 2018) and melanoma cell motility in sentinel lymph nodes from a melanoma xenograft model (Olmeda et al., 2017). MPLSM has also been utilized in understanding immune cell trafficking *in vivo*, including intravital imaging of T cells in the lymph node (Miller et al., 2003; Okada et al., 2016), as well as the motility of osteoclast precursor monocytes (Ishii et al., 2009).

The benefit of MPLSM is the use of near-infrared (NIR) radiation with femtosecond lasers to excite fluorophores, allowing for deep imaging of living samples, diminished photodamage and phototoxicity, and the excitation of multiple fluorophores simultaneously (Stutzmann and Parker, 2005). Second harmonic generation (SHG), which can be used to image ordered structures, such as type I collagen fibers *in vivo*, coupled to MPLSM has demonstrated the effects of the extracellular matrix topology on tumor cell migration (Clark and Vignjevic, 2015). The vasculature, which also affects tumor cell motility and intravasation, can be labeled with fluorophore-conjugated dextrans and imaged with MPLSM to provide a more complete picture of how the microenvironment regulates tumor cell motility *in vivo* (Harney et al., 2015).

Third harmonic generation (THG) has been used to image infiltrating leukocytes (Rehberg et al., 2011) and the microenvironment surrounding melanoma cell invasion (Weigelin et al., 2012). In contrast to SHG, THG requires the simultaneous arrival of three photons and the emission of a single photon at one-third the wavelength and triple the energy (Weigelin et al., 2016). THG occurs at cell and tissue interfaces (e.g. between lipid-rich structures and aqueous fluids, or protein-rich structures and aqueous fluids), including cell and nuclear membranes, intracellular and extracellular vesicles, adipocytes, blood vessels, erythrocytes, nerve fibers, and muscle fibers (Weigelin et al., 2016). Both SHG and THG allow for label-free imaging of cell-matrix interactions during cell motility *in vivo*.

To track the motility of tumor or immune cells, cells can be engineered with a photoactivatable protein (e.g. Dendra2) that can be photoconverted using multiphoton systems and tracked in a spatiotemporal manner (Kedrin et al., 2008; Okada et al., 2016; Pignatelli et al., 2016). Custom ImageJ plugins can be used to analyze four-dimensional (4D) (x, y, z, t) datasets (Entenberg et al., 2011). A disadvantage of intravital imaging with MPLSM is that it is limited by the availability of fluorescent probes and mouse models. One type of correlative microscopy that combines the advantages of cryosection labeling and intravital microscopy is CLIM. Using a multiphoton laser, a 3D autofluorescent fiducial mark, or "photo-tattoo," is created in the living mouse. Tissues are then frozen, sectioned with a cryostat, fluorescently-labeled, and imaged; the images from intravital microscopy are then correlated with the photo-tattoo. This method has been used to link how T cells affect the migration of mouse invasive lobular carcinoma cells (Ritsma et al., 2013).

Bioengineered devices that release biomolecules and *in vivo* microscopy have also been combined to develop technology that allows for the manipulation of tumor microenvironments i.e. by implanting nanodevices inside tumors, as the recently described iNANIVID (Raja et al., 2010; Williams et al., 2016; Fluegen et al., 2017).

13.2.3 *In Vivo* Imaging and FRET Biosensors

More recently, *in vivo* two-photon microscopy has been combined with FRET biosensors. Johnsson et al. (2014) developed a Rac-FRET mouse to study the spatiotemporal activation of Rac in different tumor types. In this study, the authors utilize FLIM-FRET imaging to monitor changes in Rac activation in depth with high resolution. They crossed the Rac-FRET reporter mouse strain with several mouse models of disease to study colorectal, pancreatic, and mammary tumors. Their results show changes in Rac activation *in vivo* during disease progression, revealing important insights into the spatiotemporal regulation of Rac in development and tumorigenesis. Nobis et al. (2017) generated a RhoA-FRET mouse and studied RhoA spatiotemporal activation during development and disease progression. In this study, they developed a conditional RhoA-OFF mouse model to control the expression of the biosensor. RhoA biosensor expression was prevented by a floxed transcriptional stop sequence; upon Cre-mediated recombination, the RhoA biosensor was ubiquitously expressed. After crossing the RhoA-OFF mouse to various mouse models of cancer, they utilized elegant intravital imaging, imaging

windows, and FRET-FLIM analysis to show changes in RhoA activity in response to anticancer drugs (Nobis et al., 2018).

In addition to Rho GTPase biosensors, other sensors for signaling molecules such as PKA have been used in mouse models to study the behavior of tumor cells *in vivo*. Recently, the Matsuda group used a PKA mouse biosensor to study changes in endothelial barrier function in tumors (Yamauchi et al., 2016).

13.3 Imaging Cell Adhesion

Cell adhesion is a biological process by which cells create contacts either with each other, or with the extracellular matrix (ECM). These adhesions are mediated via cell adhesion molecules (CAM), such as integrins, selectins, and cadherins. The interaction of a cell with a neighboring cell is mediated by adherens junctions, tight junctions, and desmosomes, while the interaction of a cell with the underlying ECM components, such as collagen, fibronectin, and laminin, occurs through focal adhesions, fibrillar adhesions, and hemidesmosomes (Niessen, 2007; Parsons et al., 2010).

Adhesion is an essential process in cell communication, development, disease progression, and in maintaining the architecture of tissues. In addition, adhesions play a role in many cellular processes including cell division, differentiation, migration, ECM degradation, and survival. Different pathologies have been assigned to defects in adhesion formation, such as arthritis, osteoporosis, atherosclerosis, and genetic diseases, such as leucocyte adhesion deficiency-I (Harris et al., 2012); autoimmune diseases, such as pemphigus (Galichet et al., 2014); and cancer (Bendas and Borsig, 2012). Given the complexity of these structures and their dynamics, different imaging approaches have been used to study their function and regulation. In this part of the chapter, we will review different microscopy techniques that have revealed important aspects of adhesion dynamics both *in vitro* and *in vivo*.

13.3.1 Imaging Cellular Adhesion *In Vitro*

Microscopy has played a major role in the identification of focal adhesions and its different components. In 1961, work by Dr. A.C. Taylor (1961) described the attachment and spreading of cells on glass using phase contrast light microscopy. Studies using interference reflection microscopy (IRM) first observed the close contact between the ventral surface of cells and the substratum (Curtis, 1964). IRM also showed localization of adhesion molecules in focal adhesion structures and allowed for the study of adhesion dynamics (Horwitz, 2012). Heath and Dunn. (1978) showed that coupling IRM with electron microscopy was able to explore further the adhesion complex, showing that, indeed, the complexes link the extracellular matrix with the actin cytoskeleton. Early work using scanning force microscopy and scanning electron microscopy provided high-resolution images of the adhesion structures (Rajaraman et al., 1974).

Total internal reflection fluorescence microscopy (TIRF) and total internal reflection aqueous fluorescence microscopy (TIRAF) allow for the visualization of the adhesion between cells and the substratum; one caveat is that the samples must remain transparent and not be too thick. This technique can be used to image a region only of 100 microns close to the glass surface used to cultivate cells (Fish, 2009). Acquisition of time-lapse movies by using wide-field microscopy or spinning disk confocal microscopy allows for the study the focal adhesion dynamics and focal adhesion protein turnover following adhesion assembly and disassembly. Spinning disk confocal is preferred for long-term live-cell imaging because of its fast acquisition time and low photobleaching (Worth and Parsons, 2010). High content image analysis tools have also been implemented to quantitatively analyze the changes induced in focal adhesions when interfering with molecules involved in cell migration and to identify new regulators of focal adhesion dynamics (Bravo-Cordero et al., 2016).

Photoactivation and photobleaching are imaging methods used to study protein dynamics and turnover over time. A few examples of the application of this technology to study key focal adhesion molecules are listed here: Himmel et al. (2009) defined the kinetics of talin in focal adhesions using fluorescence recovery after photobleaching (FRAP); Hamadi et al. (2009) used fluorescence loss in photobleaching (FLIP) to investigate the dynamics of Src in focal adhesions.

Other imaging tools to study focal adhesion biology include traction force microscopy, atomic force microscopy (AFM), AFM-based force spectroscopy, fluorescence microscopy and fluid force microscopy-based single-cell force spectroscopy, and photonic crystal enhanced microscopy (PCEM). Traction force microscopy is a technique that is particularly useful for measuring cell traction forces in generating mechanical signals driving adhesion dynamics and allows for the analysis of both adhesion molecule dynamics and ECM deformation by the cell contact (Wang and Li, 2009). Atomic force microscopy (AFM) for example, was used in a study in which the adhesion forces of cervical carcinoma cells in tissue culture were measured by using the manipulation force microscope, a novel AFM (Sagvolden et al., 1999). AFM-based force spectroscopy has been used to quantify single-molecule adhesion forces in living amoeboid cells (Eibl and Benoit, 2004). Kashef and Franz (2015) used a combination of TIRF and AFM to study adhesion and to better understand the initial adhesion events when cells engage contact with the ECM. Sankaran et al. (2017) used fluorescence microscopy and fluid force microscopy-based single-cell force spectroscopy to show that actin filaments, focal adhesions, adhesion forces, and cell contractility are comparable between cells adhering to covalent and noncovalent surfaces. Photonic crystal enhanced microscopy (PCEM) is a novel biosensor-based microscopy technique that allows for the movement of cellular materials at the plasma membrane of individual living cells to be monitored dynamically and imaged quantitatively (Zhuo et al., 2016). This is a high sensitivity label-free live-cell imaging technique allowing for the profiling of dynamic adhesions of single cells. This technique has been used to investigate the adhesion process in early stages of stem cell migration.

13.3.1.1 Super-Resolution Microscopy

New microscopy techniques that have improved resolution have also been successfully applied to the study of focal adhesions. Super-resolution imaging increases spatial resolution compared to confocal microscopy. The development of techniques such as structured illumination microscopy (SIM), photoactivated localization microscopy (PALM), stochastic optical reconstruction microscopy (STORM), and stimulated emission depletion (STED) enable the characterization of subdomains of focal adhesions and revealed their dynamic architecture with a single-molecule localization scale (Schwartz, 2011; Sahl et al., 2017). All these technologies have contributed to the improvement and understanding of adhesions as well as the establishment of novel models that are shedding light into how cells organize focal adhesions at the molecular level. For example, PALM has revealed that signaling and cytoskeletal proteins reside at specific vertical distances between the plasma membrane and F-actin (Schwartz, 2011).

FRET techniques have also been used to study focal adhesions. Seong et al. used fluorescence resonance energy transfer (FRET) to detect focal adhesion kinase activation at membrane microdomains (Seong et al., 2011). FRET biosensors for RhoGTPases have also been used to study the activation of GTPases during adhesion (Martin et al., 2016). By using FRET techniques, Grashoff et al. (2010) developed a sensor that measures forces across specific molecules. They applied this technology to focal adhesion components such as vinculin and talin (Kumar et al., 2016) to measure the tension across these molecules at focal adhesions.

13.3.2 Adhesion *In Vivo*

Studying focal adhesions in a physiological context has been a challenge in the cell biology field. Work by Cukierman et al. (2001, 2002) showed that in 3D matrices, focal adhesions are morphologically different than ones in 2D, and also have a different composition. Drosophila embryos, as well as *D. rerio* and *C. elegans* models have been used to study cell adhesion *in vivo*. Because of their size, low thickness, and transparency, cell adhesion can be studied in these models with high resolution using confocal spinning disk microscopy. Several groups have developed imaging technologies to study adhesion *in vivo* by using different model organisms. David Sherwood and his team examined the cell biological aspects of cell–cell and cell-basement membrane establishment during uterine-vulval attachment in *C. elegans* using confocal spinning disk microscopy (Ihara et al., 2011). Shen et al. (2011) developed a method of single-cell adhesion force measurements based on nanorobotic manipulation systems inside an environment scanning electron microscope (ESEM). By using this method, they evaluated the adhesion of single yeast cells to ITO substrates. An elegant study using Drosophila embryos and larvae analyzed the dynamics of adhesions *in vivo*. The authors used fluorescence recovery after photobleaching (FRAP) to measure integrin dynamics in adhesion structures showing the importance of protein turnover inside a stable structure for proper function of adhesions (Yuan et al., 2010). Xu et al. (2012) showed that the motility of chemokine-guided germ cells within the zebrafish embryo requires the function of small RhoGTPases Rac1 and RhoA, as well as E-cadherin-mediated cell–cell adhesion by using FRET imaging.

13.4 Conclusions and Perspectives

Microscopy has proven to be a powerful tool to uncover mechanisms mediating cell migration and adhesion. State-of-the-art imaging approaches are being used to understand the process of cell migration from the single-cell to the single-molecule level. The innovative technologies being developed in the cell biology field will uncover new aspects of motility and adhesion that will impact our understanding of development and disease.

13.5 Acknowledgments

This work was supported by a National Cancer Institute (NCI) Career Transition Award (K22CA196750, to J.J.B.C), a Susan G. Komen for the Cure Career Catalyst Award (CCR18547848, to J.J.B.C), an NCI National Research Service Award Institutional Training Grant (T32) (5T32CA078207-19, to C.M), an NCI Cancer Center Support Grant (P30-CA196521), and a Schneider-Lesser Fellowship Award (to J.J.B.C).

REFERENCES

Abercrombie, M., Heaysman, J. E., and Pegrum, S. M. (1971). The locomotion of fibroblasts in culture. IV. Electron microscopy of the leading lamella. *Exp Cell Res, 67*: 359–367.

Allazetta, S., Cosson, S., and Lutolf, M. P. (2011). Programmable microfluidic patterning of protein gradients on hydrogels. *Chem Commun, 47*: 191–193.

Aoki, K., and Matsuda, M. (2009). Visualization of small GTPase activity with fluorescence resonance energy transfer-based biosensors. *Nat Protoc, 4*: 1623–1631.

Bendas, G., and Borsig, L. (2012). Cancer cell adhesion and metastasis: Selectins, integrins, and the inhibitory potential of heparins. *Int J Cell Biol, 2012*: 676731.

Boyden, S. (1962). The chemotactic effect of mixtures of antibody and antigen on polymorphonuclear leucocytes. *J Exp Med, 115*: 453–466.

Bravo-Cordero, J. J. et al. (2016). A novel high-content analysis tool reveals Rab8-driven cytoskeletal reorganization through Rho GTPases, calpain and MT1-MMP. *J Cell Sci, 129*: 1734–1749.

Bravo-Cordero, J. J., Hodgson, L., and Condeelis, J. (2012). Directed cell invasion and migration during metastasis. *Curr Opin Cell Biol, 24*: 277–283.

Bravo-Cordero, J. J., Hodgson, L., and Condeelis, J. S. (2014). Spatial regulation of tumor cell protrusions by RhoC. *Cell Adh Migr, 8*: 263–267.

Bravo-Cordero, J. J., Magalhaes, M. A. O., Eddy, R. J., Hodgson, L., and Condeelis, J. (2013a). Functions of cofilin in cell locomotion and invasion. *Nat Rev Mol Cell Biol, 14*: 405–415.

Bravo-Cordero, J. J., Moshfegh, Y., Condeelis, J., and Hodgson, L. (2013b). Live cell imaging of RhoGTPase biosensors in tumor cells. *Methods Mol Biol, 1046*: 359–370.

Bravo-Cordero, J. J., Oser, M., Chen, X., Eddy, R., Hodgson, L., and Condeelis, J. (2011). A novel spatiotemporal RhoC activation pathway locally regulates cofilin activity at invadopodia. *Curr Biol, 21*: 635–644.

Bravo-Cordero, J. J. et al. (2013). Spatial regulation of RhoC activity defines protrusion formation in migrating cells. *J Cell Sci, 126*(15): 3356–3369.

Burridge, K. (2017). Focal adhesions: A personal perspective on a half century of progress. *FEBS J 284*: 3355–3361.

Chen, B.-C. et al. (2014). Lattice light-sheet microscopy: Imaging molecules to embryos at high spatiotemporal resolution. *Science, 346*: 1257998.

Cheng, S.-Y., Heilman, S., Wasserman, M., Archer, S., Shuler, M. L., and Wu, M. (2007). A hydrogel-based microfluidic device for the studies of directed cell migration. *Lab Chip, 7*: 763–769.

Clark, A. G., and Vignjevic, D. M. (2015). Modes of cancer cell invasion and the role of the microenvironment. *Curr Opin Cell Biol, 36*: 13–22.

Cukierman, E., Pankov, R., Stevens, D. R., and Yamada, K. M. (2001). Taking cell-matrix adhesions to the third dimension. *Science, 294*: 1708–1712.

Cukierman, E., Pankov, R., and Yamada, K. M. (2002), Cell interactions with three-dimensional matrices. *Curr Opin Cell Biol, 14*: 633–639.

Curtis, A. S. (1964). The mechanism of adhesion of cells to glass. A study by interference reflection microscopy. *J Cell Biol, 20*: 199–215.

Donnelly, S. K., Bravo-Cordero, J. J., and Hodgson, L. (2014), Rho GTPase isoforms in cell motility: Don't fret, we have FRET. *Cell Adh Migr, 8*: 526–534.

Doyle, A. D., Wang, F. W., Matsumoto, K., and Yamada, K. M. (2009). One-dimensional topography underlies three-dimensional fibrillar cell migration. *J Cell Biol, 184*: 481–490.

Dunn, G. A., and Jones, G. E. (2004). Cell motility under the microscope: Vorsprung durch Technik. *Nat Rev Mol Cell Biol, 5*: 667–672.

Eibl, R. H., and Benoit, M. (2004). Molecular resolution of cell adhesion forces. *IEE Proc Nanobiotechnol, 151*: 128–132.

Ellenbroek, S. I. J., and van Rheenen, J. (2014). Imaging hallmarks of cancer in living mice. *Nat Rev Cancer, 14*: 406–418.

Entenberg, D. et al. (2018). A permanent window for the murine lung enables high-resolution imaging of cancer metastasis. *Nat Methods, 15*: 73–80.

Entenberg, D. et al. (2011). Setup and use of a two-laser multiphoton microscope for multichannel intravital fluorescence imaging. *Nat Protoc, 6*: 1500–1520.

Fish, K. N. (2009). Total internal reflection fluorescence (TIRF) microscopy. *Curr Protoc Cytom, 50*(1): 12–18.

Fluegen, G. et al. (2017). Phenotypic heterogeneity of disseminated tumour cells is preset by primary tumour hypoxic microenvironments. *Nat Cell Biol. 19*: 120–132.

Follain, G. et al. (2018a) Hemodynamic forces tune the arrest, adhesion, and extravasation of circulating tumor cells. *Dev Cell, 45*: 33–52.e12.

Follain, G., Osmani, N., Fuchs, C., Allio, G., Harlepp, S., and Goetz, J. G. (2018b). Using the zebrafish embryo to dissect the early steps of the metastasis cascade. *Methods Mol Biol, 1749*: 195–211.

Franz, C. M., Jones, G. E., and Ridley, A. J. (2002). Cell migration in development and disease. *Dev Cell, 2*: 153–158.

Galichet, A., Borradori, L., and Muller, E. J. (2014), A new light on an old disease: Adhesion signaling in pemphigus vulgaris. *J Invest Dermatol, 134*: 8–10.

Gerisch, G., and Keller, H. U. (1981). Chemotactic reorientation of granulocytes stimulated with micropipettes containing fMet-Leu-Phe. *J Cell Sci, 52*: 1–10.

Grashoff, C. et al. (2010). Measuring mechanical tension across vinculin reveals regulation of focal adhesion dynamics. *Nature, 466*: 263–266.

Haessler, U., Kalinin, Y., Swartz, M. A., and Wu, M. (2009). An agarose-based microfluidic platform with a gradient buffer for 3D chemotaxis studies. *Biomed Microdevices, 11*: 827–835.

Hamadi, A., Deramaudt, T. B., Takeda, K., and Ronde, P. (2009). Src activation and translocation from focal adhesions to membrane ruffles contribute to formation of new adhesion sites. *Cell Mol Life Sci, 66*: 324–338.

Hanna, S., Miskolci, V., Cox, D., and Hodgson, L. (2014). A new genetically encoded single-chain biosensor for Cdc42 based on FRET, useful for live-cell imaging. *PLoS One, 9*: e96469.

Harney, A. S. et al. (2015). Real-time imaging reveals local, transient vascular permeability, and tumor cell intravasation stimulated by TIE2[hi] macrophage–derived VEGFA. *Cancer Discov, 5*: 932–943.

Harris, E. S., Smith, T. L., Springett, G. M., Weyrich, A. S., and Zimmerman, G. A. (2012). Leukocyte adhesion deficiency-I variant syndrome (LAD-Iv, LAD-III): Molecular characterization of the defect in an index family. *Am J Hematol, 87*: 311–313.

Heath, J. P., and Dunn, G. A. (1978). Cell to substratum contacts of chick fibroblasts and their relation to the microfilament system. A correlated interference-reflexion and high-voltage electron-microscope study. *J Cell Sci, 29*: 197 LP – 212.

Heuze, M. L., Collin, O., Terriac, E., Lennon-Dumenil, A.-M., and Piel, M. (2011). Cell migration in confinement: A microchannel-based assay. *Methods Mol Biol, 769*: 415–434.

Himmel, M. et al. (2009). Control of high affinity interactions in the talin C terminus: How talin domains coordinate protein dynamics in cell adhesions. *J Biol Chem, 284*: 13832–13842.

Horwitz, A. R. (2012). The origins of the molecular era of adhesion research. *Nat Rev Mol Cell Biol, 13*: 805–811.

Ihara, S. et al. (2011). Basement membrane sliding and targeted adhesion remodels tissue boundaries during uterine-vulval attachment in Caenorhabditis elegans. *Nat Cell Biol, 13*: 641–651.

Ishii, M. et al. (2009). Sphingosine-1-phosphate mobilizes osteoclast precursors and regulates bone homeostasis. *Nature, 458*: 524–528.

Johnsson, A.-K. E. et al. (2014). The Rac-FRET mouse reveals tight spatiotemporal control of Rac activity in primary cells and tissues. *Cell Rep, 6*: 1153–1164.

Kashef, J., and Franz, C. M. (2015). Quantitative methods for analyzing cell–cell adhesion in development. *Dev Biol, 401*: 165–174.

Kedrin, D. et al. (2008). Intravital imaging of metastatic behavior through a mammary imaging window. *Nat Method, 5*: 1019–1021.

Keenan, T. M., and Folch, A. (2008). Biomolecular gradients in cell culture systems. *Lab Chip, 8*: 34–57.

Kim, B. J., and Wu, M. (2012). Microfluidics for mammalian cell chemotaxis. *Ann Biomed Eng, 40*: 1316–1327.

Kim, H.-D., and Peyton, S. R. (2012). Bio-inspired materials for parsing matrix physicochemical control of cell migration: A review. *Integr Biol, 4*: 37–52.

Komatsu, N. et al. (2011). Development of an optimized backbone of FRET biosensors for kinases and GTPases. *Mol Biol Cell, 22*: 4647–4656.

Kumar, A. et al. (2016). Talin tension sensor reveals novel features of focal adhesion force transmission and mechanosensitivity. *J Cell Biol, 213*: 371 LP – 383.

Kurokawa, K., and Matsuda, M. (2005). Localized RhoA activation as a requirement for the induction of membrane ruffling. *Mol Biol Cell, 16*: 4294–4303.

Lane, N. (2015). The unseen world: Reflections on leeuwenhoek (1677) 'concerning little animals.' *Philos Trans R Soc B Biol Sci, 370*: 20140344.

Lautenschlager, F., and Piel, M. (2013). Microfabricated devices for cell biology: All for one and one for all. *Curr Opin Cell Biol, 25*: 116–124.

Leung, E. et al. (2017), Blood vessel endothelium-directed tumor cell streaming in breast tumors requires the HGF/C-Met signaling pathway. *Oncogene, 36*: 2680–2692.

Lin, B. et al. (2012). Synthetic spatially graded Rac activation drives cell polarization and movement. *Proc Natl Acad Sci U S A, 109*: E3668– E3677.

Liu, T. L. et al. (2018). Observing the cell in its native state: Imaging subcellular dynamics in multicellular organisms. *Science (80-), 360* (6386): eaaq1392.

Lohmer, L. L., Kelley, L. C., Hagedorn, E. J., and Sherwood, D. R. (2014). Invadopodia and basement membrane invasion in vivo. *Cell Adh Migr, 8*: 246–255.

Lühmann, T., Hänseler, P., Grant, B., and Hall, H. (2009). The induction of cell alignment by covalently immobilized gradients of the 6th Ig-like domain of cell adhesion molecule L1 in 3D-fibrin matrices. *Biomaterials, 30*: 4503–4512.

Martin, K., Reimann, A., Fritz, R. D., Ryu, H., Jeon, N. L., and Pertz, O. (2016). Spatio-temporal co-ordination of RhoA, Rac1 and Cdc42 activation during prototypical edge protrusion and retraction dynamics. *Sci Rep, 6*: 21901.

Medwig, T. N., Matus, D. Q. (2017). Breaking down barriers: The evolution of cell invasion. *Curr Opin Genet Dev, 47*:33–40.

Miller, M. J., Wei, S. H., Cahalan, M. D., and Parker, I. (2003). Autonomous T cell trafficking examined in vivo with intravital two-photon microscopy. *Proc Natl Acad Sci U S A, 100*: 2604–2609.

Miskolci, V., Wu, B., Moshfegh, Y., Cox, D., and Hodgson, L. (2016). Optical tools to study the isoform-specific roles of small GTPases in immune cells. *J Immunol, 196*: 3479–3493.

Moshfegh, Y., Bravo-Cordero, J. J., Miskolci, V., Condeelis, J., and Hodgson, L. (2014). A Trio-Rac1-PAK1 signaling axis drives invadopodia disassembly. *Nat Cell Biol, 16*: 574–586.

Nelson, R. D., Quie, P. G., and Simmons, R. L. (1975). Chemotaxis under agarose: A new and simple method for measuring chemotaxis and spontaneous migration of human polymorphonuclear leukocytes and monocytes. *J Immunol, 115*: 1650–1656.

Niessen, C. M. (2007). Tight junctions/adherens junctions: Basic structure and function. *J Invest Dermatol, 127*: 2525–2532.

Nobis, M. et al. (2017). A RhoA-FRET biosensor mouse for intravital imaging in normal tissue homeostasis and disease contexts. *Cell Rep, 21*: 274–288.

Nobis, M. et al. (2018). Shedding new light on RhoA signalling as a drug target in vivo using a novel RhoA-FRET biosensor mouse. *Small GTPases*: 1–8.

Okada, T., Takahashi, S., Ishida, A., and Ishigame, H. (2016). In vivo multiphoton imaging of immune cell dynamics. *Pflugers Arch*, 468: 1793–1801.

Olmeda, D. et al. (2017). Whole-body imaging of lymphovascular niches identifies pre-metastatic roles of midkine. *Nature*, 546: 676–680.

Paluch, E. K., and Raz, E. (2013). The role and regulation of blebs in cell migration. *Curr Opin Cell Biol*, 25: 582–590.

Parsons, J. T., Horwitz, A. R., and Schwartz, M. A. (2010). Cell adhesion: Integrating cytoskeletal dynamics and cellular tension. *Nat Rev Mol Cell Biol*, 11: 633–643.

Pertz, O., Hodgson, L., Klemke, R. L., and Hahn, K. M. (2006). Spatiotemporal dynamics of RhoA activity in migrating cells. *Nature*, 440: 1069–1072.

Pignatelli, J. et al. (2016). Macrophage-dependent tumor cell transendothelial migration is mediated by Notch1/Mena(INV)-initiated invadopodium formation. *Sci Rep*, 6: 37874.

Pujic, Z., Mortimer, D., Feldner, J., and Goodhill, G. J. (2009). Assays for eukaryotic cell chemotaxis. *Comb Chem High Throughput Screen*, 12: 580–588.

Raja, W. K., Gligorijevic, B., Wyckoff, J., Condeelis, J. S., and Castracane, J. (2010). A new chemotaxis device for cell migration studies. *Integr Biol*, 2: 696–706.

Rajaraman, R., Rounds, D. E., Yen, S. P., and Rembaum, A. (1974). A scanning electron microscope study of cell adhesion and spreading in vitro. *Exp Cell Res*, 88: 327–339.

Rehberg, M., Krombach, F., Pohl, U., and Dietzel, S. (2011). Label-free 3D visualization of cellular and tissue structures in intact muscle with second and third harmonic generation microscopy. *PLoS One*, 6: e28237.

Ricoult, S. G., Kennedy, T. E., and Juncker, D. (2015). Substrate-bound protein gradients to study haptotaxis. *Front Bioeng Biotechnol*, 3: 40.

Ritsma, L., Vrisekoop, N., and van Rheenen, J. (2013). In vivo imaging and histochemistry are combined in the cryo-section labelling and intravital microscopy technique. *Nat Commun*, 4: 2366.

Roh-Johnson, M. et al. (2014). Macrophage contact induces RhoA GTPase signaling to trigger tumor cell intravasation. *Oncogene*, 33: 4203–4212.

Sagvolden, G., Giaever, I., Pettersen, E. O., and Feder, J. (1999). Cell adhesion force microscopy. *Proc Natl Acad Sci*, 96: 471 LP – 476.

Sahl, S. J., Hell, S. W., and Jakobs, S. (2017). Fluorescence nanoscopy in cell biology. *Nat Rev Mol Cell Biol*, 18: 685–701.

Sankaran, S., Jaatinen, L., Brinkmann, J., Zambelli, T., Voros, J., and Jonkheijm, P. (2017). Cell adhesion on dynamic supramolecular surfaces probed by fluid force microscopy-based single-cell force spectroscopy. *ACS Nano*, 11: 3867–3874.

Schwartz, M. A. (2011). Super-resolution microscopy: A new dimension in focal adhesions. *Curr Biol*, 21: R115–R116.

Seong, J. et al. (2011). Detection of focal adhesion kinase activation at membrane microdomains by fluorescence resonance energy transfer. *Nat Commun*, 2: 406.

Sharma, V. P. et al. (2012). Reconstitution of in vivo macrophage-tumor cell pairing and streaming motility on one-dimensional micro-patterned substrates. *Intravital*, 1: 77–85.

Shen, Y., Ahmad, M. R., Nakajima, M., Kojima, S., Homma, M., and Fukuda, T. (2011). Evaluation of the single yeast cell's adhesion to ITO substrates with various surface energies via ESEM nanorobotic manipulation system. *IEEE Trans Nanobioscience*, 10: 217–224.

Shrestha, D., Jenei, A., Nagy, P., Vereb, G., and Szöllősi, J. (2015). Understanding FRET as a research tool for cellular studies. *Int J Mol Sci*, 16: 6718–6756.

Spiering, D., Bravo-Cordero, J. J., Moshfegh, Y., Miskolci, V., and Hodgson, L. (2013). Quantitative ratiometric imaging of FRET-biosensors in living cells. *Methods Cell Biol*, 114: 593–609.

Stutzmann, G. E., and Parker, I. (2005). Dynamic multiphoton imaging: A live view from cells to systems. *Physiology*, 20: 15–21.

Taylor, A. C. (1961). Attachment and spreading of cells in culture. *Exp Cell Res Suppl*, 8: 154–173.

Taylor, D. L., and Condeelis, J. S. (1979). Cytoplasmic structure and contractility in amoeboid cells. *Int Rev Cytol*, 56: 57–144.

Wang, J. H.-C., and Li, B. (2009). Application of cell traction force microscopy for cell biology research. *Methods Mol Biol*, 586: 301–313.

Weigelin B, Bakker, G.-J., and Friedl, P. (2012). Intravital third harmonic generation microscopy of collective melanoma cell invasion: Principles of interface guidance and microvesicle dynamics. *Intravital*, 1: 32–43.

Weigelin, B., Bakker, G.-J., and Friedl, P. (2016). Third harmonic generation microscopy of cells and tissue organization. *J Cell Sci*, 129: 245–255.

Williams, J. K. et al. (2016). Validation of a device for the active manipulation of the tumor microenvironment during intravital imaging. *Intravital*, 5.

Worth, D. C., and Parsons, M. (2010). Advances in imaging cell–matrix adhesions. *J Cell Sci*, 123:3629 LP – 3638.

Wu, B. et al. (2015). Synonymous modification results in high-fidelity gene expression of repetitive protein and nucleotide sequences. *Genes Dev*, 29: 876–886.

Xu, H., Kardash, E., Chen, S., Raz, E., and Lin, F. (2012). Gbetagamma signaling controls the polarization of zebrafish primordial germ cells by regulating Rac activity. *Development*, 139: 57–62.

Yamauchi, F., Kamioka, Y., Yano, T., and Matsuda, M. (2016). In Vivo FRET Imaging of tumor endothelial cells highlights a role of low PKA activity in vascular hyperpermeability. *Cancer Res*, 76: 5266–5276.

Yoon, D. et al. (2016). Study on chemotaxis and chemokinesis of bone marrow-derived mesenchymal stem cells in hydrogel-based 3D microfluidic devices. *Biomater Res*, 20: 25.

Yuan, L., Fairchild, M. J., Perkins, A. D., and Tanentzapf, G. (2010). Analysis of integrin turnover in fly myotendinous junctions. *J Cell Sci*, 123: 939–946.

Zhuo, Y., Choi, J. S., Marin, T., Yu, H., Harley, B. A., and Cunningham, B. T. (2016). Quantitative imaging of cell membrane-associated effective mass density using photonic crystal enhanced microscopy (PCEM). *Prog Quantum Electron*, 50: 1–18.

Zicha, D., Dunn, G. A., and Brown, A. F. (1991). A new direct-viewing chemotaxis chamber. *J Cell Sci*, 99 (Pt 4): 769–775.

Zigmond, S. H. (1977). Ability of polymorphonuclear leukocytes to orient in gradients of chemotactic factors. *J Cell Biol*, 75: 606–616.

14

Imaging the Living Eye

Brian T. Soetikno, Lisa Beckmann, and Hao F. Zhang

CONTENTS

Since the dawn of photography in the 1800s, scientists and physicians have pursued techniques to capture images from inside the living eye. Capturing the fundus, or the interior surface of the retina, initially proved to be challenging. Indeed, early fundus photographs suffered from motion artifacts, overlapping corneal reflections, and long exposure times (Jackman and Webster, 1886). Since these early explorations, however, significant advances in eye imaging have emerged, owing to innovative modalities, including the fundus camera, scanning laser ophthalmoscope (SLO), and optical coherence tomography (OCT). Of note, OCT has enabled three-dimensional (3D) imaging of the living retina and provided images that closely resembled the quality of *ex vivo* histology. More recently, techniques such as photoacoustic ophthalmoscopy (PAOM) and adaptive optics (AO) have made an additional impact by providing new optical absorption imaging contrast and aberration correction for much improved spatial resolution, respectively. Each of these technologies has provided new avenues toward better understanding of vision and eye diseases.

Understanding the fundamentals of ophthalmic imaging is important to a wide variety of scientists and clinicians. For scientists, ophthalmic imaging can serve as a critical tool in studies that seek to answer questions about both physiological and pathophysiological processes in the eye. For example, studies have combined animal models of human ocular diseases with advanced imaging tools to improve our understanding of disease pathogenesis and develop new therapies. Ophthalmic imaging has also revealed mechanisms of neural circuitry, as the eye is developmentally an extension of the brain (London et al., 2013). For clinicians, many ophthalmic imaging devices are routinely used in patina care, and it can be very beneficial to understand the fundamentals and applications of these imaging modalities. This chapter will begin by introducing the ocular anatomy and then follow with discussion of the most important ophthalmic imaging technologies. The operating principles underlying each technology will be outlined, and their applications to ophthalmic research will be discussed.

14.1 Ocular Anatomy

An understanding of the ocular anatomy is required to grasp the principles of ophthalmic imaging. Refractive elements of the eye focus parallel rays of light to the light-sensitive posterior layers of the eye, and, therefore, the optical system of the eye must be considered when designing an ocular imaging system. Figure 14.1a illustrates the human eye, which can be divided into anterior and posterior segments. The anterior segment contains the curved surfaces of the cornea and lens, which focus light onto the retina. In addition, the iris dilates and constricts, controlling the amount of light reaching the retina. The posterior segment consists of the vitreous, retina,

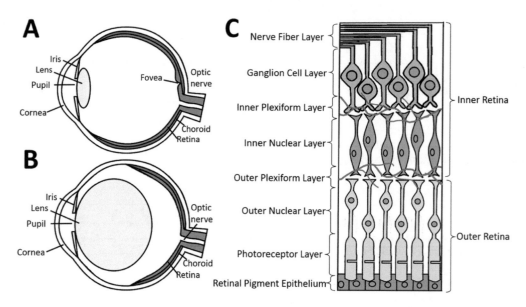

FIGURE 14.1   Anatomical structures of human eye, mouse eye, and retina. A. Schematic of a human eye; B. Schematic of a mouse eye; C. Schematic of the cross-sectional view of the cellular structures of the retina. The intermediate and deep capillary plexi are shown as well.

choroid, and optic nerve head. Most of the ophthalmic imaging modalities focus on capturing the posterior segment and, more specifically, the retina.

A significant portion of current ophthalmic research is performed in rodents, in part, because of their simple handling and housing, and their similarities in development, morphology, and function to the human eye. In addition, genetic manipulations in mice are simpler than in other mammals (Levkovitch-Verbin, 2004; Geng et al., 2012). These genetic manipulations have enabled the creation of animal models of human diseases, the investigation of disease mechanisms, and the study of therapeutic targets. As shown in Figure 14.1B, the anatomy of the human eye and the mouse eye are similar but have some important distinctions (Veleri et al., 2015). In particular, the rodent eye lacks a fovea and macula, an area responsible for sharp vision in humans that mainly consists of cone photoreceptors. Thus, rod photoreceptors predominate the rodent retina. Notably, the rodent lens takes up a significant portion of the ocular volume (Remtulla and Hallett, 1985). Because the focal length of the rodent eye is smaller to that of the human eye, the numerical aperture (NA) in mice (NA~0.5) is larger than in humans (NA~0.18) (Geng et al., 2011). This chapter will focus on ophthalmic imaging in rodents, but many of the principles can be translated to humans.

As shown in Figure 14.1C, the retina is composed of eight distinct layers grouped into the outer and inner retinae. The outer retina consists of photoreceptors and the retinal pigment epithelium (RPE). Photoreceptors are further divided into rod and cone photoreceptors, which are responsible for low-light vision and color perception, respectively. The RPE layer contains various pigments, including melanin, that aid in absorbing irradiating light energy among other physiological roles. In addition, the RPE also transports nutrients between the photoreceptors and the choriocapillaris, participates in the visual cycle, removes shed photoreceptor membranes by

phagocytosis, and secretes a variety of growth factors (Strauss, 2005). The inner retina consists of four classes of neurons: ganglion cells, amacrine cells, horizontal cells, and bipolar cells. The combination and connections of these cells form small neuronal networks that amplify, process, and filter information from the photoreceptors (Wassle, 2004). Eventually, the ganglion cell bodies collect the partially processed information, and their axons, which together form the optic nerve, transmit the information to the visual cortex of the brain for further higher-level information processing.

The retina possesses an intricate circulatory system that provides the nutrients necessary to meet the high metabolic demand of the tissue (Wangsa-Wirawan and Linsenmeier, 2003; Linsenmeier and Zhang, 2017). Unlike any other tissue in the body, retina has two independent circulations: the anterior, termed the retinal circulation, and the posterior, termed the choroidal circulation. Each is distinct by anatomy and by autoregulatory control. The retinal circulation is divided into a trilaminar network (Kornfield and Newman, 2014). Arterioles exit at the optic nerve head and spread outward at the level of the nerve–fiber layer and the ganglion-cell layer (Figure 14.1c). Branches dive deeper into the retina and form an intermediate capillary plexus at the boundary between the inner nuclear layer and the inner plexiform layer. A dense, deep capillary plexus is also formed at the boundary of the outer nuclear layer and the outer plexiform layer. The capillaries vertically ascend and drain to large veins that exit at the optic nerve head. The choroidal circulation consists of large vessels that ultimately branch to form the choriocapillaris, a dense mesh of capillaries beneath the RPE layer. The retinal circulation is autoregulated, while the choroidal circulation is not. Thus, the flow rate of the retinal circulation is controlled and varies depending on the metabolic demands of the eye. On the other hand, the flow rate of the choroidal circulation remains mostly constant with one of the highest flow rates in the body (Kur et al., 2012).

14.2 Fundus Camera

The fundus camera magnifies and photographs the retinal fundus, providing two-dimensional (2D) anatomical information. Images can be used to determine the progression of retinal diseases, especially those that involve retinal vasculature or retinal lesions. With monochromatic filters, the fundus camera can also enhance its contrast, for example, of hemoglobin in imaging vasculature or other pigmented chromophores in the retina, such as drusen in age-related macular degeneration (AMD). Fluorescent dyes can also be imaged by the fundus camera by using filters designed to pass specific wavelength ranges of fluorescence emission. This technique is primarily used for fluorescein angiography, a method to map the retinal vasculature after intravenous injection of the fluorescein dye. Fundus cameras can be integrated with other imaging modalities, such as photoacoustic ophthalmoscopy (Liu et al., 2013) or optical coherence tomography (Song et al., 2012) to guide these techniques by providing a large, real-time field-of-view for the operator. Multiwavelength and hyperspectral fundus cameras, which can obtain spectroscopic measurements of hemoglobin absorption inside the vasculature, have also been tested for quantifying the oxygen saturation in the retinal blood vessels (Hardarson et al., 2006; Li et al., 2017).

Figure 14.2a illustrates the principle of the fundus camera. There are two light paths in a fundus camera: the observation light path (solid lines) and the illumination light path (dashed lines). The observation light path is similar to that of an indirect ophthalmoscope, in which a lens placed in front of the eye focuses parallel rays exiting the eye to form an aerial image. As shown in Figure 14.2a, the first objective lens (L_{obj}), in conjunction with the subject's own cornea and lens, forms an aerial image of the fundus. The aerial image is then imaged through a 50/50 beam splitter to a camera's charged coupled device (CCD) sensor by the combination of a second objective lens (L_{obj2}) and a zoom lens.

The illumination light path takes advantage of the Gullstrand principle, which was first introduced in 1910 by Alvar Gullstrand (1910). Gullstrand realized that to eliminate the strong corneal reflections in ophthalmoscopy, the illumination and observation light paths should be separated at the pupil plane. To implement the Gullstrand principle, the light source illuminates an annular mask. The light annulus is then projected to the subject's outer pupil plane by L_{obj} and L3. The outer rim of the annulus serves as the illumination light, while the center dark portion of the annulus allows observation light to pass through (Figure 14.1b). Fundus cameras typically use a noncoherent, incandescent source (e.g. halogen or xenon lamp). In the design shown in Figure 14.2A, a beam splitter was used so that the illumination and observation light paths were concentric. Alternative designs have incorporated a mirror with a pinhole to achieve a similar effect (Link et al., 2011). At the pupil plane, the inner diameter of the annulus must be small enough so that enough light can enter the eye. If the subject's pupil cuts off the annular illumination, the retina will not be uniformly illuminated.

Mouse and rat eyes have much higher chromatic aberration than human eyes (Bedford and Wyszecki, 1957; Chaudhuri et al., 1983; Geng et al., 2011), ranging from 6 to 10 diopters

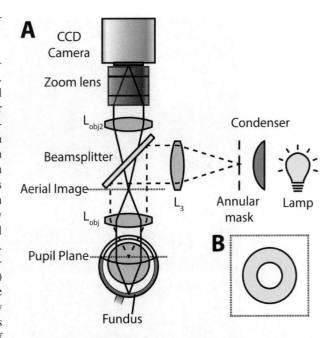

FIGURE 14.2 Principles of fundus photography. A. Schematic of a fundus camera. CCD: charge-coupled device camera; solid lines: fundus photography observation light path; dashed lines: illumination light path; B. Illustration of the Gullstrand principle. An annulus illumination is projected at the pupil plane.

in the visible spectral range. Additionally, eyes from albino rodent strains have more severe chromatic aberration than those from pigmented animals. Without correcting the chromatic aberration, broadband fundus photographs in rodents are blurry and provide low lateral resolution (Li et al., 2015). By limiting the illumination bandwidth, the lateral resolution can be optimized. Using a combination of numerical simulation and experimental studies, Li et al. (2015) demonstrated the highest resolution images that can be achieved when the light source bandwidth is between 10 nm and 20 nm.

Figure 14.3 shows fundus camera images for six different spectral bandwidths. The filter bandwidths were 1: 19 nm, 2: 30 nm, 3: 89 nm, 4: 143 nm, 5: 162 nm, 6: 270 nm, respectively (Figure 14.3a–f). The image resolution degrades with increasing bandwidth, which can be best observed in the magnified views of a blood vessel in Figures 14.3g–m. The best resolution achieved was ~10 μm using the 19 nm bandwidth illumination, which is consistent with the expected fundus camera performance given the large chromatic aberration in the rodent eye.

Other solutions for performing fundus photography have been reported. Paques et al. (2007) developed a low-cost, direct-contact solution by combining an illuminating endoscope with a digital camera. The eye is still illuminated with annular or crescent illumination to minimize specular corneal reflections. Standard laboratory microscopes can also be used to visualize the fundus by incorporating an additional objective lens or by applanation of the cornea with a contact lens or a microscope cover-slip (Hawes et al., 1999; Cohan et al., 2003). Finally, a handheld smartphone's flash light and camera for illumination and observation, with the help of a standard indirect ophthalmoscope lens, was also demonstrated (Haddock et al., 2013).

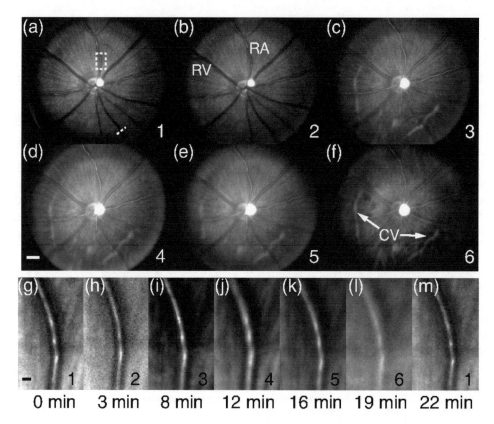

FIGURE 14.3 Fundus photographs from a pigmented adult rat. The chromatic aberration in rodent eyes is severe. Each photograph shows the same animal, but with different illumination bandwidth. (a–f) As the bandwidth increases, the resolution decreases. Scale bar: 500 µm; (g–i) Magnified view of the white dashed box in (a–f), illustrating the change in resolution with increasing illumination bandwidth. Filters were changed at the times indicated below each image. 1–6 indicate different filters that were used. (m) The filter for 1 was replaced and the image quality was similar to (g), indicating that the effects of eye quality were minimal. Scale bar: 50 µm. Adapted with permission from Li et al. (2015).

14.3 Confocal Scanning Laser Ophthalmoscope

The scanning laser ophthalmoscope (SLO) was first described in 1980 by Webb et al. (1980). Here we discuss the confocal SLO (cSLO) as it predominates most research and clinical SLO technology today. In the cSLO design, the focus at the retina and the focus at the detector are conjugate to each other (Webb et al., 1987). Similar to a standard confocal microscope, a pinhole, positioned at a plane conjugate to the retina (usually in front of the detector), rejects off-axis scattered light, which increases the image contrast, improves the point spread function, and allows for depth-sectioning. Contrast in cSLO is derived from retinal reflectance, autofluorescence, or fluorescence from exogenous dyes. In genetically modified animal models, fluorescent proteins in specific retinal cells can also be imaged (Zhang et al., 2015b).

Figure 14.4 shows the schematic of a cSLO that is designed for fluorescent imaging of rose bengal dye injected in the vasculature (Soetikno et al., 2017). In contrast to fundus camera, a coherent light source is typically used, such as the 532 nm continuous-wave laser shown in this design. The beam is reflected off a dichroic mirror (DC) (cutoff wavelength: 560 nm) and then resized by a telescope (L_3 and L_4). Two galvanometer mirrors direct the beam and are placed perpendicularly to steer the beam along orthogonal axes. Tilting the mirrors translates to linear motions of the focused beam across the retina. After the mirrors,

a Keplerian telescope (L_1 and L_2) images the mirror plane to the pupil plane of the rodent's eye, making planes conjugate and creating a stationary entrance point at the pupil plane. The cornea and lens of the eye collectively act as a low NA objective lens, which focuses the illumination light onto the retina.

Unlike the fundus camera, cSLO has the advantage that it does not require geometric pupil separation of the illumination and detection light to avoid crosstalk. In fact, the cSLO's illumination and detection light paths are interchanged with those of the fundus camera. The central portion of the pupil is used for illumination, while the entire pupil is used to collect light for detection. The same Keplerian telescope both delivers and collects light from the eye. The collected light is passed through the DC, where it is focused by a lens (L_5) through a pinhole (PH). At each scan position of the gavlo mirrors, the signal is recorded by a photodetector, such as a photomultiplier tube (PMT) or an avalanche photodetector (APD, not shown in Figure 14.4), resulting in a 2D image.

The field of view (FOV) in angular units is determined by two parameters: the maximum scanning angle of the mirrors, θ, and the magnification of the telescope, *M* (LaRocca et al., 2013). This can be written as:

$$FOV = \theta \times M. \quad (14.1)$$

The FOV can be converted to metric units with an approximate eyeball length and simple geometry. For example, approximate

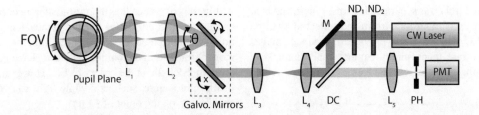

FIGURE 14.4 Schematic of a cSLO. A pinhole (PH) in front of the detector rejects off-axis scattered light. The magnification of the Keplerian telescope made by L1 and L2 and the maximum scanning angle of the galvanometer mirrors determine the field of view (FOV). CW: continuous wave. ND: neutral density filters. DC: dichroic mirror. PMT: photomultiplier tube.

mouse eye and rat eyeball lengths are 3.3 mm and 6.4 mm, respectively (Hughes, 1979; Remtulla and Hallett, 1985).

The lateral and axial resolution of cSLO can be approximated by examining the point spread function (PSF) (Roorda, 2010; Zhang et al., 2015a):

$$PSF_{cSLO} = PSF_{in} \times \left(PSF_{out} \otimes PSF_{pinhole} \right), \quad (14.2)$$

where PSF_{in} and PSF_{out} denote the PSFs in and out of the eye, which are determined by the cornea and lens. The $PSF_{pinhole}$ is convolved with the PSF_{out}. If the pinhole is adequately small (<0.25 airy unit), the PSF_{cSLO} approaches the square of the PSF of the eye; therefore, smaller lateral and axial resolutions can potentially be reached. However, in cases when the pinhole is large (>1 airy unit), its diffraction effects can be ignored, and, in this scenario, the lateral and axial resolutions are approximated by:

$$FWHM_{lateral} = \frac{0.51\,\lambda_{exc}}{NA}, \text{ and} \quad (14.3)$$

$$FWHM_{axial} = \frac{1.67\,\lambda_{exc}}{NA^2}, \quad (14.4)$$

where λ_{exc} is the excitation wavelength (Wilhelm et al., 2003; Park et al., 2004). The illumination NA can be calculated as:

$$NA = n\sin\left(\arctan\left(\frac{d}{2f} \right) \right) \approx \frac{nd}{2f}, \quad (14.5)$$

where n is the refractive index; d is the beam diameter; and f is the focal length of the eye (Liu et al., 2015b; Roorda and Duncan, 2015a). Increasing the magnification of the Keplerian telescope will result in increasing the FOV, but will also decrease d, which will in turn decrease the lateral resolution and depth of focus. Therefore, a tradeoff generally exists between FOV, lateral resolution, and depth of focus. Studies have reported ideal lateral resolutions of ~3 μm at 530 nm in mice and ~7 μm at 840 nm in humans (LaRocca et al., 2013; Zhang et al., 2015b). These equations serve as rough approximations, as the natural aberrations of the eye will prevent achievement of diffraction-limited resolution. AO technology, discussed in Section 14.6, can help counteract the eye's aberrations and achieve improved image resolution. A variant of fluorescence cSLO, called oblique-scanning cSLO, can overcome the tradeoffs between FOV and lateral resolution by breaking the coaxial alignment of excitation and emission detection (Zhang et al., 2017, 2018).

The pinhole of the cSLO reduces stray light, which results in improved contrast. In addition, the pinhole also enables optical sectioning. In rodents, the high NA of the eye (NA~0.5) allows adjusting the image z-plane. By consecutively taking z-slices at retinal depth positions, a 3D image can be created. When the pinhole is large, the depth discrimination is primarily determined by the emission-side diffraction pattern and the geometric optics effects of the pinhole (Wilhelm et al., 2003).

SLO's can be designed with different illumination light sources and collection filters, enabling high-contrast imaging of fluorescent dyes in the living eye. Common dyes used in ophthalmic research include fluorescein, indocyanine green, rose bengal, and acridine orange. These dyes are often administered via tail injection or intraperitoneal injection. Fluorescein (excitation peak: 494 nm, emission peak: 512 nm) is primarily used for fluorescein angiography (FA) and labeling the vasculature. In rodents, shifting of the confocal plane allows the observation of the three distinct capillary plexi (Paques et al., 2006). In certain disease models, such as choroidal neovascularization in AMD, the vasculature is immature, and fluorescein leaks from the vasculature, which can also be observed with FA (Liu et al., 2015b). The dye rose bengal (excitation peak: 560 nm, emission peak: 570 nm) is a fluorescein derivative that generates singlet oxygen radicals upon high-intensity laser excitation. The singlet oxygen radicals activate the clotting cascade. The clotting process can be used to model retinal occlusive diseases in rodents, such as branch retinal vein occlusion (Zhang et al., 2005, 2008; Dominguez et al., 2015; Ebneter et al., 2015; Soetikno et al., 2017). Acridine orange (excitation peak: 500 nm; emission peak: 526 nm) can label the nuclear DNA of leukocytes within the bloodstream (Nishiwaki et al., 1996; Miyahara et al., 2004; Cahoon et al., 2014). Acridine orange fluorography can be used to assess the level of inflammatory response in models of diabetic retinopathy (DR) (Miyamoto et al., 1999; Miyamoto et al., 2000).

In addition to the single-channel design discussed above, multichannel setups for cSLO have also been explored. Manivannan et al. (2001) combined red (semiconductor diode, 670 nm), green (diode pumped YAG, 532 nm), and blue (argon, 488 nm) lasers into a single fiber and sequentially acquired each channel's reflectance using a single detector. The resulting images appear similar to fundus photographs but with higher contrast. Each laser has a narrow spectral linewidth that samples only a small portion of the retinal reflectance spectrum. LaRocca et al. (2014) proposed a "true-color" SLO setup that used a visible-range supercontinuum light source with broadband illumination, similar to that of a fundus camera.

The backscattered light was separated into red, green, and blue channels by a series of dichroic mirrors and collected by three PMTs. Such SLO designs could provide reflectance images with higher contrast than fundus cameras and potentially reveal subtle retinal pathological changes.

14.4 Optical Coherence Tomography

Optical coherence tomography (OCT) is one of the most successful ophthalmic imaging technologies of the past two decades. First reported by Huang et al. (1991), OCT revolutionized both fundamental investigation and clinical care of a wide variety of ocular diseases. Here, we briefly review the basic concepts of OCT before discussing more recent technology development, such as OCT angiography and visible-light OCT.

The objective of OCT is to recover the time-of-flight of backscattered light echoes from tissue. An OCT A-line is a one-dimensional (1D) depth-resolved signal (Figure 14.5A). By raster scanning the focused illumination along a linear direction across the sample, a set of A-lines can be acquired to form a 2D cross section, called a B-scan (Figure 14.5B). Multiple B-scans form a 3D OCT volume data (Figure 14.5C). Early work with femtosecond optics and ultrahigh speed photography showed that light echoes could be time-gated in exact analogy to ultrasound imaging. However, the poor scalability of these systems led to an alternative technique to measure light echoes (Fujimoto and Swanson, 2016).

Instead of time-gating echoes directly, OCT uses the temporal coherence properties of light. Temporal coherence refers to the ability of light to interfere with a time-delayed copy of itself. Delayed light beams are said to be *temporally* coherent, when their combined detection results in a high-contrast constructive or destructive interference pattern. Ideally, a perfectly sinusoidal light field will have perfect temporal coherence, meaning it will always be able to interfere with a time-delayed version of itself. In reality, however, light fields fluctuate randomly in time, and therefore, interference will depend on the statistical temporal similarity of interfering light beams. One measure of the statistical temporal similarity is the coherence length. Light beams that have temporal delay within the coherence length are statistically correlated enough to interfere. Importantly, the coherence length is inversely proportional to the bandwidth of the light source, $\Delta\lambda$. Thus, a monochromatic light source, such as a HeNe laser, has a long coherence length on the order of ~1 m. However, a broad bandwidth light source, such as a white light source, has a coherence length on the order of ~1 μm.

OCT is based on the Michelson interferometer, a device that measures the temporal interference between two light beams emitted from the same source. Light from a broadband light source is directed to a 50/50 beam splitter, sending the light into two arms. One arm contains the sample, while the other contains a reference mirror. The same beam splitter combines the back-reflected light from the sample and reference arms to a detector. The detector signal can be expressed as

$$I(\tau) = \langle I_S(t) \rangle + \langle I_R(t) \rangle + 2Re\left[\Gamma_{SR}(\tau)\right], \quad (14.6)$$

where $\langle I_S(t) \rangle$ and $\langle I_R(t) \rangle$ denote the collected light intensity (amplitude-squared of the electric field) from the sample and reference arm, respectively; $\langle \ \rangle$ denotes time averaging; and τ denotes the time delay between the two arms. The first-order cross-correlation function $\Gamma_{SR}(\tau)$ represents the interference term and can be written as

$$\Gamma_{SR}(\tau) = \langle E_S(t) E_R^*(t+\tau) \rangle = \Gamma_{source}(\tau) \circledast h(\tau), \quad (14.7)$$

where $E_S(t)$ and $E_R(t)$ are the sample and reference arm electric field amplitudes and the superscript * denotes the complex conjugate; $\Gamma_{source}(\tau)$ is the source coherence function; $h(\tau)$ is the reflectivity function of the sample; and $\circledast$ denotes convolution. $\Gamma_{SR}(\tau)$ is the OCT A-line itself. If the difference in path length between the two arms is within the coherence length of the source, an interference signal will be observed at the detector. In other words, the short coherence length acts as a coherence gate; reflections within the coherence length are accepted and reflections outside are rejected. This forms the basis of depth-resolved imaging with OCT.

In the original OCT design, the reference mirror was translated back and forth to detect interference signals at

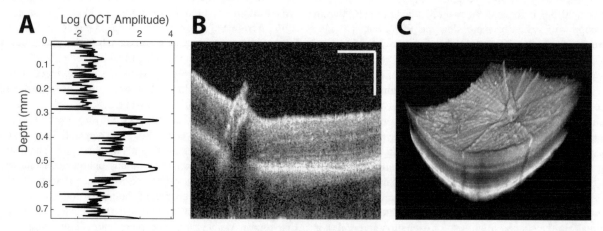

FIGURE 14.5 OCT image formation in a wild-type C57BL/6 mouse retina. A. One-dimensional OCT A-line; B. Cross-sectional B-scan image through the murine optic nerve head. Horizontal scale bar: 500 μm. Vertical scale bar: 200 μm; C. OCT volume rendering of the retina.

different depths to generate an OCT A-line, which is known as time-domain OCT (TD-OCT). TD-OCT varies τ such that $\Gamma_{SR}(\tau)$ (the OCT A-line) can be reconstructed. Unfortunately, TD-OCT suffers from slow imaging speeds and low sensitivity. An alternative to obtain OCT A-line, termed Fourier-domain OCT (FD-OCT), measures a spectral interferogram to improve imaging speeds and increase sensitivity (de Boer et al., 2017b). When the Fourier transform of Equation 14.6 is taken, the spectral density function is obtained (Fercher, 2010) as

$$S(\omega,\tau) = \langle S_S(\omega) \rangle + \langle S_R(\omega) \rangle + 2Re\left[W_{SR}(\omega)\right]\cos 2\pi\omega\tau, \quad (14.8)$$

where $W_{SR}(\omega)$ is the cross spectral density given by:

$$W_{SR}(\omega) = S_{Source}(\omega)H(\omega) \quad (14.9)$$

and $S(\omega,\tau)$ can be measured by using a spectrometer or a swept-source laser. Note that while $\omega \leftrightarrow t$ are a Fourier transform pair, they can also be converted to $k \leftrightarrow z$ by multiplying by the phase velocity of light c/n. Spectral domain OCT (SD-OCT) measures $S(\omega,\tau)$ by incorporating a diffractive element in the detection arm. The diffracted light (spectrum) is then focused onto and recorded by a 1D array detector. In this technique, $S(\omega,\tau)$ is obtained in a single shot, i.e. all wavelengths are acquired simultaneously. In swept-source OCT (SS-OCT), a swept-source laser emits only narrow band light and sweeps ω over time, and a single-element photodetector in the detection arm collects $S(\omega,\tau)$. The sweep rate of the laser can be made very high, resulting in MHz A-line rate. In both cases, taking the inverse Fourier transform of $S(\omega,\tau)$ allows reconstruction of $\Gamma_{SR}(\tau)$, and the reference arm is stationary during imaging.

Figure 14.6 shows an example of a near-infrared (NIR), fiber-based SD-OCT system (Soetikno et al., 2017). The light source is a superluminescent light emitting diode (SLED) with a center wavelength of 840 nm and a bandwidth of 95 nm (full width at half maximum). A 50/50 fiber coupler collects the light from the SLED and splits it into the sample and reference arms. The sample arm is steered by two galvanometer mirrors in the same fashion described in the SLO section. The *x-y* galvanometer mirrors deflect the sample arm beam, while a Keplerian telescope creates a point conjugate from the scanning mirrors to the rodent's pupil plane. The reference arm beam is back-reflected by a silver mirror (M_1). Several BK7 glass plates are used in the reference arm for dispersion compensation. The back-reflected sample beam recombines and interferes with the backscattered light from the sample. A spectrometer, consisting of a collimating lens, a transmission diffraction grating, a focusing lens, and a line CCD camera detects the spectral interferogram.

A complete mathematical framework for the image reconstruction process of FD-OCT is beyond the scope of this chapter (Fercher et al., 2003; Izatt et al., 2015; Kalkman, 2017); however, the key principles can be grasped with knowledge of Fourier transform (FT) theory. Figure 14.7 depicts several important concepts related to the FD-OCT image reconstruction process. There are several key parameters to characterize the performance of an OCT system, which include the maximum imaging depth, axial resolution, and sensitivity rolloff. These parameters are explained below.

1. **Frequency encoding (Figure 14.7A).** From Equation 14.8, the spectral interferogram includes a sinusoidal term, which depends on both τ and ω. In FD-OCT, a spectral interferogram is acquired as a function of ω (or k, if the coordinates are changed from optical frequency to wavenumber). Figure 14.7a shows the sinusoidal interferogram in k-space. Taking the Fourier transform of the sinusoid gives the sample reflectivity function: two delta functions positioned a distance away from zero. The frequency of the interferogram with respect to k, which depends on the depth of the sample, determines the position of the delta functions away from the zero-depth position. In other words, the k values in the spectral interferogram encodes the depth information z. Usually the amplitude and the frequency of the interferogram is

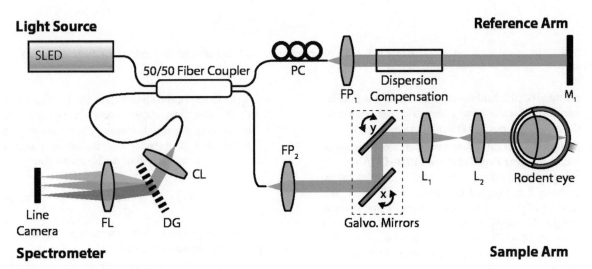

FIGURE 14.6 Schematic of a spectral-domain OCT. SLED: superluminescent light emitting diode. M: mirror. L: lens. PC: polarization controller. FP: fiber port. CL: collimating lens. FL: focusing lens. DG: diffraction grating.

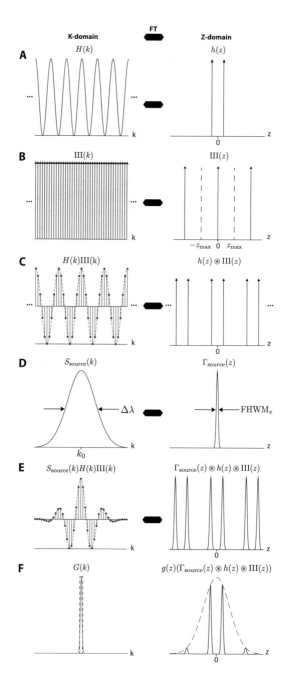

FIGURE 14.7 Image reconstruction process in FD-OCT. A. Frequency encoding. B.–C. Sampling of the interferogram. D.–E. Axial resolution. F. Sensitivity rolloff. ⊛ denotes the convolution operator.

recorded but not its absolute phase, leading to Fourier transformed result being symmetric with respect to the zero-depth position (Hermitian symmetry).

2. **Sampling interval in *k*-space determines the maximum imaging depth (Figure 14.7B–C).** Although the spectral interferogram is ideally continuous, in practice, it is sampled in *k*-space. In SD-OCT, the individual pixels of the spectrometer's line camera sample the spatially distributed spectral interferogram. In SS-OCT, the evenly spaced wavenumbers are recorded and act as the sampling function. The sampling process is modeled by multiplying the spectral interferogram with a Dirac comb function, III(*k*)

(Hu et al., 2007), shown in Figure 14.7B. Because of the finite sampling, aliasing occurs in the *z*-domain. The sampling step size in *k*-space, δk, determines the Nyquist limit in *z*-space. The maximum imaging depth corresponds to half of the Nyquist value:

$$z_{max} = \frac{\pi}{2\delta k} \qquad (14.10)$$

Aliasing also means that images are repeated in *z*-space, as shown in Figure 14.7C.

3. **Coherence length of the light source primarily determines the axial resolution (Figure 14.7D–E).**

In OCT, the bandwidth of the low-coherence light source is the dominant factor that determines the axial resolution. The spectral interferogram is modulated by the power spectral density of the light source, which is depicted as a Gaussian density function in Figure 14.7d. The full width at half maximum (FWHM) in z-space is typically assumed as the axial resolution, which is given by the following equation.

$$FWHM_z = \frac{2ln2}{\pi n} \frac{\lambda_0^2}{\Delta\lambda}, \qquad (14.11)$$

where λ_0 is the center wavelength of the light source; $\Delta\lambda$ is the FWHM of the light source power spectrum; and n is the refractive index of the sample. Figure 14.7E shows the sampled spectral interferogram, modulated by the Gaussian power spectral density and its respective Fourier transformed result.

4. **Sensitivity Roll-off is related to the spectral resolution (Figure 14.7F).** In FD-OCT, portions of the cross-spectral density are integrated in k-space. In spectrometer-based systems, this is due to the square or rectangular pixel shape. In addition, the spectrometer has a limited spewctral resolution because of optical aberrations of the internal lens elements (Hu et al., 2007). As a result, the size of the PSF of the spectrometer's optical system does not always match the individual pixel dimensions across the full length of the 1D CCD. In swept-source systems, the instantaneous linewidth of the source determines the range of k integrated at each wavelength. In either scenario, this results in the cross-spectral density being convolved with an integration function $G(k)$. In z-space, this leads to multiplication with a sensitivity roll-off function $g(z)$. $g(z)$ usually has both sinc and Gaussian behavior. Therefore, OCT signals further away from the zero delay location have lower amplitude than those closer to the zero delay.

14.4.1 OCT Angiography

OCT angiography (OCTA) extends OCT imaging by providing label-free, high-contrast, 3D imaging of the retinal vasculature (Gao et al., 2016; Chen and Wang, 2017). FD-OCT led to much improved OCT imaging speed, making OCTA possible. In 2006, Makita et al. (2006) demonstrated volumetric OCTA by utilizing the Doppler phase shift. Since then, multiple algorithms have been studied and developed to perform OCTA. In general, the algorithms can be categorized based on the components of the OCT signal utilized, which include changes in Doppler shift, OCT amplitude, OCT phase, or a combination thereof (Gao et al., 2016; Gorczynska et al., 2016). Spectral, temporal, or depth-dependent filtering is often incorporated to reduce dependence of the OCTA signal on subject motion. For example, the split-spectrum amplitude decorrelation algorithm (SSADA) uses spectral splitting to generate multiple OCT volumes with reduced axial resolution, and the average of those volumes is calculated (Jia et al., 2012). By reducing the axial resolution, subject motion has less impact on the SSADA signal and image contrast may be improved (Gao et al., 2015). In comparison, the optical microangiography algorithm (OMAG) uses the full complex OCT signal, instead of amplitude alone, and B-scans are averaged temporally (An and Wang, 2008).

Regardless of the algorithm used, a common acquisition protocol exists between OCTA techniques. As depicted in Figure 14.8A, OCTA uses repeated B-scans colocalized on the retina. The raster scanning pattern of OCT (solid lines) and OCTA (dashed lines) are drawn at low density for illustration. The number of B-scans can vary between techniques, but at least two are needed to achieve OCTA contrast. In this illustration, five OCT B-scans were repeated at the same location, as illustrated by the group of B-scans over time in Figure 14.8B. In ultrafast OCT systems, in which the A-line acquisition rate exceeds MHz speeds, OCT volumes can also be repeated at the same location to achieve angiography contrast (Zhi et al., 2015). By applying an OCTA algorithm to these B-scans, OCT signal changes due to blood cell movement can be highlighted while eliminating signals from static tissue (Figure 14.8C). The final 3D OCTA volume can be displayed as a depth colored *en face* OCTA, as shown in Figure 14.8D.

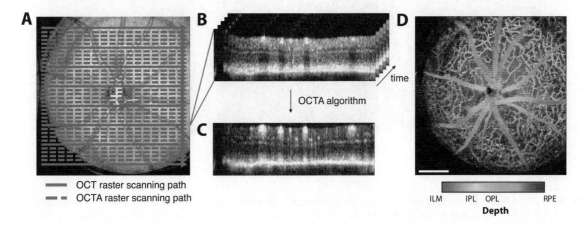

FIGURE 14.8 OCTA scanning protocol and image formation. A. *En face* OCT of a mouse retinal fundus. OCT and OCTA raster scanning paths are illustrated by the solid and dashed arrows, respectively. B. OCT B-scans acquired from the OCTA raster scanning path. C. After processing by an OCTA algorithm, an OCTA B-scan is created. D. Depth-colored *en face* OCTA. Scale bar: 500 μm.

Figure 14.9 shows some of the applications of an OCTA algorithm toward studying rodent models of retinal diseases. These images were constructed using a complex OCT values difference-based algorithm that utilizes spectral-splitting and time averaging (Yi et al., 2014). Choroidal neovascularization (CNV) is a hallmark of wet AMD. CNV refers to abnormal vessels that grow upward from the choroid, through the RPE layer, and into the retina. To investigate CNV in rodents, a laser-induced model has been described, in which high-power laser shots are delivered to areas of the retina (Shah et al., 2015; Park et al., 2016). The laser damages the RPE and Bruch's membrane, enabling CNV formation. Figure 14.9A shows the *en face* projection of the OCT volume near the CNV lesion acquired with a visible-light OCT (vis-OCT) system. OCTA of the same area shows abnormal vasculature (green) above the choroid (blue) (Figure 14.9B). By numerically removing the inner retinal vasculature (red) from the OCTA volume, the CNV lesion can be separated for further analysis (Figure 14.9C). OCTA can also be integrated with fluorescein angiography to compare the differences between the two techniques when studying this animal model. OCTA and fluorescein angiogram matched well; however, OCTA cannot assess dye leakage from the vasculature (Liu et al., 2015b).

Branch retinal vein occlusion (BRVO) is an ischemic disease of the retina that results from decreased blood flow in one or more major retinal veins. Figures 14.9D–F show OCTA images of a BRVO model created by imaging-guided photocoagulation, taken with a NIR OCT system (Soetikno et al., 2017). Figure 14.9D shows an *en face* OCTA image of the vasculature before occlusion. The major arteries and veins are labeled as an *a* and *v*, respectively. The vein branch, indicated by the white circle, was occluded. The OCTA image after occlusion (Figure 14.9E) shows capillary nonperfusion in the area (white-dashed area) surrounding the occlusion site. This area increased on Day 1 as detected by the OCTA image shown in Figure 14.9F.

While OCTA holds promise for answering translational research questions, as highlighted by the previous examples, OCTA is also being adopted rapidly for answering clinical questions (Kashani et al., 2017). There are several commercial OCT systems for imaging the retina in patients, with the ability to perform an OCTA scan pattern (Tan et al., 2018). Each device includes its own OCTA

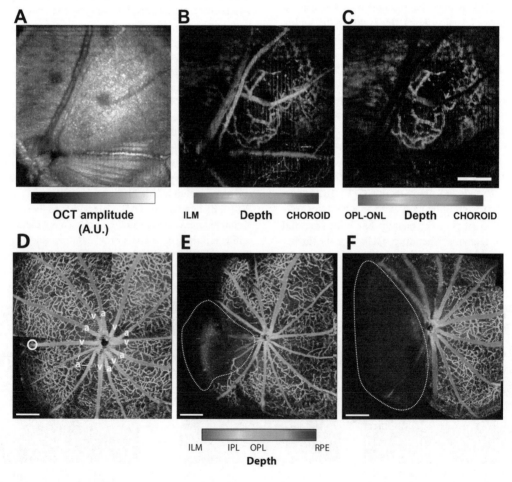

FIGURE 14.9 OCTA in animal models of age-related macular degeneration and retinal vein occlusion. A. *En face* vis-OCT of the area encompassing a CNV lesion. The CNV lesion is not visible on standard OCT. A.U.: arbitrary units. B. *En face* vis-OCTA of the same area, revealing the CNV lesion. Image color-coded by depth. C. Final postprocessed outer retina *en face* angiogram, color-coded by depth. Scale bar: 200 μm D. Montage of 9 OCTA images before vein occlusion. Arteries and veins denoted by a and v, respectively. White circle denotes the targeted site of occlusion on the 9 o' clock vein. E. Montage after vein occlusion. F. Montage on Day 1 after the vein occlusion. White dotted regions in E. and F. denote areas of capillary nonperfusion. Scale bars: 500 μm. Adapted with permission from Shah et al. (2016); Soetikno et al. (2017).

algorithm, as discussed previously, and automated segmentation software to segment each of the retinal vascular networks. Because OCTA is still considered an emerging technology, studies using OCTA in clinics have compared OCTA images with that revealed by accepted gold-standard methods of fundus fluorescein angiography or indocyanine green angiography (Ang et al., 2016; Inoue et al., 2016; Tanaka et al., 2017; Abucham-Neto et al., 2018). Thus far, OCTA studies in patients have found promise in a wide variety of retinal diseases, such as diabetic retinopathy, AMD, vascular occlusion, and choroidal neovascularization. OCTA imaging could also have potential to study the anterior segment for diseases, such as glaucoma or corneal neovascularization. OCTA imaging will likely become a key complementary imaging technique for monitoring the healing process after a corneal transplantation, or the effect of anti-VEGF treatments in a patient with diabetic retinopathy (Hagag et al., 2017).

14.4.2 Visible-Light OCT

14.4.2.1 Retinal Oximetry

Visible-light OCT (vis-OCT) is a functional extension of OCT, which uses a light source in the visible spectrum (instead of NIR light). By using shorter wavelengths, vis-OCT typically provides a higher axial resolution than most NIR counterparts (Pi et al., 2017). The most promising application of vis-OCT being demonstrated so far, however, is its potential to measure oxygenation within the retinal vasculature (Yi et al., 2013). Because several retinal diseases exhibit dysfunctions in oxygenation, vis-OCT is especially attractive for its potential to provide quantitative measurements of oxygen metabolism, which could provide insight about several retinal functions. In this section, we will describe the process of extracting oxygen saturation of hemoglobin (sO_2) from within the blood using vis-OCT. In addition, we will discuss combining these measurements with Doppler OCT to derive the metabolic rate of oxygen in the retinal circulation.

Figure 14.10 illustrates the process of retinal oximetry with vis-OCT. A vis-OCT image of a healthy rat fundus was acquired and reconstructed (Figure 14.10a). The supercontinuum light-source spectrum was centered at 585 nm and

had a bandwidth of 85 nm. Further details of this system can be found in previous studies (Chen et al., 2015a). The full-spectrum vis-OCT B-scan in Figure 14.10b, corresponding to the white-dashed line in Figure 14.10a, shows several blood vessel cross sections. For each A-line within every vessel, the spectral interferogram was split into 14 Gaussian bands using STFT. The bandwidth of each band was ~0.32 μm⁻¹ (FWHM) corresponding to an axial resolution of ~8.9 μm in air. Figure 14.10c shows the split-spectrum images of a selected vessel in the white-dashed box of Figure 14.10b. For each of the wavelength bands, the OCT signal intensity, $I(\lambda)$, at the bottom of each of the circles labeled in Figure 14.10c, was measured.

The Beer–Lambert law was used to model light attenuation after passing through the blood vessel. The model can be written as

$$I(\lambda) = I_0 R_0 r \exp\left(-2nd\left(\mu_{HbO_2}(\lambda)sO_2 + \mu_{Hb}(\lambda)(1-sO_2)\right)\right), \quad (14.12)$$

where I_0 is the incident light intensity; R_0 is the reference arm reflectance; r is the reflectivity of the vessel wall; n is the refractive index; d is the thickness of the blood vessel; and μ_{HbO_2} and μ_{Hb} are the attenuation coefficients of oxygenated and deoxygenated hemoglobin, respectively (Yi et al., 2013). The factor of 2 in the exponential expression is due to round-trip optical attenuation. The reflectance of the vessel wall can be modeled as a power law with a first-order Born approximation, as $r(\lambda) = A\lambda^{-\alpha}$.

Taking the logarithm of Equation 14.12 converts it into a linear equation:

$$\ln\left(\frac{I(\lambda)}{I_0(\lambda)}\right) = -nd\left(sO_2\mu_{HbO_2}(\lambda) + (1-sO_2)\mu_{Hb}(\lambda)\right)$$
$$-\frac{1}{2}\alpha\ln(\lambda) + \frac{1}{2}\ln AR_0. \quad (14.13)$$

With the measured $I(\lambda)$, at the bottom of the vessel wall, and experimental values of μ_{HbO_2} and μ_{Hb} spectra (Faber et al., 2004), a least-squares fitting can be performed to derive sO_2. Figure 14.10d shows the measured sO_2 for the major retinal arteries and veins color-coded on the *en face* OCT.

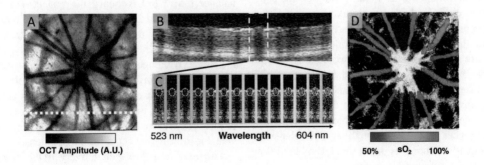

FIGURE 14.10 Retinal oximetry using vis-OCT. A. *En face* OCT image of a rat retinal fundus. B. OCT B-scan at the location of the white dashed line in A. C. Short-time Fourier transforms split the OCT image into 14 spectral bands that are used for sO_2 fitting. D. sO_2 measurements in the major retinal arteries and veins. Adapted with permission from Nesper et al. (2017).

14.4.2.2 Doppler vis-OCT

In addition to performing retinal oximetry, vis-OCT can also measure the volumetric blood flow rate using established Doppler OCT methods (Wang et al., 2007; Yimin Wang et al., 2008). To measure the volumetric blood flow rate (in µL/min), the blood vessel diameter (in µm) and blood flow velocity (in mm/s) must be measured. The vessel diameter can be measured by calculating the distance between the vessel walls on OCT B-scans, while the blood flow velocity is measured by analyzing the phase of the OCT signal. When there is a small time difference between successive OCT A-lines, blood flow imparts a linear phase shift on the OCT signal (Nam et al., 2014). This phase difference between two successive A-lines can be calculated from the phase of the complex OCT signal after Fourier transform of the real-valued spectral interferogram. The phase difference gives information about subresolution offsets within the coherence gate (Uttam and Liu, 2015). The phase offset between two temporally delayed A-scans can be expressed as:

$$\Delta\phi' = \tan^{-1}\left(\frac{\mathrm{Im}(I_{t_2})}{\mathrm{Re}(I_{t_2})}\right) - \tan^{-1}\left(\frac{\mathrm{Im}(I_{t_1})}{\mathrm{Re}(I_{t_1})}\right), \qquad (14.14)$$

where t_1 and t_2 are the time points separated by $\Delta T = t_2 - t_1$. Animal motion is inevitable for *in vivo* flow measurements. Movement along the direction of the probing beam imparts a bulk phase shift, $\Delta\phi_{bulk}$, on top of the phase shift from blood flow, $\Delta\phi_{flow}$. Therefore, the total measured phase difference is:

$$\Delta\phi = \Delta\phi_{bulk} + \Delta\phi_{flow}. \qquad (14.15)$$

The bulk phase shift can be determined by estimating the phase difference for the tissue outside the blood vessel. For example, average-shifted histograms can determine the mode of the phase shifts in a given A-line (Makita et al., 2006). After determining $\Delta\phi_{bulk}$, it can be subtracted from $\Delta\phi$ to obtain $\Delta\phi_{flow}$, which is proportional to the flow speed (White et al., 2003).

Since $\Delta\phi$ is only sensitive to motion along the direction parallel to the sample beam, the measured velocity must be corrected for the Doppler angle θ, the angle between the probing beam and the blood vessel direction. Dual-circle Doppler OCT is one method being reported to obtain the Doppler angle and measure the actual volumetric flow rate (Wang et al., 2007, 2008). In the dual-circle method, two circular scans of different diameters around the optic nerve head are acquired. In the original method, the coordinates of the blood vessel in the two circular scans are used to compute θ. However, this method relies on prior knowledge of the eyeball's axial length, which is unknown and varies between subjects. Liu et al. (2015a) proposed an alternative calculation of θ from dual-circle scans, which is independent of the eyeball's axial length. The calculated phase differences for a vessel from the two circular scans are used to compute θ. Once θ is determined, the blood flow speed can be calculated with $\Delta\phi_{flow}$ using

$$v = \frac{f_s \lambda_0 \Delta\phi_{flow}}{4\pi n \cos\theta} = v_{max} \frac{\Delta\phi_{flow}}{\pi}, \qquad (14.16)$$

where f_s is the A-line rate; λ_0 is the center wavelength of the light source; and n is the refractive index of the sample (Wang et al., 2007). Note that $\Delta\phi_{flow}$ is limited to values between $\left[-\pi, \pi\right]$ and the maximum speed $v_{max} = \frac{f_s \lambda_0}{4n\cos\theta}$. To obtain velocity values greater than v_{max}, $\Delta\phi_{flow}$ must be phase-unwrapped to extend its range to $\left[-2\pi, 2\pi\right]$ (Hendargo et al., 2011).

Figure 14.11 shows flow measurements using dual-circle Doppler vis-OCT. Figure 14.11a shows an *en face* vis-OCT image of a healthy rat retina (Liu et al., 2015a). The white-dashed circles illustrate the paths of the inner and outer circular scans. Each circular scan must have high scanning density, such that the scanning step size is smaller than the beam diameter on the retina. The high scanning density maintains the assumption that adjacent A-lines are acquired at the same location. In this example, the 4096 A-lines per circular scan were collected. Figure 14.11b shows the average phase difference B-scan, from the inner circular scan, after bulk-motion correction. The red and blue arrows indicate an artery and vein, respectively. The phase-differences were averaged over multiple points in the cardiac cycle. Figure 14.11c shows the average phase value with each cumulative circular B-scan for the phase measurements from the small and large circular scans. Figure 14.11d shows the measured Doppler angles for each of the 10 blood vessels. Figure 14.11e shows the recorded electrocardiogram that was compared to pulsatile velocity profiles measured by phase-resolved vis-OCT and shown in Figure 14.11f–g. The vis-OCT pulsatile profiles correlate well with the cardiac cycle measured by electrocardiogram. To estimate blood flow from velocity, the velocity was multiplied by the cross-sectional area of the vessel, with the assumption that the flow was parabolic. The cross-sectional area was approximated as a circle, in which the diameter was measured from the OCT B-scan. Figure 14.11h shows the repeatability of flow measurements over several days of testing.

14.4.2.3 Calculating the Inner Retinal Metabolic Rate of Oxygen

The inner retinal metabolic rate of oxygen (irMRO$_2$) and delivery rate of oxygen (irDO$_2$) are two functional markers that provide absolute measurements of oxygen consumption in the inner retina. According to the Fick principle, which is based on mass balance, the irMRO$_2$ is equal to the product of the arterial-venous sO$_2$ difference and the volumetric blood flow rate. irMRO$_2$ measurements assume that the difference in oxygen mass, between the arterial and venous circulations, is the oxygen consumed by the tissue. Although this is true in normal tissue circulations, the retina has contributions from both the retinal and choroidal circulation. Because choroidal circulation is currently difficult to assess, the true metabolic rate of the entire retina is difficult to measure *in vivo*. Nevertheless, the irMRO$_2$ can still provide valuable information about retinal functions.

The inner retinal oxygen delivery, irDO$_2$, in nl min^{-1} can be expressed as

$$irDO_2 = \varepsilon \times C_{Hb} \times F_{Total} \times \overline{s_a O_2}, \qquad (14.17)$$

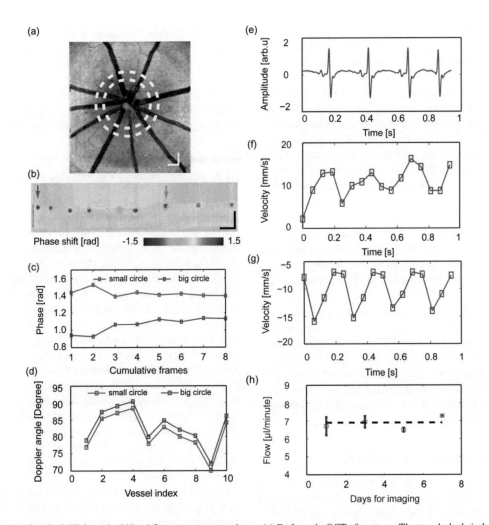

FIGURE 14.11 Doppler vis-OCT for retinal blood flow measurements in rat. (a) *En face* vis-OCT of a rat eye. The two dashed circles indicate where the dual-ring scans were performed. (b) The phase image from the inner-circle scan. (c) Phase stability across a different number of cumulative frames for a sample vessel, indicated by the left arrow in (b). (d) Estimated Doppler angles for all the vessels from both small and large scanning locations. (e) Recorded rat electrocardiogram during imaging. (f) Measured pulsatile flow from one retinal artery, highlighted by the left arrow in (b). (g) Measured pulsatile flow from one retinal vein, highlighted by the right arrow in (b). (h) Repeatability of flow imaging for the same rodent subject over seven days. Scale bar: 200 µm. Adapted with permission from Liu et al. (2015a).

where ε is the oxygen-binding capacity of hemoglobin (1.36 ml O_2/g of Hb) (Hall and Guyton, 2011); C_{Hb} is the total hemoglobin concentration (150 g of Hb/L of blood); $\overline{s_aO_2}$ is the average arterial oxygenation; and F_{Total} (in ml/min) is the total blood flow entering and leaving the inner retinal circulation, measured with Doppler OCT. If blood flow is measured in all the blood vessels, the total flow rate can be calculated as the sum of flow rates from all the vessels. However, in practice, measurements may be incomplete. In this situation, the total blood flow can be approximated by using an average value for the arterial and venous blood flows:

$$F_{Total} = \frac{N}{2} \frac{\left(\overline{F_a} + \overline{F_v} \right)}{2}, \quad (14.18)$$

where $\overline{F_a}$ and $\overline{F_v}$ are the average arterial and venous blood flows, respectively; and N indicates the total number of vessels. The irMRO$_2$ can be calculated with an equation similar to Equation 14.17

$$irMRO_2 = \varepsilon \times C_{Hb} \times F_{Total} \times \left(\overline{s_aO_2} - \overline{s_vO_2} \right), \quad (14.19)$$

where $\overline{s_aO_2}$ and $\overline{s_vO_2}$ are the average arterial and venous sO$_2$ percentage values, respectively.

Another measure of oxygen consumption is the oxygen extraction fraction (OEF), which is defined by

$$OEF = \frac{\overline{s_aO_2} - \overline{s_vO_2}}{\overline{s_aO_2}}. \quad (14.20)$$

The OEF is a *relative* marker of oxygenation consumption by the retina because it does not include a blood flow term. Therefore, OEF can be difficult to interpret in circumstances when blood flow is also changing. Nevertheless, it is a commonly used measurement in metabolic imaging.

14.4.2.4 Validation of Metabolic Imaging Using vis-OCT

The gold standard for oxygenation measurements in the retina is electrode-based oxygen measurements. However, these measurements are difficult to perform experimentally, and a study comparing vis-OCT to this gold standard has yet to be performed. However, the accuracy of vis-OCT for retinal oximetry has been studied *in silico*, *in vitro*, and *in vivo* studies.

1. ***In silico* verification:** Chen et al. (2015b) modeled visible light propagation through a retinal blood vessel using Monte Carlo simulations. The parameters used to simulate the vis-OCT measurements matched the current state-of-the-art systems. The major finding was that OCT in the visible range can accurately quantify retinal oxygenation, while NIR OCT failed to do so in practice because of scattering over dominating absorption.

2. ***In vitro* verification:** Yi et al. used vis-OCT to quantify sO_2 of bovine blood at different oxygen saturations values in capillary tubes. Measurements agreed well with those from a blood-gas analyzer (Yi et al., 2015).

3. ***In vivo* verification:** Pulse oximetry measurements from the peripheral arteries in rats have also been compared with that of vis-OCT sO_2 readings (Chen et al., 2015a; Soetikno et al., 2015). The level of inhaled oxygen was modulated to change the arterial oxygenation. Vis-OCT measured sO_2 from the retinal arteries correlated well with measurements made from the peripheral arteries.

14.4.2.5 Animal Studies with vis-OCT

Vis-OCT has been used to study diabetic retinopathy (DR) in animal models. With respect to early stage diabetes, vis-OCT has been used to quantify the $irMRO_2$ in the Akita[+]/TSP[−/−] mouse. The Akita[+] mouse develops type 1 diabetes and early signs of mild diabetic retinopathy (Sorenson et al., 2013). Liu et al. (2017) found that the $irMRO_2$ increased from Week 6 to Week 13, while microvascular pericyte numbers did not change. This suggested that functional measurements with vis-OCT could provide an early biomarker of DR before major vascular changes are detectable.

Vis-OCT has also been used to study an animal model of late stage DR. The oxygen-induced retinopathy (OIR) model in rats was originally developed as a model of retinopathy of prematurity, but has become an accepted model of late stage DR because it possesses retinal avascularity and proliferative

neovascularization. Soetikno et al. (2015) used vis-OCT to measure $irMRO_2$ on postnatal Day 18. The $irDO_2$ and $irMRO_2$ were significantly decreased in rats with OIR compared to healthy controls, which correlated with decreased retinal thickness in the OIR rats compared to controls. It is hypothesized that the decrease in retinal thickness may contribute to the observed decreased in retinal metabolic demand.

14.5 Photoacoustic Ophthalmoscopy

Photoacoustic ophthalmoscopy (PAOM) relies on the photoacoustic effect to produce 3D images of optical absorbers in the eye (de la Zerda et al., 2010; Hu et al., 2010; Jiao et al., 2010). PAOM is a variation of photoacoustic microscopy (PAM) (Zhang et al., 2006; Hu et al., 2011; Yeh et al., 2014, 2015), with the objective lens replaced by the natural refractive surfaces of the eye. In PAM, nanosecond pulsed laser illuminates the sample. When the pulses reach an optical absorber, such as hemoglobin within a blood vessel, a transient temperature rise is generated (Wang and Hu, 2012). Part of that energy is converted into mechanical energy, which manifests as spherical, broadband ultrasonic waves, known as photoacoustic waves. The photoacoustic waves are detected by placing an ultrasonic needle transducer on the surface of the eye (Figure 14.12A). The time-gated arrival of the waves gives the depth information of the absorber. The axial resolution of PAOM is primarily determined by the bandwidth of the ultrasonic detector. Scanning the illumination beam allows for volumetric imaging.

Unlike OCT that is sensitive to optical scattering and absorption, PAOM is primarily sensitive to optical absorption, as the initial pressure rise is directly proportional to the absorption coefficient of the absorber. In the eye, optical absorbers include hemoglobin in blood vessels and melanin in the RPE cells. Hemoglobin imaging allows visualization of vasculature, and applications include imaging of the retinal fundus vessels, corneal neovascularization, and vessels of the iris (Hu et al., 2010; Liu et al., 2014; Song et al., 2014). With multiwavelength PAOM, absorption spectroscopy of hemoglobin can be performed, and the sO_2 of major retinal blood vessels can be calculated (Zhang et al., 2006; Song et al., 2014). When combined with Doppler OCT for blood flow, PAOM can be used to measure the $irMRO_2$ (Song et al., 2014). Melanin imaging with

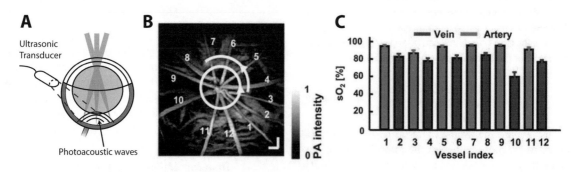

FIGURE 14.12 PAOM imaging of retinal sO_2. A. The basic schematic of PAOM showing the incoming laser beam scanned across the retina as well as the resulting photoacoustic waves and ultrasonic transducer. B. A PAOM fundus image of the rat retina. C. sO_2 values calculated from the image in Panel B. Adapted with permission from Song et al. (2014).

PAOM may provide a noninvasive screening tool to evaluate the health of the RPE, which has been implicated in several retinal diseases including AMD (Shu et al., 2015, 2017).

Figure 14.12B shows an example PAOM image acquired from an adult Sprague-Dawley rat. The light source was a tunable dye laser, pumped by a pulsed laser, at wavelengths of 578 nm, 580 nm, and 588 nm. The lateral and axial resolutions were 20 μm and 23 μm, respectively. The ultrasonic bandwidth was 30 MHz (Jiao et al., 2010; Song et al., 2014). By fitting the photoacoustic signals to the oxygenated and deoxygenated hemoglobin absorption values at 578 nm, 580 nm, and 588 nm, the sO_2 was calculated for each vessel. Figure 14.12C shows the sO_2 measurements for the vessels labeled in Figure 14.12b.

PAOM is uniquely sensitive to optical absorption. However, current systems possess relatively low axial resolution, especially when compared with OCT. High bandwidth detectors using microring resonators have been demonstrated that can provide higher axial resolution (Dong et al., 2014; Li et al., 2014). Unfortunately, PAOM requires direct-contact onto the surface of the eye, which can be cumbersome when working with the sensitive tissues of the eye. PAOM has not yet demonstrated the capability of measuring retinal blood flow, and therefore, PAOM is likely required to be coupled with OCT or ultrasound to measure metabolic rate (de la Zerda et al., 2010; Song et al., 2014; Shu et al., 2016; Dong et al., 2017).

14.6 Adaptive Optics

AO improves the lateral resolution of scanning-based ophthalmic imaging technologies by correcting for the natural aberrations of the eye. Irregularities in the lens and cornea cause high-order aberrations that distort the regular structure of laser beam wavefronts entering the eye. Ultimately, this distortion of the normal diffraction pattern of the near Gaussian beam leads to a larger and distorted spot shape on the retina and even larger aberrations at the signal collection. These aberrations prevent retinal imaging devices from reaching the optical diffraction limit performance and reduce overall signal-to-noise ratio (SNR) and quality of the images (Lombardo et al., 2013). For example, the diffraction-limited lateral resolution of an OCT system with center wavelength 840 nm in a fully dilated mouse eye with NA~0.5 is ~0.6 μm, but the actual lateral resolution without aberration correction is ~15 μm. To correct for the aberrations, AO systems use technology originally developed by astronomers to image objects in space, such as stars and planets, where the real image of the object is corrupted after passing through the earth's scattering atmosphere.

To apply AO correction, a sensor must first measure the wavefront aberrations introduced by the eye. Then, a deformable mirror predistorts the wavefront in an opposite manner to compensate for the aberrations of the eye. Predistortion of the wavefront of the illumination light allows achievement of a regular, focused Gaussian beam wavefront at the retina. Ultimately, near diffraction-limited performance of the imaging system can be obtained in an arbitrary eye.

The Shack-Hartmann wavefront sensor (SHWS) is the most commonly used wavefront sensor in AO retinal imaging (Roorda and Duncan, 2015b). Figure 14.13A shows the principle of the SHWS. A collimated light source is reflected by a beam splitter and then focused by the anterior eye's refractive system onto the retina. The wavefront of the reflected light from the eye is distorted by the eye's natural aberrations. The distorted wavefront passes through the beam splitter onto a lenslet array, which consists of hundreds of lenslets. These lenslets focus the light onto a 2D detector, which creates a spot diagram of the wavefront. Considering a plane wave without aberrations, the 2D detector should detect evenly distributed, well-focused spot diagram with comparable brightness. With aberrations, however, the spatial distribution, brightness, and axial focus of the detected spot diagram will not be highly uniform, which is used to calculate an approximation of the wavefront correction.

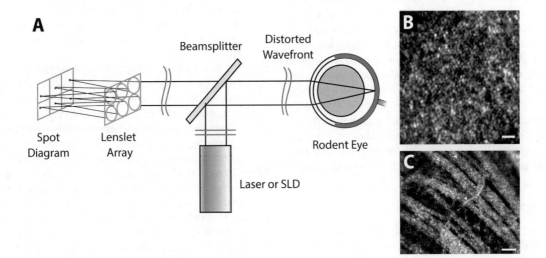

FIGURE 14.13 Principle of wavefront sensing in AO retinal imaging. A. The basic schematic of a Shack–Hartmann wavefront sensor. Light from a laser or superluminescent diode is reflected at a beam splitter and focused onto the back of the eye. The returning distorted light is transmitted through the beam splitter and passes through a lenslet array that focus the light onto a 2D camera to form a spot diagram. This spot diagram is used for wavefront correction. B. Images of the photoreceptors and C. nerve fiber layer taken by an AOSLO system. Adapted with permission from Geng et al. (2012).

Once the aberrations are sensed, predistortions can be introduced to the illumination light using a deformable mirror. The deformable mirror is a 2D mirror array in which each mirror can be individually translated to adjust the optical pathlength for a portion of the illumination beam reflected by the particular mirror. Collectively, the 2D mirror away can predistort the illumination beam into the opposite wavefront comparing with the detected aberrations by the SHWS. Such correction process is usually iterative, and near diffraction-limited resolution can be achieved in retinal imaging.

AO techniques can be combined with several ophthalmic imaging systems to improve lateral resolution, including fundus, SLO, and OCT (Jian et al., 2013; Roorda and Duncan, 2015b). AO enhanced imaging modalities have been used to study photoreceptors, the nerve fiber layer (NFL), and the RPE. As shown in Figure 14.13B, adaptive-optics SLO or AOSLO can image the mosaics of distinct photoreceptors. The best reported images of photoreceptors were obtained using confocal AOSLO (Roorda and Duncan, 2015b), which also image the entire mosaic of photoreceptors in the living eyes for the first time (Roorda and Duncan, 2015b). AO imaging of photoreceptors may potentially be used to analyze the health of photoreceptors. Figure 14.13C shows a detailed image of the NFL taken in an adult mouse. Individual nerve fiber bundles are visible and small capillaries (arrowhead) are also visible within the field of view. The NFL is made up almost entirely of the axons of retinal ganglion cells (RGC). RGC loss has been shown to be an important indicator of glaucoma progression.

14.7 Summary

In conclusion, this chapter has introduced the major ophthalmic imaging techniques used in biomedical research and clinical practice. While the fundus camera, cSLO, and OCT have become well established in vision science and ophthalmology, newer technologies, such as OCTA, vis-OCT, PAOM, and AO, are still ongoing research efforts and finding clinical utilities. Some imaging technologies were not discussed, including ultrasound imaging and other functional variants of OCT (Silverman, 2009), including polarization-sensitive OCT (de Boer et al., 2017a). In many of the discussed techniques, ideas were borrowed from microscopy, chemistry, physics, and astronomy. Therefore, by continuing to look at these fields for new ideas, we may realize yet undiscovered ophthalmic technologies, bringing with them further understanding of the eye and its function in health and disease. Ultimately, new ophthalmic imaging technologies provide powerful tools for diagnostics and pave the way for improved therapeutic strategies.

REFERENCES

Abucham-Neto JZ, Torricelli AAM, Lui ACF, Guimaraes SN, Nascimento H, Regatieri CV (2018) Comparison between optical coherence tomography angiography and fluorescein angiography findings in retinal vasculitis. *Int J Retina Vitreous* 4:15.

An L, Wang RK (2008) In vivo volumetric imaging of vascular perfusion within human retina and choroids with optical micro-angiography. *Opt Express* 16:11438–11452.

Ang M, Cai Y, MacPhee B, Sim DA, Keane PA, Sng CC, Egan CA, Tufail A, Larkin DF, Wilkins MR (2016) Optical coherence tomography angiography and indocyanine green angiography for corneal vascularisation. *Br J Ophthalmol* 100:1557–1563.

Bedford RE, Wyszecki G (1957) Axial chromatic aberration of the human eye. *J Opt Soc Am* 47:564–565.

Cahoon JM, Olson PR, Nielson S, Miya TR, Bankhead P, McGeown JG, Curtis TM, Ambati BK (2014) Acridine orange leukocyte fluorography in mice. *Exp Eye Res* 120:15–19.

Chaudhuri A, Hallett PE, Parker JA (1983) Aspheric curvatures, refractive indices and chromatic aberration for the rat eye. *Vision Res* 23:1351–1363.

Chen CL, Wang RK (2017) Optical coherence tomography based angiography [Invited]. *Biomed Opt Express* 8:1056–1082.

Chen S, Yi J, Zhang HF (2015a) Measuring oxygen saturation in retinal and choroidal circulations in rats using visible light optical coherence tomography angiography. *Biomed Opt Express* 6:2840–2853.

Chen S, Yi J, Liu W, Backman V, Zhang H (2015b) Monte Carlo investigation of optical coherence tomography retinal oximetry. *IEEE Trans Bio-Med Eng* 62:2308–15.

Cohan BE, Pearch AC, Jokelainen PT, Bohr DF (2003) Optic disc imaging in conscious rats and mice. *Invest Ophthalmol Vis Sci* 44:160–163.

de Boer JF, Hitzenberger CK, Yasuno Y (2017a) Polarization sensitive optical coherence tomography - a review [Invited]. *Biomed Opt Express* 8:1838–1873.

de Boer JF, Leitgeb R, Wojtkowski M (2017b) Twenty-five years of optical coherence tomography: the paradigm shift in sensitivity and speed provided by Fourier domain OCT [Invited]. *Biomed Opt Express* 8:3248–3280.

de la Zerda A, Paulus YM, Teed R, Bodapati S, Dollberg Y, Khuri-Yakub BT, Blumenkranz MS, Moshfeghi DM, Gambhir SS (2010) Photoacoustic ocular imaging. *Opt Lett* 35:270–272.

Dominguez E, Raoul W, Calippe B, Sahel JA, Guillonneau X, Paques M, Sennlaub F (2015) Experimental branch retinal vein occlusion induces upstream pericyte loss and vascular destabilization. *PLoS One* 10:e0132644.

Dong B, Sun C, Zhang HF (2017) Optical detection of ultrasound in photoacoustic imaging. *IEEE Trans Biomed Eng* 64:4–15.

Dong B, Chen S, Zhang Z, Sun C, Zhang HF (2014) Photoacoustic probe using a microring resonator ultrasonic sensor for endoscopic applications. *Opt Lett* 39:4372–4375.

Ebneter A, Agca C, Dysli C, Zinkernagel MS (2015) Investigation of retinal morphology alterations using spectral domain optical coherence tomography in a mouse model of retinal branch and central retinal vein occlusion. *PLoS One* 10:e0119046.

Faber DJ, Aalders MC, Mik EG, Hooper BA, van Gemert MJ, van Leeuwen TG (2004) Oxygen saturation-dependent absorption and scattering of blood. *Phys Rev Lett* 93:028102.

Fercher AF (2010) Optical coherence tomography–Development, principles, applications. *Z Med Phys* 20:251–276.

Fercher AF, Drexler W, Hitzenberger CK, Lasser T (2003) Optical coherence tomography–Principles and applications. *Rep Prog Phy* 66:239–303.

Fujimoto J, Swanson E (2016) The development, commercialization, and impact of optical coherence tomography. *Invest Ophthalmol Vis Sci* 57:OCT1–OCT13.

Gao SS, Liu G, Huang D, Jia Y (2015) Optimization of the split-spectrum amplitude-decorrelation angiography algorithm on a spectral optical coherence tomography system. *Opt Lett* 40:2305–2308.

Gao SS, Jia Y, Zhang M, Su JP, Liu G, Hwang TS, Bailey ST, Huang D (2016) Optical coherence tomography angiography. *Invest Ophthalmol Vis Sci* 57:OCT27–OCT36.

Geng Y, Dubra A, Yin L, Merigan WH, Sharma R, Libby RT, Williams DR (2012) Adaptive optics retinal imaging in the living mouse eye. *Biomed Opt Express* 3:715–734.

Geng Y, Schery LA, Sharma R, Dubra A, Ahmad K, Libby RT, Williams DR (2011) Optical properties of the mouse eye. *Biomed Opt Express* 2:717–738.

Gorczynska I, Migacz JV, Zawadzki RJ, Capps AG, Werner JS (2016) Comparison of amplitude-decorrelation, speckle-variance and phase-variance OCT angiography methods for imaging the human retina and choroid. *Biomed Opt Express* 7:911–942.

Gullstrand A (1910) Neue methoden der reflexlosen ophthalmoskopie. *Ber Dtsch Ophthalmol Ges* 36:326.

Haddock LJ, Kim DY, Mukai S (2013) Simple, inexpensive technique for high-quality smartphone fundus photography in human and animal eyes. *J Ophthalmol* 2013:518479.

Hagag AM, Gao SS, Jia Y, Huang D (2017) Optical coherence tomography angiography: Technical principles and clinical applications in ophthalmology. *Taiwan J Ophthalmol* 7:115–129.

Hall JE, Guyton AC (2011) *Textbook of Medical Physiology.* Saunders.

Hardarson SH, Harris A, Karlsson RA, Halldorsson GH, Kagemann L, Rechtman E, Zoega GM, Eysteinsson T, Benediktsson JA, Thorsteinsson A, Jensen PK, Beach J, Stefansson E (2006) Automatic retinal oximetry. *Invest Ophthalmol Vis Sci* 47:5011–5016.

Hawes NL, Smith RS, Chang B, Davisson M, Heckenlively JR, John SW (1999) Mouse fundus photography and angiography: A catalogue of normal and mutant phenotypes. *Mol Vis* 5:22.

Hendargo HC, McNabb RP, Dhalla AH, Shepherd N, Izatt JA (2011) Doppler velocity detection limitations in spectrometer-based versus swept-source optical coherence tomography. *Biomed Opt Express* 2:2175–2188.

Hu S, Maslov K, Wang LV (2011) Second-generation optical-resolution photoacoustic microscopy with improved sensitivity and speed. *Opt Lett* 36:1134–1136.

Hu Z, Pan Y, Rollins AM (2007) Analytical model of spectrometer-based two-beam spectral interferometry. *Appl Opt* 46:8499–8505.

Hu S, Rao B, Maslov K, Wang LV (2010) Label-free photoacoustic ophthalmic angiography. *Opt Lett* 35:1–3.

Huang D, Swanson EA, Lin CP, Schuman JS, Stinson WG, Chang W, Hee MR, Flotte T, Gregory K, Puliafito CA (1991) Optical coherence tomography. *Science* 254:1178–1181.

Hughes A (1979) A schematic eye for the rat. *Vision Res* 19:569–588.

Inoue M, Jung JJ, Balaratnasingam C, Dansingani KK, Dhrami-Gavazi E, Suzuki M, de Carlo TE, Shahlaee A, Klufas MA, El Maftouhi A, Duker JS, Ho AC, Quaranta-El Maftouhi M, Sarraf D, Freund KB, Grp C-S (2016) A comparison between optical coherence tomography angiography and fluorescein angiography for the imaging of type 1 neovascularization. *Investig Ophthalmol Vis Sci* 57:Oct314–Oct323.

Izatt JA, Choma MA, Dhalla A-H (2015) Theory of optical coherence tomography. In *Optical Coherence Tomography: Technology and Applications*, Drexler W, Fujimoto JG (Eds), pp. 65–94. Springer.

Jackman W, Webster J (1886) On photographing the retina of the living human eye. *Phila Photograph* 23:275–276.

Jia Y, Tan O, Tokayer J, Potsaid B, Wang Y, Liu JJ, Kraus MF, Subhash H, Fujimoto JG, Hornegger J, Huang D (2012) Split-spectrum amplitude-decorrelation angiography with optical coherence tomography. *Opt Express* 20:4710–4725.

Jian Y, Zawadzki RJ, Sarunic MV (2013) Adaptive optics optical coherence tomography for in vivo mouse retinal imaging. *J Biomed Opt* 18:056007.

Jiao S, Jiang M, Hu J, Fawzi A, Zhou Q, Shung KK, Puliafito CA, Zhang HF (2010) Photoacoustic ophthalmoscopy for in vivo retinal imaging. *Opt Express* 18:3967–3972.

Kalkman J (2017) Fourier-domain optical coherence tomography signal analysis and numerical modeling. *Int J Opt* 2017:16.

Kashani AH, Chen CL, Gahm JK, Zheng F, Richter GM, Rosenfeld PJ, Shi Y, Wang RK (2017) Optical coherence tomography angiography: A comprehensive review of current methods and clinical applications. *Prog Retin Eye Res* 60:66–100.

Kornfield TE, Newman EA (2014) Regulation of blood flow in the retinal trilaminar vascular network. *J Neurosci* 34:11504–11513.

Kur J, Newman EA, Chan-Ling T (2012) Cellular and physiological mechanisms underlying blood flow regulation in the retina and choroid in health and disease. *Prog Retin Eye Res* 31:377–406.

LaRocca F, Dhalla AH, Kelly MP, Farsiu S, Izatt JA (2013) Optimization of confocal scanning laser ophthalmoscope design. *J Biomed Opt* 18:076015.

LaRocca F, Nankivil D, Farsiu S, Izatt JA (2014) True color scanning laser ophthalmoscopy and optical coherence tomography handheld probe. *Biomed Opt Express* 5:3204–3216.

Levkovitch-Verbin H (2004) Animal models of optic nerve diseases. *Eye* 18:1066.

Li H, Dong B, Zhang Z, Zhang HF, Sun C (2014) A transparent broadband ultrasonic detector based on an optical micro-ring resonator for photoacoustic microscopy. *Sci Rep* 4:4496.

Li H, Liu W, Dong B, Kaluzny JV, Fawzi AA, Zhang HF (2017) Snapshot hyperspectral retinal imaging using compact spectral resolving detector array. *J Biophotonics* 10:830–839.

Li H, Liu W, Zhang HF (2015) Investigating the influence of chromatic aberration and optical illumination bandwidth on fundus imaging in rats. *J Biomed Opt* 20:106010.

Link D, Strohmaier C, Seifert BU, Riemer T, Reitsamer HA, Haueisen J, Vilser W (2011) Novel non-contact retina camera for the rat and its application to dynamic retinal vessel analysis. *Biomed Opt Express* 2:3094–3108.

Linsenmeier RA, Zhang HF (2017) Retinal oxygen: From animals to humans. *Prog Retin Eye Res* 58:115–151.

Liu W, Li H, Shah RS, Shu X, Linsenmeier RA, Fawzi AA, Zhang HF (2015b) Simultaneous optical coherence tomography angiography and fluorescein angiography in rodents with normal retina and laser-induced choroidal neovascularization. *Opt Lett* 40:5782–5785.

Liu T, Li H, Song W, Jiao S, Zhang HF (2013) Fundus camera guided photoacoustic ophthalmoscopy. *Curr Eye Res* 38:1229–1234.

Liu W, Schultz KM, Zhang K, Sasman A, Gao F, Kume T, Zhang HF (2014) In vivo corneal neovascularization imaging by optical-resolution photoacoustic microscopy. *Photoacoustics* 2:81–86.

Liu W, Wang S, Soetikno B, Yi J, Zhang K, Chen S, Linsenmeier RA, Sorenson CM, Sheibani N, Zhang HF (2017) Increased retinal oxygen metabolism precedes microvascular alterations in type 1 diabetic mice. *Invest Ophthalmol Vis Sci* 58:981–989.

Liu W, Yi J, Chen S, Jiao S, Zhang HF (2015a) Measuring retinal blood flow in rats using Doppler optical coherence tomography without knowing eyeball axial length. *Med Phys* 42:5356–5362.

Lombardo M, Serrao S, Devaney N, Parravano M, Lombardo G (2013) Adaptive optics technology for high-resolution retinal imaging. *Sensors* 13:334–366.

London A, Benhar I, Schwartz M (2013) The retina as a window to the brain-from eye research to CNS disorders. *Nat Rev Neurol* 9:44–53.

Makita S, Hong Y, Yamanari M, Yatagai T, Yasuno Y (2006) Optical coherence angiography. *Opt Express* 14:7821–7840.

Manivannan A, Van der Hoek J, Vieira P, Farrow A, Olson J, Sharp PF, Forrester JV (2001) Clinical investigation of a true color scanning laser ophthalmoscope. *Arch Ophthalmol* 119:819–824.

Miyahara S, Kiryu J, Miyamoto K, Katsuta H, Hirose F, Tamura H, Musashi K, Honda Y, Yoshimura N (2004) In vivo three-dimensional evaluation of leukocyte behavior in retinal microcirculation of mice. *Invest Ophthalmol Vis Sci* 45:4197–4201.

Miyamoto K, Khosrof S, Bursell SE, Moromizato Y, Aiello LP, Ogura Y, Adamis AP (2000) Vascular endothelial growth factor (VEGF)-induced retinal vascular permeability is mediated by intercellular adhesion molecule-1 (ICAM-1). *Am J Path* 156:1733–1739.

Miyamoto K, Khosrof S, Bursell SE, Rohan R, Murata T, Clermont AC, Aiello LP, Ogura Y, Adamis AP (1999) Prevention of leukostasis and vascular leakage in streptozotocin-induced diabetic retinopathy via intercellular adhesion molecule-1 inhibition. *Proc Nat Acad Sci U S A* 96:10836–10841.

Nam AS, Chico-Calero I, Vakoc BJ (2014) Complex differential variance algorithm for optical coherence tomography angiography. *Biomed Opt Express* 5:3822–3832.

Nesper PL, Soetikno BT, Zhang HF, Fawzi AA (2017) OCT angiography and visible-light OCT in diabetic retinopathy. *Vision Res* 139:191–203.

Nishiwaki H, Ogura Y, Kimura H, Kiryu J, Miyamoto K, Matsuda N (1996) Visualization and quantitative analysis of leukocyte dynamics in retinal microcirculation of rats. *Invest Ophthalmol Vis Sci* 37:1341–1347.

Paques M, Guyomard JL, Simonutti M, Roux MJ, Picaud S, Legargasson JF, Sahel JA (2007) Panretinal, high-resolution color photography of the mouse fundus. *Invest Ophthalmol Vis Sci* 48:2769–2774.

Paques M, Simonutti M, Roux MJ, Picaud S, Levavasseur E, Bellman C, Sahel JA (2006) High resolution fundus imaging by confocal scanning laser ophthalmoscopy in the mouse. *Vision Res* 46:1336–1345.

Park JR, Choi W, Hong HK, Kim Y, Jun Park S, Hwang Y, Kim P, Joon Woo S, Hyung Park K, Oh WY (2016) Imaging laser-induced choroidal neovascularization in the rodent retina using optical coherence tomography angiography. *Invest Ophthalmol Vis Sci* 57:OCT331–OCT340.

Park JS, Choi CK, Kihm KD (2004) Optically sliced micro-PIV using confocal laser scanning microscopy (CLSM). *Exp Fluids* 37:105–119.

Pi S, Camino A, Zhang M, Cepurna W, Liu G, Huang D, Morrison J, Jia Y (2017) Angiographic and structural imaging using high axial resolution fiber-based visible-light OCT. *Biomed Opt Express* 8:4595–4608.

Remtulla S, Hallett PE (1985) A schematic eye for the mouse, and comparisons with the rat. *Vision Res* 25:21–31.

Roorda A (2010) Applications of adaptive optics scanning laser ophthalmoscopy. *Optom Vis Sci* 87:260–268.

Roorda A, Duncan JL (2015) Adaptive optics ophthalmoscopy. *Annu Rev Vis Sci* 1:19–50.

Shah RS, Soetikno BT, Lajko M, Fawzi AA (2015) A mouse model for laser-induced choroidal neovascularization. *J Vis Exp*, 106:e53502.

Shah RS, Soetikno BT, Yi J, Liu W, Skondra D, Zhang HF, Fawzi AA (2016) Visible-light optical coherence tomography angiography for monitoring laser-induced choroidal neovascularization in mice. *Invest Ophthalmol Vis Sci* 57:OCT86–OCT95.

Shu X, Bondu M, Dong B, Podoleanu A, Leick L, Zhang HF (2016) Single all-fiber-based nanosecond-pulsed supercontinuum source for multispectral photoacoustic microscopy and optical coherence tomography. *Opt Lett* 41:2743–2746.

Shu X, Li H, Dong B, Sun C, Zhang HF (2017) Quantifying melanin concentration in retinal pigment epithelium using broadband photoacoustic microscopy. *Biomed Opt Express* 8:2851–2865.

Shu X, Liu W, Zhang HF (2015) Monte Carlo investigation on quantifying the retinal pigment epithelium melanin concentration by photoacoustic ophthalmoscopy. *J Biomed Opt* 20:106005.

Silverman RH (2009) High-resolution ultrasound imaging of the eye–A review. *Clin Exp Ophthalmol* 37:54–67.

Soetikno BT, Yi J, Shah R, Liu W, Purta P, Zhang HF, Fawzi AA (2015) Inner retinal oxygen metabolism in the 50/10 oxygen-induced retinopathy model. *Sci Rep* 5:16752.

Soetikno BT, Shu X, Liu Q, Liu W, Chen S, Beckmann L, Fawzi AA, Zhang HF (2017) Optical coherence tomography angiography of retinal vascular occlusions produced by imaging-guided laser photocoagulation. *Biomed Opt Express* 8:3571–3582.

Song W, Wei Q, Liu T, Kuai D, Burke JM, Jiao S, Zhang HF (2012) Integrating photoacoustic ophthalmoscopy with scanning laser ophthalmoscopy, optical coherence tomography, and fluorescein angiography for a multimodal retinal imaging platform. *J Biomed Opt* 17:061206.

Song W, Wei Q, Liu W, Liu T, Yi J, Sheibani N, Fawzi AA, Linsenmeier RA, Jiao S, Zhang HF (2014) A combined method to quantify the retinal metabolic rate of oxygen using photoacoustic ophthalmoscopy and optical coherence tomography. *Sci Rep* 4:6525.

Sorenson CM, Wang S, Gendron R, Paradis H, Sheibani N (2013) Thrombospondin-1 deficiency exacerbates the pathogenesis of diabetic retinopathy. *J Diabetes Metab* (Suppl 12).

Strauss O (2005) The retinal pigment epithelium in visual function. *Physiol Rev* 85:845–881.

Tan ACS, Tan GS, Denniston AK, Keane PA, Ang M, Milea D, Chakravarthy U, Cheung CMG (2018) An overview of the clinical applications of optical coherence tomography angiography. *Eye* 32:262–286.

Tanaka K, Mori R, Kawamura A, Nakashizuka H, Wakatsuki Y, Yuzawa M (2017) Comparison of OCT angiography and indocyanine green angiographic findings with subtypes of polypoidal choroidal vasculopathy. *Br J Ophthalmol* 101:51–55.

Uttam S, Liu Y (2015) Fourier phase in Fourier-domain optical coherence tomography. *J Opt Soc Am A Opt Image Sci Vis* 32:2286–2306.

Veleri S, Lazar CH, Chang B, Sieving PA, Banin E, Swaroop A (2015) Biology and therapy of inherited retinal degenerative disease: Insights from mouse models. *Dis Mod Mech* 8:109–129.

Wang Y, Bower BA, Izatt JA, Tan O, Huang D (2008) Retinal blood flow measurement by circumpapillary Fourier domain Doppler optical coherence tomography. *J Biomed Opt* 13:064003.

Wang Y, Bower BA, Izatt JA, Tan O, Huang D (2007) In vivo total retinal blood flow measurement by Fourier domain Doppler optical coherence tomography. *J Biomed Opt* 12:041215.

Wang LV, Hu S (2012) Photoacoustic tomography: in vivo imaging from organelles to organs. *Science* 335:1458–1462.

Wangsa-Wirawan ND, Linsenmeier RA (2003) Retinal oxygen: Fundamental and clinical aspects. *Arch Ophthalmol* 121:547–557.

Wassle H (2004) Parallel processing in the mammalian retina. *Nat Rev Neurosci* 5:747–757.

Webb RH, Hughes GW, Delori FC (1987) Confocal scanning laser ophthalmoscope. *Appl Opt* 26:1492–1499.

Webb RH, Hughes GW, Pomerantzeff O (1980) Flying spot TV ophthalmoscope. *Appl Opt* 19:2991–2997.

White BR, Pierce MC, Nassif N, Cense B, Park BH, Tearney GJ, Bouma BE, Chen TC, de Boer JF (2003) In vivo dynamic human retinal blood flow imaging using ultra-high-speed spectral domain optical Doppler tomography. *Opt Express* 11:3490–3497.

Wilhelm S, Grobler B, Gluch M, Heinz H (2003) Confocal laser scanning microscopy. Principles. Microscopy from Carl Zeiss, microspecial.

Yeh C, Soetikno B, Hu S, Maslov KI, Wang LV (2014) Microvascular quantification based on contour-scanning photoacoustic microscopy. *J Biomed Opt* 19:96011.

Yeh C, Soetikno B, Hu S, Maslov KI, Wang LV (2015) Three-dimensional arbitrary trajectory scanning photoacoustic microscopy. *J Biophotonics* 8:303–308.

Yi J, Chen S, Backman V, Zhang HF (2014) In vivo functional microangiography by visible-light optical coherence tomography. *Biomed Opt Express* 5:3603.

Yi, J., Liu, W., Chen, S. *et al.* (2015). Visible light optical coherence tomography measures retinal oxygen metabolic response to systemic oxygenation. *Light Sci Appl* 4:e334

Yi J, Wei Q, Liu W, Backman V, Zhang HF (2013) Visible-light optical coherence tomography for retinal oximetry. *Opt Lett* 38:1796–1798.

Zhang L, Capilla A, Song W, Mostoslavsky G, Yi J (2017) Oblique scanning laser microscopy for simultaneously volumetric structural and molecular imaging using only one raster scan. *Sci Rep* 7:8591.

Zhang Y, Cho CH, Atchaneeyasakul LO, McFarland T, Appukuttan B, Stout JT (2005) Activation of the mitochondrial apoptotic pathway in a rat model of central retinal artery occlusion. *Invest Ophthalmol Vis Sci* 46:2133–2139.

Zhang Y, Fortune B, Atchaneeyasakul LO, McFarland T, Mose K, Wallace P, Main J, Wilson D, Appukuttan B, Stout JT (2008) Natural history and histology in a rat model of laser-induced photothrombotic retinal vein occlusion. *Curr Eye Res* 33:365–376.

Zhang P, Goswami M, Zam A, Pugh EN, Zawadzki RJ (2015a) Effect of scanning beam size on the lateral resolution of mouse retinal imaging with SLO. *Optics Lett* 40:5830–5833.

Zhang HF, Maslov K, Stoica G, Wang LV (2006) Functional photoacoustic microscopy for high-resolution and noninvasive in vivo imaging. *Nat Biotechnol* 24:848–851.

Zhang L, Song W, Shao D, Zhang S, Desai M, Ness S, Roy S, Yi J (2018) Volumetric fluorescence retinal imaging in vivo over a 30-degree field of view by oblique scanning laser ophthalmoscopy (oSLO). *Biomed Opt Express* 9:25–40.

Zhang P, Zam A, Jian Y, Wang X, Li Y, Lam KS, Burns ME, Sarunic MV, Pugh EN, Jr., Zawadzki RJ (2015b) In vivo wide-field multispectral scanning laser ophthalmoscopy-optical coherence tomography mouse retinal imager: longitudinal imaging of ganglion cells, microglia, and Muller glia, and mapping of the mouse retinal and choroidal vasculature. *J Biomed Opt* 20:126005.

Zhi Z, Qin W, Wang J, Wei W, Wang RK (2015) 4D optical coherence tomography-based micro-angiography achieved by 1.6-MHz FDML swept source. *Optics Lett* 40:1779–1782.

Section III

Whole-Organ and Whole-Organism Imaging

15

Heart Imaging

Leonardo Sacconi and Claudia Crocini

CONTENTS

15.1 Echocardiography of Mammalian Hearts

In vivo assessment of cardiac function is indispensable for diagnosis and follow-up in patients, as well as for assessment of cardiovascular structure and function in animal models. Echocardiography (ECG) represents a gold standard for reliable *in vivo* cardiac assessment, thanks to its noninvasive nature that overcomes the limits provided by the anatomical location of the heart. ECG is widely used in cardiac research to assess morphological and hemodynamic features of the heart in a variety of animal models. Ideally, ECG should be performed with awake and cooperative animals, as anesthetics induce a direct or indirect depression of hemodynamic values and cardiac functional parameters. However, performing ECG in conscious animals requires extensive animal training to avoid excitement that can result in enhanced sympathetic tone and heart rate. Thus, imaging acquisition under mild anesthesia is recommended, but the type and the dose of the anesthesia has to be carefully evaluated depending on the animal model and when comparing data from different labs.

15.1.1 Echocardiographic Modes

An echocardiogram produces a real-time image of the heart, exploiting different acoustic impedances of tissues to high-frequency sound beams: the myocardium appears white because it reflects more ultrasound (hyperechoic) than blood (hypoechoic), which appears black. Three principal imaging formats can be obtained to assess cardiac function: brightness mode (B-mode), motion mode (M-mode), and Doppler imaging.

The most basic of the echo modes is the B-mode, which produces a real-time black and white image of the targeted site. B-mode images provide two-dimensional (2D) views of the heart and vasculature (Coatney, 2001), allowing for non-quantitative assessment of cardiac phenotype, chamber dimensions, and heart function. The B-mode can serve as a guidance for further evaluation using other imaging formats, such as M-mode and Doppler imaging.

M-mode images are obtained by a rapid sequence of B-mode scans along a single line displayed over time. The beam of sound reflected from the moving cardiac walls is converted into continuous waves showing myocardium motion during systole and diastole. M-mode images allow for assessment of left ventricle (LV) functional parameters, including ejection fraction (EF), fractional shortening (FS), cardiac output, and stroke volume, as well as wall and interventricular diameters, abnormal segmental wall contraction, and LV mass (Collins et al., 2001; Kiatchoosakun et al., 2002).

Doppler imaging exploits the Doppler shift principle of a moving target to determine blood flow velocity and direction. Using a pulsed-wave, it is possible to determine blood flow velocity and calculate peak velocities, ejection time (ET), and velocity time intervals. Transmittal flow velocity profiles can provide useful insights of diastolic function, including isovolumic contraction and relaxation times, ratio of early (E)-to-late (atrial, A) ventricular filling velocities (E/A), and deceleration of E wave. Fusion of E and A waves may be an indicator of diastolic dysfunction. However, the rapid heart rate of rodents (especially mice and rats) may also result in partial or complete fusion of E and A waves.

15.1.2 Echocardiographic Measurements

From M-mode images, it is possible to obtain a variety of measurements relative to the LV structure, such as interventricular septal thicknesses (IVS), LV internal dimensions (LVID), and posterior wall thicknesses (PW), both at diastole and systole (IVSd, LVIDd, PWd, and IVSs, LVIDs, PWs). From those values, LV EF and LV FS can be calculated using the following formulas (Gardin et al., 1995; Syed et al., 2005; Tsujita et al., 2005):

$$EF(\%) = ((LVIDd^{3} - LVIDs)^{3} / LVIDd^{3}) \times 100$$

$$FS(\%) = \frac{LVIDd - LVIDs}{LVIDd} \times 100$$

LV EF and LV FS are related in the healthy heart. However, upon changes of LV geometry, EF calculated by the simple cubic assumption of LV volume may not be accurate.

Using pulse-wave Doppler imaging, the velocity of circumferential fiber shortening (Vcf), a preload independent measurement of LV systolic function, can be calculated as $Vfc = FS/ET$, where ET is the ejection time of the LV. It should be noted that ET is heart rate dependent and thus requires correction by dividing it by the square root of the R–R interval (Odley et al., 2004). This correction is relevant in both conscious and anesthetized mice. Finally, LV mass can be calculated from end-diastolic linear measurements of LV diameter and wall thickness using M-mode images. To convert the volume to mass, the myocardial density (approximately 1.05 g/mL) can be used. LV mass is an important predictor of cardiovascular events (Verdecchia et al., 2001; Verma et al., 2008).

$$LV\ mass = 1.05 \times \left[\left(IVSd + LVIDd + PWd \right)^{3} - LVIDd^{3} \right]$$

15.2 Optical Mapping of Mammalian Hearts

15.2.1 Langendorff Heart

Despite the numerous advantages of ECG, specific manipulation of cardiac function in the intact animal is frequently complicated and can be affected by extracardiac factors, such as vascular tone, circulating factors, and anesthetic interventions. In this respect, the isolated perfused heart preparation offers unique advantages for studying cardiac function.

In 1898, Oscar Langendorff described the isolated mammalian heart preparation, in which an excised intact heart (from cats) could be kept alive for a prolonged period of time by delivery of blood through a cannula attached to the ascending aorta (Zimmer, 1998; Skrzypiec-Spring et al., 2007). The basic idea of this preparation is still frequently used in cardiac research and is called retrograde Langendorff's perfusion. Today, the perfusion is not performed using blood but a crystalloid buffer solution whose composition is strictly controlled by the investigator. The perfusion solution often includes a contraction uncoupler (e.g. 2,3 butanedione monoxime or blebbistatin) for maintaining the heart at rest. This approach is particularly indicated for microscopy. Perfusion via the coronary circulation does not allow the heart to perform pressure-volume work, as no fluid is present in the left ventricle (Merx and Schrader, 2009). An alternative strategy consists of cannulating the left side of the heart and introducing the perfusion solution into the left atrium, and eventually into the left ventricle (Neely et al., 1967; Merx and Schrader, 2009). Simultaneous measurement of pressure and volume is considered the most effective mean for experimental characterization of intact heart function (Kass, 1992; Georgakopoulos et al., 1998), as it allows assessing load-independent indices of contractility and relaxation. Measurement of volume and ventricular chamber dimensions obtained by ECG offer information about the pump function of the heart, though this measurement averages steady state images over several seconds and, therefore, does not provide load-independent assessment of contractility. In contrast, the pressure-conductance catheter technique provides an instantaneous and continuous signal for pressure and volume. Occlusion maneuvers performed during conductance catheterization permit assessment of end systolic and end diastolic pressure volume relationships, respectively, allowing characterization of the intrinsic contractile properties of the ventricle and providing useful information for both clinical and basic research (Burkhoff et al., 2005). In general, Langendorff's heart preparations and its variants have proven to be an invaluable analytic tool to assess cardiac function (e.g. action potential propagation, calcium transients, contraction, metabolism) at the whole heart level.

15.2.2 Voltage Sensitive Dyes and Calcium Indicators

Electrical activity may be considered the most peculiar characteristic of excitable cells, and several efforts have been made to create sensors capable of following membrane potential changes. There are some important technical problems that need to be overcome in order to measure a membrane potential variation. For instance:

1. The electrical field decreases exponentially with distance from the cell membrane, meaning that the voltage sensor must be either inside or physically attached to the membrane in order to probe the phenomenon. Moreover, the cell membrane is an active cellular component, and its electrical properties may vary with the insertion or attachment of chromophores.

2. The membrane that borders the cellular environment (plasmalemma) is not the only membrane present in the cell. It is necessary to exclusively localize fluorescent molecules at the plasmalemma. Because intracellular membranes are not involved in electrical activity, any dye attached to them would only increase background noise.

3. Compared to the bulk cytoplasm, the membrane is very thin and the amount of chromophores in that volume is limited. Thus, only few photons can be emitted providing a low signal to noise ratio (SNR). Increasing the light-source power or the time and area of illumination would improve the signal but could also exert photodamage and eventually jeopardize the measurement.

Once these problems are overcome, effective voltage imaging can be achieved by exploiting different physicochemical features of molecules. One of the most used methods to generate voltage-sensitive dyes (VSD) is the synthesis of organic chromophores that bind to the outer cellular membrane. A lot of effort has been spent in the past decades to develop synthetized organic chromophores. These approaches rely on several different mechanisms of voltage sensing that are common both to absorption and fluorescence. The most important classes of VSDs are:

1. Nernstian dyes: These dyes are characterized by a mechanism called redistribution, whereby the change in the electric field causes the chromophore to move into or out of the cell, either completely or partially, changing the absolute concentration of the fluorophore in the cell, and hence the fluorescence it can produce. The differences in chemical environment between membrane and cytoplasm alter the relative stabilities and energies of the ground and excited states of the chromophore, changing its spectroscopic properties. They are called Nernstian dyes because they redistribute according to Nernstian equilibrium, or, alternatively, slow dyes because their insertion on detachment from the membrane is a relatively slow ($\sim$ s) equilibrium process compared to the other mechanisms. These dyes are thus not suitable to detect action potential propagation.

2. Reorientation-based dyes: The chromophore lies in or on the membrane with a particular orientation, determined by the sum of the interaction forces on the chromophore. Changes in the electric field act on the chromophore's dipole moment, producing a torque that alters the orientation angle of the chromophore. The change in terms of orientation generally implies altered fluorescent spectra and quantum yield. Reorientation can be fast because it does not involve a significant movement of the chromophore.

3. Electrochromic dyes: These are the dyes that have received the most interest and are based on electrochromism i.e. the direct electrical modulation of the electronic structure, and thus the spectra, of a chromophore. Chromophores that exhibit strong electrochromism typically have remarkable differences in dipole moment between the ground and excited states, so they are easily polarizable with strong induced dipoles. These changes in electronic structure lead to changes in both the excitation and emission spectra, which are subsequently transduced as variations in terms of absorption and re-emission. The electrochromic effect, also known as Stark effect, is fast because it involves only intramolecular charge redistribution without chromophore movement. By generating spectral differences, it offers a convenient method to monitor changes in membrane potential by monitoring optical signals at selected wavelengths. Hemicyanine dyes attracted attention because they exhibit electrochromism as a voltage-sensing mechanism. The electrochromic mechanism is a direct interaction of the electric field with the dye and does not require any movement of the chromophore. This means that the dye rapidly responds to membrane potential variations. Because it is a main actor in the excitation contraction coupling, Ca^{2+} imaging in cardiac cells is of great importance, too. Over the years, many Ca^{2+} indicators have been developed with different Ca^{2+} sensitivities and excitation/emission spectra to better meet researcher necessities. Ca^{2+} indicators have been used for a long time in the form of bioluminescent Ca^{2+}-binding photoproteins. Great progress was made in the early '80s with the development of Ca^{2+} fluorescent indicators that, in one molecule, combine the features of Ca^{2+} chelators with fluorescent chromophores. Some indicators of this first generation are still widely used, such as fura-2. It allows ratiometric quantitative measurements of the intracellular calcium concentration $[Ca^{2+}]_i$ to be performed.

15.2.3 Optical Mapping

The basic concept of a fluorescence microscope is to irradiate the sample and to detect the fluorescence emitted from the fluorescent indicators (VSD and Ca^{2+} sensors for instance). Wide-field fluorescence microscopy is a largely used technique to obtain both topographical and dynamic information. It is based on a simultaneous illumination of the whole sample (Figure 15.1). Mercury lamps or LEDs are generally used as a source of light. The beam of excitation light uniformly illuminates the heart surface while an objective is used to

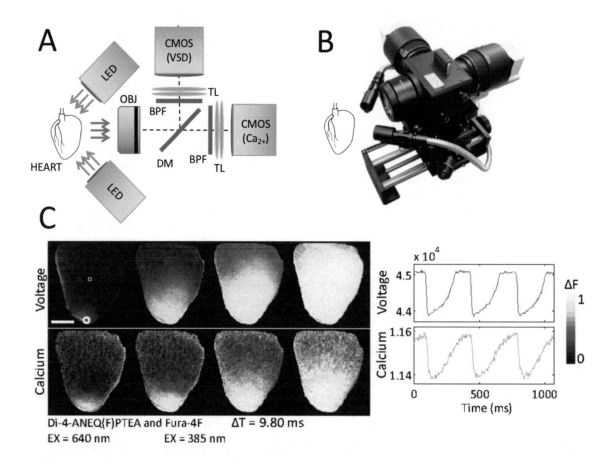

FIGURE 15.1 Optical mapping. A. Scheme of a multiparametric optical mapping system. A Langendorff-perfused isolated heart is illuminated using two independent LEDs. The fluorescence signal (voltage and/or calcium) is collected using a low magnification objective (OBJ). A dichroic mirror (DM) is used to split the emission spectra of the Ca2+ probe and VSD. The fluorescence signals are finally detected by two independent fast CMOS cameras. B. An example of a commercial fluorescence microscope specifically developed for imaging membrane potential propagation of Langendorff-perfused isolated heart preparations. C. Normalized fluorescence intensity maps of voltage and calcium at four different time points. The heart was electrical stimulated of the apex (at the site of the white circle). Raw VSD and Ca2+ probe signals are shown from a region of interest on the ventricle. Color bar shows normalized fluorescence intensity maps. Modified from Prigge et al. (2012).

collect the fluorescence signal. A high-speed imaging system is needed to capture and visualize small changes in fluorescence intensity from the heart stained with fluorescent probes. Considering the temporal resolution required to follow action potential propagation (1 ms) and the minimum sampling frequency necessary according to the Nyquist theorem, the wide-field microscope needs to operate at least at a frame rate of around 2 kHz, corresponding to a frame exposure time of 500 μs. This can be achieved with high-speed complementary metal oxide semiconductor (CMOS) cameras with a frame rate > 2 kHz. Furthermore, considering the small fluorescent variation associated with action potential or calcium release, a very low read-out noise, high quantum efficiency and well depth are required. Dedicated imaging acquisition systems are commercially available (see for example Sci Media and RedShirt). An alternative and cheaper solution is provided by using a multipurpose scientific CMOS (sCMOS) camera, such as the Hamamatsu OrcaFlash4.0. The maximum number of pixel lines that OrcaFlash4.0 sCMOS camera can support in free-run mode at this approximate frame rate is 128 lines when using camera link. In this case, considering a typical mouse heart of approximately 10 mm × 10 mm, the optical system

must demagnify the whole heart to fit into the central portion of the sensor.

A multimodal acquisition can be achieved using spectral separation both in excitation and in detection. A dichroic mirror is generally used to split the two spectral components of the fluorescence signal, the VSD, and the calcium sensor emission light. The fluorescence signal can be detected by two independent CMOS cameras or even using a single sensor. In the last case, different LEDs are turned on and off sequentially and in a fixed pattern during synchronized image acquisition (Lee et al., 2011).

While many electrophysiological properties can be quantified mapping the heart with a single view, there are some that necessitate a multiview imaging acquisition to map the whole heart surface (Figure 15.2). In arrhythmias for example, meandering re-entrant rotors can easily span across the heart surface, and a panoramic optical mapping based on multiview imaging is very useful in visualizing the entirety of the heart surface. Different strategies can be applied for panoramic imaging using multiple charge-coupled device (CCD) camera sensors (Lin and Wikswo, 1999) or multiple photodetectors arrays (Qu et al., 2007) or even employing mirrors to project

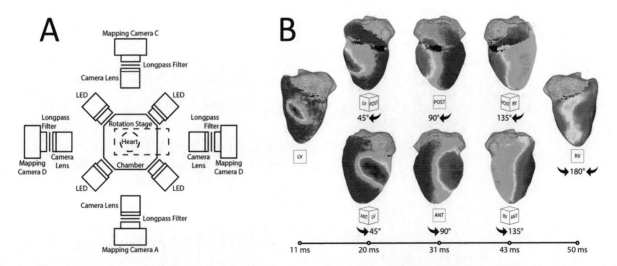

FIGURE 15.2 Cardiac panoramic optical mapping. A. Schematic of an optical mapping system. A Langendorff-perfused isolated heart is imaged from four different perspectives using four independent cameras. B. Representative rabbit panoramic optical data. A paced beat starts on the left ventricular free wall, and the divergent wavefronts are tracked over time around both the posterior and anterior sides to where they meet on the right ventricular free wall. Modified from Gloschat et al. (2018).

multiple views onto a single CCD (Lee et al., 2017). Recently, an open-source imaging toolkit for cardiac panoramic optical mapping has been developed by Efimov's lab (https://github.com/optocardiography). This open-source toolkit includes instructions for the 3D printing of experimental components and software for data acquisition, processing, and analysis.

15.3 Optical Manipulation of Heart Electrical Activity

15.3.1 Optogenetics in Cardiovascular Research

In addition to optically observing cardiac function, more recently, light has been employed to modulate cardiac activity, too. Specifically, photosensitive ion channels or pumps can be expressed on the plasmalemma of cardiac myocytes for manipulation of electrical excitability. Several microbial rhodopsins are available to generate both depolarizing and hyperpolarizing currents upon expression in mammalian cells (Miesenbock, 2009; Deisseroth, 2015). Currently, the most widely used optogenetic tool is ChR2, a light-gated cation channel cloned from algae (*Chlamydomonas reinhardtii*) (Nagel et al., 2003). The spectral response of ChR2 peaks at 470 nm, and, upon interaction with a photon, the all-trans retinal isomerizes to 13-cis retinal causing ChR2 to open. ChR2 is a promiscuous channel that preferentially conducts protons (H^+), then sodium cations (Na^+), and thirdly potassium (K^+) (Lin et al., 2009). Therefore, at the resting membrane potential (about –70 mV in mammalian excitable cells), a depolarizing inward current, referred as "photocurrent," is generated. As extensively reviewed by Schneider et al. (2015), ChR2 biophysical properties, such as channel gating, photocurrent amplitude, and activation/inactivation properties, can be modulated by the intensity and duration of the light stimulus. Several mutants of ChR2 have been generated to improve diverse features, such as enhanced ion selectivity (Kleinlogel et al., 2011;

Pan et al., 2014; Wietek et al., 2014), faster or slower photocycle kinetics (Lin et al., 2009; Bamann et al., 2010; Berndt et al., 2011; Klapoetke et al., 2014), higher retinal affinity (Nagel et al., 2005; Berndt et al., 2011), or modulation of absorption spectrum (Yizhar et al., 2011; Prigge et al., 2012; Govorunova et al., 2013; Lin et al., 2013; Klapoetke et al., 2014). Among those, the most commonly used mutant is the ChR2-(H134R), characterized by higher retinal affinity (Nagel et al., 2005) and by increased channel conductance to Na^+ than the wild-type (Lin et al., 2009). Among the hyperpolarizing tools, the chloride pump halorhodopsin (Schobert and Lanyi, 1982) and the proton pump archaerhodopsin (Chow et al., 2010) have been successfully used in mammalian cells to mediate an inhibitory effect. In cardiomyocytes, the hyperpolarizing current of halorhodopsin is not particularly large (Park et al., 2014) as compared to the magnitude of ChR2 current; however, a new family of hyperpolarizing ion channel rhodopsin have recently been described to produce significantly higher photocurrents (Govorunova et al., 2015, 2017).

In 2010, two simultaneous studies performed in zebrafish and mice (Arrenberg et al., 2010; Bruegmann et al., 2010) have employed optogenetics for cardiac research for the first time. In the zebrafish model, the transgenic expression of ChR2 and halorhodopsin was employed to study the origin of cardiac rhythmicity, combining optogenetics and light sheet microscopy to spatially map pacemaker activity during zebrafish development. Importantly, the authors demonstrated the ability of optogenetic approaches to precisely reverse rhythm disorders. Bruegmann et al. (2010) developed a transgenic mouse embryonic stem cell (ESC) line expressing ChR2-(H134R) to study its function both *in vitro* and *in vivo*. Illumination of ESC-derived cardiomyocytes and cardiac tissue from transgenic mice generated from ChR2-expressing ESC resulted in induced depolarizations and Ca^{2+} releases, perturbations of the sinus rhythm, and generation of extra beats. More recently, Zaglia et al. (2015) developed mouse lines with cell-specific expression of ChR2 and dissected the factors triggering

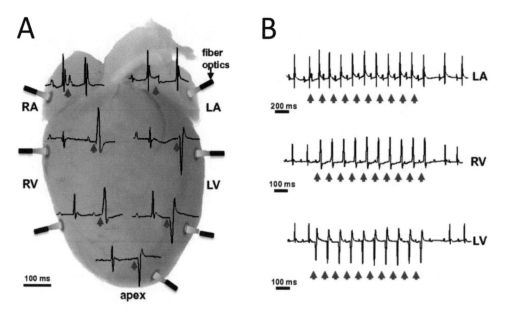

FIGURE 15.3   Cardiac optogenetics. A. Cardiac optogenetics allows noninvasive localized heart pacing *in vivo*. right atrium (RA), left atrium (LA) right ventricle (RV), left ventricle (LV), and apex. B. Representative ECG traces of ectopic beats originated by blue light stimulation of different regions: Modified from Zaglia et al. (2015).

arrhythmic beats by stimulating the heart with high spatial and temporal resolution (see Figure 15.3). Since then, a number of original studies and reviews have attempted to validate optogenetics as a tool for clinical applications.

15.3.2 Optogenetic Resynchronization and Defibrillation

The first therapeutic approach based on optogenetic stimulation *in vivo* was reported in 2015 (Nussinovitch and Gepstein, 2015). Rats were infected using intramyocardium injection of adeno-associated virus (AAV) vectors carrying ChR2-(H134R) and a fiber-coupled monochromic LED (centered at 450 nm) was used to deliver flashes of focused blue light to the site of ChR2-viral expression. Optogenetic pacing was effective both in open-chest and Langerdorff's perfused-experiments. Moreover, optical resynchronization therapy was achieved in hearts, in which multiple sites were infected with the virus. For each type of pacing (electrical, optogenetic), the authors studied the total ventricular activation times and standard deviation and compared them to the sinus rhythm parameters. Multisite optogenetic pacing restored both parameters to values similar to sinus rhythm, demonstrating the capability of optogenetics to rescue electrical desynchrony. Interestingly, the work is supported by optical mapping measurements of electrical propagation using a red-shifted voltage-sensitive dye (Matiukas et al., 2007). Recently, one of the more expected therapeutic applications of optogenetics has been explored by three independent and simultaneous studies (Bruegmann et al., 2016; Crocini et al., 2016c; Nyns et al., 2016). The works of Bruegmann et al. (2016) and Nyns et al. (2016) have demonstrated optogenetic termination of ventricular tachycardias (VTs) in mice and rats, respectively. They both employed a k_{ATP} channel opener

for inducing monomorphic (in mice), as well as, polymorphic (in rats) VTs. Bruegmann et al. (2016) employed both transgenic mice expressing ChR2-(H134R), generated by the same group, and mice after systemic infection of AAV9-ChR2-(H134R) virus, also previously reported by the same authors (Vogt et al., 2015). Similarly, Nyns et al. (2016) employed systemic infection of AAV9 in adult rats, but their vectors encoded the red-activatable channelrhodopsin (ReaChR). The excitation spectrum of ReaChR has a peak around 590 nm (Lin et al., 2013) and the red-shifted activation could be exploited for deeper optogenetic excitation. However, in the work by Nyns et al., ReaChR is excited with a LED centered at 470 nm and, therefore, without taking full advantage of the red-shifted rhodopsin. For VT termination, both groups chose to illuminate relatively wide regions of the ventricular myocardium (143 mm² for mice and 125 mm² for rats), but with different illumination intensity. Bruegmann et al. (2016) employed a protocol of four light pulses lasting 1 second each with intensity of 0.4 mW/mm², while Nyns et al. (2016) used a protocol consisting of one single light pulse of 1 second with intensity of 2.97 mW/mm². In both cases, an optogenetic defibrillation efficacy of ~100 % was achieved. The remarkably higher light intensity needed in the work by Nyns et al. might be due to the one-pulse protocol (instead of four) to the different animal model or to the out-of-peak ReaChR excitation. Additionally, the work by Bruegmann et al. (2016) explored the possibility of optogenetic defibrillation in human hearts with an elegant *in silico* approach. They established important requirements for human hearts: 1) the necessity of transmural penetration realizable with red light, and 2) the 5-fold increase of light intensity for pericardium defibrillation.

The work by Crocini et al. (2016c) explored, instead, optogenetics as a tool for designing novel strategies for defibrillation (Crocini et al., 2016b, 2016c, 2016d). In order to make a direct comparison with currently employed electrical leads of implantable cardioverter defibrillators (ICDs), the authors deliberately chose to use ChR2-(H134R) expressed via transgenesis in mice and used advanced imaging to learn how VTs develop in the mouse heart. In fact, a wide-field mesoscope, operating at 2,000 frames per second, was developed to map the action potential propagation in Langendorff horizontally perfused mouse hearts loaded with a red-shifted VSD (di-4-ANBDQPQ) (Matiukas et al., 2007). Acousto-optical deflectors (AOD) (see Section 15.4.2) (Reddy and Saggau, 2005; Prigge et al., 2012; Sacconi et al., 2012; Crocini et al., 2014a, 2016b, 2016e), were used to draw arbitrary ChR2 stimulation patterns. As a result, the optical toolkit proposed by Crocini et al., allows for full optical control at sub-ms temporal resolution and enables the assessment of cardiac arrhythmias and novel optogenetic defibrillation strategies in whole isolated mouse hearts. VTs were induced by reducing glucose and oxygen to mimic ischemic conditions, and the wide-field mesoscope was used to study the features of the induced functional re-entry circuits. Consistently with conduction velocity and dimensions of mouse hearts, monomorphic VTs were generated, as a re-entrant circuit could self-sustain only by invading the entire left ventricular free wall. Based on optical mapping studies and thanks to the full-optical control provided by the imaging system, the authors designed optogenetic protocols to terminate VTs based on the re-entry mechanisms and compared those with a generalized intervention (whole-heart illumination). Based on the kinetics of ChR2, the maximum capture rate of ChR2 stimulation in myocardium is 5–15 Hz (Nussinovitch and Gepstein, 2015; Zaglia et al., 2015; Crocini et al., 2016c), and defibrillation protocols featured 10 pulses of illumination lasting 5 or 10 ms each at 10 Hz. The authors demonstrated that three lines of conduction block perpendicular to the re-entry circuit (*triple barrier*) interrupted VTs (see Figure 15.4) as effectively as whole-heart illumination but with far lower energy requirements

than the works by Bruegmann et al., (2016) and Nyns et al. (2016) Most importantly, the authors demonstrated that the cardioversion efficiency was stringently dependent on the design based on the specific mechanism of VTs. In fact, by using three lines with the same total irradiation energy, but positioned regardless of the re-entry wavefront, defibrillation efficacy dropped dramatically. This result redefines the concept that a critical mass of myocardium is always needed for defibrillation (Zipes et al., 1975).

Despite its exciting potential, optogenetic applications have a long way to go before they are used in the clinic. Gene therapy in humans entails evident critical elements. The safety of ChR2 expression has been repeatedly demonstrated in mammalian cardiomyocytes (Bruegmann et al., 2010; Ambrosi and Entcheva, 2014; Vogt et al., 2015; Klimas et al., 2016), and it does not represent an issue for cardiac function, but targeting cardiac tissue instead could be quite challenging. Even though primordial human cardiac progenitor cells have been recently identified (Bu et al., 2009), the regenerative potential of the heart is extremely low, and thus cardiac gene therapy must be performed on terminally differentiated cells. Cardiomyocytes represent about 20% (2–3 billion) of the total number of cells in the human heart, but 70–80% of the entire cardiac mass (Dow et al., 1981). Taking advantage of the tight electrical coupling of cardiac cells, gene therapy could be effective even if just a portion of cardiomyocytes is expressing ChR2. However, it is still an enormous number of cells to be genetically manipulated. Gene delivery vehicles belong to two main categories: nonviral or viral gene delivery vectors. The safety profile and minimum immunogenicity are attractive features of the nonviral vehicles, but because of their low transfection efficiency, nonviral vectors have limited *in vivo* applications. Viral vectors are efficient and represent the most concrete alternative for *in vivo* human therapy. However, there are many unresolved issues related to viral transduction in humans, mainly because of the pathogenicity of viral vectors and the uncertain long-lasting effects of transduction.

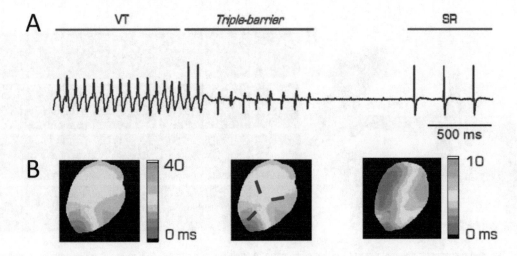

FIGURE 15.4 Optogenetic defibrillation of a mouse heart. A. Example of ECG recording showing a cardioversion. After the onset of the ventricular tachycardia (VT), the triple-barrier pattern was applied for 1 s at 10 Hz. The heart stops showing VT and restores its own sinus rhythm. B. Activation maps relative to the ECG recording reported in Panel A. Activation map depicting the VT (left), the triple-barrier intervention on the arrhythmic heart (middle), and the restored sinus rhythm (right). Modified from Crocini et al. (2016c).

The safest option for human clinical applications is represented by AAVs, which are a diverse collection of nonpathogenic, naturally replication-deficient parvoviruses (Blacklow et al., 1967). In fact, gene therapy via AAV delivery has been already attempted in human hearts (Jaski et al., 2009; Jessup et al., 2011; Greenberg et al., 2016) and may allow for effective ChR2 expression, too. Another infrequently considered issue is how the light is going to be delivered in humans. In fact, the device would need to be an integrated, multifunctional, biocompatible, and elastic system, custom-formed to match the shape of each patient heart. It would need to address minimally invasive access, motion artifacts, and photon scattering/absorption as well as provide optogenetic therapy with spatiotemporal resolution of physiological parameters for feedback control. In animal models, conformable devices that can wrap the heart have been already tested (Xu et al., 2014; Roche et al., 2017). Particularly, Xu et al. have proposed to shape 3D elastic membranes to precisely match the epicardium of the heart. This would allow the membrane to completely envelop the heart in a form fitting manner and support for deformable arrays of multifunctional sensors and actuators, including optical stimulators. Implantation of such devices in humans would have an enormous impact, as infections, coagulation, and repairs are occasionally needed in patients with implanted devices (Joy et al., 2017), hence the complete and rapid removal of those devices is crucial in the surgery room (Margey et al., 2010).

15.3.3 All Optical Platform to Probe and Control Heart Activity

The complex spatiotemporal dynamics of excitable cells require high-resolution actuation in combination with optical mapping (reviewed by (Entcheva and Bub, 2016; Crocini et al., 2017)). To date, the cardiac research community has produced a limited number of works that employ advanced optical methods for both stimulation and detection (Arrenberg et al.,

2010; Mickoleit et al., 2014; Burton et al., 2015; Crocini et al., 2016c). Two features of optogenetics are of the greatest importance: the high spatiotemporal resolution and the specificity. The main advantage of the optical setup proposed by Crocini et al. over the others employed for defibrillation is indeed the high spatiotemporal resolution of the stimulation patterns. As mentioned, AODs allow for an ultrafast manipulation of light (Crocini et al., 2016c) but generate patterns serially and are complicated tools to integrate in an optical system. Recently, a high-resolution all-optical approach has been developed using computer-controlled digital micromirror device (DMD), realizing true parallel optical manipulation. (Burton et al., 2015). In this work, the authors combined dye-free optical imaging with optogenetic actuation to achieve dynamic control of cardiac excitation waves. The dye-free imaging system is based on off-axis oblique illumination, a common technique for enhancing contrast in microscopic samples. Wave propagation was imaged with high contrast and high resolution, relying on changes in the optical properties of the tissue induced by cellular contraction. ChR2-(H134R) expression in cardiomyocytes was achieved by adenoviral infection. Another important achievement in pattern designing has been realized recently in the form of real-time closed-loop feedback. In fact, even though Crocini and colleagues have successfully designed mechanistically based patterns for defibrillation, the authors relied on the knowledge acquired from optical mapping recordings of several VTs. This offline strategy was possible because mouse hearts exhibit stereotypical monomorphic VTs, comprising the whole left ventricle. As also shown by Nyns et al. (2016), larger animals would likely display more complex VTs that would need fine-tuned interventions. Although a real-time stimulation approach requires an enormous effort in terms of engineering and programming, it represents a completely new method for the investigation of arrhythmia and will provide fundamental insights in cardiac diseases. Recently, Scardigli et al. (2018b) developed an all-optical platform (see Figure 15.5) to monitor and control electrical activity in real-time.

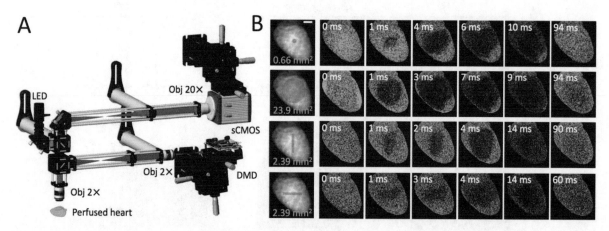

FIGURE 15.5 Targeted optogenetic manipulation of cardiac conduction. A. Scheme of a wide-field fluorescence mesoscope. A commercial light steering solution based on a digital micromirror device (DMD) is coupled with the mesoscope using a high numerical aperture relay system and a dichroic beam splitter. B. Optical mapping during four patterns of optogenetic stimulation: single-point, whole ventricle, vertical line, and horizontal line, designed as reported by the blue traits on the fluorescence image (F_0) of the heart. Six representative frames of optical mapping ($\Delta F/F_0$) showing the electrical activation in red and the baseline in cyan. Scale bar of 2 mm in white. Modified from Scardigli et al. (2018b).

15.4 Nonlinear Imaging

15.4.1 Two-Photon Fluorescence Microscopy

A significant implementation in laser scanning microscopy (such as confocal microscopy) is introduced by nonlinear microscopy. The most used nonlinear microscopy technique in biology is two-photon excitation (TPE) microscopy. TPE provides 3D optical sectioning without absorption (which would lead to photobleaching and phototoxicity) above and below the plane of focus. The phenomenon of TPE arises from the quasisimultaneous absorption of two photons in a single quantized event (within $\approx 10^{-16}$s). Because the energy of a photon is inversely proportional to its wavelength, the two absorbed photons must have a wavelength about twice that required for one-photon excitation. If another photon strikes within this time window, the energy of the two photons can be summed to allow an energy state transition. Halving the energy by two implies that λ must be doubled (from the photon energy equation), and, therefore, the wavelengths used are redder compared to single photon excitation (less phototoxic events). Moreover, the multiple-photons requirement for fluorescence excitation implies that the generated fluorescence depends nonlinearly on the number of photons per time and cm^2 i.e. $P \propto In$, where P is the probability, I is the intensity of the photon flux, and n is the number of photons required for excitation. Thus, for TPE: $P \propto I2$. This means that to generate sufficient signal, the excitation light has to be concentrated in space and time. Concentration in the time domain requires the use of lasers that emit short pulses (typically 100 fs at a repetition rate of 80–100 MHz) with high peak intensities. Then, by focusing a laser beam through a high numerical aperture (NA) objective, it is possible to spatially limit TPE probability to the focal plane (Denk et al., 1990). Away from the focus, TPE probability drops off rapidly so that, elsewhere in the sample, TPE is a very unlikely event. The localization of the excitation volume is the greatest advantage of nonlinear microscopy compared to conventional microscopy techniques. In fact, such localization is also maintained in scattering tissues, as the excitation photons diffused by the sample do not produce a density of energy that induces a nonlinear transition. Furthermore, the near-infrared light used in TPE penetrates deeper into biological samples with respect to the visible light generally used in confocal microscopy. For all of these advantages, TPE is used for experiments that require deep penetration into living tissue or intact animal specimens in preference to other techniques. Imaging methods can be coupled with different scanning systems, depending on the phenomenon of interest. One of these methods is random access microscopy, which is suitable for fast dynamic phenomena.

15.4.2 Functional Imaging at Subcellular Level

The major limiting factor of nonlinear microscopes is the scanning time. For this reason, the optical recording of fast physiological events is possible only in a single position of the cell by using a line scanning procedure. In principle, the optical measurement of time-dependent processes does not require the production of images at all. Instead, more time could be spent collecting as many photons as possible from selective positions,

where the image plane intersects the biological objects of interest. Using this approach, fast physiological processes, such as action potentials in cardiac cells, can be recorded at a sampling frequency of more than 1 kHz. This cannot be achieved with a standard galvanometer mirror because about 1 ms is required to reach and stabilize a new position. Scanning a set of points within a plane at high speed is possible with two orthogonal AODs. In an AOD, a propagating ultrasonic wave establishes a grating that diffracts a laser beam at a precise angle that can be changed within a few microseconds. For that reason, this scanning modality is called *random access*. In particular, AODs consist of a crystal attached to a piezo-electrical transducer. An oscillating electric signal drives the transducer to vibrate, which creates sound waves in the crystal (usually of radio frequency), and this generates a refractive index wave that behaves like a sinusoidal grating. An incident laser beam passing through this grating will be diffracted into several orders. With an appropriate design, the first order beam has the highest efficiency. It predicts that the angular position is linearly proportional to the acoustic frequency, so the higher the frequency the larger the diffracted angle. Based on the velocity of sound propagation in the crystal and the dimension of the laser beam, the commutation time can be estimated around 5 µs for multisite measurements in multiplexed modality. This scanning modality can be applied to standard fluorescence microscopy, single photon, or multiple photons excitation. A random access two photon microscope has been recently developed to study multiple subcellular sites simultaneously within isolated cardiomyocytes (Sacconi et al., 2012). Cardiomyocytes are characterized by an extensive system of deep invaginations of the cell membrane called *T-tubules*. T-tubules mediate the rapid spread of the action potential within the cell, from the surface to the cell core. As a result, Ca^{2+} release from the sarcoplasmic reticulum is triggered synchronously throughout the cell. Then, calcium ions bind proteins of the sarcomere, initiating the contraction. Given the limited accessibility of T-tubules within cardiomyocytes, the function of these subcellular structures has been deducted indirectly for a long time. Light-based imaging technologies have opened countless opportunities to assess T-tubules' role in cardiac physiology and pathology (Crocini et al., 2014a). In 2012, the action potential propagation has been measured at single T-tubular level (see Figure 15.6), unveiling the presence of electrical abnormalities in a rat model of heart failure (Sacconi et al., 2012).

Specifically, isolated cardiomyocytes were labeled with a fluorinated voltage sensitive dye, and action potentials were measured simultaneously in multiple T-tubules using a random access two-photon microscope. This work has demonstrated that, even though a cardiomyocyte exhibits action potentials, single T-tubules may fail to propagate the electrical activation in diseases. Given the high number of T-tubules in each cardiomyocyte, one might think that the consequences of single electrical failure on cellular function are small. The impact of local T-tubular electrical abnormalities has been investigated in a follow-up work that employed an enhanced random access microscope to simultaneously record action potential and Ca^{2+} release (Crocini et al., 2014b). The presence of electrically failing T-tubules dramatically compromises local Ca^{2+} release, generating sites of slower and faster activation. This result clearly

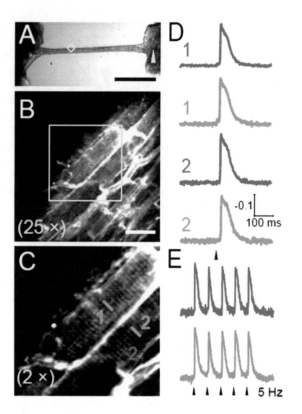

FIGURE 15.6 Action potential propagation intact cardiac sample. A. Bright-field image of a rat ventricular trabecula. The yellow arrowhead marks the stimulation site and the yellow diamond encompasses the recording area. (Scale bar = 1 mm.) B. TPF image of the area highlighted in yellow in Panel A (Scale bar = 20 μm.) C. The region in the yellow box of B. shows two adjacent myocytes magnified. D. Normalized fluorescence traces from the scanned lines indicated in C. Action potentials are elicited in correspondence to the black arrowhead. E. Normalized fluorescence traces recorded from SS and TT in cell 1 during high-frequency stimulation. Modified from Sacconi et al. (2012).

highlights the importance of single T-tubular elements in determining local function within cardiomyocytes. Additionally, electrically failing T-tubules can blunt β-adrenergic signaling in heart failure (Crocini et al., 2016a) and are found in other animal models, such as hypertrophic cardiomyopathy with compromised cardiac contraction (Crocini et al., 2016e) or spontaneously hypertensive rats (Scardigli et al., 2017, 2018a). Multisite optical recordings of action potential have also been performed in multicellular preparations like ventricular trabeculae in which cell-to-cell conduction occurs through gap junctions (Sacconi et al., 2012). These experiments have confirmed the tight electrical coupling between surface membrane and T-tubules in a tissue context as well as between different cells within trabeculae in healthy preparation. Subcellular and cellular investigations can inform on the cardiac function at the organ level and should be integrated with whole heart imaging assessments.

15.4.3 Intravital Imaging of Cardiac Function at the Single-Cell Level

Combining Langendorff's preparations with fluorescent dyes and two-photon imaging allows for deeper high-resolution imaging in intact hearts. Such approaches can be applied for both voltage and calcium imaging as reported by the group led by Prof. Godfrey Smith at the University of Glasgow (Ghouri et al., 2015). They measured voltage and intracellular calcium in intact heart preparations by ratiometric two-photon microscopy in a near simultaneous fashion. Such an approach is particularly interesting because it allows for subepicardial measurements (Kelly et al., 2013). In fact, action potential shape and propagation differ within the cardiac muscle. Action potential rise time within the myocardium is significantly slower than at the epicardial surface, an effect accentuated at higher pacing rates. Accessing information deeper in cardiac tissue is extremely interesting to better understand cardiac electrophysiology in healthy and diseased hearts as well as to study drug efficacy. To perform this kind of imaging experiment, uncoupling contraction from excitation and Ca^{2+} release is often necessary to avoid beating artifacts. Recently, an intravital optical microscope that compensates for motion of the contracting heart has been developed (see Figure 15.7) allowing for measurement of contractile function at the single-cell level (Lee et al., 2012; Aguirre et al., 2014; Vinegoni et al., 2015). There are multiple challenges in measuring single cells in the beating heart with microscopy, but it is a powerful way to study individual cardiomyocyte physiology and cell-to-cell heterogeneity. This method enables quantitative analysis of dynamic events in the myocardium and microvasculature, including arrhythmia, calcium signaling, leukocyte trafficking (Li et al., 2012), cell-to-cell communication, and investigation of cellular responses to injury and pharmacologic intervention.

15.5 Whole Heart Imaging in Zebrafish Larva

15.5.1 Single Plane Illumination Microscopy

Optical sectioning in light sheet microscopy is achieved by confining illumination to the focal plane of the objective lens. In particular, the optical paths of excitation and detection light are decoupled, guaranteeing optical sectioning also with low-NA optics. The sample is illuminated with a thin sheet of light, and the fluorescence emission is observed from an axis perpendicular to the illumination plane (see Figure 15.8). Confinement of excitation to fluorophores, which are actually observed, dramatically reduces photobleaching with comparison to confocal microscopy (up to three orders of magnitude (Keller et al., 2008)). As the detection architecture is that of wide-field microscopy, light sheet illumination (LSI) can afford the same frame rate, ensuring optical sectioning over large fields of view.

The simpler strategy to shape a laser beam into a sheet of light is to put a cylindrical lens in its path (Voie et al., 1993; Huisken et al., 2004). In this way, the beam, which is supposed to be axially symmetric before entering the lens, is focused only in one direction. The shape of the beam at the focal plane is thus a highly elongated ellipse, which is a reasonable approximation of a light sheet. To obtain a more homogeneous light sheet, a symmetric Gaussian beam can be scanned inside the specimen (Keller et al., 2008). While the radial resolution of an LSI microscope is the same as a wide-field microscope,

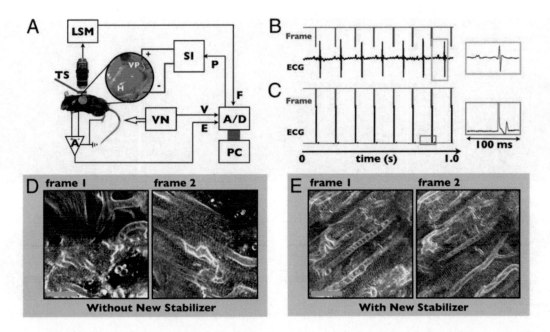

FIGURE 15.7 Real-time prospective cardiac gating for intravital microscopy. A. System schematic. Laser scanning microscope (LSM) acquisition is synchronized to the cardiac cycle using pacing. B. and C. Timing waveforms demonstrate asynchronous acquisition B. and synchronization of the acquisition with the cardiac cycle using pacing C. The insets illustrate ventricular capture with widening of the paced electrocardiogram (ECG) complex C., compared with the native ECG B. D. and E. Sequential two-photon microscopy frames without stabilizer D. have significant frame-to-frame variation, while the image stabilizer allows tracking of individual myocytes E. Modified from Aguirre et al. (2014).

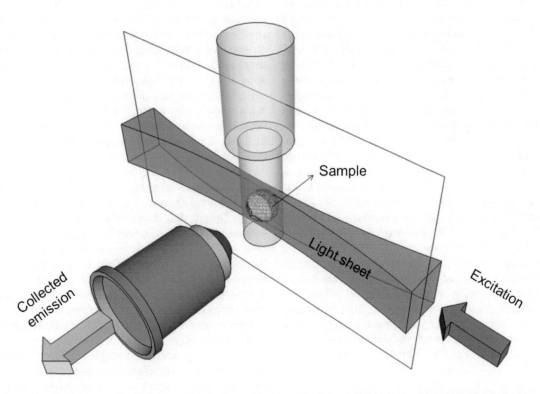

FIGURE 15.8 Simplified scheme of a light sheet illumination microscope. The light sheet lies in the focal plane of the detection objective. Fluorescence and excitation light are indicated in green and blue, respectively. From Olarte et al. (2018).

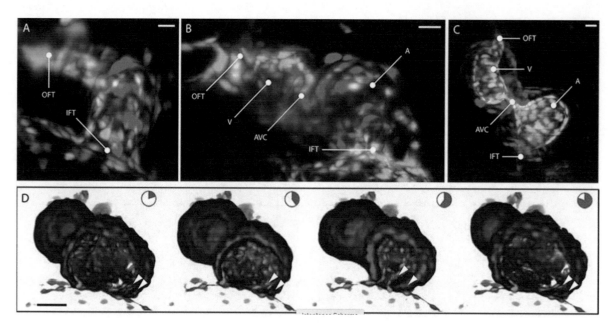

FIGURE 15.9   3D recordings of the embryonic zebrafish heart. (a–c) Maximum intensity projections of image stacks recorded from hearts of 1–3 dpf transgenic Tg(kdrl:eGFP,myl7:dsRed, gata1a:dsRed) zebrafish. Red, myocardium; cyan, vasculature. Inflow tract (IFT), atrium (A), atrioventricular canal (AVC), ventricle (V), outflow tract (OFT). Scale bar = 20 µm. (d) 3D rendering of synchronized movie stack cut open in the atrium to reveal inflow tract and impact of cardiac jelly on pumping efficiency (arrowheads). Red, myocardium; cyan, vasculature. Modified from Weber and Huisken, (2015).

axial sectioning is instead related to the thickness of the light sheet, which, in turn, depends on the characteristics of illumination optics. Although LSI was introduced in microscopy more than a century ago, it has been applied to fluorescence imaging of biological specimens only in recent years. After the seminal paper of Voie et al. (1993), describing a technique called orthogonal-plane fluorescence optical sectioning (OPFOS), light sheet fluorescence microscopy has known a true renaissance, with the development of many techniques as selective plane illumination microscopy (SPIM) (Huisken et al., 2004), ultramicroscopy (Dodt et al., 2007), multidirectional SPIM (mSPIM) (Huisken and Stainier, 2007), multiple-views SPIM (Verveer et al., 2007), digital scanned laser light sheet microscopy (DSLM) (Keller et al., 2008), oblique plane microscopy (OPM) (Dunsby, 2008), objective coupled planar illumination microscopy (OCPI) (Holekamp et al., 2008), highly inclined and laminated optical sheet microscopy (HILO) (Tokunaga et al., 2008), and swept confocally aligned planar excitation (SCAPE) microscopy (Bouchard et al., 2015). The flourishing of laboratories involved in the development of such methods is justified by the inherent advantages of light sheet illumination among the other optical sectioning approaches: intrinsic optical sectioning, high frame rate, high SNR, and low phototoxicity and photobleaching. These features of LSI make it the ideal candidate to image organs inside transparent specimen like zebrafish embryos.

15.5.2 High-Resolution Reconstruction of the Beating Zebrafish Heart

The zebrafish (*Danio rerio*) is a particularly powerful animal model to study development and organogenesis. In fact, zebrafish embryos are accessible for light microscopy studies

from the earliest stages. The cardiac development of the zebrafish shows remarkable similarities to that of humans. Less than 2 days after fertilization, the zebrafish heart develops from progenitor cells into a two-chambered organ that sits dorsally between the head and trunk. The two chambers, atrium and ventricle, are connected by the atrioventricular canal and consist of two main cell layers: the inner endocardium and the outer myocardium. A double-walled sac, the pericardium, contains the heart and the roots of the great vessels while also fixing the heart to the thorax, providing lubrication and protecting against infection. In contrast to those of mammals and amphibians, the zebrafish heart does not progress to septation and retains a simpler structure. However, there are broad similarities between zebrafish and mammals with respect to the genetic determinants of heart tube and chamber formation (Lawson and Weinstein, 2002). Despite the comparatively simple structure of the zebrafish heart, its electrocardiogram and the overall shape of its action potentials are very similar to those of mammalian hearts (Nemtsas et al., 2010). Overall, the zebrafish has unique features that make it attractive and complementary to existing model systems. The instantaneous optical sectioning capabilities of light sheet microscopy are particularly valuable for *in vivo* imaging of zebrafish hearts. Selected regions of the heart can be captured *in vivo* with great detail using light sheet microscopy. Although high-speed movies of a single plane are sufficient to describe and quantify numerous cardiac properties, they lack the depth information needed to reconstruct the entire beating heart.

However, its continuous motion in all three dimensions makes it challenging to record high-resolution images of the entire organ. One option could be to perform a postacquisition synchronization (Taylor et al., 2011, 2012; Taylor, 2014). The light sheet microscope captures a z-stack of movies, with each movie covering at least one cardiac cycle (see Figure 15.9). After the recording is finished, one 3D cardiac cycle is reconstructed by synchronizing the movies in time (Mickoleit et al., 2014). In this fashion, the complex movement of the myocardium during systole and diastole and the propagation of cardiac contraction from inflow to outflow tract are visualized by tracking individual heart cells. Recently, Sacconi et al. (2020) have introduced a multi-plane imaging method that measures membrane voltage in tissue volumes on a millisecond time scale. The microscope was used to measure action potential propagation through a zebrafish heart in three dimensions, which is something that has not been previously achieved using any experimental method.

REFERENCES

Aguirre AD, Vinegoni C, Sebas M, Weissleder R (2014) Intravital imaging of cardiac function at the single-cell level. *Proceedings of the National Academy of Sciences of the United States of America* 111:11257–11262.

Ambrosi CM, Entcheva E (2014) Optogenetic control of cardiomyocytes via viral delivery. *Methods in Molecular Biology* 1181:215–228.

Arrenberg AB, Stainier DY, Baier H, Huisken J (2010) Optogenetic control of cardiac function. *Science* 330:971–974.

Bamann C, Gueta R, Kleinlogel S, Nagel G, Bamberg E (2010) Structural guidance of the photocycle of channelrhodopsin-2 by an interhelical hydrogen bond. *Biochemistry* 49:267–278.

Berndt A et al. (2011) High-efficiency channelrhodopsins for fast neuronal stimulation at low light levels. *Proceedings of the National Academy of Sciences of the United States of America* 108:7595–7600.

Blacklow NR, Hoggan MD, Rowe WP (1967) Isolation of adenovirus-associated viruses from man. *Proceedings of the National Academy of Sciences of the United States of America* 58:1410–1415.

Bouchard MB et al. (2015) Swept confocally-aligned planar excitation (SCAPE) microscopy for high speed volumetric imaging of behaving organisms. *Nature Photonics* 9:113–119.

Bruegmann T et al. (2010) Optogenetic control of heart muscle in vitro and in vivo. *Nature Methods* 7:897–900.

Bruegmann T et al. (2016) Optogenetic defibrillation terminates ventricular arrhythmia in mouse hearts and human simulations. *The Journal of Clinical Investigation* 126:3894–3904.

Bu L et al. (2009) Human ISL1 heart progenitors generate diverse multipotent cardiovascular cell lineages. *Nature* 460:113–117.

Burkhoff D, Mirsky I, Suga H (2005) Assessment of systolic and diastolic ventricular properties via pressure-volume analysis: A guide for clinical, translational, and basic researchers. *American Journal of Physiology Heart and Circulatory Physiology* 289:H501–512.

Burton RA et al. (2015) Optical control of excitation waves in cardiac tissue. *Nature Photonics* 9:813–816.

Chow BY et al. (2010) High-performance genetically targetable optical neural silencing by light-driven proton pumps. *Nature* 463:98–102.

Coatney RW (2001) Ultrasound imaging: Principles and applications in rodent research. *ILAR Journal* 42:233–247.

Collins KA et al. (2001) Accuracy of echocardiographic estimates of left ventricular mass in mice. *American Journal of Physiology Heart and Circulatory Physiology* 280:H1954–1962.

Crocini C, Coppini R, Ferrantini C, Pavone FS, Sacconi L (2014a) Functional cardiac imaging by random access microscopy. *Frontiers in Physiology* 5:403.

Crocini C, Ferrantini C, Pavone FS, Sacconi L (2017) Optogenetics gets to the heart: A guiding light beyond defibrillation. *Progress in Biophysics and Molecular Biology* 130:132–139.

Crocini C et al. (2014b) Defects in T-tubular electrical activity underlie local alterations of calcium release in heart failure. *Proceedings of the National Academy of Sciences of the United States of America* 111:15196–15201.

Crocini C et al. (2016a) T-Tubular electrical defects contribute to blunted beta-adrenergic response in heart failure. *International Journal of Molecular Sciences* 17(9):1471.

Crocini C et al. (2016b) Optogenetics to rethink ICD technology. *European Heart Journal* 37:696–696.

Crocini C et al. (2016c) Optogenetics design of mechanistically-based stimulation patterns for cardiac defibrillation. *Scientific Reports* 6:35628.

Crocini C et al. (2016d) Optical treatment of cardiac arrhythmias. *Cardiovascular Research* 111:S90–S91.

Crocini C et al. (2016e) Novel insights on the relationship between T-tubular defects and contractile dysfunction in a mouse model of hypertrophic cardiomyopathy. *Journal of Molecular and Cellular Cardiology* 91:42–51.

Deisseroth K (2015) Optogenetics: 10 years of microbial opsins in neuroscience. *Nature Neuroscience* 18:1213–1225.

Denk W, Strickler JH, Webb WW (1990) Two-photon laser scanning fluorescence microscopy. *Science* 248:73–76.

Dodt HU et al. (2007) Ultramicroscopy: Three-dimensional visualization of neuronal networks in the whole mouse brain. *Nature Methods* 4:331–336.

Dow JW, Harding NG, Powell T (1981) Isolated cardiac myocytes. I. Preparation of adult myocytes and their homology with the intact tissue. *Cardiovascular Research* 15:483–514.

Dunsby C (2008) Optically sectioned imaging by oblique plane microscopy. *Optics Express* 16:20306–20316.

Entcheva E, Bub G (2016) All-optical control of cardiac excitation: Combined high-resolution optogenetic actuation and optical mapping. *The Journal of Physiology* 594:2503–2510.

Gardin JM, Siri FM, Kitsis RN, Edwards JG, Leinwand LA (1995) Echocardiographic assessment of left ventricular mass and systolic function in mice. *Circulation Research* 76:907–914.

Georgakopoulos D et al. (1998) In vivo murine left ventricular pressure-volume relations by miniaturized conductance micromanometry. *The American Journal of Physiology* 274:H1416–1422.

Ghouri IA, Kelly A, Burton FL, Smith GL, Kemi OJ (2015) 2-Photon excitation fluorescence microscopy enables deeper high-resolution imaging of voltage and Ca(2+) in intact mice, rat, and rabbit hearts. *Journal of Biophotonics* 8:112–123.

Gloschat C et al. (2018) RHYTHM: An open source imaging toolkit for cardiac panoramic optical mapping. *Scientific Reports* 8:2921.

Govorunova EG, Sineshchekov OA, Janz R, Liu X, Spudich JL (2015) NEUROSCIENCE. Natural light-gated anion channels: A family of microbial rhodopsins for advanced optogenetics. *Science* 349:647–650.

Govorunova EG, Sineshchekov OA, Li H, Janz R, Spudich JL (2013) Characterization of a highly efficient blue-shifted channelrhodopsin from the marine alga Platymonas subcordiformis. *The Journal of Biological Chemistry* 288:29911–29922.

Govorunova EG et al. (2017) The expanding family of natural anion channelrhodopsins reveals large variations in kinetics, conductance, and spectral sensitivity. *Scientific Reports* 7:43358.

Greenberg B et al. (2016) Calcium upregulation by percutaneous administration of gene therapy in patients with cardiac disease (CUPID 2): A randomised, multinational, double-blind, placebo-controlled, phase 2b trial. *Lancet* 387:1178–1186.

Holekamp TF, Turaga D, Holy TE (2008) Fast three-dimensional fluorescence imaging of activity in neural populations by objective-coupled planar illumination microscopy. *Neuron* 57:661–672.

Huisken J, Stainier DY (2007) Even fluorescence excitation by multidirectional selective plane illumination microscopy (mSPIM). *Optics Letters* 32:2608–2610.

Huisken J, Swoger J, Del Bene F, Wittbrodt J, Stelzer EH (2004) Optical sectioning deep inside live embryos by selective plane illumination microscopy. *Science* 305:1007–1009.

Jaski BE et al. (2009) Calcium upregulation by percutaneous administration of gene therapy in cardiac disease (CUPID Trial), a first-in-human phase 1/2 clinical trial. *Journal of Cardiac Failure* 15:171–181.

Jessup M et al. (2011) Calcium upregulation by percutaneous administration of gene therapy in cardiac disease (CUPID): A phase 2 trial of intracoronary gene therapy of sarcoplasmic reticulum Ca2+-ATPase in patients with advanced heart failure. *Circulation* 124:304–313.

Joy PS, Kumar G, Poole JE, London B, Olshansky B (2017) Cardiac implantable electronic device infections: Who is at greatest risk? *Heart Rhythm* 14(6):839–45.

Kass DA (1992) Clinical evaluation of left heart function by conductance catheter technique. *European Heart Journal* 13 (Suppl E):57–64.

Keller PJ, Schmidt AD, Wittbrodt J, Stelzer EH (2008) Reconstruction of zebrafish early embryonic development by scanned light sheet microscopy. *Science* 322:1065–1069.

Kelly A et al. (2013) Subepicardial action potential characteristics are a function of depth and activation sequence in isolated rabbit hearts. *Circulation Arrhythmia and Electrophysiology* 6:809–817.

Kiatchoosakun S, Restivo J, Kirkpatrick D, Hoit BD (2002) Assessment of left ventricular mass in mice: Comparison between two-dimensional and m-mode echocardiography. *Echocardiography* 19:199–205.

Klapoetke NC et al. (2014) Independent optical excitation of distinct neural populations. *Nature Methods* 11:338–346.

Kleinlogel S et al. (2011) Ultra light-sensitive and fast neuronal activation with the Ca(2)+-permeable channelrhodopsin CatCh. *Nature Neuroscience* 14:513–518.

Klimas A, Ambrosi CM, Yu J, Williams JC, Bien H, Entcheva E (2016) OptoDyCE as an automated system for high-throughput all-optical dynamic cardiac electrophysiology. *Nature Communications* 7:11542.

Lawson ND, Weinstein BM (2002) In vivo imaging of embryonic vascular development using transgenic zebrafish. *Developmental Biology* 248:307–318.

Lee P, Bollensdorff C, Quinn TA, Wuskell JP, Loew LM, Kohl P (2011) Single-sensor system for spatially resolved, continuous, and multiparametric optical mapping of cardiac tissue. *Heart Rhythm* 8:1482–1491.

Lee P et al. (2017) Low-cost optical mapping systems for panoramic imaging of complex arrhythmias and drug-action in translational heart models. *Scientific Reports* 7:43217.

Lee S et al. (2012) Real-time in vivo imaging of the beating mouse heart at microscopic resolution. *Nature Communications* 3:1054.

Li W et al. (2012) Intravital 2-photon imaging of leukocyte trafficking in beating heart. *The Journal of Clinical Investigation* 122:2499–2508.

Lin JY, Lin MZ, Steinbach P, Tsien RY (2009) Characterization of engineered channelrhodopsin variants with improved properties and kinetics. *Biophysical Journal* 96:1803–1814.

Lin JY, Knutsen PM, Muller A, Kleinfeld D, Tsien RY (2013) ReaChR: A red-shifted variant of channelrhodopsin enables deep transcranial optogenetic excitation. *Nature Neuroscience* 16:1499–1508.

Lin SF, Wikswo JP (1999) Panoramic optical imaging of electrical propagation in isolated heart. *Journal of Biomedical Optics* 4:200–207.

Margey R et al. (2010) Contemporary management of and outcomes from cardiac device related infections. *Europace: European Pacing, Arrhythmias, and Cardiac Electrophysiology: Journal of the Working Groups Cardiac Pacing, Arrhythmias, and Cardiac Cellular Electrophysiology of the European Society of Cardiology* 12:64–70.

Matiukas A et al. (2007) Near-infrared voltage-sensitive fluorescent dyes optimized for optical mapping in blood-perfused myocardium. *Heart Rhythm* 4:1441–1451.

Merx MW, Schrader Jp- (2009) *The Working Heart*. Springer Berlin Heidelberg Practical Methods in Cardiovascular Research, pp. 173–189.

Mickoleit M et al. (2014) High-resolution reconstruction of the beating zebrafish heart. *Nature Methods* 11:919–922.

Miesenbock G (2009) The optogenetic catechism. *Science* 326:395–399.

Nagel G, Brauner M, Liewald JF, Adeishvili N, Bamberg E, Gottschalk A (2005) Light activation of channelrhodopsin-2 in excitable cells of *Caenorhabditis elegans* triggers rapid behavioral responses. *Current Biology CB* 15:2279–2284.

Nagel G et al. (2003) Channelrhodopsin-2, a directly light-gated cation-selective membrane channel. *Proceedings of the National Academy of Sciences of the United States of America* 100:13940–13945.

Neely JR, Liebermeister H, Battersby EJ, Morgan HE (1967) Effect of pressure development on oxygen consumption by isolated rat heart. *The American Journal of Physiology* 212:804–814.

Nemtsas P, Wettwer E, Christ T, Weidinger G, Ravens U (2010) Adult zebrafish heart as a model for human heart? An electrophysiological study. *Journal of Molecular and Cellular Cardiology* 48:161–171.

Nussinovitch U, Gepstein L (2015) Optogenetics for in vivo cardiac pacing and resynchronization therapies. *Nature Biotechnology* 33:750–754.

Nyns EC et al. (2016) Optogenetic termination of ventricular arrhythmias in the whole heart: Towards biological cardiac rhythm management. *European Heart Journal*.

Odley A et al. (2004) Regulation of cardiac contractility by Rab4-modulated beta2-adrenergic receptor recycling. *Proceedings of the National Academy of Sciences of the United States of America* 101:7082–7087.

Olarte OE, Andilla J, Gualda EJ, Loza-Alvarez P (2018) Light-sheet microscopy: A tutorial. *Advances in Optics and Photonics* 10:111–179.

Pan ZH, Ganjawala TH, Lu Q, Ivanova E, Zhang Z (2014) ChR2 mutants at L132 and T159 with improved operational light sensitivity for vision restoration. *PloS One* 9:e98924.

Park SA, Lee SR, Tung L, Yue DT (2014) Optical mapping of optogenetically shaped cardiac action potentials. *Scientific Reports* 4:6125.

Prigge M et al. (2012) Color-tuned channelrhodopsins for multiwavelength optogenetics. *The Journal of Biological Chemistry* 287:31804–31812.

Qu F, Ripplinger CM, Nikolski VP, Grimm C, Efimov IR (2007) Three-dimensional panoramic imaging of cardiac arrhythmias in rabbit heart. *Journal of Biomedical Optics* 12:044019.

Reddy GD, Saggau P (2005) Fast three-dimensional laser scanning scheme using acousto-optic deflectors. *Journal of Biomedical Optics* 10:064038.

Roche ET et al. (2017) Soft robotic sleeve supports heart function. *Science Translational Medicine* 9(373):eaaf3925.

Sacconi L et al. (2012) Action potential propagation in transverse-axial tubular system is impaired in heart failure. *Proceedings of the National Academy of Sciences of the United States of America* 109:5815–5819.

Sacconi L., et al. (2020) KHz-rate volumetric voltage imaging of the whole zebrafish heart. *bioRxiv* 2020.07.13.196063.

Scardigli M, Ferrantini C, Crocini C, Pavone FS, Sacconi L (2018a) Interplay between sub-cellular alterations of calcium release and T-Tubular defects in cardiac diseases. *Frontiers in Physiology* 9:1474.

Scardigli M et al. (2017) Quantitative assessment of passive electrical properties of the cardiac T-tubular system by FRAP microscopy. *Proceedings of the National Academy of Sciences of the United States of America* 114:5737–5742.

Scardigli M et al. (2018b) Real-time optical manipulation of cardiac conduction in intact hearts. *The Journal of Physiology* 596:3841–3858.

Schneider F, Grimm C, Hegemann P (2015) Biophysics of channelrhodopsin. *Annual Review of Biophysics* 44:167–186.

Schobert B, Lanyi JK (1982) Halorhodopsin is a light-driven chloride pump. *The Journal of Biological Chemistry* 257:10306–10313.

Skrzypiec-Spring M, Grotthus B, Szelag A, Schulz R (2007) Isolated heart perfusion according to Langendorff---still viable in the new millennium. *Journal of Pharmacological and Toxicological Methods* 55:113–126.

Syed F, Diwan A, Hahn HS (2005) Murine echocardiography: A practical approach for phenotyping genetically manipulated and surgically modeled mice. *Journal of the American Society of Echocardiography: Official Publication of the American Society of Echocardiography* 18:982–990.

Taylor JM (2014) Optically gated beating-heart imaging. *Frontiers in Physiology* 5:481.

Taylor JM, Girkin JM, Love GD (2012) High-resolution 3D optical microscopy inside the beating zebrafish heart using prospective optical gating. *Biomedical Optics Express* 3:3043–3053.

Taylor JM, Saunter CD, Love GD, Girkin JM, Henderson DJ, Chaudhry B (2011) Real-time optical gating for three-dimensional beating heart imaging. *Journal of Biomedical Optics* 16:116021.

Tokunaga M, Imamoto N, Sakata-Sogawa K (2008) Highly inclined thin illumination enables clear single-molecule imaging in cells. *Nature Methods* 5:159–161.

Tsujita Y, Kato T, Sussman MA (2005) Evaluation of left ventricular function in cardiomyopathic mice by tissue Doppler and color M-mode Doppler echocardiography. *Echocardiography* 22:245–253.

Verdecchia P et al. (2001) Left ventricular mass and cardiovascular morbidity in essential hypertension: The MAVI study. *Journal of the American College of Cardiology* 38:1829–1835.

Verma A et al. (2008) Prognostic implications of left ventricular mass and geometry following myocardial infarction: The VALIANT (valsartan in acute myocardial infarction) echocardiographic study. *JACC Cardiovascular Imaging* 1:582–591.

Verveer PJ, Swoger J, Pampaloni F, Greger K, Marcello M, Stelzer EH (2007) High-resolution three-dimensional imaging of large specimens with light sheet-based microscopy. *Nature Methods* 4:311–313.

Vinegoni C, Aguirre AD, Lee S, Weissleder R (2015) Imaging the beating heart in the mouse using intravital microscopy techniques. *Nature Protocols* 10:1802–1819.

Vogt CC et al. (2015) Systemic gene transfer enables optogenetic pacing of mouse hearts. *Cardiovascular Research* 106:338–343.

Voie AH, Burns DH, Spelman FA (1993) Orthogonal-plane fluorescence optical sectioning: Three-dimensional imaging of macroscopic biological specimens. *Journal of Microscopy* 170:229–236.

Weber M, Huisken J (2015) In vivo imaging of cardiac development and function in zebrafish using light sheet microscopy. *Swiss Medical Weekly* 145:w14227.

Wietek J et al. (2014) Conversion of channelrhodopsin into a light-gated chloride channel. *Science* 344:409–412.

Xu L et al. (2014) 3D multifunctional integumentary membranes for spatiotemporal cardiac measurements and stimulation across the entire epicardium. *Nature Communications* 5:3329.

Yizhar O et al. (2011) Neocortical excitation/inhibition balance in information processing and social dysfunction. *Nature* 477:171–178.

Zaglia T et al. (2015) Optogenetic determination of the myocardial requirements for extrasystoles by cell type-specific targeting of channelrhodopsin-2. *Proceedings of the National Academy of Sciences of the United States of America* 112:E4495–4504.

Zimmer HG (1998) The isolated perfused heart and its pioneers. *News in Physiological Sciences: An International Journal of Physiology Produced Jointly by the International Union of Physiological Sciences and the American Physiological Society* 13:203–210.

Zipes DP, Fischer J, King RM, Nicoll Ad, Jolly WW (1975) Termination of ventricular fibrillation in dogs by depolarizing a critical amount of myocardium. *The American Journal of Cardiology* 36:37–44.

16

Visualizing Hepatic Immunity through the Eyes of Intravital Microscopy

Maria Alice Freitas-Lopes, Maísa Mota Antunes, Raquel Carvalho-Gontijo, Érika de Carvalho, and Gustavo Batista Menezes

CONTENTS

16.1 Introduction

The liver is one of the main, important organs in the body, performing indispensable metabolic and immune functions. The liver is strategically positioned in the abdominal cavity and is in contact with the intestine and the systemic circulation. Because of its location, the liver is continually exposed to nutritional antigens, products from the intestinal microbiota, and toxic substances. Hepatocytes are the major functional constituents of the hepatic lobes and perform most of the liver's secretory and synthesizing functions, although other groups of cells sustain the vitality of the organ: the hepatic immune cells. Hepatic leukocytes play a fundamental role in host immune responses and exquisite mechanisms are necessary to govern the density and the location of these cells. Given the importance of such cell populations to both metabolism and immune systems, major efforts have been employed to visualize these cells in their native context using different *in vivo* imaging approaches. In this chapter, we will discuss how intravital microscopy (IVM) has enhanced our knowledge on the different fields of hepatic immunity and biology.

16.2 Hepatic Morphology

16.2.1 Structure and Organization

After the skin, the liver is the second largest organ in the human body. The liver plays key functions including metabolism, synthesis, and secretion of several substances, such as bile, enzymes, and different proteins. In addition, the liver is an important immune organ, acting as a filter to clear blood-borne bacteria (Balmer et al., 2014; Kubes and Jenne, 2018). In mammals, it is subdivided into two lobes: a thin layer of connective tissue, called Glisson's capsule, covers the left – the largest – and the right. The liver is located in the abdominal cavity, below the diaphragm, and is directly connected to the systemic circulation through the hepatic arteries and veins and to the gastrointestinal tract system, more specifically to the intestine through the portal vein.

The liver consists mainly of hepatocytes (liver parenchymal cells). These cells are juxtaposed, forming strings that are separated from each other by the sinusoids (formed by endothelial cells), where the blood circulates. It also has bile ducts and hepatic blood vessels. In addition, a range of nonparenchymal cells has been described as resident cells of this organ, for example macrophages and dendritic cells, as well as Ito cells, natural killers, and T and B-lymphocytes (Freitas-Lopes et al., 2017).

16.2.1.1 Ex Vivo × In Vivo *Images*

Histological techniques allow us to study this organ's structure with great precision, especially the identification of its morphology. For many years, studies using fixed specimens, such as histological slides by conventional light microscopy or immunohistochemistry, using specific antibodies to identify different structures, were the only way to reinforce the hypotheses on how different processes occur in a tissue, such as organ formation, cellular interaction, modification of the structure to an injury, etc. However, these techniques give us a static perspective and do not allow for observation of how these processes occur. So, these techniques provide us what is called '*Ex Vivo* Images' With the evolution of technological processes, new microscopes have been developed, allowing us to delve deeper into the studies of phenomena that we previously only believed to exist, providing the possibility to visualize, cells and tissues from the inside of the animal; for example, *in vivo* cellular interactions through the technique of IVM.

Using IVM, scientists are able to see not only a static image, but also the cells interacting with each other. A clear example of the difference among IVM and histological techniques can be observed, for instance, in a cell-migration framework. Histology shows us the rearrangement of tissue structure, such as the formation of pores. It is possible to see cells statically at different stages of migration. Through the use of IVM, scientists are able to observe phenomena, in real time, that previously have not been possible, such as blood flow inside the vessels, interactions of microorganisms with phagocytic cells, and metabolic and immunological processes (Antunes et al., 2018a).

These techniques complement each other; structural details of tissues and cells are favored when analyzed by histological technique, and visualizing processes of cellular interaction and real-time mechanisms are only possible using *in vivo* imaging techniques. On a technological evolutionary scale, we can compare comic books with cartoons: both tell us a story, but in different ways (Figure 16.1).

16.2.2 Cells *In Vivo*: Location and Distribution

Because the liver is strategically located in the abdominal cavity, it is in intrinsic contact with the systemic circulation; blood from the spleen, pancreas, and gastrointestinal circulation enters the liver via the portal vein, and the arterial input reaches from the hepatic artery. Inside the liver, blood flows through microvessels – the sinusoids – to circulate into the lobules through the hepatic portal spaces, later drained by centrilobular vein out of the liver via the hepatic vein. There are special transporting structures, such as channels, vesicles, diaphragms, and fenestrae, that contribute to the high permeability of capillary endothelium to water and solutes (Bismuth, 2014). Not only macrostructures are involved in hepatic functions from this singular hemodynamic scheme, but the hepatic microenvironment also participates in important metabolic and immune functions (Eckert et al., 2015; Antunes et al., 2018b).

The hepatic environment harbors a singular endothelial cell population, called liver sinusoidal endothelial cells (LSECs). LSECs are found between the hepatic microcirculation and hepatocytes. Inside the vessels, on the luminal side, LSECs continuously survey blood from the gastrointestinal tract. On the extravascular side, LSECs interact with hepatic stellate cells (or Ito cells) and hepatocytes. This is crucial for the hepatic metabolism because LSECs are a permeable barrier that mediate the exchange of active uptake and degradation of circulating molecules (Poisson et al., 2017). In order to visualize the LSECS *in vivo*, a murine model of IVM can be performed (Marques et al., 2015), along with an immunostaining-using

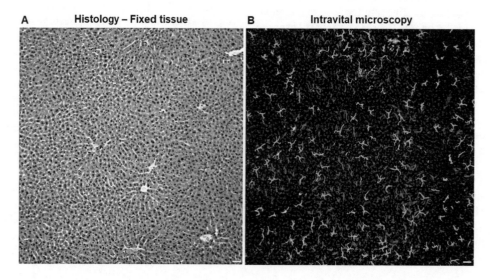

FIGURE 16.1   *Ex vivo versus in vivo* images**.** Structure of an adult mouse liver. A. Liver histological preparation, stained for hematoxylin and eosin, evidencing the hepatic structure: hepatocytes (in dark purple), hepatic cord (in light purple) and large vessels. B. Liver intravital microscopy evidencing the hepatic structure through antibody staining: hepatocyte nucleus (in blue), stained with DAPI; Kupffer cells (in magenta), stained with anti-F4/80; and dendritic cells (in green); green fluorescent protein positive (GFP+) cells from a reporter mouse. Scale bars in A., B. 35 μm.

antibody, which also enables the visualization of the hepatic vessels (Figure 16.2a).

Hepatocytes are the major functional constituents of the hepatic lobes and perform most of the liver's secretory and synthesizing functions. Other groups of cells have important roles in sustaining the organ: the hepatic immune cells. The liver harbors different immune cell populations that are established during the embryonic period of life and are one of the largest populations of immune cells in the body (Nakagaki et al., 2018b). Subsets of these cells, resident phagocytes and other leukocytes, can be found within the liver in homeostatic conditions, and through IVM they can be visualized and studied in different research fields. The role of hepatic leukocytes and resident phagocytes is still unknown in distinct biological contexts. *In vivo* imaging of the liver compartments are a practical and more realistic method to elucidate biological phenomena in the liver, once it is continually exposed to nutritional products, infectious microorganisms from the intestinal microbiota, and to toxic substances (Medzhitov et al., 2012; Nakagaki et al., 2018a).

Hepatic macrophages, known as Kupffer cells (KCs), originate in the yolk sac during embryogenesis and, unlike other organs in which the resident macrophages are located in the parenchyma, in the liver, these cells are inside the sinusoids and in contact with blood circulation (Naito et al., 1997; Kinoshita et al., 2010). KCs are adhered to the endothelium, and they emit extensions into the extravascular space among blood and components of liver parenchyma (Figure 16.2b, d). Moreover, the KCs form the first line of defense against particles and

immunorreactive material passing from the gastrointestinal tract via the portal circulation. These cells act as an intravascular and natural hepatic barrier against hazardous substances from the circulation. This process of catching circulating bacteria (Zeng et al., 2016) can be imaged through IVM. In situations in which damage has occurred and there is death of hepatocytes, KCs act by recruiting circulating monocytes into the injured tissue, and these cells will originate monocyte-derived macrophages that can substitute the embryonic KCs (Auffray et al., 2009).

Another resident cell population in the liver is the dendritic cells (DCs). DCs are antigen-presenting cells (APCs), able to induce immune and tolerogenic responses in lymphoid and nonlymphoid organs, such as the liver. Their location within the liver has long been researched, although it is well established that hepatic DCs represent a heterogeneous and large population within this organ (Ardavín, 2003; Lian et al., 2003). The hepatic DCs are located underneath the liver capsule and around large vessels. They are extravascular cells in steady-state conditions and morphologically different amongst them; *in vivo* imaging shows that the subcapsular DCs are larger and have more dendrites (David et al., 2016) (Figure 16.2c, d). Unlike KCs, DCs are rarely distributed within the parenchyma. They play a role in capturing, processing, and presenting antigens by interacting with T cells, being the key to initiate immune responses.

Neutrophils are polymorphonuclear leukocytes and are essential for controlling infections and sterile inflammation (Wang et al., 2017). These cells are developed from

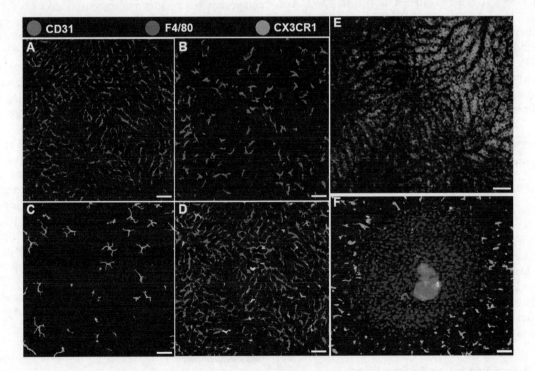

FIGURE 16.2 Liver cells *in vivo*. Intravital microscopy enables the visualization and distinction of liver cells. A. Liver sinusoidal endothelial cells (LSECs), or liver sinusoids, here stained with andi-CD31 PECAM-1 (in red). B. Kupffer cells, an intravascular phagocyte, here stained with anti-F4/80 (in magenta). C. There is an exclusively extravascular population of dendritic cells (GFP+ cells in green) that inhabit the liver surface. This image was obtained from a reporter mouse. D. Merge of the three channels. E. Hepatocytes and hepatic cord evidenced by their autofluorescence. F. Circulating cells (in red) are recruited to the lesion focus (in blue) caused by sterile damage in the liver, and GFP+ cells (in green) do not migrate to the lesion site. Scale bars in A–F = 35 μm.

hematopoietic stem cells in bone marrow, a process called "granulopoiesis," and are found patrolling the hepatic sinusoids. During inflammation in the liver, the neutrophils are recruited to lesion sites. These granulocytes play an important role migrating from blood vessels to the inflammation focus, a process driven by chemokines and chemotactic agents (Ramaiah and Jaeschke, 2007). Once attracted, neutrophils accumulate within the hepatic microvasculature before the transmigration process (Figure 16.2f).

Eosinophils are granulocytes characterized by cytoplasmic granules originated in the bone marrow and are predominantly tissue cells that migrate from the blood into tissues as a consequence of correlated events that involve adhesion pathways and chemoattractants (Weller and Spencer, 2017). The recruitment of eosinophils to the injured liver is regulated by several events involving cytokines and chemokines released by other eosinophils and T lymphocytes; the location and the consequences of eosinophilic infiltration in the liver depends on the disease condition (Zimmermann et al., 2008).

The liver also harbors large populations of natural killer (NK) cells, natural killer T (NKT) cells, and other lymphocytes, such as T and B cells (Richards et al., 2015). NK and NKT cells can proliferate under certain experimental or pathological conditions. The normal liver contains a large number of lymphocytes CD4 and CD8 T cells and B cells. In inflammatory conditions, the number of lymphocytes in the liver increases, and they form infiltrating cells that will determine the nature of the inflammation. Under healthy conditions, the human liver contains significant numbers of T lymphocytes in the portal tracts and scattered through the parenchyma (Lalor et al., 2002). Under homeostasis, both CD4 and CD8 T cells are found in portal tracts at low numbers, and a population of cells is found in association with biliary epithelium (Probert et al., 1997). B-lymphocytes perform production of antibodies, antigen presentation, secretion of multiple cytokines, and regulation of immune responses, but little is known about the functional biology of liver B cells by *in vivo* imaging.

16.3 Immunostaining for *In Vivo* Imaging

16.3.1 Common Cell Markers under Homeostasis

The liver consists of many different cell types, and their functions and interactions make this organ a pivotal agent in the maintenance of the body's homeostasis. In the face of a range of cell types, different tools have been proposed to identify these cells in a precise and specific way in order to study deeply the location and function of liver cells by imaging. Below, we separate the main cellular types of the liver and suggest labeling strategies to identify these cells according to their specific phenotypic characteristics.

16.3.1.1 Hepatocytes Can Be Identified by Their Autofluorescence

As mentioned in Section 16.1.2, the hepatocytes are the most abundant hepatic cells and represent about 80% of the liver mass. One of the functions of the hepatocytes is the synthesis of

albumin – the human version is human serum albumin – and it normally constitutes about 50% of human plasma protein. Serum albumin has intrinsic fluorescence properties and, when excited at 280 nm, exhibits fluorescence with maximum emission in the region of 350 nm associated with the L-amino acids tryptophan, tyrosine, and phenylalanine. Thus, the identification of hepatocytes can be made based on the wavelength emitted by the albumin when excited by a laser, without the need of antibodies coupled to fluorophores (David et al., 2016) (Figure 16.2e).

16.3.1.2 Liver Vessels Labeling Strategy

The blood vessels' architecture can be visualized labeling endothelial cells with anti-CD31 or PECAM-1 monoclonal antibody coupled to a fluorochrome, which is an adhesion molecule expressed at the endothelial cells' surface (Figure 16.2a). By staining these cells, it is possible to identify the location of intra- and extravascular cells and to monitor the rolling, migration, and arrival of cells during the inflammatory process.

16.3.1.3 Liver Immune Cells Can Be Labeled by Their Surface Antigen Expression

The liver harbors several immune cells that play an important role in the clearance of injurious microorganisms that enter into the bloodstream. Several studies have sought to discover specific biomarkers capable of identifying hepatic immune cells and assisting researchers to understand more clearly these cells' role in health and disease. Coupling antibodies to fluorescent compounds has been one of the most used tools for labeling cells *in vitro* and *in vivo* and identifying them through imaging strategies. Another strategy generates genetically modified animals that express fluorescent proteins on the surface of liver immune cells under the control of a promoter gene, allowing the identification of these cells using confocal microscopy.

Many immune cell markers are named following the clusters of differentiation (CD) nomenclature, which aims to provide targets for cell immunophenotyping. The majority of these surface antigens is not only specific to a unique cell type. A classic example is the CD11b, an integrin family member, expressed on the surface of many leukocytes including monocytes, neutrophils, NK cells, granulocytes, and macrophages. To better identify a specific cell type, it is important to search for a marker expressed exclusively for this lineage or cell. Additionally, imaging experiments allow the evaluation of morphological aspects of the cells, as well as the size and presence of dendrites, for example, which facilitates their identification jointly to the marking.

The main immune cell surface markers are listed in Table 16.1 to help identify liver immune cells by imaging experiments.

16.4 Liver Imaging Acquisition

Visualization of the hepatic environment in real time through the use of a confocal microscope allows us to observe varied phenomena, from the migration of inflammatory cells to a lesion site (Marques et al., 2015; McDonald et al., 2010) to

TABLE 16.1

Liver Immune Cells and Their Main Markers

Liver Immune Cells	Location in Homeostasis	Main Surface Markers
Kupffer Cells	Inside the sinusoids, adhered to the endothelium	F4/80
Dendritic Cells	Underneath hepatic capsule; around large vessels	CD19- CD11c+; CD8a+ B22C CD11b-(lymphoid); CD8a- B2 CD11b+ (myeloid); B22C+ CD1 (plasmacytoid)
Monocytes		CD11b^{hi}CD115^{hi}Gr1^{lo}
Neutrophils		Ly6G+CD11b+F4/80-
Eosinophils	Inside the sinusoids as patrolling cells	CD11b+CD193+Siglec F+
Natural Killer Cells		CD3-NK1.1+
T lymphocytes		CD3+
B lymphocytes		CD19+

phagocytosis of bacteria or toxic products (David et al., 2016; Wiesner et al., 2010), all at the exact moment they occur, using live animals.

There are many alternative imaging techniques to study hepatic phenomena, which may be both *ex vivo* and *in vivo*. As examples of techniques in which tissues extracted from animals are used, i.e. *ex vivo*, histology with specific staining for the cells and/or phenomena is appreciated, as well as immunofluorescence techniques, in which specific fluorescent antibodies are used in order to visualize events. Among the techniques in which it is possible to perform imaging with live animals, we can mention magnetic resonance (MRI), positron emission tomography (PET), and fluorescence endomicroscopy. These methods are of great importance for basic research, especially as they are applicable to humans. In addition, MRI and PET are completely noninvasive imaging methods that offer a wealth of information. Although they have been adapted to small animals, such as experimental animals, these techniques have limited image resolution, giving a further advantage to the confocal IVM technique.

The use of live animals and a microscope explain the term "intravital microscopy," and it means exactly this: the visualization and imaging inside a living organism. It is possible to visualize many structures and cells in the liver using histologic slides, for example. However, the use of *in vivo* imaging allows us to explore, with detail, the hepatic vascular arrangement, allowing a detailed study, rich in images and videos, and even the visualization of hepatic blood flow and its distribution throughout the organ.

The IVM technique is relatively simple, however, much care must be taken, as animals are used for viewing and acquiring images. Care involves everything from the handling and the content of the animals, to the maintenance of the living animal, ensuring that it is free of any kind of pain or suffering. Here, we will describe in detail the procedures necessary for the visualization of multiple areas of the liver, ensuring the acquisition of static images and videos capable of assessing the movement of cells and blood flow within the organ.

16.4.1 The Procedure

As in any procedure involving experimental animals, IVM requires the careful management of mice, as well as the

approval of ethical regulations and adequate anesthesia. The procedure for liver exposure includes a laparotomy followed by static placement of the organ (Marques et al., 2015). In this way, prior preparation and practice by the researchers are important.

The first step is the containment and administration of anesthesia in the animal. Here, we will deal with techniques and doses related to mice weight, which are, today, one of the most used species in animal experimentation. Anesthetic solution should be freshly made, and it should be protected from light. For the preparation, mix 10% (wt/vol) ketamine solution (2.5 mL), 2% (wt/vol) xylazine solution (2.0 mL), and 1× PBS (0.01M) (5.5 mL). Inject 6 µl intraperitoneal per g of body weight (ketamine: 60–80 mg/kg and xylazine: 8–15 mg/kg) (Marques et al., 2015).

Throughout the procedure, it is essential that the animal remains warm, hydrated, and sedated. After complete sedation, attach the mouse on the surgical stage using a piece of adhesive tape (Figure 16.3A). After fixation, the animal will undergo laparotomy to expose the liver. To do this, apply mineral oil to the abdominal skin to avoid imaging interference by small hairs in the imaging field. Alternatively, the mouse may be shaved before the experiment. The next step is laparotomy. Using surgical scissors and forceps, perform the midline incision in the abdomen from the pubis to the xiphoid process (Figure 16.3B). Dissociate the abdominal musculature from the skin and cauterize the vessels in the field of view (Figure 16.3C, D). The correct and complete cauterization of the vessels is a critical step because it will prevent bleeding.

After cauterization and making sure the animal is not bleeding, remove the skin and abdominal muscle along the costal margin to the midaxillary line to expose the liver (Figure 16.3E, F, G). Use the cauterizer when removing the muscle. This procedure aims to minimize liver movement and to allow visualization of other organs sequentially. It is important to note that the cauterizer can reach very high temperatures, which can seriously injure the researcher. Still, it is advisable to avoid long exposures of the animal's organs, especially the liver, preventing lesions in the organ that will be the object of study. After that, remove any hairs from the incision with sterile saline, and immediately cauterize the blood vessels that have been cut with the cauterizer.

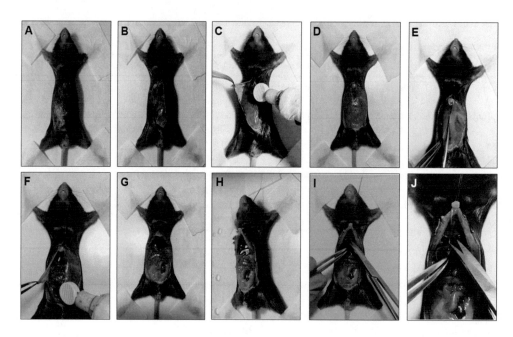

FIGURE 16.3 Detailed liver exposure surgery for confocal intravital microscopy. A. Attach the anesthetized mouse in the surgical stage using tape. B. Perform midline laparotomy. C. Cauterize all vessels in the skin and D. remove the abdominal skin using scissors. E. Perform midline laparotomy. F. Remove the abdominal muscle using the cauterizer G. to expose the abdominal cavity. H. Make a knot and hold the xiphoid process. I., J. Cut the falciform ligament between the liver and the diaphragm. Appropriate Institutional Regulatory Board permission from CEUA/UFMG (Comissão de Ética no Uso de Animais – UFMG) for animal use was obtained.

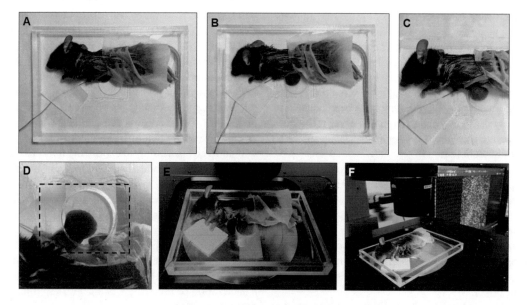

FIGURE 16.4 Positioning the animal for image acquisition. A., B. Move the mouse to the imaging stage and gently expose the right lobe. C. Cover the exposed lobe with a piece of Kimwipes. D. Bottom view of liver lobe placed on the surgical stage. E., F. Position the mouse on the microscope stage. Appropriate Institutional Regulatory Board permission from CEUA/UFMG (Comissão de Ética no Uso de Animais – UFMG) for animal use was obtained.

The last step of the surgical procedure is to neutralize the respiratory movement of the animal, thus avoiding movements during the acquisition of the images. To do this, hold the xiphoid process with a knot made of suture thread in order to facilitate access to the falciform ligament between the liver and the diaphragm. This may be done by bending the tied xiphoid cranially and taping the suture thread behind the mouse's head (Figure 16.3H). Cut the falciform ligament carefully to separate the liver from the diaphragm (Figure 16.3I, J). Cutting

the falciform ligament requires the utmost care and attention, as the lower perforation of the diaphragm is lethal to the animal. The surgery for liver exposure is complete, and the liver can be exposed and positioned for imaging. For this, place the mouse in the right lateral position and gently exteriorize the right lobe in a custom-made acrylic platform (Figure 16.4A). Move the liver and the guts apart carefully, using a wet cotton swab in order to isolate the liver lobe from the internal organs. Remember, never touch the liver using sharp instruments

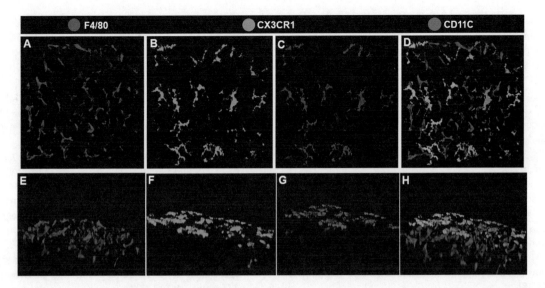

FIGURE 16.5 Three-dimensional rendering of liver phagocytes. Liver confocal intravital microscopy imaging through the capsule evidencing the distribution of: A. F4/80+ macrophages (magenta), B. CX3CR1gfp (green) and C. CD11c-YFP (red) cells. D. Cells merged. Liver confocal intravital microscopy imaging through an oblique plan: E. F4/80+ macrophages (magenta), F. CX3CR1gfp (green) and G. CD11c-YFP (red) cells. H. Cells merged. Note that the macrophages are abundantly found in different extensions of the liver, while the other cell types are located in the liver capsule.

(i.e. forceps). The liver is very fragile, and it will bleed even if the operator roughly touches it with a cotton swab. As mentioned earlier, it is critical to keep the animal hydrated throughout the imaging period. This is done by involving the exposed peritoneal organs with a piece of Kimwipes, moving the organs closer to the mouse's body. Keep Kimwipes soaked with warm saline or 1× PBS (0.01M). This step, in addition to keeping the animal hydrated, keeps the intestine away from the liver, thus preventing peristaltic movements from interfering with the acquisition of the images or a 3D reconstruction (z-series) (Figure 16.5A–H).

After removal of the intestine, the liver lobe will be isolated. Cut a small square piece of a Kimwipe (~ 3 × 3 cm), and gently cover the liver lobe (Figure 16.4B). This piece of Kimwipes will touch the moist liver surface, causing relatively strong adhesion, allowing the operator to make small movements with the liver without directly touching the organ, which will avoid unwanted injuries. After covering the liver lobe, gently pull the Kimwipes together with the liver in order to place the liver in a "stretch" mode. The moist Kimwipes will then adhere to the Plexiglas surface, thus stabilizing the liver (Figure 16.4C, D). This will be the key to prevent liver movement related to breathing and peristalsis.

After all surgical procedures and proper positioning in the acrylic stage, the animal is ready to be positioned under the microscope (Figure 16.4E). Check the general condition of the mouse: breathing rate, bleeding, hydration, temperature, and sedation. Keep all exposed tissues moistened with saline-soaked Kimwipes to prevent dehydration during imaging. For the duration of all experiments, the liver should be continuously perfused with physiological saline buffer. In order to keep animal adequately heated, body temperature should be maintained with an infrared heat lamp or a heating blanket. The distance between the lamp and the mouse will depend on both the lamp power and the room temperature. If there is imaging interference when using the IR lamp, cover the mouse

with a piece of aluminum foil. The animal is ready to be taken under the microscope, and the images acquired (Figure 16.4F).

16.4.2 Experimental Design and Probes Administration

IVM liver-imaging protocol, as well as visualization of its structures and cells, encompasses procedures and care for the animal and the surgical procedure for exposure of the organ, the administration of fluorescent probes that will allow the visualization of these structures of interest, and the administration of specific probes or antibodies, which will generate reliable and consistent results. It is important to emphasize that during the performance of an experiment in which the IVM will be used, all the experimental groups must undergo the same treatment. In addition, control groups must be used, and the use of fluorescent isotype antibodies is also mandatory when IVM studies are performed.

Antibodies and fluorescent probes should be administered intravenously in the animal after sedation, 10 min before imaging to allow efficient distribution and labeling. The most used routes are the retro-orbital sinus and the tail vein. It is possible for the administration of more than one antibody and/ or probe to be performed at the same time by mixing them. The solutions should be freshly prepared, and they should be protected from light before injection.

REFERENCES

Antunes, M. M., Araújo, A. M., Diniz, A. B., Pereira, R. V. S., Alvarenga, D. M., David, B. A., Rocha, R. M., Lopes, M. A. F., Marchesi, S. C., Nakagaki, B. N., Carvalho, É., Marques, P. E., Ryffel, B., Quesniaux, V., Guabiraba Brito, R., Filho, J. C. A., Cara, D. C., Rezende, R. M., and Menezes, G. B. (2018b). IL-33 signalling in liver immune cells enhances drug-induced liver injury and inflammation. *Inflamm Res*, *67*: 77–88.

Antunes, M. M., Carvalho, É., and Menezes, G. B. (2018a). DIY: "Do imaging yourself" - Conventional microscopes as powerful tools for in vivo investigation. *Int J Biochem Cell Biol*, *94*: 1–5.

Ardavín, C. (2003). Origin, precursors and differentiation of mouse dendritic cells. *Nat Rev Immunol*, *3*: 582–590.

Auffray, C., Sieweke, M. H., and Geissmann, F. (2009). Blood monocytes: Development, heterogeneity, and relationship with dendritic cells. *Annu Rev Immunol*, *27*: 669–692.

Balmer, M. L., Slack, E., de Gottardi, A., Lawson, M. A., Hapfelmeier, S., Miele, L., Grieco, A., Van Vlierberghe, H., Fahrner, R., Patuto, N., Bernsmeier, C., Ronchi, F., Wyss, M., Stroka, D., Dickgreber, N., Heim, M. H., McCoy, K. D., and Macpherson, A. J. (2014). The liver may act as a firewall mediating mutualism between the host and its gut commensal microbiota. *Sci Transl Med*, *6*: 237ra266.

Bismuth, H. (2014). A new look on liver anatomy: Needs and means to go beyond the Couinaud scheme. *J Hepatol*, *60*: 480–481.

David, B. A. Rezende, R. M., Antunes, M. M., Santos, M. M., Lopes, M. A., Diniz, A. B., Pereira, R. V., Marchesi, S. C., Alvarenga, D. M., Nakagaki, B. N., and Araújo, A. M. (2016). Combination of mass cytometry and imaging analysis reveals origin, location, and functional repopulation of liver myeloid cells in mice. *Gastroenterology*, *151*: 1176–1191.

Eckert, C., Klein, N., Kornek, M., and Lukacs-Kornek, V. (2015). The complex myeloid network of the liver with diverse functional capacity at steady state and in inflammation. *Front Immunol*, *6*.

Freitas-Lopes, M. A., Mafra, K., David, B. A., Carvalho-Gontijo, R., and Menezes, G. B. (2017). Differential location and distribution of hepatic immune cells. *Cells*, *6*.

Kinoshita, M., Uchida, T., Sato, A., Nakashima, M., Nakashima, H., Shono, S., Habu, Y., Miyazaki, H., Hiroi, S., and Seki, S. (2010). Characterization of two F4/80-positive Kupffer cell subsets by their function and phenotype in mice. *J Hepatol*, *53*: 903–910.

Kubes, P., and Jenne, C. (2018). Immune responses in the liver. *Annu Rev Immunol*, *36*: 247–277.

Lalor, P. F., Shields, P., Grant, A., and Adams, D. H. (2002). Recruitment of lymphocytes to the human liver. *Immunol Cell Biol*, *80*: 52–64.

Lian, Z. X., Okada, T., He, X. S., Kita, H., Liu, Y. J., Ansari, A. A., Kikuchi, K., Ikehara, S., and Gershwin, M. E. (2003). Heterogeneity of dendritic cells in the mouse liver: Identification and characterization of four distinct populations. *J Immunol*, *170*: 2323–2330.

Marques, P. E., Antunes, M. M., David, B. A., Pereira, R. V., Teixeira, M. M., and Menezes, G. B. (2015). Imaging liver biology in vivo using conventional confocal microscopy. *Nat Protoc*, *10*: 258–268.

McDonald, B., Pittman, K., Menezes, G. B., Hirota, S. A., Slaba, I., Waterhouse, C. C., Beck, P. L., Muruve, D. A., and Kubes, P. (2010). Intravascular danger signals guide neutrophils to sites of sterile inflammation. *Science*, *330*: 362–366.

Medzhitov, R., Schneider, D. S., and Soares, M. P. (2012). Disease tolerance as a defense strategy. *Science*, *335*: 936–941.

Naito, M., Hasegawa, G., and Takahashi, K. (1997). Development, differentiation, and maturation of Kupffer cells. *Microsc Res Tech*, *39*: 350–364.

Nakagaki, B. N., Vieira, A. T., Rezende, R. M., David, B. A., and Menezes, G. B. (2018a). Tissue macrophages as mediators of a healthy relationship with gut commensal microbiota. *Cell Immunol*, *330*: 16–26.

Nakagaki, B. N., Mafra, K., de Carvalho, É., Lopes, M. E., Carvalho-Gontijo, R., de Castro-Oliveira, H. M., Campolina-Silva, G. H., de Miranda, C. D., Antunes, M. M., Silva, A. C., and Diniz, A. B. (2018b). Immune and metabolic shifts during neonatal development reprogram liver identity and function. *J Hepatol*, *69*:1294–1307.

Poisson, J., Lemoinne, S., Boulanger, C., Durand, F., Moreau, R., Valla, D., and Rautou, P. E. (2017). Liver sinusoidal endothelial cells: Physiology and role in liver diseases. *J Hepatol*, *66*: 212–227.

Probert, C. S., Christ, A. D., Saubermann, L. J., Turner, J. R., Chott, A., Carr-Locke, D., Balk, S. P., and Blumberg, R. S. (1997). Analysis of human common bile duct-associated T cells: Evidence for oligoclonality, T cell clonal persistence, and epithelial cell recognition. *J Immunol*, *158*: 1941–1948.

Ramaiah, S. K., and Jaeschke, H. (2007). Role of neutrophils in the pathogenesis of acute inflammatory liver injury. *Toxicol Pathol*, *35*: 757–766.

Richards, J. A., Bucsaiova, M., Hesketh, E. E., Ventre, C., Henderson, N. C., Simpson, K., Bellamy, C. O., Howie, S. E., Anderton, S. M., Hughes, J., and Wigmore, S. J. (2015). Acute liver injury is independent of B cells or immunoglobulin M. *PLoS One*, *10*: e0138688.

Wang, J., Hossain, M., Thanabalasuriar, A., Gunzer, M., Meininger, C., and Kubes, P. (2017). Visualizing the function and fate of neutrophils in sterile injury and repair. *Science*, *358*: 111–116.

Weller, P. F., and Spencer, L. A. (2017). Functions of tissue-resident eosinophils. *Nat Rev Immunol*, *17*: 746–760.

Wiesner, P., Choi, S. H., Almazan, F., Benner, C., Huang, W., Diehl, C. J., Gonen, A., Butler, S., Witztum, J. L., Glass, C. K., and Miller, Y. I. (2010). Low doses of lipopolysaccharide and minimally oxidized low-density lipoprotein cooperatively activate macrophages via nuclear factor kappa B and activator protein-1: Possible mechanism for acceleration of atherosclerosis by subclinical endotoxemia. *Circ Res*, *107*: 56–65.

Zeng, Z., Surewaard, B. G., Wong, C. H., Geoghegan, J. A., Jenne, C. N., and Kubes, P. (2016). CRIg functions as a macrophage pattern recognition receptor to directly bind and capture blood-borne gram-positive bacteria. *Cell Host Microbe*, *20*: 99–106.

Zimmermann, N., McBride, M. L., Yamada, Y., Hudson, S. A., Jones, C., Cromie, K. D., Crocker, P. R., Rothenberg, M. E., and Bochner, B. S. (2008). Siglec-F antibody administration to mice selectively reduces blood and tissue eosinophils. *Allergy*, *63*: 1156–1163.

17

Optical Imaging of the Mammalian Oviduct In Vivo

Shang Wang and Irina V. Larina

CONTENTS

17.1 Introduction

17.1.1 Why We Need to Image the Oviduct

The mammalian oviduct (or fallopian tube) is a tubular reproductive organ essential for fertilization and preimplantation embryonic development *in vivo* (Coy et al., 2012; Li and Winuthayanon, 2017). In a normal pregnancy, the ovary releases oocytes surrounded by cumulus cells, which are transferred to the site of fertilization within the oviduct. Ejaculated sperm, which survived journey through the uterus, are guided and regulated by the oviduct environment toward the fertilization. After fertilization takes place, the oviduct transports embryos to the uterus for implantation, and at the same time, creates the microenvironment for the embryos to progress through preimplantation development. Anatomically, the oviduct has two major portions, the ampulla where fertilization occurs and the isthmus that hosts the major part of early development of the new life. The ampulla connects to the ovary through the infundibulum, and the isthmus ends with the uterotubal junction in connection with the uterus. An illustration of the mouse oviduct is shown in Figure 17.1. Functionally, there are two sources of force in the oviduct, the contraction

of smooth muscle surrounding the tube and the beat of motile cilia lining on the luminal epithelium. It is believed that the flow of secretory fluid inside the oviduct lumen, induced by these functional activities, is essential for delivering the cumulus-oocyte complex to the fertilization site, guiding the sperm toward the oocytes and transferring the embryos for implantation (Lyons et al., 2006). However, the specific roles of the muscle contraction and cilia beating in these reproductive events are currently unclear (Coy et al., 2012; Li and Winuthayanon, 2017; Lyons et al., 2006). In fact, the oviduct is less studied than other reproductive organs, e.g. the ovary and the uterus (Li and Winuthayanon, 2017). As a result, the reproductive process taking place natively in the oviduct has yet to be visualized, and thus remains largely unknown. In the cases of reproductive disorders, such as infertility and ectopic pregnancy, whose causes are not entirely clear, whether the oviduct plays a role in them and what particular part of the reproductive process is disrupted in the oviduct are poorly understood. This significantly restricts effective treatment options for human reproductive diseases and prevents improvement of the assisted reproductive technology (ART), where the environmental factors for early embryonic development require

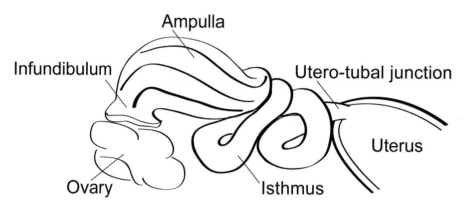

FIGURE 17.1 Illustration of the mouse oviduct, which consists of the infundibulum, the ampulla, the isthmus, and the uterotubal junction, in the ovary-to-uterus direction. Reproduced from Wang et al., 2015.

further optimization (Lucas, 2013; Kondapalli and Perales-Puchalt, 2013). This very limited understanding of the mammalian oviduct is largely due to the lack of proper imaging access to study its dynamics and functions. Given that the recently enriched biochemical tools, especially in the mouse model, allow advanced manipulations of the genetics and molecular pathways associated with the oviduct [Pazeli et al., 2004; Georgiou et al., 2005; Mondéjar et al., 2012), phenotypic analysis through imaging has become increasingly critical to delineate the tissue and cellular responses, which, as a whole, will help to elucidate the functional roles of the oviduct in mammalian reproduction and reproductive disorders.

17.1.2 *In Vitro* and *Ex Vivo* Imaging Approaches

For a long time, studies on the mammalian oviduct have been heavily relying on *in vitro* and *ex vivo* imaging methods, primarily including electron microscopy, histology, bright-field and fluorescence microscopies. Electron microscopy uses electron beams that have a much shorter wavelength than light, thus providing a spatial resolvability at subnanometer level. The scanning and transmission electron microscopy techniques have been mainly utilized to image the topography of the oviduct luminal epithelium (Dirksen and Satir, 1972; Abe and Oikawa, 1992; Sharma et al., 2015), providing a clear understanding of the structure, the morphology, and the distribution of motile cilia and secretory cells in the mammalian oviduct. Histology applies staining or fluorescent labeling of tissue slices to form molecular specific contrast that can be probed by light microscopy. Histological analysis enabled studies of the oviduct at the molecular level (Li et al., 2017;Ghosh et al., 2017; Tanwar et al., 2012)), bringing essential discoveries on regulatory molecules and signaling pathways in the oviductal physiology and pathology. With these *in vitro* imaging approaches, researchers have gained great insights into the structure and molecular function of the mammalian oviduct; however, neither of these methods are able to probe the dynamics of tissues and cells in the oviduct because of the requirement of a vacuum environment or slicing the sample. Bright-field microscopy, equipped with a high-speed camera, was employed to address this demand for live tissue and cell imaging under *ex vivo* experimental settings, with the

oviduct dissected out and placed in a dish (Kölle, 2012; Shi et al., 2014). With this setup and bright-field illuminations, the movements of the cumulus-oocyte complex and the sperm can be conveniently observed (Kölle et al., 2009). Notably, with a high-frame-rate data acquisition, this method has become a standard technique to measure the cilia beat frequency (CBF) in the *ex vivo* oviduct (Lyons et al., 2002; Shi et al., 2011; Zhao et al., 2011), enabling investigations concerning the influences of chemical (e.g. hormones) and physical (e.g. flows) factors on ciliary dynamics [Andrade et al., 2005; Bloodgood, 2010; Bylander et al., 2010). Although it only requires simple procedures and has been widely used, its limited imaging depth requires the oviduct to be opened up for better access to the cilia. This could significantly alter the original environment of the ciliated cells, complicating the interpretation of measurement results. Similar to the bright-field microscopy approach, fluorescence imaging, which employs fluorescence protein labeling and wavelength-specific illumination and detection, can be utilized to improve imaging contrast, as well as to detect specific cells and molecular components (La Sina et al., 2016; Hino et al., 2016).

17.1.3 The Demand for *In Vivo* Imaging

Ex vivo imaging methods have brought exciting findings on the activities of oviductal cilia, the movements of the cumulus-oocyte complex, and the behaviors of sperm. However, the oviduct naturally is a highly dynamic organ with a very complex and fast-changing biochemical and biomechanical environment, so *ex vivo* imaging results cannot fully represent the *in vivo* cases, where tissues and cells are in their native state. Specifically, with the oviduct removed from the reproductive system, although immediately imaged, the fluid mechanics inside the lumen are expected to change (Fauci and Dillon, 2006), and the signaling activities, such as the molecular response to the presence of gametes (Georgiou et al., 2005), are likely lost. These could largely contribute to the alteration of oviductal functions, including smooth muscle contraction and cilia beating, which in return affect the fluid flow and gamete movement inside the lumen through a possible looping effect. Thus, *ex vivo* imaging is not sufficient to further advance research and our understanding of the mammalian oviduct. A number of hypothesized mechanisms of mammalian fertilization, as well as the currently believed roles of the oviduct in this fascinating process, are required to be tested *in vivo* (Coy et al., 2012; Li and Winuthayanon, 2017; Lyons et al., 2006). These bring up an urgent demand for *in vivo* imaging and the phenotypic analysis of oviductal structures and functions in the mammalian model.

17.1.4 An Overview of This Chapter

In recognition of the significance of imaging the oviduct in its native condition, we review in detail the most recent advancements (in the past 6 years) in achieving *in vivo* imaging and assessment of the mammalian oviduct and the cellular dynamics inside the oviduct lumen. Among these, fluorescence microscopy provides two-dimensional (2D) localization of the labeled sperm as they migrate through the mouse oviduct isthmus, while optical coherence tomography (OCT) enables label-free three-dimensional (3D) structural and functional imaging of the mouse oviduct, including the mucosa fold morphology, cumulus-oocyte complex, smooth muscle contraction, and CBF. As an overview, in Section 2, we describe the technical aspect and the experimental results obtained from *in vivo* fluorescence microscopy of the sperm migration in the isthmus. In Section 3, we first introduce OCT and its endogenous tissue contrast, then present the *in vivo* mouse imaging setup with a recent advancement in intravital imaging through a dorsal imaging window, and finally review the specific methods and representative results from three major applications covering the oviductal structures and functions. In Section 4, we comment on several possible technical improvements of these methods and foresee exciting studies these *in vivo* approaches could bring for improving our understanding of the mammalian oviduct. In Section 5, we conclude this review with an anticipation for a bright future for *in vivo* optical imaging in biomedical research on mammalian reproduction.

17.2 Fluorescence Microscopy

17.2.1 Fluorescent Labeling and Imaging Setup

Taking one step further from the *ex vivo* fluorescence imaging, *in vivo* imaging of labeled sperm in the isthmus of the mouse oviduct was recently demonstrated (Isikawa et al., 2016; Muro et al., 2016). Male transgenic mice were utilized to produce sperm, with the acrosome expressing green fluorescent proteins (EGFP or GFP) and the mitochondria expressing red fluorescent proteins (DsRed2 or RFP). This labeling strategy not only allowed for the localization of sperm, but also made it possible to monitor the acrosome reaction (Muro et al., 2016). To conduct *in vivo* imaging, with the postcoitus female anesthetized, one side of the reproductive organs (ovary, oviduct, and uterus) were exposed through a midlateral incision and placed in a glass-bottomed dish filled with phosphate-buffered saline (PBS). A plate heater and a lens heater were used to keep the mouse and the dish warm during imaging. Regular fluorescence microscopes with video-recording cameras were employed to provide the location of sperm in the oviduct and the visualization of their acrosome statuses.

17.2.2 *In Vivo* Sperm Imaging in the Isthmus

With this *in vivo* imaging capability, two studies have so far been reported, both focusing on sperm imaging in the isthmus. Ishikawa et al. (2016)applied this method to investigate the influence of muscle contraction in sperm migration through the oviduct. The study successfully located the sperm and its assemblage in the isthmus of the oviduct, and with padrin (prifinium bromide) administrated to suppress the contraction of smooth muscle *in vivo*, the study found that oviductal muscle contraction plays an important role in the formation of sperm assemblage and the directional transfer of this assemblage to the middle part of the oviduct (Ishikawa et al., 2016). In another work, Muro et al. (2016) documented sperm attachment to the epithelium in both lower and upper portions of the isthmus. These studies provided *in vivo* evidence of sperm aggregation and reservoir in the isthmus. The simplicity of the experimental procedure and the wide availability of the transgenic mouse model make it greatly expected to produce more findings based on this *in vivo* imaging approach.

17.3 Optical Coherence Tomography

17.3.1 OCT and Its Imaging Contrast

As a low-coherence optical imaging technique, OCT provides a unique imaging scale, combining microlevel spatial resolution with millimeter-level imaging depth, which perfectly fills in the gap between confocal microscopy and ultrasonic imaging (Fercher et al., 2003). Since its invention in 1991 (Huang et al., 1991), OCT has experienced a fast and dramatic evolution from the original time-domain scheme, where the coherence gating is directly utilized by scanning the reference-arm mirror, to the current Fourier-domain configuration, where an inverse Fourier transform is applied to obtain depth-resolved spatial information (Leitgeb et al., 2003). Relying on the short coherence length from a broadband laser source, OCT is capable of providing an axial resolution ranging from ~1 μm to ~15 μm; taking advantage of the near-infrared light that lies in the low-absorption regime of biological tissue, penetration depth of up to several millimeters in highly scattering tissues can be achieved (Fercher et al., 2003). The state-of-the-art OCT system can reach a data acquisition speed of up to MHz for depth-resolved A-scans. The 3D OCT imaging is obtained based on 2D transverse scanning of the imaging beam across a sample, and recently, volumetric imaging at a rate over tens of Hz has been demonstrated (Wang et al., 2015).

Figure 17.2 shows the setup of a representative spectral-domain OCT system. A fiber-based Michaelson interferometer is used for interference of light from the reference arm and the sample arm, and a spectrometer is utilized to spatially resolve the interference fringes. Synchronization between the A-scan acquisition and the 2D laser beam scanning is implemented for various ways of acquiring the OCT data. The OCT structural imaging contrast relies on the mismatch of the refractive index inside the tissue that generates different magnitudes of backscattering of light; the spatial mapping of such light intensities forms the OCT images. Thus, OCT does not require any exogenous contrast to

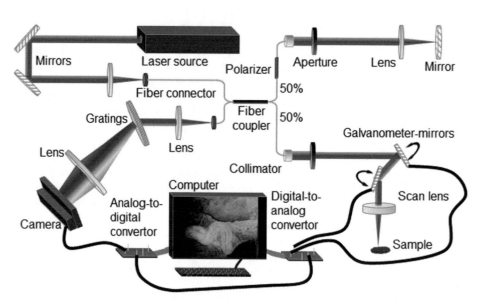

FIGURE 17.2 Schematic of a representative spectral-domain OCT system that is based on a broadband laser source, a fiber-based Michaelson interferometer, a sample arm, a reference arm, a line-field spectrometer, and a computer with signal converters.

be introduced to tissues or cells. In addition to structural imaging, a number of functional OCT imaging techniques are widely available based on the endogenous tissue and cell dynamics, organization, and properties to form the desired imaging contrast (Kim et al., 2015). As some of the best-known examples, these functional techniques include speckle variance OCT (Mahmud et al., 2013) and OCT angiography (Chen and Wang, 2017) to map the perfused blood vessels, Doppler OCT (Leitgeb et al., 2014) to measure blood flow velocity, OCT elastography (Larin and Sampson, 2017) to assess tissue stiffness, spectroscopic OCT (Morgner et al., 2000) to extract blood oxygen saturation, and polarization-sensitive OCT (de Boer et al., 2017) to reveal tissue birefringence. The nonlabeling high-resolution structural imaging and the different functional derivatives of OCT have enabled numerous applications across a variety of biomedical fields, particularly in ophthalmology (Adhi and Duker, 2013), cardiology (Bourma et al., 2017), oncology (Vakoc et al., 2012), neurology (Lamirel et al., 2009), dermatology (Sattler et al., 2013), and developmental biology (Men et al., 2016). As to the field of mammalian reproduction, OCT was first employed for imaging the human and bovine oviduct *in vitro* and *ex vivo* (Herrmann et al., 1998; Trottmann et al., 2016), demonstrating its feasibility to distinguish the detailed tissue structures and cell complex noninvasively (Burton et al., 2015; Burton et al., 2016; Rodriguez et al., 2016). Recently, our group has introduced OCT into live studies of mammalian reproduction in the oviduct, developed *in vivo* approach for OCT imaging of the mouse oviduct, and demonstrated its structural, cellular, and functional imaging capabilities (Wang et al., 2015; Burton et al., 2015; Wang et al., 2018).

17.3.2 *In Vivo* Imaging Setup

The initial setup for *in vivo* OCT imaging of the mouse oviduct adopted the surgical procedure routinely used for zygote injection in transgenic mouse production (Wang et al., 2015; Burton et al., 2015). As shown in Figure 17.3, incisions were made in the skin and muscle layers on the right dorsal side of

the anesthetized mouse, and the ovary, oviduct, and part of the uterus were gently pulled outside the body cavity for imaging. During OCT data acquisition, a pair of clamps were utilized to hold the fat pad associated with the ovary for stabilizing the tissues and minimizing the influence from motions induced by breathing. The handles of the clamps were firmly fixed on the optical table. Because OCT offers a relatively large working distance at the level of centimeters, this setup can be conveniently placed under the OCT imaging beam. Although this method is very efficient for experiments, because the reproductive organs are exposed to the air, this setup is not suitable for prolonged imaging sessions or longitudinal studies. To address this limitation, we have optimized a dorsal imaging window for OCT imaging of the mouse oviduct (Wang et al., 2018).

The original design of the window was adopted from a study in which intravital confocal imaging was performed on the mouse ovary to probe cancer cell migration (Bochner et al., 2015). For OCT imaging, because a larger working distance is available, we have increased the space to hold and fit the reproductive organs (ovary and oviduct), which provided minimal physical pressure on the tissues, thus helping to preserve oviductal functions. Also, instead of using titanium as the material, we utilized a high-resolution liquid 3D printer to produce the window with lower cost and higher efficiency. Specifically, the Form 2 printer from Formlabs (Massachusetts, United States) and the grey resin were used. Figure 17.4 shows the dorsal imaging window, its implantation on the mouse, and the stabilization method during OCT imaging. The optimized dorsal imaging window featured a clear aperture with a diameter of 10 mm, two extensions from the window frame for convenience of stabilizing the setup, and a pair of prongs as tissue holders. A total of 14 eyelets were evenly distributed close to the edge of the frame for suturing purposes. The details of the window dimension can be found in (Wang et al., 2018).

Implantation of this window was conducted on the right dorsal side of the mouse, as the left ovary is adjacent to the spleen. Routine procedures for survival surgeries on mice should be

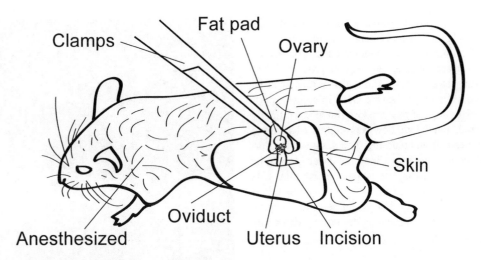

FIGURE 17.3 Illustration of the *in vivo* imaging method with the ovary, oviduct, and part of the uterus exposed through an incision. Reproduced from Wang et al., 2015.

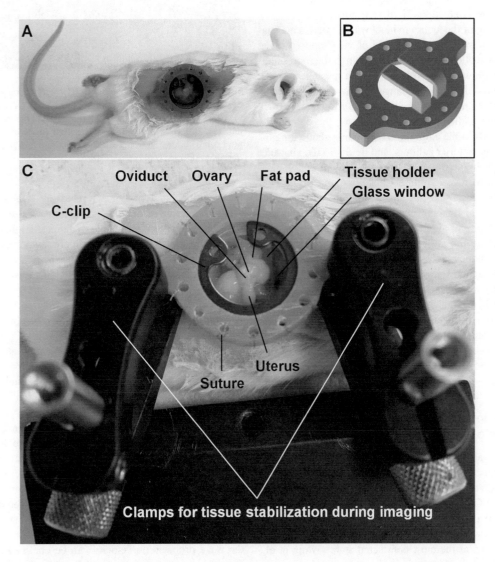

FIGURE 17.4 Setup for *in vivo* OCT imaging of the mouse oviduct through a dorsal imaging window. A. Implantation of the intravital window on the right dorsal side of the mouse. B. The model of the dorsal window for 3D printing. C. *In vivo* imaging setup with stabilization of the window by clamps. Reproduced from Wang et al., 2018.

followed, including sterile environment, proper anesthesia, treatment with analgesic, such as buprenorphine and non-steroidal anti-inflammatory drug, application of ophthalmic ointment, and postoperative care. The detailed procedures are presented in (Wang et al., 2018). The entire surgery can be completed within 45 minutes. During OCT imaging experiments, the two extensions of the window frame were tightly clamped, stabilized, and the window was slightly lifted up to avoid motion influence from breathing. The design of the dorsal window and the *in vivo* setup allows the cover glass to be replaced if needed while the mouse is under anesthesia. This makes it possible to conduct injections or topical applications of desired substances when necessary, and permits changing of the cover glass in case of scratching or staining that could affect the imaging quality.

17.3.3 Structural Imaging of Tissues and Cells

Structural imaging helps us to understand the oviduct morphology and locate the specific features inside the lumen. With endogenous contrast, OCT directly provides 3D depth-resolved imaging of the mouse oviduct with a microscale spatial resolution, a millimeter-level field of view, and covering the entire cross section of the tube, as shown in the representative images in Figure 17.5. With a ~4–5 μm resolution in 3D, we have observed the gradual change of the mucosa fold morphology from the ampulla to the isthmus (Burton et al., 2015). Specifically, the folding patterns transit from the continuous longitudinal folds in the ampulla to the nodule-shape longitudinal folds in the anterior isthmus, and finally to the ring-like transverse folds in the posterior isthmus. This morphological change is hard to reveal from 2D histology and bright-field microscopic images. With the dorsal imaging window, longitudinal structural imaging of the oviduct morphology can be obtained (Wang et al., 2018), which is potentially useful for understanding the structural changes of the oviduct in response to hormone regulation over the estrous cycle.

The OCT structural imaging can also capture the cumulus-oocyte complex with a cellular resolvability in the ampulla (Burton et al., 2015). Individual cumulus cells surrounding the oocytes were well distinguished in 3D (Burton et al., 2015). This accurate localization of the oocytes inside the oviductal lumen provides the feasibility to track their movements volumetrically, thus enabling studies of the dynamics of oocytes and cumulus-oocyte complexes being transferred through the oviduct *in vivo*.

17.3.4 Functional Assessment of the Muscle Contraction

As one of the major sources of force supporting the oviductal transporting function, the muscle contraction can be directly imaged, visualized in 3D, and quantitatively characterized with time-lapse OCT imaging (Wang et al., 2018). For example, Figure 17.6A shows the 3D OCT imaging of the muscle contraction in the ampulla, where the lumen was found to remain open even at the most contracted state (Wang et al., 2018). In contrast, in the isthmus, the oviductal lumen was fully closed when most contracted (Wang et al., 2018). This was caused by a thicker muscle layer in the isthmus than in

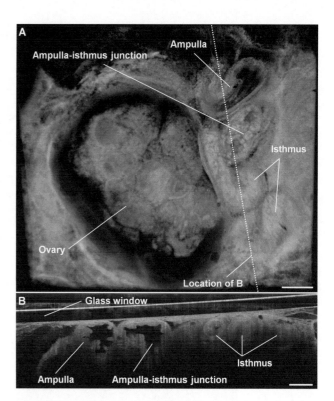

FIGURE 17.5 Representative 3D OCT structural image of the mouse oviduct through the dorsal imaging window. A. 3D OCT image covering the ovary and the oviduct, including the ampulla, the isthmus, and their junction. B. Cross-sectional, depth-resolved image showing the structural details of the oviduct. Scale bar is 500 μm in A. and 300 μm in B. Reproduced from Wang et al., 2018.

the ampulla, which produced a relatively greater contractility. The volume rate of 3D data acquisition was set as ~0.67 Hz, sufficient to capture the whole dynamics. Higher volume rate is possible, which could provide more temporal details of the oviduct wall movement. The representative videos showing the dynamic visualizations of the oviduct muscle contraction can be found in (Wang et al., 2018).

In addition to visualizing the contraction, the contraction wave propagation along the oviduct tube can be quantitatively measured and analyzed (Wang et al., 2018). Cross sections in 2D were selected perpendicular to the tube at five different locations in the ampulla, as shown in Figure 17.6A, and the luminal area for all the locations was measured over time. By normalizing the area from each position into the range between 0 and 1 (0 for smallest and 1 for largest), as shown in Figure 17.6B, the propagation of the luminal area change or wall displacement can be clearly seen in the direction from the proximal end to the distal end with respect to the ovary, indicating the contraction wave propagation, which has the same propagating direction as the conduction wave (Dixon et al., 2011). Through calculating the time difference (or delay) between the most contracted states at different locations and measuring the distance between these measurement positions, the wave propagation velocity can be obtained based on the linear regression of the time delay data versus the distance data (Wang et al., 2018), potentially allowing for comparisons across estrous stages or different animals.

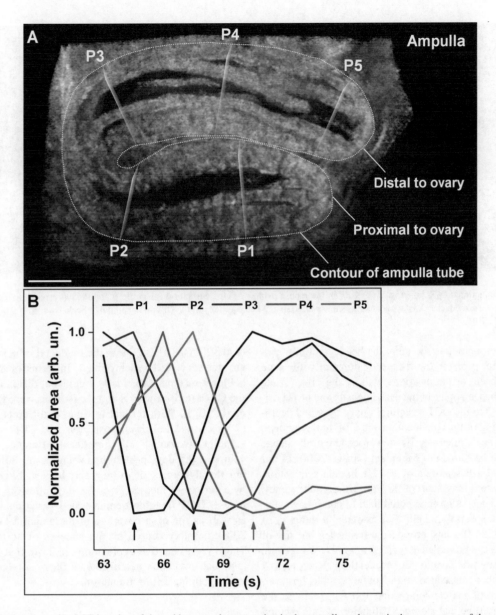

FIGURE 17.6 Time-lapse 3D OCT imaging of the oviduct muscle contraction in the ampulla and quantitative assessment of the contraction wave propagation. A. Time-lapse 3D imaging of the oviduct shows muscle contraction along the ampulla tube. B. Based on the measurement of the luminal area at five locations along the oviduct, the profile of the normalized area over time indicates the contraction wave propagation. Arrows point at the most contracted states at the five different locations. Scale bar is 300 μm. Reproduced from Wang et al., 2018.

17.3.5 Functional Imaging of the Ciliary Dynamics

The motile cilia in the mouse oviduct have a length of ~5–10 μm and a diameter of ~300 nm. Because of their small size, the oviductal cilia cannot be directly resolved based on general OCT imaging. We have developed a speckle-based OCT imaging method to functionally assess the ciliary dynamics in the oviduct *in vivo*. As a low-coherence imaging modality, OCT generates speckles in its structural images, which are sensitive to the movement of scatters in the tissue. A small displacement of the scatter is possible to change the status of interference between constructive state and destructive state, which, respectively, increases and decreases the intensity of the speckle. The beat of motile cilia can introduce this type of change to the OCT speckles (Oldenburg et al., 2012). We took advantage of the periodic cilia beating that produces

periodic intensity change in the OCT structural image and reconstructed the mapping of cilia location and CBF through frequency domain analyses (Wang, et al., 2015). This method allows for measurement of CBF in the oviduct, for the first time, without opening of the oviduct lumen, thus preserving the native biochemical and biomechanical state of the ciliated cells.

Specifically, at a selected location on the oviduct tube, 2D B-scan OCT images were taken over time to capture the OCT intensity fluctuations caused by the ciliary dynamics. The intensity profile from each pixel of the B-scan was then analyzed in spectral domain with a fast Fourier transform. To ensure this analysis was only conducted on the oviduct tissue, a binary image created from the averaged B-scans was used for eliminating the background. Within the B-scan, at the

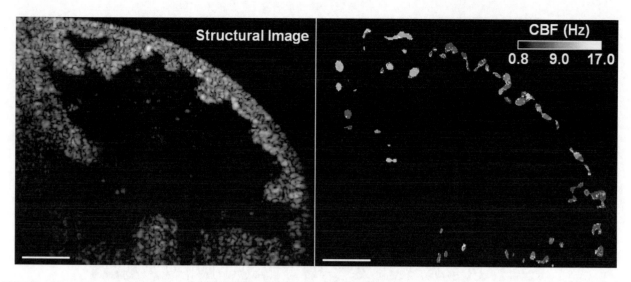

FIGURE 17.7   Functional OCT imaging of the CBF in the mouse oviduct *in vivo*. Structural image (left) and the corresponding mapping of CBF (right) from the mouse oviduct at 0.5 days postcoitus show a large spatial heterogeneity of the CBF in the ampulla. Scale bars are 100 μm. Reproduced from Wang et al., 2015.

positions corresponding to the cilia, the beating activity generated dominant peaks in the frequency domain, while such peaks were absent at the positions without the cilia (Wang et al., 2016). Thus, mapping of the peak intensity and peak position (frequency) to the OCT structural image provided depth-resolved imaging of the cilia location and CBF in the oviduct, respectively. The detailed step-by-step procedure of the image processing can be found in (Wang and Larina, 2018). In our studies, we had a B-scan rate of 100 Hz in data acquisition, corresponding to a maximum of 50 Hz resolvable frequency. A longer data acquisition time could help to obtain a smaller bin size in the spectral domain. For example, a duration of 2.56 seconds (256 B-scans) produced a frequency bin size of ~0.4 Hz, which can be sufficient for CBF analysis (Wang et al., 2018). Increasing this duration is beneficial for detecting low frequency components, which might not be possible, however, in the presence of muscle contraction and will increase the data processing time and storage. Validation of this method was conducted with the existing standard techniques. For the mapping of cilia location, we performed immunostaining histology on the same oviduct imaged by OCT, revealing a good agreement of the cilia coverage between the two methods (Wang et al., 2015). Also, we employed the bright-field video microscopy on the same region of the ampulla tissue and confirmed that the OCT-based approach is capable of providing CBF measurement and mapping (Wang et al., 2015).

This functional OCT method has been applied to a number of studies on the oviductal ciliary activities *in vivo*, including spatial, temporal, and longitudinal analyses of CBF (Wang et al., 2015; Wang et al., 2018). In particular, we have found that there exists a huge spatial heterogeneity of the CBF in the mouse oviduct, in both the ampulla and the isthmus. An example is shown in Figure 17.7. The CBF values generally spread over a large range covering as high as ~20 Hz (Wang et al., 2015; Wang et al., 2018). From the same mucosa fold, the CBF also spreads out (Wang et al., 2018), and even the adjacent cilia groups can beat at distinct frequencies (Wang

et al., 2015). This suggests the oviductal cilia in the mouse are not very well synchronized. In terms of the anatomical locations, no statistically significant difference between the CBFs in the ampulla and the isthmus were found (Wang et al., 2015). With respect to time, at different time points (0.5 days and 2.5 days) postcoitus, the CBF values in the ampulla have statistically significant difference, with higher values at 0.5 days postcoitus (Wang et al., 2015), suggesting the dynamics of ciliary activities is likely regulated by a variety of factors. Over a 3-day follow up of the same mouse, OCT imaging demonstrated variations of the CBF as well as the cilia coverage in the ampulla (Wang et al., 2018), possibly caused by the changes of hormone levels. These representative applications indicate that this *in vivo* approach could be a useful tool for the cilia-related research in the mammalian reproduction.

17.4 Discussions

17.4.1 Current Challenges and Technical Improvements

The described optical *in vivo* imaging approaches are promising and powerful tools for various reproductive studies in the mouse model. While optimistic, we would like to note the current challenges for these approaches and share our views for the potential technical improvements.

For the fluorescence imaging approach, the experimental procedure involved exposing the oviduct and associated reproductive organs in the dish filled with PBS. Currently, only imaging sessions lasting for a relatively short period were reported (Muro et al., 2016). Although the tissue was kept hydrated and the temperature was controlled, with this setup, prolonged data acquisition or longitudinal analysis are difficult to achieve. Potentially, the implantable dorsal imaging window can be implemented to address this limitation. As shown in

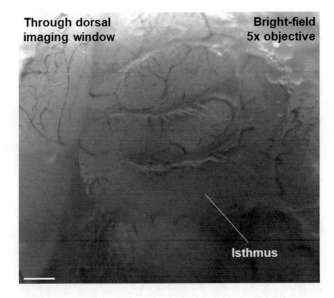

FIGURE 17.8 Bright-field microscopy of the mouse oviduct through the dorsal imaging window. Scale bar is 200 μm.

Figure 17.8, bright-field microscopy of the oviduct is possible through the window, suggesting the feasibility of fluorescence imaging. The relatively large window size in combination with a low-magnification objective lens will allow for covering a large field of view. Thus, an extended imaging session may enable a number of new studies, including further investigation of how ejaculated fluorescently labeled sperm migrate throughout the mouse oviduct.

For OCT imaging through the dorsal window, so far, only 2D depth-resolved mapping of the CBF was reported. With a single cross section, temporal and longitudinal analyses of ciliary behavior are less reliable, because the 2D location under investigation could change over time as a result of the muscle contraction. Mapping of the CBF in 3D is potentially possible through a C-scan of the repeated B-scan imaging and combining the 2D CBF maps into a volume (Wang et al., 2015). When performing time-lapse 3D imaging, a balance between the volume rate, the spatial sampling interval, and the transverse field of view is required. Specifically, the volume rate needs to be sufficiently high to capture the dynamics, the sampling interval should be small enough to not sacrifice the spatial resolution, and the field of view is expected to cover the region of interest. The current A-scan rate of the OCT system limits a further optimization of this balance. Significantly increasing the OCT A-scan rate, which is possible through implementation of recently developed MHz swept laser sources, will be highly beneficial for volumetric analysis of cilia beat frequency and other functional parameters.

17.4.2 Future Applications

We anticipate that these *in vivo* approaches will open a variety of new opportunities for studies of fertilization and preimplantation pregnancy. The dynamic aspects of the oviduct, including muscle contraction, ciliary activity, and gamete movement, can now be analyzed under different experimental conditions.

Studies on the sperm in the ampulla, so far, have been very limited (Ardon et al., 2016; Chang and Suarez, 2012), resulting in a gap of knowledge on the sperm activities at the site of fertilization (Suarez, 2016). Through implementation of the *in vivo* fluorescence microscopy approach to track fluorescently labeled sperm from transgenic males [Ishikawa et al., 2016; Muro et al., 2016], one can study interactions of sperm with the ampulla epithelium and the cumulus-oocyte complex. Prolonged imaging sessions may be needed to capture such events. Although only 2D information is available with this approach, these studies could provide valuable information about fertilization *in vivo*, such as the timing, location, and efficiency, to complement previous *ex vivo* or *in vitro* studies.

In vivo volumetric OCT imaging will likely reveal dynamic visualizations of multiple aspects of the reproductive process in the near future, which have been previously inaccessible, including ovulation, fertilization, and oocyte/embryo transport within the oviduct. These observations will reveal how the mammalian oviduct natively drives the oocytes and embryos for a successful pregnancy, and how genetic, pharmacological, and environmental factors influence these events. This will likely bring new and important advancements in understanding human reproductive disorders, such as infertility and ectopic pregnancy, and will contribute to further optimization of ARTs and contraception.

17.5 Conclusions

The function of the oviduct is a significant and attractive research topic, as the oviduct hosts the start of a new life, yet many unknows remain to be addressed in this field. *In vivo* high-resolution imaging of the mouse oviduct is promising to bring new discoveries about the role of the oviduct in the reproductive processes, including gamete transport, fertilization, and embryo transfer, taking our understanding of the oviductal physiology and pathology to a new level. In this chapter, we have reviewed the most recent advancements in developing and establishing *in vivo* structural and functional imaging of the mouse oviduct and the cellular dynamics in the oviduct lumen. We also discussed potential technological developments of these approaches and future applications. The fluorescence microscopy and the OCT imaging approaches are complementary and allow for characterization of nearly a full range of dynamic tissues and cells associated with the oviduct. These include the mucosal fold morphology, smooth muscle contraction, and cilia beating, as well as the dynamics of sperm and the cumulus-oocyte complexes. Given the robust capabilities of the described methods, we envision that more studies with these *in vivo* optical approaches will be carried out in the future to help elucidate the specific functions of the mammalian oviduct in fertilization and pregnancy.

Acknowledgments

This work was supported by the National Institutes of Health with grants R21EB028409 to SW, R01HD096335, R01EB027099, and R01HD099026 to IVL, as well as the Start-Up funding from Stevens Institute of Technology to SW.

REFERENCES

Abe, H., and Oikawa, T. (1992). Examination by scanning electron microscopy of oviductal epithelium of the prolific Chinese Meishan pig at follicular and luteal phases. *The Anatomical Record*, 233: 399–408.

Adhi, M. and Duker, J. S. (2013). Optical coherence tomography – Current and future applications. *Current Opinion in Ophthalmology*, 24: 213–221.

Andrade, Y. N., Fernandes, J., Vázquez, E., Fernández-Fernández, J. M., Arniges, M., Sánchez, T. M., Villalón, M., and Valverde, M. A. (2005). TRPV4 channel is involved in the coupling of fluid viscosity changes to epithelial ciliary activity. *The Journal of Cell Biology*, 168: 869–874.

Ardon, F., Markello, R. D., Hu, L., Deutsch, Z. I., Tung, C.-K., Wu, M., and Suarez, S. S. (2016). Dynamics of bovine sperm interaction with epithelium differ between oviductal isthmus and ampulla. *Biology of Reproduction*, 95: 90.

Bochner, F., Fellus-Alyagor, L., Kalchenko, V., Shinar, S., and Neeman, M. (2015). A novel intravital imaging window for longitudinal microscopy of the mouse ovary. *Scientific Reports*, 5: 12446.

Bouma, B. E., Villiger, M., Otsuka, K., and Oh, W.-Y. (2017). Intravascular optical coherence tomography [Invited]. *Biomedical Optics Express*, 8: 2660–2686.

Bloodgood, R. A. (2010). Sensory reception is an attribute of both primary cilia and motile cilia. *Journal of Cell Science*, 123: 505–509.

Burton, J. C., Wang, S., Behringer, R. R., and Larina, I. V. (2016). Three-dimensional imaging of the developing mouse female reproductive organs with optical coherence tomography. In *Proceedings of SPIE*, p. 97160E. SPIE.

Burton, J. C., Wang, S., and Larina, I. V. (2015). Dynamic imaging of preimplantation embryos in the murine oviduct. In *Proceedings of SPIE*, p. 933409. SPIE.

Burton, J. C., Wang, S., Stewart, C. A., Behringer, R. R., and Larina, I. V. (2015). High-resolution three-dimensional in vivo imaging of mouse oviduct using optical coherence tomography. *Biomedical Optics Express*, 6: 2713–2723.

Bylander, A., Nutu, M., Wellander, R., Goksör, M., Billig, H., and Larsson, D. G. J. (2010). Rapid effects of progesterone on ciliary beat frequency in the mouse fallopian tube. *Reproductive Biology and Endocrinology*, 8: 48.

Chang, H. and Suarez, S. S. (2012). Unexpected flagellar movement patterns and epithelial binding behavior of mouse sperm in the oviduct. *Biology of Reproduction*, 86, 140, 141-148-140, 141–148.

Chen, C.-L. and Wang, R. K. (2017). Optical coherence tomography based angiography [Invited]. *Biomedical Optics Express*, 8: 1056–1082.

Coy, P., García-Vázquez, F. A., Visconti, P. E., and Avilés, M. (2012). Roles of the oviduct in mammalian fertilization. *Reproduction*, 144: 649–660.

de Boer, J. F., Hitzenberger, C. K., and Yasuno, Y. (2017). Polarization sensitive optical coherence tomography - a review [Invited]. *Biomedical Optics Express*, 8: 1838–1873.

Dirksen, E. R. and Satir, P. (1972) Ciliary activity in the mouse oviduct as studied by transmission and scanning electron microscopy. *Tissue and Cell*, 4: 389–403.

Dixon, R. E., Britton, F. C., Baker, S. A., Hennig, G. W., Rollings, C. M., Sanders, K. M., and Ward, S. M. (2011). Electrical slow waves in the mouse oviduct are dependent on extracellular and intracellular calcium sources. *American Journal of Physiology. Cell Physiology*, 301: C1458–C1469.

Fauci, L. J. and Dillon, R. (2006). Biofluidmechanics of reproduction. *Annual Review of Fluid Mechanics*, 38: 371–394.

Fazeli, A., Affara, N. A., Hubank, M., and Holt, W. V. (2004). Sperm-induced modification of the oviductal gene expression profile after natural insemination in mice. *Biology of Reproduction*, 71: 60–65.

Fercher, A. F., Drexler, W., Hitzenberger, C. K., and Lasser, T. (2003). Optical coherence tomography - Principles and applications. *Reports on Progress in Physics*, 66: 239.

Georgiou, A. S., Sostaric, E., Wong, C. H., Snijders, A. P. L., Wright, P. C., Moore, H. D., and Fazeli, A. (2005). Gametes alter the oviductal secretory proteome. *Molecular & Cellular Proteomics*, 4: 1785–1796.

Ghosh, A., Syed, S. M., and Tanwar, P. S. (2017). In vivo genetic cell lineage tracing reveals that oviductal secretory cells self-renew and give rise to ciliated cells. *Development*, 144: 3031–3041.

Herrmann, J. M., Brezinski, M. E., Bouma, B. E., Boppart, S. A., Pitris, C., Southern, J. F., and Fujimoto, J. G. (1998). Two- and three-dimensional high-resolution imaging of the human oviduct with optical coherence tomography. *Fertility and Sterility*, 70: 155–158.

Hino, T., Muro, Y., Tamura-Nakano, M., Okabe, M., Tateno, H., and Yanagimachi, R. (2016). The behavior and acrosomal status of mouse spermatozoa in vitro, and within the oviduct during fertilization after natural mating. *Biology of Reproduction*, 95: 50.

Huang, D., Swanson, E. A., Lin, C. P., Schuman, J. S., Stinson, W. G., Chang, W., Hee, M. R., Flotte, T., Gregory, K., and Puliafito, C. A. (1991). Optical coherence tomography. *Science*, 254: 1178–1181.

Ishikawa, Y., Usui, T., Yamashita, M., Kanemori, Y., and Baba, T. (2016). Surfing and swimming of ejaculated sperm in the mouse oviduct. *Biology of Reproduction*, 94: 89.

Kim, J., Brown, W., Maher, J. R., Levinson, H., and Wax, A. (2015). Functional optical coherence tomography: Principles and progress. *Physics in Medicine and Biology*, 60: R211–R237.

Kölle, S. (2012). Chapter twenty-one - Live cell imaging of the oviduct. In P. M. Conn (Ed.), *Methods in enzymology*, pp. 415–423. Academic Press.

Kölle, S., Dubielzig, S., Reese, S., Wehrend, A., König, P., and Kummer, W. (2009). Ciliary transport, gamete interaction, and effects of the early embryo in the oviduct: Ex vivo analyses using a new digital videomicroscopic system in the cow. *Biology of Reproduction*, 81: 267–274.

Kondapalli, L. A. and Perales-Puchalt, A. (2013). Low birth weight: Is it related to assisted reproductive technology or underlying infertility? *Fertility and Sterility*, 99: 303–310.

Lamirel, C., Newman, N., and Biousse, V. (2009). The use of optical coherence tomography in neurology. *Reviews in Neurological Diseases*, 6: E105–E120.

Larin, K. V. and Sampson, D. D. (2017). Optical coherence elastography - OCT at work in tissue biomechanics [Invited]. *Biomedical Optics Express*, 8: 1172–1202.

La Spina, F. A., Puga Molina, L. C., Romarowski, A., Vitale, A. M., Falzone, T. L., Krapf, D., Hirohashi, N., and Buffone, M. G. (2016). Mouse sperm begin to undergo acrosomal exocytosis in the upper isthmus of the oviduct. *Developmental Biology, 411*: 172–182.

Leitgeb, R., Hitzenberger, C. K., and Fercher, A. F. (2003). Performance of fourier domain vs. time domain optical coherence tomography. *Optics Express, 11*: 889–894.

Leitgeb, R. A., Werkmeister, R. M., Blatter, C., and Schmetterer, L. (2014). Doppler optical coherence tomography. *Progress in Retinal and Eye Research, 41*: 26–43.

Li, S., O'Neill, S. R. S., Zhang, Y., Holtzman, M. J., Takemaru, K.-I., Korach, K. S., and Winuthayanon, W. (2017). Estrogen receptor α is required for oviductal transport of embryos. *The FASEB Journal, 31*: 1595–1607.

Li, S. and Winuthayanon, W. (2017). Oviduct: Roles in fertilization and early embryo development. *Journal of Endocrinology 232*: R1–R26.

Lucas, E. (2013). Epigenetic effects on the embryo as a result of periconceptional environment and assisted reproduction technology. *Reproductive BioMedicine Online, 27*: 477–485.

Lyons, R. A., Djahanbakhch, O., Mahmood, T., Saridogan, E., Sattar, S., Sheaff, M. T., Naftalin, A. A., and Chenoy, R. (2002). Fallopian tube ciliary beat frequency in relation to the stage of menstrual cycle and anatomical site. *Human Reproduction, 17*: 584–588.

Lyons, R. A., Saridogan, E., and Djahanbakhch, O. (2006). The reproductive significance of human fallopian tube cilia. *Human Reproduction Update, 12*: 363–372.

Mahmud, M. S., Cadotte, D. W., Vuong, B., Sun, C., Luk, T. W., Mariampillai, A., and Yang, V. X. (2013). Review of speckle and phase variance optical coherence tomography to visualize microvascular networks. *Journal of Biomedical Optics, 18*: 50901.

Men, J., Huang, Y., Solanki, J., Zeng, X., Alex, A., Jerwick, J., Zhang, Z., Tanzi, R. E., Li, A., and Zhou, C. (2016). Optical coherence tomography for brain imaging and developmental biology. *IEEE Journal of Selected Topics in Quantum Electronics, 22*: 120–132.

Mondéjar, I., Acuña, O. S., Izquierdo-Rico, M. J., Coy, P., and Avilés, M. (2012). The oviduct: functional genomic and proteomic approach. *Reproduction in Domestic Animals, 47*: 22–29.

Morgner, U., Drexler, W., Kärtner, F. X., Li, X. D., Pitris, C., Ippen, E. P., and Fujimoto, J. G. (2000). Spectroscopic optical coherence tomography. *Optics Letters, 25*: 111–113.

Muro, Y., Hasuwa, H., Isotani, A., Miyata, H., Yamagata, K., Ikawa, M., Yanagimachi, R., and Okabe, M. (2016). Behavior of mouse spermatozoa in the female reproductive tract from soon after mating to the beginning of fertilization. *Biology of Reproduction, 94*: 80.

Oldenburg, A. L., Chhetri, R. K., Hill, D. B., and Button, B. (2012). Monitoring airway mucus flow and ciliary activity with optical coherence tomography. *Biomedical Optics Express, 3*: 1978–1992.

Rodriguez, A., Tripurani, S. K., Burton, J. C., Clementi, C., Larina, I., and Pangas, S. A. (2016). SMAD signaling is required for structural integrity of the female reproductive tract and uterine function during early pregnancy in mice. *Biology of Reproduction, 95*: 44.

Sattler, E., Kastle, R., and Welzel, J. (2013). Optical coherence tomography in dermatology. *Journal of Biomedical Optics, 18*: 061224.

Sharma, R. K., Singh, R., and Bhardwaj, J. K. (2015). Scanning and transmission electron microscopic analysis of ampullary segment of oviduct during estrous cycle in caprines. *Scanning, 37*: 36–41.

Shi, D., Komatsu, K., Hirao, M., Toyooka, Y., Koyama, H., Tissir, F., Goffinet, A. M., Uemura, T., and Fujimori, T. (2014). Celsr1 is required for the generation of polarity at multiple levels of the mouse oviduct. *Development, 141*: 4558–4568.

Shi, D., Komatsu, K., Uemura, T., and Fujimori, T. (2011). Analysis of ciliary beat frequency and ovum transport ability in the mouse oviduct. *Genes to Cells, 16*: 282–290.

Suarez, S. S. (2016). Mammalian sperm interactions with the female reproductive tract. *Cell and Tissue Research, 363*: 185–194.

Tanwar, P. S., Kaneko-Tarui, T., Zhang, L., Tanaka, Y., Crum, C. P., and Teixeira, J. M. (2012). Stromal liver kinase B1 [STK11] signaling loss induces oviductal adenomas and endometrial cancer by activating mammalian target of rapamycin complex 1. *PLOS Genetics, 8*: e1002906.

Trottmann, M., Kölle, S., Leeb, R., Doering, D., Reese, S., Stief, C. G., Dulohery, K., Leavy, M., Kuznetsova, J., Homann, C., and Sroka, R. (2016). Ex vivo investigations on the potential of optical coherence tomography (OCT) as a diagnostic tool for reproductive medicine in a bovine model. *Journal of Biophotonics, 9*: 129–137.

Vakoc, B. J., Fukumura, D., Jain, R. K., and Bouma, B. E. (2012). Cancer imaging by optical coherence tomography: Preclinical progress and clinical potential. *Nature Reviews. Cancer, 12*: 363–368.

Wang, S., Burton, J. C., Behringer, R. R., and Larina, I. V. (2015). In vivo micro-scale tomography of ciliary behavior in the mammalian oviduct. *Scientific Reports, 5*: 13216.

Wang, S., Burton, J. C., Behringer, R. R., and Larina, I. V. (2016). Functional optical coherence tomography for high-resolution mapping of cilia beat frequency in the mouse oviduct in vivo. In *Proceedings of SPIE*, p. 96893R. SPIE.

Wang, S. and Larina, I. V., "In vivo imaging of the mouse reproductive organs, embryo transfer, and oviduct cilia dynamics using optical coherence tomography, In *Mouse Embryogenesis: Methods and Protocols*. Edited by P. D. Olguin (Humana Press, Springer Nature, 2018).

Wang, S., Singh, M., Lopez, A. L., Wu, C., Raghunathan, R., Schill, A., Li, J., Larin, K. V., and Larina, I. V. (2015). Direct four-dimensional structural and functional imaging of cardiovascular dynamics in mouse embryos with 1.5 MHz optical coherence tomography. *Optics Express, 40*, 4791–4794.

Wang, S., Syed, R., Grishina, O. A., and Larina, I. V. (2018). Prolonged in vivo functional assessment of the mouse oviduct using optical coherence tomography through a dorsal imaging window. *Journal of Biophotonics, 11*: e201700316.

Zhao, W., Zhu, Q., Yan, M., Li, C., Yuan, J., Qin, G., and Zhang, J. (2015). Levonorgestrel decreases cilia beat frequency of human fallopian tubes and rat oviducts without changing morphological structure. *Clinical and Experimental Pharmacology and Physiology, 42*: 171–178.

18

Immune System Imaging

Michael J. Hickey and M. Ursula Norman

CONTENTS

18.1 Introduction

The anatomical components of the immune system play host to many key events in the lives of immune cells, including their generation, maturation, activation, and elimination. The major sites for these events are the bone marrow, thymus, lymph nodes, peripheral lymphatics, and spleen. Advanced *in vivo* imaging has been critical in understanding what takes place beneath the surfaces of these organs. In this chapter we will discuss the approaches used and lessons learned from intravital microscopy studies of each of these organs.

18.2 Imaging Bone Marrow

The bone marrow is the cellular factory for the immune system. In addition to erythrocytes and platelets, the bone marrow is responsible for the generation of all circulating immune cell subsets. The maintenance of the appropriate circulating numbers of all of these cells is dependent on the proper functioning of the bone marrow and continual release of enormous numbers of cells into the bloodstream. Imaging has provided key insights into the physiology and pathophysiology of these processes. Two main approaches have been used for these studies: two-photon microscopy- (2PM) based imaging of the bone marrow in the mouse skull (calvarial) bone and imaging the bone marrow in long bones, such as the tibia, after mechanical bone thinning (Lo Celso et al., 2009; Kohler et al., 2011; Devi et al., 2013; Beck et al., 2014). Where similar experiments have used both approaches, there has been, in general, good agreement in the data generated from these two structurally distinct locations. These experiments typically involve labeling the immune/stem cell of interest via a genetically encoded or exogenously added fluorochrome and detecting the bone marrow microvasculature via an alternate fluorochrome, while using the inherent signal derived from second harmonic generation to detect bone and collagen (Lo Celso et al., 2009). In the sections to follow, we describe studies that have examined the nature of the haematopoietic niche and the behavior of haematopoietic stem/precursor cells (HSPCs) within it, the mechanisms of mobilisation of neutrophils, monocytes and B lymphocytes from the bone marrow, and the nature of platelet generation by megakaryocytes.

18.2.1 The Hematopoietic Stem Cell Niche

The hematopoietic stem cell niche is the name given to the complex multicellular and molecular microenvironment within the bone marrow that enables HSPCs to perform their functions of progenitor cell generation and self-renewal (Wei and Frenette, 2018). Numerous cell types resident in the bone marrow are integral to development and maintenance of a niche suitable for HSPC function (Figure 18.1). These include mesenchymal stem progenitor cells (MSPCs), which are the dominant producers of factors that promote HSPC activity, such as CXCL12 and stem cell factor. Additionally, cells of the osteoblast lineage (cells responsible for bone formation), endothelial cells, and adrenergic neurons, as well as the progeny of the HSPCs themselves, have roles in this process (Wei and Frenette, 2018).

Imaging of the bone marrow has been critical to the development of this understanding. Early studies of the bone marrow of the mouse skull demonstrated the existence of "hot spots" of expression of chemokine CXCL12 and vascular adhesion molecule E-selectin in sinusoids (bone marrow blood vessels) forming vascular niches that supported cell trafficking from the bloodstream (Sipkins et al., 2005). Lo Celso et al. (2009) used 2PM to investigate the cellular composition of the HSPC niche in the bone marrow of the skull, focusing on the role of the osteoblast (Lo Celso and Scadden, 2011). Examination of a reporter mouse expressing green fluorescent protein (GFP) driven by an osteoblast-specific promoter revealed that most osteoblasts were located within 20 μm of a sinusoid. To determine the locations of HSPCs within the niche, labeled HSPCs were transferred into W/Wv mice in which endogenous hematopoietic stem cell (HSC) generation is compromised and the niche is underpopulated, leaving it available for population by donor cells. Transferred HSPCs localized ~25 μm from the endosteum (inner lining of the bone) and, over the subsequent days, underwent proliferation, forming clusters derived from a single cell. The more differentiated cell types of multipotent progenitor-enriched and committed progenitor-enriched HSCs were found to reside further away from the endosteum. These observations demonstrated that the niche is located in

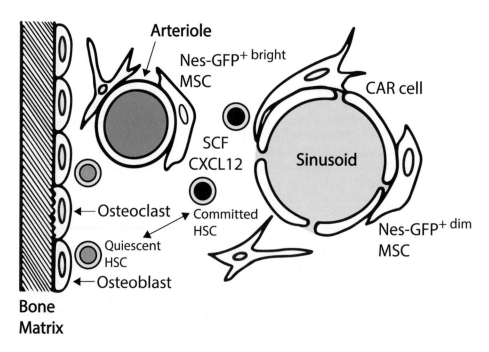

FIGURE 18.1 Simplified schematic of the hematopoietic stem cell niche in the bone marrow. Osteoblasts (bone-forming cells) and osteoclasts (bone-resorbing cells) line the interface of the bone and the bone marrow (BM). Osteoblasts, together with endothelial cells and various BM stromal cells, are thought to be important in the BM hematopoietic stem cell (HSC) niche. BM stromal cells, particularly those identified via Nestin (Nes) expression, as well as CXCL12-abundant reticular (CAR) cells and mesenchymal stem progenitor cells (MSPCs) are the main producers of CXCL12 and SCF, factors important in HSC retention in the BM. These cell types are found in close proximity to arterioles and sinusoids. Quiescent HSC (light grey) tend to be located close to osteoblasts, while more committed progenitor HSCs (dark grey) are typically located closer to sinusoids. At the appropriate time in their development, immune cells leave the BM via the sinusoids. Adapted from Wei and Frenette, 2018.

a perivascular/periendosteal position and that localization of HSPCs within the niche is determined by their differentiation status.

Using a similar approach, Mendez-Ferrer et al. (2010) found that nestin-expressing MSPCs form perivascular HSPC-containing niches in the bone marrow that facilitate HSPC homing. This finding was supported by studies using 3D confocal imaging of fixed bone marrow samples in specific reporter mice demonstrating the existence of two subsets of these MSPCs: an abundant low nestin-expressing population adjacent to bone marrow sinusoids and a rare population expressing high levels of nestin located adjacent to arterioles (Kunisaki et al., 2013) (Figure 18.1). Imaging studies have also demonstrated further cellular complexity in these HSPCs niches such as inclusion of regulatory T cells (Tregs) that act to support HSPC function by reducing oxidative stress in the niche (Fujisaki et al., 2011; Hirata et al., 2018).

To investigate the biology of HSPCs in more detail, Vandoorne et al. (2018) isolated stem cells from donor bone marrow, fluorescently labeled them, and transferred them into recipient mice for imaging. Using this approach to visualize HSPCs specifically, they found that systemic inflammatory activation with bacterial lipopolysaccharide induced an increase in HSPC proliferation, a response associated with increased neutrophil egress from the bone marrow. In addition, recent work using innovative long-term longitudinal intravital bone marrow imaging has revealed that the structure of the bone marrow microvasculature is not static – in contrast, it is highly dynamic undergoing continual structural reorganisation under homeostatic conditions (Reismann et al., 2017).

Transplantation of allogeneic HSPCs is a widely used therapeutic strategy in hematological malignancy, made effective by the long-term persistence of allogeneic cells in the hematopoietic stem cell niche. Removal of immune-suppressive Tregs was shown to promote elimination of transplanted allograft HSPCs, providing evidence that Tregs facilitate persistence of these allograft cells. Fujisaki et al. (2011) used imaging to examine this phenomenon, demonstrating that Tregs colocalized with transferred allogeneic HSPCs in bone marrow niches, adjacent to the endosteum. This colocalization during the prolonged persistence of allograft HSPCs within the bone marrow suggested that, along with their function in maintaining oxidative conditions within the niche, Tregs contribute to the generation of an immune-suppressive environment that enables allogeneic HSPCs to avoid rejection by the host immune system.

18.2.2 Neutrophil and Monocyte and Mobilization

Neutrophils and monocytes play essential roles in the innate immune system. Because of their short half-lives in the circulation, large numbers of both cell types are generated in the bone marrow and released into the circulation continuously. Additionally, in response to infection or sterile injury, the circulating numbers can rapidly increase. *In vivo* imaging has been critical in developing our understanding of the regulation of neutrophil and monocyte release from the bone marrow.

Granulocyte colony-stimulating factor (G-CSF) and the CXCR4 antagonist plerixafor are therapeutic agents used to restore circulating neutrophil numbers in neutropenic patients (Devi et al., 2013). 2PM has been used to investigate neutrophil behavior in the bone marrow and the effects of these therapeutic agents, making use of LysM-GFP mice as neutrophil reporter mice because of the high expression of GFP in neutrophils in these animals (Kohler et al., 2011; Devi et al., 2013). Under resting conditions, the majority of neutrophils in the bone marrow were found to be nonmotile, with ~30% undergoing slow random migration. Egress from the bone marrow via transmigration into the sinusoidal vasculature was seen only rarely, although neutrophils adjacent to these vessels actively probed their surroundings. However, G-CSF administration induced marked changes in neutrophil behavior within 1–2 hours, increasing the proportion of actively migratory cells as well as the average migration velocity. In response to G-CSF, neutrophils migrated toward the blood vessels and were seen exiting into the bloodstream. This response was dependent on CXCR2, consistent with a role for induction of chemokine ligands including CXCL1 and CXCL2. Notably, CXCR4 antagonism via plerixafor, which also increases circulating neutrophil numbers, did not modulate neutrophil behavior in the bone marrow, indicating that these agents work via distinct mechanisms. These studies also revealed that neutrophils move from the circulation back into the bone marrow space, undergoing recruitment from the bone marrow microvasculature, and that this behavior is blocked by CXCR4 inhibition.

Monocyte release from the bone marrow is mediated by different mechanisms from those used by neutrophils, with CCR2 being central to this response (Serbina and Pamer, 2006). To examine monocyte behavior in the bone marrow, various myeloid cell reporter mice have been examined. *Cx3Cr1gfp/+* mice express GFP in monocytes, macrophages and other myeloid lineages. GFP-labeled cells in the bone marrow of these mice adopt both spherical and dendritic morphologies, consistent with GFP labeling monocytes and their progenitors, as well as macrophages and dendritic cells (DCs) (Evrard et al., 2015). To detect monocyte progenitors/immature monocytes more selectively, MacBlue mice (expressing cyan fluorescent protein/CFP in monocytes) have been used (Jacquelin et al., 2013). In these mice, CFP is more monocyte-restricted than GFP is in *Cx3Cr1gfp/+* mice, and CFP+ cells in the bone marrow are predominantly spherical in nature and, therefore, more likely to represent immature monocytes. In 2PM imaging studies of the bone marrow of MacBlue mice, CFP+ cells displayed variable motility. On average, monocytes were slowly migratory, although cells adjacent to the vasculature moved more rapidly and in a more directed fashion, even being occasionally observed migrating on the endothelial surface of the sinusoidal vasculature (Jacquelin et al., 2013). In studies of bone marrow repopulation following chemotherapy-mediated ablation, CFP+ cells repopulated the parenchyma over 3–5 days, initially as single, blastic cells, which developed into clusters by Day 5. At this stage, highly motile monocytes were detected in the perivascular regions, and monocytes were seen to transmigrate into the sinusoidal vasculature. In a more acute model, LPS stimulation was seen to stimulate monocyte migration within 1 hour and induce egress into the sinusoidal vasculature within 2 hours (Evrard et al., 2015).

18.2.3 Lymphocyte Dynamics in the Bone Marrow

Imaging studies of B cells in the bone marrow revealed a mechanistically distinct mechanism for leukocyte egress from the bone marrow (Beck et al., 2014). Beck et al. (2014) imaged the bone marrow of *Rag1*[GFP/+] mice, in which >99% of GFP[+] cells are of the B cell lineage. GFP[+] B lineage cells underwent constitutive migration, at ~ 2.4 µm/min. As for other cell types, CXCR4 antagonism promoted B cell egress from the bone marrow. Shortly after administration of a CXCR4 antagonist or anti-CXCL12, B cell migration was markedly reduced (Beck et al., 2014), demonstrating the key role of the CXCR4/CXCL12 pathway in promoting migration of immature B cells in the bone marrow. However, given the ability of CXCR4 inhibition to promote B cell egress from the bone marrow, this finding was unexpected. How do B cells leave the bone marrow parenchyma when migration is inhibited? Further investigation of the mechanisms of B cell migration revealed that inhibition of all chemokine-driven migration increased B cell egress from the bone marrow (Beck et al., 2014). These findings led to the hypothesis that inhibition of B cell migration "trapped" these cells adjacent to the sinusoidal vessels. Imaging perisinusoidal B cells under conditions of inhibition of all chemoattractant receptors revealed that these cells did not undergo ameboid movement typical of migratory leukocytes but, nevertheless, entered the sinusoids and were swept away in the bloodstream. This phenomenon was associated with a highly leaky endothelial barrier indicating that it was facilitated by the fenestrated nature of the sinusoidal endothelium (Beck et al., 2014). In the absence of CXCR4 inhibition, i.e. under homeostatic conditions, maturing B cells also lose the ability to respond to CXCL12 by progressively downregulating CXCR4, thereby facilitating their egress from the bone marrow. Inhibition of chemoattractant receptors also promoted egress of natural killer (NK) cells, monocytes, and granulocytes into the bloodstream. Together these findings indicate that chemokine-driven migration in the bone marrow acts to inhibit leukocyte mobilization into the bloodstream. Intriguingly, they raise the possibility that leukocyte entry into the sinusoids is driven by passive forces external to the leukocyte, such as fluid flow into the sinusoid. Such a phenomenon would explain how poorly motile erythrocytes are able to be delivered into the bloodstream.

18.2.4 Thrombopoiesis

Platelets are the most abundant cell in the circulation, being generated in large numbers on a daily basis. Bone marrow-resident megakaryocytes are responsible for platelet generation, although prior to the application of imaging it was unclear how platelets derived from this large cell located in the bone marrow stroma were delivered into the bloodstream. Junt et al. (2007a) addressed this making use of a CD41-EYFP reporter mouse in which yellow fluorescent protein (YFP), expressed under control of a megakaryocyte-specific promoter, labels megakaryocytes and platelets. They used 2PM to image the bone marrow in the skull of these animals, noting that megakaryocytes were immotile, but predominantly located adjacent to bone marrow microvessels. Dynamic imaging revealed that

megakaryocytes extend long proplatelet-bearing processes into the bone marrow microvasculature, exposing them to the shear forces of flowing blood. This leads to the detachment of heterogeneous fragments of these processes into the circulation. This was the first description of *in vivo* generation of proplatelets, platelet precursors thought to be converted into mature platelets via intravascular shear forces at other locations in the circulation.

More recent imaging studies have challenged these findings. Using advanced intravital and correlative electron microscopy, Brown et al. (2018) revealed that the predominant mechanism of megakaryocyte exposure to the bloodstream is via extrusion of numerous, large cellular protrusions into the bone marrow sinusoids. These large structures were found to anchor to the sinusoidal endothelium where they formed the major initial substrate for platelet biogenesis in the circulation.

18.3 Imaging the Thymus

18.3.1 Introduction

T lymphocytes play a critical role in the adaptive immune response, providing long-term specific immunity against microbial pathogens, while restricting inappropriate autoimmune responses. T cells initially arise from progenitors in the bone marrow. However, in contrast to many other immune cells, T cells must complete their development outside the bone marrow, in the thymus (Bhandoola et al., 2007). T cell progenitors enter the thymus from the bloodstream, where they undergo a series of selection events, while migrating through the different regions of the thymus, interacting with stromal cells and other immune cells. Only ~2% of the T cell progenitors that enter the thymus survive these stringent processes, going on to enter the circulation as naïve T cells (Sprent et al., 1996). The key step in intrathymic T cell development is recombination of T cell receptor (TCR) genes to generate a TCR. Thymocytes that express a functional TCR (capable of recognizing antigenic peptide presented on major histocompatibility complex (MHC) molecules) are allowed to survive ("positive selection") and undergo further development. Subsequently, potentially autoreactive T cells are eliminated ("negative selection"). Thymic stromal cells play essential roles in both positive and negative selection. The thymus also supports development of cell types including γδ T cells, invariant natural killer T (iNKT cells) cells and B cells (Bhandoola et al., 2007; Klein Wolterink et al., 2010; Cowan et al., 2015; Perera and Huang, 2015; Male and Brady, 2017). However, the majority of cells that exit the adult thymus are conventional CD4[+] and CD8[+] αβ T cells (Akashi et al., 2000; Klein Wolterink et al., 2010; Cowan et al., 2015). As such, this section will focus on how imaging techniques have advanced our understanding of differentiation αβ T cells in this organ.

18.3.2 Overview of Thymocyte Development

The thymus is structurally divided into four compartments based on the types of thymic epithelial cells (TEC) present: 1) the subcapsular zone (SCZ), 2) the cortex, 3)

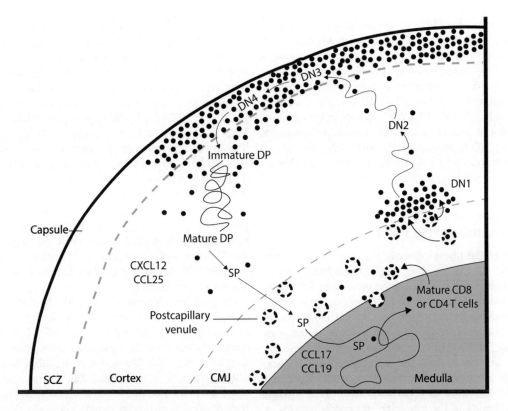

FIGURE 18.2 Structure of the mouse thymus. The mouse thymus is comprised of two lobes encased in a thin connective tissue capsule. Multipotent progenitor cells enter the thymus at the highly vascular corticomedullary junction (CMJ). This region borders the inner (medulla) and outer (cortex) regions of the thymus. Early T cell progenitors that have recently exited vessels in the CMJ remain confined within this area for approximately 9–10 days and undergo proliferation. These cells do not express the coreceptors CD4 or CD8 and are termed double negative (DN) cells. DN cells are further subdivided into four groups based on differential expression of surface markers. DN cells migrate in a slow random manner toward the outer regions of the cortex until they reach the outermost region, the subcapsular zone (SCZ). During their migration toward the SCZ, thymocytes undergo rearrangement of the TCRβ, γ and δ gene loci, and upon reaching the SCZ they rearrange the TCRα gene loci and upregulate both CD4 and CD8 to become double positive (DP) cells. Substantial proliferation of DP thymocytes occurs within the SCZ, and immature DP cells undergo slow random migration through the cortex interacting with thymic epithelial cells to undergo positive selection. After undergoing successful positive selection, mature DP thymocytes migrate to the medulla in a rapid, nonrandom manner and differentiate into single positive (SP) T cells expressing either CD8 or CD4. CD4 or CD8 SP T cells with high avidity for peripheral tissue-restricted self-antigen undergo negative selection within the medulla. Mature T cells that survive the negative selection process are released into the circulation at the CMJ. In the cortex and medulla, region-specific expression of the chemokines shown contributes to thymocyte migration.

the corticomedullary junction (CMJ), and 4) the medulla (Figure 18.2) (Petrie and Zuniga-Pflucker, 2007). The outer compartments (cortex and SCZ) are densely packed with immature thymocytes with fewer cortical thymic epithelial cells (cTECs) and macrophages. By comparison, the inner compartment (medulla) is more loosely packed and contains mature thymocytes with more prominent medullary epithelial cells (mTECs) as well as DCs and macrophages (Pearse, 2006). The CMJ is the highly vascularized region where progenitor cells enter, and mature thymocytes leave, the thymus (Lind et al., 2001; Mori et al., 2007).

Prior to studies using 2PM, the movement of thymocytes throughout the thymus could only be inferred through snapshot views of fixed tissue (Petrie and Zuniga-Pflucker, 2007; Dzhagalov and Phee, 2012). These studies revealed that thymocytes moved throughout the thymus, with developmental milestones occurring within distinct areas (Figure 18.2). Briefly, T cell progenitors that have just entered the thymus in the CMJ undergo proliferation and remain confined in this area for ~9–10 days (Lind et al., 2001; Porritt et al., 2003). As these cells

do not express CD4 or CD8, they are termed double negative (DN) cells. DN cells undergo various migration and maturation events through the inner, mid, and outer cortex (differentiating from DN1 to DN4 cells), before becoming committed to the T cell lineage via rearrangement of the TCRβ, γ, and δ gene loci. In the SCZ, they upregulate expression of CD4 and CD8, becoming double positive (DP) cells, and rearrange their TCRα gene loci to enable expression of a surface TCR (Lind et al., 2001; Porritt et al., 2003; Petrie and Zuniga-Pflucker, 2007; Bunting et al., 2011). In the SCZ, DP thymocytes undergo substantial proliferation before migrating toward the medulla. Survival of DP thymocytes is dependent on intermediate affinity TCR-peptide MHC interactions (positive selection) (Klein et al., 2009). Positively selected DP thymocytes migrate to the medulla and become single positive (SP) T cells (either CD8+ or CD4+). These cells interact with mTECs and DCs that express peripheral tissue-restricted self-antigens to undergo negative selection or differentiation into Foxp3+ regulatory T cells (Nitta and Suzuki, 2016). Mature T cells are then exported into the circulation at the CMJ (Figure 18.2).

In regard to the mechanisms of this complex choreography of intrathymic thymocyte migration, immunohistochemical and flow cytometric studies have highlighted roles for stage-specific adhesion molecule and chemokine receptor expression. Similarly, compartmentalized expression of specific chemokines within the cortex (CCL25, CXCL12) and medulla (CCL19, CCL21, CCL17, CCL22) provides evidence of a role for these chemokines in directing thymocyte migration (Misslitz et al., 2006; Bunting et al., 2011; Dzhagalov et al., 2012). However, only via 2PM imaging has it been possible to test the hypotheses arising from these studies.

18.3.3 Imaging the Thymic Microenvironment via 2PM

The location of the thymus directly above the heart makes it susceptible to cardiac and respiration-induced movements, and renders it problematic to image *in vivo*. Therefore, researchers have turned to imaging thymic explants maintained under physiological conditions. Studies using whole-organ explants limit observations to the cortex as imaging can only penetrate 200–300 μm beneath the capsule (Dzhagalov et al., 2012). In order to study the deeper medullary regions, thymic slices are used (Dzhagalov et al., 2012; Kurd and Robey, 2016; Lancaster and Ehrlich, 2017). This method allows investigation of the migration of specific subsets of thymocytes, as purified thymocyte subsets can be layered on top of thymic slices, allowing the cells to migrate into the tissue and their behavior in the intact thymic microenvironment to be visualized (Lancaster and Ehrlich, 2017). Chemotactic and differentiation signals are preserved in these preparations as thymocytes migrate to their normal location according to their developmental stage and can successfully undergo differentiation within the slice (Ehrlich et al., 2009; Melichar et al., 2013; Kurd and Robey, 2016).

18.3.4 Thymocyte Motility During Positive and Negative Selection

Early 2PM studies examined thymocyte interactions with stromal cells in reaggregated thymic organ cultures (RTOC), where DP thymocytes with a fixed TCR were cocultured with a mix of thymic stromal cells (Bousso et al., 2002). These studies revealed that thymocytes are highly mobile and undergo active crawling on the surface of stromal cells. In intact thymus and thymic slice cultures, 2PM revealed that DP thymocytes undergo two types of behavior during positive selection (Bhakta et al., 2005; Witt et al., 2005). The majority move at a relatively slow rate (3–8 μm/min) in a nondirected fashion (random walk) while a relatively small population move directionally toward the medulla at a faster velocity (> 10 μm/min) (Witt et al., 2005). It is hypothesized that these faster, more directed cells represent DP cells that have survived positive selection. Interestingly, DN cells also migrate in a slow random walk but are restricted to the cortex (Ehrlich et al., 2009). In a 2PM study, in which all thymic stromal cells were labeled with GFP, thymocytes were shown to be in constant contact with 3D mesh-like structures formed by the thymic stromal cells within the cortex and medulla (Sanos et al., 2011). The random

walk behavior exhibited by thymocytes may be explained by these cells using this 3D network as a substrate for migration.

2PM examination of preselection DP thymocyte behavior reveals that both MHC-I and MHC-II/TCR interactions induce brief elevations in intracellular calcium (Ca^{2+}) concentration, an effect associated with a reduction in thymocyte motility (Bhakta et al., 2005; Melichar et al., 2013; Ross et al., 2014). Cells undergoing positive selection via an MHC-II-restricted TCR have longer periods of immobility (~ 30 mins) (Bhakta et al., 2005) compared with selection on an MHC-I-restricted TCR (5 mins) (Ross et al., 2014). As selection progresses, the basal level of Ca^{2+} is increased while thymocytes still experience rapid Ca^{2+} fluctuations, albeit for short durations accompanied by shorter periods of thymocyte immobility (Ross et al., 2014). These studies have been interpreted to indicate that DP thymocytes scan a range of stromal cells, integrating multiple TCR signals to reach a threshold for positive selection (Lancaster and Ehrlich, 2017).

Postpositive selection DP and SP thymocytes within or near the medulla are characterized by high motility, whereas negative selection, induced by high avidity ligand-binding, is marked by sustained Ca^{2+} elevation and migratory arrest (Le Borgne et al., 2009; Dzhagalov et al., 2013; Melichar et al., 2013). While it was believed that negative selection was restricted to the medulla, negative selection of self-reactive DP thymocytes has also been found to occur in the cortex via a mechanism involving thymocyte association with cortical DCs (McCaughtry et al., 2008). In support of this concept, 2PM studies have shown that DP thymocytes extensively interact with cortical DCs (Ladi et al., 2008). These observations indicate that selection events might not be as compartmentalized as previously thought. After negative selection, thymocytes show reduced motility, but do retain a degree of movement for hours after the initial TCR signaling, until they encounter a phagocytic cell. Thymocyte death ultimately occurs in contact with or enclosed by this phagocyte (Dzhagalov et al., 2013).

18.3.5 Chemokine Control of Thymocyte Movement

The role of chemokines in directing the migration of thymocytes within the thymus is undoubtedly complex. Imaging thymocytes after their application to thymic slices reveals that DN or DP cells segregate to the cortex with DP cells preferentially localizing to the CMJ, while SP thymocytes migrate to the medulla (Ehrlich et al., 2009; Dzhagalov et al., 2012). These observations show that thymocyte migration differs according to the differentiation state of the cell, but what are the underlying mechanisms?

Leukocyte migration is typically driven by interaction of chemokines with leukocyte-expressed chemokine receptors. In the cortex, chemokine expression is relatively low and uniform with CCL25 and CXCL12 contributing to DN thymocyte accumulation at the SCZ (Plotkin et al., 2003; Bunting et al., 2011; Dzhagalov and Phee, 2012). In the medulla, both the CCR4 ligands (CCL17, CCL22) and the CCR7 ligands (CCL19, CCL21) are highly expressed by medullary DCs, setting up a chemotactic gradient to attract cells expressing the appropriate receptors. After positive selection, thymocytes upregulate CCR4 and CCR7 allowing these cells to migrate into the medulla in response to this chemotactic gradient (Misslitz et al., 2006; Petrie and Zuniga-Pflucker, 2007). 2PM studies show that this migration is blocked by an inhibitor of chemokine receptor signaling. Similarly, positively selected thymocytes deficient in CCR7 showed reduced entry into the medulla following positive selection (Kurobe et al., 2006; Ehrlich et al., 2009), while CCR4 deficiency also reduced the ability of postpositive selection DP and SP CD4 thymocytes to accumulate in the medulla and to interact with medullary DCs (Hu et al., 2015). Together these observations support a role for chemokine-driven migration in these responses.

However, chemokine-directed migration does not explain all thymocyte behavior. For example, while DN and DP thymocytes express the same chemokine receptors (CCR9 and CXCR4), they undergo different patterns of migration, indicating that differential expression of chemokine receptors is insufficient to explain differences in migration between these populations (Bunting et al., 2011; Dzhagalov and Phee, 2012). Passive chemokine-independent processes may also influence thymocyte localization. For example, the slow inward movement of DP thymocytes prior to positive selection may be driven passively by intense cell proliferation at the SCZ. Together, these data demonstrate the complex and dynamic nature of the mechanisms controlling the positioning of thymocyte subsets within the thymus. In the absence of 2PM imaging, unraveling this complexity would have presented an extremely challenging problem.

18.3.6 Imaging the Thymus – Summary

Imaging has aided immunologists to understand how thymocytes are directed to enter the medulla and interact with different APC populations. This is important as even subtle changes in thymocyte migration within the medulla can influence negative selection and central tolerance. This was highlighted recently in 2PM studies showing that the minor contribution of the chemokine receptor Epstein-Barr virus- induced receptor 2 (EBI2) to thymocyte migration speed is sufficient to alter negative selection (Ki et al., 2017). This ability to study small but important changes in thymocyte migration within a 3D environment highlights the power of intravital imaging techniques. Future studies examining the contribution of chemokines and adhesion molecules to thymocyte interactions with APCs across the thymus will be illuminating, for example in studying the movement and stromal interactions of rarer populations of progenitor cells including γδ T cells, iNKT cells and B cells.

18.4 Imaging the Lymph Node

The body is exposed to countless pathogenic microorganisms and other challenges on a daily basis. To counter this, the adaptive immune system develops the capacity to recognize and respond to an enormous diversity of molecular antigens. The exquisite specificity of the adaptive immune response, conferred via the capacity of T cells and antibodies to recognize unique molecular patterns, plays a key role in combating invading pathogens and malignancy. The number of different antigen specificities recognized by the body's CD4+ T cells for example is believed to be at least in the $10^5–10^6$ range (Moon et al., 2007). The corollary of this, however, is that the total number of T cells specific for any one antigenic peptide is extremely low under resting conditions, being estimated for CD4+ T cells in mice as being in the range of 20–200 (Moon et al., 2007). As such, the challenge for the immune system during pathogen infection for example, is to enable the small number of T cells with the antigenic specificity appropriate for the pathogen to be in the right place at the right time to detect it and respond before it overwhelms the host. How does the immune system overcome this daunting challenge?

18.4.1 Lymph Node Structure

The lymph nodes serve as key sites for development of cellular and humoral adaptive immune responses. It has been long-established that lymph nodes play host to a large number of transient and resident immune cell populations, as well as serving as a collection point for antigens and pathogen-derived particles draining from adjacent organs (Gowans and Knight, 1964). The outermost regions of the lymph node comprise the capsule, and the subcapsular sinus through which percolate lymph and cells draining into the node (Figure 18.3). The lymph node cortex immediately subjacent to this structure is comprised of distinct regions characterized by compartmentalization of lymphocyte subsets: B cells are concentrated in follicles, whereas T cells are located in the interfollicular T cell zones and deeper paracortex (Lindquist et al., 2004). These areas are also populated by DCs and supporting fibroblastic reticular cells (FRCs), and supplied by a microvasculature characterized by the highly specialized lymph node high endothelial venules (HEV).

18.4.2 Lymphocyte Interactions in High Endothelial Venules

Early intravital microscopy studies examined the mechanisms whereby T cells entered the lymph node from the bloodstream via interactions in HEV. Endothelial cells in HEV express an adhesion molecule signature that facilitates interactions of circulating naïve lymphocytes, enabling these cells to enter the lymph node via a sequence of interactions, culminating in transmigration. Using epifluorescence intravital microscopy to visualize the lymph node, Von Andrian and Warnock (von Andrian, 1996; Warnock et al., 1998) demonstrated that lymphocytes rolled and adhered in paracortical HEV, with lymphocyte rolling being dependent on L-selectin and adhesion on LFA-1. Subsequent experiments revealed a role for the CCR7 ligand CCL21 in mediating arrest of rolling lymphocytes, demonstrating the importance of this pathway in delivery of naïve lymphocytes into the lymph node (Stein et al., 2000).

While these experiments revealed the basis of entry of blood-borne lymphocytes into the lymph, the subsequent actions of these cells remained a mystery. Indeed, prior to the advent of 2PM-based *in vivo* imaging, how immune cells in the lymph node worked together in the promulgation of immune responses could only be inferred from histology and flow cytometry-based studies. However, over the past 15 years,

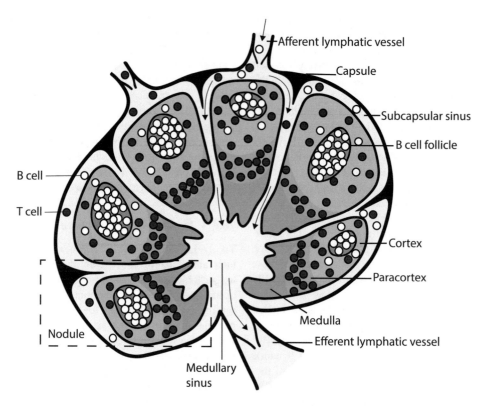

FIGURE 18.3 Structure of the mouse lymph node. The lymph node is composed of several nodules surrounded by a thick connective tissue layer called the capsule. Most nodules are serviced by a single afferent lymphatic vessel that delivers lymph containing lymphocytes and antigen-presenting cells from peripheral tissues. A constant stream of lymph enters the supcapsular sinus, drains around the lobules into the medullary sinus, and exits the lymph node via a single efferent lymphatic vessel. The nodule cortex is divided into the outer cortex that contains B cells localized into follicles and the inner cortex (paracortex) that mainly contains T cells. The deeper medulla areas are populated by plasma cells and macrophages. This area also contains large blood vessels that branch to supply each lobule (not shown). High endothelial venules (not pictured) are mainly found in the paracortical areas and represent the main port of lymphocyte entry into the lymph nodes from the bloodstream.

2PM imaging has enabled investigation of the dynamics of immune cells up to 300 μm below the surface of lymph nodes. This approach has revealed cellular behaviors in T cell zones and B cell follicles and described the incredibly complex choreography of immune cell migration, cell–cell interactions, signaling, activation, and proliferation that occurs during these responses. Two main approaches have been used for 2PM-based investigation of the lymph node: *ex vivo* analysis of isolated lymph nodes in organ baths that are maintained in appropriately oxygenated buffers at 37°C (Miller et al., 2002; Stoll et al., 2002). Alternatively, lymph nodes have been examined *in vivo*, i.e. *in situ* in the animal, ensuring that normal blood and lymphatic flow are maintained (Mempel et al., 2004; Miller et al., 2004). In general, there has been good agreement between the findings generated using these two approaches (Germain et al., 2012). In the sections below we will describe findings generated using these approaches.

18.4.3 Dynamics of T Cells, B Cells, and Dendritic Cells in the Lymph Node

Initial studies focused on characterization of T and B cell motility in lymph nodes (Cahalan et al., 2002; Miller et al., 2002). In these studies, T and B cells from donor mice were purified and differentially labeled with fluorochromes before

intravenous transfer into recipient mice. After allowing time for the lymphocytes to home to the nodes, lymph nodes were removed and imaged in organ baths. Consistent with previous homing analyses, B cells migrated to follicles while T cells were found in interfollicular and paracortical T cell zones. T cells underwent rapid, random migration interspersed with changes of direction and periods of reduced migration or stasis. B cells were also spontaneously migratory, although at slower velocities. Use of lymphocytes from TCR transgenic mice, in which most T cells recognize the same antigen, enabled assessment of the effect of antigen recognition on this behavior. Antigen-challenged cells were typically larger and, in many cases, relatively immobile, while some cells remained freely motile. Evidence of proliferation of T cells in the node, assessed by analysis of reduction of dye staining intensity, was apparent 5 days after antigen stimulation.

It was known that T cell responses to antigen were driven by DCs presenting antigen for recognition by lymphocytes. Therefore, imaging experiments were devised to enable simultaneous visualization of differentially labeled lymphocytes and DCs (Stoll et al., 2002; Mempel et al., 2004). T cells were transferred intravenously, while DCs were injected subcutaneously, enabling them to migrate via the lymphatic vasculature to the local lymph node. Eighteen hours later, lymph nodes were removed and imaged via 2PM. When DCs were transferred with antigen recognized by the T cells, T cells underwent sustained stable interactions with DCs in T cell zones, a response associated with activation of the lymphocyte. T cells also interacted with nonantigen-bearing DCs, although these interactions were less prolonged, and T cells detached and moved away. Examination 37 hours after T cell transfer revealed that T cells were now highly migratory.

In vivo analyses of T cell and DC behavior in the lymph node revealed similar staging in the development of the antigen-specific T cell response (Mempel et al., 2004; Miller et al., 2004). Soon after their entry into lymph nodes, T cells were highly migratory, making contact with numerous DCs as they migrated randomly within the node. The behavior of DCs differed from that of T cells, with these cells being less actively migratory but rapidly moving cellular processes around the surrounding area. DCs made short-lived contacts with numerous T cells in a short time span, often with multiple T cells simultaneously (Miller et al., 2004). Notably, DCs that migrated into the lymph node were often found adjacent to HEVs within the paracortex, sites appropriate for encountering T cells immediately after they enter the node from the vasculature. Under conditions in which the DCs were loaded with antigen recognized by the T cells, 8-12 hours after T cell transfer interactions between T cells and DCs became significantly longer. During these interactions, DCs migrated slowly, bringing the attached T cell with them. At the end of this phase, T cells demonstrated expression of IL-2 and interferon-γ and upregulation of the IL-2 receptor CD25, indicating that these prolonged interactions were central to T cell activation. By 24 hours, behavior had reverted to more short-lived T cell/DC contacts, with T cell proliferation becoming detected ~1.5 days after T cell transfer (Stoll et al., 2002). More recent imaging work compared the actions of migratory DCs entering lymph nodes from the periphery with those of lymph node-resident DCs, demonstrating that migratory DCs can disseminate antigen from the periphery to the highly abundant lymph node-resident DCs (Gurevich et al., 2017). This increases the number of cells able to present antigen to T cells, thereby increasing the efficiency of T cell activation.

The mechanisms underlying lymphocyte migration in the lymph node subsequently became a major focus of investigations. Studies soon revealed that T and B cells migrate in the paracortex on a network formed by FRCs, cells that also act as a substrate for DC migration (Bajenoff et al., 2006). These were innovative experiments in that they made use of mice expressing GFP in all nucleated cells, but replaced their bone marrow and, therefore, lymphocytes with nonfluorescent cells, leaving only stromal cells expressing GFP. T cells labeled with an alternative red fluorochrome were transferred into these mice, enabling differentiation of T cells from FRCs. In B cell follicles, follicular DCs were found to play a similar role in supporting and directing B cell migration (Bajenoff et al., 2006). Together, these studies revealed that lymphocytes, supported by lymph node stromal cells, adopt a random search strategy to maximize their opportunities for encountering multiple DCs. At later stages of the response, T cells undergo prolonged interactions with antigen-bearing DCs that support T cell activation and, eventually, proliferation.

After undergoing antigen-mediated activation within lymph nodes, T cells leave the lymph nodes via the lymphatic vasculature, eventually entering the bloodstream to undergo recirculation. Imaging has also revealed the mechanisms underlying lymphocyte egress from the lymph node, demonstrating that T cells migrate toward and within the lymph node sinus at which stage they are able to be captured by lymph flow (Grigorova et al., 2009). This process is driven by interaction of sphingosine 1-phosphate (S1P), a chemoattractant abundant in lymph, via interaction with T cell-expressed sphingosine 1-phosphate receptor type 1 ($S1P_1$). The latter finding is notable for the fact that prevention of T cell egress from the lymph node via $S1P_1$ inhibition forms the basis of action of Fingolimod (FTY720), an immunosuppressive agent used in the treatment of multiple sclerosis (Calabresi et al., 2014).

18.4.4 B Cells and Humoral Responses

One of the major responses that takes place in the lymph node is development of antibody-mediated humoral immunity via antigen-dependent activation and maturation of B cells. Studies with intravenously delivered antigens have revealed that following uptake of target antigen, B cells migrate preferentially toward the border between T and B cell zones (Okada et al., 2005). In this location, they are able to undergo interactions with CD4+ T cells, forming mobile B cell-T cell conjugates. These cells then migrate back into the follicle and initiate the germinal center reaction, the ultimate outcome of which is plasma cell development and generation of high affinity antibodies (Okada et al., 2005).

Soluble antigens, immune complexes, or microbial particles can also enter the node via the lymphatics, draining into the subcasular sinus of the lymph node. Here, viral particles can be captured by subcapsular sinus macrophages, one of the functions of which is to restrict viral spread (Gonzalez et al., 2010). Alternatively, this antigenic material can be passed from the macrophages onto B cells that carry the antigen into the follicle or passively drain farther into the lymph node and be captured by lymph node-resident DCs for initiation of B cell-dependent immune responses (Junt et al., 2007b; Phan et al., 2007; Gonzalez et al., 2010).

18.4.5 Immune Regulation in the Lymph Node

The process of immune regulation ensures that immune responses do not proceed to inappropriate levels of activation. Tregs acting within lymph nodes are central to this response. However, given the complex multicellular nature of immune responses and the variety of inhibitory mechanisms Tregs have at their disposal (Vignali et al., 2008), the mechanisms underlying this effect were unclear. To address this, 2PM experiments were performed in which Tregs were visualized in the lymph node, along with effector lymphocytes and DCs (Mempel et al., 2006; Tang et al., 2006). Tregs were found to inhibit the function of CD8+ T cells, although this occurred in the absence of direct contact between these cells. In contrast, Tregs were more likely to interact closely with DCs and reduce the capacity of DCs to undergo prolonged interactions with effector CD8+ T cells. These studies provided evidence that

Tregs modulate effector lymphocyte function indirectly via effects on DC function.

18.4.6 Imaging the Lymphatic Vasculature

Immune cell migration from peripheral locations into secondary lymphoid organs is a key process whereby the immune system monitors the periphery. Immune cells in peripheral organs, particularly DCs and lymphocytes, migrate into lymphatic vessels and these vessels serve as a conduit to local lymph nodes. This has been studied in most detail via the examination of DCs entering skin-draining lymph vessels, where preexisting discontinuities in the basement membrane supporting the lymphatic vasculature serve as portals for immune cell entry (Pflicke and Sixt, 2009; Sen et al., 2010). This process does not require proteolysis or integrin-mediated adhesive contacts but is driven by chemokine gradients guiding the immune cells toward the vessel (Lammermann et al., 2008; Pflicke and Sixt, 2009; Russo et al., 2016). After entering the vessel, DCs then crawl intraluminally along the lymphatic endothelial surface before detaching and being transported passively via lymph flow (Nitschke et al., 2012; Russo et al., 2016). Similar findings have been made for T cells in that they transmigrate into lymphatic microvessels and then migrate intraluminally within these vessels until they reach collecting vessels where they are swept away toward the lymph node by lymphatic flow (Teijeira et al., 2017).

18.4.7 Lymph Node and Lymphatic Imaging – Summary

The lymph node is central to many of the key events in immunology, playing host to the comings and goings of numerous cell types, antigens and immune stimuli. Here, we have attempted to illustrate how the power of advanced *in vivo* imaging has been indispensable in revealing how this complexity comes together to result in the generation of effective and appropriate immune responses. It should be noted that there is considerably more intricacy than described here, as exemplified by studies examining the effects of variation in antigen load, thresholds for lymphocyte activation, and complexity in T cell:B cell interactions in development of the humoral response (Okada et al., 2005; Henrickson et al., 2008; Qi et al., 2008). Undoubtedly, our understanding of the workings of immunity in this vital organ will continue to grow in the years to come.

18.5 Imaging the Spleen

18.5.1 Introduction

The spleen is the body's largest secondary lymphoid organ, containing approximately one-fourth of the body's lymphocytes. Its primary functions are two-fold: 1) surveying the blood for foreign material/infectious agents and mounting immune responses to captured foreign antigens, and 2) removal of old or defective red blood cells (RBCs) from circulation. These functions are carried out in morphologically distinct compartments termed the white pulp (immune regulation) and the red pulp (filtering of aging RBCs), with these regions being separated by an interface called the marginal zone (MZ) (Figure 18.4). Different leukocyte populations in these compartments are specialized to carry out specific functions. For example, the red pulp consists of a network of reticular fibers interspersed with macrophages specialized in phagocytosis and recycling components of defective RBCs, while in the MZ, different populations of macrophages are specialized in removal of blood-borne pathogens. This section will highlight how imaging techniques have advanced our understanding of leukocyte behavior within different compartments of the spleen under steady-state conditions and during immune responses.

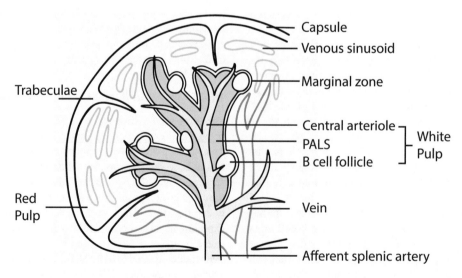

FIGURE 18.4　Structure of the mouse spleen. The spleen is divided into red pulp and white pulp areas surrounded by a fibrous capsule. The white pulp consists of T cell zones (periarteriolar lymphoid sheath, PALS) and B cell follicles that encircle arterial branches termed central arterioles. The white pulp is surrounded by the marginal zone that contains several types of macrophages, specialized B cells, and dendritic cells. The circulation of the spleen is open ended with arterioles emptying into the red pulp and marginal zone areas. Blood from the red pulp runs into venous sinusoids that collect into veins and exit the spleen via the efferent splenic vein.

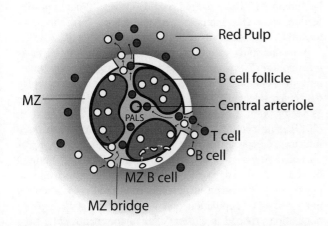

FIGURE 18.5 Schematic view of T and B cell movement throughout the white pulp. Fibroblast reticular cells (FRCs) form an extensive 3D network within the PALS of the white pulp. This network extends across areas of the marginal zone (MZ), forming bridges that connect the PALS to the red pulp (MZ bridges). T cells and follicular B cells use MZ bridges to enter and exit the white pulp. Specialized B cells within the MZ (MZ B cells) also continuously migrate between B cell follicles and the MZ performing the function of antigen capture and delivery to the follicles. Adapted from den Haan et al., 2012.

18.5.2 Spleen Structure

The spleen is surrounded by a fibrous capsule that extends protrusions (trabeculae) into the splenic tissue to provide support for its vast branching vasculature (Figure 18.4) (Mebius and Kraal, 2005). As a filter of the blood, it has a unique open circulation in that capillaries empty blood out into either the marginal sinus (the inner junction between the white pulp and the MZ), the MZ, or directly into the red pulp (Cesta, 2006). Blood freely moves through the stromal tissue of these areas and is recollected into specialized sinuses in the red pulp for venous drainage. The smallest artery branches (central arterioles) are surrounded by lymphoid tissue (white pulp) that is structurally organized into T cell (periarteriolar lymphoid sheath [PALS]) and B cell zones (B-cell follicles) (Figure 18.5), paralleling the layout found in lymph nodes. The MZ separating the white and red pulp represents an important transit area for the cells leaving the circulation. This zone contains specialized macrophages, DCs, and a resident population of B cells (MZ B cells) that have distinct features relative to B cells in splenic follicles. Access to the white pulp through the marginal zone is restricted to B cells, T cells, and DCs that are attracted to their respective zones by specific chemokines.

18.5.3 Imaging Lymphocytes in the Spleen

Early intravital laser scanning confocal microscopy studies tracked fluorescently labeled dextran and T cells within spleens (Grayson et al., 2001, 2003), revealing that T cells adhere in channels just distal to the terminal buds of splenic capillaries. The shear forces within these channels are relatively weak, being similar to that found in bone marrow sinusoids. Selectin-mediated interactions do not play a role in lymphocyte accumulation in the white pulp, suggesting that these vessels do not support leukocyte rolling (Nolte et al., 2002; Lo et al., 2003).

In contrast, the movement of lymphocytes across the MZ into the white pulp involves the actions of chemokines and integrins (LFA-1 and $\alpha_4\beta_1$), molecules typically involved in adhesion and migration (Lo et al., 2003).

Intravital laser scanning confocal microscopy studies were hampered by the inability to image deep into the spleen. Some studies have overcome this by visualising spleen slices, maintained in warm oxygenated medium (Bajenoff et al., 2008). Alternatively, the use of 2PM rather than confocal microscopy enables structures 200–300 µm below the surface of the spleen to be visualized, including the most superficial white pulp areas in intact spleens (Aoshi et al., 2008; Arnon et al., 2013).

Using a technique previously used to examine T and B cell migration along stromal cells within the lymph nodes, Bajenoff et al. (2006) showed that T cells migrate within the PALs in a random stop-go manner. As seen in the lymph nodes, T cells actively crawled and migrated along FRCs that form an extensive 3D network within the PALS of the splenic white pulp (Bajenoff et al., 2008). The FRC network was found to extend throughout the PALS and interrupt the MZ in certain areas, forming bridging channels (MZBCs) directly connecting the red pulp to the T cell zones (Figure 18.5). T cells predominantly entered the PALS using these bridging channels. Some B cells also used the MZBCs to access the white pulp using the outer edge of the MZBC closest to the B cell follicles (Bajenoff et al., 2008). Studies using confocal microscopy to examine the white pulp CD8⁺ T cell response to *Listeria monocytogenes* infection revealed that both activated CD8⁺ T cells from the initial response to *Listeria* and restimulated memory CD8⁺ T cells exit the PALS via MZBCs (Khanna et al., 2007).

Though the lymphocyte response to antigens in the spleen has been thought to resemble that in the lymph nodes, recently Barinov et al. (2017) used 2PM to reveal that differences exist. Examination of a naïve T cell response to endogenously expressed antigen revealed that, unlike in the lymph nodes, initial T cell/DC interactions can be of long duration in the red pulp of the spleen. A striking difference between CD8⁺ T cell and CD4⁺ T cell interactions with DCs was observed, with CD8⁺ T cells forming stable interactions with antigen-positive (Ag⁺) DCs, while all initial CD4⁺ T cell/Ag⁺ DC interactions were short-lived (Barinov et al., 2017). The authors propose that this was a reflection of low expression of MHC-II on this relatively immature DC population. CD8⁺ T cells did not stop moving upon DC interaction. Instead these two cells migrated as a group (Barinov et al., 2017). In this setting, CD8⁺ T cells were shown to recruit CD4⁺ T cells to Ag⁺ DC to form ternary complexes, highlighting the fact that CD8⁺ T cells can also modulate CD4⁺ T cell immune responses and vice versa.

The importance of T cell motility to CD8⁺ T cell-mediated killing of infected cells was highlighted by 2PM studies examining T cell responses to virus in the spleen. The responses to two types of virus were examined, one that is cleared quickly and one that persists and results in T cell exhaustion (Zinselmeyer et al., 2013). These studies showed that viral persistence results in PD-1/PDL-1-dependent reduced motility of CD4⁺ and CD8⁺ T cell movement within the MZ and red pulp (sites of highest viral infection). This T cell paralysis was associated with impaired antiviral T cell immunity, a response thought to represent an important aspect of immune exhaustion during persistent infection.

18.5.4 Imaging B Cell Behavior in the Spleen

The spleen plays host to two major populations of B cells: MZ B cells and follicular B cells. MZ B cells have unique properties that cross boundaries between innate and adaptive immune systems, including expressing polyreactive BCRs, high levels of innate receptors (TLRs) and the capacity to readily recognize opsonized antigens (reviewed in Cerutti et al., 2013). These specialized B cells are positioned with the MZ and enmeshed with macrophages, neutrophils, and DCs, all of which are strategically located to interact with antigens entering from the bloodstream. Intravital 2PM was used to compare the behavior of MZ and follicular B cells (Arnon et al., 2013). Fluorescently labeled immune complexes were administered to label MZ macrophages and provide a clear delineation between the MZ and follicles. These experiments revealed that MZ B cells are highly migratory and continually shuttle between the MZ and the follicles, explaining how they rapidly deliver antigens from the blood to the follicles (Cinamon et al., 2008). Similarly, follicular B cells can transit from the follicles to the MZ, a process dependent on S1P$_1$-driven migration. However, they are released subsequently from the MZ into the red pulp (Arnon et al., 2013).

18.5.5 Visualizing Responses to Pathogens in the Spleen

2PM studies of immune response to live pathogens have illustrated that immune responses in the spleen are complex and heterogeneous and differ according to the type of challenge and location of the response. For example, immunohistochemical studies of *Listeria* infection revealed that a variety of phagocytic cells within the MZ, including MZ macrophages, CD11b$^+$ cells, and CD11c$^+$ cells, can uptake *Listeria* and enter the PALS after *Listeria* infection (Aoshi et al., 2008). Studying this response via 2PM showed that CD11c$^+$ DCs move randomly in and out of the MZ and PALS, while MZ macrophages were more sessile (Aoshi et al., 2008). Indeed, it was the migration of infected MZ DCs into the PALS that underpinned initiation of the CD8-dependent response to *Listeria* infection. Although the majority of *Listeria* is captured via phagocytic cells of the MZ, a minority of this pathogen can establish foci of infection in the red pulp. Imaging the latter response revealed that *Listeria* that penetrate the red pulp rapidly infect local CD11c$^+$ DCs. This population of DC is nonmotile but highly active in extending and retracting cell processes (Waite et al., 2011). The other immune populations involved in surveilling this region are myelomonocytic cells (neutrophils and monocytes), which constitutively migrate through the red pulp and form transient interactions with DCs. After *Listeria* infection, infected CD11c$^+$ DCs become surrounded by a swarm of myelomonocytes causing restriction of regional blood flow. Antigen-specific CD8$^+$ T cells are recruited to these foci at later time points, where they engage with infected cells (Waite et al., 2011).

The behavior of red pulp DCs has also been found to differ depending upon the invading pathogen. 2PM imaging studies of the spleen after infection with the *Plasmodium chabaudi* parasite revealed that infected RBCs are rapidly phagocytosed by nonmotile DCs in the red pulp. However, 12 hours postinfection, these DCs increase their migration in order to move toward CD4$^+$ T cell-rich areas (Borges da Silva et al., 2015).

A recent study examining the response to *Streptococcus pneumoniae (S. pneumoniae)* in the spleen used intravital spinning disk confocal and 2PM imaging to reveal the interplay between resident macrophages, neutrophils, and MZ B cells in *S. pneumoniae* clearance (Deniset et al., 2017). Deniset et al. (2017) examined the initial innate response to *S. pneumoniae* infection, demonstrating that early after infection, a large proportion of *S. pneumoniae* flows across the MZ and is captured by red pulp macrophages. These studies characterized two populations of resident neutrophils in the red pulp: a mobile mature Ly6G^{hi} population and an immature static Ly6G^{int} population grouped in clusters. The mature neutrophils were found to capture and phagocytose *S. pneumoniae* off the surface of red pulp macrophages. At later time points, neutrophils recruited directly from the circulation were found to associate with MZ macrophages and MZ B cells where they contributed to the subsequent MZ B cell response. Interestingly, the clusters of immature immobile neutrophils were capable of a low level of replication and a proportion underwent differentiation into mature Ly6G^{hi} cells after infection. This suggested that these cells serve as a neutrophil reservoir in the spleen that can be mobilized to action during local immune responses. This has parallels in earlier observations of a reservoir of immature monocytes in the red pulp that, in response to injury, can be stimulated to exit the spleen and contribute to systemic responses (Swirski et al., 2009).

18.6 Concluding Remarks

Recent advancements in intravital imaging have allowed researchers to study different leukocyte subsets and their complex behaviors in the organs where immune cells are generated, mature, and respond to immune challenge. 2PM studies using a wide range of visualization techniques and reporter strategies have revealed the dynamic interplay between cells of the adaptive (lymphocytes) and innate (DCs and macrophages) immune systems and stromal cells. These studies have been integral in enabling us to understand how the immune system overcomes the daunting physical and numerical challenges of providing immune surveillance against the countless threats faced on a daily basis. So what does the future hold for *in vivo* imaging of the immune system? Investigators are now moving beyond descriptions of behaviors and cell–cell interactions to the examination of intracellular signaling in immune and other cells (Yoshikawa et al., 2016; Nobis et al., 2017). These studies are beginning to reveal information about the dynamics of the intracellular responses of cells *in vivo* to their immediate environment at a single-cell level. This work requires creation of novel reporter mice hand-in-hand with development of novel image analysis algorithms for processing of complex multidimensional data sets. However, work of this nature will ensure that *in vivo* imaging of the immune system continues to reward researchers via generation of novel and important information for many years to come.

REFERENCES

Akashi, K., Richie, L. I., Miyamoto, T., Carr, W. H., and Weissman, I. L. (2000). B lymphopoiesis in the thymus. *J Immunol, 164*: 5221–5226.

Aoshi, T. et al. (2008). Bacterial entry to the splenic white pulp initiates antigen presentation to CD8+ T cells. *Immunity, 29*: 476–486.

Arnon, T. I., Horton, R. M., Grigorova, I. L., and Cyster, J. G. (2013). Visualization of splenic marginal zone B-cell shuttling and follicular B-cell egress. *Nature, 493*: 684–688.

Bajenoff, M., Glaichenhaus, N., and Germain, R. N. (2008). Fibroblastic reticular cells guide T lymphocyte entry into and migration within the splenic T cell zone. *J Immunol, 181*: 3947–3954.

Bajenoff, M. et al. (2006). Stromal cell networks regulate lymphocyte entry, migration, and territoriality in lymph nodes. *Immunity, 25*: 989–1001.

Barinov, A., Galgano, A., Krenn, G., Tanchot, C., Vasseur, F., and Rocha, B. (2017). CD4/CD8/Dendritic cell complexes in the spleen: CD8+ T cells can directly bind CD4+ T cells and modulate their response. *PLoS One, 12*: e0180644.

Beck, T. C., Gomes, A. C., Cyster, J. G., and Pereira, J. P. (2014). CXCR4 and a cell-extrinsic mechanism control immature B lymphocyte egress from bone marrow. *J Exp Med, 211*: 2567–2581.

Bhakta, N. R., Oh, D. Y., and Lewis, R. S. (2005). Calcium oscillations regulate thymocyte motility during positive selection in the three-dimensional thymic environment. *Nat Immunol, 6*: 143–151.

Bhandoola, A., von Boehmer, H., Petrie, H. T., and Zuniga-Pflucker, J. C. (2007). Commitment and developmental potential of extrathymic and intrathymic T cell precursors: Plenty to choose from. *Immunity, 26*: 678–689.

Borges da Silva, H. et al. (2015). In vivo approaches reveal a key role for DCs in CD4+ T cell activation and parasite clearance during the acute phase of experimental blood-stage malaria. *PLoS Pathog, 11*: e1004598.

Bousso, P., Bhakta, N. R., Lewis, R. S., and Robey, E. (2002). Dynamics of thymocyte-stromal cell interactions visualized by two-photon microscopy. *Science, 296*: 1876–1880.

Brown, E., Carlin, L. M., Nerlov, C., Lo Celso, C., and Poole, A. W. (2018). Multiple membrane extrusion sites drive megakaryocyte migration into bone marrow blood vessels. *Life Sci Alliance, 1*(2): pii: e201800061.

Bunting, M. D., Comerford, I., and McColl, S. R. (2011). Finding their niche: Chemokines directing cell migration in the thymus. *Immunol Cell Biol, 89*: 185–196.

Cahalan, M. D., Parker, I., Wei, S. H., and Miller, M. J. (2002). Two-photon tissue imaging: Seeing the immune system in a fresh light. *Nat Rev Immunol, 2*: 872–880.

Calabresi, P. A. et al. (2014). Safety and efficacy of fingolimod in patients with relapsing-remitting multiple sclerosis (FREEDOMS II): A double-blind, randomised, placebo-controlled, phase 3 trial. *Lancet Neurol, 13*: 545–556.

Cerutti, A., Cols, M., Puga, I. (2013). Marginal zone B cells: Virtues of innate-like antibody-producing lymphocytes. *Nat Rev Immunol, 13*: 118–132.

Cesta, M. F. (2006). Normal structure, function, and histology of the spleen. *Toxicol Pathol, 34*: 455–465.

Cinamon, G., Zachariah, M. A., Lam, O. M., Foss, F. W., Jr., and Cyster, J. G. (2008). Follicular shuttling of marginal zone B cells facilitates antigen transport. *Nat Immunol, 9*: 54–62.

Cowan, J. E., Jenkinson, W. E., and Anderson, G. (2015). Thymus medulla fosters generation of natural Treg cells, invariant gammadelta T cells, and invariant NKT cells: What we learn from intrathymic migration. *Eur J Immunol, 45*: 652–660.

den Haan, J. M., Mebius, R. E., and Kraal, G. (2012). Stromal cells of the mouse spleen. *Front Immunol, 3*: 201.

Deniset, J. F., Surewaard, B. G., Lee, W. Y., and Kubes, P. (2017). Splenic Ly6G(high) mature and Ly6G(int) immature neutrophils contribute to eradication of S. pneumoniae. *J Exp Med, 214*: 1333–1350.

Devi, S. et al. (2013). Neutrophil mobilization via plerixafor-mediated CXCR4 inhibition arises from lung demargination and blockade of neutrophil homing to the bone marrow. *J Exp Med, 210*: 2321–2336.

Dzhagalov, I., and Phee, H. (2012). How to find your way through the thymus: A practical guide for aspiring T cells. *Cell Mol Life Sci, 69*: 663–682.

Dzhagalov, I. L., Chen, K. G., Herzmark, P., and Robey, E. A. (2013). Elimination of self-reactive T cells in the thymus: A timeline for negative selection. *PLoS Biol, 11*: e1001566.

Dzhagalov, I. L., Melichar, H. J., Ross, J. O., Herzmark, P., and Robey, E. A. (2012). Two-photon imaging of the immune system. *Curr Protoc Cytom, 60*: 12–26.

Ehrlich, L. I., Oh, D. Y., Weissman, I. L., and Lewis, R. S. (2009). Differential contribution of chemotaxis and substrate restriction to segregation of immature and mature thymocytes. *Immunity, 31*: 986–998.

Evrard, M. et al. (2015). Visualization of bone marrow monocyte mobilization using Cx3cr1gfp/+Flt3L-/- reporter mouse by multiphoton intravital microscopy. *J Leukoc Biol, 97*: 611–619.

Fujisaki, J. et al. (2011). In vivo imaging of Treg cells providing immune privilege to the haematopoietic stem-cell niche. *Nature, 474*: 216–219.

Germain, R. N., Robey, E. A., and Cahalan, M. D. (2012). A decade of imaging cellular motility and interaction dynamics in the immune system. *Science, 336*: 1676–1681.

Gonzalez, S. F. et al. (2010). Capture of influenza by medullary dendritic cells via SIGN-R1 is essential for humoral immunity in draining lymph nodes. *Nat Immunol, 11*: 427–434.

Gowans, J. L., Knight, E. J. (1964). The route of re-circulation of lymphocytes in the rat. *Proc R Soc Lond B Biol Sci, 159*: 257–282.

Grayson, M. H., Chaplin, D. D., Karl, I. E., and Hotchkiss, R. S. (2001). Confocal fluorescent intravital microscopy of the murine spleen. *J Immunol Methods, 256*: 55–63.

Grayson, M. H., Hotchkiss, R. S., Karl, I. E., Holtzman, M. J., and Chaplin, D. D. (2003). Intravital microscopy comparing T lymphocyte trafficking to the spleen and the mesenteric lymph node. *Am J Physiol Heart Circ Physiol, 284*: H2213–2226.

Grigorova, I. L., Schwab, S. R., Phan, T. G., Pham, T. H., Okada, T., and Cyster, J. G. (2009). Cortical sinus probing, S1P1-dependent entry and flow-based capture of egressing T cells. *Nat Immunol, 10*: 58–65.

Gurevich, I. et al. (2017). Active dissemination of cellular antigens by DCs facilitates CD8(+) T-cell priming in lymph nodes. *Eur J Immunol, 47*: 1802–1818.

Henrickson, S. E. et al. (2008). T cell sensing of antigen dose governs interactive behavior with dendritic cells and sets a threshold for T cell activation. *Nat Immunol, 9*: 282–291.

Hirata, Y. et al. (2018). CD150(high) Bone marrow tregs maintain hematopoietic stem cell quiescence and immune privilege via adenosine. *Cell Stem Cell, 22*: 445–453 e445.

Hu, Z., Lancaster, J. N., Sasiponganan, C., and Ehrlich, L. I. (2015). CCR4 promotes medullary entry and thymocyte-dendritic cell interactions required for central tolerance. *J Exp Med, 212*: 1947–1965.

Jacquelin, S. et al. (2013). CX3CR1 reduces Ly6Chigh-monocyte motility within and release from the bone marrow after chemotherapy in mice. *Blood, 122*: 674–683.

Junt, T. et al. (2007a). Dynamic visualization of thrombopoiesis within bone marrow. *Science, 317*: 1767–1770.

Junt, T. et al. (2007b). Subcapsular sinus macrophages in lymph nodes clear lymph-borne viruses and present them to antiviral B cells. *Nature, 450*: 110–114.

Khanna, K. M., McNamara, J. T., and Lefrancois, L. (2007) In situ imaging of the endogenous CD8 T cell response to infection. *Science, 318*: 116–120.

Ki, S., Thyagarajan, H. M., Hu, Z., Lancaster, J. N., and Ehrlich, L. I. R. (2017). EBI2 contributes to the induction of thymic central tolerance in mice by promoting rapid motility of medullary thymocytes. *Eur J Immunol, 47*: 1906–1917.

Klein, L., Hinterberger, M., Wirnsberger, G., and Kyewski, B. (2009). Antigen presentation in the thymus for positive selection and central tolerance induction. *Nat Rev Immunol, 9*: 833–844.

Klein Wolterink, R. G., Garcia-Ojeda, M. E., Vosshenrich, C. A., Hendriks, R. W., and Di Santo, J. P. (2010). The intrathymic crossroads of T and NK cell differentiation. *Immunol Rev, 238*: 126–137.

Kohler, A. et al. (2011). G-CSF-mediated thrombopoietin release triggers neutrophil motility and mobilization from bone marrow via induction of Cxcr2 ligands. *Blood, 117*: 4349–4357.

Kunisaki, Y. et al. (2013). Arteriolar niches maintain haematopoietic stem cell quiescence. *Nature, 502*: 637–643.

Kurd, N., and Robey, E. A. (2016). T-cell selection in the thymus: A spatial and temporal perspective. *Immunol Rev, 271*: 114–126.

Kurobe, H. et al. (2006). CCR7-dependent cortex-to-medulla migration of positively selected thymocytes is essential for establishing central tolerance. *Immunity, 24*: 165–177.

Ladi, E. et al. (2008). Thymocyte-dendritic cell interactions near sources of CCR7 ligands in the thymic cortex. *J Immunol, 181*: 7014–7023.

Lammermann, T. et al. (2008). Rapid leukocyte migration by integrin-independent flowing and squeezing. *Nature, 453*: 51–55.

Lancaster, J. N., and Ehrlich, L. I. (2017). Analysis of thymocyte migration, cellular interactions, and activation by multiphoton fluorescence microscopy of live thymic slices. *Methods Mol Biol, 1591*: 9–25.

Le Borgne, M. et al. (2009). The impact of negative selection on thymocyte migration in the medulla. *Nat Immunol, 10*: 823–830.

Lind, E. F., Prockop, S. E., Porritt, H. E., and Petrie, H. T. (2001). Mapping precursor movement through the postnatal thymus reveals specific microenvironments supporting defined stages of early lymphoid development. *J Exp Med, 194*: 127–134.

Lindquist, R. L. et al. (2004). Visualizing dendritic cell networks in vivo. *Nat Immunol, 5*: 1243–1250.

Lo Celso, C., and Scadden, D. T. (2011). The haematopoietic stem cell niche at a glance. *J Cell Sci, 124*:3529–3535.

Lo Celso, C. et al. (2009). Live-animal tracking of individual haematopoietic stem/progenitor cells in their niche. *Nature, 457*: 92–96.

Lo, C. G., Lu, T. T., and Cyster, J. G. (2003). Integrin-dependence of lymphocyte entry into the splenic white pulp. *J Exp Med, 197*: 353–361.

Male, V., and Brady, H. J. M. (2017). Murine thymic NK cells: A case of identity. *Eur J Immunol, 47*: 797–799.

McCaughtry, T. M., Baldwin, T. A., Wilken, M. S., and Hogquist, K. A. (2008). Clonal deletion of thymocytes can occur in the cortex with no involvement of the medulla. *J Exp Med, 205*: 2575–2584.

Mebius, R. E., and Kraal, G. (2005). Structure and function of the spleen. *Nat Rev Immunol, 5*: 606–616.

Melichar, H. J., Ross, J. O., Herzmark, P., Hogquist, K. A., and Robey, E. A. (2013). Distinct temporal patterns of T cell receptor signaling during positive versus negative selection in situ. *Sci Signal, 6*: ra92.

Mempel, T. R., Henrickson, S. E., and von Andrian, U. H. (2004). T-cell priming by dendritic cells in lymph nodes occurs in three distinct phases. *Nature, 427*: 154–159.

Mempel, T. R. et al. (2006). Regulatory T cells reversibly suppress cytotoxic T cell function independent of effector differentiation. *Immunity, 25*: 129–141.

Mendez-Ferrer, S. et al. (2010). Mesenchymal and haematopoietic stem cells form a unique bone marrow niche. *Nature, 466*: 829–834.

Miller, M. J., Wei, S. H., Parker, I., and Cahalan, M. D. (2002), Two-photon imaging of lymphocyte motility and antigen response in intact lymph node. *Science, 296*: 1869–1873.

Miller, M. J., Hejazi, A. S., Wei, S. H., Cahalan, M. D., and Parker, I. (2004). T cell repertoire scanning is promoted by dynamic dendritic cell behavior and random T cell motility in the lymph node. *Proc Natl Acad Sci U S A, 101*: 998–1003.

Misslitz, A., Bernhardt, G., and Forster, R. (2006). Trafficking on serpentines: Molecular insight on how maturing T cells find their winding paths in the thymus. *Immunol Rev, 209*: 115–128.

Moon, J. J. et al. (2007). Naive CD4(+) T cell frequency varies for different epitopes and predicts repertoire diversity and response magnitude. *Immunity, 27*: 203–213.

Mori, K., Itoi, M., Tsukamoto, N., Kubo, H., and Amagai, T. (2007). The perivascular space as a path of hematopoietic progenitor cells and mature T cells between the blood circulation and the thymic parenchyma. *Int Immunol, 19*: 745–753.

Nitschke, M. et al. (2012). Differential requirement for ROCK in dendritic cell migration within lymphatic capillaries in steady-state and inflammation. *Blood, 120*: 2249–2258.

Nitta, T., and Suzuki, H. (2016). Thymic stromal cell subsets for T cell development. *Cell Mol Life Sci, 73*: 1021–1037.

Nobis, M. et al. (2017). A RhoA-FRET Biosensor mouse for intravital imaging in normal tissue homeostasis and disease contexts. *Cell Rep, 21*: 274–288.

Nolte, M. A., Hamann, A., Kraal, G., and Mebius, R. E. (2002). The strict regulation of lymphocyte migration to splenic white pulp does not involve common homing receptors. *Immunology, 106*: 299–307.

Okada, T. et al. (2005). Antigen-engaged B cells undergo chemotaxis toward the T zone and form motile conjugates with helper T cells. *PLoS Biol, 3*: e150.

Pearse, G. (2006). Normal structure, function and histology of the thymus. *Toxicol Pathol, 34*: 504–514.

Perera, J., and Huang, H. (2015). The development and function of thymic B cells. *Cell Mol Life Sci, 72*: 2657–2663.

Petrie, H. T., and Zuniga-Pflucker, J. C. (2007). Zoned out: Functional mapping of stromal signaling microenvironments in the thymus. *Annu Rev Immunol, 25*: 649–679.

Pflicke, H., and Sixt, M. (2009). Preformed portals facilitate dendritic cell entry into afferent lymphatic vessels. *J Exp Med, 206*: 2925–2935.

Phan, T. G., Grigorova, I., Okada, T., and Cyster, J. G. (2007). Subcapsular encounter and complement-dependent transport of immune complexes by lymph node B cells. *Nat Immunol, 8*: 992–1000.

Plotkin, J., Prockop, S. E., Lepique, A., and Petrie, H. T. (2003). Critical role for CXCR4 signaling in progenitor localization and T cell differentiation in the postnatal thymus. *J Immunol, 171*: 4521–4527.

Porritt, H. E., Gordon, K., and Petrie, H. T. (2003). Kinetics of steady-state differentiation and mapping of intrathymic-signaling environments by stem cell transplantation in non-irradiated mice. *J Exp Med, 198*: 957–962.

Qi, H., Cannons, J. L., Klauschen, F., Schwartzberg, P. L., and Germain, R. N. (2008). SAP-controlled T-B cell interactions underlie germinal centre formation. *Nature, 455*: 764–769.

Reismann, D. et al. (2017). Longitudinal intravital imaging of the femoral bone marrow reveals plasticity within marrow vasculature. *Nat Commun, 8*: 2153.

Ross, J. O., Melichar, H. J., Au-Yeung, B. B., Herzmark, P., Weiss, A., and Robey, E. A. (2014). Distinct phases in the positive selection of CD8+ T cells distinguished by intrathymic migration and T-cell receptor signaling patterns. *Proc Natl Acad Sci U S A, 111*: E2550–2558.

Russo, E. et al. (2016). Intralymphatic CCL21 promotes tissue egress of dendritic cells through afferent lymphatic vessels. *Cell Rep, 14*: 1723–1734.

Sanos, S. L., Nowak, J., Fallet, M., and Bajenoff, M. (2011). Stromal cell networks regulate thymocyte migration and dendritic cell behavior in the thymus. *J Immunol, 186*: 2835–2841.

Sen, D., Forrest, L., Kepler, T. B., Parker, I., and Cahalan, M. D. (2010). Selective and site-specific mobilization of dermal dendritic cells and Langerhans cells by Th1- and Th2-polarizing adjuvants. *Proc Natl Acad Sci U S A, 107*: 8334–8339.

Serbina, N. V., and Pamer, E. G. (2006). Monocyte emigration from bone marrow during bacterial infection requires signals mediated by chemokine receptor CCR2. *Nat Immunol, 7*: 311–317.

Sipkins, D. A. et al. (2005). In vivo imaging of specialized bone marrow endothelial microdomains for tumour engraftment. *Nature, 435*: 969–973.

Sprent, J. et al. (1996). The thymus and T cell death. *Adv Exp Med Biol, 406*: 191–198.

Stein, J. V. et al. (2000). The CC chemokine thymus-derived chemotactic agent 4 (TCA-4, secondary lymphoid tissue chemokine, 6Ckine, exodus-2) triggers lymphocyte function-associated antigen 1-mediated arrest of rolling T lymphocytes in peripheral lymph node high endothelial venules. *J Exp Med, 191*: 61–76.

Stoll, S., Delon, J., Brotz, T. M., and Germain, R. N. (2002). Dynamic imaging of T cell-dendritic cell interactions in lymph nodes. *Science, 296*: 1873–1876.

Swirski, F. K. et al. (2009). Identification of splenic reservoir monocytes and their deployment to inflammatory sites. *Science, 325*: 612–616.

Tang, Q. et al. (2006). Visualizing regulatory T cell control of autoimmune responses in nonobese diabetic mice. *Nat Immunol, 7*: 83–92.

Teijeira, A. et al. (2017). T cell migration from inflamed skin to draining lymph nodes requires intralymphatic crawling supported by ICAM-1/LFA-1 interactions. *Cell Rep 18*: 857–865.

Vandoorne, K. et al. (2018). Imaging the vascular bone marrow niche during inflammatory stress. *Circ Res, 123*: 415–427.

Vignali, D. A., Collison, L. W., and Workman, C. J. (2008). How regulatory T cells work. *Nat Rev Immunol, 8*: 523–532.

von Andrian, U. H. (1996). Intravital microscopy of the peripheral lymph node microcirculation in mice. *Microcirculation, 3*: 287–300.

Waite, J. C. et al. (2011). Dynamic imaging of the effector immune response to listeria infection in vivo. *PLoS Pathog, 7*: e1001326.

Warnock, R. A., Askari, S., Butcher, E. C., and von Andrian, U. H. (1998). Molecular mechanisms of lymphocyte homing to peripheral lymph nodes. *J Exp Med, 187*: 205–216.

Wei, Q., and Frenette, P. S. (2018). Niches for hematopoietic stem cells and their progeny. *Immunity, 48*: 632–648.

Witt, C. M., Raychaudhuri, S., Schaefer, B., Chakraborty, A. K., and Robey, E. A. (2005). Directed migration of positively selected thymocytes visualized in real time. *PLoS Biol, 3*: e160.

Yoshikawa, S. et al. (2016). Intravital imaging of Ca(2+) signals in lymphocytes of Ca(2+) biosensor transgenic mice: Indication of autoimmune diseases before the pathological onset. *Sci Rep, 6*: 18738.

Zinselmeyer, B. H. et al. (2013). PD-1 promotes immune exhaustion by inducing antiviral T cell motility paralysis. *J Exp Med, 210*: 757–774.

19

Imaging Living Organisms

Masahiro Fukuda, Katsuya Ozawa, and Hajime Hirase

CONTENTS

19.1 Background

One of the major goals of neuroscience is to understand how the brain works. Two-photon *in vivo* imaging is a suitable method for monitoring neuronal function and circuits over long timescales, as it has high spatial and temporal resolution (Svoboda and Yasuda, 2006). Confocal microscopy has been used for *in vivo* imaging; however, its application was limited in the past because of phototoxicity, insufficient signals, light scattering, and poor penetration depth (Fetcho and O'Malley, 1997). Two-photon microscopy (2PM), which was first reported in 1990 by Webb, Strickler, and Denk (1990), could overcome many of these problems owing to the excitation light of near-infrared wavelengths.

2PM requires that the observed target generates fluorescence. As such, the method of labeling targets with fluorescence must be considered. Initially, researchers introduced fluorophores to targets using the whole cell patch-clamp technique (Denk et al., 1995). The first paper using 2PM *in vivo* imaging in rodents was reported 7 years after the development of 2PM, largely due to more challenging *in vivo* configurations (Svoboda et al., 1997). Over time, multiple technical developments, such as generation of transgenic animals expressing GFP or related fluorescent proteins in neurons (Feng et al., 2000), refinement of techniques to monitor Ca²⁺ signals from multiple cells using bolus loading of AM-esters (Stosiek et al., 2003), and evolution of gene delivery by electroporation (Fukuchi-Shimogori and Grove, 2001; Tabata and Nakajima, 2001) or viral vectors (Kammesheidt et al., 1996; Hermens and Verhaagen, 1998; Dittgen et al., 2004) have greatly advanced the *in vivo* imaging field.

One of the shortcomings of *in vitro* brain slice imaging is that many afferent and efferent connections, including long-range connections or neuromodulatory inputs (Lee and Dan, 2012), are severed. *In vivo* 2PM imaging has been utilized to investigate actual connections and interregional interactions, which cannot be addressed with *in vitro* techniques. For example, 2PM functional brain imaging (Kerr and Denk, 2008) has been adopted to address several issues that cannot be addressed by other *in vivo* techniques, such as intrinsic optical signal imaging or electrode arrays because of insufficient spatial and/or temporal resolution. Ohki et al. (2005, 2006) reported on the cellular-resolution structure of the pinwheel center of orientation columns in the cat primary visual cortex using *in vivo* 2PM Ca²⁺ imaging. *In vivo* 2PM imaging can be applied not only to neurons but also to any fluorescently labeled structure, including astrocytes (Hirase et al., 2004), microglia (Nimmerjahn et al., 2005) and blood vessels (Shih et al., 2012).

Certain experiments can only be performed on live animals. For example, the activity of place cells (O'Keefe and Dostrovsky, 1971) in animals actively navigating a virtual reality environment, which inherently requires live-animal imaging, has been reported (Figure 19.1) (Dombeck et al., 2010). Another example is the mechanism of transcranial direct current stimulation (tDCS). tDCS is widely applied to human patients for treating depression (Nitsche et al., 2009), although its exact mechanism of action has remained unclear. Surprisingly, not neurons but astrocytes were responsible in the antidepressant effect of tDCS (Figure 19.2) (Monai et al., 2016).

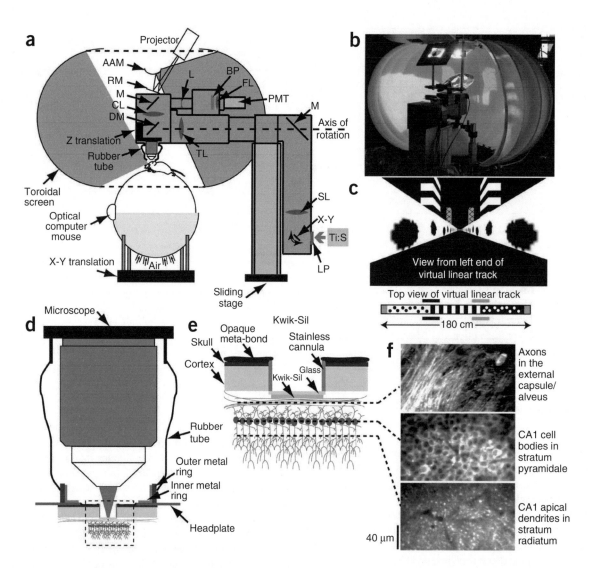

FIGURE 19.1 Experimental setup. (a) The experimental apparatus, consisting of a spherical treadmill, a virtual reality apparatus (projector, reflecting mirror [RM], angular amplification mirror [AAM], toroidal screen, and optical computer mouse to record ball rotation) and a custom two-photon microscope (Ti:sapphire laser [Ti:S], long-pass filter [LP], galvanometers [x-y], scan lens [SL], mirror [M], tube lens [TL], dichroic mirror [DM], collection lens [CL], biconcave lens [L], bandpass filter [BP], focusing lens [FL], photomultiplier tube [PMT], sliding stage [used to move microscope for treadmill access], x-y translation [moves treadmill and mouse], Z-translation [objective focus control], and rubber tube [shown in cross-section, for light shielding]). (b) Photograph of experimental setup. (c) Top, view from one end of the virtual linear track. Bottom, top view of the linear track. (d) View of materials used to block background light from entering the microscope objective hole. Hippocampal imaging window can also be seen. (e) Detailed view of hippocampal imaging window (from boxed region in d). (f) *In vivo* two-photon images at different depths through the hippocampal window. *From Figure 1 of* Dombeck, D.A., Harvey, C. D., Tian, L., Looger, L. L., Tank, D. W. (2010). Functional imaging of hippocampal place cells at cellular resolution during virtual navigation. *Nat Neurosci., 18(11): 1433–40.* doi: 10.1038/nn.2648.

19.1.1 Technical Developments

The development of *in vivo* 2PM imaging has progressed owing to multiple lines of technical advances, including improvements in microscopic optics, improvements in fluorophores and indicators, and refinement of experimental techniques. Initially, a pair of Galvano scanners were utilized for imaging live animals. However, imaging was vulnerable to drifts due to slow scanning speed, and images tended to be distorted. To overcome image distortions, multiple approaches were proposed, such as electrocardiogram- (ECG) triggered imaging (Paukert and Bergles, 2012). However, advances in

scanning methods (Kim et al., 1999; Lechleiter et al., 2002; Leybaert et al., 2005) and improvements in signal detectors from multialkali PMTs (with quantum efficiency of approximately 20%) to more efficient GaAsP PMTs (with quantum efficiency of approximately 40%) resulted in greatly improved signal-to-noise ratio. These technical advances have made it possible to acquire images with much greater speed.

In addition, the layout of microscopes has been modified extensively, especially for imaging in awake animals. For example, Dombeck, et al. constructed their own custom 2PM microscopy to incorporate apparatuses for observing awake-behaving animals in virtual reality (Dombeck et al., 2010)

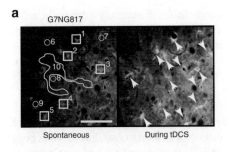
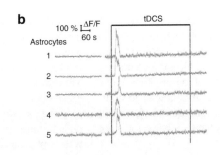
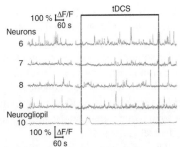

FIGURE 19.2 tDCS induces Ca²⁺ surges in layer 2/3 astrocytes, whereas neurons do not show obvious activity changes in urethane-anaesthetized mice. (a) Intracranial two-photon imaging of primary visual cortex layer 2/3 during tDCS. Astrocytes are labeled with SR101 (red). Arrows point to astrocytes that had Ca²⁺ elevations during tDCS. Numbers correspond to the cells and neurogliopil region plotted in b. Scale bar = 100 μm. (b) Fluorescent intensity (ΔF/F) traces of astrocytes (orange), neurons (green) and neurogliopil (brown). *From Figure 2 of* Monai, H., Ohkura, M., Tanaka, M., Oe, Y., Konno, A., Hirai, H., Mikoshiba, K., Itohara, S., Nakai, J., Iwai, Y., Hirase, H. (2016). Calcium imaging reveals glial involvement in transcranial direct current stimulation-induced plasticity in mouse brain. *Nat Commun., 70*: 11100. doi: 10.1038/ncomms11100.

because commercially available upright microscopes at that time could not accommodate such apparatuses. Currently, microscopes with large clearance under objective lens for behavioral apparatuses and large animals are commercially available. In addition, to observe awake animals, it is preferable to avoid keeping them in abnormal postures. Instead, it is preferable to rotate the microscopes themselves. As of 2020, rotatable microscopes and tilting nosepieces have become commercially available.

Over the past three decades, multiple fluorophores and indicators have been developed. In particular, the development of genetically encoded calcium indicators (GECIs) has greatly enhanced neuroscience research. The first protein-based FRET Ca²⁺ indicator was reported in 1997 (Miyawaki et al., 1997; Romoser et al., 1997). The first monochrome indicator, Camgaroo, was developed in 1999 (Baird et al., 1999), and the founder of the GCaMP family, GCaMP, was reported in 2001 (Nakai et al., 2001). Since then, GCaMP variants have been progressively improved, and GCaMP3 has enabled stable *in vivo* Ca²⁺ imaging (Tian et al., 2009). GCaMP6 could even resolve a single action potential (Chen et al., 2013), and its performance has exceeded that of Oregon Green BAPTA-1 (OGB-1), a widely used organic Ca²⁺ indicator. Through continuous refinements, GECIs in multiple colors have been developed. Red GECI is expected to have great potential for imaging deep structures, as its longer emission wavelength has greater penetrability than does green fluorescence (Zhao et al., 2011; Akerboom et al., 2013; Dana et al., 2016). In addition, a series of genetically encoded voltage indicators (GEVIs), which reflect membrane potential directly, has been recently developed (Baker et al., 2008; Antic et al., 2016). GEVIs have allowed researchers to record signals resembling electrophysiological signals and sparked a new field known as all optical electrophysiology (Kannan et al., 2018).

19.2 *In Vivo* 2PM Imaging

19.2.1 Vital Points in Animal Surgeries

One of the benefits of chronic *in vivo* imaging is that changes in neuronal circuits can be tracked over time in the same animal

(Huber et al., 2012). Animal preparation is thus of utmost importance for high quality *in vivo* imaging, and many protocols for *in vivo* 2PM imaging have been published (Mostany and Portera-Cailliau, 2008; Holtmaat et al., 2009). The following points will be explained in this chapter: methods of craniotomy, aseptic manipulation, damage-free surgery, control of bleeding, pain/stress-free operation, and keeping craniotomy clear over time.

First, optical access is needed to observe brains using 2PM. There are two major methods of craniotomy: glass window (Trachtenberg et al., 2002) and thin skull (Christie et al., 2001; Drew et al., 2010; Yang et al., 2010). Both methods utilize a glass coverslip, which can be substituted with alternative materials such as polydimethylsiloxane (PDMS) (Heo et al., 2016). The glass-window method might induce astrocytic and microglial activation (Holtmaat et al., 2009), while the thin-skull method is considered more physiological, as the brain is left untouched. It has been debated whether the preparation for imaging itself affects experimental results (Xu et al., 2007). This is mainly due to the fact that the glass window could not be maintained over time with traditional methods; however, it is now possible to keep the cranial window clear for prolonged periods. Thus, sufficient recovery time is possible, and this issue is no longer a significant problem.

Aseptic manipulation is the most basic and important aspect of cranial window preparation. All surgical apparatus must be sterilized with 70% ethanol, UV light, and/or glass bead sterilizer. In addition, disinfection of hands is equally important. During the operation, the experimenter may contact objects in the lab without being aware of those that are not disinfected. Thus, it should thus be a priority to frequently sterilize the hands. The surgical area has to be disinfected with ethanol and povidone-iodine to prevent germ transmission and infection from mouse hair and skin. With adequate aseptic manipulation, antibiotics are not necessary, even in immunodeficient mice.

It is important to avoid too much damage to brain parenchyma. Heating by drilling should be avoided, and drilling speed should be reduced to avoid subdural bleeding. Pressure on the skull should be avoided to prevent glial activation.

Animals should be permitted sufficient recovery time after surgery to avoid adverse effects of inflammation. It is vital to control bleeding by washing the surgical area with extracellular solution or using a gelatin sponge. If bleeding continues, the cranial window may be filled with blood or blood clots, which will hamper subsequent imaging. It is crucial to minimize pain and use analgesics, not only for animal welfare, but also to reduce stress on the animals so that their behavior is not adversely affected.

The main cause of cranial window clouding is the regeneration of tissue and bones beneath the implanted coverslip (Goldey et al., 2014). There are two steps to keep the cranial window clear over time: (1) Use dexamethasone and NSAIDSs, and (2) gently depress the coverslip onto the brain so that opaque regenerative tissue will not be formed. Agarose was placed between the brain and coverslip to reduce drifts; however, this may result in regenerative tissue formation within the agarose over time. Additionally, the craniotomy can be kept clear only with a narcotic (buprenorphine) (Yu et al., 2011), which suggests that pain control may be important for keeping the craniotomy clear by preventing bone regeneration (Gerner and O'Connor, 2008). If damage is adequately managed, invasive procedures such as virus injections (Huber et al., 2012; Lukasiewicz et al., 2016) can be simultaneously performed with craniotomy.

In rats or larger animals, the dura mater must be removed because of its thickness and optical properties. The dura mater of mice is thin enough to transmit light; therefore, dura removal is not necessary for observation. However, there are cases that require dura removal in mice such as virus injection. In such instances, the dura can be removed without damage (Kyweriga et al., 2017).

The attachment of custom head-gear onto the skull has to be carefully considered. The bonding between stainless steel and dental cement is relatively weak, and screws are needed to obtain stable attachment. In contrast, titanium is biocompatible and has stronger bonding with dental cement. Therefore, custom-made head gear made of titanium is preferable. By combining cyanoacrylate application after periosteum removal, dental cement and custom-made titanium gear can be firmly attached for more than 1 year.

Numerous improvements in surgical techniques have also been made. The superior sagittal sinus is the largest vein in the brain, situated in the midline. If it is damaged, bleeding may become uncontrollable, which will prevent observation or manipulation of areas close to the midline, such as the anterior cingulate or retrosplenial cortices. Encouragingly, recent studies have reported that it is possible to remove the entire skull without damage to the superior sagittal sinus (Lim et al., 2012; Kim et al., 2016).

One hallmark of the advances in *in vivo* imaging techniques is chronic microglial imaging over years (Fuger et al., 2017). Microglial proliferation has been investigated by tracking the same region in the same animals; this is only possible with chronic *in vivo* imaging.

19.2.2 Imaging Procedures

Many studies can be performed only on awake-behaving animals. In this regard, experiments in awake animals are more desirable because anesthetics directly affect brain activity,

and different anesthetics may differentially alter brain activity (Ruebhausen et al., 2012). Indeed, anesthetized brain states demonstrate substantial differences from awake states (Liang et al., 2012), which makes it challenging to interpret results. Moreover, the pharmacokinetics of anesthetics are not fully understood (Crosby et al., 2010).

In the past, *in vivo* imaging was limited to anesthetized animals because of larger drifts from heartbeat and breathing, as well as phenomena that occur in awake conditions, such as body or eyeball movements. Stronger heartbeat and respiration in awake animals result in larger image drifts, causing targets to disappear from the imaging plane, thereby rendering it difficult to apply a single-plane image registration algorithm. Second, mice are more likely to fidget when they are awake, therefore causing dislocation of imaging targets. In short, the greatest challenge for achieving awake animal imaging was how to manage image drifts. In 2007, the first report of Ca^{2+} imaging of the cerebral cortex in awake animals was published, using combined methodology that enabled head-fixation of animals without stress and post-hoc image registration techniques (Dombeck et al., 2007).

In awake animal imaging, it is vital to keep animals under head-restraint conditions without stress, as stress alters brain activity (McEwen and Sapolsky, 1995). By reducing motion force, the probability of image disruptions can be decreased. For that purpose, several options are available, such as spherical treadmill (Dombeck et al., 2007), rotating stage (Kato et al., 2013), or air-lifted platform (Kislin et al., 2014). Conversely, in awake but immobilized preparations (Ono et al., 1986; Kleinfeld et al., 2002), animals can be kept in a small tube via behavioral training (Komiyama et al., 2010; Guo et al., 2014). Through training and acclimation, animals can even learn to sleep on a spherical treadmill (Yang et al., 2014).

Most research has been limited to cellular resolution, as finer structures are more prone to drifts. Dendritic spines, which measure in the range of a few micrometers, tend to be smaller than the size of brain drifts induced by heartbeat and breathing. Therefore, imaging of dendritic spines in live animals was previously intractable. However, functional Ca^{2+} imaging of dendritic spines has been reported in 2011 in anesthetized animals (Chen et al., 2011), and in 2018 in awake animals (El-Boustani et al., 2018).

19.3 Optical Aberrations in Brain Tissue

19.3.1 Deep Imaging

The cortex has columnar organization; as such, imaging capabilities that permit scanning of the full extent of cortical thickness are desired. Only a few studies have examined cells in deep layers of the mouse cerebral cortex (Kondo and Ohki, 2014). The depth of interest used in most studies is limited to superficial layers of up to 400 µm depth, even though 2PM can theoretically obtain images up to the working distance of the objective lens (typically 2–3 mm). Imaging of deeper structures, such as the hippocampal CA1, has been achieved by surgical removal of the overlying cortex (Dombeck et al., 2010; Sakaguchi et al., 2012) or by using specialized lasers (Kobat et al., 2011; Horton et al., 2013; Kawakami et al., 2013, 2015; Perillo et al., 2016). Other options are to adopt a gradient refractive index (GRIN) lens (Jung et al., 2004), although the field of view is highly limited when imaging through GRIN lenses (Barretto et al., 2009).

In 2013, it was demonstrated that hippocampal CA1 cells through intact cortex could be observed using commercially available 2PM (Hirase and Ozawa, 2013). Despite the need for deep brain imaging, and even though Ca^{2+} imaging of CA1 cells has been reported (Kondo et al., 2017), optimal conditions to achieve deep 2PM imaging have not been explored.

19.3.2 Light Refraction and Aberrations

Light scattering by brain tissue has received considerable attention (Helmchen and Denk, 2005); however, there have been limited studies examining how brain tissue or thin skull affect the point spread function (PSF) of 2PM excitation (Helm et al., 2009; Chaigneau et al., 2011; Isshiki and Okabe, 2014). We have elucidated the necessary conditions for deep imaging with 2PM through optical simulation (Figure 19.3a–e) and *in vivo* experiments (Figure 19.3f–h). We observed that PSF deteriorates with cranial window tilt, which is often overlooked in *in vivo* 2PM imaging (Figure 19.3a, b). Using a commercially available ultrashort pulse laser, we demonstrated that a practical method to align the cranial window surface can achieve imaging of hippocampal CA1 pyramidal cells in Thy1-YFP (H-line) transgenic mice without having to remove the dura and cortex.

First, we propose the importance of creating a larger opening on the skull to enable full utilization of the numerical aperture of the objective lens. When imaging deep areas, reduction of both fluorescent collection efficiency and excitation efficiency occurs (Oheim et al., 2001). In conventional *in vivo* electrophysiology, a smaller cranial window is preferred to reduce drift of the brain due to heartbeat, breathing, and brain pressure changes. In contrast, in imaging that utilizes an objective lens with large numerical aperture and wide area, the remaining skull can be a barrier in delivering excitation light and collecting fluorescent signals if the craniotomy is small.

Second, it is vital to keep the objective lens opening and cranial window completely parallel. In both the glass-window method and thin-skull method, the interface is sealed with a thin glass coverslip, which has a distinct refractive index. The ultrashort pulse of a typical mode-locked laser used in 2PM contains a broadband optical spectrum and is subject to refraction according to Snell's law (Diels and Rudolph, 2006). Attachment of the glass window is usually performed manually by humans, which may introduce errors when trying to align the glass window in parallel to the objective lens. In addition, the dispersion caused by brain tissue can worsen PSF (Ji et al., 2010). Increased excitation power to deliver more energy to deep areas may elicit out-of-focus 2PM excitation, which deteriorates the signal-to-noise ratio. Aberrations are caused not only by tilting cover slips but also by refractive index mismatches. Lights of different wavelengths does not converge to the same point through media, which results in enlarged PSF and lower energy concentration in focus, thereby hampering deep imaging.

PSF enlargements due to optical aberrations become problematic when the sizes of observation targets are similar, as subtle changes in PSF size can greatly affect measurements, while large targets, such as cell bodies, are not affected. Moreover, PSF enlargements and optical aberrations in *in vivo*

configuration are unpredictable, and the measurement of tiny objects are greatly affected by those optical effects (Wang et al., 2015). We have noticed that axons in the corpus callosum appeared thicker in our *in vivo* imaging (Figure 19.3i). To address the effects of PSF size changes in deep regions, we quantified the diameter of individual axons in *in vivo* imaging and in lightly fixed samples of Thy1-eYFP mice. In lightly fixed samples, axonal diameters were 1.40 ± 0.0041 μm ($n = 118$), similar to shrinkage-adjusted values of corpus callosum axons reported in the literature using electron microscopy (Wang et al., 2008). By contrast, axon diameters from *in vivo* imaging were 2.59 ± 0.08 μm ($n = 80$), which indicates a significantly larger appearance of axons in *in vivo* imaging ($P = 9.3594 \times 10^{-30}$) due to optical aberrations (Figure 19.3j). Our results indicate that measurements of object size are significantly affected by optical aberrations in unpredictable manner, which may lead to erroneous size estimations.

19.4 Future Directions

Technological advancements have made it feasible to dissect neuronal circuit function in intact brains. Electron microscopy (EM) provides information on fine circuit structure but lacks dynamic information (Kasthuri et al., 2015). Using *in vivo* imaging, the dynamics of neuronal circuits can be investigated in live animals, and neuronal circuits can be manipulated by combining other techniques, such as behavioral experiments or optogenetics.

Currently, the field of view of 2PM is mostly limited to regions smaller than 1 mm^2, which restricts the ability to establish correlations between different brain regions. By conducting large field imaging, multiple areas can be observed simultaneously (Tsai et al., 2015; Sofroniew et al., 2016; Stirman et al., 2016). How different brain regions interact can be investigated, such as the relationship between the primary motor cortex and the primary sensory cortex, or information flow from the primary visual cortex to higher visual cortices. As such, future advances in 2PM should incorporate widefield imaging, bigger and faster volumetric imaging, and adaptive optics (AO).

When large volumetric imaging is required, point scanning itself can be disadvantageous because of insufficient imaging speed. The combination of resonant scanner and piezo z-drive is currently the most popular setup (Peron et al., 2015), which is not fast enough to monitor the entire cortex. To overcome issues of speed, several solutions have been proposed such as imaging cells of interest via random access scanning (Katona et al., 2012), or performing volumetric imaging with Bessel beam (Lu et al., 2017) or special devices (Har-Gil et al., 2018).

2PM excitation requires a high density of photons to achieve 2-photon absorption in focus. However, light is interfered with by media during its passage, and photons emitted simultaneously (wavefront) do not reach the focal point at the same time (Figure 19.4). AO is a technique used to correct aberrations induced by media, such as brain parenchyma using spatial light modulators and/or deformable mirrors, and to enhance the efficiency of 2PM excitation and imaging resolution (Figure 19.5) (Ji et al., 2010; Sun et al., 2016). The skull itself does not block excitation light; rather, it scatters excitation

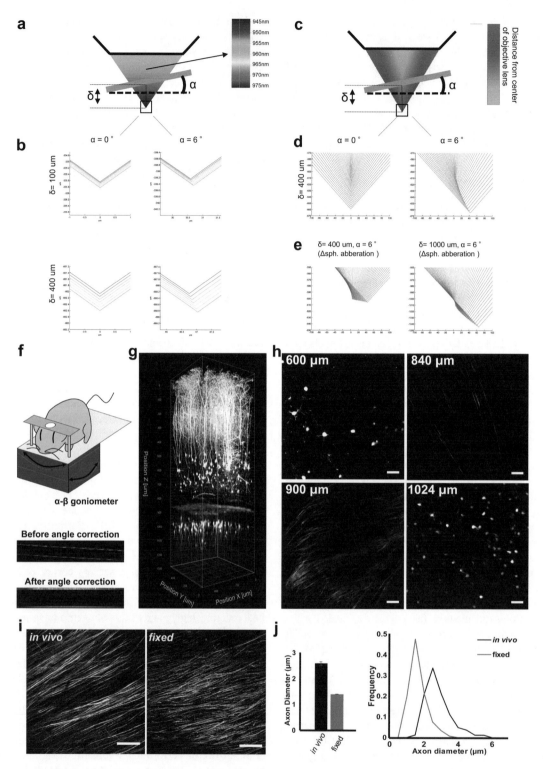

FIGURE 19.3 (a) Schematics for spectral simulation. Each color represents the light path of each wavelength. (b) Simulation of the light paths at different depth without and with 6 degrees tilting. (c) Schematics for simulation of spherical aberration. Each color represents distance from center of objective lens. (d) Without correction of spherical aberration, lights from each position do not converge (left) and the 6 degrees tilting of glass cause huge divergence vertically and horizontally (right). (e) With correction of spherical aberration, the 6 degrees tilting of glass diverges light from each position of objective lens (left) and such divergence is depth dependent (right). (f) Top: Schematics for angle correction using an α-β goniometer. The animal was attached to a custom stage that could be rotated. Bottom: Reflection from cranial window before and after angle correction. (g) 3D image of whole stack throughout intact cortex *in vivo*. (h) Representative images of layer 5 cells at 600 μm, corpus callosum at 840 μm, alveus at 900 μm, and CA1 stratum pyramidale at 1024 μm. Scale bar = 50 μm. (i) Comparison of axon thickness between *in vivo* and lightly fixed sample. representative two-photon images of corpus callosum from *in vivo* imaging (left) and fixed sample (right). Scale bar = 50 μm. (j) Axon diameters from *in vivo* imaging and lightly fixed samples were 2.59 ± 0.08 μm (*n* = 80) and 1.40 ± 0.0041um (*n* = 118), respectively. *In vivo* imaging resulted in significantly thicker appearance of axons (p = 9.3594 × 10^{-30}). Data represented in average ± s.e.m.

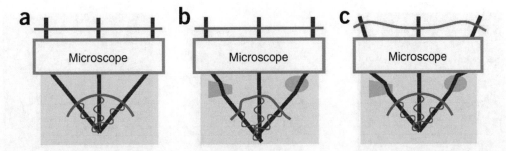

FIGURE 19.4 A simple model of optical focus formation. (a) An ideal microscope converts a planar wavefront (top red line) to a converging spherical one (bottom semicircle) in a sample of the design optical properties. Propagation vectors or "rays", defined by the direction normal to the wavefront, converge at a common point and, being in phase, constructively interfere there to create an optimal focus. Sinusoidal curves denote the phase variation along each ray. (b) Inhomogeneities in the refractive index of the sample change the directions and phases of the rays, leading to a distorted wavefront and an enlarged focal volume with lower peak intensity. (c) Controlling the input wavefront using an active optical element (not shown) can cancel these aberrations, recovering a diffraction-limited focus. *From Figure 1 of* Ji, N., Milkie, D. E., Betzig, E. (2009). Adaptive optics via pupil segmentation for high-resolution imaging in biological tissues. *Nat Methods, 39*: 141–7. doi: 10.1038/nmeth.1411.

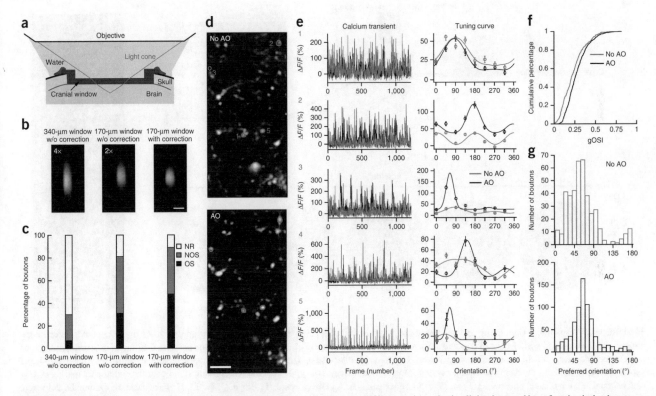

FIGURE 19.5 Adaptive optics are essential for tuning curve characterization. (a) Differences in excitation light aberrated by refractive index between water and cranial window or brain. (b) Axial images of a 2-μm bead below a 340-μm window, a 170-μm window or a 170-μm window with adaptive optics (AO) correction. Images taken without AO have 4× and 2× gain for better visibility. Scale bar = 2 μm. (c) Percentages of nonresponsive (NR), not orientation selective (NOS), and orientation selective (OS) boutons at 300–350-μm depth under the conditions in b. (d) Images of GCaMP6s+ axons at 170-μm depth measured without and with AO. Images are saturated to improve visibility of dim features. Scale bar = 10 μm. (e) Calcium transients and tuning curves for ROIs (region of interest) labeled in (d) measured without (red) and with (black) AO. Error bars represent s.e.m. (standard error of the mean) (f) Cumulative distributions of gOSI for boutons at 300–350-μm depth measured without and with AO. (g) Their preferred orientation distributions measured without and with AO. Data in (f) and (g) are the same as in (c), boutons imaged under 170-μm window without and with AO correction and analyzed with independent ROI selections. Cranial window thickness in (d)–(g) is 170 μm. Numeric aperture of the microscope objective is 1.05. *From Figure 2 of* Sun, W., Tan, Z., Mensh, B. D., Ji, N. (2015). Thalamus provides layer 4 of primary visual cortex with orientation- and direction-tuned inputs. *Nat Neurosci. 19*(2): 308–15. doi: 10.1038/nn.4196.

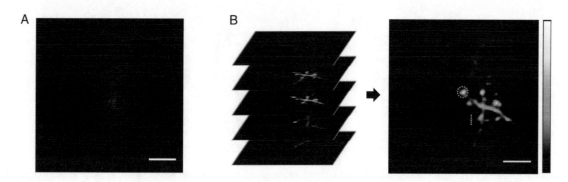

FIGURE 19.6 *In vivo* imaging of dendritic spines through the intact skull. (a) Maximum-intensity projection (MIP) of the volume stack obtained with system correction. Neurons at the superficial layer of the brain were imaged through 150 μm of intact skull. (b) MIP of the volume imaged with wavefront correction after 3D deconvolution. We merged 20 z-stacks with 1-μm axial spacing for the MIP. Submicron-size spine neck and head structures can clearly be identified. The blue dotted circle denotes the laser beam position during wavefront measurement. (a) and (b) share the same color bar and were acquired with the same laser power. (Scale bar = 5 μm.) *From Figure 4 of* Park, J. H., Sun, W., Cui, M. (2015). High-resolution *in vivo* imaging of mouse brain through the intact skull. *Proc Natl Acad Sci U S A., 112*(30): 9236-41. doi: 10.1073/pnas.1505939112

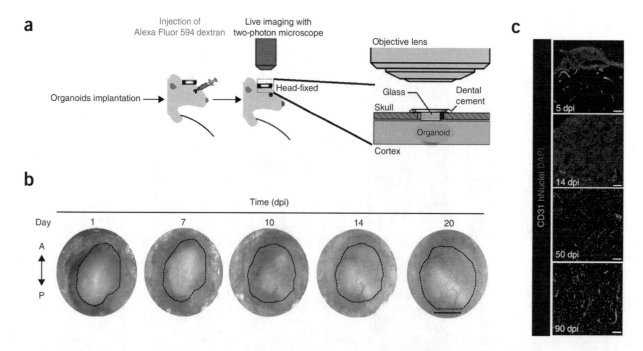

FIGURE 19.7 Functional vasculature and decreased apoptosis in grafts. (a) Illustration of the imaging approach for live two-photon microscopy imaging of the organoid graft. Image of lens and stage adapted from Sadakane, et al. Long-term two-photon calcium imaging of neuronal populations with subcellular resolution in adult nonhuman primates. *Cell Rep.* 13, 1989–1999, 2015. (b) Serial macroscopic tracking of grafts showing dynamics of blood vessel perfusion by the recipient vascular system. Dotted area indicates the graft. (c) Grafts coimmunostained for the endothelial markers CD31 and hNuclei at the indicated time points. *From Figure 4 of* Mansour, A. A., Gonçalves, J. T., Bloyd, C. W., Li, H., Fernandes, S., Quang, D., Johnston, S., Parylak, S. L., Jin, X., Gage, F. H. (2018). An *in vivo* model of functional and vascularized human brain organoids. *Nat Biotechnol., 36*(5):432-441. doi: 10.1038/nbt.4127.

light in the near infrared range, which results in aberration of wavefronts and deterioration of image quality (Helm et al., 2009; Lapchak et al., 2015; Tehrani et al., 2017). By correction of wavefronts utilizing AO, structures deeper than 500 μm from the surface can be imaged without any manipulation of the skull (Figure 19.6) (Park et al., 2015).

Finally, *in vivo* brain imaging has various applications. The craniotomy setup provides blood and oxygen supply from brain tissue and cerebrospinal fluid, mechanical protection by the skull, an immune-privileged environment, and optical access that is suitable for chronic observation. By exploiting these features, transplantation of pancreatic islet (Takahashi et al., 2014) or liver buds from induced pluripotent stem cells (Takebe et al., 2013, 2015) has been performed. Recently, it was demonstrated that human cerebral organoids transplanted into mouse brain parenchyma could be vascularized and integrated (Figure 19.7) (Mansour et al., 2018). By combining human cell transplantation and chronic *in vivo* imaging, it is possible to chronically observe cellular function and dynamics, which will be highly useful in drug development and screening (Akassoglou et al., 2017).

REFERENCES

Akassoglou, K., Merlini, M., Rafalski, V. A., Real, R., Liang, L., Jin, Y., Dougherty, S. E., De Paola, V., Linden, D. J., Misgeld, T., and Zheng, B. (2017). In vivo imaging of CNS injury and disease. *J Neurosci, 37*: 10808–10816.

Akerboom, J., Carreras Calderón, N., Tian, L., Wabnig, S., Prigge, M., Tolö, J., Gordus, A., Orger, M. B., Severi, K. E., Macklin, J. J., and Patel, R. (2013). Genetically encoded calcium indicators for multi-color neural activity imaging and combination with optogenetics. *Front Mol Neurosci, 6*: 2.

Antic, S. D., Empson, R. M., and Knopfel, T. (2016). Voltage imaging to understand connections and functions of neuronal circuits. *J Neurophysiol, 116*: 135–152.

Baird, G. S., Zacharias, D. A., and Tsien, R. Y. (1999). Circular permutation and receptor insertion within green fluorescent proteins. *Proc Natl Acad Sci U S A, 96*: 11241–11246.

Baker, B. J., Mutoh, H., Dimitrov, D., Akemann, W., Perron, A., Iwamoto, Y., Jin, L., Cohen, L. B., Isacoff, E. Y., Pieribone, V. A., Hughes, T., and Knopfel, T. (2008). Genetically encoded fluorescent sensors of membrane potential. *Brain Cell Biol, 36*: 53–67.

Barretto, R. P., Messerschmidt, B., and Schnitzer, M. J. (2009). In vivo fluorescence imaging with high-resolution microlenses. *Nat Methods, 6*: 511–512.

Chaigneau, E., Wright, A. J., Poland, S. P., Girkin, J. M., and Silver, R. A. (2011). Impact of wavefront distortion and scattering on 2-photon microscopy in mammalian brain tissue. *Opt Express, 19*: 22755–22774.

Chen, T. W., Wardill, T. J., Sun, Y., Pulver, S. R., Renninger, S. L., Baohan, A., Schreiter, E. R., Kerr, R. A., Orger, M. B., Jayaraman, V., Looger, L. L., Svoboda, K., and Kim, D. S. (2013). Ultrasensitive fluorescent proteins for imaging neuronal activity. *Nature, 499*: 295–300.

Chen, X., Leischner, U., Rochefort, N. L., Nelken, I., and Konnerth, A. (2011). Functional mapping of single spines in cortical neurons in vivo. *Nature, 475*: 501–505.

Christie, R. H., Bacskai, B. J., Zipfel, W. R., Williams, R. M., Kajdasz, S. T., Webb, W. W., and Hyman, B. T. (2001). Growth arrest of individual senile plaques in a model of Alzheimer's disease observed by in vivo multiphoton microscopy. *J Neurosci, 21*: 858–864.

Crosby, G., Culley, D. J., and Patel, P. M. (2010). At the sharp end of spines: Anesthetic effects on synaptic remodeling in the developing brain. *Anesthesiology, 112*: 521–523.

Dana, H., Mohar, B., Sun, Y., Narayan, S., Gordus, A., Hasseman, J. P., Tsegaye, G., Holt, G. T., Hu, A., Walpita, D., Patel, R., Macklin, J. J., Bargmann, C. I., Ahrens, M. B., Schreiter, E. R., Jayaraman, V., Looger, L. L., Svoboda, K., and Kim, D. S. (2016). Sensitive red protein calcium indicators for imaging neural activity. *Elife, 5*: e12727.

Denk, W., Strickler, J. H., Webb, W. W. (1990). Two-photon laser scanning fluorescence microscopy. *Science, 248*: 73–76.

Denk, W., Sugimori, M., and Llinas, R. (1995). Two types of calcium response limited to single spines in cerebellar Purkinje cells. *Proc Natl Acad Sci U S A, 92*: 8279–8282.

Diels, J.-C., and Rudolph, W. (2006). Ultrashort laser pulse phenomena: Fundamentals, techniques, and applications on a femtosecond time scale. In *Optics and Photonics* (2nd ed.), pp 1 online resource (xxi, 652 pages). Burlington, MA: Academic Press.

Dittgen, T., Nimmerjahn, A., Komai, S., Licznerski, P., Waters, J., Margrie, T. W., Helmchen, F., Denk, W., Brecht, M., and Osten, P. (2004). Lentivirus-based genetic manipulations of cortical neurons and their optical and electrophysiological monitoring in vivo. *Proc Natl Acad Sci U S A, 101*: 18206–18211.

Dombeck, D. A., Khabbaz, A. N., Collman, F., Adelman, T. L., and Tank, D. W. (2007). Imaging large-scale neural activity with cellular resolution in awake, mobile mice. *Neuron. 56*: 43–57.

Dombeck, D. A., Harvey, C. D., Tian, L., Looger, L. L., and Tank, D. W. (2010). Functional imaging of hippocampal place cells at cellular resolution during virtual navigation. *Nat Neurosci, 13*: 1433–1440.

Drew, P. J., Shih, A. Y., Driscoll, J. D., Knutsen, P. M., Blinder, P., Davalos, D., Akassoglou, K., Tsai, P. S., and Kleinfeld, D. (2010). Chronic optical access through a polished and reinforced thinned skull. *Nat Methods, 7*: 981–984.

El-Boustani, S., Ip, J. P. K., Breton-Provencher, V., Knott, G. W., Okuno, H., Bito, H., and Sur, M. (2018) Locally coordinated synaptic plasticity of visual cortex neurons in vivo. *Science, 360*: 1349–1354.

Feng, G., Mellor, R. H., Bernstein, M., Keller-Peck, C., Nguyen, Q. T., Wallace, M., Nerbonne, J. M., Lichtman, J. W., and Sanes, J. R. (2000). Imaging neuronal subsets in transgenic mice expressing multiple spectral variants of GFP. *Neuron, 28*: 41–51.

Fetcho, J. R., and O'Malley, D. M. (1997). Imaging neuronal networks in behaving animals. *Curr Opin Neurobiol, 7*: 832–838.

Fuger, P., Hefendehl, J. K., Veeraraghavalu, K., Wendeln, A. C., Schlosser, C., Obermuller, U., Wegenast-Braun, B. M., Neher, J. J., Martus, P., Kohsaka, S., Thunemann, M., Feil, R., Sisodia, S. S., Skodras, A., and Jucker, M. (2017). Microglia turnover with aging and in an Alzheimer's model via long-term in vivo single-cell imaging. *Nat Neurosci, 20*: 1371–1376.

Fukuchi-Shimogori, T., and Grove, E. A. (2001). Neocortex patterning by the secreted signaling molecule FGF8. *Science, 294*: 1071–1074.

Gerner, P., and O'Connor, J. P. (2008). Impact of analgesia on bone fracture healing. *Anesthesiology, 108*: 349–350.

Goldey, G. J., Roumis, D. K., Glickfeld, L. L., Kerlin, A. M., Reid, R. C., Bonin, V., Schafer, D. P., and Andermann, M. L. (2014). Removable cranial windows for long-term imaging in awake mice. *Nat Protoc, 9*: 2515–2538.

Guo, Z. V., Hires, S. A., Li, N., O'Connor, D. H., Komiyama, T., Ophir, E., Huber, D., Bonardi, C., Morandell, K., Gutnisky, D., Peron, S., Xu, N. L., Cox, J., and Svoboda, K. (2014). Procedures for behavioral experiments in head-fixed mice. *PLoS One, 9*: e88678.

Har-Gil, H., Golgher, L., Israel, S., Kain, D., Cheshnovsky, O., Parnas, M., and Blinder, P. (2018). PySight: Plug and play photon counting for fast continuous volumetric intravital microscopy. *Optica, 5*: 1104–1112.

Helm, P. J., Ottersen, O. P., and Nase, G. (2009). Analysis of optical properties of the mouse cranium--implications for in vivo multi photon laser scanning microscopy. *J Neurosci Methods, 178*: 316–322.

Helmchen, F., and Denk, W. (2005). Deep tissue two-photon microscopy. *Nat Methods*, 2: 932–940.

Heo, C., Park, H., Kim, Y. T., Baeg, E., Kim, Y. H., Kim, S. G., and Suh, M. (2016). A soft, transparent, freely accessible cranial window for chronic imaging and electrophysiology. *Sci Rep*, 6: 27818.

Hermens, W. T., and Verhaagen, J. (1998). Viral vectors, tools for gene transfer in the nervous system. *Prog Neurobiol*, 55: 399–432.

Hirase, H., and Ozawa, K. (2013). *Deep brain imaging - Thorlabs multiphoton microscope*. Available at: https://www.youtube.com/watch?v=kyuY7F3cvNU

Hirase, H., Qian, L., Bartho, P., and Buzsaki, G. (2004). Calcium dynamics of cortical astrocytic networks in vivo. *PLoS Biol*, 2: E96.

Holtmaat, A., Bonhoeffer, T., Chow, D. K., Chuckowree, J., De Paola, V., Hofer, S. B., Hubener, M., Keck, T., Knott, G., Lee, W. C., Mostany, R., Mrsic-Flogel, T. D., Nedivi, E., Portera-Cailliau, C., Svoboda, K., Trachtenberg, J. T., and Wilbrecht, L. (2009). Long-term, high-resolution imaging in the mouse neocortex through a chronic cranial window. *Nat Protoc*, 4: 1128–1144.

Horton, N. G., Wang, K., Kobat, D., Clark, C. G., Wise, F. W., Schaffer, C. B., and Xu, C. (2013). In vivo three-photon microscopy of subcortical structures within an intact mouse brain. *Nat Photonics*, 7(3): 205–209.

Huber, D., Gutnisky, D. A., Peron, S., O'Connor, D. H., Wiegert, J. S., Tian, L., Oertner, T. G., Looger, L. L., and Svoboda, K. (2012). Multiple dynamic representations in the motor cortex during sensorimotor learning. *Nature*, 484: 473–478.

Isshiki, M., and Okabe, S. (2014), Evaluation of cranial window types for in vivo two-photon imaging of brain microstructures. *Microscopy*, 63: 53–63.

Ji, N., Milkie, D. E., and Betzig, E. (2010). Adaptive optics via pupil segmentation for high-resolution imaging in biological tissues. *Nat Methods*, 7: 141–147.

Jung, J. C., Mehta, A. D., Aksay, E., Stepnoski, R., and Schnitzer, M. J. (2004). In vivo mammalian brain imaging using one- and two-photon fluorescence microendoscopy. *J Neurophysiol*, 92: 3121–3133.

Kammesheidt, A., Ito, K., Kato, K., Villarreal, L. P., and Sumikawa, K. (1996). Transduction of hippocampal CA1 by adenovirus in vivo. *Brain Res*, 736: 297–304.

Kannan, M., Vasan, G., Huang, C., Haziza, S., Li, J. Z., Inan, H., Schnitzer, M. J., and Pieribone, V. A. (2018). Fast, in vivo voltage imaging using a red fluorescent indicator. *Nat Methods*, 15(12): 1108–1116.

Kasthuri, N., Hayworth, K. J., Berger, D. R., Schalek, R. L., Conchello, J. A., Knowles-Barley, S., Lee, D., Vázquez-Reina, A., Kaynig, V., Jones, T. R., and Roberts, M. (2015). Saturated reconstruction of a volume of neocortex. *Cell*, 162: 648–661.

Kato, H. K., Gillet, S. N., Peters, A. J., Isaacson, J. S., and Komiyama, T. (2013). Parvalbumin-expressing interneurons linearly control olfactory bulb output. *Neuron*, 80: 1218–1231.

Katona, G., Szalay, G., Maak, P., Kaszas, A., Veress, M., Hillier, D., Chiovini, B., Vizi, E. S., Roska, B., and Rozsa, B. (2012). Fast two-photon in vivo imaging with three-dimensional random-access scanning in large tissue volumes. *Nat Methods*, 9: 201–208.

Kawakami, R., Sawada, K., Sato, A., Hibi, T., Kozawa, Y., Sato, S., Yokoyama, H., and Nemoto, T. (2013). Visualizing hippocampal neurons with in vivo two-photon microscopy using a 1030 nm picosecond pulse laser. *Sci Rep*, 3: 1014.

Kawakami, R., Sawada, K., Kusama, Y., Fang, Y. C., Kanazawa, S., Kozawa, Y., Sato, S., Yokoyama, H., and Nemoto, T. (2015). In vivo two-photon imaging of mouse hippocampal neurons in dentate gyrus using a light source based on a high-peak power gain-switched laser diode. *Biomed Opt Express*, 6: 891–901.

Kerr, J. N., and Denk, W. (2008). Imaging in vivo: Watching the brain in action. *Nat Rev Neurosci*, 9: 195–205.

Kim, K. H., Buehler, C., and So, P. T. (1999). High-speed, two-photon scanning microscope. *Appl Opt*, 38: 6004–6009.

Kim, T. H., Zhang, Y., Lecoq, J., Jung, J. C., Li, J., Zeng, H., Niell, C. M., and Schnitzer, M. J. (2016). Long-term optical access to an estimated one million neurons in the live mouse cortex. *Cell Rep*, 17: 3385–3394.

Kislin, M., Mugantseva, E., Molotkov, D., Kulesskaya, N., Khirug, S., Kirilkin, I., Pryazhnikov, E., Kolikova, J., Toptunov, D., Yuryev, M., Giniatullin, R., Voikar, V., Rivera, C., Rauvala, H., and Khiroug, L. (2014). Flat-floored air-lifted platform: A new method for combining behavior with microscopy or electrophysiology on awake freely moving rodents. *J Vis Exp*, 2014 Jun 29(88): e51869.

Kleinfeld, D., Sachdev, R. N., Merchant, L. M., Jarvis, M. R., and Ebner, F. F. (2002). Adaptive filtering of vibrissa input in motor cortex of rat. *Neuron*, 34: 1021–1034.

Kobat, D., Horton, N. G., and Xu, C. (2011). In vivo two-photon microscopy to 1.6-mm depth in mouse cortex. *J Biomed Opt*, 16: 106014.

Komiyama, T., Sato, T. R., O'Connor, D. H., Zhang, Y. X., Huber, D., Hooks, B. M., Gabitto, M., and Svoboda, K. (2010). Learning-related fine-scale specificity imaged in motor cortex circuits of behaving mice. *Nature*, 464: 1182–1186.

Kondo, M., Kobayashi, K., Ohkura, M., Nakai, J., and Matsuzaki, M. (2017). Two-photon calcium imaging of the medial prefrontal cortex and hippocampus without cortical invasion. *Elife*, 6: e26839

Kondo, S., and Ohki, K. (2014). Functional imaging of thalamic axons in mouse primary visual cortex. In *Annual Meeting of the Society For Neuroscience*.

Kyweriga, M., Sun, J., Wang, S., Kline, R., and Mohajerani, M. H. (2017). A large lateral craniotomy procedure for mesoscale wide-field optical imaging of brain activity. *J Vis Exp*. 2017 May 7(123):52642.

Lapchak, P. A., Boitano, P. D., Butte, P. V., Fisher, D. J., Holscher, T., Ley, E. J., Nuno, M., Voie, A. H., and Rajput, P. S. (2015). Transcranial near-infrared laser transmission (NILT) profiles (800 nm): Systematic comparison in four common research species. *PLoS One*, 10: e0127580.

Lechleiter, J. D., Lin, D. T., and Sieneart, I. (2002). Multi-photon laser scanning microscopy using an acoustic optical deflector. *Biophys J*, 83: 2292–2299.

Lee, S. H., and Dan, Y. (2012). Neuromodulation of brain states. *Neuron*, 76: 209–222.

Leybaert, L., de Meyer, A., Mabilde, C., and Sanderson, M. J. (2005). A simple and practical method to acquire geometrically correct images with resonant scanning-based line scanning in a custom-built video-rate laser scanning microscope. *J Microsc*, 219: 133–140.

Liang, Z., King, J., and Zhang, N. (2012), Intrinsic organization of the anesthetized brain. *J Neurosci*, 32: 10183–10191.

Lim, D. H., Mohajerani, M. H., Ledue, J., Boyd, J., Chen, S., and Murphy, T. H. (2012) In vivo large-scale cortical mapping using channelrhodopsin-2 stimulation in transgenic mice reveals asymmetric and reciprocal relationships between cortical areas. *Front Neural Circuits*, 6: 11.

Lu, R., Sun, W., Liang, Y., Kerlin, A., Bierfeld, J., Seelig, J. D., Wilson, D. E., Scholl, B., Mohar, B., Tanimoto, M., Koyama, M., Fitzpatrick, D., Orger, M. B., and Ji, N. (2017). Video-rate volumetric functional imaging of the brain at synaptic resolution. *Nat Neurosci*, 20: 620–628.

Lukasiewicz, K., Robacha, M., Bozycki, L., Radwanska, K., and Czajkowski, R. (2016). Simultaneous two-photon in vivo imaging of synaptic inputs and postsynaptic targets in the mouse retrosplenial cortex. *J Vis Exp*. 2016 Mar 13(109):53528.

Mansour, A. A., Goncalves, J. T., Bloyd, C. W., Li, H., Fernandes, S., Quang, D., Johnston, S., Parylak, S. L., Jin, X., and Gage, F. H. (2018). An in vivo model of functional and vascularized human brain organoids. *Nat Biotechnol*, 36: 432–441.

McEwen, B. S., and Sapolsky, R. M. (1995). Stress and cognitive function. *Curr Opin Neurobiol*, 5: 205–216.

Miyawaki, A., Llopis, J., Heim, R., McCaffery, J. M., Adams, J. A., Ikura, M., and Tsien, R. Y. (1997). Fluorescent indicators for Ca2+ based on green fluorescent proteins and calmodulin. *Nature*, 388: 882–887.

Monai, H., Ohkura, M., Tanaka, M., Oe, Y., Konno, A., Hirai, H., Mikoshiba, K., Itohara, S., Nakai, J., Iwai, Y., and Hirase, H. (2016). Calcium imaging reveals glial involvement in transcranial direct current stimulation-induced plasticity in mouse brain. *Nat Commun*, 7: 11100.

Mostany, R., and Portera-Cailliau, C. (2008). A craniotomy surgery procedure for chronic brain imaging. *J Vis Exp*. 2008 Feb 15(12):680.

Nakai, J., Ohkura, M., and Imoto, K. (2001). A high signal-to-noise Ca(2+) probe composed of a single green fluorescent protein. *Nat Biotechnol*. 19: 137–141.

Nimmerjahn, A., Kirchhoff, F., and Helmchen, F. (2005). Resting microglial cells are highly dynamic surveillants of brain parenchyma in vivo. *Science*, 308: 1314–1318.

Nitsche, M. A., Boggio, P. S., Fregni, F., and Pascual-Leone, A. (2009). Treatment of depression with transcranial direct current stimulation (tDCS): A review. *Exp Neurol*, 219: 14–19.

O'Keefe, J., and Dostrovsky, J. (1971). The hippocampus as a spatial map. Preliminary evidence from unit activity in the freely-moving rat. *Brain Res*, 34: 171–175.

Oheim, M., Beaurepaire, E., Chaigneau, E., Mertz, J., and Charpak, S. (2001). Two-photon microscopy in brain tissue: Parameters influencing the imaging depth. *J Neurosci Methods*, 111: 29–37.

Ohki, K., Chung, S., Ch'ng, Y. H., Kara, P., and Reid, R. C. (2005). Functional imaging with cellular resolution reveals precise micro-architecture in visual cortex. *Nature*, 433: 597–603.

Ohki, K., Chung, S., Kara, P., Hubener, M., Bonhoeffer, T., and Reid, R. C. (2006). Highly ordered arrangement of single neurons in orientation pinwheels. *Nature*, 442: 925–928.

Ono, T., Nakamura, K., Nishijo. H., and Fukuda, M. (1986). Hypothalamic neuron involvement in integration of reward, aversion, and cue signals. *J Neurophysiol*, 56: 63–79.

Park, J. H., Sun, W., and Cui, M. (2015). High-resolution in vivo imaging of mouse brain through the intact skull. *Proc Natl Acad Sci U S A*, 112: 9236–9241.

Paukert, M., and Bergles, D. E. (2012). Reduction of motion artifacts during in vivo two-photon imaging of brain through heartbeat triggered scanning. *J Physiol*, 590: 2955–2963.

Perillo, E. P., McCracken, J. E., Fernee, D. C., Goldak, J. R., Medina, F. A., Miller, D. R., Yeh, H. C., and Dunn, A. K. (2016). Deep in vivo two-photon microscopy with a low cost custom built mode-locked 1060 nm fiber laser. *Biomed Opt Express*, 7: 324–334.

Peron, S. P., Freeman, J., Iyer, V., Guo, C., and Svoboda, K. (2015). A cellular resolution map of barrel cortex activity during tactile behavior. *Neuron*, 86: 783–799.

Podgorski, K., and Ranganathan, G. (2016). Brain heating induced by near-infrared lasers during multiphoton microscopy. *J Neurophysiol*, 116: 1012–1023.

Romoser, V. A., Hinkle, P. M., and Persechini, A. (1997). Detection in living cells of Ca2+-dependent changes in the fluorescence emission of an indicator composed of two green fluorescent protein variants linked by a calmodulin-binding sequence. A new class of fluorescent indicators. *J Biol Chem*, 272: 13270–13274.

Ruebhausen, M. R., Brozoski, T. J., and Bauer, C. A. (2012). A comparison of the effects of isoflurane and ketamine anesthesia on auditory brainstem response (ABR) thresholds in rats. *Hear Res*, 287: 25–29.

Sakaguchi, T., Ishikawa, D., Nomura, H., Matsuki, N., and Ikegaya, Y. (2012). Normal learning ability of mice with a surgically exposed hippocampus. *Neuroreport*, 23: 457–461.

Shih, A. Y., Driscoll, J. D., Drew, P. J., Nishimura, N., Schaffer, C. B., and Kleinfeld, D. (2012). Two-photon microscopy as a tool to study blood flow and neurovascular coupling in the rodent brain. *J Cereb Blood Flow Metab*, 32: 1277–1309.

Sofroniew, N. J., Flickinger, D., King, J., and Svoboda, K. (2016). A large field of view two-photon mesoscope with subcellular resolution for in vivo imaging. *Elife*, 5: e14472

Stirman, J. N., Smith, I. T., Kudenov, M. W., and Smith, S. L. (2016). Wide field-of-view, multi-region, two-photon imaging of neuronal activity in the mammalian brain. *Nat Biotechnol*, 34: 857–862.

Stosiek, C., Garaschuk, O., Holthoff, K., and Konnerth, A. (2003). Thalamus provides layer 4 of primary visual cortex with orientation- and direction-tuned inputs. *Proc Natl Acad Sci U S A*. 100: 7319–7324.

Sun, W., Tan, Z., Mensh, B. D., and Ji, N. (2016). Thalamus provides layer 4 of primary visual cortex with orientation- and direction-tuned inputs. *Nat Neurosci*, 19: 308–315.

Svoboda, K., and Yasuda, R. (2006). Principles of two-photon excitation microscopy and its applications to neuroscience. *Neuron*, 50: 823–839.

Svoboda, K., Denk, W., Kleinfeld, D., and Tank, D. W. (1997). In vivo dendritic calcium dynamics in neocortical pyramidal neurons. *Nature*, 385: 161–165.

Tabata, H., and Nakajima, K. (2001). Efficient in utero gene transfer system to the developing mouse brain using electroporation: Visualization of neuronal migration in the developing cortex. *Neuroscience*, 103: 865–872.

Takahashi, Y., Takebe, T., Enomura, M., Koike, N., Lee, S., Nemeno, J. G., Sekine, K., Lee, J. I., and Taniguchi, H. (2014). High-resolution intravital imaging for monitoring the transplanted islets in mice. *Transplant Proc*, *46*: 1166–1168.

Takebe, T., Sekine, K., Enomura, M., Koike. H., Kimura, M., Ogaeri, T., Zhang, R. R., Ueno, Y., Zheng, Y. W., Koike, N., Aoyama, S., Adachi, Y., and Taniguchi, H. (2013). Vascularized and functional human liver from an iPSC-derived organ bud transplant. *Nature*, *499*: 481–484.

Takebe, T., Enomura, M., Yoshizawa, E., Kimura, M., Koike, H., Ueno, Y., Matsuzaki, T., Yamazaki, T., Toyohara, T., Osafune, K., Nakauchi, H., Yoshikawa, H. Y., and Taniguchi, H. (2015). Vascularized and complex organ buds from diverse tissues via mesenchymal cell-driven condensation. *Cell Stem Cell*, *16*: 556–565.

Tehrani, K. F., Kner, P., and Mortensen, L. J. (2017). Characterization of wavefront errors in mouse cranial bone using second-harmonic generation. *J Biomed Opt*, *22*: 36012.

Tian, L., Hires, S. A., Mao, T., Huber, D., Chiappe, M. E., Chalasani, S. H., Petreanu, L., Akerboom, J., McKinney, S. A., Schreiter, E. R., Bargmann, C. I., Jayaraman, V., Svoboda, K., and Looger, L. L. (2009). Imaging neural activity in worms, flies and mice with improved GCaMP calcium indicators. *Nat Methods*, *6*: 875–881.

Trachtenberg, J, T., Chen, B. E., Knott, G. W., Feng, G., Sanes, J. R., Welker, E., and Svoboda, K. (2002). Long-term in vivo imaging of experience-dependent synaptic plasticity in adult cortex. *Nature*, *420*: 788–794.

Tsai, P. S., Mateo, C., Field, J. J., Schaffer, C. B., Anderson, M. E., and Kleinfeld, D. (2015). Ultra-large field-of-view two-photon microscopy. *Opt Express*, *23*: 13833–13847.

Wang, K., Sun, W., Richie, C. T., Harvey, B. K., Betzig, E., and Ji, N. (2015). Direct wavefront sensing for high-resolution in vivo imaging in scattering tissue. *Nat Commun*, *6*: 7276.

Wang, S. S., Shultz, J. R., Burish, M. J., Harrison, K. H., Hof, P. R., Towns, L. C., Wagers, M. W., and Wyatt, K. D. (2008). Functional trade-offs in white matter axonal scaling. *J Neurosci*, *28*: 4047–4056.

Xu, H. T., Pan, F., Yang, G., and Gan, W. B. (2007). Choice of cranial window type for in vivo imaging affects dendritic spine turnover in the cortex. *Nat Neurosci*, *10*: 549–551.

Yang, G., Pan, F., Parkhurst, C. N., Grutzendler, J., and Gan, W. B. (2010). Thinned-skull cranial window technique for long-term imaging of the cortex in live mice. *Nat Protoc*, *5*: 201–208.

Yang, G., Lai, C. S., Cichon, J., Ma, L., Li, W., and Gan, W. B. (2014). Sleep promotes branch-specific formation of dendritic spines after learning. *Science*, *344*: 1173–1178.

Yu, H., Majewska, A. K., and Sur, M. (2011). Rapid experience-dependent plasticity of synapse function and structure in ferret visual cortex in vivo. *Proc Natl Acad Sci U S A*, *108*: 21235–21240.

Zhao, Y., Araki, S., Wu, J., Teramoto, T., Chang, Y. F., Nakano, M., Abdelfattah, A. S., Fujiwara, M., Ishihara, T., Nagai, T., and Campbell, R. E. (2011). An expanded palette of genetically encoded Ca(2)(+) indicators. *Science*, *333*: 1888–1891.

20

Live Imaging of Zebrafish

Yinan Wan, Philipp J. Keller, and Burkhard Höckendorf

CONTENTS

20.1 Introduction – Why and When to Use Zebrafish for *In Vivo* Imaging

From the making of a protein molecule to the division of a cell, from the generation of an action potential to the motion of an animal, all biological processes intrinsically have a dynamic nature. Capturing such dynamics in intact organisms is a powerful approach for understanding the underlying biological mechanisms. Over the past three decades, the zebrafish (*Danio rerio*) has emerged in this context as an excellent model organism for vertebrate biology, forming a strong synergy with the concurrent, rapid advancement of *in vivo* imaging technologies.

Compared to other vertebrate model organisms, the zebrafish is highly accessible by microscopy because of its optical transparency and external development. The life cycle of zebrafish is illustrated in Figure 20.1A. From the single-cell stage, embryos develop outside the mother and early embryogenesis – e.g. cleavage, gastrulation, and early organogenesis – can be visualized without prior surgery procedures (Figure 20.1b). By the end of the first day of development, the embryo has already established a characteristic fish-like form (Figure 20.1c) with most of its tissues and organ primordia in

place (Kimmel et al., 1995). Within three days, the zebrafish larvae hatch and start to exhibit various types of behaviors, including swimming and foraging. The transparency of the brain also allows optical studies and photoablations/perturbations of neural circuits to be carried out in a minimally invasive way (Fetcho and Liu, 1998).

The zebrafish is an outstanding model organism for genetic analyses and manipulation due to the ease of maintenance, high fecundity, and the short generation time. A single female can produce up to 200 eggs in a single spawning session and it takes 3–4 months for the fish to reach sexual maturity (Figure 20.1A). Since the first introduction of zebrafish for genetic studies by George Streisinger in the 1970s, the field has experienced tremendous success in large-scale screening (Granato et al., 1996; Haffter et al., 1996; Driever et al., 1996), establishment of permanent transgenic lines (Udvadia and Linney, 2003), and more recently gene editing techniques (Hisano et al., 2014). The transgenic techniques made it possible to visualize the expression of different genes using fluorescent proteins (FPs) driven by the corresponding promoters and to study their function in wild type or mutant animals. The zebrafish genome has been sequenced and assembled, and the ongoing annotation revealed evidence for more than 26,000

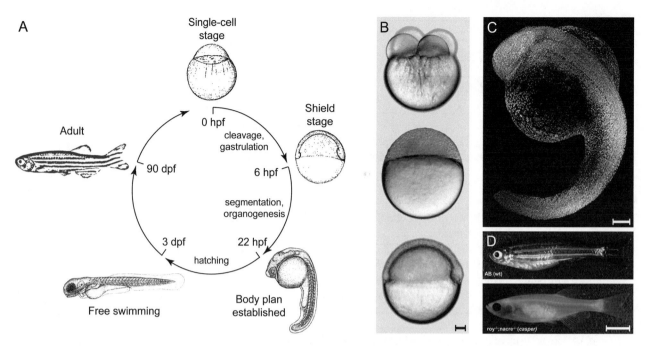

FIGURE 20.1 Zebrafish life cycle. A. Life cycle of the zebrafish. The zebrafish embryo develops from a single cell sitting on top of the yolk. It develops rapidly outside the mother, and by the end of the first day after fertilization, acquires the form of a tiny fish with major organs visible. After hatching, larval zebrafish exhibit swimming and foraging behavior and continuously grow in size. Adult zebrafish become sexually mature by ~3 months after fertilization and can be used for breeding the next generation. B. Zebrafish embryo at cleavage (top), blastula (middle), and gastrula (bottom) stages. The images were captured at 1, 4, and 6 hours after fertilization, respectively, to serve as a reference for normal embryonic development (Haffter et al., 1996). C. Nuclei-labeled zebrafish embryo at 22 hours postfertilization. The 3D image stack was acquired with a simultaneous multiview light sheet microscope (SiMView). Gradient indicates depth in the image (Keller, 2013). D. Adult zebrafish with wild-type (top) or mutated pigmentation (bottom) pattern. The largely transparent mutant *casper* was created by crossing *roy obison* and *nacre* mutants and exhibits a combination of melanocyte and iridophore loss (White et al., 2008). Parts of A. have been reprinted from Kimmel et al. (2005) with permission from the publisher John Wiley and Sons. B. has been reprinted from Haffter et al. (1996) with permission from the publisher The Company of Biologists Ltd. C. has been reprinted from Keller (2013) with permission from the publisher AAAS. D. has been reprinted from White et al. (2008) with permission from the publisher Elsevier. Scale bars = 100 μm (b), 100 μm (c), 5 mm (d).

protein-coding genes (Collins et al., 2012; Howe et al., 2013b). To date, more than 14,000 genes have been described in publications, according to statistics from the Zebrafish Information Network (http://zfin.org) (Howe et al., 2013a).

Although there are other established model organisms such as *Drosophila melanogaster* and *Caenorhabditis elegans* that are also small and amenable to large-scale genetic screening, they cannot be utilized to address questions that are specific to vertebrate development and function. Invertebrates lack certain organ systems, such as kidney and the multichambered heart, and thus are limited in studying their normal and abnormal functions. The zebrafish nervous system is also similar to that of mammals, with representative brain anatomy (fore-, mid- and hind-brain) and conserved structure and function of many peripheral sensory organs. For example, the zebrafish retina preserves the laminar organization and stereotypic location of cell types compared to higher vertebrates (Agathocleous and Harris, 2009). About 70% of human genes have at least one orthologue in the zebrafish genome (Howe et al., 2013b). Over the years, many mutations have been identified in zebrafish as reminiscent of human clinical disorders, including cancer, infectious disease, inflammation, and immunological disorders. Conserved pathophysiological features have been found at the molecular, cellular or histopathological level and provided valuable insights into the genetic pathways controlling normal development and/

or disease-generating states (reviewed in Dooley and Zon, 2000; Lieschke and Currie, 2007; Goldsmith and Jobin, 2012).

Despite these advantages, there are certain limitations of using zebrafish as a model organism. Certain organs in zebrafish are not directly equivalent to the mammalian counterparts. For example, although the zebrafish skin is also of ectodermal origin, it lacks the mammalian appendages, such as hair follicles and sebaceous glands. The telencephalon in fish has only a rudimentary cortex, and their behavior and cognitive functions are much more simplified or abstracted compared with humans. From around 24 hours post fertilization, zebrafish start to form pigment cells and become less transparent over development, making optical studies more challenging. Treating the embryos with N-phenylthiourea (PTU) inhibits the formation of melanophores, and the *casper* mutants (mutations in genes *nacre* and *mpv17*) lead to transparent embryos that can be imaged up to late larval or even adult stages (White et al., 2008; D'Agati et al., 2017), which partially alleviates the challenges encountered when imaging through the skin (Figure 20.1d).

20.2 Approaches for Labeling Zebrafish

Although the inhomogeneous properties of zebrafish tissues can provide a certain level of optical contrast in label-free embryos, modern microscopy relies heavily on fluorescent labeling techniques and has benefited tremendously from the development of fluorescent dyes and proteins. Here, we review some fluorescent labeling strategies that are commonly used in zebrafish in the order of decreasing sparseness, focusing on their respective strengths and weaknesses for live imaging. From ubiquitous labeling to marking a single cell, approaches can be selected as needed for a given application.

20.2.1 Ubiquitous Labeling

Ubiquitous labeling is suitable for *in toto* imaging of the entire zebrafish embryo or specific tissues at early developmental stages. Common strategies for labeling all or nearly all the cells include chemical dye staining, mRNA injection, and the use of transgenic lines. Vital dyes (e.g. DiI, BODYPY, Rhodamine, etc.) can be applied by bath incubation or by pressure injection into one-cell stage embryos (Godinho, 2011b). Alternatively, mRNAs that encode various fluorescent proteins can be transcribed *in vitro* and injected into one-cell stage embryos (Godinho, 2011a). Both dye and RNA injection label the cells transiently. Fluorescent dyes are visible immediately after application but will eventually dilute out after multiple rounds of cell divisions after about 30 hours. For RNA injection, fluorescent proteins will not be visible until they are properly translated and folded, introducing a delay of several hours after injection. However, the florescence from RNA injection is amplified through multiple rounds of translation and can last for up to 5–10 days. Cells can be permanently labeled in a transgenic line that expresses FPs driven by ubiquitous gene promoters (e.g. beta-actin) (see [Burket et al., 2008] for a comprehensive review).

Dyes and FPs can be delivered to cytoplasm, but for all three methods, there are also options available for targeting specific subcellular structures. For example, nuclei can be specifically stained by Hoechst-33342 dye or can be labeled by cells expressing a fusion protein of the FP and a nuclear localization signal.

20.2.2 Labeling a Defined Population of Cells

Labeling cells effectively and reliably in a tissue-restricted and stage-specific manner is critical if we wish to observe cell behavior in a wild-type or perturbed context. The development of genetic techniques in zebrafish has led to many powerful tools for establishing stable transgenic lines through traditional or modern genetic methods (Udvadia and Linney, 2003; Hisano et al., 2014). *Tol2* transposon-mediated transgenesis is commonly used to integrate external genes and reporters in the zebrafish genome. Kwan et al. (2007) established the Tol2kit system, where the promoter, coding sequence and 3' tag can be modularly engineered into individual entry clones and assembled into a Tol2 transposon backbone using multisite Gateway technology. When the construct is coinjected into the single-cell stage embryos with Tol2 mRNA, the transgene can be randomly inserted into the fish genome through transposition (Kwan et al., 2007). For example, ubiquitous labeling of nuclei in red and membrane in green can be achieved by combining the *bactin2* promoter, nlsmCherry sequence, and

the IRES-EGFPCAAX-pA tag (Figure 20.2a). Over the past two decades, the field has been constantly moving toward the goal of controlling the spatiotemporal pattern of gene expression with higher precision and more flexibility. In addition to the lines generated during the studies of specific gene functions, cell-type specific transgenic lines have also been generated through large-scale enhancer trap or gene trap screens (Scott et al., 2007; Asakawa et al., 2008; Ogura et al., 2009; Otsuna et al., 2015; Kimura et al., 2014). Transcriptional control can be further refined by a combination of Gal4, Cre, and other regulatory systems. Temporally inducible systems are also available and use heat (Adam et al., 2000; Le et al., 2007) or chemicals (Knopf et al., 2010; Hans et al., 2009) to trigger transcription *in vivo*. FPs have been engineered to have higher brightness and photostability as well as to cover a broader range in the spectrum. The recently developed modular protein tagging systems (e.g. HaloTag or SNAP-tag) (Los et al., 2008; Keppler et al., 2003) employ synthetic ligands that covalently bind to the protein tags, providing more photostability and flexibility in excitation and emission spectra without a need for re-creating transgenic lines (Xue et al., 2015; Grimm et al., 2017).

20.2.3 Mosaic Labeling

Dense labeling of all cells in the entire embryo or in specific tissues can sometimes be problematic if information (e.g. morphology or molecular identity) needs to be extracted from individual cells. This problem can be circumvented by several mosaic labeling strategies. Because cell fates are not determined in early cleavage stages, dye or mRNA injection into individual blastomeres in this early stage will lead to transient random labeling of a subset of cells. Injection of DNA into the embryos at the single-cell stage will cause the injected genes to randomly integrate into the genome of a subset of the cells – usually in a sparser manner than for mRNA injection – and their descendants, thus permanently labeling a cell population in a mosaic manner. Similarly, inducible systems will also exhibit some level of randomness with a short duration of heat shock or ligand treatment. Mosaic analysis in zebrafish (MAZe) is a system that uses transient heat activation to drive Cre-mediated recombination in a random subset of cells, in which inheritable genetic change causes the activation of reporter genes via the Gal4/UAS system (Figure 20.2B) (Collins et al., 2010). Zebrabow, a more sophisticated reporter system, has been developed to generate mosaic labeling with multiple colors through a random combination of spectrally distinct fluorescent proteins (Figure 20.2c) (Pan et al., 2013).

20.2.4 Clonal and Single-Cell Labeling

Cell lineaging and single-cell physiology studies require precise labeling of single cells and/or their descendants. Fluorescent dyes, DNA, or RNA can be delivered to a small number of cells or even individual cells through focal electroporation (Tawk et al., 2009). Individual cells can be transplanted from one fluorescent embryo (donor) to another embryo (host) that expresses different fluorescent reporters or no reporters at all. Donor cells can be placed in desired

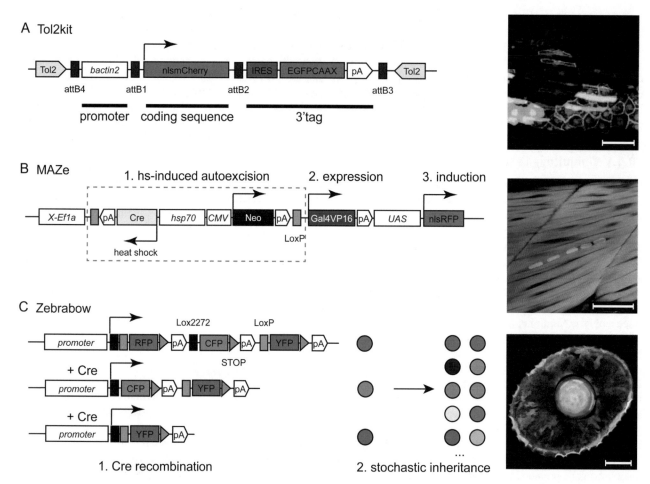

FIGURE 20.2 Genetic labeling techniques. A. Ubiquitous labeling of cell nuclei and membranes using *Tol2kit*. Driven under a ubiquitous *beta-actin* promoter, the nuclear-localized mCherry and membrane-targeted EGFP are connected with an internal ribosome entry sequence (IRES) and engineered into the destination vector such that the two genes are expressed in tandem. The construction kit is based on the multisite gateway system to assemble different modules of promoters, coding sequences and 3' tags to efficiently target the intended population of cells. The panel on the right shows the trunk of an embryo injected with the construct DNA at the one-cell stage, and imaged at 24 hpf, with mCherry in red and EGFP in green (Kwan et al., 2007). B.Clonal labeling using mosaic analysis in zebrafish (MAZe). Transient activation of a heat-shock promoter triggers the Cre recombinase to remove itself from the genome, leading to expression of the reporter gene via the Gal4/UAS system. The panel on the right shows MAZe-labeled cells in nuclei of a muscle fiber of a 5 dpf embryo that was heat-shocked at 8 hpf (Collins et al., 2010). C. Stochastic multicolor labeling using Zebrabow. Fluorescent proteins RFP, CFP, and YFP are arranged with interspersed LoxP and Lox2272 sites. After Cre-mediated recombination, cells acquire multiple inserts of genes encoding the different fluorescent proteins and exhibit a spectrum of different colors as an outcome of the stochastic combination of RFP, CFP, and YFP. The panel on the right shows the eye of a 2 dpf embryo constructed with ubiquitous Zebrabow insertion and broad injection of Cre protein Pan et al. (2013). The image included in Panel A has been reprinted from Kwan et al. (2007) with permission from the publisher John Wiley and Sons. The image included in Panel B has been reprinted from Collins et al. (2010) with permission from the publisher Springer Nature. The image included in Panel C has been reprinted from Pan et al. (2013) with permission from the publisher The Company of Biologists Ltd. Scale bars = 50 µm (A, B and C).

progenitor domains depending on the precision of the micromanipulator. Transplantation is a commonly used in embryology and can also be used to study cell–cell signaling if the donor and host embryos are of different genetic backgrounds.

With the advances in understanding of chromophore photochemistry, a range of photomodulatable fluorescent proteins have become available, which, in combination with precise optical manipulation, can be used to efficiently label cells with high spatial and temporal resolution. Upon exposure to light at a certain wavelength, proteins switch from a dark/dim to a bright fluorescent state (photoactivatable fluorescent proteins), or convert from one color to another color (photoconvertible fluorescent proteins) (for a review, see Lukyanov et al.,

2005). The activation or conversion process is irreversible, but the light-induced protein form will be eventually diluted and replaced by the unconverted color with the synthesis and turnover of the proteins. Alternatively, focused light or the heat generated from it can be used to trigger gene recombination in the targeted cells, leading to permanent labeling of cells and their descendants (Ando et al., 2001; Halloran et al., 2000).

20.2.5 Biosensors in Cells

The reporter systems for *in vivo* imaging apply far beyond visualization of cell location and shape. Fluorescent biosensors have been designed and engineered for studying the dynamic processes of signaling molecules, ions, and metabolites, the changes of which are physiologically relevant to the organism (for review see Okumoto et al., 2012). A typical example is the use of genetically encoded calcium indicators (GCaMP), which is widely used in neuroscience to monitor population neural activity in live imaging applications (reviewed in Broussard et al., 2014). A number of transgenic fish have also been generated recently to express *in vivo* reporter proteins under the control of signaling pathways such as Wnt, BMP, FGF, Shh and Notch (reviewed in Moro et al., 2013).

Various methods are also well established for detecting transcripts or proteins of interest in zebrafish embryos or larvae after fixation (Gross-Thebing et al., 2014; Thisse and Thisse, 2008). These methods can be valuable if applied after a live imaging experiment for correlative analysis but are beyond the scope of this chapter and thus not described in detail here.

20.3 Imaging Techniques

As discussed above, zebrafish are ideal for live imaging applications because of their transparency and the broad availability of various labeling techniques for fluorescence microscopy. In this section, we will first discuss the general principles and challenges in the design of light-based live imaging methods, and then introduce several microscopy techniques that are commonly used and actively researched for zebrafish studies.

20.3.1 Basic Principles in Light-Based Live Imaging Methods

Excessive amounts of excitation light can cause photobleaching and phototoxicity that are detrimental to the cells. The affordable light exposure and resulting levels of phototoxicity are dependent on the biological questions, samples and imaging modalities, and there is no universal approach to their evaluation (for a detailed review see Laissue et al., 2017). However, all light-based live imaging methods are fundamentally constrained by the "photon budget" – the number of photons that can be used for imaging before the natural biological processes are perturbed, and the data can no longer be trusted. To obtain optimal result in a given experiment, the photon budget needs to be carefully allocated in space and time to balance different imaging parameters, including but not limited to 1) spatial resolution, 2) temporal resolution, 3) length of observation and 4) signal-to-noise ratio. With a fixed photon budget, improvement of a single parameter is often achieved at the expense of others. For example, if imaging speed needs to be increased while the signal-to-noise ratio is kept constant, one may need to reduce the length of the time-lapse experiment or reduce spatial sampling/resolution during image acquisition. As different imaging methods favor different regimes of this parameter space, the desired range of imaging parameters should be considered first when choosing between different imaging modalities.

However, this situation is typically further complicated by the fact that the photon budget cannot always be perfectly converted into fluorescence that effectively contributes to meaningful image contrast. Fluorescence emitted by fluorophores outside the focus of the detection optics can contribute

to image formation and lead to a decrease of image contrast for in-focus features. As light penetrates deeper into the tissue, contrast is usually further degraded because of aberrations and light scattering. Many new imaging modalities and improvements are designed with the motivation of converting the photon budget more effectively and efficiently into meaningful signal. Moreover, imaging parameters are sometimes limited by microscope hardware or software, rather than the photon budget itself. With the rapid advancement of relevant technology and computational hardware, those imaging methods that utilize the photon budget more efficiently in space and time will thus have more room for improvements in the future.

With the concept of photon budget and imaging performance in mind, we will now introduce several light microscopy techniques that are currently used for live imaging of zebrafish.

20.3.2 Wide-Field Light Microscopy

Records of time-lapse imaging of zebrafish embryos can be dated back as early as 1942 by Warren Lewis using bright-field light microscopy (Lewis, 1942) and 1957 by Hisaoka and colleagues with the aid of phase contrast microscopy (Hisaoka et al., 1957). Wide-field microscopes with Nomarski differential interference contrast optics contributed substantially to early zebrafish embryology and lineage-tracing studies and are still commonly used in laboratories worldwide. When equipped with an illuminating lamp and the appropriate excitation and emission filters, the microscope can be used to visualize common fluorescent dyes or proteins in zebrafish embryos or larvae, and time-lapse movies can be captured with a computer-controlled charge-coupled device (CCD) camera at a relatively low cost. Although objectives with high numerical aperture (NA) can be used to achieve high spatial resolution, the limitation of wide-field light microscopy is obvious: because the entire field is simultaneously illuminated throughout its full width and depth, fluorescence from outside the objective's depth of field contributes to the background. Deconvolution methods are available to partially address this problem (Sibarita, 2005) but cannot handle light scattering in deep tissues. Thus, wide-field microscopy is most suitable for two-dimensional (2D) imaging of samples with sparsely labeled cells that are approximately distributed within a thin plane.

20.3.3 Confocal and Multiphoton Laser Scanning Microscopy

Confocal and multiphoton laser scanning microscopy overcome the main limitation of wide-field microscopy by introducing optical sectioning capability. Confocal laser scanning microscopy (CLSM) uses a pinhole to prevent out of focus light from reaching the detector. In two-photon laser scanning microscopy (2PLSM), only the fluorophores within the focal plane are excited. In both imaging methods, volumetric imaging is achieved by scanning the focal spot three dimensionally (3D), which constrains acquisition speed. The highest imaging speed is achieved using galvanometer scanners capable of resonant scanning, or alternatively by spinning disk confocal microscopy, which further improves speed by utilizing multiple excitation beams in parallel. Jonkman and Brown (2015) provide a practical guide to selecting an instrument from the pool of commercially available confocal microscopes based on the user's specific imaging applications.

The main drawback of CLSM is that it suffers from high rates of photodamage and photobleaching because, for the acquisition of each point in the 3D volume, the excitation light cone exposes a large section of the sample to laser light, including the respective out-of-focus regions from which fluorescence is later rejected by the pinhole. Moreover, in scattering tissues, less light (and less signal) can pass through the pinhole, leading to rapid loss in signal brightness and contrast with increasing penetration depth. 2PLSM reduces light scattering in the excitation process, decreases phototoxicity and photobleaching, and improves penetration depth compared to single-photon CLSM because of localized fluorescence excitation and the use of longer wavelengths. However, extra care should be taken when using multiphoton microscopy to image pigmented specimens, as the absorption of high-power infrared pulsed laser beams typically results in severe photodamage in pigment-producing cells. All point-scanning techniques are intrinsically constrained by the lack of parallelization of signal collection in time. For samples with sparse labeling along the axial direction, a recently developed optical module can collect 2D projections of the 3D volume without a need for z-scanning, using an axially extended Bessel beam. The module can be easily integrated into standard 2PLSM and has been demonstrated to provide video-rate functional imaging in the brain of larval zebrafish (Lu et al., 2017).

20.3.4 Light Sheet Fluorescence Microscopy

Light sheet fluorescence microscopy (LSFM) (also known as selective plane illumination microscopy, or ultramicroscopy) tackles the shortcomings of point-scanning techniques by uncoupling the illumination and detection light paths. In a basic LSFM setup, the sample is illuminated by a thin sheet of light perpendicular to the detection objective, such that the light sheet and the focal plane of the detection objective overlap. Instead of collecting the signal point by point, the fluorescence image at any given plane can be rapidly captured with a wide-field detector, such as a CCD or sCMOS camera, providing imaging speeds several orders of magnitude higher than point-scanning techniques. Because excitation is confined to the focal plane, photobleaching and photodamage throughout the sample volume is greatly reduced compared to CLSM. High speed and low phototoxicity thus make LSFM an ideal system for live imaging over a long period of time. Light sheet can be generated through a cylindrical lens (Huisken et al., 2004; Voie et al., 1993) or by rapid scanning of a weakly focused Gaussian beam (Keller et al., 2008; Keller et al., 2010). 3D volumetric imaging is achieved either by moving the sample through a stationary light sheet, or through synchronized movement of the light sheet and the detection focal plane if higher speed is desired (Ahrens et al., 2013; Lemon et al., 2015). Alternative approaches to this scheme have also been explored and extend the depth-of-field of the detection optics with a cubic phase mask (Quirin et al., 2016) or by introducing spherical aberrations (Tomer et al., 2015) in the detection path,

so the detection objective can remain stationary during high-speed imaging. However, these approaches also intrinsically sacrifice axial resolution and photon collection efficiency.

The orthogonal arrangement of objectives employed by LSFM requires the sample to be optically accessible along multiple directions. This can be challenging for samples with geometrical constraints (e.g. internally embedded tissues), but is ideal for specimens like zebrafish embryos for which light-scattering yolk is located in the center and covered by cells that are easily accessible from the outside along different angles. In this scenario, the concept of multiview imaging (acquiring multiple views of the sample from different directions) can furthermore increase high-resolution physical coverage of the sample and facilitate reconstruction of images with isotropic resolution. Multi-view images can be acquired by physical rotation of the specimen, an approach that is easy to implement but sacrifices imaging speed. By contrast, simultaneous multi-view light sheet microscopy (SiMView, also known as MuVi-SPIM) uses two opposing illumination objectives to generate two light sheets, and two opposing detection objectives to acquire images from opposite directions simultaneously. The respective four views generated from each light sheet camera combination can be fused by image postprocessing to achieve *in toto* physical coverage of specimens that can only be partially imaged in high quality from a single view (Tomer et al., 2012; Krzic et al., 2012). This approach has been used to image zebrafish embryonic development with high resolution across the embryo (Amat et al., 2014). Isotropic multiview light sheet microscopy (IsoView) improves on the SiMView design in that each of the microscope's four optical arms can be used for illumination and detection sequentially or simultaneously, making it possible to perform zebrafish whole-brain functional imaging with near-isotropic spatial resolution (Chhetri et al., 2015).

In the absence of multiview imaging capabilities, the axial resolution of LSMF is determined by the depth of focus of the detection objective and the thickness of light sheet. However, due to the intrinsic diffraction of Gaussian beams, a thinner light sheet would inevitably lead to a smaller usable field of view. This problem can be overcome by performing light sheet imaging of fluorescent signal produced by the thin, central intensity peak of Bessel beams (Planchon et al., 2011) or by constructing ultrathin 2D optical lattices (Chen et al., 2014). These methods provide substantially improved spatial resolution and, when combined with structured illumination, can be used to resolve structures beyond the diffraction limit.

If better depth penetration is needed, two-photon light sheet imaging can also be utilized. In addition to reduced scattering of excitation light at near-infrared wavelengths, two-photon light sheet fluorescence microscopy (2PLSFM) also benefits from lower background fluorescence because 2p-excitation depends quadratically on excitation power density, suppressing fluorescence in out-of-focus regions and in low-power regions along the light sheet that are far away from the beam waist. These advantages come at the expense of signal strength and imaging speed in 2PLSFM (Lemon et al., 2015). 2PLSFM has been successfully implemented to reconstruct embryonic development in *Drosophila* embryos (Tomer et al., 2012; Truong et al., 2011). In addition, 2P-LSFM is also used in functional imaging experiments where the fish needs to perform visual tasks and must not be disturbed by visible light (Wolf et al., 2015).

Optimal image quality in LSFM requires a perfect overlap between the illuminating light sheet and the focal plane of the detection objective. In optically complex specimens and in time-lapse imaging of multicellular organisms, however, light sheet and detection focal plane are usually mismatched to some extent because of their geometrical distortion within the tissue or the change of optical properties over time. To address this problem, Royer et al. (2016) developed the AutoPilot framework, which automatically evaluates and optimizes image quality throughout the sample by translation and rotation of light sheet and detection planes in a SiMView microscopy setup.

The unique strengths of LSFM for live imaging of biological specimens is recognized by many biologists, but the barrier for purchasing or building an LSFM is still often prohibitively high. Commercial setups are available on the market, and the OpenSPIM project provides an open-source guide to building a basic LSFM from off-the-shelf components (Pitrone et al., 2013). The commentary written by Reynaud et al. (2015) can help interested readers better understand the practical challenges in light sheet imaging and find resources available from the light sheet community.

20.4 Applications

20.4.1 Cell Behavior and Fate During Morphogenesis

During embryonic development, cells migrate and rearrange to shape the organism while simultaneously displaying dynamic gene expression profiles as they adopt specific fates and cell type identities. Studying these interrelated processes is a challenge, and direct observation supported by imaging technology has been an invaluable tool to this end. Zebrafish have quickly emerged as a popular model for such experiments because of their small size and relatively transparent vertebrate that develops *ex utero*.

A major event in early embryogenesis is gastrulation, during which the main body axes of an organism are shaped from an initially continuous and outwardly seemingly homogeneous sheet of cells (for review, see Solnica-Krezel and Sepich, 2012). To investigate this process in zebrafish, Kimmel et al. (1990) injected an intravital dye into cells of a blastula-stage (pregastrulation) embryo and recorded the labeled anatomical position and cell types postgastrulation. They systematically sampled spatial positions in the blastula and found that the fate of the cells inhabiting many sites is not random, but predictable to some extent. Their work thereby established the first cell fate map of the zebrafish blastoderm and provided detailed evidence for pregastrulation cell fate restrictions (Figure 20.3A–F). Note that cell behavior and lineage restrictions were not observed directly, but instead inferred from the defined pregastrulation position of the marked cells and the observed anatomical position and cell types of their progeny postgastrulation.

In a separate work, Kimmel et al. (1994) investigated the early development of the central nervous system (CNS) of zebrafish. They focused on the early CNS at a developmental time when there is less cell mixing, which enabled the direct observation of cell divisions and manual reconstruction of clonal cell lineage relationships (Figure 20.3G–M). The authors found that the labeled cells undergo a robustly conserved developmental program. Cells belonging to the same clone divide synchronized to within a few minutes of each other. Moreover, early cell divisions are predominantly oriented along the head to tail axis (anterior-posterior, AP), but then rapidly reorient by 90 degrees to a mediolateral orientation. The descendants of the labeled cells thus almost invariably form clones that are elongated along the AP axis and contribute equally to both sides of the midline. This work demonstrates how direct observation of cell behavior can reveal exiting details about the developmental programs that underlie the formation of organized tissues.

When starting their experiments, Kimmel et al. (1990) suspected that the cells in the late zebrafish blastoderm become increasingly lineage-restricted and less of a homogenous population of equivalent cells. However, they lacked the technical means to specifically define and label subpopulations of cells based on biomolecular markers, such as expressed genes. Subsequent to their work, techniques and resources that enable this have become routine and widely available. In particular, it has become easy to generate transgenic reporter lines that label specific cell types and employ powerful microscopy techniques (discussed above) that enable following cells through developmental programs. However, cells in tissues are often densely packed. This can severely hamper following single cells through a developmental program, even when using genetic driver lines that exclusively label the cells of interest. It is more challenging to spatially resolve multiple densely packed cells, especially in deeper tissues where light microscopy is increasingly limited by light attenuation and scattering. Moreover, if it is desirable to follow individual cells, the temporal resolution of the recording must be high enough to avoid ambiguities when tracking them as they move, both collectively and individually. To facilitate tracking single cells in densely packed tissues, cells can be labeled sparsely using techniques based on the ideas of Brainbow (Livet et al., 2007). Briefly, the cells of interest are made to express one out of a few options of possible labels, most often fluorescent proteins of different colors. The transgene construct is usually designed such that the decision regarding which label to express is irreversible and essentially stochastic (Figure 20.2). Thus the cells of interest will express a salt-and-pepper pattern of different colors, which greatly simplifies distinguishing individual cells from one another. The colors are passed on to all subsequent generations of daughter cells and thereby keep a record of the clonal origin of each cell (Figure 20.4a). For a detailed review of genetic multicolor labeling techniques, see Weissman and Pan (2015).

Collins et al. (2010) established the reusable MAZe toolkit to sparsely label and follow single cells and their progeny in zebrafish. They induced clones in a variety of tissues and followed myoblast fusion in developing muscle fibers. He et al. (2012) modified this system to investigate the development of the zebrafish retina. They used MAZe to sparsely express

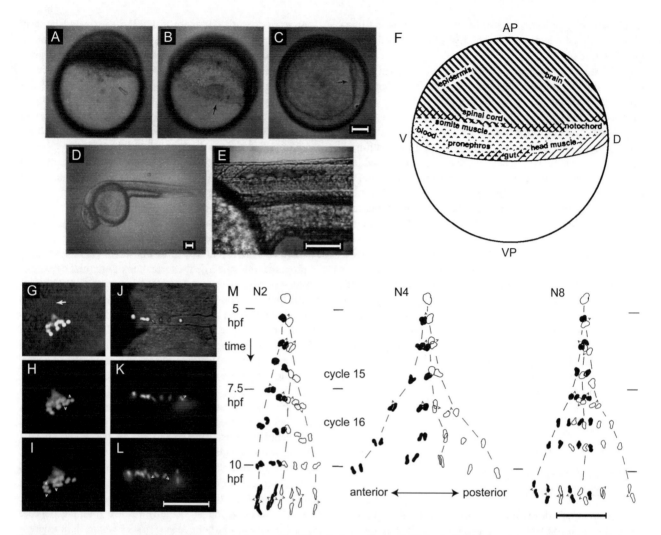

FIGURE 20.3   Endpoint analysis and direct observation reveal cell fate during embryogenesis. A.–D. Zebrafish embryo at different developmental stages, from A. blastula (pregastrulation, ~3.7 h) to D. 24 h (postgastrulation with established body axis). A single cell was injected with a vital dye, which reveals the developmental fate of its daughter cells to be dorsolateral periderm. E. Labeled cells in D., magnified to better illustrate their anatomical position. F. Fate map of the pregastrulation zebrafish embryo, generated by systematically sampling different positions in the blastula and observing their fate (as in A.–E.). Cells at different sites of the blastoderm predominantly contribute to the indicated tissues after embryonic development. G.–L. Direct observation of cell lineages in the developing early central nervous system of zebrafish reveals a tightly conserved developmental program. Displayed are labeled cells at different time points from A. late gastrula (7.5 h) to L. 11 h. The cells undergo a series of divisions (indicated by >). Up to cell cycle 15, the division axis is predominantly oriented longitudinally, which elongates the clone in anterior-posterior direction H., I. At cell cycle 16, the cell division orientation switches by ~90 degrees to mediolateral E., F. M. Extracted diagrammatic cell lineages of three independent clones. As in G.–L/, up to cell cycle 15, divisions occur in anterior-posterior orientation before switching to mediolateral in cell cycle 16. Therefore, the resulting clones are elongated in anterior-posterior direction and strikingly symmetric around the midline. Panels A–F have been reprinted from Kimmel et al. (1990) with permission from the publisher The Company of Biologists Ltd. Panels G.–M. have been reprinted from Kimmel et al. (1994) with permission from the publisher The Company of Biologists Ltd. Scale bars = 100 µm C., 100 µm D., 100 µm E., 200 µm L., 200 µm M.

a photoconvertible fluorescent protein in the densely packed retinal progenitor cells. This enabled exquisite control to mark single cells, their progeny, and even subclones within that progeny by targeted photoconversion. Following the marked cells through retina development uncovered a large variability in observed clone sizes, strongly indicating that retinal progenitor cells in zebrafish do not follow a fixed developmental program, unlike e.g. *Drosophila* neuroblasts (Gallaud et al., 2017). Instead, the authors suggest that relatively simple stochastic rules are sufficient to reconcile the observed clone size variability with the requirement to build an organ of defined shape and cell type composition. Pan et al. (2013) demonstrated that multicolor labeling can be used to study zebrafish

neuroanatomy, which is particularly challenging because it requires the accurate recording of cell identities though the extensive network of thin membrane filaments that underlie neural networks. The authors traced neurons and additionally followed the clonal lineage of postembryonic cornea cells (Figure 20.4a). Chen et al. (2016) used multicolor brainbow to barcode and follow skin cells in adult zebrafish for over 20 consecutive days. In this continuously renewing tissue, they characterized the rates of cell loss and cell gain, as well as the distinct types of local space remodeling during homeostasis and regeneration after injury.

The development of modern LSFM (Voie et al., 1993; Huisken et al., 2004; Keller et al., 2008) has enabled the

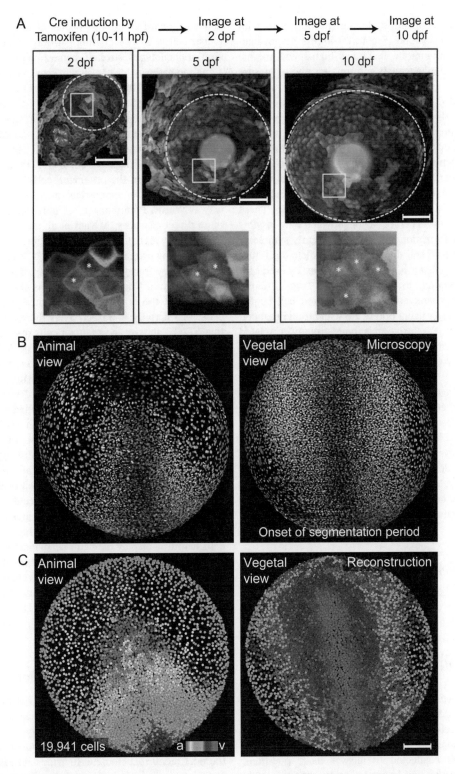

FIGURE 20.4 Multicolor labeling techniques and computational methods for observing large numbers of cells over long time periods. A. Multicolor labeling with *zebrabow*. Recombination of the zebrabow transgene insertions (see Figure 20.2C) was performed at 10–11 hpf, and images were sporadically recorded until 10 dpf. Displayed is a zebrafish eye (dashed circle), with multicolor-labeled cells in the cornea. The labels were randomly assigned at recombination time and stably propagated to all progeny, allowing to reconstruct clonal relationships. The inset shows the addition of two cells to a clone (asterisks). Note that this information is readily apparent from a few images taken over ten days. B. Two time points from time-lapse recording of zebrafish embryogenesis. In contrast to A., all cell nuclei of the embryo were labeled uniformly. In this case, reconstructing cell behavior and clonal relationships necessitates recording at high spatiotemporal resolution. The time lapse was recorded with a light sheet microscope at a rate of one complete volume per minute, for over 13 h. C. Results of automated cell tracking of the time lapse depicted in B. The color code indicates the origin of the cells in the blastoderm according to the displayed color bar (a = animal/center, v = vegetal/periphery). A. has been reprinted from Pan et al. (2013) with permission from the publisher The Company of Biologists Ltd. B. and C. have been reprinted from Amat et al. (2014) with permission from the publisher Springer Nature. Scale bars = 50 μm (a), 100 μm (c).

recording of entire zebrafish developing from embryos with essentially cellular resolution (Figure 20.4b). Its perpendicular illumination configuration facilitates high-speed volume acquisition while simultaneously minimizing photoexposure of the sample. Furthermore, relatively large and opaque samples can be recorded from multiple angles to improve coverage, as discussed above. LSFM is therefore uniquely suited to record larger samples at high temporal resolution and over extended time periods. Keller et al. (2008) recorded the entire volume of a developing zebrafish embryo during the first 24 h of its development, with essentially cellular resolution and at sufficient temporal resolution to follow single cells and characterize spatially variant cell migration behaviors. These reconstructions of *in toto* cell behavior provided new insights into mesendoderm formation and revealed an early symmetry break in the pattern of cell divisions that predicts the future body axis of the fish. Shah et al. (2017) additionally used specific transgenic reporters to distinguish cells from different germ layers. They characterized and compared cell behaviors in each germ layer using optical flow and found that cell behavior during early gastrulation is more strongly correlated with cell position within the embryo than germ layer identity. Germ layer-specific cell behaviors instead emerge during late gastrulation.

Technology and available resources have improved considerably since Kimmel et al. began to study cell fate specification in the zebrafish blastula. Today, genetic targeting of specific cells is routine, and it is becoming possible to follow them in a minimally invasive manner over the extended time periods required to cover developmental processes. The problem of reconstructing developmental programs, therefore, increasingly shifts to the development of software to aid with the analysis of larger numbers of cells and high data volumes (see below).

20.4.2 Cell Biology

As briefly noted above, the main head-to-tail body axis is shaped in a process called gastrulation during early embryogenesis. Gastrulation starts with the embryo as a fairly continuous and symmetric sheet of cells, and one of the prominent collective cell behaviors during gastrulation is convergence and extension. Cells converge laterally toward the forming body axis and pack densely by intercalation. As more cells intercalate, the forming body axis extends in head-to-tail direction (Figure 20.5A) (for review, see Roszko et al., 2009).

What instructs cells to behave in this way and how is this behavior implemented at the biomolecular level? Extensive research has been conducted to answer these questions. Directed cell migration and intercalation requires the establishment of cell polarity. Gong et al. (2004) investigate cell polarity during gastrulation by recording the orientation of the cell division axis. They find that cells in dorsal tissues display a strong bias toward dividing along the animal-vegetal axis of the embryo. They compare these observations with mutants known to have strong defects in body axis formation and show that mutant cells never establish a similar polarity bias, thereby failing to engage in convergence and extension behavior (Figure 20.5 B–G). Interestingly, multiple mutants

deficient for different components of the noncanonical Wnt/planar cell polarity signaling pathway (Wnt-PCP) show very similar phenotypes, strongly suggesting the importance of this signaling pathway for convergence and extension behavior.

Miyagi et al. (2004) investigated what activates Wnt-PCP in these cells. The authors had previously observed that STAT3, a component of separate signaling pathways, is active in organizer cells. These cells are part of a well-established signaling center coordinating several crucial aspects of gastrulation (for review, see Roszko et al., 2009). When interfering with STAT3 activity in the organizer, cells outside the organizer failed to establish appropriate polarity, much in the same way as Wnt-PCP mutant cells (Figure 20.5H.–I.). To establish more direct causality, Miyagi et al. measured the activity of RhoA, a key signaling factor influencing cytoskeletal dynamics, and a downstream target of Wnt-PCP, using Foerster resonance energy transfer (FRET, Figure 20.5J–L). Their results show increased RhoA activity in cells close to the organizer. However, if the organizer is populated by STAT3-deficient cells, this activity is absent. Thus the authors hypothesized that STAT3 activity in organizer cells results in the secretion of a molecular signal that activates Wnt-PCP in target cells. The target cells in turn activate RhoA to remodel the cytoskeleton and establish appropriate cell polarity.

This topic remains an active area of research with many open questions. Experiments that investigate the action of signaling pathways and molecular components in an intact multicellular organism are often at the limit of what is technically possible. Such experiments may require expressing labeled components of signaling pathways at appropriate levels and subcellular localization and to quantitatively record fluorescence emitted from relatively few molecules for long periods of time. Performing this in an intact organism confounds these technical challenges. However, it is during embryogenesis that cells naturally display their full behavioral and molecular repertoire. Furthermore, cells act within their native environment, which is shaped by other cells, and this enables observing the short-range and long-range cell–cell interactions that initiate and coordinate complex cell behaviors. Despite the challenges, a developing embryo is thus a compelling model to study cell biology.

20.4.3 Brain Activity and Behavior

The electrical firing of neurons occurs at rapid, millisecond timescales. Most functional brain imaging studies do not image the membrane potential directly, but instead employ sensors that provide a fluorescence readout of calcium concentration. Calcium accumulates in firing neurons and can be used as an integrating and more persisting measure of recent activity (Figure 20.6A). This relaxes the imaging speed requirements significantly and, furthermore, enables the use of well established, high signal-to-noise genetically encoded calcium indicators (GECI) (Looger and Griesbeck, 2012).

Gahtan et al. (2002) perform calcium imaging of neurons in the zebrafish hindbrain and find that escape response behavior involves activity of a widespread network of approximately 100 neurons. Notably, this finding was directly facilitated by the imaging approach. Electrophysiological recordings are

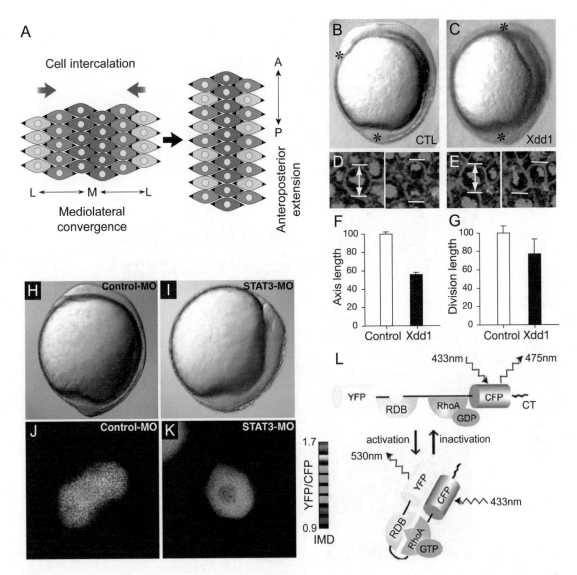

FIGURE 20.5 Cell biology during convergence and extension. A. During vertebrate gastrulation, cells engage in convergence and extension behavior. They converge mediolaterally toward the forming body axis and intercalate to extend the body axis in anterior-posterior direction. B.–C. Comparison of body axis length of wild type B. and mutant C. zebrafish embryos. The mutant embryo expressed an inactive form of disheveled (Xdd1), a component of the Wnt signaling pathway. Note that the body axis of the mutant is significantly shorter than that of the wild type embryo (body axis limits indicated by asterisks). This phenotype suggests a possible defect of convergence and extension behavior. D.–E. For convergence and extension to occur, cells must establish appropriate polarity. Orientation of cell division can be used as an indicator of cell polarity. Depicted are cell divisions of wild type D. and Xdd1-expressing cells E., aligned relative to the anterior-posterior axis of the embryo (top-bottom). Note that the mutant cell division was oblique relative to that axis. F. Quantification of body axis lengths in wild type and Xdd1-expressing embryos. Wild type axis length (open bar) was normalized to 100. In comparison, Xdd1-expressing embryos reach an axis length of less than 60% of that of the wild type fish (solid bar). G. Relative contribution of cell division orientation to body axis elongation in wild type and Xdd1-expressing embryos. The displacement distance of the daughter cells along the anterior-posterior axis (top-bottom) was quantified (white lines in D., E.). Average cell displacement in this direction is markedly reduced in the mutants because of the increased variability in division orientation. This quantification links the axis elongation and cell polarity defects. H.–I. Comparison of wild type H. and STAT3-depleted I. zebrafish embryos. STAT3-depletion was achieved by injecting a morpholino and resulted in shortened body axis length. Note that this phenotype is very similar to that of the Xdd1-expressing embryo in C., suggesting that both experimental perturbations may affect related processes. J.–K. Measurements of RhoA activity during convergence and extension. Depicted are cells from a wild type embryo J. and a STAT3-depleted embryo. The STAT3-depleted embryo shows a pronounced reduction in RhoA activity. RhoA has well established roles in coordinating cytoskeleton and cell polarity, suggesting a link to the observed cell polarity defects. L. Design of the Raichu-RhoA FRET sensor used to measure RhoA activity. FRET depends on spectral overlap and spatial distance. The emission spectrum of CFP overlaps with the excitation spectrum of YFP, which enables CFP to sensitize YFP across very short distances. In the Raichu sensors, the FPs are linked in the same peptide chain, but separated by RhoA itself, and a RhoA binding domain (RBD). RhoA naturally acts as a switch by cycling through inactive (bound to GDP – guanosine diphosphate) and active (bound to GTP – guanosine triphosphate) states. The RBD binds exclusively to active RhoA, thereby folding the peptide chain and shortening the average distance between the FPs. The shorter distance significantly improves FRET efficiency, which can be measured as an increase in YFP emission during CFP excitation. Panel A has been reprinted from Tada et al. (2012) with permission from the publisher The Company of Biologists Ltd. B.–G. have been reprinted from Gong et al. (2004) with permission from the publisher Springer Nature. H.–L. have been reprinted from Miyagi et al. (2004) with permission from the publisher Rockefeller University Press.

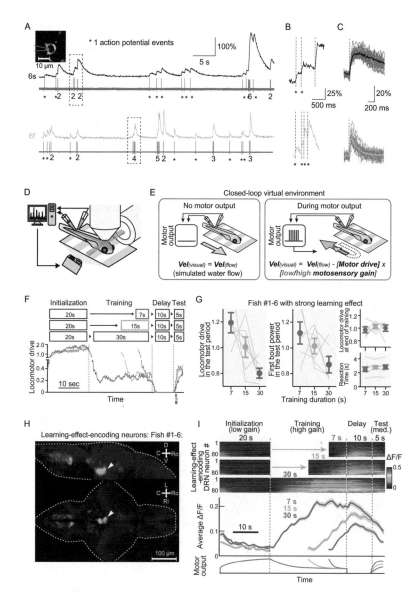

FIGURE 20.6 Recording and analysis of brain activity and behavior. A. Imaging and electrophysiology in the mouse visual cortex. Displayed are calcium imaging traces and electrophysiological readout from the same cell, recorded with two different Ca-indicators (GCaMP6s, top; GCaMP6f, bottom). Action potentials are indicated below the as asterisk (single-action potential) or number (multiple-action potentials). B. Ca sensors provide a relatively persistent measure of spiking activity and integrate over multiple action potentials occurring in rapid succession. Note that the response kinetics vary for specific indicators (compare top and bottom). C. Responses to single spikes (gray) and averaged response (black) of the two presented Ca-indicators. D. Whole-brain volumetric Ca-imaging with simultaneous behavioral readout in larval zebrafish. A light sheet microscope images the brain while electrodes record fictive swimming from motor neuron axons in the tail. A one-dimensional visual scene is displayed below the fish and updated in response to fictive swims (see E.). E. Closed-loop virtual environment. When the fish does not fictively swim (left), a visual scene moves forward slowly to simulate the effect of water current. This elicits swimming as the fish strives to maintain a constant position. Fictive swim signals (right, inset) move the visual scene backward to mimic the effect of forward swimming. The speed of visual feedback is controlled by the motosensory gain. For the same swim bout, high gain yields more visual flow (magenta arrow) than low gain (gray arrow). F. Experimental design to test motor learning (top) and representative fish behavior (bottom). During the initialization phase, the fish is exposed to low motosensory gain, resulting in an increase in locomotor drive because the fish must swim harder to achieve the required displacement. The gain was then increased for a variable period of time (no time passes between the displayed arrows). In this training phase, the fish adjusts to the higher gain by decreasing its locomotor drive. After a delay phase, during which the visual scene was kept stationary, the fish was exposed to intermediate motosensory gain. During this test period, fish attenuate their locomotor drive dependent on how long they were trained, with longer training resulting in a stronger attenuation. Colored lines below the x axis of the test period, mean ± SD of reaction times. G. Behavior across six fish with strong learning effect. Left: locomotor drive in the test period is more attenuated after longer training. Right: locomotor drive at the end of training (top, last 5 s, normalized across conditions) and reaction time to the test stimulus (bottom) are similar across training durations. Gray lines indicate data from individual fish. Error bars indicate SEM (standard error of the mean) across fish. H. Functional brain map of neurons associated with motor learning (cyan for positive correlation to the learning effect; magenta for negative). The neurons were selected because their activity during the delay period depended on the duration of the training period and were predictive of the locomotor drive in the test period. Note the density of selected neurons in the dorsal raphe nucleus (DRN) indicated by the white arrow. I. Dynamics of learning-effect-encoding DRN neurons during the motor learning paradigm. Top: trial-averaged ΔF/F from 80 individual DRN neurons in a representative fish. Middle: ΔF/F traces averaged over these neurons. Shadows indicate SEM across neurons. Bottom: schematic of locomotor drive. Panels A–C were reprinted from Chen et al. (2014) with permission from the publisher Springer Nature. D.–I. were reprinted from Kawashima et al. (2016) with permission from the publisher Elsevier. Scale bars = 10 μm (a), 100 μm (h).

more direct and accurate, but they do not scale to enable the simultaneous recording from hundreds of neurons required to screen for the ones involved in a specific behavior. Ahrens et al. (2012) combined recording of brain activity from a paralyzed fish with a virtual reality setup in which visual stimuli abstractly representing a visual scene are updated in response to fictive swim patterns. Fictive swims were recorded with an electrode on motor nerve axons in the tail to minimize the latency of visual scene updates and to acquire a more direct and accurate readout of motor behavior. Intriguingly, the swimming strength of the fish increased and decreased when the gain coupling electrophysiological swimming readout and visual stimulus update was altered, demonstrating motor learning behavior in zebrafish.

Both studies employ point-scanning techniques to acquire images pixel-wise. As discussed above, compared to newer technologies such as LSFM, point scanning is at a severe disadvantage with respect to both speed and photobleaching. In two new studies, Ahrens et al. (Ahrens et al., 2013) and Panier et al. (2013) thus applied LSFM to the functional brain imaging of zebrafish. These earliest applications of LSFM to functional brain imaging were still limited by a low volumetric rate relative to the timescales of neural computations, but this was rapidly improved. Vladimirov et al. (2014) combined LSFM with a virtual reality setup much like Ahrens et al. in their 2012 study and achieved a volumetric rate of 3 Hz while minimizing direct illumination of the fishes' eyes by laser light. Kawashima et al. (2016)used this approach to follow-up on the motor gain adaptation experiments described above. Recording the entire brain enabled mining the data systematically for neuronal populations with activity profiles expected for gain adaptation (Figure 20.6B–I). The authors identified the dorsal raphe nucleus as a center for short-term locomotor learning and used both cell ablation and optogenetic stimulation to experimentally validate their results.

Extracting the approximate activity of a brain cell or region requires analyzing its fluorescence changes over time. To facilitate this, the fish is usually paralyzed to prevent rapid nonlinear motion and aid with subsequent image registration. However, this experimental setup constrains the behaviors that can be studied. The gold standard would be to image brain activity and behavior of freely behaving fish. Kim et al. (2017) and Cong et al. (2017) developed microscope hardware and software that maintains almost freely behaving fish inside the field of view of a microscope. The fishes' position is tracked in real time, and its movements are compensated by sample stage position updates. As the motion of the fish occurs at rapid timescales and can involve nonlinear anatomical deformations, volumes can no longer be acquired plane-wise. Instead, Kim et al. developed differential illumination focal filtering (DIFF), which achieves optical sectioning with a pair of structured and complementary illumination patterns. By contrast, Cong et al. used additional optics to combine multifocal and light field microscopy and acquired 3D information from a single camera image.

20.4.4 High Content Screening

High content screening is usually performed with the aim to identify molecular effectors that induce, confound, or mitigate a given measurable phenotype. To this end, biological samples are exposed to large libraries of small molecules likely to be bioactive. As the expected number of hits is low, it is often desirable to screen large substance libraries. Hence, the screening procedure must be designed to be efficient and inexpensive. Zebrafish can routinely produce large quantities of offspring, and their housing is inexpensive compared to other vertebrate models. Furthermore, being vertebrates, zebrafish can adequately recapitulate many biomedically relevant conditions (Lieschke and Currie, 2007). Their small size, rapid *ex utero* development, and translucent skin enables direct identification and scoring of even subtle phenotypes (Williams and Hong, 2016). For these reasons, zebrafish are a long-standing model for high-throughput toxicity and drug screening.

As an example, Peal et al. (2010) screened for therapeutic drug candidates targeting Long QT syndrome, a life-threatening heart disorder characterized by pathologically prolonged cardiac repolarization. They used a zebrafish mutant line recapitulating this disorder and identified two compounds from a 1,200-compound library that restore essentially normal heartbeat in the mutant fishes. An important readout for this screen was the optical mapping of voltage in the cardiac muscle through the use of a voltage indicator dye.

The feasibility of simultaneously recording the behavior of multiple zebrafish in multiwell plates over extended time periods enabled Rihel et al. (2010) to screen a library of more than 5,000 compounds for modulators of general restfulness and wakefulness. Their library contained substances with known therapeutic effects on mammals, such as substances used to treat attention deficit hyperactivity disorder (ADHD), and their effects on zebrafish behavior are consistent with their known function in mammals. This work is one of the first examples of behavioral screening in zebrafish, and it suggests the possibility to perform screens for modulators of basic neuropsychiatric conditions.

Most of the screening in zebrafish uses imaging technology to collect data. To support this, larger-scale screening often requires the development of custom hardware and software, especially with respect to sample handling and positioning. For instance, Pulak (2016) introduced a combined hard- and software solution to automate sequential collection, positioning, and bright-field as well as fluorescence imaging of up to 7-day old zebrafish. For these applications, it is highly desirable to not only automate data collection but also data processing, as manual scoring would not be feasible. Thus, considerable effort is invested into developing robust and automated readouts, i.e. phenotype scores, based on image data, sometimes in combination with reporters of physiological state. In the following sections, we will discuss a few approaches to common and relevant biological image processing.

20.5 Data Analysis

The advent of quantitative imaging has led to a paradigm shift, where imaging is used not only to observe, but to measure. However, this results in a novel set of challenges that necessitate an interdisciplinary effort involving biologists, engineers, and computer scientists.

As described above, zebrafish are a compelling model to record morphogenetic dynamics with cellular resolution. However, making full use of this level of detail requires as a prerequisite the detection of cell positions (segmentation) and the tracking of their movements and divisions over time. Manual approaches, such as the massive multiview racker (MaMuT) (Wolff et al., 2017), although extremely useful for smaller-scale applications, do not scale well to large cell populations. Hence, it is desirable to develop largely unsupervised and automated software that yields high-quality results as a starting point for final manual curation. The Ilastik framework (Sommer et al., 2011) aims to provide users with a user-friendly way to train generalizable machine learning algorithms on their data. The initial focus of the software has been image segmentation, but tracking functionality has been added more recently (Schiegg et al., 2013; Haubold et al., 2016). Their elegant approach builds on a probabilistic graphical model that accounts for segmentation errors and can be solved in a globally optimal manner. Zebrafish are relatively large specimens, which poses the practical problem of scalability of algorithms and their implementations. For cell detection and tracking, Amat et al. (2014) addressed this by first performing a coarse but fast segmentation step that subdivides the image volume into a set of supervoxels. As the number of supervoxels is small compared to the total number of voxels, they form a greatly simplified input for a more accurate and slow subsequent detection and tracking step, resulting in good runtime performance. As multiple conceptual approaches are developed and implemented, it becomes important to compare them carefully under realistic and standardized conditions. The Cell Tracking Challenge (http://celltrackingchallenge.net) aims to address this need. Ulman et al. (2017) summarized their results obtained with 21 competing algorithms applied to 13 different data sets covering a range of target scenarios and data modalities. Their work provides a great overview of the currently available options and their most appropriate applications.

Detecting and tracking cells is most often performed with a cell nucleus label, as it facilitates this task by providing relatively sparse point-like sources. On the other hand, this does not include detailed information about cell morphology. For this purpose, it is necessary to characterize the shape of the outer-cell membrane. Stegmaier et al. (2016)developed a fast membrane segmentation approach to this end, which has been applied to confocal as well as light sheet data sets of zebrafish embryos.

In addition to imaging cell behavior, zebrafish are becoming an increasingly popular model to observe brain activity. Fast imaging methods are required to keep up with the speed of dynamic processes occurring in the nervous system, and the resulting data rate is challenging to handle. A particular problem is that images are acquired one volume per time point basis, whereas the analysis requires pixel-wise time traces. The required transformation can be problematic, as loading the time trace of a single pixel necessitates loading all time points, which is a slow process. Therefore, it is desirable to keep all time points in memory, but data set sizes are frequently prohibitively large. To aid with this, Freeman et al. (2014) developed Thunder, a library for image and time-series analysis based on the Apache Spark distributed memory cluster computing engine. With Thunder, entire large data sets can be kept in the collective memory of multiple machines and includes a set of routine processing algorithms implemented and optimized for this architecture.

Finally, a general challenge is to combine data from multiple specimens. In many cases, this can involve anatomical information. Ronneberger et al. (2012) and Randlett et al. (2015) developed atlases of the zebrafish larval brain, along with standardized staining and registration protocols that are used by the community to curate data and gene expression patterns.

20.6 Future Directions

Although very appealing and suitable for imaging, zebrafish have been lagging behind other model systems such as *Drosophila* and mice in the available genetic tools. With the emergence of novel methods, in particular CRISPR/Cas9 (Albadri et al., 2017), this gap is slowly closing. This new technique is enabling targeted mutagenesis to study gene function and the construction of knock-in alleles (Auer et al., 2014). Notably, there are about 3,600 known single-allelic genetic conditions in humans (Lander, 2015), and fish can be an appealing complementary model to study these conditions as well as to rapidly screen candidate drugs.

Imaging technology will also continue to improve with the availability of better hardware and the development of new software approaches to maximize the utility of the recorded data. As noted above, live imaging is constrained by an exquisite photon budget, and an experimental approach must balance the available photons in lieu of sample physiology and lifetime and the required spatial and temporal resolution. In the future, computational approaches could help address this problem from an entirely new angle. For example, a recently developed software uses machine learning approaches to restore, improve, and denoise images acquired with very low light exposure (Weigert et al., 2017). Solutions such as this will facilitate recording faster or longer time-lapse data sets with better volume coverage and less phototoxicity.

REFERENCES

Adam, A., Bartfai, R., Lele, Z., Krone, P. H., and Orban, L. (2000). Heat-inducible expression of a reporter gene detected by transient assay in zebrafish. *Exp Cell Res*, 256: 282–90.

Agathocleous, M., and Harris, W. A. (2009). From progenitors to differentiated cells in the vertebrate retina. *Annu Rev Cell Dev Biol*, 25: 45–69.

Ahrens, M. B., Li, J. M., Orger, M. B., Robson, D. N., Schier, A. F., Engert, F., and Portugues, R. (2012). Brain-wide neuronal dynamics during motor adaptation in zebrafish. *Nature*, 485: 471–77.

Ahrens, M. B., Orger, M. B., Robson, D. N., Li, J. M., and Keller, P. J. (2013). Whole-brain functional imaging at cellular resolution using light-sheet microscopy. *Nat Methods*, 10: 413–20.

Albadri, S., Del Bene, F. and Revenu, C. (2017). Genome editing using CRISPR/Cas9-based knock-in approaches in zebrafish. *Methods*, 121–122: 77–85.

Amat, F., Lemon, W., Mossing, D. P., Mcdole, K., Wan, Y., Branson, K., Myers, E. W., and Keller, P. J. (2014). Fast, accurate reconstruction of cell lineages from large-scale fluorescence microscopy data. *Nat Methods*, 11: 951–8.

Ando, H., Furuta, T., Tsien, R. Y., and Okamoto, H. (2001). Photo-mediated gene activation using caged RNA/DNA in zebrafish embryos. *Nat Genet, 28*: 317–25.

Asakawa, K., Suster, M. L., Mizusawa, K., Nagayoshi, S., Kotani, T., Urasaki, A., Kishimoto, Y., Hibi, M., and Kawakami, K. (2008). Genetic dissection of neural circuits by Tol2 transposon-mediated Gal4 gene and enhancer trapping in zebrafish. *Proc Natl Acad Sci U S A, 105*: 1255–60.

Auer, T. O., Duroure, K., De Cian, A., Concordet, J. P., and Del Bene, F. (2014). Highly efficient CRISPR/Cas9-mediated knock-in in zebrafish by homology-independent DNA repair. *Genome Res, 24*: 142–53.

Broussard, G. J., Liang, R., and Tian, L. (2014). Monitoring activity in neural circuits with genetically encoded indicators. *Front Mol Neurosci, 7*: 97.

Burket, C. T., Montgomery, J. E., Thummel, R., Kassen, S. C., Lafave, M. C., Langenau, D. M., Zon, L. I., and Hyde, D. R. (2008). Generation and characterization of transgenic zebrafish lines using different ubiquitous promoters. *Transgenic Res, 17*: 265–79.

Chen, B. C., Legant, W. R., Wang, K., Shao, L., Milkie, D. E., Davidson, M. W., Janetopoulos, C., Wu, X. S., Hammer, J. A., 3rd, Liu, Z., English, B. P., Mimori-Kiyosue, Y., Romero, D. P., Ritter, A. T., Lippincott-Schwartz, J., Fritz-Laylin, L., Mullins, R. D., Mitchell, D. M., Bembenek, J. N., Reymann, A. C., Bohme, R., Grill, S. W., Wang, J. T., Seydoux, G., Tulu, U. S., Kiehart, D. P. and Betzig, E. (2014). Lattice light-sheet microscopy: Imaging molecules to embryos at high spatiotemporal resolution. *Science, 346*: 1257998.

Chen, C.-H., Puliafito, A., Cox, B. D., Primo, L., Fang, Y., Di Talia, S., and Poss, K. D. (2016). Multicolor cell barcoding technology for long-term surveillance of epithelial regeneration in zebrafish. *Dev Cell, 36*: 668–80.

Chhetri, R. K., Amat, F., Wan, Y., Hockendorf, B., Lemon, W. C. and Keller, P. J. (2015). Whole-animal functional and developmental imaging with isotropic spatial resolution. *Nat Methods, 12*: 1171–8.

Collins, R. T., Linker, C., and Lewis, J. (2010). MAZe: A tool for mosaic analysis of gene function in zebrafish. *Nat Methods, 7*: 219–23.

Collins, J. E., White, S., Searle, S. M., and Stemple, D. L. (2012). Incorporating RNA-seq data into the zebrafish Ensembl genebuild. *Genome Res, 22*: 2067–78.

Cong, L., Wang, Z., Chai, Y., Hang, W., Shang, C., Yang, W., Bai, L., Du, J., Wang, K., and Wen, Q. (2017). Rapid whole brain imaging of neural activity in freely behaving larval zebrafish (*Danio rerio*). *Elife, 6*:e28158.

D'agati, G., Beltre, R., Sessa, A., Burger, A., Zhou, Y., Mosimann, C., and White, R. M. (2017). A defect in the mitochondrial protein Mpv17 underlies the transparent casper zebrafish. *Dev Biol, 430*: 11–17.

Dooley, K., and Zon, L. I. (2000). Zebrafish: A model system for the study of human disease. *Curr Opin Genet Dev, 10*: 252–6.

Driever, W., Solnica-Krezel, L., Schier, A. F., Neuhauss, S. C., Malicki, J., Stemple, D. L., Stainier, D. Y., Zwartkruis, F., Abdelilah, S., Rangini, Z., Belak, J., and Boggs, C. (1996). A genetic screen for mutations affecting embryogenesis in zebrafish. *Development, 123*: 37–46.

Fetcho, J. R., and Liu, K. S. (1998). Zebrafish as a model system for studying neuronal circuits and behavior. *Ann N Y Acad Sci, 860*: 333–45.

Freeman, J., Vladimirov, N., Kawashima, T., Mu, Y., Sofroniew, N. J., Bennett, D. V., Rosen, J., Yang, C.-T., Looger, L. L., and Ahrens, M. B. (2014). Mapping brain activity at scale with cluster computing. *Nat Methods, 11*: 941–50.

Gahtan, E., Sankrithi, N., Campos, J. B., and O'malley, D. M. (2002). Evidence for a widespread brain stem escape network in larval zebrafish. *J Neurophysiol, 87*: 608–14.

Gallaud, E., Pham, T., and Cabernard, C. (2017). *Drosophila melanogaster* neuroblasts: A model for asymmetric stem cell divisions. *Results Probl Cell Differ, 61*: 183–210.

Godinho, L. (2011a). Injecting zebrafish with DNA or RNA constructs encoding fluorescent protein reporters. *Cold Spring Harb Protoc, 2011*: 871–4.

Godinho, L. (2011b). Using intravital dyes to ubiquitously label embryonic zebrafish. *Cold Spring Harb Protoc, 2011*: 877–8.

Goldsmith, J. R., and Jobin, C. (2012). Think small: Zebrafish as a model system of human pathology. *J Biomed Biotechnol, 2012*: 817341.

Gong, Y., Mo, C., and Fraser, S. E. (2004). Planar cell polarity signalling controls cell division orientation during zebrafish gastrulation. *Nature, 430*: 689–93.

Granato, M., Van Eeden, F. J., Schach, U., Trowe, T., Brand, M., Furutani-Seiki, M., Haffter, P., Hammerschmidt, M., Heisenberg, C. P., Jiang, Y. J., Kane, D. A., Kelsh, R. N., Mullins, M. C., Odenthal, J. and Nusslein-Volhard, C. (1996). Genes controlling and mediating locomotion behavior of the zebrafish embryo and larva. *Development, 123*: 399–413.

Grimm, J. B., Muthusamy, A. K., Liang, Y., Brown, T. A., Lemon, W. C., Patel, R., Lu, R., Macklin, J. J., Keller, P. J., Ji, N., and Lavis, L. D. (2017). A general method to fine-tune fluorophores for live-cell and in vivo imaging. *Nat Methods, 14*: 987–94.

Gross-Thebing, T., Paksa, A., and Raz, E. (2014). Simultaneous high-resolution detection of multiple transcripts combined with localization of proteins in whole-mount embryos. *Bmc Biol, 12*: 55.

Haffter, P., Granato, M., Brand, M., Mullins, M. C., Hammerschmidt, M., Kane, D. A., Odenthal, J., Van Eeden, F. J., Jiang, Y. J., Heisenberg, C. P., Kelsh, R. N., Furutani-Seiki, M., Vogelsang, E., Beuchle, D., Schach, U., Fabian, C., and Nusslein-Volhard, C. (1996). The identification of genes with unique and essential functions in the development of the zebrafish, Danio rerio. *Development, 123*: 1–36.

Halloran, M. C., Sato-Maeda, M., Warren, J. T., Su, F., Lele, Z., Krone, P. H., Kuwada, J. Y., and Shoji, W. (2000). Laser-induced gene expression in specific cells of transgenic zebrafish. *Development, 127*, 1953–60.

Hans, S., Kaslin, J., Freudenreich, D., and Brand, M. (2009). Temporally-controlled site-specific recombination in zebrafish. *PLoS One, 4*: e4640.

Haubold, C., Schiegg, M., Kreshuk, A., Berg, S., Koethe, U., and Hamprecht, F. A. (2016). Segmenting and tracking multiple dividing targets using Ilastik. *Adv Anat Embryol Cell Biol, 219*: 199–229.

He, J., Zhang, G., Almeida, A. D., Cayouette, M., Simons, B. D., and Harris, W. A. (2012). How variable clones build an invariant retina. *Neuron, 75*: 786–98.

Hisano, Y., Ota, S., and Kawahara, A. (2014). Genome editing using artificial site-specific nucleases in zebrafish. *Dev Growth Differ, 56*: 26–33.

Hisaoka, K., Ott, J., and Marchese, A. (1957). *Time lapse studies on the embryology of the zebra fish, Brachydanio-rerio. Anatomical record.* New York, NY: Wiley-Liss Div John Wiley and Sons Inc, pp. 565–5.

Howe, D. G., Bradford, Y. M., Conlin, T., Eagle, A. E., Fashena, D., Frazer, K., Knight, J., Mani, P., Martin, R., Moxon, S. A., Paddock, H., Pich, C., Ramachandran, S., Ruef, B. J., Ruzicka, L., Schaper, K., Shao, X., Singer, A., Sprunger, B., Van Slyke, C. E., and Westerfield, M. (2013a). Zfin, the zebrafish model organism database: Increased support for mutants and transgenics. *Nucleic Acids Res, 41*: D854–60.

Howe, K., Clark, M. D., Torroja, C. F., Torrance, J., Berthelot, C., Muffato, M., Collins, J. E., Humphray, S., Mclaren, K., Matthews, L., Mclaren, S., Sealy, I., Caccamo, M., Churcher, C., Scott, C., Barrett, J. C., Koch, R., Rauch, G. J., White, S., Chow, W., Kilian, B., Quintais, L. T., Guerra-Assuncao, J. A., Zhou, Y., Gu, Y., Yen, J., Vogel, J. H., Eyre, T., Redmond, S., Banerjee, R., Chi, J., Fu, B., Langley, E., Maguire, S. F., Laird, G. K., Lloyd, D., Kenyon, E., Donaldson, S., Sehra, H., Almeida-King, J., Loveland, J., Trevanion, S., Jones, M., Quail, M., Willey, D., Hunt, A., Burton, J., Sims, S., Mclay, K., Plumb, B., Davis, J., Clee, C., Oliver, K., Clark, R., Riddle, C., Elliot, D., Threadgold, G., Harden, G., Ware, D., Begum, S., Mortimore, B., Kerry, G., Heath, P., Phillimore, B., Tracey, A., Corby, N., Dunn, M., Johnson, C., Wood, J., Clark, S., Pelan, S., Griffiths, G., Smith, M., Glithero, R., Howden, P., Barker, N., Lloyd, C., Stevens, C., Harley, J., Holt, K., Panagiotidis, G., Lovell, J., Beasley, H., Henderson, C., Gordon, D., Auger, K., Wright, D., Collins, J., Raisen, C., Dyer, L., Leung, K., Robertson, L., Ambridge, K., Leongamornlert, D., Mcguire, S., Gilderthorp, R., Griffiths, C., Manthravadi, D., Nichol, S., and Barker, G. (2013b). The zebrafish reference genome sequence and its relationship to the human genome. *Nature, 496*: 498–503.

Huisken, J., Swoger, J., Del Bene, F., Wittbrodt, J., and Stelzer, E. H. K. (2004). Optical sectioning deep inside live embryos by selective plane illumination microscopy. *Science, 305*: 1007–9.

Jonkman, J., and Brown, C. M. (2015). Any way you slice it-A comparison of confocal microscopy techniques. *J Biomol Tech, 26*: 54–65.

Kawashima, T., Zwart, M. F., Yang, C.-T., Mensh, B. D., and Ahrens, M. B. (2016). The serotonergic system tracks the outcomes of actions to mediate short-term motor learning. *Cell, 167*: 933–946.e20.

Keller, P.J. (2013). Imaging morphogenesis: technological advances and biological insights. *Science, 340*(6137): 1234168.

Keller, P. J., Schmidt, A. D., Santella, A., Khairy, K., Bao, Z., Wittbrodt, J., and Stelzer, E. H. (2010). Fast, high-contrast imaging of animal development with scanned light sheet-based structured-illumination microscopy. *Nat Methods, 7*: 637–42.

Keller, P. J., Schmidt, A. D., Wittbrodt, J., and Stelzer, E. H. K. (2008). Reconstruction of zebrafish early embryonic development by scanned light sheet microscopy. *Science, 322*: 1065–9.

Keppler, A., Gendreizig, S., Gronemeyer, T., Pick, H., Vogel, H., and Johnsson, K. (2003). A general method for the covalent labeling of fusion proteins with small molecules in vivo. *Nat Biotechnol, 21*: 86–9.

Kim, D. H., Kim, J., Marques, J. C., Grama, A., Hildebrand, D. G. C., Gu, W., Li, J. M., and Robson, D. N. (2017). Pan-neuronal calcium imaging with cellular resolution in freely swimming zebrafish. *Nat Methods, 14*: 1107–14.

Kimmel, C. B., Ballard, W. W., Kimmel, S. R., Ullmann, B., and Schilling, T. F. (1995). Stages of embryonic development of the zebrafish. *Dev Dyn, 203*: 253–310.

Kimmel, C. B., Warga, R. M. and Kane, D. A. (1994). Cell cycles and clonal strings during formation of the zebrafish central nervous system. *Development, 120*: 265–76.

Kimmel, C. B., Warga, R. M., and Schilling, T. F. (1990). Origin and organization of the zebrafish fate map. *Development, 108*: 581–94.

Kimura, Y., Hisano, Y., Kawahara, A., and Higashijima, S. (2014). Efficient generation of knock-in transgenic zebrafish carrying reporter/driver genes by CRISPR/Cas9-mediated genome engineering. *Sci Rep, 4*: 6545.

Knopf, F., Schnabel, K., Haase, C., Pfeifer, K., Anastassiadis, K. and Weidinger, G. (2010). Dually inducible TetON systems for tissue-specific conditional gene expression in zebrafish. *Proc Natl Acad Sci U S A, 107*: 19933–8.

Krzic, U., Gunther, S., Saunders, T. E., Streichan, S. J., and Hufnagel, L. (2012). Multiview light-sheet microscope for rapid in toto imaging. *Nat Methods, 9*: 730–3.

Kwan, K. M., Fujimoto, E., Grabher, C., Mangum, B. D., Hardy, M. E., Campbell, D. S., Parant, J. M., Yost, H. J., Kanki, J. P., and Chien, C. B. (2007). The Tol2kit: A multisite gateway-based construction kit for Tol2 transposon transgenesis constructs. *Dev Dyn, 236*: 3088–99.

Laissue, P. P., Alghamdi, R. A., Tomancak, P., Reynaud, E. G., and Shroff, H. (2017). Assessing phototoxicity in live fluorescence imaging. *Nat Methods, 14*: 657–61.

Lander, E. S. (2015). Brave new genome. *N Engl J Med, 373*: 5–8.

Le, X., Langenau, D. M., Keefe, M. D., Kutok, J. L., Neuberg, D. S. and Zon, L. I. (2007). Heat shock-inducible Cre/Lox approaches to induce diverse types of tumors and hyperplasia in transgenic zebrafish. *Proc Natl Acad Sci U S A, 104*: 9410–5.

Lemon, W. C., Pulver, S. R., Hockendorf, B., Mcdole, K., Branson, K., Freeman, J., and Keller, P. J. (2015). Whole-central nervous system functional imaging in larval Drosophila. *Nat Commun, 6*: 7924.

Lewis, W. H. (1942). The formation of the blastodisc in the egg of the zebrafish, Brachydanio rerio, illustrated with motion pictures. *Anat Rec, 84*: 463–4.

Lieschke, G. J., and Currie, P. D. (2007). Animal models of human disease: Zebrafish swim into view. *Nat Rev Genet, 8*: 353–67.

Livet, J., Weissman, T. A., Kang, H., Draft, R. W., Lu, J., Bennis, R. A., Sanes, J. R., and Lichtman, J. W. (2007). Transgenic strategies for combinatorial expression of fluorescent proteins in the nervous system. *Nature, 450*: 56–62.

Looger, L. L., and Griesbeck, O. (2012). Genetically encoded neural activity indicators. *Curr Opin Neurobiol, 22*: 18–23.

Los, G. V., Encell, L. P., Mcdougall, M. G., Hartzell, D. D., Karassina, N., Zimprich, C., Wood, M. G., Learish, R., Ohana, R. F., Urh, M., Simpson, D., Mendez, J., Zimmerman, K., Otto, P., Vidugiris, G., Zhu, J., Darzins, A., Klaubert, D. H., Bulleit, R. F., and Wood, K. V. (2008). HaloTag: A novel protein labeling technology for cell imaging and protein analysis. *Acs Chem Biol, 3*: 373–82.

Lu, R., Sun, W., Liang, Y., Kerlin, A., Bierfeld, J., Seelig, J. D., Wilson, D. E., Scholl, B., Mohar, B., Tanimoto, M., Koyama, M., Fitzpatrick, D., Orger, M. B., and Ji, N. (2017). Video-rate volumetric functional imaging of the brain at synaptic resolution. *Nat Neurosci, 20*: 620–8.

Lukyanov, K. A., Chudakov, D. M., Lukyanov, S., and Verkhusha, V. V. (2005). Innovation: Photoactivatable fluorescent proteins. *Nat Rev Mol Cell Biol, 6*: 885–91.

Miyagi, C., Yamashita, S., Ohba, Y., Yoshizaki, H., Matsuda, M. and Hirano, T. (2004). STAT3 noncell-autonomously controls planar cell polarity during zebrafish convergence and extension. *J Cell Biol, 166*: 975–81.

Moro, E., Vettori, A., Porazzi, P., Schiavone, M., Rampazzo, E., Casari, A., Ek, O., Facchinello, N., Astone, M., Zancan, I., Milanetto, M., Tiso, N., and Argenton, F. (2013). Generation and application of signaling pathway reporter lines in zebrafish. *Mol Genet Genomics, 288*: 231–42.

Ogura, E., Okuda, Y., Kondoh, H., and Kamachi, Y. (2009). Adaptation of GAL4 activators for GAL4 enhancer trapping in zebrafish. *Dev Dyn, 238*: 641–55.

Okumoto, S., Jones, A., and Frommer, W. B. (2012). Quantitative imaging with fluorescent biosensors. *Annu Rev Plant Biol, 63*: 663–706.

Otsuna, H., Hutcheson, D. A., Duncan, R. N., Mcpherson, A. D., Scoresby, A. N., Gaynes, B. F., Tong, Z., Fujimoto, E., Kwan, K. M., Chien, C. B., and Dorsky, R. I. (2015). High-resolution analysis of central nervous system expression patterns in zebrafish Gal4 enhancer-trap lines. *Dev Dyn, 244*: 785–96.

Pan, Y. A., Freundlich, T., Weissman, T. A., Schoppik, D., Wang, X. C., Zimmerman, S., Ciruna, B., Sanes, J. R., Lichtman, J. W., and Schier, A. F. (2013). Zebrabow: Multispectral cell labeling for cell tracing and lineage analysis in zebrafish. *Development, 140*: 2835–46.

Panier, T., Romano, S. A., Olive, R., Pietri, T., Sumbre, G., Candelier, R. and Debrégeas, G. (2013). Fast functional imaging of multiple brain regions in intact zebrafish larvae using selective plane illumination microscopy. *Front Neural Circuits, 7*: 65.

Peal, D. S., Mills, R. W., Lynch, S. N., Mosley, J. M., Lim, E., Ellinor, P. T., January, C. T., Peterson, R. T., and Milan, D. J. (2010). Novel chemical suppressors of long Qt syndrome identified by an in vivo functional screen. *Circulation, 123.* 23–30.

Pitrone, P. G., Schindelin, J., Stuyvenberg, L., Preibisch, S., Weber, M., Eliceiri, K. W., Huisken, J., and Tomancak, P. (2013). OpenSPIM: An open-access light-sheet microscopy platform. *Nat Methods, 10*: 598–9.

Planchon, T. A., Gao, L., Milkie, D. E., Davidson, M. W., Galbraith, J. A., Galbraith, C. G. and Betzig, E. (2011). Rapid three-dimensional isotropic imaging of living cells using Bessel beam plane illumination. *Nat Methods, 8*: 417–23.

Pulak, R. (2016). Tools for automating the imaging of zebrafish larvae. *Methods, 96*: 118–126.

Quirin, S., Vladimirov, N., Yang, C. T., Peterka, D. S., Yuste, R., and Ahrens, M. B. (2016). Calcium imaging of neural circuits with extended depth-of-field light-sheet microscopy. *Opt Lett, 41*: 855–8.

Randlett, O., Wee, C. L., Naumann, E. A., Nnaemeka, O., Schoppik, D., Fitzgerald, J. E., Portugues, R., Lacoste, A. M. B., Riegler, C., Engert, F., and Schier, A. F. (2015). Whole-brain activity mapping onto a zebrafish brain atlas. *Nat Methods, 12*: 1039–1046.

Reynaud, E. G., Peychl, J., Huisken, J., and Tomancak, P. (2015). Guide to light-sheet microscopy for adventurous biologists. *Nat Methods, 12*: 30–4.

Rihel, J., Prober, D. A., Arvanites, A., Lam, K., Zimmerman, S., Jang, S., Haggarty, S. J., Kokel, D., Rubin, L. L., Peterson, R. T., and Schier, A. F. (2010). Zebrafish behavioral profiling links drugs to biological targets and rest/wake regulation. *Science, 327*: 348–51.

Ronneberger, O., Liu, K., Rath, M., RUEB, D., Mueller, T., Skibbe, H., Drayer, B., Schmidt, T., Filippi, A., Nitschke, R., Brox, T., Burkhardt, H., and Driever, W. (2012). ViBE-Z: A framework for 3D virtual colocalization analysis in zebrafish larval brains. *Nat Methods, 9*: 735–42.

Roszko, I., Sawada, A., and Solnica-Krezel, L. (2009). Regulation of convergence and extension movements during vertebrate gastrulation by the Wnt/PCP pathway. *Semin Cell Dev Biol, 20*: 986–97.

Royer, L. A., Lemon, W. C., Chhetri, R. K., Wan, Y., Coleman, M., Myers, E. W., and Keller, P. J. (2016). Adaptive light-sheet microscopy for long-term, high-resolution imaging in living organisms. *Nat Biotechnol, 34*: 1267–1278.

Schiegg, M., Hanslovsky, P., Kausler, B. X., Hufnagel, L., and Hamprecht, F. A. (2013). Conservation tracking. In *2013 IEEE International Conference on Computer Vision.*

Scott, E. K., Mason, L., Arrenberg, A. B., Ziv, L., Gosse, N. J., Xiao, T., Chi, N. C., Asakawa, K., Kawakami, K., and Baier, H. (2007). Targeting neural circuitry in zebrafish using GAL4 enhancer trapping. *Nat Methods, 4*: 323–6.

Shah, G., Thierbach, K., Schmid, B., Reade, A., Roeder, I., Scherf, N., and Huisken, J. (2019). Pan-embryo cell dynamics of germlayer formation in zebrafish. *Nat Commun, 10*:5753.

Sibarita, J. B. (2005). Deconvolution microscopy. *Adv Biochem Eng Biotechnol, 95*, 201–43.

Solnica-Krezel, L., and Sepich, D. S. (2012). Gastrulation: Making and shaping germ layers. *Annu Rev Cell Dev Biol, 28*: 687–717.

Sommer, C., Straehle, C., Kothe, U., and Hamprecht, F. A. (2011). Ilastik: Interactive learning and segmentation toolkit. In *IEEE International Symposium on Biomedical Imaging: From Nano to Macro*, Chicago, IL, 230–233.

Stegmaier, J., Amat, F., Lemon, W. C., Mcdole, K., Wan, Y., Teodoro, G., Mikut, R., and Keller, P. J. (2016). Real-time three-dimensional cell segmentation in large-scale microscopy data of developing embryos. *Dev Cell, 36*: 225–240.

Tada, M., and Heisenberg, C.P. (2012). Convergent extension: using collective cell migration and cell intercalation to shape embryos. *Development, 139*(21): 3897–3904.

Tawk, M., Bianco, I. H., and Clarke, J. D. (2009). Focal electroporation in zebrafish embryos and larvae. *Methods Mol Biol, 546*: 145–51.

Thisse, C., and Thisse, B. (2008). High-resolution in situ hybridization to whole-mount zebrafish embryos. *Nat Protoc, 3*: 59–69.

Tomer, R., Khairy, K., Amat, F. and Keller, P. J. (2012). Quantitative high-speed imaging of entire developing embryos with simultaneous multiview light-sheet microscopy. *Nat Methods, 9*: 755–63.

Tomer, R., Lovett-Barron, M., Kauvar, I., Andalman, A., Burns, V. M., Sankaran, S., Grosenick, L., Broxton, M., Yang, S., and Deisseroth, K. (2015). SPED light sheet microscopy: Fast mapping of biological system structure and function. *Cell, 163*: 1796–806.

Truong, T. V., Supatto, W., Koos, D. S., Choi, J. M., and Fraser, S. E. (2011). Deep and fast live imaging with two-photon scanned light-sheet microscopy. *Nat Methods*, 8: 757–60.

Udvadia, A. J., and Linney, E. (2003). Windows into development: Historic, current, and future perspectives on transgenic zebrafish. *Dev Biol*, 256: 1–17.

Ulman, V., Maška, M., Magnusson, K. E. G., Ronneberger, O., Haubold, C., Harder, N., Matula, P., Matula, P., Svoboda, D., Radojevic, M., Smal, I., Rohr, K., Jaldén, J., Blau, H. M., Dzyubachyk, O., Lelieveldt, B., Xiao, P., Li, Y., Cho, S.-Y., Dufour, A. C., Olivo-Marin, J.-C., Reyes-Aldasoro, C. C., Solis-Lemus, J. A., Bensch, R., Brox, T., Stegmaier, J., Mikut, R., Wolf, S., Hamprecht, F. A., Esteves, T., Quelhas, P., Demirel, Ö., Malmström, L., Jug, F., Tomancak, P., Meijering, E., Muñoz-Barrutia, A., Kozubek, M., and Ortiz-De-Solorzano, C. (2017). An objective comparison of cell-tracking algorithms. *Nat Methods*, 14: 1141–52.

Vladimirov, N., Mu, Y., Kawashima, T., Bennett, D. V., Yang, C.-T., Looger, L. L., Keller, P. J., Freeman, J., and Ahrens, M. B. (2014). Light-sheet functional imaging in fictively behaving zebrafish. *Nat Methods*, 11: 883–4.

Voie, A. H., Burns, D. H., and Spelman, F. A. (1993). Orthogonal-plane fluorescence optical sectioning: Three-dimensional imaging of macroscopic biological specimens. *J Microsc*, 170: 229–36.

Weigert, M., Schmidt, U., Boothe, T., Müller, A., Dibrov, A., Jain, A., Wilhelm, B., Schmidt, D., Broaddus, C., Culley, S., Rocha-Martins, M., Segovia-Miranda, F., Norden, C., Henriques, R., Zerial, M., Solimena, M., Rink, J., Tomancak, P., Royer, L., Jug, F., and Myers, E. W. (2018). Content-aware image restoration: Pushing the limits of fluorescence microscopy. *Nat Methods, 15*: 1090–1097.

Weissman, T. A., and Pan, Y. A. (2015). Brainbow: New resources and emerging biological applications for multicolor genetic labeling and analysis. *Genetics, 199*: 293–306.

White, R. M., Sessa, A., Burke, C., Bowman, T., Leblanc, J., Ceol, C., Bourque, C., Dovey, M., Goessling, W., Burns, C. E., and Zon, L. I. (2008). Transparent adult zebrafish as a tool for in vivo transplantation analysis. *Cell Stem Cell, 2*: 183–9.

Williams, C. H., and Hong, C. C. (2016). Zebrafish small molecule screens: Taking the phenotypic plunge. *Comput Struct Biotechnol J, 14*: 350–6.

Wolf, S., Supatto, W., Debregeas, G., Mahou, P., Kruglik, S. G., Sintes, J. M., Beaurepaire, E., and Candelier, R. (2015). Whole-brain functional imaging with two-photon light-sheet microscopy. *Nat Methods, 12*: 379–80.

Wolff, C., Tinevez, J.-Y., Pietzsch, T., Stamataki, E., Harich, B., Preibisch, S., Shorte, S., Keller, P. J., Tomancak, P., and Pavlopoulos, A. (2018). Reconstruction of cell lineages and behaviors underlying arthropod limb outgrowth with multi-view light-sheet imaging and tracking. *Elife, 7*:e34410.

Xue, L., Karpenko, I. A., Hiblot, J., and Johnsson, K. (2015). Imaging and manipulating proteins in live cells through covalent labeling. *Nat Chem Biol, 11*: 917–23.

21

Whole-Body Fluorescence Imaging in Cancer Research

Marina V. Shirmanova, Diana V. Yuzhakova, Maria M. Lukina, Alena I. Gavrina,
Aleksandra V. Meleshina, Ilya V. Turchin, and Elena V. Zagaynova

CONTENTS

21.1 Introduction

Fluorescence whole-body imaging of small animals is a powerful *in vivo* technology with many biomedical applications, among which cancer research occupies a central position. Modern fluorescence-based macroscopic approaches provide a unique opportunity to observe a whole organism or a whole tumor in a way that is crucial to appreciating tumor heterogeneity and for identifying the specifics of tumor-host interactions and drug distribution in individual animals.

Whole-body animal studies on mice using fluorescent agents can be performed using different imaging approaches: superficial methods, transillumination, and fluorescence tomography (Ntziachristos, 2005; Turchin, 2016).

The epi-illumination technique, one of the superficial methods, is the most common for this type of study. It is based on the excitation of fluorescence in the investigated area by a broad light beam and the detection of fluorescence using a digital camera with an optical filter that passes the fluorescence emission while blocking the wavelengths of the excitation light. Another superficial method is raster scanned point-source imaging, which is able to provide better spatial resolution than the broad beam geometries (Pogue et al., 2004). The principle of raster scanning is similar to laser scanning microscopy: the use of point scanning by a narrow light beam with simultaneous reception of the fluorescence caused in a given region. This approach reduces the multiple scattered light component coming from neighboring areas at each measuring point, thereby providing more accurate quantification. Because of the high contrast of fluorescent agents, the superficial methods can provide a qualitative image of the fluorophore distribution in the tissue but do not allow resolution of the depth distribution of the fluorescent agent.

Fluorescence tomography techniques use multiple projections of the object under study when it is irradiated from a fluorescence-exciting source, followed by registration of the emissions for subsequent reconstruction of the fluorophore distribution in the tissue. Fluorescence diffuse tomography (FDT) detects multiple scattered photons and, therefore, faces a number of additional complications, i.e. high sensitivity to boundary conditions, ill-posed inverse problem, and the limitations of the current analytical models of light propagation in tissues. In order to reconstruct the fluorophore distribution correctly, the signal-to-background ratio must be sufficiently high. A high signal-to-background ratio can be achieved with the use of markers, which are able to accumulate very selectively in the zone of interest or may be specifically expressed by pathological cells, and have excitation and emission spectra within the therapeutic transparency window. The greatest number of papers on FDT is associated with the use of red and far-red fluorescent proteins as tumor markers. Nevertheless, even in the near-infrared (NIR) region, the level of autofluorescence (intrinsic fluorescence of tissues) is quite high, thus limiting the quality of reconstruction (Darne et al., 2014). The quality of reconstruction in FDT can be improved by the spectral unmixing technique. The spectral characteristics of exogenous agents are well known, and several measurements with different excitation and emission bands can be used to separate the useful signals from those of the endogenous fluorophores (Chaudhari et al., 2009; Kleshnin et al., 2013).

Because superficial methods do not allow deep tissue imaging (beyond a few millimeters) and tomographic techniques are too complicated and time-consuming, in some cases it is reasonable to use an intermediate approach, e.g. the transillumination raster-scanning technique. This method reveals the two-dimensional (2D) fluorophore distribution, while being less sensitive to depth, compared with the superficial methods, and, in contrast to tomographic approaches, does not need any reconstruction procedure (only Born normalization is applied to reduce the influence of optical property variations and the boundaries of the object). The projection method has proven itself valuable for the study of tumors labeled with far-red proteins (Kleshnin et al., 2015).

The use of time-resolved techniques allows registration of both the fluorescence intensity and the fluorescence lifetime. Lifetime measurements are usually provided by a short (subnanosecond) laser pulse and time-resolved fluorescence detection by using a time-correlated single photon counting (TCSPC) technique (Becker, 2005) or a camera with a fast-switchable multichannel photocathode (Nothdurft et al., 2009). Another approach for measuring the lifetime of a fluorophore is based on measuring the phase shift between the excitation radiation intensity and the received fluorescent signal, made possible when the excitation is modulated at a high frequency (about 50 MHz). Because the lifetime of fluorescence greatly depends on the molecular environment, fluorescence lifetime imaging (FLIM) opens up the possibility of investigating the functional characteristics (e.g. metabolism and protein–protein interactions) and various physical and chemical parameters of cells and tissues (e.g. pH, viscosity, ion concentration).

This chapter gives an overview of the major applications of fluorescence whole-body imaging in living mice with a focus on cancer studies.

21.2 *In Vivo* Imaging of Biodistribution of Novel Anticancer Agents

The pharmacokinetics of new drugs are routinely investigated using radiolabeling methods and scintillation counting of the radioactivity in harvested tissue samples, or by PET and SPECT imaging of experimental animals (Ding and Wu, 2012; Hillyar et al., 2015). Although these techniques give accurate data on drug concentrations in tissues, they are complicated and time-consuming and require the radioisotope to be incorporated into an appropriate biologically active compound.

An alternative approach for the examination of pharmacokinetics is the use of fluorescence-based imaging. Being technically easy to implement and simple in operation, fluorescence imaging techniques enable noninvasive visualization and quantification of the biodistribution of fluorescent agents over time at the whole-body level of a small animal. In preclinical studies on tumor-bearing mice, whole animal fluorescence imaging has been employed extensively to obtain information regarding the selectivity of potential drugs to particular tumors (via quantification of the tumor-to-normal tissue fluorescence ratio), the kinetics of drug uptake by tumors (via quantification of the fluorescence in the tumor area) and the details of their distribution throughout the animal body. Typically, longitudinal monitoring of the same living animals is also performed, with the initial time point being taken immediately post injection of the agent, while subsequent time points capture information on the maximum uptake in the tumor and the kinetics of washout from normal tissues.

Among anticancer agents, photosensitizers for photodynamic therapy (PDT) are a major class that has been extensively investigated in preclinical models using fluorescence imaging, as they can be readily detected by their characteristic fluorescence.

PDT is a cancer treatment approach based on the light activation of phototoxic dyes called photosensitizers, resulting in the generation of reactive oxygen species (ROS) that destroy neoplastic tissue (Agostinis et al., 2011; Dolmans et al., 2003). Most photosensitizers are fluorescent molecules, so this enables clinical diagnosis of malignancies, and preclinical pharmacokinetics studies in animals. The development of photosensitizers showing preferential accumulation in tumors and with minimum uptake by the skin and mucosae is crucial for successful treatment and for avoiding side effects.

First studies of the fluorescence kinetics of photosensitizers in animal tumor models *in vivo* date to the late 1990s. Using direct visualization of fluorescence in ultraviolet (UV) light (Major et al., 1997) or with imaging systems relying on charged-coupled device (CCD) cameras to record fluorescence in either intensity (Harada et al., 1998) or lifetime modes (Cubeddu et al., 1997), it was possible to demonstrate stronger fluorescence and longer retention time of photosensitizers in tumors than in the surrounding normal areas, allowing the real-time identification of neoplastic tissue. Further development of imaging systems has allowed the difficulties associated with animal motion artifacts and strong autofluorescence backgrounds to be overcome and enabled the fast acquisition of fluorescence images over large areas, as well as the three-dimensional (3D) reconstruction of fluorophore distribution. Our study demonstrated the possibility of *in vivo* investigation of the pharmacokinetics of various photosensitizers in mice by means of transillumination imaging (Shirmanova et al., 2010). This is preferable for photosensitizer detection because of its higher sensitivity to deep-seated fluorophores. It was shown that different photosensitizers have different kinetics of tumor uptake and that tumor fluorescence linearly increased with increasing drug dose. Ozturk et al. (Ozturk et al., 2014) and Mo et al. (Mo et al., 2012) reported the development of a technique of fluorescence tomographic imaging to assess the photosensitizer distribution in deep tissues and demonstrated the heterogeneous distribution of 2-[1-hexyloxyethyl]-2 devinyl pyropheophorbide-a (HPPH) in cancer models. Meanwhile, 2D epi-fluorescence (back-reflectance) imaging is a rapid and easy to use technology and is most widely exploited in biodistribution studies of photosensitizers and other drugs.

Examples of novel agents for PDT investigated with the use of fluorescence whole-body imaging, include newly developed substances (bacteriochlorin analogues (Patel et al., 2016), pheophorbide a (Ahn et al., 2017), ytterbium cyanoporphyrazine complex (Klapshina et al., 2010; Shirmanova et al., 2011), butadiyne-linked conjugated porphyrin dimer (Khurana et al., 2012), modified forms of classical photosensitizers (phthalocyanines conjugated with folate and triglycerol substitutions) (Li et al., 2015), chlorin e6 complexes with amphiphilic polymers (Shirmanova et al., 2014; Gavrina et al., 2018) and a range of multifunctional probes that involve a photosensitizer and an magnetic resonance imaging (MRI) contrast agent for image-guided cancer therapy (Wolfbeis, 2015; Bechet et al., 2012).

All these studies prove that *in vivo* fluorescence imaging can be the method of choice to determine the substances with the highest selectivity for tumor tissue and to determine, at the preclinical stage, the optimal time for conducting PDT.

A typical biodistribution of a passively delivered photosensitizer in a tumor-bearing mouse is shown in Figure 21.1. In that study, we investigated the effect of amphiphilic polymers on the tissue distribution of chlorin e6 (Ce6) in mice with subcutaneously transplanted cervical carcinomas. Fluorescence images were acquired *in vivo* before (control) and at different time points after intravenous injection of the Ce6 encapsulated in Pluronic F127 micelles (Shirmanova et al., 2014). Within 30 min postinjection, an intense fluorescence was detected across the whole animal body due to the photosensitizer circulating in the blood stream and its uptake in the skin and abdominal organs. Accumulation of the photosensitizer in the tumor and faster elimination from the surrounding normal tissues allowed the tumor to be distinguished by fluorescence after 1 h. The maximum tumor-to-normal ratio of 4:1 was observed 8 h after injection, but the fluorescence intensity of the tumor (and consequently, the concentration of the photosensitizer in it) was quite low. A tumor-to-normal ratio between 2:1 and 3:1, typical of many chemically synthesized photosensitizers, was detected in the period 2–6 h post-injection. Because PDT success is, in part, a compromise between drug selectivity and concentration, this time period may be recommended for PDT with this drug formulation. At 24 h, the photosensitizer remained predominantly in the tumor, but by 96 h had been completely eliminated from the body. As can be seen from whole-body fluorescence images, the distribution patterns of the photosensitizer changes with time, but verification of its presence in specific organs is not possible within living mice. Therefore, for detailed biodistribution analysis fluorescence from the excised organs and tumor is usually recorded after the completion of the *in vivo* study or for a separate group of animals. Quantification of the fluorescence intensity from the harvested tissues revealed that the highest content of the Ce6-F127 complex was in the organs responsible for excretion of the drug from the body – the colon and liver. The lowest concentrations were in the brain, heart, spleen, and muscles.

A number of contrast agents for fluorescence-based cancer diagnosis have been evaluated in mice using fluorescence imaging. Among them, NIR fluorescence (NIRF) probes (650–900 nm) are receiving increasing attention because of the greater transparency of animal tissues for NIR light, which provides centimeter-depth visualization. Besides excellent optical characteristics and suitable biocompatibility, the ability to accumulate selectively at tumor sites is an important requirement for these probes in terms of their potential clinical application. Biodistribution studies of the NIRF organic dyes and nanoparticles, delivered to the tumors using the EPR (enhanced permeability and retention) effect, or actively targeted by specific ligands, validate NIR fluorescence as a novel strategy for the accurate detection of malignancies (Jiang et al., 2010; Nam et al., 2010; Harrison et al., 2015; Fei-Peng et al., 2016).

Potential agents for chemotherapy, one of the major treatment approaches for most cancers, can also be characterized *in vivo* using fluorescence assessments. because most therapeutic molecules are nonfluorescent, labeling with a fluorophore is

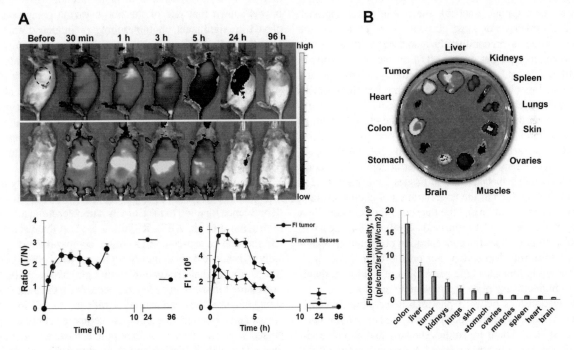

FIGURE 21.1 *In vivo* monitoring of the biodistribution of the photosensitizer chlorin e6 in mice with subcutaneous mouse cervical carcinomas. Chlorin e6 was encapsulated in Pluronic F127 micelles and injected intravenously via the tail vein at 10 mg/kg. A. Fluorescence whole-body images acquired from the left side (upper row) and ventral side (bottom row) at different time points using an IVIS-Spectrum system (Caliper Life Sciences, USA), with excitation at 640/30 nm and detection at 720/20 nm. Tumor is shown by the dashed circle. Kinetics of fluorescence intensity (FI) in tumors and the tumor-to-nuormal (T/N) tissue fluorescence ratio. B. Fluorescence of the excised tumor and organs at 6 hours after injection. Mean ± SEM, *n* = 3 mice.

required to visualize their localization within the whole mouse body. This is a frequently used approach in preclinical studies of nanoparticle-based formulations, when a fluorescent dye can be easily incorporated in the nanoparticle along with the therapeutic component (Kim et al., 2009; Sailor and Park, 2012). Several chemotherapeutic drugs, such as doxorubicin, paclitaxel, and bleomycin, do possess their own fluorescent properties (Motlagh et al., 2016); however, because of their low fluorescence quantum yield and the blue-green emission/excitation spectrum range, this feature is not suitable for *in vivo* visualization. Whole-body fluorescence imaging of chemotherapeutics by inherent fluorescence is now limited to studies of doxorubicin as this emits at ~560 nm. For example, Kanno et al. (2015)monitored the accumulation of doxorubicin in non-small cell lung cancer xenografts with multidrug resistance phenotypes and reported on higher uptake of the drug in tumors that had been exposed to radiotherapy.

Therefore, fluorescence imaging *in vivo* enables noninvasive, rapid, and cost-effective monitoring of the tumor accumulation and clearance of novel agents for cancer diagnostics and therapy in small animal models. A number of studies *in vivo* demonstrate the possibility of fluorescence whole-body imaging to assess the pharmacokinetics and biodistribution of inherently fluorescent compounds (e.g. photosensitizers for PDT) and therapeutic agents labeled with fluorescent markers.

21.3 Assessment of Photobleaching of Photosensitizer during PDT

Photosensitizer photobleaching, as measured by fluorescence, is considered a simple, practical, and cost-effective approach in PDT dosimetry. For most classical photosensitizers, the photobleaching is predominantly associated with an attack by ROS (mainly singlet oxygen, $^1O_2$) on the photosensitizer molecules, followed by their photochemical destruction and irreversible loss of fluorescence. However, other photobleaching mechanisms are also possible (e.g. some photosensitizers can bleach due to photoreduction of the chromophore as a result of its direct interaction with cell components) (Jarvi et al., 2012; Dysart et al., 2005; Sheng et al., 2007; Anbil et al., 2012). In fact, many chemical photosensitizers exhibit complicated bleaching mechanisms in tissues, and these depend on many factors, such as the concentration of photosensitizer, the oxygen concentration and the fluence of the delivered light (Huang et al., 2012; Vollet-Filho et al., 2009). On the one hand, photobleaching reduces the concentration of photoactive photosensitizer during irradiation, but on the other hand, high photobleaching indicates high production of ROS and, therefore, the high efficacy of its photodynamic reaction.

The relationships between photosensitizer photobleaching and the PDT responses have been investigated *in vivo* for many photosensitizers. There are studies showing that photobleaching and PDT response are correlated. Vollet-Filho et al. (2009) observed a high degree of correlation between the photodegradation of Photogem, measured by the surface fluorescence signal during PDT and the depth of necrosis in rat liver *in vivo* for light doses lower than 250 J/cm². In a transplanted tumor model, Photofrin fluorescence measurements *in vivo* correlated

with the PDT response (Wilson et al., 1997). A correlation was found between the bleaching kinetics of 5-ALA-PpIX and PDT-induced edema in rodent esophagus – a rapid photobleaching occurred when there was an ample oxygen supply but was not present when the oxygen was limited (Sheng et al., 2007). The ability of a photobleaching-based metric to predict the clinical phototoxic response (erythema) resulting from 5-ALA PDT was demonstrated by Mallidi et al. (2015). Ascencio et al. (2008) showed, *in vivo*, that the fluorescence photobleaching of PpIX is useful to predict tumor responses to hexaminolevulinate PDT. In a clinical study, it was found that low photobleaching of Photodithazine in nonmelanoma skin malignancies can predict tumor recurrence (Gamayunov et al., 2016).

However, photobleaching is not always predictive of the PDT response. Rezzoug et al. (1998) noted higher photodynamic activity of Foscan in tumor-bearing mice at lower irradiance, while the photobleaching was the same for both low and high fluence rates. Using fluorescence imaging *in vivo*, we have demonstrated a weak correlation ($r = 0.386$) between Photodithazine photobleaching and tumor growth inhibition in mice (Sirotkina et al., 2017).

Nevertheless, *in vivo* measurement of photosensitizer photobleaching has proven useful in preclinical settings, as it indicates, at least, that a photodynamic reaction has occurred. This helps in evaluating the photodynamic properties of new agents and in selecting the optimal light dose needed to achieve the maximal therapeutic effect. For example, we performed preliminary assessments on the photodynamic activity of two novel Gd(III) cation-porphyrazine chelates by the detection of their *in vivo* photobleaching in mouse tumors (Figure 21.2). It was shown that one of the porphyrazine complexes displayed no significant photobleaching or influence on tumor growth, while PDT with the other compound resulted in a ~20% decrease in fluorescence intensity and a moderate inhibition of tumor growth (Yuzhakova et al., 2017). In a recent study of a new formulation of chlorin e6-polyvinyl alcohol (Ce6-PVA) on a mouse tumor model, we observed its greater photobleaching rate in comparison with Ce6 alone and with Photodithazine, and that this correlated with fast, reproducible tumor regression and more advanced necrosis after PDT (Gavrina et al., 2018).

In our studies, we assessed the *in vivo* phototoxicity of the first genetically encoded photosensitizer, KillerRed, and used fluorescence imaging to measure its photobleaching after PDT (Shirmanova et al., 2013). KillerRed is a red fluorescent protein that can be expressed in a chosen compartment of genetically transduced cells to induce highly targeted photodamage. It was found that irradiation with light of the KillerRed-expressing tumors in mice led to a decrease in the fluorescence signal (Figure 21.3) and caused significant destruction of the cancer cells. In a separate study, we compared continuous and pulsed laser regimens in an attempt to achieve a better antitumor effect with KillerRed and showed that the pulsed mode provided a higher rate of photobleaching and induced more pronounced histopathological changes and inhibition of tumor growth (Shirmanova et al., 2015). Possible explanations for this are that the higher light dose was delivered to the tissue in the pulsed mode but without any temperature increase on

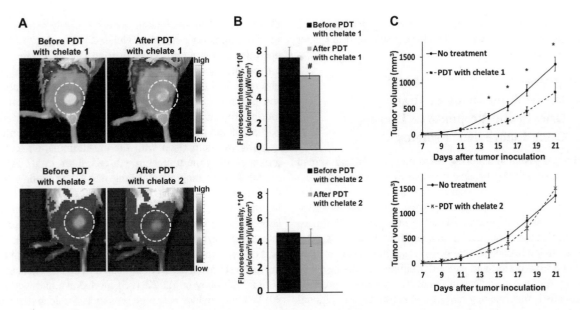

FIGURE 21.2 Photobleaching of Gd(III) cation-porphyrazine chelates in CT26 tumor. A. *In vivo* fluorescence images of CT26 tumors before and after laser irradiation. The chelates 1 and 2 were injected into mice intravenously 3 hours prior to PDT. Fluorescence images were captured on an IVIS-Spectrum system (Caliper Life Sciences, USA) using excitation at 605/30 nm and detection at 660/20 nm. Tumors are shown by the white circles. B. Quantification of fluorescence intensity in the tumors is based on the *in vivo* images. Mean ± SD ($n = 3$ tumors). #, $P=0.034$ from the control before PDT. C. Tumor growth curves for the PDT treated and untreated groups. *, $P<0.027$ from the untreated control.

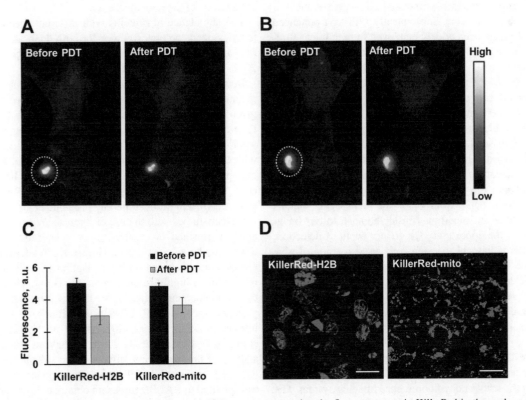

FIGURE 21.3 *In vivo* fluorescence imaging of mice with HeLa tumors expressing the fluorescent protein KillerRed in the nucleus A. or in the mitochondria B. Fluorescence whole-body images before and immediately after irradiation at 593 nm, 85 mW/cm², and at 30 min are shown. Tumors are indicated by the yellow circles. Imaging was performed using a home-built back-reflectance imaging system (Institute of Applied Physics RAS, Russia), excitation at 585 nm, detection at 645–730 nm. C. Quantification of fluorescence intensity in the tumors based on the *in vivo* images. D. *Ex vivo* verification of the KillerRed expression in tumor cells by fluorescence laser scanning microscopy on an LSM 510 (Carl Zeiss, Germany), excitation at 543 nm, detection at 597–661 nm. Scale bar is 20 μm.

the tumor surface, and that the high-energy pulse initiated a more effective photochemical reaction, resulting in a higher photobleaching rate.

21.4 *In Vivo* Monitoring of Tumor Growth and Response to Therapy Using Fluorescent Proteins

Continuous visual monitoring of tumor growth and spread within intact animals became possible by the use of fluorescent proteins of the green fluorescent protein (GFP) family. The fluorescent protein encoding gene can be easily introduced into cultured cancer cells using transgenic techniques, meaning that the tumor model obtained from these cells, after transplantation to rodents, stably expresses the protein over an indefinite time period and can be visualized by fluorescence imaging. A major advantage of fluorescent protein-expressing tumor cells is that imaging requires no substrate or additional contrast agents and therefore, is uniquely suited for *in vivo* studies (Hoffman, 2002; Kaijzel et al., 2007).

The pioneering use of GFP for *in vivo* whole-body fluorescence imaging of tumors in mice was reported in 2000 by Yang et al. (2000). They established stable GFP-expressing mouse melanoma and human colon cancer cells and demonstrated the possibility of fluorescent visualization of primary tumors and metastatic lesions, together with quantitative measurement of tumor growth in real time. Subsequently, GFP and enhanced GFP (EGFP) have been widely employed in preclinical studies for the *in vivo* monitoring of solid tumors (Yamaoka et al., 2010; Castano et al., 2006; Hoffman, 2005) and metastases (Hoffman, 2014; Hoffman, 2002; Yamamoto et al., 2011).

The spectral properties of currently available GFP-like proteins and their mutants cover nearly the entire visible spectrum from blue to far-red. However, the far-red and NIR ranges are preferable for live-animal imaging because tissue autofluorescence and light absorption in those ranges are lower than in the blue/green region, and, therefore, the red light can pass deeper into the tissues (Chudakov et al., 2010; Luker et al., 2015). The depth from which the fluorescence signals are detectable is an important consideration in tumor imaging. Owing to the opportunity for greater depth of detection, far-red proteins have been extensively used for noninvasive fluorescence imaging of deeply located tumors in live animals. For example, Bouvet et al. (2005) developed an RFP-expressing orthotopic model of human pancreatic cancer and demonstrated its whole-body fluorescence visualization in nude mice showing strong correlations of the fluorescent areas and tumor volumes to those measured by magnetic resonance imaging (MRI). Yu et al. (2009) showed that micrometastases of RFP-labeled pancreatic cancer could be detected in early stage using the whole-body fluorescence imaging system. The noninvasive optical tracking of cancer cells expressing the red fluorescent protein tdTomato in a model of human metastatic breast cancer in SCID mice was shown by Winnard et al. (2006). 3D image reconstructions that indicated the detection of tdTomato fluorescence extended to approximately 1 cm below the surface. In a study by Yamaoka et al., (2010) mCherry-expressing peritoneally disseminated and orthotopic

pleural mesothelioma models were sensitively detected using *in vivo* fluorescence imaging. Nunez-Cruz et al. (2010) developed an orthotopic model of ovarian cancer, expressing the far-red fluorescent protein Katushka, which permitted *in vivo* observation of the growth for at least 3 weeks using a noninvasive procedure. In a previous study, we established orthotopic murine colon carcinoma expressing the red fluorescent protein KillerRed and demonstrated the efficacy of transillumination imaging in comparison with the epi-illumination technique for monitoring the tumor growth (Kleshnin et al., 2015). In subcutaneous xenograft and lung metastasis models in rats, Christensen et al. (2015) detected breast cancer cells expressing the far-red fluorescence protein E2-Crimson by means of fluorescence tomography and showed that the fluorescence intensities correlated with tumor volume and weight. To the best of our knowledge, currently, the fluorescent protein with the furthest far-red (excitation > 650 nm) and useful brightness is iRFP (Filonov et al., 2011; Shcherbakova and Verkhusha, 2013). This provides very high sensitivity for deep tissue imaging (Lai et al., 2016).

An example of fluorescent visualization of lung metastasis of human breast adenocarcinoma stably expressing the red fluorescent protein Turbo FP650 is presented in Figure 21.4. Metastasis development in nude mice was monitored over 8 weeks by whole-body epi-fluorescence imaging *in vivo*, and confirmed by fluorescence imaging *ex vivo,* and by histological analysis of the lung thereafter.

The abundance of proteins with different spectral characteristics opens up opportunities for both dual- and multicolor *in vivo* whole-body fluorescence imaging of cancer (Hoffman, 2014; Yang et al., 2004; Tran Cao et al., 2009).

Because fluorescence imaging allows for reliable and robust quantification of the tumor growth, special attention can be given to monitoring tumor responses to anticancer therapy. Real-time evaluation of therapeutic efficacy using noninvasive whole-body imaging and tumor models genetically labeled with fluorescent proteins has been demonstrated for chemotherapy (Katz et al., 2003), PDT (Castano et al., 2006; Mallidi et al., 2015), targeted therapy (Zdobnova et al., 2015; Kimura et al., 2010), and immunotherapy (Leblond et al., 2010).

Cancer studies with the use of fluorescent proteins are generally performed on human tumor xenografts transplanted into immunodeficient mice. However, the development of tumor therapies based on the activation of antitumor immunity requires highly immunogenic tumor models growing in the immunocompetent animals. It is known that fluorescent proteins, such as GFP, EGFP, and KillerRed, possess such immunogenic properties (Stripecke et al., 1999; Steinbauer et al., 2003; Castano et al., 2006; Yuzhakova et al., 2015). In a study of the therapeutic effects of the soluble OX40 ligand (OX40L), we used EGFP-expressing murine colon cancer transplanted to BALB/c mice and obtained an effective tumor rejection and the development of immunological memory in the treated animals (Serebrovskaya et al., 2016). In our tumor model, the OX40L was genetically encoded in the cancer cells and coexpressed with EGFP; therefore, fluorescence imaging of the tumors enabled not only monitoring of the tumor regression but also the intravital control of OX40L production. Figure 21.5 illustrates the regression of a tumor coexpressing

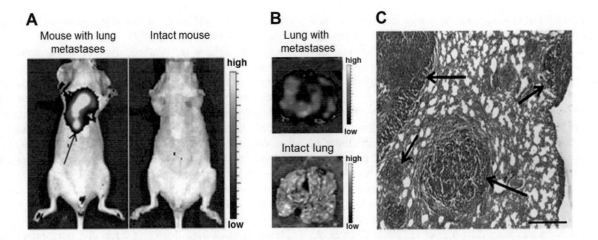

FIGURE 21.4 Fluorescence imaging of metastases in mice. A. *In vivo* fluorescence images of a nude mouse with metastases in the lungs and of an intact mouse. The metastases were induced by intravenous injection of human breast cancer cells MDA-MB-231 expressing the red fluorescent protein Turbo FP650. Fluorescence images were captured on an IVIS-Spectrum system (Caliper Life Sciences, USA) using excitation at 605/30 nm and detection at 620/20 nm over the 8 weeks postinjection of the cancer cells. B. Verification of fluorescent metastases in the excised lungs. C. H&E histology of lung tissue with metastases (shown by arrows). Scale bar is 40 μm.

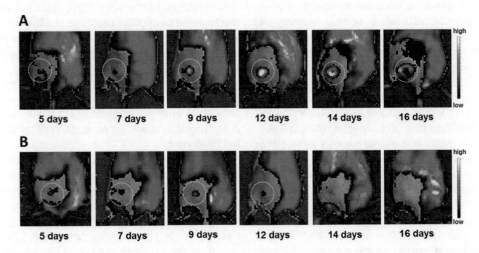

FIGURE 21.5 *In vivo* fluorescence imaging of a mouse with CT26 tumor expressing EGFP A. or coexpressing EGFP and OX40L B. Tumor is shown by the yellow circle. Days after tumor inoculation are indicated under the images. The skin around the tumor was shaved. Fluorescence images were captured on an IVIS-Spectrum system (Caliper Life Sciences, USA) using excitation at 462 nm and detection at 520 nm.

EGFP and OX40L and the progression of tumors expressing EGFP alone.

It is apparent that tumor models genetically labeled with fluorescent proteins, in combination with fluorescence imaging technologies, represent a valuable instrument for *in vivo* investigations of tumor progression and for the evaluation of novel therapeutic agents in small animals.

21.5 Functional Fluorescence Imaging of Tumors *In Vivo*

The development of high-sensitivity fluorescent imaging techniques and fluorescent sensors offers unique opportunities to study different functional processes in cancer cells *in vivo* at the whole-body level.

The sensor may be represented by a fluorescent protein (or a construct of two proteins) genetically introduced into the cells, or by a chemical fluorescent dye that is administered to the tumor tissue exogenously. Genetically encoded indicators have the advantages of their stable expression in any particular compartment within the cells, and this opens up possibilities for continuous monitoring of the cellular activities in tumors. The shortcomings of chemical sensors include problems with their delivery to the cancer cells, their redistribution of the dyes and leakage from the cells, their interactions with other molecules, and their own toxicity. Nevertheless, the use of both types of sensors is realistic for *in vivo* applications in cancer biology.

Optical detection of tumor-associated proteases (enzymes with the common ability to hydrolyze peptide bonds) has been demonstrated, including the activity of cathepsins (Drake et al., 2011; Blum et al., 2009), matrix metalloproteinases (Lee et al., 2012; Shimizu et al., 2014; Zhu et al., 2011) and caspases (Edgington et al., 2009; Schellenberger et al., 2003; Goryashchenko et al., 2015; Zherdeva et al., 2018). Sensing is generally achieved by the separation of a fluorophore-quencher pair or a FRET (Förster resonance energy transfer) pair on a sensor resulting from the proteolytic activity, which potentiates an increase in fluorescence of the probe. Because proteases play pivotal roles in tumor progression, their *in vivo* noninvasive imaging is of great importance to improve cancer diagnosis and therapy.

Targeting of fluorescently labeled ligands to tumor-specific receptors allows for *in vivo* fluorescence visualization of molecular processes in cancer models. The feasibility of this approach to enable quantitative imaging of the interaction of receptor-ligand complexes in tumors has been shown, for example, for vascular endothelial growth factor (VEGF) (Fukumura et al., 1998; Zhang et al., 2004), human epidermal growth factor 2 (HER2) (Ardeshirpour et al., 2012), cell adhesion molecules (integrins) (Cai et al., 2006; Chen et al., 2004; Hsu et al., 2006), and the iron-binding glycoprotein, transferrin (Rudkouskaya et al., 2017; Abe et al., 2013; Zhao et al., 2014).

Dysregulated pH, with increased intracellular and decreased extracellular values compared with normal tissues is recognized as a hallmark of solid tumors and an important determinant of carcinogenesis. Therefore, *in vivo* imaging and measurement of tumor pH is of great interest. In general, a map of pH can be obtained from the ratio of the fluorescence signals of the pH probe at different excitation or emission wavelengths. In contrast to intensiometric registration, ratiometric measurements are relatively independent of such factors as optical path length, focusing, excitation intensities, emission collection efficiencies, probe concentration, and photobleaching, so this makes ratiometric probes a more robust class of fluorescent probes for *in vivo* pH mapping. Evaluating extracellular pH in tumors has been performed with the use of synthetic pH-sensitive fluorescence dyes, such as 5,6-CF (Mordon et al., 1994), SNARF-1 (Gatenby et al., 2006; Lin et al., 2012; Estrella et al., 2013; Leung et al., 2013), or with nanoparticles combining pH-sensitive and pH-insensitive dyes (Tsai et al., 2013). Ratiometric fluorescence probes have been designed for simultaneous visualization of protease activities

and sensing the pH of the tumor microenvironment *in vivo* (Hou et al., 2015; Ma et al., 2017).

Although some of the synthetic dyes are able to enter cells *in vitro*, their use for the mapping of intracellular pH in solid tumors remains problematic. For this, genetically encoded pH-probes based on GFPs are more appropriate. Matlashov et al. (2015) have reported on a pH-sensitive ratiometric indicator, SypHer2, based on the cpYFP fluorophore, for intracellular pH measurements. Using the novel pH-sensor, SypHer2, we have demonstrated, for the first time, the possibility of ratiometric (dual excitation) imaging of cytoplasmic pH in tumors *in vivo*. Tumor xenografts were generated of HeLa cells stably expressing SypHer2 in the cytosol and observed *in vivo* using whole-body fluorescence imaging via opening a reversible skin flap (Figure 21.6). A high degree of inter- and intratumoral heterogeneity of the pH was detected in HeLa tumors in mice. Comparison of the fluorescence ratio map with pathomorphology and hypoxia staining in the tumors revealed a correspondence of the zones with higher cytoplasmic pH to the necrotic and hypoxic areas (Shirmanova et al., 2015). Furthermore, we looked into the relationships between the intracellular pH and the response of cervical cancer to cisplatin, a cytotoxic drug widely used in clinical chemotherapy, and detected a lowering of the cytoplasmic pH in treatment-responsive tumors *in vivo* (Shirmanova et al., 2017).

Introduction of FLIM into whole-body scale visualization expands the possibilities for functional characterization of tumors. *In vivo* macroscopic FLIM has been exploited as a direct tool to evaluate, for example, cells' metabolism or drug internalization.

Probing of the metabolic status of cells can be performed by measuring fluorescence from the endogenous metabolic cofactors reduced nicotinamide adenine dinucleotide (NADH) and oxidized flavin adenine dinucleotide (FAD) (Heikal, 2010; Georgakoudi et al., 2012). Metabolic changes resulting in changes of the concentration and/or state (free or protein-bound) of the cofactors can be readily detected by the ratio of intensities of the NADH and FAD emissions (redox-ratio)

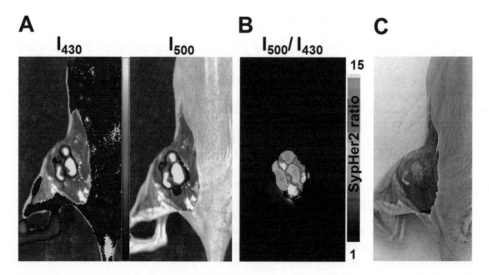

FIGURE 21.6 *In vivo* intracellular pH mapping of HeLa tumors expressing the genetically encoded pH-sensor, SypHer2. A. Fluorescence images with excitation at 430 nm and 500 nm (detection at 540 nm); B. images of SypHer2 ratio (I500/I430) from tumor *in vivo*; C photograph of the corresponding tumor. The data were recorded on Day 14 of tumor growth. The HeLa tumor expressing SypHer2 was imaged with the skin flap opened.

and their fluorescence lifetimes, respectively. A prevalence of glycolysis over oxidative glucose metabolism, even under normal oxygenation, is the best characterized metabolic feature of cancer cells, and this is manifested as a decreased redox-ratio, FAD/NADH, and an increased contribution of the short lifetime component of NADH. Typically, metabolic imaging, including *in vivo* studies on animal tumors, is implemented at a cellular level by two-photon fluorescence microscopy (Skala et al., 2007; Shah et al., 2015; Shirmanova et al., 2017). Our recent experiments show the possibility for interrogation of tumor metabolism on the macroscale using a novel confocal one-photon macro-FLIM system (Shcheslavskiy et al., 2018). Using this system, we have obtained a picture of metabolic heterogeneity in subcutaneously transplanted mouse colorectal tumors *in vivo*.

In a recent study by Rudkouskaya et al. (2018) whole-body macroscopic FLIM/FRET imaging was used to quantify intracellular transferrin receptor-ligand binding in live animals bearing tumor xenografts, which shows the prospects of time-resolved technique for guiding the development of targeted drug therapy.

Therefore, using fluorescence-based whole-body imaging, various physiological and physicochemical parameters can be interrogated in a tumor within a live small animal model. The studies of molecular and metabolic processes in cancer are crucial for our understanding of the mechanisms of tumor development and responses to treatment.

21.6 Conclusions

In summary, the examples provided here demonstrate that imaging of fluorescence at the whole-body level of small animals has been a highly successful approach in cancer studies. Because it is nondestructive and can be performed over long periods, it offers an ideal solution to visualize drug distributions throughout the mouse body and to assess drug efficacy without the need to sacrifice the animal. While the majority of studies have used fluorescence intensity mapping, recent achievements in time-resolved techniques open up the possibility of assessing the functional characteristics of tumors, although this has yet to be realized at the macro-level. In general, the development of whole-body optical imaging has led to a paradigm shift in how cancer can be studied, especially taking into account its complex biological behavior.

Acknowledgments

This work was financially supported by the Russian Foundation for Basic Research (project # 18-29-01022) and the Ministry of Education and Science of the Russian Federation (grant 14.W03.31.0005). The authors are grateful to Dr. Marina A. Sirotkina (NNSMA), Dr. Ludmila B. Snopova (NNSMA), Dr. Varvara V. Dudenkova (NNSMA), Irina N. Druzhkova (NNSMA), Anastasia Shumilova (NNSMA), Dr. Vladislav I. Shcheslavskiy (Becker & Hickl GmbH), Dr. Michail Kleshnin (IAP RAS), Vladimir I. Plekhanov (IAP RAS), Dr. Irina V. Balalaeva (NNSU), Prof. Konstantin A. Lukyanov (IBC RAS), Prof. Vsevolod V. Belousov (IBC RAS), Dr. Ekaterina O. Serebrovskaya (IBC RAS), Dr. Larisa G. Klapshina (IMC RAS), Dr. Anna B. Solovieva (ICP RAS), Dr. Nadeshda A. Aksenova (ICP RAS), and many others who made important contributions to the experiments mentioned in this chapter.

REFERENCES

Abe, K., Zhao, L., Periasamy, A., Intes, X., and Barroso, M. (2013). Non-invasive in vivo imaging of near infrared-labeled transferrin in breast cancer cells and tumors using fluorescence lifetime FRET. *PLoS One, 8*: e80269.

Agostinis, P. et al. (2011). Photodynamic therapy of cancer: An update. *CA Cancer J Clin., 61*(4): 250–281.

Ahn, M. Y., Yoon, H. E., Moon, S. Y., Kim, Y. C., and Yoon, J. H. (2017). Intratumoral photodynamic therapy with newly synthesized pheophorbide a in murine oral cancer. *Oncol Res., 25*(2): 295–304.

Anbil, S., Rizvi, I., Celli, J. P., Alagic, N., and Hasan, T. (2012). A photobleaching-based PDT dose metric predicts PDT efficacy over certain BPD concentration ranges in a three-dimensional model of ovarian cancer. *Proc SPIE, 8568*: 85680S.

Ardeshirpour, Y. et al. (2012). In vivo fluorescence lifetime imaging monitors binding of specific probes to cancer biomarkers. *PLoS ONE, 7*(2): e31881.

Ascencio, M., Collinet, P., Farine, M. O., and Mordon, S. (2008). In vivo PpIX fluorescence photobleaching is useful to predict the tissue response to HAL-PDT. *Lesers Surg Med, 40*(5): 332–341.

Bechet, D., Frochot, C., Vanderesse, R., and Barberi-Heyob, M. (2012). Innovations of photodynamic therapy for brain tumors: Potential of multifunctional nanoparticles. *J Carcinogene Mutagene, S8*: 001.

Becker, W. (2005). *Advanced Time-Correlated Single-Photon Counting Techniques.* New York: Springer.

Blum, G., Weimer, R. M., Edgington, L. E., Adams, W., and Bogyo, M. (2009). Comparative assessment of substrates and activity based probes as tools for non-invasive optical imaging of cysteine protease activity. *PLoS One, 4*(7): e6374

Bouvet, M. et al. (2005). High correlation of whole-body red fluorescent protein imaging and magnetic resonance imaging on an orthotopic model of pancreatic cancer. *Cancer Res., 65*(21): 9829–9833.

Cai, W. et al. (2006). Peptide-labeled near-infrared quantum dots for imaging tumor vasculature in living subjects. *Nano Lett., 6*: 669–676.

Castano, A. P., Liu, Q., and Hamblin, M. R. (2006). A green fluorescent protein-expressing murine tumour but not its wild-type counterpart is cured by photodynamic therapy. *Br J Cancer, 94*(3): 391–397.

Chaudhari, A. J. et al. (2009). Excitation spectroscopy in multispectral optical fluorescence tomography: Methodology, feasibility, and computer simulation studies. *Med Biol., 54*(15): 4687–4704.

Chen, X. et al. (2004). In vivo near-infrared fluorescence imaging of integrin avh3 in brain tumor xenografts. *Cancer Res., 64*: 8009–8014.

Christensen, J., Vonwil, D., and Shastri, V. P. (2015). Non-invasive *in vivo* imaging and quantification of tumor growth and metastasis in rats using cells expressing far-red fluorescence protein. *PLoS One, 10*(7): e0132725.

Chudakov, D. M., Matz, M. V., Lukyanov, S., and Lukyanov, K. A. (2010). Fluorescent proteins ad their applications in imaging living cells and tissues. *Physiol Rev.*, *90*(3): 1103–1163.

Cubeddu, R., Canti, G., Pifferi, A., Taroni, P., and Valentini, G. (1997). Fluorescence lifetime imaging of experimental tumors in hematoporphyrin derivative-sensitized mice. *Photochem Photobiol.*, *66*(2): 229–236.

Darne, C., Lu, Y., and Sevick-Muraca, E. M. (2014). Small animal fluorescence and bioluminescence tomography: A review of approaches, algorithms and technology update. *Phys Med Biol.*, *59*(1): R1–R64.

Ding, H., and Wu, F. et al. (2012). Image guided biodistribution and pharmacokinetic studies of theranostics. *Theranostics*, *2*(11): 1040–1053.

Dolmans, D. E., Fukumura, D., and Jain, R. K. (2003) Photodynamic therapy for cancer. *Nat Rev Cancer*, *3*: 380–387.

Drake, C. R., Miller, D. C., and Jones, E. F. (2011). Activatable optical probes for the detection of enzymes. *Curr Org Synth.*, *8*: 498–520.

Dysart, J. S., Singh, G., and Patterson, M. S. (2005). Calculation of singlet oxygen dose from photosensitizer fluorescence and photobleaching during mTHPC photodynamic therapy of MLL cells. *Photochem Photobiol.*, *81*: 196–205.

Edgington, L. E. et al. (2009). Non-invasive optical imaging of apoptosis using caspase-targeted activity based probes. *Nat Med.*, *15*(8): 967–973.

Estrella, V. et al. (2013). Acidity generated by the tumor microenvironment drives local invasion. *Cancer Res.*, *73*(5): 1524–1535.

Fei-Peng, Z. et al. (2016). Dual-modality imaging probes with high magnetic relaxivity and near-infrared fluorescence based highly aminated mesoporous silica nanoparticles. *J Nanomat*, *2016*: 1–9.

Filonov, G. S. et al. (2011). Bright and stable near-infrared fluorescent protein for in vivo imaging. *Nat Biotechnol.*, *29*: 757–761.

Fukumura, D. et al. (1998). Tumor induction of VEGF promoter activity in stromal cells. *Cell*, *94*: 715–725.

Gamayunov, S. et al. (2016). Fluorescence imaging for photodynamic therapy of non-melanoma skin malignancies – A retrospective clinical study. *Photonics Lasers Med*, *5*(2): 101–111.

Gambotto, A. et al. (2000). Immunogenicity of enhanced green fluorescent protein (EGFP) in BALB/c mice: Identification of an H2-Kdrestricted CTL epitope. *Gene Ther.*, *7*(23): 2036–2040.

Gatenby, R. A., Gawlinski, E. T., Gmitro, A. F., Kaylor, B., and Gillies, R. J. (2006). Acid-mediated tumor invasion: A multidisciplinary study. *Cancer Res.*, *66*(10): 5216–5223.

Gavrina, A. I. et al. (2018). Photodynamic therapy of mouse tumor model using chlorin e6- polyvinyl alcohol complex. *J Photochem Photobiol B*, *178*: 614–622.

Georgakoudi, I. et al. (2012). Optical imaging using endogenous contrast to assess metabolic state. *Annu Rev Biomed Eng.*, *14*: 351–367.

Goryashchenko, A. S. et al. (2015). Genetically encoded FRET-sensor based on terbium chelate and red fluorescent protein for detection of caspase-3 activity. *Int J Mol Sci.*, *16*(7): 16642–16654.

Harada, M., Aizawa, K., Okunaka, T., and Kato, H. (1998). In vivo fluorescence kinetics of mono-l-aspartyl chlorin e6 (NPe6) and influence of angiogenesis in fibrosarcoma-bearing mice. *Int J Clin Oncol.*, *3*(4): 209–215.

Harrison, V. S., Carney, C. E., MacRenaris, K. W., Waters, E. A., and Meade, T. J. (2015). Multimeric near IR–MR contrast agent for multimodal *in vivo* imaging. *J Am Chem Soc.*, *137*(28): 9108–9116.

Heikal, A. A. (2010). Intracellular coenzymes as natural biomarkers for metabolic activities and mitochondrial anomalies. *Biomark Med.*, *4*(2): 241–63.

Hillyar, C. R., Knight, J. C., Vallis, K. A., and Cornelissen, B. (2015). PET and SPECT Imaging for the acceleration of anti-cancer drug development. *Curr Drug Targets*, *16*(6): 582–591.

Hoffman, R. M. (2002). In vivo imaging of metastatic cancer with fluorescent proteins. *Cell Death Differ.*, *9*: 786–789.

Hoffman, R. M. (2005). The multiple uses of fluorescent proteins to visualize cancer in vivo. *Nat Rev Cancer*, *5*(10): 796–806.

Hoffman, R. M. (2014). Imaging metastatic cell trafficking at the cellular level in vivo with fluorescent proteins. *Methods Mol Biol.*, *1070*: 171–179.

Hoffman, R. M. (2015). Application of GFP imaging in cancer. *Lab Investig.*, *95*(4): 432–452.

Hou, Y. et al. (2015). Protease-activated ratiometric fluorescent probe for pH mapping of malignant tumors. *ACS Nano*, *9*(3): 3199–3205.

Hsu, A. R. et al. (2006). In vivo near-infrared fluorescence imaging of integrin a(v)h(3) in an orthotopic glioblastoma model. *Mol Imaging Biol.*, *8*: 315–323.

Huang, K., Chen, L., Lv, S., and Xiong, J. (2012). Protoporphyrin IX photobleaching of subcellular distributed sites of leukemic HL60 cells based on ALA-PDT in vitro. *Biomed Sci Eng.*, *5*: 548–555.

Jarvi, M. T., Patterson, M. S., and Wilson, B. C. (2012). Insights into photodynamic therapy dosimetry: Simultaneous singlet oxygen luminescence and photosensitizer photobleaching measurements. *Biophys J.*, *102*(3): 661–671.

Jiang, S., Gnanasammandhan, M. K., and Zhang, Y. (2010). Optical imaging-guided cancer therapy with fluorescent nanoparticles. *J R Soc Interface*, *7*(42): 3–18.

Kaijzel, E. L., van der Pluijm, G., and Löwik, C. W. (2007). Whole-body optical imaging in animal models to assess cancer development and progression. *Clin Cancer Res.*, *13*(12): 3490–3497.

Kanno, S., Utsunomiya, K., Kono, Y., Tanigawa, N., and Sawada, S. (2015). The effect of radiation exposure on multidrug resistance: In vitro and in vivo studies using non-small lung cancer cells. *EJNMMI Res*, *5*: 11.

Katz, M. H. et al. (2003). A novel red fluorescent protein orthotopic pancreatic cancer model for the preclinical evaluation of chemotherapeutics. *J Surg Res.*, *113*(1): 151–160.

Kim, D., Gao, Z. G., Lee, E. S., and Bae, Y. H. (2009). In vivo evaluation of doxorubicin-loaded polymeric micelles targeting folate receptors and early endosomal pH in drug-resistant ovarian cancer. *Mol Pharm.*, *6*(5): 1353–1362.

Khurana, M. et al. (2012). Biodistribution and pharmacokinetic studies of a porphyrin dimer photosensitizer (Oxdime) by fluorescence imaging and spectroscopy in mice bearing xenograft tumors. *Photochem Photobiol.*, *88*(6): 1531–1538.

Kleshnin, M. S. et al. (2013). Fluorescence diffuse tomography technique with autofluorescence removal based on dispersion of biotissue optical properties. *Laser Phys Lett.*, *10*(7): 075601.

Kleshnin, M. S. et al. (2015). Trans-illumination fluorescence imaging of deep-seated tumors in small animals. *Photon Lasers Med.*, *4*(1): 85–92.

Klapshina, L. G. et al. (2010). Novel PEG-organized biocompatible fluorescent nanoparticles doped with an ytterbium cyanoporphyrazine complex for biophotonic applications. *Chem Commun.*, *46*: 8398–8400.

Kimura, H. et al. (2010). Targeted therapy of spinal cord glioma with a genetically-modified Salmonella typhimurium. *Cell Proliferation*, *43*(1): 41–48.

Lai, C. W. et al. (2016). Using dual fluorescence reporting genes to establish an in vivo imaging model of orthotopic lung adenocarcinoma in mice. *Mol Imaging Biol.*, *18*(6): 849–859.

Lin, Y., Wu, T.-Y., and Gmitro, A. F. (2012). Error analysis of ratiometric imaging of extracellular pH in a window chamber model. *J Biomed Opt.*, *17*(4): 046004.

Li, Y., Wang, J., Zhang, X., Guo, W., Li, F., Yu, M., Kong, X., Wu, W., and Hong, Z. (2015). Highly water-soluble and tumor-targeted photosensitizers for photodynamic therapy. *Org Biomol Chem.*, *13*(28): 7681–7694.

Leblond, F., Davis, S. C., Valdés, P. A., and Pogue, B. W. (2010). Preclinical whole-body fluorescence imaging: Review of instruments, methods and applications. *J Photochem Photobiol B*, *98*(1): 77–94.

Lee, C. M. et al. (2012). Optical imaging of MMP expression and cancer progression in an inflammation-induced colon cancer model. *Int J Cancer*, *131*(8): 1846–1853.

Leung, H. M., Schafer, R., Pagel, M. M., Robey, I. F., and Gmitro, A. F. (2013). Multimodality pH imaging in a mouse dorsal skin fold window chamber model. *Proc SPIE Int Soc Opt Eng.*, *8574*: 85740L.

Luker, K. E. et al. (2015). Comparative study reveals better far-red fluorescent protein for whole body imaging. *Sci Rep.*, *5*: 10332.

Mallidi, S., Spring, B. Q., and Hasan, T. (2015). Optical imaging, photodynamic therapy and optically-triggered combination treatments. *Cancer J.*, *21*(3): 194–205.

Major, A. L. et al. (1997). In vivo fluorescence detection of ovarian cancer in the NuTu-19 epithelial ovarian cancer animal model using 5-aminolevulinic acid (ALA). *Gynecol Oncol.*, *66*(1): 122–132.

Matlashov, M. E. et al. (2015). Fluorescent ratiometric pH indicator SypHer2: Applications in neuroscience and regenerative biology. *Biochim Biophys Acta*, *1850*(11): 2318–2328.

Ma, T. et al. (2017). Dual-ratiometric target-triggered fluorescent probe for simultaneous quantitative visualization of tumor microenvironment protease activity and pH *in vivo*. *J Am Chem Soc.*, *140*(1): 211–218.

Mordon, S., Devoisselle, M., and Maunoury, V. (1994). In vivo pH measurement and imaging of tumor tissue using a pH-sensitive fluorescent probe (5,6-carboxyfluorescein): Instrumental and experimental studies. *Photochem Photobiol.*, *60*(3): 274–279.

Motlagh, N. S., Parvin, P., Ghasemi, F., and Atyabi, F. (2016), Fluorescence properties of several chemotherapy drugs: Doxorubicin, paclitaxel and bleomycin. *Biomed Opt Express*, *7*(6): 2400–2406.

Mo, W., Rohrbach, D., and Sunar, U. (2012). Imaging a photodynamic therapy photosensitizer in vivo with a time-gated fluorescence tomography system. *J Biomed Opt.*, *17*(7): 071306.

Nam, T. et al. (2010). Tumor targeting chitosan nanoparticles for dual-modality optical/MR cancer imaging. *Bioconjug Chem.*, *21*(4): 578–582.

Nothdurft, R. E. et al. (2009). In vivo fluorescence lifetime tomography. *J Biomed Opt.*, *14*(2): 024004.

Ntziachristos, V. (2005). Looking and listening to light: The evolution of whole-body photonic imaging. *Nat Biotechnol.*, *23*: 313–320.

Nunez-Cruz, S., Connolly, D. C., and Scholler, N. (2010). An orthotopic model of serous ovarian cancer in immunocompetent mice for in vivo tumor imaging and monitoring of tumor immune responses. *J Visual Exp.*, (*45*): 2146.

Ozturk, M. S., Rohrbach, D., Sunar, U., and Intes, X. (2014). Mesoscopic fluorescence tomography of a photosensitizer (HPPH) 3D biodistribution in skin cancer. *Acad Radiol.*, *21*(2): 271–280.

Patel, N. et al. (2016). Highly effective dual-function near-infrared (NIR) photosensitizer for fluorescence imaging and photodynamic therapy (PDT) of cancer. *J Med Chem.*, *59*(21): 9774–9787.

Pogue, B. W., Gibbs, S. L., Chen, B., and Savellano, M. (2004). Fluorescence imaging in vivo: Raster scanned point-source imaging provides more accurate quantification than broad beam geometries. *Technol Cancer Res Treat.*, *3*(1): 15–21.

Rezzoug, H., Bezdetnaya, L., A'amar, O., Merlin, L., and Guillemin, F. (1998), Parameters affecting photodynamic activity of foscan® or meta-tetra(hydroxyphenyl)chlorin (mTHPC) *in vitro* and *in vivo*. *Lasers Med Sci.*, *13*:119–125.

Rudkouskaya, A., Sinsuebphon, N., Ward, J., Tubbesing, K., Intes, X., and Barroso, M. (2018). Quantitative imaging of receptor-ligand engagement in intact live animals. *J Controlled Release*, *286*: 451–459.

Rudkouskaya, A., Sinsuebphon, N., Intes, X., Mazurkiewicz, J. E., and Barroso, M. (2017). Fluorescence lifetime FRET imaging of receptor-ligand complexes in tumor cells in vitro and in vivo. *Proc SPIE 10069, Multiphoton Microscopy in the Biomedical Sciences XVII*, 1006917.

Sailor, M. J., and Park, J.-H. (2012). Hybrid nanoparticles for detection and treatment of cancer. *Adv Mat.*, *24*(28): 3779–3802.

Schellenberger, E. A. et al. (2003). Optical imaging of apoptosis as a biomarker of tumor response to chemotherapy. *Neoplasia*, *5*(3): 187–192.

Serebrovskaya, E. O. et al. (2016). Soluble OX40L favors tumor rejection in CT26 colon carcinoma model. *Cytokine*, *84*: 10–16.

Shah, A. T., Diggins, K. E., Walsh, A. J., Irish, J. M., and Skala, M. C. (2015). In vivo autofluorescence imaging of tumor heterogeneity in response to treatment. *Neoplasia*, *17*: 862–870.

Shcherbakova, D. M., and Verkhusha, V. V. (2013). Near-infrared fluorescent proteins for multicolor in vivo imaging. *Nat Methods*, *10*: 751–754.

Shcheslavskiy, V. et al. (2018). Fluorescence time-resolved macroimaging. *Opt Lett.*, *43*(13): 3152–3155.

Sheng, C., Hoopes, P. J., Hasan, T., and Pogue, B. W. (2007). Photobleaching-based dosimetry predicts deposited dose in ALA-PpIX PDT of rodent esophagus. *Photochem Photobiol.*, *83*(3): 738–748.

Shimizu, Y. et al. (2014). In vivo imaging of membrane type-1 matrix metalloproteinase with a novel activatable near-infrared fluorescence probe. *Cancer Sci.*, 105: 1056–1062.

Shirmanova, M. V. et al. (2010). In vivo study of photosensitizer pharmacokinetics by fluorescence transillumination imaging. *J Biomed Opt.*, *15*(4): 048004.

Shirmanova, M. V. et al. (2011). Design and testing of a new photosensitizer based on an ytterbium porphyrazine complex. *Biophysics*, *56*(6): 1083–1087.

Shirmanova, M. V. et al. (2013). Phototoxic effects of fluorescent protein KillerRed on tumor cells in mice. *J Biophoton*, *6*(3): 283–290.

Shirmanova, M. V. et al. (2014). Comparative study of tissue distribution of chlorin e6 complexes with amphiphilic polymers in mice with cervical carcinoma. *J Anal Bioanal Tech. S*, *1*: 008.

Shirmanova, M. V. et al. (2015). Intracellular pH imaging in cancer cells in vitro and tumors in vivo using the new genetically encoded sensor SypHer2. *Biochim Biophys Acta*, *1850*(9): 1905–1911.

Shirmanova, M. V. et al. (2015). Towards PDT with genetically encoded photosensitizer KillerRed: A comparison of continuous and pulsed laser regimens in an animal tumor model. *PLoS ONE*, *10*(12): e0144617.

Shirmanova, M. V. et al. (2017). Chemotherapy with cisplatin: Insights into intracellular pH and metabolic landscape of cancer cells in vitro and in vivo. *Sci Rep.*, *7*(1): 8911.

Sirotkina, M. A. et al. (2017). Photodynamic therapy monitoring with optical coherence angiography. *Sci Rep.*, 7:41506.

Skala, M. C. et al. (2007). In vivo multiphoton microscopy of NADH and FAD redox states, fluorescence lifetimes, and cellular morphology in precancerous epithelia. *Proc Natl Acad Sci USA.*, *104*(49): 19494–19499.

Steinbauer, M. et al. (2003). GFP-transfected tumor cells are useful in examining early metastasis in vivo, but immune reaction precludes long-term tumor development studies in immunocompetent mice. *Clin Exp Metastasis*, *20*(2): 135–141.

Stripecke, R. et al. (1999). Immune response to green fluorescent protein implications for gene therapy. *Gene Ther.*, *6*(7): 1305–1312.

Tran Cao, H. S. et al. (2009). Development of the transgenic cyan fluorescent protein (CFP)-expressing nude mouse for "Technicolor" cancer imaging. *J Cell Biochem.*, *107*(2): 328–334.

Tsai, Y. T. et al. (2013). Real-time noninvasive monitoring of in vivo inflammatory responses using a pH ratiometric fluorescence imaging probe. *Adv Healthcare Mater.*, *3*(2): 221–229.

Turchin, I. V. (2016). Methods of biomedical optical imaging: From subcellular structures to tissues and organs. *Phys Usp.*, *59*: 487–501.

Vollet-Filho, J. D. et al. (2009), Possibility for a full optical determination of photodynamic therapy outcome. *J Appl Phys.*, *105*:102038.

Wilson, B. C., Patterson, M. S., and Lilge, L. (1997). Implicit and explicit dosimetry in photodynamic therapy: A new paradigm. *Lasers Med Sci.*, *12*: 182–199.

Winnard, P. T. Jr., Kluth, J. B., and Raman, V. (2006). Noninvasive optical tracking of red fluorescent protein-expressing cancer cells in a model of metastatic breast cancer. *Neoplasia*, *8*(10): 796–806.

Wolfbeis, O. S. (2015). An overview of nanoparticles commonly used in fluorescent bioimaging. *Chem Soc Rev.*, *44*(14): 4743–4768.

Yamamoto, N., Tsuchiya, H., and Hoffman, R. M. (2011). Tumor imaging with multicolor fluorescent protein expression. *Int J Clin Oncol.*, *16*(2): 84–91.

Yamaoka, N. et al. (2010). Establishment of in vivo fluorescence imaging in mouse models of malignant mesothelioma. *Int J Oncol.*, *37*(2): 273–279.

Yang, M. et al. (2000). Whole-body optical imaging of green fluorescent protein-expressing tumors and metastases. *Proc Natl Acad Sci USA*, *97*: 1206–1211.

Yang, M., Reynoso, J., Jiang, P., Li, L., Moossa, A. R., and Hoffman, R. M. (2004). Transgenic nude mouse with ubiquitous green fluorescent protein expression as a host for human tumors. *Cancer Res. 64*(23): 8651–8656.

Yuzhakova, D. V. et al. (2015). CT26 murine colon carcinoma expressing the red fluorescent protein KillerRed as a highly immunogenic tumor model. *J Biomed Opt.*, *20*(8): 088002.

Yuzhakova, D. V. et al. (2017). In vivo multimodal tumor imaging and photodynamic therapy with novel theranostic agents based on the porphyrazine framework-chelated gadolinium (III) cation. *Biochim Biophys Acta*, 1861(12): 3120–3130.

Yu, Z. Q. et al. (2009). Establishment of red fluorescent protein orthotopic transplantation nude mice metastasis model of pancreatic cancer and whole-body fluorescent imaging. *Zhonghua Wai Ke Za Zhi*, *47*(14): 1092–1095.

Zdobnova, T. et al. (2015). A novel far-red fluorescent xenograft model of ovarian carcinoma for preclinical evaluation of HER2-targeted immunotoxins. *Oncotarget*, *6* (31): 30919–30928.

Zhang, N., Fang, Z., Contag, P. R., Purchio, A. F., and West, D. B. (2004). Tracking angiogenesis induced by skin wounding and contact hypersensitivity using a Vegfr2-luciferase transgenic mouse. *Blood*, 103: 617–626.

Zhao, L., Abe, K., Rajoria, S., Pian, Q., Barroso, M., and Intes, X. (2014). Spatial light modulator based active wide-field illumination for ex vivo and in vivo quantitative NIR FRET imaging. *Biomed Opt Express*, 5: 944.

Zherdeva, V. et al. (2018). Long-term fluorescence lifetime imaging of a genetically encoded sensor for caspase-3 activity in mouse tumor xenografts. *J Biomed Opt.*, *23*(3): 1–11.

Zhu, L. et al. (2011) In vivo optical imaging of membrane-type matrix metalloproteinase (MT-MMP) *Activity Mol Pharm.*, *8*(6): 2331–2338.

22

Large-Scale Fluorescence Imaging in Neuroscience

Robert Prevedel

CONTENTS

22.1 Introduction

A major challenge in the field of neuroscience is to understand how the orchestrated activity of millions of neurons leads to cognitive function and behavior. This endeavor depends critically on appropriate tools that are capable of recording the activity of all the neurons that comprise a functional network. Yet, it is often technically challenging to sample neuronal activity at large-scale, i.e. over large volumes, with sufficient spatial and temporal resolution that is relevant to neuronal circuit function. This challenge has fueled the development of microscopy technologies that enable recording of neuronal activity across large brain volumes. Optically imaging these activities through calcium-sensitive and genetically encoded reporters such as GCaMP (Akerboom et al. 2013; Chen et al. 2013) seems very promising in this respect, as light microscopy is a noninvasive tool with high spatial resolution, and, in contrast to electrode-based recordings, offers straightforward scalability to comprehensively record from hundreds or even thousands of neurons in parallel. Furthermore, imaging calcium-mediated activity with fluorescent proteins enables targeted recording of genetically defined cell types, while chronic imaging windows allow for long-term monitoring of the same neuronal populations over months. Current calcium indicators are based on modified versions of common fluorescent proteins (such as GFP in the case of GCaMP), which include structural subunits that, when bound to calcium, undergo conformational changes that result in an increase of the absorption cross section and, therefore, fluorescence emission. Hence, the intensity of the calcium-indicator expressing neuron's fluorescence is a direct proxy for its neuronal activity. From such intensity time traces, action potential sequences can often be detected by using deconvolution methods (Yaksi and Friedrich 2006). On the technical imaging side, multiphoton microscopy (MPM) is often the method of choice because of the highly scattering nature of most (mammalian) brain tissues. In this chapter, we will give an introduction to the underlying principles of MPM and discuss their technical realization. Here, we have chosen to pay special consideration to current and emerging trends in multiphoton excitation-based modalities. In this respect, we will provide an overview of the current research in this field, especially in the pursuit of imaging fast, deep, and with multiple colors in the context of neuroscience. Light-sculpting for arbitrary engineering of the point-spread function is an interesting concept in this domain, which we chose to highlight and discuss in more detail, accompanied with coverage of recent experiments that utilized this technique to image and record neuronal activity at high temporal resolution on a large scale in brains of two different model organisms.

22.2 Multiphoton Microscopy

As discussed in Chapter 1, confocal microscopy and related one-photon absorption-based imaging methods achieve three-dimensional (3D) resolution (optical sectioning) by rejecting light that does not seem to originate from the focus. This is important, as most biological tissues are highly scattering, i.e.

cellular inhomogeneities in refractive index cause photons to stray away from their straight, "ballistic" propagation direction. This severely limits the usefulness of these one-photon techniques to thin sections or cultured cells *in vitro* and superficial layers of tissue *in vivo*. MPM on the other hand relies on a nonlinear light-matter interaction to generate the fluorescence signal, thus making it less sensitive to scattering. Over the past three decades, such nonlinear optical microscopy techniques have been continuously refined and engineered for improved performance. Since the seminal paper by Denk et al. (1990), two-photon microscopy in combination with *in vivo* labeling strategies have led to a rapidly expanding field with vastly different applications in live tissue and animal imaging. In the context of neuroscience, the development of calcium indicators (Stosiek et al. 2003; Kerr and Denk 2008) has helped MPM become the technique of choice for imaging neuronal activity in small animals, most importantly the mouse. Besides neuroscience, typical imaging applications of MPM include mouse skin (Laiho et al. 2005), heart (Vinegoni et al. 2015), kidney (Dunn, Sutton, and Sandoval 2012) and lymph nodes (Liu et al. 2015). In the following, we briefly introduce the concept and technical requirements for multiphoton microscopy, discuss current implementations and highlight some important parameters that influence its performance within scattering and live specimen.

22.2.1 Physical Principle

The fact that more than one photon contributes to the signal generation in MPM leads to qualitatively new properties, because of the "higher order" light matter interaction, in which the fluorophore excitation depends on the square (or cube, etc.) of the incident light intensity. This effectively results in a tighter confinement of the signal (fluorescence) generation and practically leads to focal volumes on the order of femtoliters, close to the optical diffraction limit (see Figure 22.1b). In order to excite a molecule, two or more photons need to impinge near simultaneously (within ~0.5 fs; see Figure 22.1a); therefore, the efficiency of multiphoton absorptions depends crucially on the physical properties of the molecule as well as the spatial and temporal density of the excitation light. In general, the multiphoton excitation efficiency can be approximated by n_a, the number of photons absorbed per fluorophore per unit time (Denk et al., 1990),

$$n_a \approx \frac{p_0^N \delta_N NA^{2N-4}}{(\tau_p f_p)^{N-1} \lambda^{2N-3}},$$

where p_0 is the average laser power, τ_p is the pulse duration, λ is the excitation wavelength, δ_N is the n-photon absorption cross-section, f_p is the laser's repetition rate, and NA is the numerical

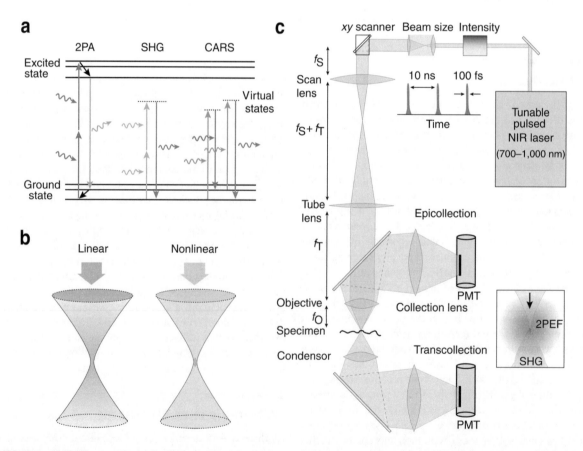

FIGURE 22.1 (a) Nonlinear excitation leads to a more confined region of fluorescence generation. In one-photon based imaging methods, such as confocal imaging, fluorescence is excited along the entire cone of the focus, while in nonlinear microscopy, the higher-order dependence leads to an effectively smaller excitation spot size. (b) Schematic Jablonski diagram, showing two-photon absorption (2PA) as well as second-harmonic generation (SHG). (c) Basic scheme of nonlinear laser-scanning microscope. $f_{S,T,O}$ denote focal lengths of respective lenses. See text for more details. Figure modified and reprinted by permission from Helmchen and Denk, 2005.

aperture of the focusing objective. Note that we have neglected nonrelevant prefactors and parameters for simplicity here. Most typical fluorophores and dyes have "absorption cross sections" that are many orders of magnitude smaller than in the one-photon case (for two-photon absorption, a typical unit is 1 GM = 10^{-50} cm^4 s, named after Maria Göppert-Mayer who, in 1931, predicted this quantum-mechanical effect). This necessitates highly confined light pulses in both space and time, through high numerical aperture (NA) objectives, as well as through ultrashort laser pulses ($10^{-14} - 10^{-12}$ s), which in turn achieve high peak intensities. Lasers typically used in MPM feature τ~100 fs pulses at f~100 MHz. Thus, compared to continuous-wave lasers, these femtosecond pulsed lasers offer an "enhancement" of $\left(\tau_p f_p\right)^{1-N}$, which for two-photon microscopy (N=2) would imply roughly a ~300-fold reduction in the necessary average power. Another advantage of MPM is that nonlinear contrast can also be created based on organization and orientation of nanostructures in the specimen, and, therefore, provide an alternative to active labeling for studying cellular structures and functions inside living animals. This includes second and third harmonic generations (SHG and THG) (see Figure 22.1a), which can be described as instantaneous signals, as they do not involve relaxation processes inside the molecule (Friedl et al. 2007). Biological materials that possess large, ordered, noncentrosymmetric structures such as collagen, microtubules, and muscle myosin can produce SHG signals, while THG is present particularly at phase transitions, such as water-lipid interfaces, and has been used to visualize lipids (Mahou et al. 2012) as well as myosin (Chu et al. 2004) in live tissue.

The advantages of multiphoton excitation for imaging in scattering tissues and small, living animals are manifold: (1) The higher nonlinearity strictly confines the excitation volume, resulting in comparable resolution to confocal techniques; (2) yet, at the same time, the excitation wavelength is markedly increased, which lowers scattering as well as photobleaching and related phototoxic events; (3) increased detection efficiency because the entire fluorescence collected by the objective, i.e. both ballistic as well as scattered photons, contribute to the signal. Therefore, no photons have to be rejected as is the case in confocal imaging. Also, there is no overlap between excitation and emission spectra. All these advantages have led to the rapid adoption of MPM in biomedical research fields and have made it an indispensable tool for 3D live-tissue studies.

22.2.2 Realization

An MPM requires more advanced components compared to other fluorescence microscopy techniques, in particular a pulsed laser, such as the widely used titanium-sapphire oscillator that features ~80 MHz repetition rate and ~100 femtosecond long pulses, and whose wavelength can be tuned (~650–1050 nm) to match the two-photon excitation spectra of most fluorophores used in biology. Other crucial components in the excitation path include intensity control (often through a fast electro-optic modulator), as well as a scanning module that allows to steer the beam in the lateral xy-plane. Typically, this is achieved by a pair of galvanometric scan mirrors.

Subsequently, a suitable combination of scan and tube lenses magnifies the beam diameter such that it fills the back aperture of the microscope objective, thus achieving high NA and hence resolution. The excited fluorescence is normally detected in an epidirection with photomultiplier tubes (PMT) that detect the fluorescence that is reflected by a dichroic mirror, although transdetection is also possible (see Figure 22.1c). In order to maximize the signal from scattering tissues, especially when imaging deep, multiple parameters have to be carefully considered and optimized (Helmchen and Denk, 2005; Oheim et al. 2001). In particular, low-magnification, high-NA objectives, together with large area PMTs positioned as closely as possible to the objective, maximize the collection of scattered fluorescence photons. Furthermore, the excitation beam should not overfill the objective to prevent unnecessary beam-clipping and loss of excitation power. Finally, pulse length broadening due to dispersion should be minimized or compensated by appropriate prechirping, such as not to decrease the excitation efficiency.

22.2.3 Recent Advances in Multiphoton Microscopy

A significant portion of this Ssction is devoted to some more exciting and upcoming techniques in MPM, especially regarding imaging deep in living tissues and improving the imaging speed and capacity to distinguish multiple fluorophores. Therefore, we will highlight and review recent literature below that have achieved impressive progress in these directions.

22.2.3.1 Imaging Depth

Most multiphoton imaging experiments that deal with live and scattering tissue will likely be limited at some point by the attainable image depth or at least will face rapidly decreasing signal and contrast when trying to image deeper. In the following, we want to highlight some more general design parameters and discuss promising strategies as well as recent experiments that have aimed to overcome these limitations.

The main problem that limits imaging depth is light attenuation caused by absorption and scattering, both of which reduce intensity exponentially with depth. Most tissues, while absorbing quite well in the blue and green light spectrum because of the blood's hemoglobin, do not actually absorb in the near-infrared (NIR) region, where most two-photon excitation wavelengths fall. Except for a peak in water absorption ~1400–1600 nm, most often the dominant attenuation source therefore becomes scattering; however, it significantly decreases with wavelength, pointing to a viable route for deeper imaging if red-shifted fluorescent dyes can be employed (Kobat, Horton, and Xu, 2011). Unfortunately, establishing these in live animal experiments has been hindered by low expression levels as well as long development time of e.g. transgenic mouse lines. Therefore, the majority of currently used and well-established fluorescent dyes still emit in a narrow visible region (~450–650nm). Nevertheless, the lower attenuation properties of longer-wavelength NIR light can be exploited through higher-order multiphoton absorption, in particular three-photon (3P) excitation. Especially, the spectral windows

~1300 and 1700 nm that are marked by low attenuation in mammalian tissues, conveniently overlap with the 3P excitation wavelength of green and red fluorophores, such as GFP and RFP, respectively (see Figure 22.2a). An advantage of higher-order excitation is the fact that out-of-focus excitation of (surface) fluorescence is markedly decreased, paving the road to imaging beyond depths of 1 mm, where traditional 2P schemes face a hard limit (Theer and Denk, 2006) (see Figure 22.2b). Pioneering experiments performed by the Xu group (Horton et al., 2013; Ouzounov et al., 2017) have demonstrated impressive imaging depths by exploiting lasers with high pulse energies (at low repetition rate) at these long NIR wavelengths through 3P excitation (3PE). Demonstrations include imaging of subcortical neurons at over 1mm imaging depth (Horton et al., 2013) and with sufficient speed to resolve calcium dynamics (Ouzounov et al., 2017). Hopefully, future experiments will reveal the extent to which this approach can also be useful in noncerebral tissues in live animals.

In addition to attenuation, the other main physical factor that negatively influences imaging depth is optical aberration, arising mostly due to refractive index inhomogeneities inside the tissue that distort the optical wavefront. This sample-induced aberration normally increases with imaging depth, leading to the common observation that the image becomes less "sharp," which is a consequence of an enlarged focal spot and lowered excitation intensity. But aberrations not only originate from the complex, biological tissue, but also from imperfections in the excitation path and its optics, as well as because of the mismatch of immersion media, cover glass and/or tissue indices. The good news here is that at least conceptually, these aberrations can be measured and corrected by active optical components, such as deformable mirrors, thereby actively shaping the wavefront such as to cancel out tissue-induced aberrations and recovering ideal, diffraction-limited imaging quality in the process (see Figure 22.2c). These "adaptive optics" techniques (Booth, 2014; Wang et al., 2014, 2015) have recently found widespread adoption and applications in biology with demonstration also in the context of small-animal imaging. The drawback of adaptive optics techniques is that the complexity of aberrations that can be corrected are rather low. Active optical elements that give access to higher order corrections such as spatial light modulators (SLM) in principle allow more advanced "wavefront shaping" (Katz et al., 2014; Judkewitz et al., 2013) that can also counter the effects of light scattering, which, although it appears random, is essentially a deterministic process within tissue. Currently, wavefront-shaping demonstrations are still mostly confined to optics lab and have not been successfully used to their full advantage in live animal imaging, which is predominantly due to the intrinsic motion of these tissues that leads to rapidly changing wavefront aberrations (on the order of milliseconds). Future work will show whether innovative approaches can be found to circumvent these limitations that currently prevent practical applications (Horstmeyer, Ruan, and Yang, 2015).

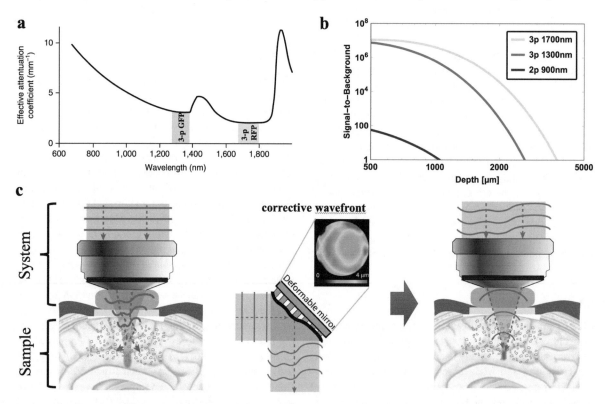

FIGURE 22.2 (a) The dependence of the effective attenuation coefficient on wavelength indicates that optimal excitation wavelength windows are near 1300 and 1700 nm. (b) The intrinsic signal-to-background ratio is higher in 3PE at 1300 and 1700 nm, compared to 2PE, and hence enables deeper imaging (calculation here assumes mouse brain tissue optical properties). (c) Left: Wavefront aberration can originate from imaging system imperfections, such as refractive index mismatch between immersion media, coverslip, and sample, as well as from biological tissue inhomogeneities. Middle: A deformable mirror allows wavefront correction at high speeds. Right: Precompensating aberrations lead to increased excitation efficiency and recovers ideal, diffraction limited resolution.

22.2.3.2 *Imaging Speed*

Improving imaging speed is maybe the single most important parameter for most life scientists who work with small animal models. This can be due to the need for large, 3D volumes to be imaged in a reasonable time to resolve dynamic events, such as neuronal activity, or, on a more practical level, because animal motion during the imaging necessitates fast frame rates for adequate motion correction in real time or during postprocessing. There exist several strategies to push imaging speed, which can be loosely grouped in methods that utilize fast remote scanning, spatial as well as temporal multiplexing, and sophisticated sampling strategies, respectively. Scanning of the multiphoton laser focus in the sample's lateral plane is typically achieved by galvanometric mirrors. Much faster deflection speeds are offered by resonant scanners (Fan et al., 1999) that oscillate at a fixed, but much higher (>10kHz) frequency compared to galvanometers. Efficient scanning along the axial direction is especially challenging because of the large weight and high inertia of the microscope objective. Therefore, alternative approaches have been realized in which e.g. the wavefront of the light is shaped such that the beam divergence at the objective's back focal plane is modulated, resulting in an axial shift in the sample. Technically this can be realized with variable-focus lenses (Grewe et al., 2011; Kong et al., 2015); wavefront shaping devices, such as deformable mirrors (Žurauskas et al., 2017); or SLMs. An optically elegant method is to perform remote focusing through a lightweight mirror and an additional objective (Botcherby et al., 2008), thereby leading to aberration-free axial scanning over a few hundred μm (Rupprecht et al., 2016). An alternative approach to mechanical laser scanning is to use acousto-optic deflectors (AODs) that offer the advantage of faster beam steering, as no movable mechanical parts are involved. With AODs, it becomes possible to access only selected locations in a 3D volume for so-called random-access scanning, with μs scan times, thereby avoiding imaging positions that e.g. do not contain labeled sample features. Utilizing this approach, calcium transients triggered by back-propagating action potentials along neuronal axons have been visualized over large cortical volumes (Duemani et al., 2008; Grewe et al., 2010; Katona et al., 2012). Another way to increase imaging speed is by using multiple excitation spots in parallel, e.g. by scanning a 2D array of foci across the sample plane (Bewersdorf, Pick, and Hell, 1998; Buist et al., 1998). Imaging depth is severely limited by this approach, as the signal from the multiple foci must be captured by an array detector and scattering can lead to significant crosstalk (Kim et al., 2007). However, temporally separating, i.e. interleaving the excitation of multiple laser pulses (foci) on fast (ns) time scales can make these resilient to scattering. Chen et al. (2013) demonstrated simultaneous recording of four different axial planes in the mouse cortex by using four temporally multiplexed (i.e. interleaved) beams of an 80MHz laser with a fast calcium sensor whose lifetime of ~1 ns was shorter than the time between successive laser pulses (~3 ns). In the case where the structure of interest (or labeling) is spatially very sparse, faster volumetric imaging can also be performed by imaging with an axially extended focus, as, in this case, the chance of overlapping structures along the

z-direction is fairly small. This can be achieved by generating a Bessel-beam (Fahrbach, Simon, and Rohrbach, 2010) whose extended depth of field can thus reduce the problem of scanning a 3D volume to a 2D frame only (Lu et al., 2017; Thériault et al., 2014). A final consideration when attempting to extract as much useful information from a sample in a minimal time is to optimize the sampling strategy. Optimized scanning trajectories and engineering of the point-spread function can enable rapid volume imaging even inside scattering tissues. A simple approach is to switch from simple raster-scanning to an arbitrary line scan, by minimizing the travel time (i.e. dead time) between sample points in 3D. This targeted scanning is easily implemented by the abovementioned AODs (Katona et al., 2012; Duemani et al., 2008) but can also be achieved by more traditional scanners. When optimizing imaging parameters, care must be taken to ensure that sampling time on each imaging voxel is long enough to ensure good signal-to-noise ratio (SNR). In practice, the achievable SNR is determined by the total number of photons that can be collected from each region of interest. It depends on the concentration, brightness, and photostability of the fluorophore, as well as the excitation and collection efficiencies of the imaging system. A final factor that constrains the imaging speed, but also attainable depth, is tissue heating and damage because of limited probe brightness or lower efficiency of higher order multiphoton absorption cross section. Naturally, care must always be taken to ensure that the physiological events under investigation are not affected by the pursuit of increased imaging depth or speed (Podgorski and Ranganathan, 2016). Therefore, it becomes especially crucial to optimize the practical photon budget, i.e., the maximal number of photons extractable before photon- or probe-induced artifacts make the biological organism under observation behave nonphysiological.

22.2.3.3 *Multicolor Imaging*

Being able to image multiple fluorophores or "colors" simultaneously is becoming increasingly important in biology for a variety of purposes, such as visualizing different specific molecules in a subcellular context, recording activity and signaling events, but also for tracking cell anatomy, movement, and lineage. However, multicolor, multiphoton tissue imaging methods that enable one to resolve more than two or three colors in thick or live samples are challenging to achieve, as the laser's wavelength typically has to be retuned for sequential multicolor measurements. Therefore, one very promising approach has been to simultaneously use multiple wavelengths laser pulses and to coalign them spatially as well as temporally in the focus (see Figure 22.3b) (Mahou et al., 2012). This adds an effective additional "virtual" excitation wavelength, stemming from two photons originating from separate wavelength pulses and gives access to two-beam processes, such as sum-frequency generation (SFG) and two-color, two-photon excited fluorescence (2c-2PEF) (see Figure 22.3a,c). A key advantage of this scheme is that it achieves simultaneous excitation of blue, green, and red fluorophores and that all signals are generated simultaneously, resulting in faster acquisition speeds compared to sequential single-channel imaging and, therefore, is also amendable to recording dynamic multiple-color

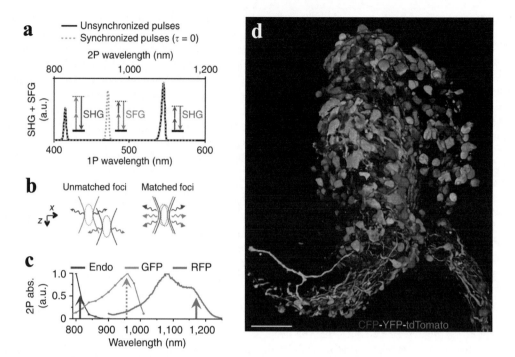

FIGURE 22.3 Multicolor two-photon imaging: (a) Excitation with unsynchronized pulses or unmatched foci can produce blue and red two-photon excited fluorescence (2PEF) only. (b) Coalignment in space and time additionally gives access to two-beam processes such as sum-frequency generation (SFG) and two-color two-photon-excited fluorescence (2c-2PEF). (c) 2P excitation and emission spectra of endogenous fluorescence ("Endo"), GFP and RFP. Arrows indicate the effective excitation wavelengths. (d) Three-color and THG imaging of live, Brainbow-labeled mouse embryo tissue (spinal cord at E3.5). Figure adapted and reprinted by permission from Mahou et al., 2012.

events, such as rapid morphogenic movements during embryo development in *Drosophila* (Mahou et al., 2012). Notably, the simultaneous excitation and acquisition of multiple fluorophores and colors also permits spectral unmixing if wavelength and power-matched reference spectra are available, so that, in principle, much more than 2–3 colors can be discriminated (Zimmermann et al., 2014). Together with multicolor genetic labeling strategies, such as the Brainbow mice (Livet et al., 2007), this provides a new avenue for neuronal-connectivity tracing, cell-lineage tracking, and related bioimaging problems that benefit from spectrally separating the otherwise spatially nonresolvable, dense cell assemblies (see Figure 22.3d). Furthermore, it could enable imaging of the activity of different neuronal cell types in parallel, each labeled with a Ca^{2+}-indicator of a different color, e.g. in order to separate the role of excitatory and inhibitory neurons within a particular network.

In the next section, we will introduce an imaging approach that offers a new "dimension" in multiphoton imaging that can be used to optimize the ever-present trade-offs in imaging regarding spatial and temporal resolution, imaging field-of-view and indirectly, photodamage.

22.3 Temporal Focusing – Light-Sculpting an Arbitrary PSF

Conventional multiphoton imaging approaches are based on the spatial focusing of Gaussian beams. This, however, also leads to an intrinsic trade-off between lateral and axial

resolution, as these two properties scale with the objective's numerical aperture as $\sim NA^{-1}$, and $\sim NA^{-2}$, respectively. To obtain high resolution, in particular along the axial direction, the excitation spots are typically focused to the diffraction limit, i.e. ~500 nm in the lateral, as well as ~1–2 μm in the axial dimension (see Figure 22.4a). Although this high resolution is often desired, it also comes at a price for volumetric imaging, as this increases the overall number of voxels to be imaged from a given volume. In the following, we describe an approach that effectively decouples optical parameters that govern the lateral and axial resolution and allows the spatial extent of an excitation PSF within a multiphoton microscope to be shaped arbitrarily. This so-called "light-sculpting" approach is based on an ultrafast laser physics technique known as temporal focusing (Oron, Tal, and Silberberg, 2005; Zhu et al., 2005) (see Figure 22.4f). In temporal focusing, the frequency spectrum of an ultrafast (femtosecond) laser pulse is spatially dispersed by a grating, and the spot on the grating is imaged inside the sample via a relay lens and objective (see Figure 22.4g). The spectral components of the laser pulse will only spatially overlap in the focus region of the objective where the original, Fourier-limited pulse duration will be restored. Outside the focal region, the pulses are spatially dispersed and temporally chirped (i.e. stretched) to pico-second lengths. Because the probability to excite a fluorescent sample scales inversely with laser pulse length, $\sim \tau^{-1}$, this leads to a confinement of the excitation in a relatively small axial region. Experimentally, axial resolutions as high as ~1μm have been achieved (Andrasfalvy et al., 2010), and the flexible PSF engineering of temporal focusing has been most often utilized

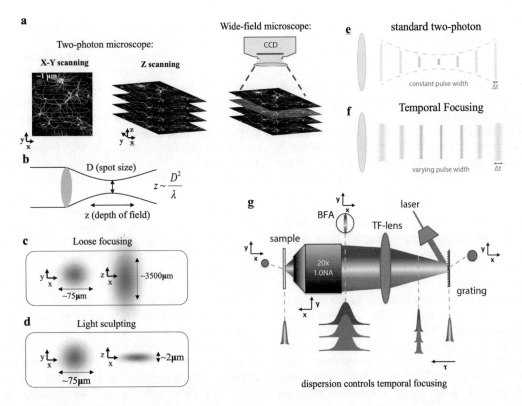

FIGURE 22.4 (a) In multiphoton microscopy, a diffraction-limited laser spot is scanned in the lateral as well as axial dimension in order to capture the entire volume. Faster acquisition can be achieved by recording an entire plane concurrently, e.g. by using a wide-field detector, but this requires the excitation light to be axially confined to a single section (red plane). (b) In Gaussian optics, lateral (spot size) and axial (depth of field) resolution are interlinked, prohibiting (c) wide focal spots with good axial resolution. (d) This can be circumvented by utilizing light-sculpting. (e) While a standard two-photon microscope achieves focusing in space only, (f) temporal focusing modulates the pulse duration in an axially dependent fashion. (g) Schematic illustration of temporal focusing setup, illustrating spatial dispersion and pulse width broadening outside the focal (sample) region.

in the quest to increasing the lateral extent of the excitation spot while retaining good optical sectioning (Vaziri et al., 2008; Chang et al., 2017). This can be useful in the context of efficient excitation of optogenetic actuators inside neurons (Andrasfalvy et al., 2010; Papagiakoumou et al., 2010; Rickgauer, Deisseroth, and Tank, 2014) but also in order to increase recording speed in volumetric imaging, as we will see in the following section. Most notably, temporal focusing also maintains the advantage of high depth penetration because of the longer wavelengths employed, and its axial confinement has also been shown to be highly resilient to scattering (Sela, Dana, and Shoham, 2013) in contrast to conventional multiphoton microscopy, in which the axial extent of the PSF deteriorates with increasing imaging depth.

22.4 Applications of Light-Sculpting in *In Vivo* Neuronal Activity Imaging

Many current questions in systems biology and neuroscience, such as understanding brain dynamics of neuronal networks, require the ability to capture functional dynamics of large cellular populations with single-cell resolution at high speed. As we have seen so far in this chapter, MPM and especially two-photon laser-scanning microscopy (TPLSM) is often the method of choice because of the highly scattering nature of

most (mammalian) brain tissues. However, as discussed in the above sections, standard TPLSM is fairly slow when it comes to imaging across large (brain) volumes. Yet, for understanding integrative brain function, access to all neuronal cells in a given circuitry is often required, as only their combined dynamics can reveal correlations or other statistical properties crucial to signal processing in these networks. In the following, we will highlight some recent work from our group in which we exploited the intrinsic advantages of light-sculpting to enable large-volume brain activity imaging. In particular, we will discuss its application to visualizing the dynamics of the entire brain of the nematode, *C. elegans,* and show how this approach can be extended to record from large neuronal populations in the mouse cortex.

22.4.1 Whole-Brain Imaging of *C. elegans*

The simple nematode, *C. elegans*, is an interesting model organism for studying brain function for several reasons. Its nervous system is comprised of only 302 neurons and ~8000 synaptic connections, a tractable number compared to many other organisms, yet, it features a repertoire of complex behaviors. It is also the only animal for which a complete nervous system has been anatomically mapped (White et al. 1986); however, this detailed anatomical knowledge alone has not been sufficient to predict all functional connections underlying behavior. Thus, it is clear that in order to understand how complex sensory inputs are represented and processed by the nervous system to generate behavior, anatomical as well as precise functional knowledge is required. In essence, this can be referred to as information about how these neuronal populations engage in dynamic activity patterns that encode for distinct behavioral states.

Recording of these dynamics in *C. elegans* have been notoriously difficult because of the small size and high density of its neurons, which are surrounded by neuropil, and has limited targeted recordings to a handful of individual neurons only. This led us to develop a pan-neuronal but nuclear-confined version of the Ca²⁺-indicator GCaMP5K, which resulted in reliable segmentations of individual neurons, even in the densely packed head-ganglia of the nematode's brain, where about half of *C. elegans* neurons are concentrated in a volume comprising ~70 × 50 × 30 µm. In order to record the calcium dynamics from this volume in a rapid fashion, we utilized light-sculpting based on temporal focusing. In this particular case, we tailored the PSF of the excitation light to cover the entire lateral extent of the *C. elegans'* brain, i.e. 70 × 70 µm, while, at the same time, retaining a high axial confinement of ~2 µm (see Figure 22.5a). This simultaneous excitation over a large area, which was achieved by using a low repetition rate laser amplifier, permitted the use of a wide-field detection

scheme, in particular a scientific complementary metal-oxide semiconductor (sCMOS) camera attached to a high-gain image intensifier. This resulted in high signal-to-noise images, even under low light conditions (see Figure 22.5b). Fast volumetric imaging was hence reduced to scanning in the axial direction only, drastically speeding up the total acquisition time. This technique, termed wide-field temporal focusing (WF-TeFo), allowed us to simultaneously record the activity of ~100 neurons in the *C. elegans'* brain with high temporal and single-neuron resolution and to study global brain dynamics that are characterized by patterns of synchronized, i.e. correlated, activity and antagonisms (see Figure 22.5c) (Schrödel et al., 2013). The fact that, in *C. elegans*, individual neuron positions and identities are stereotypic further allowed us to link functionally correlated neurons to anatomical links known from the wiring map. Recently, our approach to brain-wide imaging has been complemented by behavioral as well as single-neuron recordings in freely-moving nematodes by our collaborators. Their studies revealed how globally coordinated dynamics of neuronal subpopulations encode motor commands and action sequences during locomotion in *C. elegans*. In their work, Kato et al. (2015) showed that these dynamics occupied a lower state manifold and that all behavioral sequences and transitions can be explained by trajectories within this phase space. This shows how "simple," anatomically well-defined

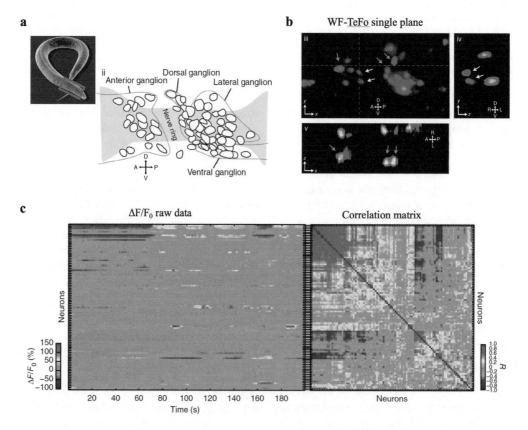

FIGURE 22.5 (a) Micrograph of the *C. elegans* nematode and schematic of the head ganglia. Black lines outline neuronal nuclei, gray lines outline the ganglia. Green indicates the pharynx. (b) Single z plane (z = 2 µm) acquired with the WF-TeFo method. Dashed lines indicate y–z and x–z cross sections shown in (iv) and (v), respectively. Arrows in (iii–v) indicate example neurons each seen in two projections. (c) Activity of 99 neurons from a volumetric recording at 5 Hz. Each row shows a time-series heat map of NLS-GCaMP5K fluorescence changes (ΔF/F0). A matrix showing correlation coefficient (R) calculated from all time series is also shown; neurons are grouped by agglomerative hierarchical clustering. Figure modified and reprinted by permission from Schrödel et al., 2013.

nervous systems, such as *C. elegans*, together with sophisticated, brain-wide imaging methods can provide mechanistic insights into the underlying principles of neural coding and information processing and hypotheses that can be further tested in higher animals with more complex brains.

22.4.2 Large-Volume Cortical Imaging in Awake Mice

Although whole-brain imaging has been demonstrated in small and transparent model organisms such as *C. elegans* (Schrödel et al., 2013) or the zebrafish larvae (Ahrens et al., 2013; Prevedel et al., 2014), recording neuronal dynamics in a similarly large-scale and high-speed fashion has been notoriously difficult to achieve in the mammalian brain, e.g. in awake-behaving mice. While TPLSM has become the gold standard in mammalian neuroimaging because of its high resolution and large penetration depth, TPLSM suffers from low temporal resolution, which makes imaging of cortical brain regions >100 × 100 × 100µm at adequate speed (>3–5 Hz) challenging. Yet, neuronal networks in the mammalian brain are known to extend several hundred micrometers in size. In fact, the smallest computational "subunit" in the mouse brain, also known as a cortical column (Mountcastle 1956), comprises at least ~500 × 500 × 500 µm, and is a region with unique sensory receptive fields and, therefore, encode for similar features in the brain (see Figure 22.6a). Understanding how sensory information is processed and stored in the complex,

highly interconnected mammalian brain requires tools that are capable of recording the activity of all the neurons that comprise a functional network, i.e. at least over an entire cortical column. The reason why imaging such an extended cortical volume with current TPLSM methods has remained so challenging is because of a combination of several technological and biological bottlenecks: The sheer number of voxels (~10^9) that need to be scanned with a diffraction-limited excitation spot place unrealistic demands on laser output (average power, pulse energy, and repetition rate) as well as on scanning (~10^7 Hz) and detection (~GHz) hardware, all of which currently offer performance at least an order of magnitude below the required parameters. Even in absence of any hardware limitations, the obtainable signal from each voxel is limited by the fluorescence lifetime (~10^{-9} s), which prohibits the required subnanosecond voxel dwell times. On the other hand, heating and photodamage place another upper limit on the laser power that can be used for imaging, typically on the order of 250 mW (Podgorski and Ranganathan, 2016). It is thus obvious that new conceptual approaches are required for functional imaging of neuronal activity in mammalian brains using MPM.

Recently, we have investigated both theoretically as well as experimentally a range of different light-sculpting strategies in order to optimize the typical trade-offs present when attempting to tailor the excitation PSF to maximize the attainable imaging volume and speed (Rupprecht et al., 2015). This has led us to put forward a novel light-sculpting approach that is suited for fast volumetric calcium imaging in scattering

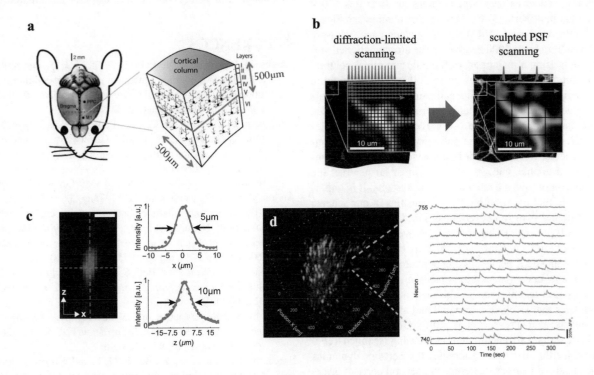

FIGURE 22.6 (a) Cartoon depicting the extent of a cortical column as well as its cortical layers in the mouse brain. Red line indicates maximum depth achieved during imaging. (b) Conceptual illustration of scanning with diffraction-limited and a sculpted, enlarged PSF over a neuronal cell. The number of voxels is significantly decreased in the latter case. (c) Measured PSF in scanned-TeFo, indicating 5 µm lateral as well as 10 µm axial resolution. (d) Fast volumetric Ca²⁺-imaging across a cortical column in the *in vivo* mouse at 3Hz volume rate. Left: 3D rendering of imaged volume (500 × 500 × 500 µm), individual neurons are clearly resolved. Right: Example Ca²⁺ signal (ΔF/F0) of GCaMP6 fluorescence extracted from the data. Time traces of 16 neurons out of a total of ~4000 are shown. Panels C, D, and B modified and reprinted by permission from Prevedel et al., 2016; Weisenburger, Prevedel, and Vaziri, 2017, respectively.

brain tissues (Prevedel et al., 2016; Weisenburger, Prevedel, and Vaziri, 2017). Our approach is based on the synergetic combination of three key principles: (1) to excite the minimally required number of voxels per unit volume necessary to resolve the structure of interest, (2) to maximize the obtainable fluorescence signal-to-noise from each voxel by optimizing the excitation PSF, and (3) the light source's parameters. It is important to realize in this context that the structures of interest in our case, i.e. neuronal somata, are relatively large (10–20μm (Kandel, Schwartz, and Jessell, 2012), hence diffraction-limited resolution is actually not beneficial (see Figure 22.6b). Our excitation scheme involves light-sculpting based on temporal focusing to shape the 3D PSF of our microscope near-isotropically (~5 × 5 × 10 μm) so as to match it to the typical size of neuronal cell bodies in the mouse cortex (see Figure 22.6c). This lowers the number of voxels that need to be scanned and effectively increases the volume acquisition rate. Quite interestingly, the enlarged PSF also leads to a higher total signal-to-noise ratio from each calcium indicator-expressing neuron, because larger voxels yield higher overall fluorescence from each voxel, and the use of lower repetition rate lasers (at the same average power) further lead to increased signal levels (lower f_p). In fact, our approach only utilizes a single excitation laser pulse per imaging voxel, in contrast to standard TPLSM, which utilize ten to hundreds of pulses per voxel (Weisenburger, Prevedel, and Vaziri, 2017). We have demonstrated the capabilities of our new imaging approach by performing *in vivo* calcium imaging of the mouse neocortex over large volumes (up to 500 × 500 × 500 μm) at multihertz (3–160 Hz) time resolution (see Figure 22.6d) (Prevedel et al., 2016). Using our approach, we were able to capture up to 4000 neurons in the aforementioned volume, comprising a major part of the mouse cortical column, and monitor the dynamics of up to 3000 active neurons in our experiments. Crucial to the realization of our experiments was the development of a suitable laser source, which offered the necessary repetition rate (~4 MHz) and pulse energies (~500 nJ) required for our one pulse per enlarged voxel excitation scheme. Recently, such lasers have also become commercially available, and we expect them to further fuel the development of this and related imaging systems. It is important to note that the resulting power densities in our scheme are not higher than in conventional TPLSM (<5.5 mW/μm² or <2.5 nJ/(μm² s)). To ensure that the overall average laser power (<220 mW) in our *in vivo* imaging experiments would not lead to any detrimental heating or photodamage, we monitored *in situ* brain temperatures using a small temperature sensor as well as ensured through postmortem immunohistochemistry that levels of microglial and astrocyte responses were not elevated (Prevedel et al., 2016). Our novel approach, which we termed scanned-temporal focusing (scanned-TeFo) allowed for the first time to monitor the network dynamics of thousands of neurons that comprise several cortical layers in the mouse. In the mammalian neocortex, these advances provide the opportunity to gain new insights into the computational principles that underlie information processing and to experimentally verify or discard a wide range of theoretical models of cortical computation that have been proposed in the past (Sompolinsky, 2014; Buonomano and Maass, 2009).

22.5 Conclusion and Outlook

In this chapter, we have introduced the reader to the underlying principles of MPM and discussed the technical requirements as well as the latest, emerging trends of this field. In particular, we have highlighted recent work that aims to increase the current imaging field-of-view, speed, and ability to image multiple labels or colors concurrently. A significant part of our chapter was devoted to the concept of light-sculpting for arbitrary engineering of the PSF in more detail, as it provides an interesting avenue for unconventional optimization of imaging parameters. In this respect, we have covered some recent demonstrations that focused on imaging neuronal activity in different model organisms in order to elucidate basic principles in neuronal coding and information processing.

The future is bright for multiphoton fluorescence microscopy, and the next years will certainly bring new and exciting concepts and developments. Brighter and more red-shifted fluorescence dyes and proteins combined with technological progress will ensure that instrument developers in close collaboration with biologists will continue to push the envelope of what is possible in fluorescence (brain) imaging, especially in respect to recording dynamic (neuronal) events in deep mammalian tissues at high spatial and temporal resolution. *In vivo* MPM is specially poised for success in this domain, as it allows longitudinal studies of structural and functional changes on a subcellular level over long periods of time, while providing the ability to put cellular behavior in a large context, both spatially and temporally.

REFERENCES

Ahrens, M. B., Orger, M. B., Robson, D, N., Li, J. M., and Keller, P. J. 2013. Whole-brain functional imaging at cellular resolution using light-sheet microscopy. *Nature Methods*, *10*(5): 413–20. doi: 10.1038/nmeth.2434

Akerboom, J., Calderón, N. C., Lin, T., Wabnig, S., Prigge, M., Tolö, J., Gordus, A., et al. 2013. Genetically encoded calcium indicators for multi-color neural activity imaging and combination with optogenetics. *Frontiers in Molecular Neuroscience*, *6*(March): 2. doi: 10.3389/fnmol.2013.00002

Andrasfalvy, B. K., Zemelman, B. V., Tang, J., and Vaziri, A. 2010. Two-photon single-cell optogenetic control of neuronal activity by sculpted light. *Proceedings of the National Academy of Sciences of the United States of America*, *107*(26): 11981–86. doi: 10.1073/pnas.1006620107

Bewersdorf, J., Pick, R., and Hell, S. W. 1998. Multifocal multiphoton microscopy. *Optics Letters*, *23*(9): 655. doi: 10.1364/OL.23.000655

Booth, M. J. 2014. Adaptive optical microscopy: The ongoing quest for a perfect image. *Light: Science & Applications*, *3*(4): e165. doi: 10.1038/lsa.2014.46

Botcherby, E. J., Juškaitis, R., Booth, M. J., and Wilson, T. 2008. An optical technique for remote focusing in microscopy. *Optics Communications*, *281*(4): 880–87. doi: 10.1016/j.optcom.2007.10.007

Buist, A. H., Müller, M., Squier, J., and Brakenhoff, G. J. 1998. Real time two-photon absorption microscopy using multi point excitation. *Journal of Microscopy*, *192*(2): 217–26. doi: 10.1046/j.1365-2818.1998.00431.x

Buonomano, D. V., and Maass, W. 2009. State-dependent computations: Spatiotemporal processing in cortical networks. *Nature Reviews Neuroscience, 10*(2): 113–25. doi: 10.1038/nrn2558

Chang, C. Y., Lin, C. H., Lin, C. Y., Da Sie, Y., Hu, Y. Y., Tsai, S. F., and Chen, S. J. 2017. Temporal focusing-based widefield multiphoton microscopy with spatially modulated illumination for biotissue imaging. *Journal of Biophotonics,* 10: 1–11. doi: 10.1002/jbio.201600287

Chen, T.-W., Wardill, T. J., Sun, Y., Pulver, S. R., Renninger, S. L., Baohan, A., Schreiter, E. R., et al. 2013. Ultrasensitive fluorescent proteins for imaging neuronal activity. *Nature, 499*(7458): 295–300. doi: 10.1038/nature12354

Chu, S.-W., Chen, S.-Y., Chern, G.-W., Tsai, T.-H., Chen, Y.-C., Lin, B.-L., and Sun, C.-K. 2004. Studies of $\chi(2)/\chi(3)$ tensors in submicron-scaled bio-tissues by polarization harmonics optical microscopy. *Biophysical Journal, 86*(6): 3914–22. doi: 10.1529/biophysj.103.034595

Denk, W., Strickler, J. H., and Webb, W. W. 1990. Two-photon laser scanning fluorescence microscopy. *Science, 248*(4951): 73–76.

Duemani, R. G., Kelleher, K., Fink, R., and Saggau, P. 2008. Three-dimensional random access multiphoton microscopy for functional imaging of neuronal activity. *Nature Neuroscience, 11*(6): 713–20. doi: 10.1038/nn.2116

Dunn, K. W., Sutton, T. A., and Sandoval, R. M. 2012. Live-animal imaging of renal function by multiphoton microscopy. *Current Protocols in Cytometry, 347:* 12.9.1–18.

Fahrbach, F. O., Simon, P., and Rohrbach, A. 2010. Microscopy with self-reconstructing beams. *Nature Photonics, 4*(11): 780–85. doi: 10.1038/nphoton.2010.204

Fan, G. Y., Fujisaki, H., Miyawaki, A., Tsay, R. K., Tsien, R. Y., and Ellisman, M. H. 1999. Video-rate scanning two-photon excitation fluorescence microscopy and ratio imaging with cameleons. *Biophysical Journal, 76* (5): 2412–20.

Friedl, P., Wolf, K., Von Andrian, U. H., and Harms, G. 2007. Biological Second and Third Harmonic Generation Microscopy. *Current Protocols in Cell Biology,* 1–21.

Grewe, B. F., Voigt, F. F., van 't Hoff, M., and Helmchen, F. 2011. Fast two-layer two-photon imaging of neuronal cell populations using an electrically tunable lens. *Biomedical Optics Express, 2*(7): 2035. doi: 10.1364/BOE.2.002035

Grewe, B. F., Langer, D., Kasper, H., Kampa, B. M., and Helmchen, F. 2010. High-speed in vivo calcium imaging reveals neuronal network activity with near-millisecond precision. *Nature Methods, 7*(5): 399–405. doi: 10.1038/nmeth.1453

Helmchen, F., and Denk, W. 2005. Deep tissue two-photon microscopy. *Nature Methods, 2*(12): 932–40. doi: 10.1038/nmeth818

Horstmeyer, R., Ruan, H., and Yang, C. 2015. Guidestar-assisted wavefront-shaping methods for focusing light into biological tissue. *Nature Photon, 9*(9): 563–71. doi: 10.1038/nphoton.2015.140

Horton, N. G., Wang, K., Wang, C. C., and Xu, C. 2013. In vivo three-photon imaging of subcortical structures of an intact mouse brain using quantum dots. *2013 Conference on Lasers and Electro-Optics Europe and International Quantum Electronics Conference, CLEO/Europe-IQEC 2013, 7*(3): 205–9.

Judkewitz, B., Wang, Y. M., Horstmeyer, R., Mathy, A., and Yang, C. 2013. Speckle-scale focusing in the diffusive regime with time-reversal of variance-encoded light (TROVE). *Nature Photonics, 7*(4): 300–5. doi: 10.1038/nphoton.2013.31

Kandel, E. R., Schwartz, J. H., and Jessell, T. M. 2012. Neural networks. *Principles of Neural Science,* 1581–600.

Kato, S., Kaplan, H. S., Schrödel, T., Skora, S., Lindsay, T. H., Yemini, E., Lockery, S., and Zimmer, M. 2015. Global brain dynamics embed the motor command sequence of *Caenorhabditis Elegans. Cell, 163*(3): 656–69. doi: 10.1016/j.cell.2015.09.034

Katona, G., Szalay, G., Maák, P., Kaszás, A., Veress, M., Hillier, D., Chiovini, B., Vizi, E. S., Roska, B., and Rózsa, B. 2012. Fast two-photon in vivo imaging with three-dimensional random-access scanning in large tissue volumes. *Nature Methods, 9*(2): 201–8. doi: 10.1038/nmeth.1851

Katz, O., Heidmann, P., Fink, M., and Gigan, S. 2014. Non-invasive real-time imaging through scattering layers and around corners via speckle correlations. *Nature Photonics, 8*(10): 784–90. doi: 10.1038/nphoton.2014.189

Kerr, J. N. D., and Denk, W. 2008. Imaging in vivo: Watching the brain in action. *Nature Reviews. Neuroscience, 9*(3): 195–205. doi: 10.1038/nrn2338

Kim, K. H., Buehler, C., Bahlmann, K., Ragan, T., Lee, W.-C. A., Nedivi, E., Heffer, E. L., Fantini, S., and So, P. T. C. 2007. Multifocal multiphoton microscopy based on multianode photomultiplier tubes. *Optics Express 15*(18): 11658–78. doi: 10.1364/OE.15.011658

Kobat, D., Horton, N. G., and Xu, C. 2011. In vivo two-photon microscopy to 1.6-mm depth in mouse cortex. *Journal of Biomedical Optics, 16*(10). doi: 10.1117/1.3646209.

Kong, L., Tang, J., Little, J. P., Yu, Y., Lämmermann, T., Lin, C. P., Germain, R. N., and Cui, M. 2015. Continuous volumetric imaging via an optical phase-locked ultrasound lens. *Nature Methods, 12*(8): 759–62. doi: 10.1038/nmeth.3476.

Laiho, L. H., Pelet, S., Hancewicz, T. M., Kaplan, P. D., and So, P. T. C. 2005. Two-photon 3-D mapping of ex vivo human skin endogenous fluorescence species based on fluorescence emission spectra. *Journal of Biomedical Optics, 10*(2). doi:10.1117/1.1891370.

Liu, Z., Gerner, M. Y., Van Panhuys, N., Levine, A. G., Rudensky, A. Y., and Germain, R. N. 2015. Immune Homeostasis Enforced by Co-Localized Effector and Regulatory T Cells. *Nature, 528*(7581): 225–30. doi: 10.1038/nature16169

Livet, J., Weissman, T. A., Kang, H., Draft, R. W., Lu, J., Bennis, R. A., Sanes, J. R., and Lichtman, J. W. 2007. Transgenic strategies for combinatorial expression of fluorescent proteins in the nervous system. *Nature, 450*(7166): 56–62. doi: 10.1038/nature06293

Lu, R., Sun, W., Liang, Y., Kerlin, A., Bierfeld, J., Seelig, J. D., Wilson, D. E., et al. 2017. Video-rate volumetric functional imaging of the brain at synaptic resolution. *Nature Neuroscience, 20*(4): 620–28. doi: 10.1038/nn.4516

Mahou, P., Zimmerley, M., Loulier, K., Matho, K. S., Labroille, G., Morin, X., Supatto, W., Livet, J., Débarre, D., and Beaurepaire, E. 2012. Multicolor two-photon tissue imaging by wavelength mixing. *Nature Methods, 9*(8): 815–18. doi: 10.1038/nmeth.2098

Mountcastle, V. 1957. Modality and topographic properties neurons of cat' s somatic sensory cortex. *Journal of Neurophysiology, 20*(4): 408–34.

Oheim, M., Beaurepaire, E., Chaigneau, E., Mertz, J., and Charpak, S. 2001. Two-photon microscopy in brain tissue: Parameters influencing the imaging depth. *Journal of Neuroscience Methods, 111*(1): 29–37. doi: 10.1016/S0165-0270(01)00438-1

Oron, D., Tal, E., and Silberberg, Y. 2005. Scanningless depth-resolved microscopy. *Optics Express*, *13*(5): 1468–76. doi: 10.1364/OPEX.13.001468

Ouzounov, D. G., Wang, T., Wang, M., Feng, D. D., Horton, N. G., Cruz-Hernandez, J. C., Cheng, Y.-T., et al. 2017. In vivo three-photon imaging of activity of GCaMP6-labeled neurons deep in intact mouse brain. *Nature Methods*, *14*(4): 388–90. doi: 10.1038/nmeth.4183

Papagiakoumou, E., Anselmi, F., Bègue, A., de Sars, V., Glückstad, J., Isacoff, E. Y., and Emiliani, V. 2010. Scanless two-photon excitation of channelrhodopsin-2. *Nature Methods*, *7*: 848–54. doi: 10.1038/nmeth.1505

Podgorski, K., and Ranganathan, G. N. 2016. Brain heating induced by near infrared lasers during multi-photon microscopy. *Journal of Neurophysiology*, *116*: 1012–23.

Prevedel, R., Verhoef, A. J., Pernía-Andrade, A. J., Weisenburger, S., Huang, B. S., Nöbauer, T., Fernández, A., et al. 2016. Fast volumetric calcium imaging across multiple cortical layers using sculpted light. *Nature Methods*, *13*(12): 1021–8. doi: 10.1038/nmeth.4040

Prevedel, R., Yoon, Y.-G., Hoffmann, M., Pak, N., Wetzstein, G., Kato, S., Schrödel, T., et al. 2014. Simultaneous whole-animal 3D imaging of neuronal activity using light-field microscopy. *Nature Methods*, *11*(7): 727–30. doi: 10.1038/nmeth.2964

Rickgauer, J. Peter, Deisseroth, K., and Tank, D. W. 2014. Simultaneous cellular-resolution optical perturbation and imaging of place cell firing fields. *Nature Neuroscience*, *17*(12): 1816–24. doi: 10.1038/nn.3866

Rupprecht, P., Prendergast, A., Wyart, C., and Friedrich, R. W. 2016. Remote z-scanning with a macroscopic voice coil motor for fast 3D multiphoton laser scanning microscopy. *Biomedical Optics Express*, *7*(5): 1656. doi: 10.1364/BOE.7.001656

Rupprecht, P., Prevedel, R., Groessl, F., Haubensak, W. E., and Vaziri, A. 2015. Optimizing and extending light-sculpting microscopy for fast functional imaging in neuroscience." *Biomedical Optics Express*, *6*(2): 353. doi: 10.1364/BOE.6.000353

Schrödel, T., Prevedel, R., Aumayr, K., Zimmer, M., and Vaziri, A. 2013. Brain-wide 3D imaging of neuronal activity in *Caenorhabditis elegans* with sculpted light. *Nature Methods*, *10*(10): 1013–20. Doi: 10.1038/nmeth.2637

Sela, G., Dana, H., and Shoham, S. 2013. *Ultra-deep penetration of temporally-focused two-photon excitation*. 8588: doi:10.1117/12.2005402.

Sompolinsky, H. 2014. Computational neuroscience: Beyond the local circuit. *Current Opinion in Neurobiology*, *25*: 13–8. doi: 10.1016/j.conb.2014.02.002

Stosiek, C., Garaschuk, O., Holthoff, K., and Konnerth, A. 2003. In vivo two-photon calcium imaging of neuronal networks. *Proceedings of the National Academy of Sciences of the United States of America*, *100*(12): 7319–24. doi: 10.1073/pnas.1232232100

Theer, P., and Denk, W. 2006. On the fundamental imaging-depth limit in two-photon microscopy. *Journal of the Optical Society of America A*, *23*(12): 3139–49. doi: 10.1364/JOSAA.23.003139

Thériault, G., Cottet, M., Castonguay, A., McCarthy, N., and De Koninck, Y. 2014. Extended two-photon microscopy in live samples with bessel beams: steadier focus, faster volume scans, and simpler stereoscopic imaging. *Frontiers in Cellular Neuroscience*, *8*(May): 139. doi: 10.3389/fncel.2014.00139

Vaziri, A., Tang, J., Shroff, H., and Shank, C. V. 2008. Multilayer three-dimensional super resolution imaging of thick biological samples. *Proceedings of the National Academy of Sciences*, *105*(51): 20221–6. doi: 10.1073/pnas.0810636105

Vinegoni, C., Lee, S., Aguirre, A. D., and Weissleder, R. 2015. New techniques for motion-artifact-free in vivo cardiac microscopy. *Frontiers in Physiology*, *6*(May): 147. doi: 10.3389/fphys.2015.00147

Wang, K., Milkie, D. E., Saxena, A, Engerer, P., Misgeld, T., Bronner, M. E., Mumm, J., and Betzig, E. 2014. Rapid adaptive optical recovery of optimal resolution over large volumes. *Nature Methods*, *11* (6): 625–28. doi: 10.1038/nmeth.2925

Wang, K., Sun, W., Richie, C. T., Harvey, B. K., Betzig, E., and Ji, N. 2015. Direct wavefront sensing for high-resolution in vivo imaging in scattering tissue. *Nature Communications*, *6*: 7276. doi: 10.1038/ncomms8276

Weisenburger, S., Prevedel, R., and Vaziri, A. 2017. *Quantitative evaluation of two-photon calcium imaging modalities for high-speed volumetric calcium imaging in scattering brain tissue*. *Doi.Org*, 115659. doi: 10.1101/115659

White, J. G., Southgate, E., Thomson, J. N., and Brenner, S. 1986. The structure of the nervous system of the nematode *Caenorhabditis elegans*. *Philosophical Transactions of the Royal Society B: Biological Sciences*, *314*(1165): 1–340. doi: 10.1098/rstb.1986.0056

Yaksi, E., and Friedrich, R. W. 2006. Reconstruction of firing rate changes across neuronal populations by temporally deconvolved Ca2+ imaging. *Nature Methods*, *3*(5): 377–83. doi: 10.1038/nmeth874

Zhu, G., van Howe, J., Durst, M., Zipfel, W., and Xu, C. 2005. Simultaneous spatial and temporal focusing of femtosecond pulses. *Optics Express*, *13*(6): 2153–59. doi: 10.1109/CLEO.2005.202174

Zimmermann, T., Marrison, J., Hogg, K., and Toole, P. O. 2014. Confocal microscopy, *1075*: 129–48. doi: 10.1007/978-1-60761-847-8

Žurauskas M., Barnstedt, O., Frade-Rodriguez, M., Waddell, S., and Booth, M. J. 2017. Rapid sensing of volumetric neural activity through adaptive remote focusing. *BioRxiv* im. doi: 10.1101/125070

Index

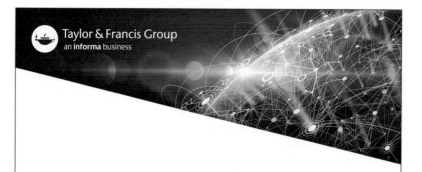

For Product Safety Concerns and Information please contact our
EU representative GPSR@taylorandfrancis.com Taylor & Francis
Verlag GmbH, Kaufingerstraße 24, 80331 München, Germany